AF443576

Geometric Measure Theory
and the Calculus of Variations

PROCEEDINGS OF SYMPOSIA
IN PURE MATHEMATICS
Volume 44

Geometric Measure Theory and the Calculus of Variations

William K. Allard and
Frederick J. Almgren, Jr., Editors

AMERICAN MATHEMATICAL SOCIETY
PROVIDENCE, RHODE ISLAND

PROCEEDINGS OF THE SUMMER INSTITUTE ON GEOMETRIC MEASURE THEORY
AND THE CALCULUS OF VARIATIONS
HELD AT HUMBOLDT STATE UNIVERSITY
ARCATA, CALIFORNIA
JULY 16–AUGUST 3, 1984
with support from the National Science Foundation, Grant DMS-8317890

1980 *Mathematics Subject Classification.* Primary 49F22, 49F20, 53A10, 53C42;
Secondary 49F10, 53C20, 35J65, 58-XX.

Library of Congress Cataloging-in-Publication Data
Main entry under title:

Geometric measure theory and the calculus of variations.
 (Proceedings of symposia in pure mathematics; v. 44)
 Includes bibliographies.
 1. Geometric measure theory—Addresses, essays, lectures. I. Allard,
William K., 1941–   . II. Almgren, Frederick J. III. Series.
QA312.G394   1985   516.3′6   85-18641
ISBN 0-8218-1470-2 (alk. paper)

COPYING AND REPRINTING. Individual readers of this publication, and nonprofit libraries
acting for them are permitted to make fair use of the material, such as to copy an article for use in
teaching or research. Permission is granted to quote brief passages from this publication in reviews
provided the customary acknowledgement of the source is given.

Republication, systematic copying, or multiple reproduction of any material in this publication
(including abstracts) is permitted only under license from the American Mathematical Society.
Requests for such permission should be addressed to the Executive Director, American Mathematical Society, P.O. Box 6248, Providence, Rhode Island 02940.

The appearance of the code on the first page of an article in this book indicates the copyright
owner's consent for copying beyond that permitted by Sections 107 or 108 of the U.S. Copyright
Law, provided that the fee of $1.00 plus $.25 per page for each copy be paid directly to the Copyright Clearance Center, Inc., 21 Congress Street, Salem, Massachusetts 01970. This consent does
not extend to other kinds of copying such as copying for general distribution, for advertising or
promotional purposes, for creating new collective works, or for resale.

Copyright © 1986 by the American Mathematical Society. All rights reserved.
Printed in the United States of America
*The American Mathematical Society retains all rights except
those granted to the United States government.*
The paper used in this book is acid-free and falls within the guidelines
established to ensure permanence and durability.

Contents

Preface

The papers in these Proceedings are associated with lectures delivered at the Thirty-Second Summer Research Institute of the American Mathematical Society devoted to Geometric Measure Theory and the Calculus of Variations. The Institute was held at Humboldt State University in Arcata, California from July 16 to August 3, 1984 and was financed by the National Science Foundation.

The Institute served as a forum for the large collection of new (and older) ideas and techniques comprising modern geometric measure theory and its applications in the calculus of variations. A major theme of the Institute was the introduction and application of multiple-valued function techniques as a basic new tool in geometric analysis; Almgren's basic paper on this subject appears in these Proceedings. In addition, major new results were announced and discussed during the Institute. Allard announced his integrality and regularity theorems for surfaces which are stationary with respect to general elliptic integrands. Scheffer announced the first example of a singular solution to the Navier–Stokes equations for fluid flow with an opposing force; this follows his earlier pioneering work on the possible size of such singular sets. Hutchinson introduced his new definition of the second fundamental form of a general varifold and gave various applications including a multiple-valued function regularity theorem. Papers on these topics are included in these Proceedings.

A session on open problems was also held during the Institute and a list of eighty such problems appears in these Proceedings.

The organizing committee for the Institute consisted of: William K. Allard, Frederick J. Almgren, Jr. (co-chairmen); Enrico Bombieri, Robert M. Hardt, H. Blaine Lawson, Jr., Jon T. Pitts, Richard Schoen, and William P. Ziemer.

The editors would like to thank the many people who cooperated to make the Institute and this volume possible. Of special direct help were Dottie Smith (Conference Secretary), Marie D. Byrnes, Susan Hautala and Elaine Wellman Becker.

William K. Allard
Durham, North Carolina
May 6, 1985

List of Speakers and Titles of Their Talks

(1) Basic Techniques of Geometric Measure Theory (7 lectures)

Examples, currents, and flat chains and basic properties, deformation theorems and their consequences, by Frederick J. Almgren, Jr.

Hausdorff measures and area and co-area formulas, Besicovitch type covering theorems, first variation of area and an isoperimetric inequality, by Enrico Bombieri

Lipschitz multiple-valued functions and approximations of currents, by Frederick J. Almgren, Jr.

Compactness theorems for rectifiable currents from multiple-valued approximations, by Frederick J. Almgren, Jr.

Applications of isoperimetric inequalities, continuity and lower semicontinuity of elliptic parametric functionals, positive currents, "monotonicity", by Enrico Bombieri

First variation distributions, mean curvature, varifolds, and formulas for first and second variations, by William K. Allard

Coflat forms and flat currents, by F. Reese Harvey

(2) Regularity Theory (5 lectures)

Single-valued regularity theory for elliptic variational problems with constraints, by Frederick J. Almgren, Jr.

Q-valued functions minimizing Dirichlet's integral, by Frederick J. Almgren, Jr.

Existence of minimal surfaces on Riemannian manifolds, by Jon T. Pitts

Regularity of minimal hypersurfaces on Riemannian manifolds, by Jon T. Pitts

Strong approximation of nearly flat area minimizing currents, by Frederick J. Almgren, Jr.

(3) Applications of Geometric Measure Theory to Modelling Physical Phenomena (4 lectures)

Flow by mean curvature of convex surfaces into spheres, by Gerhard Huisken

Some regularity results in plasticity, by Robert M. Hardt

Crystals, by Jean E. Taylor

An explicit solution to the Navier–Stokes inequality with an integral singularity, by Vladimir Scheffer

(4) Partial Differential Equations and Geometric Measure Theory (5 lectures)

Compactification of minimal submanifolds in E^n by the Gauss map, by Michael T. Anderson

PDE methods in the study of geometric variational problems, by Leon Simon (2 lectures)

Regularity properties of quasi-minima, by William P. Ziemer

Geometry of minimal surfaces, by Hyeong Choi

(5) Riemannian Curvature and Minimal Surfaces (1 lecture)

Riemannian curvature and minimal surfaces, by William K. Allard

(6) Calibrated Geometries and the Geometry of the Unit Co-mass Ball (2 lectures)

Calibrated geometries, by F. Reese Harvey

Faces of the Grassmannian and area-minimization, by Frank Morgan

(7) Least Area Mappings and Spaces of Minimal Surfaces (1 lecture)

Properties of minimal surfaces with generic boundaries, and of minimal submanifolds of generic Riemannian manifolds, area minimizing currents and area minimizing mappings, by Brian White

(8) Geometric Measure Theory over the Complex Numbers (2 lectures)

Complex analytic geometry and measure theory, by H. Blaine Lawson

Basic techniques, by Enrico Bombieri

SHORT TALKS

Allard, William K.	Toward a regularity theorem for varifolds which are stationary with respect to an elliptic integrand
	The stability inequality for general integrands
Bindschadler, David	Symmetries of area minimizing integral currents
Brothers, John E.	Stability of minimal orbits
Cook, Edith	Free boundary regularity for surfaces minimizing $\mathrm{Area}(S) + c\,\mathrm{Area}(\partial S)$
Dolbeault, Pierre	On holomorphic chains with a given boundary
Emmer, Michele	Spincasting contact lenses
Fu, Joseph	Tubular neighborhoods of planar sets
Gulliver, Robert	Removable singularities of stable minimal varieties
Huang, Wu-Hsiung	When is a soap bubble sitting on a table convex?
Hutchinson, John	Second fundamental form for varifolds
Jost, Juergen	Embedded minimal surfaces with free boundaries
Lawson, H. Blaine	Topologically nonsingular minimal cones
Margerin, Christophe	Hamilton's flow as a tool to inquire about manifolds carrying a positive curvature operator

Mattila, Pertti — Hausdorff dimension and capacities of sets in n-space

Michelsohn, Marie-Louise — A new class of metrics on complex manifolds

Miranda, Mario — MACSYMA and minimal surfaces

Morgan, Frank — On finiteness of the number of stable minimal hypersurfaces

Nance, Dana — Multiplicity theorems for projections of 2- and 3-dimensional hypersurfaces

Nishikawa, Seiki — Deformation of Riemannian metrics and manifolds with bounded curvature ratios

Parks, Harold — Elliptic isoperimetric problems

Pincus, Joel D. — Principal currents

Smith, Penny — Higher regularity for a nonelliptic, nondifferentiable functional

Taylor, Jean E. — Catalog of F-minimizing embedded F-crystalline cones

Wei, S. Walter — Strongly unstable manifolds and weakly stable currents

White, Brian — Regularity of minimizing hypersurfaces mod p

List of Participants

AFGHAHI, Mohammad	National University of Iran, Tehran, Iran
ALLARD, William K.	Duke University, Durham, NC
ALMGREN, Frederick J., Jr.	Princeton University, Princeton, NJ
ANDERSON, Michael T.	Rice University, Houston, TX
BINDSCHADLER, David E.	Wayne State University, Detroit, MI
BOMBIERI, Enrico	Institute for Advanced Study, Princeton, NJ
BROTHERS, John E.	Indiana University, Bloomington, IN
BRULOIS, Frederic	San Diego State University, San Diego, CA
CAREY, Richard W.	University of Kentucky, Lexington, KY
CHANG, Der-Chen	Princeton University, Princeton, NJ
CHANG, Sheldon	Princeton University, Princeton, NJ
CHI, Quo-Shin	Stanford University, Stanford, CA
CHOI, Hyeong	University of Chicago, Chicago, IL
COOK, Edith	Wellesley, College, Wellesley, MA
DADOK, Jiri	Indiana University, Bloomington, IN
DOLBEAULT, Pierre	University of Paris VI, Paris, France
EMMER, Michele	Princeton University, Princeton, NJ
EVANS, L. Craig	University of Maryland, College Park, MD
FU, Joseph H. G.	Massachusetts Inst. of Tech., Cambridge, MA
GULLIVER, Robert	University of Minnesota, Minneapolis, MN
HARDT, Robert M.	University of Minnesota, Minneapolis, MN
HARVEY, F. Reese	Rice University, Houston, TX
HORTON, Clark	Indiana University, Bloomington, IN
HUANG, Wu-Hsiung	Stanford University, Stanford, CA
HUISKEN, Gerhard	Australian National University, Canberra, Aust.
HUTCHINSON, John E.	Australian National University, Canberra, Aust.
JORGE, Luquesio	University of California, Berkeley, CA
JOST, Juergen	University of Bonn, Bonn, W. Germany
KUSNER, Robert B.	University of California, Berkeley, CA
LAU, Chi-Ping	University of Minnesota, Minneapolis, MN
LAWSON, H. Blaine, Jr.	SUNY at Stony Brook, Stony Brook, NY
LIN, Fang-Hua	University of Minnesota, Minneapolis, MN
LIN, Tzuemn-Renn	University of Notre Dame, Notre Dame, IN
MARGERIN, Christophe	Ecole Polytechnique, Palaiseau, France
MATTILA, Pertti	University of Minnesota, Minneapolis, MN
MEEKS, William	Rice University, Houston, TX

MICHELSOHN, Marie-Louise	SUNY at Stony Brook, Stony Brook, NY
MIRANDA, Mario	University of California, Berkeley, CA
MOORE, John Douglas	University of California, Santa Barbara, CA
MORGAN, Frank	Massachusetts Inst. of Tech., Cambridge, MA
MUMMA, Charles	University of Washington, Seattle, WA
NANCE, Dana	Duke University, Durham, NC
NISHIKAWA, Seiki	Kyushu University, Fukuaka, Japan
PARKS, Harold	Oregon State University, Corvallis, OR
PAULIK, George	Indiana University, Bloomington, IN
PERRY, Kevin	Princeton University, Princeton, NJ
PINCUS, Joel D.	SUNY at Stony Brook, Stony Brook, NY
PITTS, Jon T.	Texas A & M University, College Station, TX
POLKING, John	National Science Foundation, Washington, DC
ROSS, Martin	Stanford University, Stanford, CA
SCHEFFER, Vladimir	Rutgers University, New Brunswick, NJ
SIMON, Leon	Australian National University, Canberra, Aust.
SMITH, Penny	Texas A & M University, College Station, TX
SOLOMON, Bruce	Indiana University, Bloomington, IN
STREDULINSKY, Edward	University of Wisconsin, Madison, WI
TAM, Luen-Fai	Stanford University, Stanford, CA
TAYLOR, Jean E.	Rutgers University, New Brunswick, NJ
WEI, S. Walter	University of California, Los Angeles, CA
WHITE, Brian	Stanford University, Stanford, CA
WILLIAMS, William O.	Carnegie-Mellon University, Pittsburgh, PA
ZIEMER, William P.	Indiana University, Bloomington, IN

Geometric Measure Theory
and the Calculus of Variations

Proceedings of Symposia in Pure Mathematics
Volume **44** (1986)

An Integrality Theorem and a Regularity Theorem for Surfaces whose First Variation with respect to a Parametric Elliptic Integrand is Controlled

WILLIAM K. ALLARD

0. Introduction. The following question is fundamental in the higher dimensional parametric calculus of variations: *Suppose $C = \{(x, y) \in \mathbf{R}^n \times \mathbf{R}^N: |x| < 2\}$ and V is an n-dimensional integral varifold in C with small tilt excess which has small first variation in C with respect to a parametric integrand which is nearly translation invariant. Is it true that, given an a priori bound on the n-area of V, the n-area of V inside $\{(x, y): |x| < 1\}$ is nearly an integer times the area of the unit n-disc?* We answer this question affirmatively in this paper. In fact, we show in 2.2 that *if W is the varifold projection on $\mathbf{R}^n$ of V inside $\{(x, y): |x| < 1\}$, the total variation of W minus the varifold associated to an integer multiple of the unit disc in $\mathbf{R}^n$ is small.* I call such a result an *integrality theorem.* It holds for any parametric integrand; ellipticity is not required. The key lemma used in the proof of the above integrality theorem may be roughly stated as follows: *Suppose u is a nonnegative measure on $\{x \in \mathbf{R}^n: |x| < 2\}$ whose total mass is not too large and the flat norm of the gradient of which is small. Then there is a nonnegative constant c such that the total variation of u minus c times Lebesgue measure is small on $\{x: |x| < 1\}$.* The proof of this lemma depends on the weak $(1,1)$-inequality for singular integrals as stated, for example, on p. 33 of [**ES**]. I use these results in 2.6 to prove a Lipschitz approximation theorem for general integrands which may be considered a generalization of a similar theorem in [**AW1**, 8.12]. This approximation theorem is the basis for the Regularity Theorem of 3.6 which extends the regularity results of [**AW1**] to "nearly all" elliptic integrands in codimension one. We feel that our methods should yield regularity results for all elliptic integrands in codimension one and for a large and naturally defined class of integrands in

---

1980 *Mathematics Subject Classification.* Primary 49F20; Secondary 49F22, 28A75.

© 1986 American Mathematical Society
0082-0717/86 $1.00 + $.25 per page

higher codimension. In particular, we believe the conditions 3.6(1)–(2) are unnecessary. In order to use the method of attack in [**AW1**] to prove regularity theorems it seems necessary to establish a *height bound* as in 3.5. In the present paper this is accomplished by using barriers and this forces upon us the conditions 3.6(1)–(2). We feel one ought to be able to establish such a height bound without the use of barriers. It would suffice to establish a lower bound for mass ratios as in [**AW1**, 5.1]. These remain outstanding problems.

Conversations with F. J. Almgren, Jr. were helpful in carrying out this work.

0.1. *Notation.* Throughout this paper, n and N are positive integers and $n \geq 2$. We adopt the notation of [**FE**]; note the detailed glossary in the back of this book. We adopt the notation of [**AW1**]; note the definitions of the various varifold spaces in [**AW1**, 3]. In particular, $\mathbf{G}(n + N, n)$ is the Grassmann manifold of n-dimensional linear subspaces of $\mathbf{R}^{n+N}$; as is done in [**AW1**], we frequently identify $S \in \mathbf{G}(n + N, n)$ with orthogonal projection of $\mathbf{R}^{n+N}$ onto S. We let $\mathbf{P}$ be the projection of $\mathbf{R}^{n+N} \times \mathbf{G}(n + N, n)$ on its first factor $\mathbf{R}^{n+N}$. We let $\mathbf{p}$ be projection of $\mathbf{R}^n \times \mathbf{R}^N$ on its first factor $\mathbf{R}^n$. We will frequently identify $\mathbf{R}^{n+N}$ with $\mathbf{R}^n \times \mathbf{R}^N$, writing (x, y) for a variable point in $\mathbf{R}^n \times \mathbf{R}^N$ and writing z for a variable point in $\mathbf{R}^{n+N}$. We let $\mathbf{U}(a, r) = \{x \in \mathbf{R}^n : |x - a| < r\}$ whenever $a \in \mathbf{R}^n$ and $0 < r < \infty$. We let $e_1, \ldots, e_{n+N}$ be the standard basis for $\mathbf{R}^{n+N}$ and we let $e^1, \ldots, e^{n+N}$ be the standard basis for the dual of $\mathbf{R}^{n+N}$.

1. The Strong Constancy Lemma. For $u \in \mathscr{D}(\mathbf{R}^n)$ and $x \in \mathbf{R}^n$ we set

$$
Gu(x) = \begin{cases} \dfrac{1}{n\alpha(n)} \int \log|x - y|u(y)\, d\mathscr{L}^n y & \text{if } n = 2, \\[3mm] \dfrac{2 - n}{\alpha(n)} \int |x - y|^{2-n} u(y)\, d\mathscr{L}^n y & \text{if } n > 2; \end{cases}
$$

thus, G is a linear map from $\mathscr{D}(\mathbf{R}^n)$ to $\mathscr{E}(\mathbf{R}^n)$. As is well known,

$$
(1) \qquad\qquad G\Delta u = u \quad \text{for } u \in \mathscr{D}(\mathbf{R}^n)
$$

where $\Delta = \sum_{j=1}^n D_j D_j$ is the classical Laplacian. We set

$$
Q_j u = GD_j u \quad \text{and} \quad P_{jk} u = GD_j D_k u
$$

for $u \in \mathscr{D}(\mathbf{R}^n)$ and $j, k \in \{1, \ldots, n\}$.

Suppose $j \in \{1, \ldots, n\}$ and consider Q_j. We have

$$
(2) \qquad\qquad Q_j u = GD_j u = D_j Gu = s_j * u \quad \text{for } u \in \mathscr{D}(\mathbf{R}^n),
$$

where

$$
s_j(x) = \begin{cases} \dfrac{1}{n\alpha(n)} \dfrac{x^j}{|x|^2} & \text{if } n = 2, \\[3mm] \dfrac{(2 - n)^2}{n\alpha(n)} \dfrac{x^j}{|x|^n} & \text{if } n > 2, \end{cases}
$$

for $x \in \mathbf{R}^n$. For any $\varepsilon > 0$ we may write $s_j = a_j + b_j$ where $a_j \in \mathscr{E}(\mathbf{R}^n)$ and $\int |b_j| \, d\mathscr{L}^n \leqslant \varepsilon$. It follows that if $u_1, u_2, u_3, \ldots$ is a sequence in $\mathscr{D}(\mathbf{R}^n)$ such that $\sup\{\int |u_\nu| \, d\mathscr{L}^n : \nu = 1, 2, 3, \ldots\} < \infty$ and K is a compact subset of $\mathbf{R}^n$ then there is an $\mathscr{L}^n$-summable $\mathbf{R}$-valued function l on K such that

$$\liminf_{\nu \to \infty} \int_K |Q_j u_\nu - l| \, d\mathscr{L}^n = 0.$$

Suppose $j, k \in \{1, \ldots, n\}$ and consider P_{jk}. We assert that

$$(3) \qquad \mathscr{L}^n\left\{ x \in \mathbf{R}^n : |P_{jk} u(x)| \geqslant \alpha \right\} \leqslant \frac{C}{\alpha} \int |u| \, d\mathscr{L}^n$$

whenever $u \in \mathscr{D}(\mathbf{R}^n)$ and $0 < \alpha < \infty$, and where C depends only on n. For the proof of this highly nontrivial fact we refer to [**SE**] as follows. On p. 75 of [**SE**] we see that

$$P_{jk} u(x) = cu(x) + \lim_{\varepsilon \downarrow 0} \int_{\{y \in \mathbf{R}^n : |y| \geqslant \varepsilon\}} \frac{\Omega(y)}{|y|^n} u(x - y) \, d\mathscr{L}^n y$$

whenever $u \in \mathscr{D}(\mathbf{R}^n)$ and $x \in \mathbf{R}^n$, and where $c \in \mathbf{R}$ and $\Omega \colon \mathbf{R}^n \sim \{0\} \to \mathbf{R}$ is smooth and positively homogeneous of degree 0 and has mean value zero on $\mathbf{S}^{n-1}$. An estimate like (3) is made on p. 33, line 3 under the hypotheses of Theorem 1 on p. 29. The present Ω does not satisfy the hypotheses of Theorem 1. However, as Stein points out in the two paragraphs after the statement of Theorem 1 on p. 29, the part of the hypotheses that the present Ω does not satisfy is not important and is made for short-term technical convenience. To complete the proof of (3) look at Theorem 2 on p. 35, especially at 3.4 on pp. 37–38.

(4) THE STRONG CONSTANCY LEMMA. *Suppose*
(a) $a \in \mathbf{R}^n$ and $0 < r < \infty$;
(b) $u \in \mathscr{D}'(\mathrm{U}(a, r)), 0 \leqslant M < \infty$ and

$$0 \leqslant u(\varphi) \leqslant Mr^n \sup\{\varphi(x) : x \in \mathrm{U}(a, r)\}$$

whenever $\varphi \in \mathscr{D}(\mathrm{U}(a, r))$ and $\varphi \geqslant 0$;
(c) $f_j, X_j^i \in \mathscr{D}'(\mathrm{U}(a, r))$ for $i, j \in \{1, \ldots, n\}, 0 \leqslant \delta < \infty$ and

$$\left| X_j^i(\varphi) \right| + \left| r f_j(\varphi) \right| \leqslant \delta r^n \sup\{|\varphi(x)| : x \in \mathrm{U}(a, r)\}$$

whenever $\varphi \in \mathscr{D}(\mathrm{U}(a, r))$;
(d) $X_j = (X_j^1, \ldots, X_j^n)$ for $j \in \{1, \ldots, n\}$;
(e) $D_j u = \operatorname{div} X_j + f_j$ for $j \in \{1, \ldots, n\}$ and
(f) $0 < \lambda < 1$ and $0 < \varepsilon < \infty$.
There is $\delta_1 = \delta_1(M, \lambda, \varepsilon)$ such that if $\delta \leqslant \delta_1$ then there is c with $0 \leqslant c < \infty$ such that
(g) $r^{-n} |u(\varphi) - c \int_{\mathrm{U}(a, r)} \varphi \, d\mathscr{L}^n| \leqslant \varepsilon \sup\{|\varphi(x)| : x \in \mathrm{U}(a, r)\}$
whenever $\varphi \in \mathscr{D}(\mathrm{U}(a, r))$ and $\operatorname{spt} \varphi \subset \mathrm{U}(a, \lambda r)$.

PROOF. We may assume $a = 0$ and $r = 1$. Regularizing if necessary by non-negative approximations to the identity with small supports, we see that we may assume u, X_j and f_j, $j \in \{1,\ldots,n\}$, are smooth.

Were the lemma false, there would be M, λ and ε such that $0 \leqslant M < \infty$, $0 < \lambda < 1$ and $0 < \varepsilon < \infty$; a sequence of positive real numbers μ_1, μ_2, $\mu_3,\ldots$ with limit 0; and sequences

$$u_\nu \in \mathscr{E}(\mathbf{U}(0,1)), \quad X_{\nu,j} \in \mathscr{E}(\mathbf{U}(0,1)) \quad \text{and} \quad f_{\nu,j} \in \mathscr{E}(\mathbf{U}(0,1)),$$

$\nu = 1,2,3,\ldots$, $j \in \{1,\ldots,n\}$, such that

$$(5) \qquad\qquad u_\nu \geqslant 0 \quad \text{and} \quad \int_{\mathbf{U}(0,1)} u_\nu \, d\mathscr{L}^n \leqslant M,$$

$$(6) \qquad\qquad \int_{\mathbf{U}(0,1)} |X_{\nu,j}| + |f_{\nu,j}| \, d\mathscr{L}^n \leqslant \mu_\nu$$

and

$$(7) \qquad\qquad D_j u_\nu = \operatorname{div} X_{\nu,j} + f_{\nu,j},$$

whenever $\nu = 1,2,3,\ldots$ and $j \in \{1,\ldots,n\}$ but such that for each c with $0 \leqslant c < \infty$ and each $\nu = 1,2,3,\ldots$ we have

$$(8) \qquad\qquad \int_{\mathbf{U}(0,\lambda)} |u_\nu - c| \, d\mathscr{L}^n > \varepsilon.$$

Let $\chi \in \mathscr{D}(\mathbf{U}(0,1))$ be such that $0 \leqslant \chi \leqslant 1$ and $\chi(x) = 1$ for $x \in \mathbf{U}(0,\lambda)$. For $\nu = 1,2,3,\ldots$ and $j \in \{1,\ldots,n\}$, let

$$v_\nu = \chi u_\nu, \qquad Y_{\nu,j} = \chi X_{\nu,j}$$

and

$$g_{\nu,j} = (D_j\chi)u_\nu - \operatorname{grad} \chi \cdot X_{\nu,j} + \chi f_{\nu,j}$$

on $\mathbf{U}(0,1)$ and let them be zero elsewhere. It follows from (7) that

$$D_j v_\nu = \operatorname{div} Y_{\nu,j} + g_{\nu,j}, \qquad j \in \{1,\ldots,n\},$$

so that

$$(9) \qquad\qquad \Delta v_\nu = \sum_{j=1}^{n} D_j \operatorname{div} Y_{\nu,j} + \sum_{j=1}^{n} D_j g_{\nu,j}$$

for $\nu = 1,2,3,\ldots$. Setting

$$Z_\nu = \sum_{j=1}^{n} G D_j \operatorname{div} Y_{\nu,j} \quad \text{and} \quad h_\nu = \sum_{j=1}^{n} G D_j g_{\nu,j}$$

for $\nu = 1,2,3,\ldots$, we infer from (1) and (9) that

$$(10) \qquad\qquad v_\nu = Z_\nu + h_\nu, \qquad \nu = 1,2,3,\ldots.$$

Passing to a subsequence if necessary, we infer from (5)–(7) and the constancy theorem for distributions that there is a c such that $0 \leqslant c < \infty$ and such that

$$(11) \quad \lim_{\nu \to \infty} \int_{\mathbf{U}(0,1)} u_\nu \varphi \, d\mathscr{L}^n = c \int_{\mathbf{U}(0,1)} \varphi \, d\mathscr{L}^n \quad \text{for each } \varphi \in \mathscr{D}(\mathbf{U}(0,1)).$$

It follows from (6) that

$$(12) \qquad \lim_{\nu \to \infty} \int \varphi Z_\nu \, d\mathscr{L}^n = \lim_{\nu \to \infty} - \sum_{j=1}^{n} \int \operatorname{grad} D_j G\varphi \cdot Y_{\nu, j} \, d\mathscr{L}^n$$

$$= 0 \quad \text{whenever } \varphi \in \mathscr{D}(\mathbf{R}^n).$$

Using (5) and our observations about the operators Q_j, $j \in \{1, \ldots, n\}$, and passing again to a subsequence if necessary, we obtain an $\mathscr{L}^n$-summable real-valued function l on spt χ such that

$$(13) \qquad \lim_{\nu \to \infty} \int_{\operatorname{spt} \chi} |h_\nu - l| \, d\mathscr{L}^n = 0.$$

It follows from (11)–(13) that

$$(14) \qquad l(x) = c\chi(x) \quad \text{for } \mathscr{L}^n \text{ almost all } x \in \operatorname{spt} \chi.$$

Most importantly, we infer from (3) and (6) that

$$(15) \qquad \mathscr{L}^n\{x \in \mathbf{R}^n : |Z_\nu(x)| \geq \alpha\} \leq \frac{Cn^2}{\alpha} \sum_{j=1}^{n} \int_{\mathbf{U}(0,1)} |X_{\nu, j}| \, d\mathscr{L}^n$$

$$\leq \frac{Cn^2}{\alpha} n\mu_\nu$$

for $\nu = 1, 2, 3, \ldots$ and $0 < \alpha < \infty$.

Now suppose $0 < \eta < \infty$. For each $\nu = 1, 2, 3, \ldots$ let $G_\nu = \{x \in \mathbf{R}^n : |Z_\nu(x)| < \eta\}$ and let $B_\nu = \{x \in \mathbf{R}^n : |Z_\nu(x)| \geq \eta\}$. From (15) we infer that

$$(16) \qquad \lim_{\nu \to \infty} \mathscr{L}^n(B_\nu) = 0.$$

We have

$$\int_{B_\nu} v_\nu \, d\mathscr{L}^n = \int \chi(u_\nu - c) \, d\mathscr{L}^n - \int_{G_\nu} \chi(u_\nu - c) \, d\mathscr{L}^n + c\int_{B_\nu} \chi \, d\mathscr{L}^n.$$

Using (11), (16) and the fact that $v_\nu \geq 0$ we infer that

$$(17) \qquad \limsup_{\nu \to \infty} \int_{B_\nu} |v_\nu| \leq \eta \mathscr{L}^n(\operatorname{spt} \chi).$$

Finally, we estimate

$$\int_{\mathbf{U}(0,1)} |v_\nu - c\chi| \, d\mathscr{L}^n \leq \int_{G_\nu} |Z_\nu| + |h_\nu - c\chi| \, d\mathscr{L}^n + \int_{B_\nu} |v_\nu| + c|\chi| \, d\mathscr{L}^n$$

so that, by virtue of (13), (14) and (16),

$$\limsup_{\nu \to \infty} \int_{\mathbf{U}(0,1)} |v_\nu - c\chi| \, d\mathscr{L}^n \leq 2\eta \mathscr{L}^n(\operatorname{spt} \chi).$$

Taking η so that $2\eta \mathscr{L}^n(\operatorname{spt} \chi) < \varepsilon$ we contradict (8). $\qquad \square$

2. Some integrality theorems and a Lipschitz approximation theorem. Our object here is to prove for an arbitrary parametric integrand some integrality theorems which may be thought of as analogues of some of the results of [**AW1**, 6] and a Lipschitz approximation theorem which may be thought of as an analogue of [**AW1**, 8.12]. The results cited from [**AW1**] apply to the area integrand and their proofs depend on the monotonicity results of [**AW1**, 5]. The results which follow are far more general and rest instead on the strong constancy lemma of 1. All these results are based on rather simple geometric considerations.

2.1. *The integrand.* **For the remainder of this paper we fix an open subset** Z **of** $\mathbf{R}^{n+N}$ **and a smooth map**

$$\Phi: Z \times \mathbf{G}(n + N, n) \to \{t \in \mathbf{R}: t > 0\}.$$

We define the smooth mappings

$$\Phi_{\text{trans}}: Z \times \mathbf{G}(n + N, n) \to \mathbf{R}^{n+N},$$

$$\Phi_{\text{spin}}: Z \times \mathbf{G}(n + N, n) \to \text{Hom}(\mathbf{R}^{n+N}, \mathbf{R}^{n+N})$$

at $(z, S) \in Z \times \mathbf{G}(n + N, n)$ by requiring that

$$(1) \qquad (d/dt)\Phi(z + tv, S)\big|_{t=0} = v \cdot \Phi_{\text{trans}}(z, S) \quad \text{for } v \in \mathbf{R}^{n+N}$$

and that

$$(2) \qquad \Phi_{\text{spin}}(z, S) = S^{\perp} \circ \Phi_{\text{spin}}(z, S) \circ S$$

and

(3)

$$(d/dt)\Phi\big(z, (\mathbf{1}_{\mathbf{R}^{n+N}} + tL)(S)\big)\big|_{t=0} = L \cdot \Phi_{\text{spin}}(z, S) \text{ for } L \in \text{Hom}(\mathbf{R}^{n+N}, \mathbf{R}^{n+N}).$$

That (2), (3) serve to define Φ_{spin} follows from the fact that

$$(4) \qquad (d/dt)(\mathbf{1}_{\mathbf{R}^{n+N}} + tL)(S)\big|_{t=0} = S^{\perp} \circ L \circ S + (S^{\perp} \circ L \circ S)^{*}$$

whenever $S \in \mathbf{G}(n + N, n)$ and $L \in \text{Hom}(\mathbf{R}^{n+N}, \mathbf{R}^{n+N})$.

For each $(z, S) \in Z \times \mathbf{G}(n + N, n)$ we set

$$\delta(z, S; \Phi)(v, L) = v \cdot \Phi_{\text{trans}}(z, S) + L \cdot \big[\Phi(z, S)S + \Phi_{\text{spin}}(z, S)\big]$$

$$\text{for } (v, L) \in \mathbf{R}^{n+N} \oplus \text{Hom}(\mathbf{R}^{n+N}, \mathbf{R}^{n+N}).$$

For each n-dimensional varifold $V \in \mathbf{V}_n(Z)$ we proceed as in [**AW1**, 4] and define the linear map

$$\delta(V, \Phi): \mathscr{X}(Z) \to \mathbf{R},$$

called the **first variation of** V **with respect to** Φ, by setting

$$\delta(V, \Phi)(g) = \int \delta(z, S; \Phi)(g(z), Dg(z)) \, dV(z, S)$$

for $g \in \mathscr{X}(Z)$. This definition is motivated as follows: Fix $V \in \mathbf{V}_n(Z)$ and $g \in \mathscr{X}(Z)$; suppose $0 < \varepsilon < \infty$ and $h: (-\varepsilon, \varepsilon) \times Z \to Z$ is smooth and is such that

$$h_0(z) = z \text{ and } (d/dt)h_t(z)|_{t=0} = g(z) \text{ for each } z \in Z \text{ and } h_t$$
carries Z diffeomorphically onto Z for each $t \in (-\varepsilon, \varepsilon),$

where we have set $h_t(z) = h(t, z)$ for $(t, z) \in (-\varepsilon, \varepsilon) \times Z$. Then for each compact subset K of Z with spt $g \subset K$ we have

$$\frac{d}{dt} \int_{\mathbf{P}^{-1}[h_t[K]]} \Phi \, d\left(h_{t\#}V \right) \bigg|_{t=0}$$

$$= \frac{d}{dt} \int_{\mathbf{P}^{-1}[K]} \Phi\left(h_t(z), Dh_t(z)(S) \right) \left| \wedge_n \left[Dh_t(z) | S \right] \right| dV(z, S) \big|_{t=0}$$

$$= \delta(V, \Phi)(g).$$

For each $V \in \mathbf{V}_n(Z)$ and each open subset G of Z we let

$$\|\delta(V, \Phi)\|(G) = \sup\left\{ |\delta(V, \Phi)(g)| : g \in \mathscr{X}(Z), \text{ spt } g \subset G \right.$$

$$\left. \text{and } |g(z)| \leqslant 1 \text{ for } z \in Z \right\}.$$

In case $\|\delta(V, \Phi)\|(G) < \infty$ whenever G is a precompact open subset of Z, $\|\delta(V, \Phi)\|$ extends to a Radon measure on Z and we say $\delta(V, \Phi)$ **is a measure.**

For the remainder of this paper let us assume that

$$1 = \inf\left\{ \Phi(z, S) : (z, S) \in Z \times \mathbf{G}(n + N, n) \right\},$$

$$M_{0,0} = \sup\left\{ \Phi(z, S) : (z, S) \in Z \times \mathbf{G}(n + N, n) \right\} < \infty,$$

$$M_{1,0} = \sup\left\{ |\Phi_{\text{trans}}(z, S)| : (z, S) \in Z \times \mathbf{G}(n + N, n) \right\} < \infty \quad \text{and}$$

$$M_{0,1} = \sup\left\{ |\Phi(z, S) - \Phi(z, T)|/\|S - T\| : \right.$$

$$\left. z \in Z; S, T \in \mathbf{G}(n + N, n) \text{ and } S \neq T \right\} < \infty.$$

Suppose $(z, S) \in Z \times \mathbf{G}(n + N, n)$. We have

$$(4) \qquad |L \cdot \Phi_{\text{spin}}(z, S)| = \left| \lim_{t \downarrow 0} t^{-1}\left[\Phi\left(z, (\mathbf{1}_{\mathbf{R}^{n+N}} + tL)(S) \right) - \Phi(z, S) \right] \right|$$

$$\leqslant M_{0,1} \lim_{t \downarrow 0} |t|^{-1} \left\| (\mathbf{1}_{\mathbf{R}^{n+N}} + tL)(S) - S \right\|$$

$$= M_{0,1} \| S^{\perp} \circ L \circ S \| \quad \text{for } L \in \text{Hom}(\mathbf{R}^{n+N}, \mathbf{R}^{n+N}).$$

2.2. FIRST INTEGRALITY THEOREM. *Suppose*

(1) *$l \in \text{Hom}(\mathbf{R}^n, \mathbf{R}^N)$, $U = \{(x, l(x)): x \in \mathbf{R}^n\}$, $\theta = |\wedge_n[\mathbf{p}|U]|^{-1}$ and $q(x, y)$*
$= y - l(x)$ for $(x, y) \in \mathbf{R}^n \times \mathbf{R}^N$;

(2) *$c = (a, b) \in Z$, $0 < r < \infty$, $0 < s < \infty$, $Q = \{(x, y): |q(x - a, y - b)| \leqslant s\}$ and $Q \cap \mathbf{p}^{-1}[\mathbf{U}(a, r)] \subset Z$;*

(3) *V is an n-dimensional integral varifold in Z and*

$$\text{spt}\|V\| \cap \mathbf{p}^{-1}[\mathbf{U}(a, r)] \cap Q;$$

(4) $0 \leqslant A < \infty$ and $\|l\| \leqslant A$;

(5) $0 \leqslant M < \infty$ and $r^{-n}\|V\|(Z \cap \mathbf{p}^{-1}[\mathbf{U}(a, r)]) \leqslant M$;

(6) $0 < \lambda < 1$;

(7) $0 < \varepsilon < 1$;

(8) $E = (A, M, \lambda, \varepsilon, M_{0,0}, M_{0,1})$ and

(9) $0 \leqslant \delta < \infty$,

(a) $\max\{1, M_{1,0}\}\max\{r, s\} \leqslant \delta$;

(b) $r^{-n+1}\|\delta(V, \Phi)\|(Z \cap \mathbf{p}^{-1}[\mathbf{U}(a, r)]) \leqslant \delta$ and

(c) $r^{-n}\int_{(\mathbf{p} \circ \mathbf{P})^{-1}[\mathbf{U}(a, r)]}\|S - U\|\, dV(z, S) \leqslant \delta$.

There is a real positive real number $\delta_2 = \delta_2(E)$ *such that if* $\delta \leqslant \delta_2$ *then there is a nonnegative integer* μ *such that*

(10) $(\lambda r)^{-n}|\mathbf{p}_{\#}\|V\| - \mu\theta L^n(B)| \leqslant \varepsilon$ *for any Borel subset* B *of* $\mathbf{U}(a, \lambda r)$.

REMARK. Note that $\varepsilon < \theta\alpha(n)/2$ implies μ is unique.

PROOF. For each $j = 1, \ldots, n$ let $v_j = (e_j, l(e_j)) \in U$. For each $\psi \in \mathcal{D}(\mathbf{U}(a, r))$ and each $i, j \in \{1, \ldots, n\}$ we let

$$u(\psi) = \Phi(c, U) \int \psi \circ \mathbf{p}\, d\|V\|,$$

$$X_j^i(\psi) = -\int \psi(\mathbf{p}(z))\delta_j^i[\Phi(z, S) - \Phi(c, U)]$$

$$+ \psi(\mathbf{p}(z))e^i v_j \cdot \big[(S - U)\Phi(z, S) + \Phi_{\mathrm{spin}}(z, S)\big]\, dV(z, S),$$

$$X_j = \sum_{i=1}^{n} X_j^i e_i,$$

$$f_j(\psi) = -\delta(\Phi, V)\big((\psi \circ \mathbf{p})v_j\big) + \int \psi(\mathbf{p}(z))v_j \cdot \Phi_{\mathrm{trans}}(z, S)\, dV(z, S);$$

note that $\delta(\Phi, V)((\psi \circ \mathbf{p})v_j)$ has a natural definition because $\mathrm{spt}\|V\| \cap \mathbf{p}^{-1}[\mathbf{U}(a, r)] \subset Z$. For each $\varphi \in \mathcal{D}(\mathbf{U}(a, r))$ we calculate

$$\big(D_j u - \mathrm{div}\, \mathbf{X}_j - f_j\big)(\varphi)$$

$$= -u(D_j\varphi) + \sum_{i=1}^{n} X_j^i(D_i\varphi) - f_j(\varphi)$$

$$= -\int D_j\varphi(\mathbf{p}(z))\Phi(c, U)\, dV(z, S) - \int D_j\varphi(\mathbf{p}(z))[\Phi(z, S) - \Phi(c, U)]$$

$$+ D\varphi(\mathbf{p}(z)) \circ \mathbf{p}v_j \cdot \big[(S - U)\Phi(z, S) + \Phi_{\mathrm{spin}}(z, S)\big]\, dV(z, S)$$

$$+ \delta(V, \Phi)\big((\varphi \circ \mathbf{p})v_j\big) - \int \varphi(\mathbf{p}(z))v_j \cdot \Phi_{\mathrm{trans}}(z, S)\, dV(z, S)$$

$$= 0$$

where we have used the fact that

$$D\varphi(\mathbf{p}(z)) \circ \mathbf{p}v_j \cdot U = D_j\varphi(\mathbf{p}(z))$$

for $z \in \mathbf{U}(a, r)$ and $j \in \{1, \ldots, n\}$. Given $\psi \in \mathscr{D}(\mathbf{U}(a, r))$, we set $\sigma = \sup\{|\psi(x)|: x \in \mathbf{U}(a, r)\}$ and estimate

$$r^{-n}|u(\psi)| \leqslant M_{0,0}M\sigma,$$

$$r^{-n}|X_j^i(\psi)| \leqslant (2 + A)\delta M\sigma + (1 + A^2)^{1/2}(M_{0,0} + M_{0,1})\delta\sigma \quad \text{and}$$

$$r^{-n+1}|f_j(\psi)| \leqslant (1 + A^2)^{1/2}[\delta\sigma + \delta M\sigma].$$

Thus, by making δ smaller than a positive real number depending only on E, we infer from the Strong Constancy Lemma of 1 that there is a nonnegative real number γ such that

$$(11) \qquad (\lambda r)^{-n}\left|\int \tau \, d\left(\mathbf{p}_\#\|V\| - \gamma\theta\mathscr{L}^n\right)\right| \leqslant \varepsilon/2$$

for any Borel function $\tau: \mathbf{U}(a, \lambda r) \to [-1, 1]$.

For each $x \in \mathbf{U}(a, r)$ let $X(x)$ be the set of those $z \in Z$ such that $\mathbf{p}(z) = x$, $\Theta^n(\|V\|, z)$ is a positive integer, $\mathrm{Tan}^m(\|V\|, z) \in \mathbf{G}(n + N, n)$ and $\|\mathrm{Tan}^m(\|V\|, z) - U\| \leqslant 1$; here we have used the notation of [**FE**, 3.2.16]. Let C be a positive real number such that

$$\left|\,|\Lambda_n[\mathbf{p}|S]|^{-1} - |\Lambda_n[\mathbf{p}|U]|^{-1}\,\right| \leqslant C|\Lambda_n[\mathbf{p}|S]|^{-1}\|S - U\|$$

$$\text{whenever } S \in \mathbf{G}(n + N, n) \text{ and } \|S - U\| \leqslant 1.$$

Suppose $\tau: \mathbf{U}(a, \lambda r) \to [-1, 1]$ is a Borel function. Then

$$\left|\int \tau \, d\left(\mathbf{p}_\#\|V\|\right) - \theta\int \tau(x) \sum_{z \in X(x)} \Theta^n(\|V\|, z) \, d\mathscr{L}^n x\right|$$

$$\leqslant \left|\int \tau(x) \sum_{z \in X(x)} \Theta^n(\|V\|, z)\left\{|\Lambda_n[\mathbf{p}|\mathrm{Tan}^m(\|V\|, z)]|^{-1} - |\Lambda_n[\mathbf{p}|U]|^{-1}\right\} d\mathscr{L}^n x\right|$$

$$+ \left|\int_{\{(z, S): \|S - U\| > 1\}} \tau(\mathbf{p}(z)) \, dV(z, S)\right|$$

$$\leqslant C\left|\int \tau(x) \sum_{z \in X(x)} \Theta^n(\|V\|, z)|\Lambda_n[\mathbf{p}|\mathrm{Tan}^m(\|V\|, z)]|^{-1}\right|$$

$$\|\mathrm{Tan}^m(\|V\|, z) - U\| d\mathscr{L}^n x\Big| + \left|\int_{\{(z, S): \|S - U\| > 1\}} \tau(\mathbf{p}(z)) \, dV(z, S)\right|$$

$$\leqslant \max\{C, 1\}\int |\tau(\mathbf{p}(z))|\,\|S - U\| dV(z, S)$$

$$\leqslant \max\{C, 1\}\delta r^n.$$

We complete the proof by combining this inequality with (11). $\qquad \square$

2.3. What follows here is a generalization of [**AW**, 4.10] and is preparatory to 2.4. *Suppose $V \in \mathbf{V}_n(Z)$, G is an open subset of Z and $\|\delta(V, \Phi)\|(G) < \infty$. Suppose m is a positive integer and $f = (f^1, \ldots, f^m): Z \to \mathbf{R}^m$ is smooth and $0 < \nu < \infty$.*

Let $T = \{ t \in \mathbf{R}^m : 0 < t^i \leqslant \nu \text{ for } i \in \{1,\ldots,m\} \}$ and $\Gamma = \{ \nu\beta : \beta \in \mathbf{Z}^m \}$. We will show that

(1)

$$\nu^{-m} \int_T \sum_{\gamma \in \Gamma} \left\| \delta\left(V \llcorner (f \circ \mathbf{P})^{-1}[\gamma + t + T], \Phi \right) \right\|(G) \, d\mathscr{L}^m t$$

$$\leqslant \| \delta(V, \Phi) \|(G) + 2(M_{0,0} + M_{0,1}) \nu^{-1} \sum_{i=1}^m \int |S[\operatorname{grad} f^i(z)]| \, dV(z, S).$$

For each $i \in \{1,\ldots,m\}$ let $P^i = \{ t \in T : t^i = 0 \}$ and let

$$\mu^i(B) = \int_{\{(z, S) : f(z) \in B\}} |S[\operatorname{grad} f^i(z)]| \, dV(z, S)$$

for each Borel subset B of $\mathbf{R}^m$. Suppose $u \in \mathbf{R}^m$; we have

(2)
$$\left\| \delta\left(V \llcorner (f \circ \mathbf{P})^{-1}[u + T], \Phi \right) \right\|(G)$$

$$\leqslant \| \delta(V, \Phi) \|\left(f^{-1}[u + T] \cap G \right)$$

$$+ (M_{0,0} + M_{0,1}) \sum_{i=1}^m \limsup_{h \downarrow 0} \mu^i\big[(u + he_i + T) \sim (u + T)\big]/h.$$

To prove this, we first observe that

(3)
$$\int \zeta(z) \delta(z, S; \Phi)(g(z), Dg(z)) \, dV(z, S) = \delta(V, \Phi)(\zeta g)$$

$$- \int D\zeta(z) g(z) \cdot S \, \Phi(z, S) + D\zeta(z) g(z) \cdot \Phi_{\mathrm{spin}}(z, S) \, dV(z, S)$$

whenever $\zeta : Z \to \mathbf{R}$ is smooth and $g \in \mathscr{X}(Z)$. For each $i \in \{1,\ldots,m\}$ and each h with $0 < h < \nu$ we let

$$\psi_h^i(\tau) = \begin{cases} 0 & \text{if } \tau \leqslant u^i, \\ (\tau - u^i)/h & \text{if } u^i < \tau \leqslant u^i + h, \\ 1 & \text{if } u^i + h < \tau \leqslant u^i + \nu, \\ (u^i + h - \tau)/h & \text{if } u^i + \nu < \tau \leqslant u^i + \nu + h, \\ 0 & \text{if } u^i + \nu + h < \tau. \end{cases}$$

Given h with $0 < h < \nu$, we infer that

$$\left| \int \prod_{i=1}^m \psi_h^i \circ f^i(z) \delta(z, S; \Phi)(g(z), Dg(z)) \, dV(z, S) \right|$$

$$\leqslant \int_G \sum_{i=1}^m \psi_h^i \circ f^i d\| \delta(V, \Phi) \|$$

$$+ (M_{0,0} + M_{0,1}) \sum_{i=1}^m \mu^i\big[(u + he_i + T) \sim (u + T)\big]/h$$

whenever $g \in \mathscr{X}(Z)$, spt $g \subset G$ and $|g| \leqslant 1$ by letting ζ in (3) be a suitable approximation to $\prod_{i=1}^{m} \psi_h^i \circ f^i$. Now (2) follows by letting $h \downarrow 0$.

Using (2) we obtain

$$\int_T \sum_{\gamma \in \Gamma} \left\| \delta\left(V \llcorner (f \circ \mathbf{P})^{-1}[\gamma + t + T], \Phi \right) \right\|(G) \, d\mathscr{L}^m t$$

$$= \sum_{\gamma \in \Gamma} \int_T \left\| \delta\left(V \llcorner (f \circ \mathbf{P})^{-1}[\gamma + t + T], \Phi \right) \right\|(G) \, d\mathscr{L}^m t$$

$$\leqslant \sum_{\gamma \in \Gamma} \int_T \|\delta(V, \Phi)\| \left(f^{-1}[\gamma + t + T] \cap G \right) d\mathscr{L}^m t + \left(M_{0,0} + M_{0,1} \right)$$

$$\cdot \sum_{i=1}^{m} \sum_{\gamma \in \Gamma} \int_T \limsup_{h \downarrow 0} \mu^i \left[(\gamma + t + he_i + T) \sim (\gamma + t + T) \right] / h \, d\mathscr{L}^m t.$$

But

$$\sum_{\gamma \in \Gamma} \int_T \|\delta(V, \Phi)\| \left(f^{-1}[\gamma + t + T] \cap G \right) d\mathscr{L}^m t$$

$$= \int_T \sum_{\gamma \in \Gamma} \|\delta(V, \Phi)\| \left(f^{-1}[\gamma + t + T] \cap G \right) d\mathscr{L}^m t$$

$$= \int_T \|\delta(V, \Phi)\|(G) \, d\mathscr{L}^m t$$

$$= \nu^m \|\delta(V, \Phi)\|(G)$$

and

$$\sum_{\gamma \in \Gamma} \int_T \limsup_{h \downarrow 0} \mu^i \left[(\gamma + t + he_i + T) \sim (\gamma + t + T) \right] d\mathscr{L}^m t$$

$$\leqslant \sum_{\gamma \in \Gamma} \int_{P^i} \mu^i \left[(\gamma + s + T) \cup (\gamma + s + \nu e_i + T) \right] d\mathscr{H}^{m-1} s$$

$$= \int_{P^i} \sum_{\gamma \in \Gamma} \mu^i \left[(\gamma + s + T) \cup (\gamma + s + \nu e_i + T) \right] d\mathscr{H}^{m-1} s$$

$$= \int_{P^i} 2\mu^i(\mathbf{R}^m) \, d\mathscr{H}^{m-1} = 2\nu^{m-1} \mu^i(\mathbf{R}^m),$$

so (1) is proved. $\quad\square$

2.4. **Second Integrality Theorem.** *Suppose (1)–(9) of 2.2 hold and* $0 < \eta < \infty$. *There is a positive real number* $\delta_3 = \delta_3(E, \eta)$ *such that if* $\delta \leqslant \delta_3$ *and there is a positive integer* μ *such that*

(1) $$\mu\alpha(n)\theta - \varepsilon \leqslant (\lambda r)^{-n} \|V\|\mathbf{p}^{-1}[\mathbf{U}(a, \lambda r)] \leqslant \mu\alpha(n)\theta + \varepsilon$$

then there are a positive integer K *and for each* $k \in \{1, \ldots, K\}$ *a positive integer* μ_k *and a point* $b_k \in \mathbf{R}^N$ *such that, if*

$$C_k = \bigcap_{i=1}^{N} \left\{ (x, y) : 0 < q^i(x - a, y - b_k) \leqslant \eta r \right\} \quad \text{for } k \in \{1, \ldots, K\},$$

then

(2)
$$\mu = \sum_{k=1}^{K} \mu_k,$$

(3) $\{C_1, \ldots, C_K\}$ *is disjointed*

and

(4)
$$(\lambda r)^{-n}\left\{ \sum_{k=1}^{K} \left| \|V\|(\mathbf{p}^{-1}[B] \cap C_k) - \mu_k \theta \mathscr{L}^n(B)\right| \right.$$
$$\left. + \|V\|\left(\mathbf{p}^{-1}[\mathbf{U}(a, \lambda r)] \sim \bigcup_{k=1}^{K} C_k\right)\right\} \leqslant \varepsilon$$

whenever B is a Borel subset of $\mathbf{U}(a, \lambda r)$.

PROOF. Note that $|S[\operatorname{grad} q^i(z)]| \leqslant (1 + A^2)^{1/2}\|S - U\|$ whenever $(z, S) \in \mathbf{R}^{n+N} \times \mathbf{G}(n + N, n)$ and $i \in \{1, \ldots, n\}$. Applying 2.3 with G there equal to $Z \cap \mathbf{p}^{-1}[\mathbf{U}(a, r)]$, with v there equal to ηr and with f there equal to q we secure $t \in T$ such that

$$r^{-n+1} \sum_{\gamma \in \Gamma} \left\|\delta\left(V \llcorner (q \circ \mathbf{P})^{-1}[\gamma + t + T], \Phi\right)\right\|(G)$$

$$\leqslant r^{-n+1}\|\delta(V, \Phi)\|(G)$$

$$+ 2(M_{0,0} + M_{0,1})\eta^{-1}r^{-n}(1 + A^2)^{1/2}N \int_{(\mathbf{p} \circ \mathbf{P})^{-1}[\mathbf{U}(a, r)]} \|S - U\| dV(z, S)$$

$$\leqslant \delta + 2(M_{0,0} + M_{0,1})\eta^{-1}(1 + A^2)^{1/2}N\delta.$$

Suppose $0 < \omega < \theta\alpha(n)/3$. By assuming that δ does not exceed a positive number which depends only on E, η and ω, we may use 2.2 to choose, for each nonempty subset Λ of Γ, a nonnegative integer μ_Λ such that

(5) $(\lambda r)^{-n}\left|\mathbf{p}_\# \|V_\Lambda\| - \mu_\Lambda \theta\mathscr{L}^n(B)\right| \leqslant \omega$ for each Borel subset B of $\mathbf{U}(a, \lambda r)$,

where we have set $V_\Lambda = \sum_{\gamma \in \Lambda} V \llcorner (q \circ \mathbf{P})^{-1}[\gamma + t + T]$. Let $\Upsilon = \{\gamma \in \Gamma : \mu_\gamma > 0\}$. Suppose Λ is a nonempty subset of $\Gamma \sim \Upsilon$, $\mu_\Lambda = 0$ and $\gamma \in \Gamma \sim (\Upsilon \cup \Lambda)$. We have

$$\mu_{\Lambda \cup \{\gamma\}}\theta\alpha(n) = \left|\mu_{\Lambda \cup \{\gamma\}} - \mu_\Lambda\right|\theta\alpha(n)$$

$$\leqslant (\lambda r)^{-n}\left\{\left|\mathbf{p}_\# \|V_{\Lambda \cup \{\gamma\}}\| - \mu_{\Lambda \cup \{\gamma\}}\theta\mathscr{L}^n(\mathbf{U}(a, \lambda r))\right|\right.$$

$$\left. + \mathbf{p}_\# \|V_\Lambda\|\mathbf{U}(a, \lambda r) + \mathbf{p}_\# \|V_{\{\gamma\}}\|\mathbf{U}(a, \lambda r)\right\}$$

$$\leqslant 3\omega,$$

so $\mu_{\Lambda \cup \{\gamma\}} = 0$. Thus, $\mu_\Lambda = 0$ whenever Λ is a nonempty finite subset of $\Gamma \sim \Upsilon$ from which it follows that $\mu_\Lambda = 0$ for any nonempty subset Λ of $\Gamma \sim \Upsilon$. Since

$$2\theta\alpha(n)/3 \leqslant \mu_{\{\gamma\}}\theta\alpha(n) - \omega \leqslant (\lambda r)^{-n}\|V_{\{\gamma\}}\|\mathbf{p}^{-1}[\mathbf{U}(a, \lambda r)]$$

whenever $\gamma \in \Upsilon$, we infer that

$$2\theta\alpha(n)\operatorname{card} \Upsilon/3 \leqslant \lambda^{-n}M.$$

Finally,

$$\left| \mu - \sum_{\gamma \in \Upsilon} \mu_{\{\gamma\}} \right| \theta \alpha(n)$$

$$\leq (\lambda r)^{-n} \left\{ \left| \mathbf{p}_{\#} \|V\| - \mu \theta \mathscr{L}^n(\mathbf{U}(a, \lambda r)) \right| + \mathbf{p}_{\#} \|V_{\Gamma \sim \Upsilon}\| \mathbf{U}(a, \lambda r) \right.$$

$$\left. + \sum_{\gamma \in \Upsilon} \left| \mathbf{p}_{\#} \|V_{\{\gamma\}}\| - \mu_{\{\gamma\}} \theta \mathscr{L}^n(\mathbf{U}(a, \lambda r)) \right| \right\}$$

$$\leq \varepsilon + \omega + (\operatorname{card} \Upsilon) \omega.$$

It should now be clear how to choose δ_3.

2.5. LEMMA. *Suppose* (1)–(9) *of* 2.2. *hold,* $0 \leq \xi \leq \lambda r$,

$$(1) \qquad \rho^{-n} \int_{(\mathbf{p} \circ \mathbf{P})^{-1}[\mathbf{U}(a, \rho)]} \|S - U\| dV(z, S) \leq \delta$$

and

$$(2) \qquad \rho^{-n+1} \|\delta(V, \Phi)\| (\mathbf{p}^{-1}[\mathbf{U}(a, \rho)] \cap Z) \leq \delta \quad \text{whenever } \xi < \rho \leq r.$$

There is a positive real number $\delta_4 = \delta_4(E)$ *such that if* $\delta \leq \delta_4$ *then there is a nonnegative integer* μ *such that*

$$(3) \qquad \mu \alpha(n) \theta - \varepsilon \leq \mathbf{p}_{\#} \|V\| \mathbf{U}(a, \rho) \leq \mu \alpha(n) \theta + \varepsilon \quad \text{whenever } \xi < \rho \leq \lambda r.$$

PROOF. Choose ε_0 and ω such that $0 < \varepsilon_0 < \varepsilon$, $0 < \omega < 1$ and $\varepsilon_0 + \omega^{-n}\varepsilon_0 < 1$. Set

$$\delta_4(E) = \delta_2(\tilde{E}) \quad \text{where } \tilde{E} = \left(A, \omega^{-n}(2\varepsilon_0 + \lambda^{-n}A), \lambda, M_{0,0}, M_{0,1} \right).$$

Suppose $\delta \leq \delta_4$. Since $A \leq \omega^{-n}(2\varepsilon_0 + \lambda^{-n}A)$ and $\varepsilon_0 \leq \varepsilon$ there is by 2.2 a nonnegative integer μ such that

$$(4) \qquad (\lambda r)^{-n} \left| \mathbf{p}_{\#} \|V\| - \mu \theta \mathscr{L}^n(B) \right| \leq \varepsilon_0$$

whenever B is a Borel subset of $\mathbf{U}(A, \lambda r)$. Let R be the set of those ρ such that $\xi < \rho \leq \lambda r$ and

$$\rho^{-n} \left| \mathbf{p}_{\#} \|V\| - \mu \theta \mathscr{L}^n(B) \right| \leq \varepsilon_0$$

whenever B is a Borel subset of $\mathbf{U}(a, \rho)$. Suppose

$$\rho_1 \in R, \qquad \xi < \rho_2 < \rho_1 \quad \text{and} \quad \omega \rho_1 < \rho_2.$$

We will show that $\rho_2 \in R$ and this will prove the lemma since $\lambda r \in R$ and $\varepsilon_0 \leq \varepsilon$.

It follows from (4) that $\mu \theta \alpha(n) \leq \varepsilon_0 + \lambda^{-n}A$. Since $\rho_1 \in R$ we have

$$\rho_2^{-n} \mathbf{p}_{\#} \|V\| \mathbf{U}(a, \rho_2) \leq \omega^{-n} \rho_1^{-n} \mathbf{p}_{\#} \|V\| \mathbf{U}(a, \rho_1)$$

$$\leq \omega^{-n} \left\{ \rho_1^{-n} \left| \mathbf{p}_{\#} \|V\| - \mu \theta \mathscr{L}^n \mathbf{U}(a, \rho_1) \right| + \mu \theta \alpha(n) \right\}$$

$$\leq \omega^{-n} \left\{ \varepsilon_0 + \varepsilon_0 + \lambda^{-n}A \right\}.$$

14   W. K. ALLARD

By 2.2 there is a nonnegative integer κ such that

$$\rho_2^{-n}\big|\mathbf{p}_\#\|V\| - \kappa\theta\mathscr{L}^n(B)\big| \leqslant \varepsilon_0$$

whenever B is a Borel subset of $\mathbf{U}(a, \rho_2)$. But

$$|\kappa - \mu| \leqslant \rho_2^{-n}\big|\mathbf{p}_\#\|V\| - \kappa\theta\mathscr{L}^n(\mathbf{U}(a, \rho_2))\big|$$
$$+ \omega^{-n}\rho_1^{-n}\big|\mathbf{p}_\#\|V\| - \mu\theta\mathscr{L}^n(\mathbf{U}(a, \rho_2))\big|$$
$$\leqslant \varepsilon_0 + \omega^{-n}\varepsilon_0 < 1,$$

so $\kappa = \mu$ and $\rho_2 \in R$.

2.6. **Lipschitz Approximation Theorem.** *Suppose* (1)–(9) *of* 2.2 *hold. There are positive real numbers* $\delta_5 = \delta_5(E)$ *and* $C_1 = C_1(E)$ *such that if* $\delta \leqslant \delta_5$ *and*

$$(1) \qquad\qquad \alpha(n)\theta/2 \leqslant (\lambda r)^{-n}\mathbf{p}_\#\|V\|\mathbf{U}(a, \lambda r) \leqslant 3\alpha(n)\theta/2$$

then there are a Borel subset X of $\mathbf{U}(a, \lambda r)$ and a function $f\colon \mathbf{U}(a, \lambda r) \to \mathbf{R}^N$ such that

$$(2) \qquad\qquad (x, f(x)) \in Q \quad \text{for } x \in \mathbf{U}(a, \lambda r);$$

$$(3) \quad |(f - l)(x_1) - (f - l)(x_2)| \leqslant 24\sqrt{N}\,\varepsilon|x_1 - x_2| \quad \text{for } x_1, x_2 \in \mathbf{U}(a, \lambda r);$$

$$(4) \qquad \begin{array}{l} V\llcorner\mathbf{P}^{-1}[Z \cap \mathbf{p}^{-1}[X]] \text{ is the varifold corresponding to the rectifi-}\\ \text{able set } \{(x, f(x))\colon x \in X\} \end{array}$$

and

$$(5) \quad \mathscr{L}^n + \mathbf{p}_\#\|V\|(\mathbf{U}(a, \lambda r) \sim X)$$
$$\leqslant C_1\left\{r\|\delta(V, \Phi)\|(Q) + \int_{\mathbf{P}^{-1}[Q]} \|S - U\|^2 \, dV(z, S)\right\}.$$

REMARK. One may also prove a multiple-valued Lipschitz approximation theorem *a la* [**AF2**].

PROOF. The role of ε here is different from what it was in 2.2, 2.4 and 2.5.

Suppose $\lambda < \lambda_1 < \lambda_2 < 1$ and $\lambda_3 = (\lambda_1 - \lambda)/(\lambda_2 - \lambda)$. Choose a positive real number β such that

$$(\lambda_1/\lambda)^n\beta + \theta\alpha(n)/2 < \theta\alpha(n), \quad \beta + ((\lambda_1 - \lambda)/\lambda_1)^{-n}\beta < \theta\alpha(n),$$
$$\beta < \alpha(n)/2 \quad \text{and} \quad \beta < 3^{-n}[\theta\alpha(n) - \beta].$$

Suppose $0 < \gamma < \infty$.

By assuming that δ does not exceed a positive real number depending only on A, M, λ_1, β, $M_{0,0}$ and $M_{0,1}$ we may apply 2.2 to obtain a nonnegative integer κ such that

$$(6) \qquad\qquad (\lambda_1 r)^{-n}\big|\mathbf{p}_\#\|V\| - \kappa\theta\mathscr{L}^n(B)\big| \leqslant \beta$$

whenever B is a Borel subset of $\mathbf{U}(a, \lambda_1 r)$. But (1) and (6) imply

$$|\kappa - 1|\theta\alpha(n) \leqslant (\lambda_1/\lambda)^n (\lambda r)^{-n}|\mathbf{p}_\sharp\|V\| - \kappa\theta\mathscr{L}^n(\mathbf{U}(a, \lambda r))|$$

$$+ (\lambda r)^{-n}|\mathbf{p}_\sharp\|V\| - \theta\mathscr{L}^n(\mathbf{U}(a, \lambda r))|$$

$$\leqslant (\lambda_1/\lambda)^n \beta + \theta\alpha(n)/2$$

$$< \theta\alpha(n),$$

so

$$(7) \qquad\qquad\qquad\qquad \kappa = 1.$$

By assuming that δ does not exceed a positive real number depending only on A, M, λ_1, β, $M_{0,0}$, $M_{0,1}$ and ε we may apply 2.4 to obtain $b_0 \in \mathbf{R}^N$ such that if

$$C_0 = \bigcap_{i=1}^N \{(x, y): 0 < q^i(x - a, y - b_0) \leqslant \varepsilon r\}$$

then

(8)
$$(\lambda_1 r)^{-n}\left\{\left|\|V\|(\mathbf{p}^{-1}[B] \cap C_0) - \theta\mathscr{L}^n(B)\right| + \|V\|(\mathbf{p}^{-1}[\mathbf{U}(a, \lambda_1 r)] \sim C_0)\right\} \leqslant \beta$$

whenever B is a Borel subset of $\mathbf{U}(a, \lambda_1 r)$.

Suppose $x \in \mathbf{U}(a, \lambda r)$. We let $\xi(x)$ be the infimum of the set of those t such that $0 < t < (\lambda_2 - \lambda)r$ and

$$\rho^{-n}\int_{(\mathbf{p}\circ\mathbf{P})^{-1}[\mathbf{U}(x, \rho)]} \|S - U\|dV(z, S) \leqslant \gamma$$

and

$$\rho^{-n+1}\|\delta(V, \Phi)\|(Z \cap \mathbf{p}^{-1}[\mathbf{U}(x, \rho)]) \leqslant \gamma \quad \text{whenever } t < \rho \leqslant (\lambda_2 - \lambda)r.$$

By assuming that $(\lambda_1 - \lambda)^{-n}\delta$ and $(\lambda_1 - \lambda)^{-n+1}\delta$ are less than γ, we infer that

$$(9) \qquad\qquad\qquad 0 \leqslant \xi(x) < (\lambda_1 - \lambda)r.$$

By assuming that δ does not exceed a positive real number depending only on A, $(\lambda_2 - \lambda)$, M, λ_3, β, $M_{0,0}$ and $M_{0,1}$, and using the fact that $\lambda_3(\lambda_2 - \lambda) - \lambda_1 - \lambda$ we may use 2.5 to obtain a nonnegative integer $\mu(x)$ such that

$$(10) \qquad \mu(x)\theta\alpha(n) - \beta \leqslant \rho^{-n}\mathbf{p}_\sharp\|V\|^U(x, \rho) \leqslant \mu(x)\theta\alpha(n) + \beta$$

$$\text{whenever } \xi(x) < \rho \leqslant (\lambda_1 - \lambda)r.$$

But (6), (7) and (10) imply

$$|\mu(x) - 1|\theta\alpha(n) \leqslant ((\lambda_1 - \lambda)r)^{-n}|\mathbf{p}_\#\|V\| - \mu(x)\theta\mathscr{L}^n(\mathbf{U}(x,(\lambda_1 - \lambda)r))|$$
$$+ ((\lambda_1 - \lambda)/\lambda_1)^{-n}(\lambda_1 r)^{-n}|\mathbf{p}_\#\|V\| - \theta\mathscr{L}^n(\mathbf{U}(x, (\lambda_1 - \lambda)r))|$$
$$\leqslant \beta + ((\lambda_1 - \lambda)/\lambda_1)^{-n}\beta$$
$$< \theta\alpha(n),$$

so that

$$(11) \qquad\qquad\qquad\qquad \mu(x) = 1.$$

Let $X = \{ x \in \mathbf{U}(a, \lambda r): \xi(x) = 0 \}$. By assuming that γ does not exceed a positive real number depending only on A, $\lambda_2 - \lambda$, M, β, λ_3, $M_{0,0}$, $M_{0,1}$ and ε, and using (10), (11) and the fact that $\beta < \alpha(n)\theta/2$, we are able to choose for each $w \in X$ and each ρ with $0 < \rho \leqslant (\lambda_1 - \lambda)r$ a point $g(w, \rho) \in \mathbf{R}^N$ such that if

$$C(w, \rho) = \bigcap_{i=1}^{N} \{(x, y): 0 < q^i(x - w, y - g(w, \rho)) \leqslant \varepsilon\rho\}$$

then

$$(12) \qquad \rho^{-n}\big\{\big|\|V\|(\mathbf{p}^{-1}[B] \cap C(w, \rho)) - \theta\mathscr{L}^n(B)\big|$$
$$+ \|V\|(\mathbf{p}^{-1}[\mathbf{U}(w, \rho)] \sim C(w, \rho))\big\} \leqslant \beta$$

whenever B is a Borel subset of $\mathbf{U}(w, \rho)$.

Suppose $x_1, x_2 \in X$, $|x_1 - x_2| + \rho_1 \leqslant \rho_2 \leqslant (\lambda_1 - \lambda)r$ and $\rho_1 \geqslant \rho_2/3$. Then

$$(13) \qquad |g(x, \rho_1) - l(x_1) - [g(x_2, \rho_2) - l(x_2)]| \leqslant 2\sqrt{N}\,\varepsilon\rho_2.$$

To see this, observe that the inequality (13) is implied by $C(x_2, \rho_2) \cap C(x_1, \rho_1) \neq \varnothing$; so suppose $C(x_2, \rho_2) \cap C(x_1, \rho_1) = \varnothing$. Since $\mathbf{U}(x_1, \rho_1) \subset \mathbf{U}(x_2, \rho_2)$ we may apply (12) twice to obtain

$$\beta \geqslant \rho_2^{-n}\|V\|(\mathbf{p}^{-1}[\mathbf{U}(x_1, \rho_1)] \sim C(x_2, \rho_2))$$
$$\geqslant (\rho_2/\rho_1)^{-n}\rho_1^{-n}\|V\|(\mathbf{p}^{-1}[U(x_1, \rho_1)] \cap C(x_1, \rho_1))$$
$$\geqslant (\rho_2/\rho_1)^{-n}[\theta\alpha(n) - \beta] \geqslant 3^{-n}[\theta\alpha(n) - \beta],$$

which is a contradiction.

Since $\beta < (\lambda/(\lambda_1 - \lambda))^{-n}[\theta\alpha(n) - \beta]$ we may use the same type of argument to deduce that

$$(14) \qquad |g(x, \rho) - l(x) - [b_0 - l(a)]| \leqslant 2\sqrt{N}\,\varepsilon(\lambda_1 - \lambda)r \quad \text{for } x \in X.$$

Suppose $x \in X$. It follows from (13) that

$$|g(x, \rho/2) - g(x, \rho)| \leqslant 2\sqrt{N}\,\varepsilon\rho \quad \text{whenever } 0 < \rho \leqslant (\lambda_1 - \lambda)r$$

from which we infer that there is one and only one $g(x) \in \mathbf{R}^N$ such that

$$(15) \qquad |g(x) - g(x, \rho)| \leqslant 4\sqrt{N}\,\varepsilon\rho \quad \text{whenever } 0 < \rho \leqslant (\lambda_1 - \lambda)r.$$

Using (14) and (15), we infer that

$$(16) \qquad \left| g(x) - l(x) - [b_0 - l(a)] \right| \leqslant 6\sqrt{N}\,\varepsilon(\lambda_1 - \lambda)r.$$

Suppose $x_1, x_2 \in X$ and $2|x_1 - x_2| \leqslant (\lambda_1 - \lambda)r$. Then

$$(17) \qquad \left| g(x_1) - l(x_1) - [g(x_2) - l(x_2)] \right| \leqslant 16\sqrt{N}\,\varepsilon|x_1 - x_2|;$$

use (13) and (15) with $\rho_1 = |x_1 - x_2|$ and $\rho_2 = 2|x_1 - x_2|$. Combining (16) and (17), we obtain

$$(18) \quad \left| g(x_1) - l(x_1) - [g(x_2) - l(x_2)] \right| \leqslant 24\sqrt{N}\,\varepsilon|x_1 - x_2| \text{ for all } x_1, x_2 \in X.$$

Using Kirszbraun's Theorem [**FE**, 2.10.43], we choose f to be an extension of g which satisfies (2) and (3).

Suppose $x \in X$, $z = (x, y) \in Z$ and $\Theta^n(\|V\|, z)$ is a nonnegative integer; if $y \neq f(x)$, we infer from (12) and (15) that $\Theta^n(\|V\|, z) \leqslant \beta\alpha(n)^{-1} < 1$, so $\Theta^n(\|V\|, z) = 0$. If $y = f(x)$, we infer from (12) that

$$\liminf_{\rho \downarrow 0} \rho^{-n}\|V\|\mathbf{p}^{-1}[\mathbf{U}(x, \rho)] \geqslant \theta\alpha(n) - \beta > 0.$$

Thus (4) holds.

Finally, let us suppose $x \in \mathbf{U}(a, \lambda r) \sim X$. Evidently, *either*

$$(19) \qquad \int_{(\mathbf{p}\circ\mathbf{P})^{-1}[\mathbf{U}(x,\xi(x))]} \|S - T\|dV(z, S) \geqslant \gamma\xi(x)^n$$

or

$$(20) \qquad \xi(x)\|\delta(V, \Phi)\|\big(Z \cap \mathbf{p}^{-1}[\mathbf{U}(z, \xi(x))]\big) \geqslant \gamma\xi(x)^n.$$

Furthermore, by virtue of (10)

$$(21) \qquad \xi(x)^{-n}\|V\|\big\{z \in Z\colon |\mathbf{p}(z) - x| \leqslant \xi(x)\big\} \leqslant \theta\alpha(n) + \beta$$

since

$$\big\{z \in Z\colon |\mathbf{p}(z) - x| \leqslant \xi(x)\big\} = \bigcap_{\xi(x)<\rho\leqslant(\lambda_1-\lambda)r} \big\{z \in Z\colon |\mathbf{p}(z) - x| < \rho\big\}$$

by (9); this gives

$$\big(\mathbf{p}_\#\|V\| + \mathscr{L}^n\big)\big\{w \in \mathbf{R}^n\colon |w - x| \leqslant \xi(x)\big\} \leqslant \big(\theta\alpha(n) + \beta + \alpha(n)\big)\xi(x)^n.$$

In case (19) holds, we have

$$\gamma^2\xi(x)^n \leqslant \xi(x)^{-n}\left[\int_{(\mathbf{p}\circ\mathbf{P})^{-1}[\mathbf{U}(x,\xi(x))]} \|S - U\|dV(z, S)\right]^2$$

$$\leqslant \xi(x)^{-n}\|V\|\big\{z \in Z\colon |\mathbf{p}(z) - x| \leqslant \xi(x)\big\}$$

$$\cdot \int_{(\mathbf{p}\circ\mathbf{P})^{-1}[\mathbf{U}(x,\xi(x))]} \|S - U\|^2\,dV(z, S)$$

$$\leqslant \big(\theta\alpha(n) + \beta\big)\int_{(\mathbf{p}\circ\mathbf{P})^{-1}[\mathbf{U}(x,\xi(x))]} \|S - U\|^2\,dV(z, S).$$

We may now complete the proof with the help of the Besicovitch Covering Theorem [**FE**, 2.8.14]. $\square$

3. A regularity theorem in codimension one.

3.1. **For the remainder of this paper we assume** $N = 1$. For each $u \in \mathbf{S}^n$ we let $\sigma(u) = \{ v \in \mathbf{R}^{n+1} : v \cdot u = 0 \} \in \mathbf{G}(n+1, n)$. By a simple computation we obtain

(1) $\sigma(u)^{\perp} \circ \langle v, D\sigma(u) \rangle \circ \sigma(u) = -(\omega \circ \sigma(u))u$ whenever $u \in \mathbf{S}^n$, $v \in \mathbf{R}^{n+1}$ and $w(z) = z \cdot v$ for $z \in \mathbf{R}^{n+1}$. **We define**

$$\Psi: Z \times \mathbf{R}^{n+1} \to \{ t: 0 < t < \infty \}$$

by requiring that

(2) $\Psi(z, tu) = t\Psi(z, u)$ **whenever** $0 \leqslant t < \infty$ **and** $(z, u) \in Z \times \mathbf{R}^{n+1}$

and that

(3) $\Psi(z, u) = \Phi(z, \sigma(u))$ **whenever** $(z, u) \in Z \times \mathbf{S}^n$.

We assume there is a positive real number γ such that

(4) $\Psi(z, u) - \gamma|u|$ **is a convex function of** $u \in \mathbf{R}^{n+1}$ **for each** $z \in Z$, **and we assume that** $N_{0,0}$, $N_{0,1}$, $N_{0,2}$, $N_{0,3}$, $N_{1,0}$ $N_{1,1}$ **are all** $< \infty$ **where, for each ordered pair** (k, l) **of nonnegative integers, we have set** $N_{k,l}$ **equal to the supremum of the set of numbers**

$$\left| \langle (v_1, 0) \odot \cdots \odot (v_k, 0) \odot (0, w_1) \odot \cdots \odot (0, w_l), D^{k+l}\Psi(z, u) \rangle \right|$$

corresponding to $(z, u) \in Z \times \mathbf{S}^n$ **and** $v_1, \ldots, v_k; w_1, \ldots, w_l \in \mathbf{S}^n$. (4) is an ellipticity condition; see [FE, 5.1].

3.2. For each $z \in Z$ we set $\Psi_z(u) = \Psi(z, u)$ for $u \in \mathbf{R}^{n+1}$. Obviously, Ψ_z is smooth on $\mathbf{R}^{n+1} \sim \{0\}$. From 3.1(2) we infer that

(1) $u \cdot \operatorname{grad} \Psi_z(u) = \Psi_z(u)$ and

(2) $\langle u, D(\operatorname{grad} \Psi_z)(u) \rangle = 0$ whenever $z \in Z$ and $u \in \mathbf{R}^{n+1} \sim \{0\}$. From 3.1(3) we infer that

(3) $\Psi_z(u) = \Psi_z(-u)$ whenever $(z, u) \in Z \times \mathbf{R}^{n+1}$.

By a straightforward argument using 3.1(4) we infer that

(4) $\langle v, D(\operatorname{grad} \Psi_z)(u) \rangle \cdot v \geqslant \gamma|u|^{-3}|u \wedge v|^2$ whenever $u \in \mathbf{R}^{n+1} \sim \{0\}$ and $v \in \mathbf{R}^{n+1}$.

We have

(5) $\Psi_z(v) - v \cdot \operatorname{grad} \Psi_z(u) \geqslant \gamma(1 - u \cdot v)$ whenever $z \in Z$ and $u, v \in \mathbf{S}^n$.

To prove this, we assume $v \notin \{u, -u\}$, set $g(t) = |u + t(v \cdot u)|$ for $0 \leqslant t \leqslant 1$ and use (1) and (4) to calculate and estimate

$$\Psi_z(v) - v \cdot \operatorname{grad} \Psi_z(u) = \Psi_z(u + (v - u)) - \Psi_z(u) - (v - u) \cdot \operatorname{grad} \Psi_z(u)$$

$$= \frac{1}{2} \int_0^1 \langle v - u, D(\operatorname{grad} \Psi_z)(u + t(v - u)) \rangle \cdot (v - u)\, d(t - 1)^2$$

$$\geqslant \frac{\gamma}{2} \int_0^1 |u + t(v - u)|^{-3} |u \wedge v|^2\, d(t - 1)^2 = \frac{\gamma}{2} \int_0^1 g''(t)\, d(t - 1)^2$$

$$= \gamma \left[g(1) - g(0) - g'(0) \right] = \gamma(1 - u \cdot v).$$

We have

(6) $\Psi_z(u)\Psi_z(v) - v \cdot \operatorname{grad} \Psi_z(u)u \cdot \operatorname{grad} \Psi_z(v) \geqslant \gamma(1 - |u \cdot v|)$ whenever $z \in Z$ and $u, v \in \mathbf{S}^n$.

To prove this, we choose $\varepsilon \in \{-1, 1\}$ such that $\varepsilon v \cdot \operatorname{grad} \Psi_z(u) \geqslant 0$ and use (3) and (5) to obtain that the left side of (6) equals

$$
\begin{aligned}
\Psi_z(u)\Psi_z(\varepsilon v) &- \varepsilon v \cdot \operatorname{grad} \Psi_z(u) u \cdot \operatorname{grad} \Psi_z(\varepsilon v) \\
&= \Psi_z(u)\big[\Psi_z(\varepsilon v) - \varepsilon v \cdot \operatorname{grad} \Psi_z(u)\big] \\
&\quad + \varepsilon v \cdot \operatorname{grad} \Psi_z(u)\big[\Psi_z(u) - u \cdot \operatorname{grad} \Psi_z(\varepsilon v)\big] \\
&\geqslant \gamma(1 - |u \cdot v|)
\end{aligned}
$$

Suppose $u, v \in \mathbf{S}^n$; by [**AW1**, 8.9(3)],

$$
\tag{7}
\begin{aligned}
\|\sigma(u) - \sigma(v)\|^2 &= \big\|\sigma(u) \circ \sigma(v)^{\perp}\big\|^2 = |\sigma(u)(v)|^2 \\
&= 2\big[1 - (u \cdot v)^2\big];
\end{aligned}
$$

consequently,

$$
\tag{8}
\|\sigma(u) - \sigma(v)\|^2 \leqslant 1 - |u \cdot v| \leqslant \|\sigma(u) - \sigma(v)\|^2/2
$$

and

$$
\tag{9}
\|\sigma(u) - \sigma(v)\|/\sqrt{2} \leqslant |u - v| \leqslant \|\sigma(u) - \sigma(v)\| \quad \text{if } u \cdot v \geqslant 0.
$$

PROPOSITION. *Suppose* $(z, u) \in Z \times \mathbf{S}^n$, $v \in \mathbf{R}^{n+1}$ *and* $L \in \operatorname{Hom}(\mathbf{R}^{n+1}, \mathbf{R}^{n+1})$. *Then*

$$
\tag{10}
\begin{aligned}
&\delta(z, \sigma(u); \Phi)(v, L) \\
&\quad (\operatorname{trace} L)\Psi_z(u) - u \cdot \langle \operatorname{grad} \Psi_z(u), L \rangle + \langle (v, 0), D\Psi(z, u) \rangle.
\end{aligned}
$$

PROOF. Choose $\varepsilon > 0$ and a smooth curve $\gamma: (-\varepsilon, \varepsilon) \to \mathbf{S}^n$ such that $\gamma(0) = u$ and $\sigma(\gamma(t)) = (1 + tL)(\sigma(t))$ for $t \in (-\varepsilon, \varepsilon)$. It follows that $\gamma(t) \cdot (1 + tL)(v) = 0$ for $t \in (-\varepsilon, \varepsilon)$ and $v \in \sigma(u)$ so that $\dot\gamma(0) \cdot v = -u \cdot L(v)$ for $v \in \sigma(u)$. Thus, $\dot\gamma(0) = -L^*(u) + L(u) \cdot u\, u$. Using 3.2(1), we calculate

$$
\begin{aligned}
L \cdot \Phi_{\mathrm{spin}}(z, \sigma(u)) &= (d/dt)\, \Phi(z, \sigma(\gamma(t)))\big|_{t=0} = (d/dt)\, \Psi(z, \gamma(t))\big|_{t=0} \\
&= -L^*(u) \cdot \operatorname{grad} \Psi_z(u) + L(u) \cdot u\Psi_z(u)
\end{aligned}
$$

from which (10) follows. $\square$

3.3. PROPOSITION. *Suppose* 2.2(1)–(8) *hold and* $\delta(V, \Phi)$ *is a measure. Then there is a positive real number* $C_2 = C_2(E, N_{0,2})$ *such that*

$$
\tag{1}
\begin{aligned}
\int_{(\mathbf{p} \circ \mathbf{P})^{-1}[U(a, \lambda r)]} &\|S - U\|^2 \, dV(z, S) \\
&\leqslant C_2\bigg\{ r^{-2} \int_{Q \cap \mathbf{p}^{-1}[U(a, r)]} |q(z - c)|^2 \, d\|V\|z \\
&\quad + \int_{Q \cap \mathbf{p}^{-1}[U(a, r)]} |q(z - c)| \, d\|\delta(V, \Phi)\|z + \big(N_{1,0}^2 + N_{1,1}^2\big) M r^{n+2} \bigg\}.
\end{aligned}
$$

REMARK. Compare with [**AW1**, 8.13].

PROOF. Let $w \in \mathbf{R}^n$ be such that $l(x) = x \cdot w$ for $x \in \mathbf{R}^n$, let $T = (1 + |w|^2)^{1/2}$ and let $v = T^{-1}(-w, 1) \in \mathbf{S}^n$. Let $\varphi \in \mathscr{D}(\mathbf{U}(a, r))$ be such that $0 \leqslant \varphi \leqslant 1$ and $\varphi(x) = 1$ if $x \in \mathbf{U}(a, \lambda r)$; let $\Delta = \sup\{|\text{grad } \varphi(x)|: x \in \mathbf{U}(a, r)\}$. Let

$$g(z) = \varphi(\mathbf{p}(z))^2 q(z - c)\text{grad } \Psi_z(v) \quad \text{for } z \in Z \cap \mathbf{p}^{-1}[\mathbf{U}(a, r)].$$

Let $B(z)(u) = \langle (u, 0), D(\text{grad } \Psi_z)(v)\rangle$ for $z \in Z$ and $u \in \mathbf{R}^{n+1}$.

Suppose $z \in Q \cap \mathbf{p}^{-1}[\mathbf{U}(a, r)]$, $u \in \mathbf{R}^{n+1}$, $u \cdot v \geqslant 0$ and $S = \sigma(u)$. Set

$$A_1 = 2\varphi(\mathbf{p}(z))q(z - c)\Psi_z(u)\langle \text{grad } \Psi_z(v), D\varphi(\mathbf{p}(z)) \circ \mathbf{p}\rangle,$$

$$A_2 = \varphi(\mathbf{p}(z))^2 T\Psi_z(u)\Psi_z(v),$$

$$A_3 = \varphi(\mathbf{p}(z))^2 q(z - c)\Psi_z(u)\text{trace } B(z),$$

$$A_4 = -2\varphi(\mathbf{p}(z))q(z - c)\langle \text{grad } \Psi_z(u), D\varphi(\mathbf{p}(z)) \circ \mathbf{p}\rangle u \cdot \text{grad } \Psi_z(v),$$

$$A_5 = -\varphi(\mathbf{p}(z))^2 Tu \cdot \text{grad } \Psi_z(v) v \cdot \text{grad } \Psi_z(u),$$

$$A_6 = -\varphi(\mathbf{p}(z))^2 q(z - c)u \cdot \langle \text{grad } \Psi_z(u), B(z)\rangle \quad \text{and}$$

$$A_7 = \varphi(\mathbf{p}(z))^2 q(z - c)\langle (\text{grad } \Psi_z(v), 0), D\Psi(z, u)\rangle.$$

Keeping in mind 3.2(1) and (10) we see that

$$\delta(z, S, \Phi)(g(z), Dg(z)) = \sum_{i=1}^{7} A_i,$$

so that

(2) $$A_2 + A_5 \leqslant \delta(z, S, \Phi)(g(z), Dg(z)) + |A_1 + A_4| + |A_3| + |A_6| + |A_7|.$$

Set $\alpha = \varphi(\mathbf{p}(z))$ and $\beta = |q(z - c)|$. Using 3.2(6) and (8), we estimate

$$A_2 + A_5 \geqslant T\alpha^2\gamma\|S - U\|^2/4;$$

using 3.2(1) and (9), we estimate

$$|A_1 + A_4| = |2\alpha\beta\langle \Psi_z(u)[\text{grad } \Psi_z(v) - \text{grad } \Psi_z(u)]$$
$$+ u \cdot [\text{grad } \Psi_z(u) - \text{grad } \Psi_z(v)]\text{grad } \Psi_z(u), D\varphi(\mathbf{p}(z)) \circ p\rangle|$$
$$\leqslant 2\alpha\beta\Delta(N_{0,0} + N_{0,1})\sqrt{2}\,N_{0,2}\|S - U\|;$$

evidently,

$$|A_3| \leqslant \beta N_{0,0}(n + 1)N_{1,1},$$

$$|A_6| \leqslant \beta N_{0,1}N_{1,1} \quad \text{and} \quad |A_7| \leqslant \beta N_{0,1}N_{1,0}.$$

The proof may be completed by using (2), our estimates and Cauchy's inequality. $\square$

3.4. *Fields and barriers; a minimization property.* Suppose W is an open subset of Z, $F: W \to \mathbf{R}$ is smooth and grad $F(z) \neq 0$ for $z \in W$. We define the smooth maps

$$X: W \to \mathbf{R}^{n+1} \quad \text{and} \quad \omega: W \to \wedge^n \mathbf{R}^{n+1}$$

at $z \in W$ by setting

$$X(z) = \operatorname{grad} \Psi_z(\operatorname{grad} F(z))$$

and

$$\omega(z) = (-1)^n X(z) \llcorner e^1 \wedge \cdots \wedge e^{n+1}.$$

We have

$$(1) \qquad d\omega = (\operatorname{div} X) e^1 \wedge \cdots \wedge e^{n+1}.$$

We let $v(z) = |\operatorname{grad} F(z)|^{-1} \operatorname{grad} F(z)$ for $z \in W$. For each $t \in \operatorname{rng} F$ we let

$$T_t = \left[\mathscr{H}^n \llcorner \{ z \in W : F(z) = t \} \right] \wedge * v$$

thus defining an n-dimensional locally integral current in W and we let V_t be the n-dimensional integral varifold in W corresponding to $\{ z \in W : F(z) = t \}$. Recalling the results of [AW2], we see that

$$(2) \quad \delta(V_t, \Phi)(g) = \int (g \cdot v) \operatorname{div} X \, d\|V_t\| \quad \text{whenever } t \in \operatorname{rng} F \text{ and } g \in \mathscr{X}(W).$$

Suppose $t \in \operatorname{rng} F$, K is a compact subset of W, T is an n-dimensional locally integral current in W, S is an $(n+1)$-dimensional locally integral current in W, $\operatorname{spt} S \subset K$ and $T_t = T + \partial S$. Then T_t satisfies the following minimization property:

$$(3) \qquad \int_K \Psi_z\!\left(* \vec{T}_t(z) \right) d\|T_t\| z \leqslant \int_K \Psi_z\!\left(* \vec{T}(z) \right) d\|T\| z + S(d\omega).$$

In fact, $(T_t - T)(\omega) = S(d\omega)$,

$$\langle \vec{T}_t(z), \omega(z) \rangle = \langle * v(z), \omega(z) \rangle = v(z) \cdot X(z) = \Psi_z(v(z)),$$

for $\|T_t\|$-almost all $z \in W$, and

$$\langle \vec{T}(z), \omega(z) \rangle = (-1)^n \langle * * \vec{T}(z), \omega(z) \rangle$$
$$= * \vec{T}(z) \cdot X(z) \leqslant \Psi_z\!\left(* \vec{T}(z) \right)$$

for $\|T\|$-almost all $z \in W$ by 3.2(5).

PROPOSITION. *Suppose*
(4) $0 \leqslant H < \infty$, $t \in \operatorname{rng} F$, $V \in \mathbf{V}_n(W)$, $\operatorname{spt}\|V\|$ *is a compact subset of* W, $\|\delta(V, \Phi)\| \leqslant H\|V\|$ *on* $\{ z \in W : F(z) > t \}$ *and*
(5) $\operatorname{div} X(z) - N_{0,1}[N_{1,1} + N_{1,0}] \geqslant H N_{0,1}$ *if* $F(z) > t$.
Then
(6) $\operatorname{spt}\|V\| \subset \{ z \in W : F(z) \leqslant t \}$.

PROOF. For each $\tau \in \mathbf{R}$ and $z \in W$ we let $F_\tau(z) = F(z) - \tau$, we let $g_\tau(z) = F_\tau(z) X(z)$, we let

$$B(z)(v) = (d/d\varepsilon) D(\operatorname{grad} \Psi_{z+\varepsilon v})(v(z))\big|_{\varepsilon=0} \quad \text{for } v \subset \mathbf{R}^{n+1},$$

and we let $\xi(z) = |\operatorname{grad} F(z)|$.

Suppose $z \in \mathrm{spt}\|V\|$, $\tau \in \mathbf{R}$ and $F(z) > \tau$. Suppose $u \in S^n$ and $u \cdot \nu(z) \geqslant 0$. We have

$$\mathrm{trace}\, Dg(z)\Psi_z(u) = \xi(z)\Psi_z(\nu(z))\Psi_z(u) + F_\tau(z)\mathrm{div}\, X(z);$$

$$\begin{aligned}-u \cdot \langle \mathrm{grad}\, \Psi_z(u), Dg(z)\rangle \\ = -\xi(z)\nu(z) \cdot \mathrm{grad}\, \Psi_z(u)u \cdot \mathrm{grad}\, \Psi_z(\nu(z)) \\ + F_\tau(z)\langle \mathrm{grad}\, \Psi_z(u), D\nu(z)\rangle \cdot \langle \nu(z) - u, D(\mathrm{grad}\, \Psi_z)(\nu(z))\rangle \\ - F_\tau(z)u \cdot \langle \mathrm{grad}\, \Psi_z(u), B(z)\rangle,\end{aligned}$$

where we have used 3.2(2) and the symmetry of $D(\mathrm{grad}\, \Psi_z)(\nu(z))$, and

$$\langle (g_\tau(z),0), D\Psi(z,u)\rangle = F_\tau(z)\langle (X(z),0), D\Psi(z,u)\rangle.$$

Using 3.2(6) and (10) we see that

$$\begin{aligned}\delta(z, \sigma(u); \Phi)(g_\tau(z), Dg_\tau(z)) \\ \geqslant \xi(z)\gamma[1 - u \cdot \nu(z)] + F_\tau(z)\mathrm{div}\, X(z) \\ - F_\tau(z)\zeta[1 - u \cdot \nu(z)]^{1/2} - F_\tau(z)N_{0,1}N_{1,1} - F_\tau(z)N_{1,0}N_{0,1}\end{aligned}$$

where ζ does not depend on z or τ.

Whenever $t < \tau$, we have

$$\int_{\mathbf{P}^{-1}[\{z:\, F(z)>\tau\}]} \delta(z, S; \Phi)(g_\tau(z), Dg_\tau(z))\, dV(z, S)$$

$$\leqslant H \int_{\{z:\, F(z)>\tau\}} F_\tau(z)N_{0,1}\, d\|V\|z.$$

In the event that $t < t_1 = \sup\{ F(z): z \in \mathrm{spt}\|V\|\}$, we arrive at a contradiction by choosing τ less than but sufficiently close to t_1. $\square$

PROPOSITION. *Suppose* $c \in W$, $t = F(c)$, $0 < r < \mathrm{distance}(c, \mathbf{R}^{n+1} \sim W)$, $0 \leqslant H_1 < \infty$,

$$(7) \qquad\qquad |\mathrm{div}\, X| \leqslant H_1$$

and

$$(8) \qquad\qquad \varepsilon = H_1 r/(n+1) < 1.$$

Then

$$(9) \qquad r^{-n}\|T_t\|\{z: |z - c| \leqslant r\} \geqslant n^{-n}[(1 - \varepsilon)/\mathbf{c}(N_{0,0} + \varepsilon)]^n,$$

where **c** *is a positive real number depending only on* n.

REMARK. This is just a generalization of a similar and well-known inequality in [**FE** 5.1.6].

PROOF. Suppose $0 < \rho < \mathrm{distance}(c, \mathbf{R}^{n+1} \sim W)$. Let $U = T_t \llcorner \{z: |z - c| \leqslant \rho\}$, suppose $\mathbf{N}(U) < \infty$ and use the Isoperimetric Inequality [**FE** 4.4.2] to choose an n-dimensional integral current Q such that $\mathrm{spt}\, Q \subset \{z: |z - c| \leqslant \rho\}$, $\partial Q = \partial U$ and $\mathbf{M}(Q) \leqslant \mathbf{c}\mathbf{M}(\partial U)^{n/(n-1)}$, where $\mathbf{c}$ is a constant depending only on n. Let X be the cone on $U - Q$ with vertex c. Since $\partial S = U - Q$, we infer from (3) and 3.2(5) that

$$\mathbf{M}(U) \leqslant \int \psi_z(\ast\vec{U}(z))\,d\|U\|z \leqslant \int \Psi_z(\ast\vec{Q}(z))\,d\|Q\|z + S(d\omega)$$

$$\leqslant N_{0,0}\mathbf{M}(Q) + H_1\mathbf{M}(S)$$

$$\leqslant N_{0,0}\mathbf{c}\mathbf{M}(\partial U)^{n/(n-1)} + (H_1\rho/(n+1))\left[\mathbf{M}(U) + \mathbf{c}\mathbf{M}(\partial U)^{n/(n-1)}\right].$$

We infer from the above that

$$(1-\varepsilon)\mu(\rho) \leqslant \mathbf{c}(N_{0,0} + \varepsilon)\mu'(\rho)^{n/(n-1)}$$

for $\mathscr{L}^1$-almost all $\rho \in (0, r)$, where we have set $\mu(\rho) = \|T_t\|\{z: |z - c| \leqslant \rho\}$. Consequently,

$$(\mu^{1/n})'(\rho) \geqslant (1/n)\left[(1-\varepsilon)/\mathbf{c}(N_{0,0} + \varepsilon)\right]^{(n-1)/n}$$

for $\mathscr{L}^1$-almost all $\rho \in (0, r)$ so (9) follows by integration. $\square$

3.5. *Solubility of the Dirichlet problem; a height bound.* Throughout this section we assume that $Z = D \times \mathbf{R}$ for some open subset D of $\mathbf{R}^n$ and that

$$(1) \qquad \langle (e_{n+1}, 0), D\Psi(z, u) \rangle = 0 \quad \text{for all } (z, u) \in Z \times \mathbf{S}^{n+1}.$$

PROPOSITION. *Suppose W, F and X are as in 3.4, $H_1 \in \mathbf{R}$,*

$$(2) \qquad \qquad \operatorname{div} X = H_1$$

and

$$(3) \qquad \qquad \nu_{n+1}(z) = \nu(z) \cdot e_{n+1} \quad \text{for } z \in W.$$

Then, for each $t \in \operatorname{rng} F$, $\nu_{n+1}|\{z \in W: F(z) = t\}$ does not assume a minimum or maximum value.

PROOF. We apply the strong maximum principle [**GT**, 3.2] to the Central Identity of [**AW3**, 2.3], keeping in mind that (1) holds. $\square$

PROPOSITION. *Suppose $a \in D$, $0 < \rho < \operatorname{distance}(a, \mathbf{R}^n \sim D)$, $H_1 \in \mathbf{R}$, and*

$$(4) \qquad \qquad \varphi: \{x \in \mathbf{R}^n: |x - a| = \rho\} \to \mathbf{R} \text{ is smooth.}$$

Then there is one and only one smooth function

$$f: \{x \in \mathbf{R}^n: |x - a| \leqslant \rho\} \to \mathbf{R}$$

such that

$$(5) \qquad \qquad f|\{x \in \mathbf{R}^n: |x - a| = \rho\} = \varphi$$

and

$$(6) \qquad \qquad \operatorname{div} X = H_1,$$

where $W = \{x \in \mathbf{R}^n: |x - a| < \rho\} \times \mathbf{R}$, $F(x, y) = y - f(x)$ for $(x, y) \in W$ and X is as in 3.4, provided

$$(7) \qquad \qquad \max\{H_1, N_{1,1}\}\rho \leqslant C_3,$$

where $C_3 = C_3(\gamma, N_{0,2})$.

PROOF. We use the previous proposition, the maximum principle of [GT, 9.5] and the barrier construction of [GT, 13.5] to deduce that if f is as above then $|Df|$ may be *a priori* bounded in terms of the boundary values φ. One then uses well-known arguments as laid out, for example, in [GT] to complete the proof. $\square$

A HEIGHT BOUND. *Suppose* (1)–(8) *of 2.2 hold,*
(8) $0 \leqslant H < \infty$ *and* $\|\delta(V, \Phi)\| \leqslant H\|V\|$ *on* $\mathbf{p}^{-1}[\mathbf{U}(a, r)]$,
(9) $0 \leqslant \delta < \infty$,

(a) $\max\{N_{1,0}, N_{1,1}, H\}r \leqslant \delta$ *and*

(b) $r^{-n-2}\int_{\mathbf{p}^{-1}[\mathbf{U}(a, r)]}|q(z - c)|^2 \, d\|V\|z \leqslant \delta^2$.
Then there is a positive real number $\delta_6 = \delta_6(E, \gamma, N_{0,2})$ *such that if* $\delta \leqslant \delta_6$ *then*

$$(10) \qquad |q(z - c)| \leqslant \varepsilon r \quad \text{for } z \in \mathrm{spt}\|V\| \cap \mathbf{p}^{-1}[\mathbf{U}(a, \lambda r)]$$

provided

$$(11) \qquad \mathcal{H}^n(B) = 0 \quad \text{if } B \subset \mathrm{spt}\|V\| \text{ and } \|V\|(B) = 0.$$

REMARK. Were we able to obtain a lower bound for mass ratios as in [AW1, 5] for varifolds satisfying appropriate conditions on first variation with respect to Φ, an inequality like (10) would be evident and the proviso (11) would be unnecessary.

PROOF. Choose λ_1 such that $\lambda < \lambda_1 < 1$. We let $B = \{(x, t) \in \mathbf{R}^n \times \mathbf{R}: |x - a| < r, \ b + l(x - a) \leqslant t \text{ and } t \leqslant y \text{ for some } y \text{ such that } (x, y) \in \mathrm{spt}\|V\|\}$. Note that B is closed relative to $\mathbf{p}^{-1}[\mathbf{U}(a, r)]$ and that, by virtue of (11),

$$\mathcal{L}^{n+1}(B) \leqslant \int_{\mathbf{p}^{-1}[\mathbf{U}(a, r)]} |q(z - c)| \, d\|V\|z.$$

Thus, we may choose ρ such that $\lambda_1 r < \rho < r$ and such that

$$\mathcal{H}^n\big[B \cap \{(x, y): |x - a| = \rho\}\big] \leqslant M^{1/2}\delta(1 - \lambda_1)^{-1}r^n.$$

Since B is closed, we may choose a smooth function $\varphi: \{x \in \mathbf{R}^n: |x - a| = \rho\} \to \mathbf{R}$ such that if $x \in \mathbf{R}^n$ and $|x - a| = \rho$ then $b + l(x - a) < \varphi(x)$ and $y < \varphi(x)$ if $(x, y) \in B$ and such that

$$\zeta = \int_{\{x: |x-a|=\rho\}} \varphi(x) - b - l(x - a) \, d\mathcal{H}^{n-1}x \leqslant 2M^{1/2}\delta(1 - \lambda_1)^{-1}r^n.$$

Now suppose H_1 is such that

$$H_1 - N_{0,1}[N_{1,1} + N_{0,1}] > HN_{0,1}, \qquad H_1\rho/(n + 1) < \tfrac{1}{2}$$

and

$$\max\{H_1, N_{1,1}\}\rho \leqslant C_3(\gamma, N_{0,2}).$$

Letting f be as in (5) and (6), we infer from 3.4(6) that

$$b + l(x - a) < f(x) \quad \text{whenever } x \in \mathbf{U}(a, \rho), \text{ and}$$

$$y \leqslant f(x) \quad \text{whenever } (x, y) \in \mathrm{spt}\|V\| \cap \mathbf{p}[\mathbf{U}(a, \rho)]$$

by applying 3.4(6) when possible to the varifolds $V \llcorner (\mathbf{p} \circ \mathbf{P})^{-1}[\mathbf{U}(a, \sigma)]$, $0 < \sigma < \rho$. We need to show that $f(x)$ is small if $|x - a| \leqslant \lambda r$ and δ is small. In view of 3.4(9), it will suffice to show that

$$r^{-n-1} \int_{\mathbf{U}(a, \rho)} f(x) - b - l(x - a) \, d\mathscr{L}^n x$$

is small if δ is small. Since

$$|y - b - l(x - a)| \leqslant \sqrt{1 + A^2} \operatorname{dist}((x, y), U) \quad \text{for } (x, y) \in \mathbf{R}^n \times \mathbf{R},$$

it will suffice to show that

$$\mu = \left(r^{-n-2} \int_{\{(x, f(x)):\, x \in \mathbf{U}(a, \rho)\}} \operatorname{dist}(z, U)^2 \, d\mathscr{H}^n z \right)^{1/2}$$

is small if δ is small.

Let S be the $(n + 1)$-dimensional integral current in $\mathbf{R}^{n+1}$ corresponding to integration over $\{(x, y): |x - a| < \rho \text{ and } b + l(x - a) < y < f(x)\}$. Let T_i, $i \in \{1, 2, 3\}$, be the n-dimensional integral currents in $\mathbf{R}^{n+1}$ determined by the conditions

$$\partial S = T_1 + T_2 + T_3,$$

$$\operatorname{spt} T_1 = \{(x, y): |x - a| \leqslant \rho \text{ and } y = f(x)\},$$

$$\operatorname{spt} T_2 = \{(x, y): |x - a| = \rho \text{ and } b + l(x - a) \leqslant y \leqslant \varphi(x)\} \quad \text{and}$$

$$\operatorname{spt} T_3 = \{(x, y): |x - a| \leqslant \rho \text{ and } y = b + l(x - a)\}.$$

Using the technique of [**FE** 5.3.4, pp. 569–571], we see that

$$\mu^2 \leqslant \mathbf{c} r^{-n}[\mathbf{M}(T_1) + \mathbf{M}(T_2) - \mathbf{M}(T_3)]$$

where $\mathbf{c}$ depends only on n, λ_1 and A. Using a slight variation of the argument of [**FE**, 5.1.2] in conjunction with 3.4(3) and the fact that $\mathbf{M}(T_2) = \zeta$, we complete the proof. $\square$

3.6. *The regularity theorem.*

THE BASIC REGULARITY LEMMA. *Suppose* 2.2(1)–(8) *hold*;

(1) $\mathscr{H}^n(B) = 0$ *if* $B \subset \operatorname{spt}\|V\|$ *and* $\|V\|(B) = 0$,

(2) $\langle (e_{n+1}, 0), D\Psi(z, u) \rangle = 0$ *for all* $(z, u) \in Z \times \mathbf{S}^n$,

(3) $\alpha(n)\theta/2 < (\lambda r)^{-n}\|V\|(Z \cap \mathbf{p}^{-1}[\mathbf{U}(a, \lambda r)]) \leqslant 3\alpha(n)\theta/2$,

(4) $0 \leqslant H < \infty$ *and* $\|\delta(V, \Phi)\| \leqslant H\|V\|$ *on* $Z \cap \mathbf{p}^{-1}[\mathbf{U}(a, r)]$ *and*

(5) $0 \leqslant \delta < \infty$,

(a) $\max\{H, N_{1,0}, N_{1,1}\}r \leqslant \delta$ *and*

(b) $\mu - \left(r^{-n-2} \int_{Z \cap \mathbf{p}^{-1}[\mathbf{U}(a, r)]} |q(z - c)|^2 \, d\|V\|z \right)^{1/2} \leqslant \delta$.

There are positive real numbers $\delta_7 = \delta_7(E, \gamma, N_{0,2}, N_{0,3})$ *and* $C_4 = C_4(E, \gamma, N_{0,2}, N_{0,3})$ *such that if* $\delta \leqslant \delta_7$ *and* $0 < \omega < 1$ *then* **either**

(6) $$\mu \leqslant C_4 \omega^{-(n+2)/2} r \max\{H, N_{1,0}, N_{1,1}\}$$

or *there are* $\tilde{b} \in \mathbf{R}$ *and* $\tilde{l} \in \operatorname{Hom}(\mathbf{R}^n, \mathbf{R})$ *such that*

(7) $$\max\{r^{-1}|\tilde{b} - b|, \|\tilde{l} - l\|\} \leqslant C_4 M$$

and

$$(8) \qquad \tilde{\mu} = \left((\omega r)^{-n-2} \int_{Z \cap \mathbf{p}^{-1}[\mathbf{U}(a,\,\omega r)]} |\tilde{q}(z - \tilde{c})|^2 \, d\|V\|z \right)^{1/2}$$

$$\leqslant C_4 \omega^2 \mu.$$

where we have set $\tilde{c} = (a, \tilde{b})$ *and* $\tilde{q}(x, y) = y - \tilde{l}(x)$ *for* $(x, y) \in \mathbf{R}^n \times \mathbf{R}$.

PROOF. We follow [**AW1**, 8.16] and assume the reader is familiar with that type of argument. By assuming δ is sufficiently small, we obtain f and X satisfying 2.6(2)–(5) with ε there equal to 1; note that because (2) holds we can replace "$\max\{r, s\}$" in 2.2(9)(a) by "r"; we need also to invoke 3.3. We let $g(x) = f(x) - b - l(x - a)$ for $x \in \mathbf{U}(a, \lambda r)$. For each $x \in \mathbf{U}(a, \lambda r)$ we let $\Omega(x) \in \odot^2 \mathbf{R}^n$ be such that

$$\langle \alpha^2, \Omega(x) \rangle = \int_0^1 \langle (0, 0, \alpha)^2, D^2\Psi^{\S}(a, b, l + t\,\mathrm{grad}\,g(x)) - D^2\Psi^{\S}(a, b, l) \rangle \, dt;$$

here $\Psi^{\S}$ is as in [**FE**, 5.1], so

$$\Psi^{\S}(x, y, m) = \Psi((x, y), (-m, 1)) \quad \text{for } ((x, y), m) \in Z \times \mathrm{Hom}(\mathbf{R}^n, \mathbf{R}).$$

For each $\varphi \in \mathscr{D}(\mathbf{U}(a, \lambda r))$, we set

$$T(\varphi) = \int_{\mathbf{U}(a,\,\lambda r)} \langle (0, \varphi(x), \mathrm{grad}\,\varphi(x)), D\Psi^{\S}(x, f(x), \mathrm{grad}\,f(x)) \rangle \, d\mathscr{L}^n x,$$

$$T_1(\varphi) = \int_{\mathbf{U}(a,\,\lambda r)} \langle (0, \varphi(x), 0), D\Psi^{\S}(x, f(x), \mathrm{grad}\,f(x)) \rangle \, d\mathscr{L}^n x,$$

$$T_2(\varphi) = \int_{\mathbf{U}(a,\,\lambda r)} \big\langle (0, 0, \mathrm{grad}\,\varphi(x)), D\Psi^{\S}(x, f(x), \mathrm{grad}\,f(x))$$
$$- D\Psi^{\S}(a, b, \mathrm{grad}\,f(x)) \big\rangle d\mathscr{L}^n x,$$

$$T_3(\varphi) = \int_{\mathbf{U}(a,\,\lambda r)} \langle \mathrm{grad}\,\varphi(x) \odot \mathrm{grad}\,g(x), \Omega(x) \rangle \, d\mathscr{L}^n x,$$

$$T_4(\varphi) = \int_{\mathbf{U}(a,\,\lambda r)} \langle (0, 0, \mathrm{grad}\,\varphi(x)), D\Psi^{\S}(a, b, l) \rangle \, d\mathscr{L}^n x,$$

$$T_5(\varphi) = \int_{\mathbf{U}(a,\,\lambda r)} \big\langle (0, 0, \mathrm{grad}\,\varphi(x)) \big\rangle$$
$$\odot \big\langle (0, 0, \mathrm{grad}\,g(x)), D^2\Psi^{\S}(a, b, l) \big\rangle d\mathscr{L}^n x,$$

$$U(\varphi) = \delta(V, \Phi)((\varphi \circ \mathbf{p})e_{n+1}),$$

$$U_1(\varphi) = \int_X \langle (0, \varphi(x), \mathrm{grad}\,\varphi(x)), D\Psi^{\S}(x, f(x), \mathrm{grad}\,f(x)) \rangle \, d\mathscr{L}^n x,$$

$$U_2(\varphi) = \int_{\{(z,\,S):\,\mathbf{p}(z) \in \mathbf{U}(a,\,\lambda r) \sim X\}} \delta(z, S, \Phi)(\varphi(\mathbf{p}(z))e_{n+1},$$
$$D\varphi(p(z)) \circ \mathbf{p}e_{n+1}) \, dV(z, S).$$

Evidently,

$$T = T_1 + T_2 + T_3 + T_4 + T_5 \quad \text{and} \quad U = U_1 + U_2.$$

Consequently,

$$T_5 = (T - U_1) - T_1 - T_2 - T_3 - U_2 + U$$

since $T_4 = 0$.

Suppose $0 < \eta < \infty$. Arguing as in [**AW1**, 8.16], thereby using 3.3 when required, we see that, if δ is sufficiently small and if (6) does *not* hold, there is a smooth function $h: \mathbf{U}(a, \lambda r) \to \mathbf{R}$ such that

$$r^{-n-2} \int_{\mathbf{U}(a, \lambda r)} \left| \mu^{-1} f - h \right|^2 d\mathcal{L}^n \leqslant \eta^2,$$

$$r^{-n-2} \int_{\mathbf{U}(a, \lambda r)} |h|^2 d\mathcal{L}^n \quad \text{does not exceed} \quad C = C(E, \gamma, N_{0,2}, N_{0,3})$$

and

$$\int_{\mathbf{U}(a, \lambda r)} \langle \operatorname{grad} \varphi \odot \operatorname{grad} h, \Sigma \rangle \, d\mathcal{L}^n = 0 \quad \text{for } \varphi \in \mathscr{D}(\mathbf{U}(a, \lambda r)),$$

where we have set

$$\langle \alpha \odot \beta, \Sigma \rangle = \langle (0, 0, \alpha) \odot (0, 0, \beta), D^2 \Psi^{\S}(a, b, l) \rangle$$

for $\alpha, \beta \in \operatorname{Hom}(\mathbf{R}^n, \mathbf{R})$.

Set $\tilde{b} = \mu h(a)$ and $\tilde{l} = \mu \, Dh(a)$. Let $\tilde{q}$ be as above. It follows from standard elliptic theory that

$$\limsup_{\omega \downarrow 0} \omega^{-n-4} r^{-n-2} \int_{\mathbf{U}(a, \omega r)} \left| h(x) - \mu^{-1} [\tilde{b} + \tilde{l}(x - a)] \right|^2 d\mathcal{L}^n < \infty.$$

We now estimate

$$(\omega r)^{-n-2} \int_{\mathbf{p}^{-1}[\mathbf{U}(a, \omega r)] \cap (\mathbf{X} \times \mathbf{R})} \left| \tilde{q}(z - \tilde{c}) \right|^2 d\|V\|z$$

in terms of

$$(\omega r)^{-n-2} \int_{\mathbf{U}(a, \omega r) \cap \mathbf{X}} \left| f(x) - [\tilde{b} + \tilde{l}(x - a)] \right|^2 d\mathcal{L}^n x$$

using 3.3 and estimate

$$(\omega r)^{-n-2} \int_{\mathbf{p}^{-1}[\mathbf{U}(a, \omega r)] \sim (\mathbf{X} \times \mathbf{R})} \left| \tilde{q}(z - \tilde{c}) \right|^2 d\|V\|z$$

in terms of μ^2 by using 3.3 and the height bound of 3.5. $\square$

THE REGULARITY THEOREM. *Suppose* 2.2(1)–(8) *hold,* (1)–(5) *hold and* $0 < \alpha < 1$. *There are positive real numbers* $\delta_8 = \delta_8(E, \gamma, N_{0,2}, N_{0,3}, \alpha)$ *and* $C_5 = C_5(E, \gamma, N_{0,2}, N_{0,3}, \alpha)$ *such that if* $\delta \leqslant \delta_8$ *then there is a continuously differentiable* *function f: $\mathbf{U}(a, \lambda r) \to \mathbf{R}$ such that*

$$(9) \qquad r^{-1}|f(a) - b| + |Df(a) - l| \leqslant C_5[\mu + r\max\{H, N_{1,0}, N_{1,1}\}];$$

$$(10) \qquad |Df(\tilde{x}) - Df(x)| \leqslant C_5[\mu + r\max\{H, N_{1,0}, N_{1,1}\}](|\tilde{x} - x|/r)^{\alpha}$$

whenever $\tilde{x}, x \in \mathbf{U}(a, \lambda r)$

and

$$\mathrm{spt}\|V\| \cap \mathbf{p}^{-1}[\mathbf{U}(a, \lambda r)] = f.$$

PROOF. Proceed as in [**AW1**, 8.17–19].

REMARK. We do not believe it is necessary to assume (1) and (2) above. Hypothesis (2) might be eliminated by pursuing further the work of [**SL1**] and [**SL2**]. Both (1) and (2) could be eliminated by obtaining a lower bound for mass ratios as we remarked after the height bound of 3.5.

REFERENCES

[**AW1**] W. K. Allard, *On the first variation of a varifold*, Ann. of Math. **95** (1972), 417–491.

[**AW2**] ______, *On the first and second variation of a parametric integral*, Amer. J. Math. **105** (1983), 1255–1276.

[**AW3**] ______, *An a priori estimate for the oscillation of the normal to a hypersurface whose first and second variation with respect to an elliptic integrand is controlled*, Invent. Math. **73** (1983), 287–321.

[**AF**] F. J. Almgren, Jr., *Existence and regularity almost everywhere of solutions to elliptic variational problems among surfaces of varying topological type and singularity structure*, Ann. of Math. **87** (1968), 321–391.

[**FE**] H. Federer, *Geometric measure theory*, Springer-Verlag, 1969.

[**GT**] D. Gilbarg and N. Trudinger, *Elliptic partial differential equations of second order*, Springer-Verlag, 1977.

[**SL1**] L. Simon, *Interior gradient bounds for nonuniformly elliptic equations*, Indiana Univ. Math. J. **25** (1976), 821–855.

[**SL2**] ______, *A Hölder estimate for quasiconformal maps between surfaces in Euclidean space*, Acta Math. **139** (1977), 19–51.

[**SE**] E. M. Stein, *Singular integrals and differentiability properties of functions*, Princeton Univ. Press, Princeton, N.J., 1970.

DUKE UNIVERSITY

Proceedings of Symposia in Pure Mathematics
Volume **44** (1986)

Deformations and Multiple-Valued Functions[1]

F. ALMGREN

TABLE OF CONTENTS

1980 *Mathematics Subject Classification*. Primary 49F20; Secondary 28A75, 35R99, 49F22, 53C99, 58D99.

[1] This work was supported in part by grants from the National Science Foundation.

© 1986 American Mathematical Society
0082 0717/86 $1.00 ı $.25 por page

INTRODUCTION

This paper is a contribution to the development of multiple-valued function theory[2] as a useful and fundamental tool of geometric analysis. The multiple-valued functions f of this paper are frequently associated with the slices $\langle T, \Pi, \cdot \rangle$ of an m-dimensional (real) rectifiable current T in $\mathbf{R}^{m+n}$ by an orthogonal projection $\Pi: \mathbf{R}^{m+n} \to \mathbf{R}^m$, e.g., the value of $f(x)$ at $x \in \mathbf{R}^m$ is the 0-dimensional polyhedral chain $\langle T, \Pi, x \rangle \in \mathbf{P}_0(\mathbf{R}^{m+n})$. Under appropriate assumptions about T, we can write $T = f_{\#} \mathbf{E}^m$, i.e., T is the multiple-valued image of the m-dimensional Euclidean current $\mathbf{E}^m$.

One main object of multiple-valued function theory is to obtain strong and useful representations or approximations of general or special (e.g., area minimizing) currents T by currents $f_{\#} \mathbf{E}^m$. When this can be done we are then able to analyze the surface T (possibly having highly elaborate topological or singularity structure) in terms of properties of a function defined on a fixed simple domain and taking values in a relatively simple range; branching behavior of T, for example, fits nicely into this setup.

The basic currents studied in this paper are the size bounded rectifiable currents $T = \mathbf{t}(S, \theta, \xi)$. In accordance with the terminology of Appendix B this means that $S = \mathrm{set}(T)$ is an $(\mathscr{H}^m, m)$ rectifiable and $\mathscr{H}^m$ measurable subset of $\mathbf{R}^{m+n}$, $\mathbf{S}(T) = \mathscr{H}^m(S)$ is the **size** of T, $\theta: S \to \mathbf{R}^+$ is a positive $\mathscr{H}^m \llcorner S$ summable density function, $\mathbf{M}(T) = \int_S \theta \, d\mathscr{H}^m$ is the **mass** of T, $\xi: S \to \mathbf{G}_0(m + n, m) \subset \wedge_m \mathbf{R}^{m+n}$ is a unit simple m-vector-valued orientation function, and $T(\phi) = \int_S \langle \xi, \phi \rangle \theta \, d\mathscr{H}^m$ for appropriate differential m forms ϕ on $\mathbf{R}^{m+n}$. We frequently write $\mathbf{MS}(T) = \mathbf{M}(T) + \mathbf{S}(T)$.

---

[2] The word *function* ultimately always means a "single-valued" function. The functions f of this paper take (single) values in spaces of 0-dimensional polyhedral chains. The support of a 0-dimensional polyhedral chain is an unordered collection of points. In many cases, for $x \in \mathrm{dmn}(f)$, individual points in the support of $f(x)$ arise from or will be assembled locally into distinct sheets of a higher-dimensional surface, each sheet itself being the graph or the image of a single-valued function. It is with this geometric interpretation and application in mind that we heuristically describe such an f as a multiple-valued function.

The size function and size metric introduced formally in this paper seem essential in controlling the geometry of our surfaces S when densities can be small (for integral densities, size—formerly "reduced mass"—is dominated by mass). The size function is not a norm on the vector space $\mathscr{S}_m(\mathbf{R}^{m+n})$ of size bounded (real) rectifiable currents, although the mass function is. The size function, however, permits us to state inequalities in the Isoperimetric Inequalities 2.2 and 2.3 which are invariant under scalar multiplication and similarity transformations; compare [**FF**, Corollary 6.3] or [**FH**, 4.2.10]. The size function incorporated in the definition of the **GS** metric (see Appendix B) also enables us to generate nontrivial homotopy in the vector space $\mathbf{S}_m(\mathbf{R}^{m+n}) \cap \{T: \partial T = 0 \text{ and } \operatorname{spt}(T) \subset A\}$ of cycles lying in a compact Lipschitz neighborhood retract $A \subset \mathbf{R}^{m+n}$.

Finally, and most importantly in the present paper, multiple-valued mapping of rectifiable currents (as in 3.13) seems naturally defined for $\mathbf{P}_0(\mathbf{R}^{m+n})$-valued functions which are **GS** Lipschitz.

Perhaps the basic fact about multiple-valued approximations of currents (see 3.17) is that, for T as above with $\partial T = 0$, there will exist a **GS** Lipschitz function $f: \mathbf{R}^m \to \mathbf{P}_0(\mathbf{R}^n)$ with $T = (\mathbf{1}_{\mathbf{R}^m} \bowtie f)_{\#}\mathbf{E}^m$ (see Appendix B) if and only if

$$(*) \qquad \infty > \sup\{r^{-m}\mathbf{MS}(T \llcorner \mathbf{B}^m(x, r) \times \mathbf{R}^n): x \in \mathbf{R}^m \text{ and } 0 < r < \infty\}.$$

The experienced geometric measure theorist will recognize this condition as an invitation to fix a (large) Lipschitz constant N, define a "bad set" B to consist of those points $b \in \mathbf{R}^m$ for which

$$\mathbf{MS}(T \llcorner \mathbf{B}^m(b, r) \times \mathbf{R}^n) > \alpha(m)r^m$$

for **some** $0 < r < \infty$, and estimate $\mathscr{L}^m(B)$ (to be small) using the Besicovitch–Federer Covering Theorem. In 3.16 (the "workhorse theorem") we construct such B and then use the deformation theory constructed in 1.14 and 1.15 to deform T within $B \times \mathbf{R}^n$ (leaving $T \llcorner (\mathbf{R}^m \sim B) \times \mathbf{R}^n$ fixed) to become a new current P for which condition $(*)$ does hold (with P replacing T and for all x and r). We can thus write $T = f_{\#}\mathbf{E}^m + (T - P)$; in particular, the "error current" $(T - P)$ projects by Π to lie within (the small set) B.

We can iterate this scheme for approximations with small projected errors (why this works is sheer magic!) ultimately to write $T = g_{\#}\mathbf{E}^m + R$ where g is a **GS** Lipschitz multiple-valued function and the error or remainder current R projects to small area in many directions; see 4.1. Such R is the boundary of an $(m + 1)$-dimensional current of small mass and size according to the Isoperimetric Inequality 2.5. The uniformity of our estimates enables us to prove **G** sequential compactness for spaces of mass, size, and space bounded (real) rectifiable cycles in 4.2 (easy examples show such spaces are not **GS** sequentially compact). A consequence is the **F** compactness for corresponding spaces of real rectifiable currents shown in 4.3. The same proofs with only obvious adaptations apply to integral currents. Compactness of spaces of integral currents was shown originally by H. Federer and W. H. Fleming in [**FF**] (see also [**FH**, 4.2.17]) and later by B. Solomon using a multiple-valued function approach different from the one of this paper; see [**S1**] and [**S2**]. I do not know any adaptation of these arguments for the integral case which can be used for the real coefficient case.

Our present compactness proofs seem to apply with appropriate modifications to prove compactness of spaces of flat chains modulo ν for $\nu = 2, 3, 4, \ldots$, originally proved by W. H. Fleming in [**FW**]; see also [**FH**, 4.2.26]. I have not checked all details carefully, in part because the natural generality of such a context is not yet clear to me.

The essential integral geometric idea (1.12) for the deformation theory of Chapter One appears originally in [**FF**, 5.5]; see [**FH**, 4.2.7]. The implementation of that idea in this paper is somewhat different, however. Our estimates are local rather than global (since we use fixed cubes and variable centers of projection) and are stronger in some ways (since we use a different projection technique in the final cubes); see the remark following the Cubical Approximation Theorem 2.1. Our construction also admits a straightforward intrinsic extension to "admissible" simplicial or cubical decompositions of a variety of spaces, e.g., Riemannian manifolds.

A curious twist in the deformation theory is that essentially the original geometric construction of [**FF**, 5.5], [**FH**, 4.2.1–4.2.9] (but not its estimates) is needed (in 1.16) in deforming a cubical chain (1.11) when projected area control is essential (1.17). A combination of the two types of deformations yields a general theorem (2.9) on deforming out parts of rectifiable cycles having small projected areas; the theorem is also used in the proof of 4.2.

We call attention to the Table of Contents in which the mathematical content of this paper is indicated in some detail. We also call attention to the note [**AS**] in which a number of recent applications of multiple-valued function theory are summarized, e.g., important contributions by F. Almgren, P. Mattila, D. Nance, B. Solomon, B. Super, and B. White. Various facets of multiple-valued function theory have been part of mathematics for a very long time (e.g., functions mapping coefficients of a polynomial to its roots; note [**FH**, 4.3.12]); my own list of references is long and is certainly incomplete.

Certainly a substantial portion of the present exposition has been part of the working knowledge of experienced geometric measure theorists. It seems clear that, at least for the time being, multiple-valued function techniques are indispensable in proving various central results of geometric measure theory. It seems important to set things down in a relatively systematic way and to explore the extent to which basic as well as advanced material in the discipline should be formulated within this context. Those interested in developing a more computational geometric measure theory will appreciate the explicit combinatorial constructions of §§3.1–3.12 in Chapter Three. Several intriguing new questions and conjectures have already arisen.

Finally, I wish to acknowledge the benefit of stimulating and extensive conversations with W. K. Allard, P. Mattila, D. Nance, B. Solomon, and B. White about a variety of aspects of multiple-valued function theory.

CHAPTER ONE
CUBICAL COMPLEXES AND DEFORMATIONS

1.1. Index sets for standard projections, injections, and k vectors. Throughout this paper we assume m and n are fixed positive integers. For each $k \in \{1, 2, \ldots, m + n\}$ we recall that $\Lambda(m + n, k)$ denotes the family of all increasing maps $\{1, \ldots, k\} \to \{1, \ldots, m + n\}$, and we define

$$\Pi \colon \Lambda(m + n, k) \to \mathbf{O}(m + n, k), \qquad \Pi^* \colon \Lambda(m + n, k) \to \mathbf{O}^*(m + n, k),$$

$$\mathbf{e} \colon \Lambda(m + n, k) \to \wedge_k \mathbf{R}^{m+n}$$

by the requirements that

$$\Pi(\lambda)(x) = (x_{\lambda(1)}, \ldots, x_{\lambda(k)}), \qquad \Pi(\lambda) \circ \Pi^*(\lambda) = \mathbf{1}_{\mathbf{R}^k},$$

$$\mathbf{e}(\lambda) = \mathbf{e}_{\lambda(1)} \wedge \cdots \wedge \mathbf{e}_{\lambda(k)}$$

for each $\lambda \in \Lambda(m + n, k)$ and $x \in \mathbf{R}^{m+n}$.

For $k \in \{1, \ldots, m + n\}$ we also define

$$* \colon \Lambda(m + n, k) \to \Lambda(m + n, m + n - k)$$

by requiring $\mathrm{Im}(\lambda) \cup \mathrm{Im}(*\lambda) = \{1, \ldots, m + n\}$ for each $\lambda \in \Lambda(m + n, k)$.

1.2. Standard cubes and their interiors, boundaries, dimensions, levels, directions, projection centers, and faces. (i) We denote by $\mathbf{K}(0)$ the collection of all closed unit $(m + n)$-dimensional cubes in $\mathbf{R}^{m+n}$ associated with the integer lattice. In particular, $K \in \mathbf{K}(0)$ if and only if there are $z_1, \ldots, z_{m+n} \in \mathbf{Z}$ with

$$K = \tau(z_1, \ldots, z_{m+n})[0, 1]^{m+n}.$$

For each $N \in \mathbf{Z}$ we set

$$\mathbf{K}(N) = \{\mu(2^{-N})K \colon K \in \mathbf{K}(0)\}.$$

The members of $\mathbf{K}(N)$ are called **standard cubes of level** N.

(ii) For each $N \in \mathbf{Z}$ we set

$$\mathbf{K}_0(N) = \{\mu(2^{-N})z \colon z \in \mathbf{Z}^{m+n}\}$$

and for each $K \in \mathbf{K}_0(N)$ write $\dim(K) = 0$, $\mathrm{Int}(K) = \varnothing$, and $\partial K = K$.

(iii) For each $N \in \mathbf{Z}$, each $k \in \{1, \ldots, m + n\}$, and each $\lambda \in \Lambda(m + n, k)$ we set

$$\mathbf{K}_k(N, \lambda) = \{\mu(2^{-N}) \circ \tau(z) \circ \Pi^*(\lambda)[0, 1]^k \colon z \in \mathbf{Z}^{m+n}\}$$

and for such $K = \mu(2^{-N}) \circ \tau(z) \circ \Pi^*(\lambda)[0, 1]^k$ write

$$\dim(K) = k, \qquad \mathrm{level}(K) = N, \qquad \mathrm{direction}(K) = \lambda,$$

$$\mathrm{Int}(K) = \mu(2^{-N}) \circ \tau(z) \circ \Pi^*(\lambda)(0, 1)^k, \qquad \partial K = K \sim \mathrm{Int}(K),$$

$$\mathrm{center}(K) = \mu(2^{-N}) \circ \tau(z) \circ \Pi^*(\lambda)(\tfrac{1}{2}, \ldots, \tfrac{1}{2}),$$

and

$$\mathbf{PC}(K) = \mu(2^{-N}) \circ \tau(z) \circ \Pi^*(\lambda)\left[\tfrac{1}{4}, \tfrac{3}{4}\right]^{m+n} \quad \text{(projection centers)}.$$

We also set

$$\mathbf{K}_k(N) = \bigcup_\lambda \mathbf{K}_k(N, \lambda), \qquad \mathbf{K}_* = \bigcup_{k,N} \mathbf{K}_k(N),$$

$$\mathbf{K} = \bigcup_N \mathbf{K}(N) = \bigcup_N \mathbf{K}_{m+n}(N).$$

The members of $\mathbf{K}_k(N)$ are called **standard cubes of dimension k and level N.**

We check that whenever K, $L \in \mathbf{K}$ with $\operatorname{Int}(K) \cap \operatorname{Int}(L) \neq \varnothing$ then either $K \subset L$ or $L \subset K$. Also, corresponding to each M, $N \in \mathbf{Z}$ with $M < N$ and to each $K_N \in \mathbf{K}(N)$, we readily verify the existence of a unique sequence $K_N \subset K_{N-1} \subset K_{N-2} \subset \cdots \subset K_M$ with $K_{N-1} \in \mathbf{K}_{N-1}, \ldots, K_M \in \mathbf{K}_M$.

If $K \in \mathbf{K}_k(N)$ for some $k \in \{0, \ldots, m + n\}$ and some $N \in \mathbf{Z}$ we say that J **is a face of** K if and only if $J \in \mathbf{K}_j(N)$ for some $j \in \{0, \ldots, k\}$ and $J \subset K$.

Suppose j, $k \in \{0, \ldots, m + n\}$, M, $N \in \mathbf{Z}$ with $M \leqslant N$, and $J \in \mathbf{K}_j(M)$, $K \in \mathbf{K}_k(N)$ with $J \cap K \neq \varnothing$. We readily check that $J \cap K$ is a face of K with $\dim(J \cap K) \leqslant \inf\{j, k\}$.

Finally, we check that in case $k \in \{0, \ldots, m + n\}$, M, $N \in \mathbf{Z}$ with $M \leqslant N$, $J \in \mathbf{K}_k(M)$ is a face of $K \in \mathbf{K}_{m+n}(M)$, and $L \in \mathbf{K}_{m+n}(N)$ with $L \subset K$ and $L \cap J \neq \varnothing$, then $L \cap J$ is a face of L of dimension k.

1.3. The edge metric. When dealing with families of cubes it is frequently convenient to use the **edge metric $\mathbf{EM}\colon \mathbf{R}^{m+n} \times \mathbf{R}^{m+n} \to \mathbf{R}^+$** defined by setting

$$\mathbf{EM}(x, y) = \sup\left\{|x_i - y_i|\colon i = 1, \ldots, m + n\right\}$$

for each $x, y \in \mathbf{R}^{m+n}$. Whenever $x \in \mathbf{R}^{m+n}$ and $A, B \subset \mathbf{R}^{m+n}$ we set

$$\mathbf{EM}(x, B) = \inf\{\mathbf{EM}(x, b)\colon b \in B\},$$

$$\mathbf{EM}(A, B) = \inf\{\mathbf{EM}(a, b)\colon a \in A, b \in B\}$$

so that $\mathbf{EM}(x, \varnothing) = \mathbf{EM}(A, \varnothing) = \mathbf{EM}(\varnothing, B) = \infty$.

For each $k \in \{1, \ldots, m + n\}$, $N \in \mathbf{Z}$, and $K \in \mathbf{K}_k(N)$ we note that

$$2^{-N} = 2^{-\operatorname{level}(K)} = \sup\{\mathbf{EM}(x, y)\colon x, y \in K\}.$$

Clearly, for each $x, y \in \mathbf{R}^{m+n}$,

$$(m + n)^{-1/2}|x - y| \leqslant \mathbf{EM}(x, y) \leqslant |x - y|.$$

1.4. Proposition (General position intersections of cubical complexes). *Suppose M, $N \subset \mathbf{Z}$, $j \subset \{0, 1, \ldots, m + n - 1\}$, $k \subset \{1, 2, \ldots, m + n - 1\}$, $\lambda \in \Lambda(m + n, k)$, and $*\lambda \neq \mu \in \Lambda(m + n, m + n - k)$.*

(1) For $\mathscr{L}^{m+n}$ almost every $a \in \mathbf{R}^{m+n}$,

$$0 < \mathbf{EM}\left[\tau(a)\bigcup\mathbf{K}_j(M), \bigcup\mathbf{K}_{m+n-j-1}(N)\right]$$

$$= \mathbf{EM}\left[\bigcup\mathbf{K}_j(M), \tau(-a)\bigcup\mathbf{K}_{m+n-j-1}(N)\right].$$

(2) *For $\mathcal{L}^{m+n}$ almost every $a \in \mathbf{R}^{m+n}$,*

$$0 < \mathbf{EM}\Big[\tau(a)\bigcup \mathbf{K}_k(M, \lambda), \bigcup \mathbf{K}_{m+n-k}(N, \mu)\Big]$$
$$= \mathbf{EM}\Big[\bigcup \mathbf{K}_k(M, \lambda), \tau(-a)\bigcup \mathbf{K}_{m+n-k}(N, \mu)\Big].$$

(3) *For each $a \in \mathbf{R}^{m+n}$,*

$$\varnothing \neq \Big[\tau(a)\bigcup \mathbf{K}_k(M, \lambda)\Big] \cap \Big[\bigcup \mathbf{K}_{m+n-k}(N, *\lambda)\Big]$$
$$= \tau(a)\Big\{\Big[\bigcup \mathbf{K}_k(M, \lambda)\Big] \cap \Big[\tau(-a)\bigcup \mathbf{K}_{m+n-k}(N, *\lambda)\Big]\Big\}.$$

(4) *For $\mathcal{L}^{m+n}$ almost every $a \in \mathbf{R}^{m+n}$,*

$$0 < \mathbf{EM}\Big[\tau(a)\bigcup \{\partial K : K \in \mathbf{K}_k(M)\}, \bigcup \mathbf{K}_{m+n-k}(N)\Big]$$
$$= \mathbf{EM}\Big[\bigcup \{\partial K : k \in \mathbf{K}_k(M)\}, \tau(-a)\bigcup \mathbf{K}_{m+n-k}(N)\Big].$$

1.5. Admissible families of cubes and first and second neighbors of a cube. By an **admissible family of cubes** we mean a subset $\mathcal{F}$ of $\mathbf{K}$ for which the following three conditions hold:

(i) $K, L \in \mathcal{F}$ with $K \neq L$ implies $\operatorname{Int}(K) \cap \operatorname{Int}(L) = \varnothing$,

(ii) $K, L \in \mathcal{F}$ with $K \cap L \neq \varnothing$ implies $|\operatorname{level}(K) - \operatorname{level}(L)| \leqslant 1$,

(iii) $K \in \mathcal{F}$ implies $\partial K \subset \bigcup\{L : K \neq L \in \mathcal{F}\}$.

The empty family is admissible. Any nonempty admissible family of cubes is infinite. Also, the union of all the cubes in an admissible family is an open subset of $\mathbf{R}^{m+n}$ which may or may not be bounded.

Suppose $\mathcal{F}$ is an admissible family of cubes and $K \in \mathcal{F}$. By the **first neighbors of K in $\mathcal{F}$** we mean the subfamily

$$\operatorname{nbs}(K) = \mathcal{F} \cap \{L : L \cap K \neq \varnothing\}.$$

By the **second neighbors of K in $\mathcal{F}$** we mean the subfamily

$$\operatorname{NBS}(K) = \bigcup\{\operatorname{nbs}(L) : L \in \operatorname{nbs}(K)\}.$$

We recall 1.2 and conclude that whenever $L \in \operatorname{nbs}(K)$ then $L \cap K$ is either a face of K (of some dimension) or some $H \in \mathbf{K}(\operatorname{level}(K) + 1)$ with $H \subset K$. We check additionally that, among members of $\operatorname{nbs}(K)$, L is uniquely determined by $L \cap K$; this gives one method of estimating $\operatorname{card}(\operatorname{nbs}(K))$. We set

$$M_{1.5} = \sup\{\operatorname{card}(\operatorname{nbs}(K)) : K \text{ is a member of some admissible family of cubes}\},$$

$$N_{1.5} = \sup\{\operatorname{card}(\operatorname{NBS}(K)) : K \text{ is a member of some admissible family of cubes}\};$$

both numbers are finite and can be readily estimated.

1.6. Whitney families of cubes. Corresponding to each open subset U of $\mathbf{R}^{m+n}$ we define $\mathbf{WF}(U)$ (Whitney family) to consist of those cubes $K \in \mathbf{K}$ for which the following two conditions hold (see Figure 1.6):

(i) $\mathbf{EM}(K, \mathbf{R}^{m+n} \sim U) > 2^{1 - \operatorname{level}(K)}$,

(ii) $\mathbf{EM}(L, \mathbf{R}^{m+n} \sim U) \leqslant 2^{1 - \operatorname{level}(L)}$

for that unique $L \in \mathbf{K}(\operatorname{level}(K) - 1)$ with $K \subset L$.

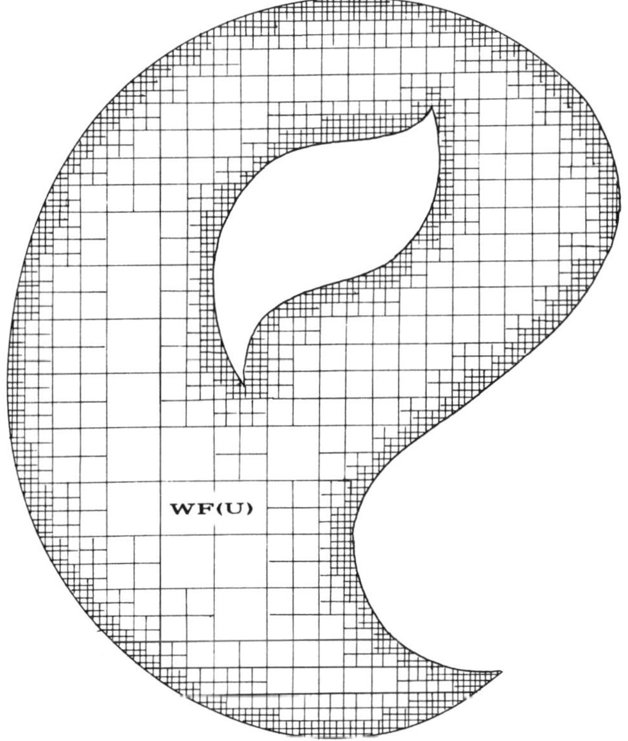

FIGURE 1.6. The Whitney family of cubes in an open set.

$\mathbf{WF}(U)$ is an admissible family of cubes with $\bigcup \mathbf{WF}(U) = U$. A straightforward estimate shows that, whenever $b \in K \in \mathbf{WF}(U)$,

$$(*) \quad \operatorname{dist}(b, \mathbf{R}^{m+n} \sim U)/6(m+n)^{1/2} \leqslant 2^{-\operatorname{level}(K)} \leqslant \operatorname{dist}(b, \mathbf{R}^{m+n} \sim U)/2.$$

Corresponding to each $k \in \{0, 1, 2, \dots\}$ and each U as above, we also set

$$\mathbf{WF}(U; k) = \mathbf{WF}(U) \cap \{K : \operatorname{level}(K) \leqslant k\} \cup \mathbf{K}(k)$$

$$\cap \{K : \operatorname{Int}(K) \cap \operatorname{Int}(L) = \varnothing \quad \text{for each } L \in \mathbf{WF}(U) \text{ with } \operatorname{level}(K) < k\}.$$

Each $\mathbf{WF}(U; k)$ is an admissible family of cubes in $\mathbf{R}^{m+n}$ with $\mathbf{R}^{m+n} = \bigcup \mathbf{WF}(U; k)$.

1.7. Refinements of admissible families of cubes. Suppose $\mathscr{F}_1$ and $\mathscr{F}_2$ are admissible families of cubes. By **the common refinement of $\mathscr{F}_1$ and $\mathscr{F}_2$** we mean the set

$$\mathscr{F} = \mathscr{F}_1 \cap \{K : K \subset L \text{ for some } L \in \mathscr{F}_2\} \cup \mathscr{F}_2 \cap \{L : L \subset K \text{ for some } K \in \mathscr{F}_1\}.$$

$\mathscr{F}$ is an admissible family of cubes with $\bigcup \mathscr{F} = \bigcup \mathscr{F}_1 \cap \bigcup \mathscr{F}_2$. Also, in case $K \in \mathscr{F}_1$, $L \in \mathscr{F}_2$ with $K \subset L$ and $\operatorname{level}(K) \geqslant \operatorname{level}(L) + 2$, then $\mathscr{F}_1 \cap \operatorname{nbs}(K) \subset \mathscr{F}$. Finally, in case U and V are open subsets of $\mathbf{R}^{m+n}$ then $\mathbf{WF}(U \cap V)$ is the common refinement of $\mathbf{WF}(U)$ and $\mathbf{WF}(V)$.

1.8. The cubical complex of an admissible family of cubes and listing functions. Suppose $\mathscr{F}$ is an admissible family of cubes. Associated with $\mathscr{F}$ is **the cubical complex of** $\mathscr{F}$, denoted $\mathbf{CX}(\mathscr{F})$, consisting of all cubes $K \in \mathbf{K}_*$ for which the following two conditions hold:

(i) K is a face (of some dimension) of some cube in $\mathscr{F}$,

(ii) in case $\dim(K) > 0$ then $\mathrm{level}(K) \geqslant \mathrm{level}(L)$ whenever L is a face of some cube in $\mathscr{F}$ with $\dim(K) = \dim(L)$ and $\mathrm{Int}(K) \cap \mathrm{Int}(L) \neq \varnothing$.

For each $k \in \{1, 2, \ldots, m + n\}$ we set

$$\mathbf{CX}_k(\mathscr{F}) = \mathbf{CX}(\mathscr{F}) \cap \{ K \colon \dim(K) = k \}.$$

For each $k \in \{1, \ldots, m + n\}$ and each $K \in \mathbf{CX}_k(\mathscr{F})$, $\partial K \subset \bigcup\{ L \colon L \in \mathbf{CX}(\mathscr{F})$ with $\dim(L) = k - 1\}$. Also, in case $\mathscr{F}_1$ and $\mathscr{F}_2$ are admissible families of cubes having $\mathscr{F}$ as common refinement, then $\bigcup \mathbf{CX}_m(\mathscr{F}_1) \cap \bigcup \mathscr{F} \subset \bigcup \mathbf{CX}_m(\mathscr{F})$.

By a **listing of** $\mathbf{CX}(\mathscr{F})$ we mean a collection of correspondences (one to one and onto mappings)

$$\zeta_k \colon \{1, 2, 3, \ldots\} \to \mathbf{CX}_k(\mathscr{F})$$

for $k = m, m + 1, \ldots, m + n$.

1.9. Index sets, associated sequences, and sources of members of cubical complexes. Suppose $\mathscr{F}$ is an admissible family of cubes, $k \in \{1, \ldots, m + n\}$, and $K_k \in \mathbf{CX}_k(\mathscr{F})$. By **the index set for** K, denoted $\mathrm{INDEX}(K)$, we mean the collection of all nested sequences

$$K_k \subset K_{k+1} \subset K_{k+2} \subset \cdots \subset K_{m+n-1} \subset K_{m+n}$$

with $K_i \in \mathbf{K}_i(\mathrm{level}(K_k))$ for each $i = k, k + 1, \ldots, m + n$. For $k \leqslant m + m - 1$ there are $2(m + n - k)$ distinct K_{k+1}'s with $K_k \subset K_{k+1} \in \mathbf{K}_{k+1}(\mathrm{level}(K_k))$ so that, for each such K_{k+1},

$$\mathrm{card}(\mathrm{INDEX}(K_k)) = 2(m + n - k)\mathrm{card}(\mathrm{INDEX}(K_{k+1}))$$

and, hence,

$$\mathrm{card}(\mathrm{INDEX}(K_k)) = 2^{m+n-k}(m + n - k)!.$$

Corresponding to each sequence $K_k \subset \cdots \subset K_{m+n}$ in $\mathrm{INDEX}(K_k)$ is a unique **associated sequence**

$$K_k \subset L_{k+1} \subset L_{k+2} \subset \cdots \subset L_{m+n-1} \subset L_{m+n}$$

characterized by the requirement that, for each $i = k + 1, \ldots, m + n$, $K_i \subset L_i \in \mathbf{CX}_i(\mathscr{F})$. It follows that $\mathrm{level}(K_k) \geqslant \mathrm{level}(L_{k+1}) \geqslant \cdots \geqslant \mathrm{level}(L_{m+n})$. Finally, we define the **source of** K_k, denoted $\mathrm{SOURCE}(K_k)$, to be the collection of all cubes $L_{m+n} \in \mathbf{CX}_{m+n}(\mathscr{F}) = \mathscr{F}$ which appear in the sequence $K_k \subset \cdots \subset L_{m+n}$ associated with some sequence $K_k \subset \cdots \subset K_{m+n}$ in $\mathrm{INDEX}(K_k)$. We compute

$$\mathrm{card}(\mathrm{SOURCE}(K_k)) = 2^{m+n-k},$$

$$\bigcup \mathrm{SOURCE}(K_k) \subset \mathbf{R}^{m+n} \cap \{ x \colon \mathbf{EM}(x, K_k) \leqslant 2^{1 - \mathrm{level}(K_k)} \}.$$

1.10. Cubical Constancy Theorem. *Suppose* $k \in \{1,\ldots,m+n\}$, $N \in \mathbf{Z}$, $z \in \mathbf{Z}^{m+n}$, $\lambda \in \Lambda(m+n,k)$, *and*

$$K = \mu(2^{-N}) \circ \tau(z) \circ \Pi^*(\lambda)[0,1]^k \in \mathbf{K}_k(N).$$

Suppose also $T \in \mathbf{N}_{k,0}(\mathbf{R}^{m+n})$ *with* $\mathrm{spt}(T) \subset K$ *and* $\mathrm{spt}(\partial T) \subset \partial K$. *Then we infer from the Constancy Theorem* [**FH**, 4.1.7] *that*

$$T = r\mathbf{t}(K,1,\mathbf{e}(\lambda)) \in \mathbf{P}_k(\mathbf{R}^{m+n})$$

for some $r \in \mathbf{R}$. *Here we use the obvious extension of terminology to denote*

$$\mathbf{e}(\lambda): K \to \{\mathbf{e}(\lambda)\} \in \bigwedge_k \mathbf{R}^{m+n};$$

we will hereafter follow this convention without additional comment.

1.11. The real and integral chains of a cubical complex. Suppose $\mathscr{F}$ is an admissible family of cubes. We say that T is a **size bounded real k chain of $\mathscr{F}$** if and only if $k \in \{0,1,\ldots,m+n\}$, $T \in \mathscr{S}_k(\mathbf{R}^{m+n})$, and there is a function r: $\mathbf{CX}_k(\mathscr{F}) \to \mathbf{R}$ such that

$$T = \sum \{r(K)[\![p(K)]\!]: \{p(K)\} = K \in \mathbf{CX}_0(\mathscr{F})\} \qquad \text{if } k = 0,$$

$$T = \sum \{r(K)\mathbf{t}(K,1,\mathbf{e}(\mathrm{direction}(K))): K \in \mathbf{CX}_k(\mathscr{F})\} \qquad \text{if } k > 0$$

with summation convergent in the **MS** metric. We further denote by $\mathscr{S}_k(\mathscr{F})$ the real vector subspace of $\mathscr{S}_k(\mathbf{R}^{m+n})$ consisting of such size bounded real k chains of $\mathscr{F}$. $\mathscr{I}_m(\mathscr{F})$, $\mathbf{S}_m(\mathscr{F})$, and $\mathbf{I}_m(\mathscr{F})$ have the obvious meanings.

The following properties are readily verified.

(1) For each $N \in \mathbf{Z}$, $\mathscr{S}_m(\mathbf{K}(N)) = \mathbf{S}_m(\mathbf{K}(N))$ and $\mathscr{I}_m(\mathbf{K}(N)) = \mathbf{I}_m(\mathbf{K}(N))$.

(2) For each $M, N \in \mathbf{Z}$ with $M < N$,

$$\mathbf{S}_m(\mathbf{K}(M)) \subset \mathbf{S}_m(\mathbf{K}(N)) \quad \text{and} \quad \mathbf{I}_m(\mathbf{K}(M)) \subset \mathbf{I}_m(\mathbf{K}(N)).$$

(3) If $N \in \mathbf{Z}$ with $0 \neq S \in \mathbf{S}_m(\mathbf{K}(N))$ and $0 \neq T \in \mathbf{I}_m(\mathbf{K}(N))$ then

$$\mathbf{S}(S) \geqslant 2^{-mN} \quad \text{and} \quad \mathbf{M}(T) \geqslant \mathbf{S}(T) \geqslant 2^{-mN}.$$

(4) If $N \in \mathbf{Z}$ and $Y \in \mathbf{S}_{m+1}(\mathbf{K}(N))$ with $0 \neq T = \partial Y \in \mathbf{I}_m(\mathbf{K}(N))$ then

$$\mathbf{M}(Y) \geqslant n^{-1}2^{-(m+1)N};$$

this is checked from the linearity of Stokes' Theorem (compare [**FH**, 4.1.8]) and the fact that for any $\lambda \in \Lambda(m+n,m)$, $n = \mathrm{card}[\Lambda(m+n,m+1) \cap \{\mu: \mathrm{Im}(\lambda) \subset \mathrm{Im}(\mu)\}]$.

1.12. Central projections within cubes. Suppose $j \in \{0,\ldots,n\}$, $K = [0,1]^{m+j} \subset \mathbf{R}^{m+j}$, $p \in \mathbf{PC}(K) = [\tfrac{1}{4},\tfrac{3}{4}]^{m+j} \subset K$, and $0 < \varepsilon < \tfrac{1}{2}$. We define

$$\sigma(P): K \sim \{p\} \to \partial K, \qquad \sigma(p,\varepsilon): K \to K, \qquad \tau(p,\varepsilon): [0,1] \times K \to K$$

by setting

$$\sigma(p)(x) = q \in \partial K \quad \text{if and only if} \quad q = p + t(x-p) \quad \text{for some } t \geqslant 1,$$

$$\sigma(p,\varepsilon)(x) = \begin{cases} \sigma(p)(x) & \text{if } |x-p| \geqslant \varepsilon, \\ p + \varepsilon^{-1}|x-p|(\sigma(p)(x)-p) & \text{if } |x-p| < \varepsilon, \end{cases}$$

$$\tau(p,\varepsilon)(t,x) = (1-t)p + t\sigma(p,\varepsilon)(x) \quad \text{for each } t, x.$$

If $k \in \{m,\ldots,m+j\}$ and $T = \mathbf{t}(S, \theta, \xi) \in \mathbf{S}_k(K)$ then it is straightforward to confirm the existence of

$$\sigma(p)_{\#}T = \lim_{\varepsilon \downarrow 0} \sigma(p, \varepsilon)_{\#}T \in \mathbf{S}_k(K) \qquad \left[\in \mathbf{S}_k(\partial K) \text{ if } k < m+j\right],$$

$$\sigma(p)_{\#}\partial T = \lim_{\varepsilon \downarrow 0} \sigma(p, \varepsilon)_{\#}\partial T \in \mathbf{S}_{k-1}(\partial K),$$

$$\tau(p)_{\#}(\llbracket 0, 1 \rrbracket \times T) = \lim_{\varepsilon \downarrow 0} \tau(p, \varepsilon)_{\#}(\llbracket 0, 1 \rrbracket \times T) \in \mathbf{S}_{k+1}(K),$$

$$\tau(p)_{\#}(\llbracket 0, 1 \rrbracket \times \partial T) = \lim_{\varepsilon \downarrow 0} \tau(p, \varepsilon)_{\#}(\llbracket 0, 1 \rrbracket \times \partial T) \in \mathbf{S}_k(K)$$

(with all limits in the **MS** metric) with $\partial \sigma(p)_{\#}T = \sigma(p)_{\#}\partial T$
and

$$\sigma(p)_{\#}T - T = \partial \tau(p)_{\#}(\llbracket 0, 1 \rrbracket \times T) + \tau(p)_{\#}(\llbracket 0, 1 \rrbracket \times \partial T)$$

(see Figure 1.12) provided

(a)
$$\int_{x \in \mathrm{set}(T) \cap \mathrm{Int}(K)} |x - p|^{-k} \, d\mathcal{H}^k x + \int_{x \in \mathrm{Int}(K)} |x - p|^{-k} \, d\|T\|x$$

is finite (if $k < m+j$) or

$$\Theta^k\left(\mathcal{H}^k \llcorner \mathrm{set}(T), p\right) \in \{0, 1\}$$

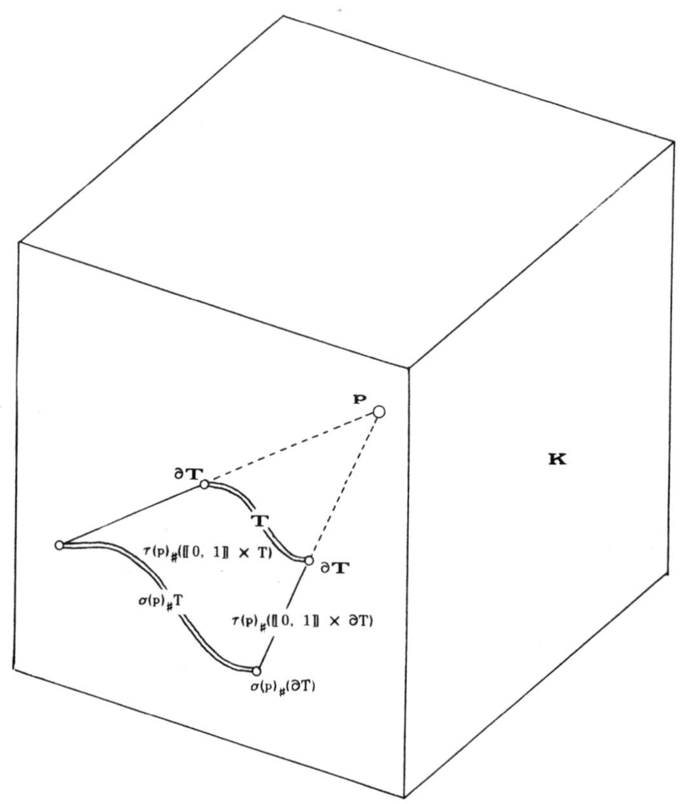

FIGURE 1.12. Deformation of a current within a cube.

and θ is approximately continuous at p (if $k = m + j$), and

(b) $\displaystyle\int_{x \in \mathrm{set}(\partial T) \cap \mathrm{Int}(K)} |x - p|^{1-k} \, d\mathscr{H}^{k-1} x + \int_{x \in \mathrm{Int}(K)} |x - p|^{1-k} \, d\|\partial T\| x$

is finite.

For a general Radon measure μ over K we check (compare [**FH**, 4.2.7])

$$\int_{p \in [1/4, 3/4]^{m+j}} \int_{x \in \mathrm{Int}(K)} |x - p|^{-k} \, d\mu x \, d\mathscr{L}^{m+j} p$$

$$= \int_{x \in \mathrm{Int}(K)} \int_{p \in [1/4, 3/4]^{m+j}} |x - p|^{-k} \, d\mathscr{L}^{m+j} p \, d\mu x \leqslant \Gamma \mu (\mathrm{Int}(K))$$

in case $k < m$; here Γ is a readily estimable absolute constant. For $k = m + j$ we have

$$\int_{p \in \mathrm{Int}(K)} \Theta^k \big(\mathscr{H}^k \, \llcorner \, \mathrm{set}(T), \, p \big) \, d\mathscr{L}^k p = \mathscr{H}^k \big[\mathrm{set}(T) \cap \mathrm{Int}(K) \big],$$

$$\int_{p \in \mathrm{set}(T) \cap \mathrm{Int}(K)} \theta(p) \, d\mathscr{L}^k p = \|T\|(\mathrm{Int}(K)).$$

We use the Hausdorff area formula to check that the integrals (or densities) in (a) and (b) dominate $\mathbf{MS}(\sigma(p)_{\#}T)$, $\mathbf{MS}(\sigma(p)_{\#}\partial T)$, $\mathbf{MS}[\tau(p)_{\#}(\llbracket 0, 1 \rrbracket \times T)]$, and $\mathbf{MS}[\tau(p)_{\#}(\llbracket 0, 1 \rrbracket \times \partial T)]$. We conclude finally the existence of a constant $\Gamma_{1.13} < \infty$ for which the following holds.

1.13. Basic central projection estimates. Corresponding to each $T \in \mathbf{S}_{k,0}(K)$ there is $p \in \mathbf{PC}(K)$ such that $\sigma(p)_{\#}T$, $\sigma(p)_{\#}\partial T$, $\tau(p)_{\#}(\llbracket 0, 1 \rrbracket \times T)$, $\tau(p)_{\#}(\llbracket 0, 1 \rrbracket \times \partial T)$ all exist with

$$\sigma(p)_{\#}T \in \mathbf{S}_{k,0}(\partial K) \quad \text{if } k < m + j,$$

$$\sigma(p)_{\#}T \in \mathbf{S}_{k,0}(K) \quad \text{and} \quad \sigma(p)_{\#}\partial T \in \mathbf{S}_{k-1,0}(\partial K) \quad \text{if } k = m + j,$$

$$\partial \sigma(p)_{\#}T = \sigma(p)_{\#}\partial T$$

(so that for $k = m + j$ the Cubical Constancy Theorem 1.10 is applicable to $\sigma(p)_{\#}T$),

$$\mathbf{M}\big(\sigma(p)_{\#}T\big) \leqslant \Gamma_{1.13}\mathbf{M}(T), \qquad \mathbf{S}\big(\sigma(p)_{\#}T\big) \leqslant \Gamma_{1.13}\mathbf{S}(T),$$

$$\mathbf{M}\big(\partial\sigma(p)_{\#}T\big) \leqslant \Gamma_{1.13}\mathbf{M}(\partial T), \qquad \mathbf{S}\big(\partial\sigma(p)_{\#}T\big) \leqslant \Gamma_{1.13}\mathbf{S}(\partial T),$$

$$\sigma(p)_{\#}(\llbracket 0, 1 \rrbracket \times T) \in \mathbf{S}_{k+1,0}(K) \quad \text{if } k < m + j,$$

$$\tau(p)_{\#}(\llbracket 0, 1 \rrbracket \times T) = 0 \quad \text{if } k = m + j,$$

$$\tau(p)_{\#}(\llbracket 0, 1 \rrbracket \times \partial T) \in \mathbf{S}_{k,0}(K),$$

$$\mathbf{M}\big(\tau(p)_{\#}(\llbracket 0, 1 \rrbracket \times T)\big) \leqslant \Gamma_{1.13}\mathbf{M}(T), \qquad \mathbf{S}\big(\tau(p)_{\#}(\llbracket 0, 1 \rrbracket \times T)\big) \leqslant \Gamma_{1.13}\mathbf{S}(T),$$

$$\mathbf{M}\big(\tau(p)_{\#}(\llbracket 0, 1 \rrbracket \times \partial T)\big) \leqslant \Gamma_{1.13}\mathbf{M}(\partial T),$$

$$\mathbf{S}\big(\tau(p)_{\#}(\llbracket 0, 1 \rrbracket \times \partial T)\big) \leqslant \Gamma_{1.13}\mathbf{S}(\partial T).$$

If $0 < r < 1$ and $L = \mu(r)K$ then maps of the form

$$\mu(r) \circ \sigma(p, \varepsilon) \circ \mu(1/r) \quad \text{and} \quad \mu(r) \circ \tau(p, \varepsilon) \circ (\mathbf{1}_\mathbf{R} \times \mu(1/r))$$

can be used to define central projections in L in the obvious way. Corresponding estimates result, e.g., for $T \in \mathbf{S}_k(L)$,

$$\mathbf{M}\big(\mu(r)_\# \circ \sigma(p)_\# \circ \mu(1/r)_\# T\big) \leqslant \Gamma_{1.13} \mathbf{M}(T),$$

$$\vdots$$

$$\mathbf{S}\big(\mu(r)_\# \circ \tau(p)_\# \circ (\mathbf{1}_\mathbf{R} \times \mu(1/r))_\#([\![0,1]\!] \times T)\big) \leqslant \Gamma_{1.13} r \mathbf{S}(T), \ldots ;$$

note the factor r.

Also, in case $\nu \in \{2, 3, 4, \ldots\}$ and $k(i) \in \{m, \ldots, m + j\}$, $T(i) \in \mathbf{S}_{k(i),0}(K)$ for $i \in \{1, \ldots, \nu\}$, then $p \in \mathbf{PC}(K)$ can be selected such that, for each $i \in \{1, \ldots, \nu\}$, $\sigma(p)_\# T(i)$, $\sigma(p)_\# \partial T(i)$, $\tau(p)_\#([\![0,1]\!] \times T(i))$, $\tau(p)_\#([\![0,1]\!] \times \partial T(i))$ all exist and the corresponding equations and inequalities above all are true provided we replace $\Gamma_{1.13}$ by $\nu \Gamma_{1.13}$.

Finally, in view of [**FH**, 2.10.43, 4.1.14, 4.1.15] we note the obvious extension of our construction (with corresponding equations and inequalities) to the case in which $\mathscr{F}$ is an admissible family of $m + n$ cubes in $\mathbf{R}^{m+n}$, $L \in \mathbf{CX}_{m+j}(\mathscr{F})$, $\nu \in \{1, 2, 3, \ldots\}$, and for each $i \in \{1, \ldots, \nu\}$ and $k(i) \in \{m, \ldots, m + j\}$, $T(i) \in \mathbf{S}_{k(i)}(\mathbf{R}^{m+n})$ with $\mathrm{spt}[T(i)] \cap \bigcup \mathscr{F} \subset \bigcup \mathbf{CX}_{m+j}(\mathscr{F})$. In such a construction, a choice of $p \in \mathbf{PC}(L)$ which yields the required equations and inequalities is called **suitable for** $T(1), \ldots, T(\nu)$.

We also adapt our terminology to give the obvious meanings to

$$\sigma(L, p)_\# T(i), \qquad \sigma(L, p)_\# \partial T(i),$$
$$\tau(L, p)_\#([\![0,1]\!] \times T(i)), \qquad \tau(L, p)_\#([\![0,1]\!] \times \partial T(i))$$

for each i; in particular,

$$\big[\sigma(L, p)_\# T(i)\big] \, \llcorner \, [\mathbf{R}^{m+n} \sim L] = T(i) \, \llcorner \, [\mathbf{R}^{m+n} \sim L]$$

and

$$\sigma(L, p)_\# T(i) - T(i) = \tau(L, p)_\#([\![0,1]\!] \times T(i)) + \tau(L, p)_\#([\![0,1]\!] \times \partial T(i)).$$

1.14. How to deform size bounded rectifiable currents onto skeletons of cubical complexes.

Assumptions. (a) $\mathscr{F}$ is admissible family of $m + n$ cubes in $\mathbf{R}^{m+n}$, $\mathbf{CX}(\mathscr{F})$ is the associated cubical complex of $\mathscr{F}$, and $\zeta_k \colon \{1, 2, 3, \ldots\} \to \mathbf{CX}_k(\mathscr{F})$, $k = m, \ldots, m + n$, is a listing of $\mathbf{CX}(\mathscr{F})$.

(b) $T = T(m + n)$, $V(m + n)$, $W(m + n)$, $X(m + n) \in \mathbf{S}_m(\mathbf{R}^{m+n})$ and $Y(m + n) \in \mathbf{S}_{m+1}(\mathbf{R}^{m+n})$ such that

$$V(m + n) - W(m + n) = X(m + n) + \partial Y(m + n)$$

and

$$\big(\|V(m + n)\| + \|W(m + n)\| + \|X(m + n)\| + \|Y(m + n)\|\big)\big(\mathbf{R}^{m+n} \sim \bigcup \mathscr{F}\big) = 0.$$

Procedures and some properties. (1) For each $k \in \{m - 1, \ldots, m + n - 1\}$ we wish to construct $V(k)$, $W(k)$, $X(k) \in \mathbf{S}_m(\mathbf{R}^{m+n})$ and $Y(k) \in \mathbf{S}_{m+1}(\mathbf{R}^{m+n})$ using iterated deformations of the type described in 1.13 applied to $V(k + 1)$, $W(k + 1)$, $X(k + 1)$, $Y(k + 1)$ respectively (except applied to $\partial Y(m)$ instead of $Y(m)$ in case $k = m - 1$) so that

(1.1) $V(k) - W(k) = X(k) + \partial Y(k)$,

(1.2) $V(m - 1), W(m - 1), X(m - 1) \in \mathbf{S}_m(\mathscr{F})$ and $Y(m - 1) \in \mathbf{S}_{m+1}(\mathscr{F})$.

Assuming inductively that $k \in \{m, \ldots, m + n - 1\}$ and $V(k + 1)$, $W(k + 1)$, $X(k + 1)$, $Y(k + 1)$ have been defined, we use 1.13 for each $K \in \mathbf{CX}_{k+1}(\mathscr{F})$ to choose $p(K) \in \mathbf{PC}(K)$ which is suitable for $V(k + 1)$, $W(k + 1)$, $X(k + 1)$, $Y(k + 1)$ in the sense of 1.13 and define

$$Z(k, 1) = \sigma\big(\zeta_{k+1}(1), p(\zeta_{k+1}(1))\big)_{\#} Z(k + 1),$$

$$Z(k, 2) = \sigma\big(\zeta_{k+1}(2), p(\zeta_{k+1}(2))\big)_{\#} Z(k, 1), \ldots,$$

$$Z(k, j) = \sigma\big(\zeta_{k+1}(j), p(\zeta_{k+1}(j))\big)_{\#} Z(k, j - 1), \ldots$$

$$Z(k) = \lim_{j \to \infty} Z(k, j)$$

for $Z = V, W, X, Y$. The limit is taken in the **MS** metrics.

We infer from the Cubical Constancy Theorem 1.10 and our construction that $Y(m) \in \mathbf{S}_{m+1}(\mathscr{F})$ and set $Y(m - 1) = Y(m)$.

For each $K \in \mathbf{CX}_m(\mathscr{F})$ we use 1.13 again to choose $p(K) \in \mathbf{PC}(K)$ which is suitable for $V(m)$, $W(m)$, $X(m)$, $\partial Y(m) = \partial Y(m - 1)$ in the sense of 1.13. We then set

$$Z(m - 1, 1) = \sigma\big(\zeta_m(1), p(\sigma_m(1))\big)_{\#} Z(m),$$

$$Z(m, 2) = \sigma\big(\zeta_m(2), p(\zeta_m(2))\big)_{\#} Z(m - 1, 1), \ldots,$$

$$Z(m - 1, j) = \sigma\big(\zeta_m(j), p(\zeta_m(j))\big)_{\#} Z(m, j - 1), \ldots$$

$$Z(m - 1) = \lim_{j \to \infty} Z(m, j)$$

for $Z = V, W, X, \partial Y$. The limit is taken in the **MS** metric. We readily check from our definitions that these deformations within m cubes leave $\partial Y(m - 1)$ unchanged. We also infer from the Cubical Constancy Theorem and our construction that $V(m - 1), W(m - 1), X(m - 1), \partial Y(m - 1) \in \mathbf{S}_m(\mathscr{F})$.

(2) For each $k \in \{m - 1, \ldots, m + n - 1\}$ we wish to construct $T(k), R(k) \in \mathbf{S}_m(\mathbf{R}^{m+n})$ and $Q(k) \in \mathbf{S}_{m+1}(\mathbf{R}^{m+n})$ using iterated deformations of the type described in 1.13 applied to $T(k + 1)$ so that

(2.1) $T(k) - T(k + 1) = \partial Q(k) + R(k)$,

(2.2) $T(k) \mathbin{\llcorner} [\mathbf{R}^{m+n} \sim \bigcup \mathscr{F}] = T(m + n) \mathbin{\llcorner} [\mathbf{R}^{m+n} \sim \bigcup \mathscr{F}]$,

(2.3) $\mathrm{spt}[T(k)] \cap \bigcup \mathscr{F} \subset \bigcup \mathbf{CX}_k(\mathscr{F})$,

(2.4) $T(m - 1) - T(m) = R(m - 1) + \partial Q(m - 1) = R(m - 1)$.

With $P - T(m - 1)$ we also wish that

(2.5) $P \mathbin{\llcorner} [\mathbf{R}^{m+n} \sim \bigcup \mathscr{F}] = T \mathbin{\llcorner} [\mathbf{R}^{m+n} \sim \bigcup \mathscr{F}]$,

(2.6) $P \mathbin{\llcorner} \bigcup \mathscr{F} \in \mathscr{S}_m(\mathscr{F})$.

Hence, in particular,

(2.7) $P - T = \partial Q + R$

provided we set

$$Q = Q(m) + \cdots + Q(m + n - 1)$$

and

$$R = R(m - 1) + R(m) + \cdots + R(m + n - 1).$$

Assuming inductively that $k \in \{m - 1, \ldots, m + n - 1\}$ and $T(k + 1)$ has been defined we use 1.13 for each $K \in \mathbf{CX}_{k+1}(\mathscr{F})$ to choose $q(K) \in \mathbf{PC}(K)$ which is suitable for $T(k + 1)$ in the sense of 1.13 and define

$$T(k, 1) = \sigma\big(\zeta_{k+1}(1), q(\zeta_{k+1}(1))\big)_\#T(k + 1),$$

$$Q(k, 1) = \tau\big(\zeta_{k+1}(1), q(\zeta_{k+1}(1))\big)_\#(\llbracket 0, 1 \rrbracket \times T(k + 1)),$$

$$R(k, 1) = \tau\big(\zeta_{k+1}(1), q(\zeta_{k+1}(1))\big)_\#(\llbracket 0, 1 \rrbracket \times \partial T(k + 1))$$

and, for each $j \in \{2, 3, 4, \ldots\}$,

$$T(k, j) = \sigma\big(\zeta_{k+1}(j - 1), q(\zeta_{k+1}(j - 1))\big)_\#T(k, j - 1),$$

$$Q(k, j) = \tau\big(\zeta_{k+1}(j - 1), q(\zeta_{k+1}(j - 1))\big)_\#(\llbracket 0, 1 \rrbracket \times T(k, j - 1)),$$

$$R(k, j) = \tau\big(\zeta_{k+1}(j - 1), q(\zeta_{k+1}(j - 1))\big)_\#(\llbracket 0, 1 \rrbracket \times \partial T(k, j - 1))$$

and, finally,

$$T(k) = \lim_{j \to \infty} T(k, j), \qquad Q(k) = \sum_{j=1}^{\infty} Q(k, j), \qquad R(k) = \sum_{j=1}^{\infty} R(k, j).$$

The limit and the sums are with respect to the **MS** metrics. Note that 1.13 implies $Q(m - 1) = 0$.

Various properties and estimates of both constructions implied by 1.13 are readily checked in view of the *a priori* bounds on the local combinatorial complexity of $\mathbf{CX}(\mathscr{F})$ implied by 1.9 in particular. The basic fact here (which is readily checked) is that, for each $K_m \in \mathbf{CX}_m(\mathscr{F})$, $P \llcorner K_m$ is determined solely by $T \llcorner \bigcup \mathrm{SOURCE}(K_m)$ and the choices of various $q(K_m)$, $q(L_{m+1}), \ldots, q(L_{m+n})$ corresponding to sequences K_m, $L_{m+1}, \ldots, L_{m+n}$ associated with members of $\mathrm{INDEX}(K_m)$. We combine this construction with the estimates and remarks of 1.13 to obtain the following general theorem.

1.15. General Deformation Theorem. *There is a constant* $1 < \Gamma_{1.15} < \infty$ *with the following properties.*

Suppose (a) $T \in \mathbf{S}_m(\mathbf{R}^{m+n})$ [*resp.* $\in \mathbf{I}_m(\mathbf{R}^{m+n})$];

(b) $\mathscr{F}$ *is an admissible family of* $m + n$ *cubes in* $\mathbf{R}^{m+n}$ *with* $N = \inf\{\mathrm{level}(K) : K \in \mathscr{F}\} > -\infty$;

(c) $V, W, X: \mathscr{F} \to \mathbf{S}_m(\mathbf{R}^{m+n})$ [*resp.* $\to \mathbf{I}_m(\mathbf{R}^{m+n})$] *and* $Y: \mathscr{F} \to \mathbf{S}_{m+1}(\mathbf{R}^{m+n})$ [*resp.* $\to \mathbf{I}_{m+1}(\mathbf{R}^{m+n})$] *such that, for each* $K \in \mathscr{F}$,

$$V(K) - W(K) = X(K) + \partial Y(K)$$

and

$$W(K) = T \llcorner U(K)$$

for some open subset $U(K)$ *of* $\mathbf{R}^{m+n}$ *with*

$$\bigcup \mathrm{nbs}(K) \subset U(K) \subset \{x : \mathbf{EM}(x, \mathrm{nbs}(K)) < 2^{-\mathrm{level}(K)-3}\}$$

(*alternatively, we could assume* V, W, X, Y *are all zero or, for example, that there are single currents* V, W, X, Y *and make appropriate adjustments to the conclusions*);

(d) *for each $K \in \mathscr{F}$,*

$$\mathrm{spt}[V(K)] \cup \mathrm{spt}[W(K)] \cup \mathrm{spt}[X(K)] \cup \mathrm{spt}[Y(K)] \subset \mathrm{NBS}(K)$$

and either $V(K) = W(K)$ with $X(K) = 0$ and $Y(K) = 0$ or

$$\varnothing = (\mathrm{spt}[\partial V(K)] \cup \mathrm{spt}[\partial W(K)] \cup \mathrm{spt}[X(K)]) \cap \bigcup \mathrm{nbs}(K);$$

(e) $K_1, K_2, K_3, \ldots \in \mathscr{F}$ *with* $N \leqslant M = \inf\{\mathrm{level}(K_i): i = 1, 2, 3, \ldots\}$.
Then

(1) *There exist* P, $R \in \mathbf{S}_m(\mathbf{R}^{m+n})$ *[resp.* $\in \mathbf{I}_m(\mathbf{R}^{m+n})$] *and* $Q \in \mathbf{S}_{m+1}(\mathbf{R}^{m+n})$
[resp. $\in \mathbf{I}_{m+1}(\mathbf{R}^{m+n})$] *such that the following conditions hold*:

(1.1) $P - T = \partial Q + R$;

(1.2) $P \llcorner [\mathbf{R}^{m+n} \sim \bigcup \mathscr{F}] = T \llcorner [\mathbf{R}^{m+n} \sim \bigcup \mathscr{F}]$;

(1.3) $P \llcorner \bigcup \mathscr{F} \in \mathscr{S}_m(\mathscr{F})$ *[resp.* $\in \mathscr{I}_m(\mathscr{F})$];

(1.4) $Q = Q \llcorner \bigcup \mathscr{F}$ *and* $R = R \llcorner \bigcup \mathscr{F}$;

(1.5) $\mathbf{M}(P \llcorner \bigcup_i K_i) \leqslant \Gamma_{1.15} \mathbf{M}(T \llcorner \bigcup_i \mathrm{nbs}(K_i))$ *and*

$$\mathbf{S}(P \llcorner \bigcup_i K_i) \leqslant \Gamma_{1.15} \mathbf{S}(T \llcorner \bigcup_i \mathrm{nbs}(K_i))$$

so that, in particular,

$$\mathbf{M}\left(P \sqcup \bigcup \mathscr{F}\right) \leqslant \Gamma_{1.15} \mathbf{M}\left(T \llcorner \bigcup \mathscr{F}\right) \quad \text{and} \quad \mathbf{S}\left(P \llcorner \bigcup \mathscr{F}\right) \leqslant \Gamma_{1.15} \mathbf{S}\left(T \llcorner \bigcup \mathscr{F}\right);$$

(1.6) $\mathbf{M}(\partial P \llcorner \bigcup_i K_i) \leqslant \Gamma_{1.15} \mathbf{M}(\partial T \llcorner \bigcup_i \mathrm{nbs}(K_i))$ *and*

$$\mathbf{S}(\partial P \llcorner \bigcup_i K_i) \leqslant \Gamma_{1.15} \mathbf{S}(\partial T \llcorner \bigcup_i \mathrm{nbs}(K_i))$$

so that, in particular,

$$\mathbf{M}\left(\partial P \llcorner \bigcup \mathscr{F}\right) \leqslant \Gamma_{1.15} \mathbf{M}\left(\partial T \llcorner \bigcup \mathscr{F}\right) \text{ and } \mathbf{S}\left(\partial P \llcorner \bigcup \mathscr{F}\right) \leqslant \Gamma_{1.15} \mathbf{S}\left(\partial T \llcorner \bigcup \mathscr{F}\right);$$

(1.7) $\qquad \mathbf{M}(Q \llcorner \bigcup_i K_i) \leqslant 2^{-M} \Gamma_{1.15} \mathbf{M}(T \llcorner \bigcup_i \mathrm{nbs}(K_i))$ *and*

$$\mathbf{S}(Q \llcorner \bigcup_i K_i) \leqslant 2^{-M} \Gamma_{1.15} \mathbf{S}(T \llcorner \bigcup_i \mathrm{nbs}(K_i))$$

so that, in particular,

$$\mathbf{M}(Q) \leqslant 2^{-N} \Gamma_{1.15} \mathbf{M}\left(T \llcorner \bigcup \mathscr{F}\right) \quad \text{and} \quad \mathbf{S}(Q) \leqslant 2^{-N} \Gamma_{1.15} \mathbf{S}\left(T \llcorner \bigcup \mathscr{F}\right);$$

(1.8) $\mathbf{M}(R \llcorner \bigcup_i K_i) \leqslant 2^{-M} \Gamma_{1.15} \mathbf{M}(\partial T \llcorner \bigcup_i \mathrm{nbs}(K_i))$ *and*

$$\mathbf{S}(R \llcorner \bigcup_i K_i) \leqslant 2^{-M} \Gamma_{1.15} \mathbf{S}(\partial T \llcorner \bigcup_i \mathrm{nbs}(K_i))$$

so that, in particular,

$$\mathbf{M}(R) \leqslant 2^{-N} \Gamma_{1.15} \mathbf{M}\left(\partial T \llcorner \bigcup \mathscr{F}\right) \quad \text{and} \quad \mathbf{S}(R) \leqslant 2^{-N} \Gamma_{1.15} \mathbf{S}\left(\partial T \llcorner \bigcup \mathscr{F}\right);$$

(1.9) *in case* $T \in \mathbf{S}_m(\mathbf{R}^{m+n})$ *with* $\partial T \in \mathbf{I}_{m-1}(\mathbf{R}^{m+n})$ *then* $\partial P \in \mathbf{I}_{m-1}(\mathbf{R}^{m+n})$ *and*
$R \in \mathbf{I}_m(\mathbf{R}^{m+n})$.

(2) *There exist* $V_*, W_*, X_* : \mathscr{F} \to \mathbf{S}_m(\mathscr{F})$ *[resp.* $\to \mathbf{I}_m(\mathscr{F})$] *and* $Y_* : \mathscr{F} \to$
$\mathbf{S}_{m+1}(\mathscr{F})$ *such that the following hold for each* $K \in \mathscr{F}$:

(2.1) $V_*(K) - W_*(K) = X_*(K) + \partial Y_*(K)$;

(2.2) $V_*(K) \llcorner K = P \llcorner K$;

(2.3) $\mathrm{spt}[V_*(K)] \cup \mathrm{spt}[W_*(K)] \cup \mathrm{spt}[X_*(K)] \cup \mathrm{spt}[Y_*(K)] \subset \bigcup \mathrm{NBS}(K)$;

(2.4) *either* $V(K) = W(K)$ *with* $X(K) = 0$ *and* $Y(K) = 0$ *or*

$$\text{spt}[\partial V_*(K)] \cup \text{spt}[\partial W_*(K)] \cup \text{spt}[X_*(K)] \subset \text{Clos}\left(\bigcup[\text{NBS}(K) \sim \text{nbs}(K)]\right);$$

(2.5) $\mathbf{M}(V_*) \leqslant \Gamma_{1.15}\mathbf{M}(V)$ *and* $\mathbf{S}(V_*) \leqslant \Gamma_{1.15}\mathbf{S}(V)$;

(2.6) $\mathbf{S}(Y_*) \leqslant \Gamma_{1.15}\mathbf{S}(Y)$;

(2.7) *in case*

$$\Gamma_{1.15}\mathbf{S}(Y) < 2^{-(m+1)(2+\text{level}(K))} \quad (\text{for each } K \in \mathscr{F})$$

then $Y_* = 0$ *and* $V_* \llcorner K = W_* \llcorner K = P \llcorner K$ *so that*

$$\mathbf{M}\left(P \llcorner \bigcup\mathscr{F}\right) \leqslant \Gamma_{1.15}\sum\{\mathbf{M}(V_*(K)): K \in \mathscr{F}\},$$

$$\mathbf{S}\left(P \llcorner \bigcup\mathscr{F}\right) \leqslant \Gamma_{1.15}\sum\{\mathbf{S}(V_*(K)): K \in \mathscr{F}\}.$$

REMARK. Suppose $\mathscr{F}_1, \mathscr{F}_2, \mathscr{F}_3,\ldots$ are admissible families of $m + n$ cubes in $\mathbf{R}^{m+n}$; $p_1, p_2, p_3,\ldots \in \mathbf{R}^{m+n}$; $r_1, r_2, r_3,\ldots \in (\frac{1}{2}, 1]$; and $\omega_1, \omega_2, \omega_3,\ldots \in \mathbf{O}(m + n)$ such that

$$\left(\tau(p_i) \circ \omega_i \circ \mu(r_i)\right) \cup \mathscr{F}_i \cap \left(\tau(p_j) \circ \omega_j \circ \mu(r_j)\right) \cup \mathscr{F}_j = \varnothing$$

whenever $i, j \in \{1, 2, 3,\ldots\}$ with $i \neq j$.

Then, since all the relevant constructions and estimates of 1.14 and 1.15 are local and adapt in an obvious way to rotated or scaled coordinate systems, our General Deformation Theorem above has a straightforward extension (with the *same* constants) to the case in which, in hypothesis (b), we set

$$\mathscr{F} = \left\{\left(\tau(p_i) \circ \omega_i \circ \mu(r_i)\right)K: i \in \{1, 2, 3,\ldots\} \text{ and } K \in \mathscr{F}_i\right\}.$$

The detailed statement of this **Extended Deformation Theorem** is left to the reader. A special case of this extended theorem will be used in proving the Strong Approximation Theorem 2.11 for size bounded real rectifiable currents.

1.16. Special central projections from exact centers within $\mathbf{K}(M)$. Suppose M, $N \in \mathbf{Z}$, $\mathscr{F} = \mathbf{K}(M)$, and $T \in \mathbf{S}_m(\mathbf{R}^{m+n})$. We basically wish to deform T into $\mathbf{UCX}_m(\mathscr{F})$ by iterations of central projections of the types discussed and utilized in 1.12, 1.13, 1.14, 1.15. Control of the mass and size of various projected currents for the basic estimates there (and here) requires advantageous positioning of centers of projection relative to the measure theoretic distribution of mass and size of the various currents being projected at each step. As an alternative to the procedure already described with variable centers within the $\mathbf{PC}(K)$'s, we can require that the center of projection $p(K)$ within K always be exactly center(K) for each K. Advantageous relative positioning of currents and centers in this case is achieved by one-time initial positioning of T by well chosen translation. This is the procedure utilized in [**FH**, 4.2.1–4.2.9]. The best estimates (essentially those of [**FH**]) which I know how to apply directly to this situation permit one iteratively to project (translated) T from centers within cubes of dimensions $m + n$, $m + n - 1,\ldots, m + 1$ to obtain a current $T(m)$ having support within $\mathbf{UCX}_m(\mathscr{F})$ but not necessarily belonging to $\mathbf{S}_m(\mathscr{F})$; more particularly, I do not see in general how to control $\Theta^m(\|T(m)\|, \text{center}(K))$ for any $K \in \mathbf{CX}_m(\mathscr{F})$. On the other hand, because of the dimensions involved, the initial positioning of T does permit

$\partial T(m)$ to be projected advantageously within the m-dimensional cubes to give a current B belonging to $\mathbf{S}_{m-1}(\mathscr{F})$ and a homotopy current $R(m-1)$ belonging to $\mathbf{S}_m(\bigcup \mathbf{CX}_m(\mathscr{F}))$.

A minor variant of the estimates of [**FH**, 4.2.1–4.2.9] combined with $1.12, 1.13, 1.14, 1.15$ guarantee the existence of a constant $1 < \Gamma_{1.16} < \infty$ depending only on m and n with the following property.

General estimates. Corresponding to each $\mathscr{L}^{m+n}$ measurable subset A of $[0, 2^{-M}]^{m+n}$ with $\mathscr{L}^{m+n}(A) \geqslant 2^{-1-M(m+n)}$ there will exist $a \in A$ and an associated current $\tau(a)_{\#}T$ for which the following is true:

(1) For each $k \in \{m, \ldots, m+n-1\}$ (but not for $k = m-1$) the deformation procedure described in 1.14(2) can be adapted in an obvious and straightforward way with $T(m+n) = \tau(a)_{\#}T$ and $q(K) = \text{center}(K)$ for each K to give currents $T(k), R(k) \in \mathbf{S}_m(\mathbf{R}^{m+n})$ and $Q(k) \in \mathbf{S}_{m+1}(\mathbf{R}^{m+n})$. Furthermore, if we set

$$Q = Q(m) + \cdots + Q(m+n-1)$$

then

(1.1) $T(m) - \tau(a)_{\#}T = \partial Q + \Sigma_j R(j)$,
(1.2) $\text{spt}[T(m)] \subset \bigcup \mathbf{CX}_m(\mathscr{F})$,
(1.3) $\mathbf{M}(T(m)) \leqslant \Gamma_{1.16}\mathbf{M}(T)$ *and* $\mathbf{S}(T(m)) \leqslant \Gamma_{1.16}\mathbf{S}(T)$,
(1.4) $\mathbf{M}(\partial T(m)) \leqslant \Gamma_{1.16}\mathbf{M}(\partial T)$ *and* $\mathbf{S}(\partial T(m)) \leqslant \Gamma_{1.16}\mathbf{S}(\partial T)$,
(1.5) $\mathbf{M}(Q) \leqslant 2^{-M}\Gamma_{1.16}\mathbf{M}(T)$ *and* $\mathbf{S}(Q) \leqslant 2^{-M}\Gamma_{1.16}\mathbf{S}(T)$,
(1.6) $\mathbf{M}(\Sigma_j R(j)) \leqslant 2^{-1-M}\Gamma_{1.16}\mathbf{M}(\partial T)$ *and* $\mathbf{S}(\Sigma_j R(j)) \leqslant 2^{-1-M}\Gamma_{1.16}\mathbf{S}(\partial T)$.

(2) For $k = m-1$ the deformation procedure of 1.13, 1.14(2) can be adapted in an obvious and straightforward way to apply to $\partial T(m)$ (but not to $T(m)$) with $q(K) = \text{center}(K)$ for each $K \in \mathbf{CX}_m(\mathscr{F})$ to give $B \in \mathbf{S}_{m-1}(\mathscr{F})$ and $R \subset \mathbf{S}_m(\bigcup \mathbf{CX}_m(\mathscr{F}))$ such that if we set

$$P = R(m-1) + T(m), \qquad R = R(m-1) + \cdots + R(m+n-1)$$

then

(2.1) $B - \partial T(m) = \partial R(m-1)$,
(2.2) $P \in \mathbf{S}_m(\mathscr{F})$,
(2.3) $P - \tau(a)_{\#}T = \partial Q + R$,
(2.4) $\mathbf{M}(B) \leqslant \Gamma_{1.16}\mathbf{M}(\partial T)$ *and* $\mathbf{S}(B) \leqslant \Gamma_{1.16}\mathbf{S}(\partial T)$,
(2.5) $\mathbf{M}(R) \leqslant 2^{-M}\Gamma_{1.16}\mathbf{M}(\partial T)$ *and* $\mathbf{S}(R) \leqslant 2^{-M}\Gamma_{1.16}\mathbf{S}(\partial T)$,
(2.6) $\mathbf{M}(P) \leqslant \Gamma_{1.16}\mathbf{M}(T) + 2^{-M}\Gamma_{1.16}\mathbf{M}(\partial T)$ *and* $\mathbf{S}(P) \leqslant \Gamma_{1.16}\mathbf{S}(T) + 2^{-M}\Gamma_{1.16}\mathbf{S}(\partial T)$.

Special estimates on $\Theta^m(\|P\|, \text{center}(K))$ *for* $K \in \mathbf{CX}_m(\mathscr{F})$ *in case* $T \in \mathbf{S}_m(\mathbf{K}(N))$. Suppose

(a) $r: \mathbf{CX}_m(\mathbf{K}(N)) \to \mathbf{R}$ such that

$$\text{card}\{K: K \in \mathbf{CX}_m(\mathbf{K}(N)) \text{ with } |r(K)| > 0\} < \infty$$

and

$$T = \sum \left\{ r(L)\mathbf{t}(L, 1, \text{direction}(L)) \colon L \in \mathbf{CX}_m(\mathbf{K}(N)) \right\},$$

(b) $\lambda \in \Lambda(m + n, m)$,

(c) $K \in \mathbf{K}_m(M, \lambda)$, e.g., $\lambda = \text{direction}(K)$,

(d) $a \in [0, 2^{-M}]^{m+n}$ is chosen in accordance with our general estimates above for some A,

(e) $c = \text{center}([0, 2^{-M}]^{m+n})$,

(f) $0 < \mathbf{EM}[\tau(a) \cup \mathbf{K}_m(N), \tau(c) \cup \mathbf{K}_{n-1}(M)]$,

(g) $0 < \mathbf{EM}[\tau(a) \cup \mathbf{K}_m(N, \lambda), \ \tau(c) \cup \mathbf{K}_n(M, \mu)]$ for each $*\lambda \neq \mu \in \Lambda(m + n, n)$,

(h) $0 < \mathbf{EM}[\tau(a) \cup \{\partial K \colon K \in \mathbf{K}_m(N)\}, \tau(c) \cup \mathbf{K}_n(M)]$.

Under these conditions a straightforward application of the Hausdorff area formula and the geometry of the iterations of central cubical projections (which are piecewise affine mappings on $\tau(a) \cup \mathbf{K}_m(N)$) shows the existence of an open neighborhood U of center(K) in $\mathbf{R}^{m+n}$ such that

$$T(m) \llcorner U = st(K \cap U, 1, \lambda)$$

where

$$s = \sum \left\{ r(L) \colon L \in \mathbf{K}_m(N, \lambda) \text{ with } \varnothing \neq \tau(a)[\text{Int}(L)] \cap \tau(c)\left[\bigcup \mathbf{K}_n(M)\right] \right.$$

$$\left. \cap \left\{ x \colon \mathbf{EM}[x, \text{center}(K)] < 2^{-1-M} \right\} \right\}.$$

Special consequences. Suppose, for each $\lambda \in \Lambda(m + n, m)$ *and each* $a \in [0, 2^{-M}]^{m+n}$,

$$f_a(\lambda) \colon \mathbf{R}^m \to \mathbf{R}^+, \qquad g_a(\lambda) \colon \mathbf{R}^m \to \mathbf{Z}^+$$

are defined by setting, for each $x \in \mathbf{R}^m$,

$$f_a(\lambda)(x) = \sum \left\{ |r(K)| \colon K \in \mathbf{K}_m(N, \lambda) \text{ with } x \in \Pi(\lambda)[\tau(a)\text{Int}(K)] \right\},$$

$$g_a(\lambda)(x) = \text{card}\left\{ K \colon K \in \mathbf{K}_m(N, \lambda) \text{ with } x \in \Pi(\lambda)[\tau(a)\text{Int}(K)] \right\}.$$

We check

(A) $\mathbf{M}(T) = \sum_{\lambda \in \Lambda(m+n, m)} \int_{\mathbf{R}^m} f_a(\lambda) \, d\mathcal{L}^m$ for each a,

(B) $\mathbf{S}(T) = \sum_{\lambda \in \Lambda(m+n, m)} \int_{\mathbf{R}^m} g_a(\lambda) \, d\mathcal{L}^m$ for each a.

For $a \in [0, 2^{-M}]^{m+n}$ chosen in accordance with our general estimates above we have, additionally,

(C) for each $K \in \mathbf{K}_m(M)$, $\sigma(K, \text{center}(K))_{\#}(T(m) \llcorner K)$ is well defined in accordance with the general provision (b) of 1.12 and equals $P \llcorner K$,

(D) $\mathbf{M}(P) \leqslant 2^{-mM} \sum \{ f_a(\lambda)(p) \colon p = 2^{-M}(z_1 + \tfrac{1}{2}, \ldots, z_m + \tfrac{1}{2}) \in \mathbf{R}^m$ for some $z_1, \ldots, z_m \in \mathbf{Z}$ and $\lambda \in \Lambda(m + n, m)\}$,

(E) $\mathbf{S}(P) \leqslant 2^{-mM} \sum \{ g_a(\lambda)(p) \colon p = 2^{-M}(z_1 + \tfrac{1}{2}, \ldots, z_m + \tfrac{1}{2}) \in \mathbf{R}^m$ for some $z_1, \ldots, z_m \in \mathbf{Z}$ and $\lambda \in \Lambda(m + n, m)\}$.

1.17. Special Deformation Theorem (Deforming out currents with small projected areas, special case).

Hypotheses. (a) $M, N \in \mathbf{Z}$ *and* $\varepsilon = 2^{-M}$,

(b) $S, T \in \mathbf{S}_m(\mathbf{K}(N))$ [*resp.* $\in \mathbf{I}_m(\mathbf{K}(N))$] *with*

$$\mathscr{L}^m(\Pi(\lambda)\mathrm{spt}[T]) \leqslant \varepsilon^m \Big/ 4\binom{m+n}{m}$$

for each $\lambda \in \Lambda(m+n, m)$,

(c) $C = \mathbf{R}^{m+n} \cap \{x: \mathbf{EM}(x, \mathrm{spt}[\partial(S+T)]) \leqslant \varepsilon\}$.

Conclusion. There exist $a \in [0, \varepsilon]^{m+n}$, $P \in \mathbf{S}_m(\mathbf{K}(M))$ [*resp.* $\in \mathbf{I}_m(\mathbf{K}(M))$], $R \in \mathbf{S}_m(\mathbf{R}^{m+n})$ [*resp.* $\in \mathbf{I}_m(\mathbf{R}^{m+n})$], *and* $Q \in \mathbf{S}_m(\mathbf{R}^{m+n})$ [*resp.* $\in \mathbf{I}_m(\mathbf{R}^{m+n})$] *with the following properties*:

(1) $\mathrm{spt}[\partial P] \cup \mathrm{spt}[R] \subset \tau(a)C$,

(2) $P - \tau(a)_{\#}(S+T) = \partial Q + R$,

(3) $\mathbf{M}(P) \leqslant 8\binom{m+n}{m}\mathbf{M}(S)$ *and* $\mathbf{S}(P) \leqslant 8\binom{m+n}{m}\mathbf{S}(S)$,

(4) $\mathbf{M}(Q) \leqslant \varepsilon\Gamma_{1.16}\mathbf{M}(S+T)$ *and* $\mathbf{S}(Q) \leqslant \varepsilon\Gamma_{1.16}\mathbf{S}(S+T)$,

(5) $\mathbf{M}(R) \leqslant \varepsilon\Gamma_{1.16}\mathbf{M}(\partial(S+T))$ *and* $\mathbf{S}(R) \leqslant \varepsilon\Gamma_{1.16}\mathbf{S}(\partial(S+T))$.

PROOF. We define $r: \mathbf{K}_m(N) \to \mathbf{R}$ by requiring

$$S = \sum \{r(K)\mathbf{t}(K, 1, \mathrm{direction}(K)): K \in \mathbf{K}_m(N)\}.$$

For each $a \in [0, \varepsilon]^{m+n}$ and each $\lambda \in \Lambda(m+n, m)$ we define

$$f_a(\lambda): \mathbf{R}^m \to \mathbf{R}^+, \qquad g_a(\lambda): \mathbf{R}^m \to \mathbf{Z}^+, \qquad h_a(\lambda): \mathbf{R}^m \to \{0,1\}$$

by setting, for each $x \in \mathbf{R}^m$,

$$f_a(\lambda)(x) = \sum \{|r(K)|: K \in \mathbf{K}_m(N) \text{ with } x \in \Pi(\lambda)[\tau(a)\,\mathrm{Int}(K)]\},$$

$$g_a(\lambda)(x) = \mathrm{card}\{K: K \in \mathbf{K}_m(N) \text{ with } x \in \Pi(\lambda)[\tau(a)\,\mathrm{Int}(K)]\},$$

$$h_a(\lambda)(x) = 1 \quad \text{if and only if} \quad x \in \Pi(\lambda)[\tau(a)\,\mathrm{spt}(T)]$$

so that, for each $x \in \mathbf{R}^m$,

$$f_0(\lambda)(x) = f_a(\lambda)(\Pi(\lambda)(a) + x), \qquad g_0(\lambda)(x) = g_a(\lambda)(\Pi(\lambda)(a) + x),$$
$$h_0(\lambda)(x) = h_a(\lambda)(\Pi(\lambda)(a) + x).$$

We estimate, for each $b' \in [0, \varepsilon]^n$,

$$\int_{b \in [0, \varepsilon]^m} \sum_{z_1, \ldots, z_m \in Z} f_{(b, b')}\left(\varepsilon\left(z_1 + \frac{1}{2}, \ldots, z_m + \frac{1}{2}\right)\right) d\mathscr{L}^m b$$

$$= \int_{b \in [0, \varepsilon]^m} \sum_{z_1, \ldots, z_m \in Z} f_0\left(-b + \varepsilon\left(z_1 + \frac{1}{2}, \ldots, z_m + \frac{1}{2}\right)\right) d\mathscr{L}^m b$$

$$= \int_{\mathbf{R}^m} f_0 \, d\mathscr{L}^m,$$

$$\int_{b \in [0, \varepsilon]^m} \sum_{z_1, \ldots, z_m \in Z} g_{(b, b')}\left(\varepsilon\left(z_1 + \frac{1}{2}, \ldots, z_m + \frac{1}{2}\right)\right) d\mathscr{L}^m b = \int_{\mathbf{R}^m} g_0 \, d\mathscr{L}^m,$$

$$\int_{b \in [0, \varepsilon]^m} \sum_{z_1, \ldots, z_m \in Z} h_{(b, b')}\left(\varepsilon\left(z_1 + \frac{1}{2}, \ldots, z_m + \frac{1}{2}\right)\right) d\mathscr{L}^m b = \int_{\mathbf{R}^m} h_0 \, d\mathscr{L}^m.$$

For such λ we denote by $B(\lambda)$ the subset of $[0, \varepsilon]^m$ consisting of those points $b \in [0, \varepsilon]^m$ for which the following conditions hold for each $b' \in [0, \varepsilon]^n$:

$$\text{(i)} \qquad \sum_{z_1,\ldots,z_m \in Z} f_{(b, b')}(\lambda)\left(\varepsilon\left(z_1 + \frac{1}{2},\ldots,z_m + \frac{1}{2}\right)\right)$$

$$\leqslant \varepsilon^{-m} 8\binom{m + n}{m} \int_{\mathbf{R}^m} f_0(\lambda)\, d\mathscr{L}^m.$$

$$\text{(ii)} \qquad \sum_{z_1,\ldots,z_m \in Z} g_{(b, b')}(\lambda)\left(\varepsilon\left(z_1 + \frac{1}{2},\ldots,z_m + \frac{1}{2}\right)\right)$$

$$\leqslant \varepsilon^{-m} 8\binom{m + n}{m} \int_{\mathbf{R}^m} g_0(\lambda)\, d\mathscr{L}^m.$$

$$\text{(iii)} \qquad h_{(b, b')}(\lambda)\left(\varepsilon\left(z_1 + \frac{1}{2},\ldots,z_m + \frac{1}{2}\right)\right) = 0 \quad \text{for each } z_1,\ldots,z_m.$$

We estimate for each λ that

$$\mathscr{L}^m[B(\lambda)] \geqslant \varepsilon^m - \varepsilon^m\Big/2\binom{m + n}{m} = \varepsilon^m\left[1 - 1\Big/2\binom{m + n}{m}\right]$$

so that

$$\mathscr{L}^{m+n}\left([0, \varepsilon]^{m+n} \cap \Pi(\lambda)^{-1}B(\lambda)\right) \geqslant \varepsilon^{m+n}\left[1 - 1\Big/2\binom{m + n}{m}\right].$$

We set

$$A = [0, \varepsilon]^{m+n} \cap \bigcap\left\{\Pi(\lambda)^{-1}B(\lambda) : \lambda \in \Lambda(m + n, m)\right\}$$

and estimate that $\mathscr{L}^{m+n}(A) \geqslant 2^{-1-M(m+n)}$ as required for the general and special estimates of 1.16. The present theorem follows readily from 1.16 and conclusions (1), (2), (3), (4) of Proposition 1.4.

1.18. Theorem (Simultaneous cubical approximation respecting small projections).

Hypotheses. (a) $S = \mathbf{t}(\text{set}(S), \theta, \xi)$, $T \in \mathscr{S}_m(\mathbf{R}^{m+n})$ [*resp.* $\in \mathscr{I}_m(\mathbf{R}^{m+n})$] *with* $S + T \in \mathbf{S}_m(\mathbf{R}^{m+n})$ [*resp.* $\in \mathbf{I}_m(\mathbf{R}^{m+n})$],

(b) $\varepsilon > 0$,

(c) $\Gamma_{1.18} = 2\Gamma_{1.15}N_{1.5}$.

Conclusion. There exist $N \in \mathbf{Z}$ together with S_*, $T_* \in \mathbf{S}_m(\mathbf{K}(N))$ [resp. $\in$ $\mathbf{I}_m(\mathbf{K}(N))$], $R \in \mathbf{S}_m(\mathbf{R}^{m+n})$ [resp. $\in \mathbf{I}_m(\mathbf{R}^{m+n})$], and $Q \in \mathbf{S}_{m+1}(\mathbf{R}^{m+n})$ [resp. $\in$ $\mathbf{I}_{m+1}(\mathbf{R}^{m+n})$] with the following properties:

(1) $(S_* + T_*) - (S + T) = \partial Q + R$,

(2) $\mathbf{M}(S_* + T_*) \leqslant \Gamma_{1.15}\mathbf{M}(S + T)$ *and* $\mathbf{S}(S_* + T_*) \leqslant \Gamma_{1.15}\mathbf{S}(S + T)$,

(3) $\mathbf{M}[\partial(S_* + T_*)] \leqslant \Gamma_{1.15}\mathbf{M}[\partial(S + T)]$ *and* $\mathbf{S}[\partial(S_* + T_*)] \leqslant \Gamma_{1.15}\mathbf{S}[\partial(S + T)]$,

(4) $\mathbf{MS}(Q) + \mathbf{MS}(R) < \varepsilon$,

(5) $\mathbf{M}(S_*) \leqslant \Gamma_{1.18}\mathbf{M}(S)$ *and* $\mathbf{S}(S_*) \leqslant \Gamma_{1.18}\mathbf{S}(S)$,

(6) $\quad \sup\{\mathscr{L}^m[\Pi(\lambda)\text{spt}(T_*)] : \lambda \in \Lambda(m + n, m)\}$
$\qquad \leqslant (1 + \varepsilon)\sup\{\mathscr{L}^m[\Pi(\lambda)\text{set}(T)] : \lambda \in \Lambda(m + n, m)\}$,

(7) $\text{spt}[\partial(S_* + T_*)] \subset \{x : \mathbf{EM}(x, \text{spt}[\partial(S + T)]) < \varepsilon\}$,

(8) *in case* $\partial(S + T) = 0$ *then* $R = 0$.

PROOF. *Step* 1. The theorem is clearly true (as a consequence of 1.15) in case $T = 0$. We henceforth assume that $T \neq 0$. We also choose and fix $0 < \varepsilon_1 < 1$ such that

$$\mathscr{H}^m\big[\operatorname{set}(S) \cap \{x: \theta(x) \leqslant \varepsilon_1\}\big]$$
$$< (\varepsilon/2)\sup\{\mathscr{L}^m[\Pi(\lambda)\operatorname{set}(T)]: \lambda \in \Lambda(m + n, m)\}$$

and set

$$S_1 = S \,\llcorner\, \big[\operatorname{set}(S) \cap \{x: \theta(x) > \varepsilon_1\}\big], \qquad T_1 = (S + T) - S_1.$$

Step 2. Associated with each $\lambda \in \Lambda(m + n, m)$ are the Radon measures $\Pi(\lambda)_{\#}\|T_1\|$ and $\Pi(\lambda)_{\#}[\mathscr{H}^m \,\llcorner\, \operatorname{set}(T_1)]$ over $\mathbf{R}^m$. In checking that these are Radon measures, we can, for example, use the formula for Hausdorff area and the fact that Hausdorff measures are not increased by orthogonal projections. Similar arguments based on [**FH**, 2.10.19(4)] readily give the following:

(i) For $\mathscr{L}^m$ almost every $x \in \mathbf{R}^m \sim \Pi(\lambda)\operatorname{set}(T_1)$,

$$0 = \Theta^m\big(\Pi(\lambda)_{\#}\|T_1\|, x\big) = \Theta^m\big(\Pi(\lambda)_{\#}[\mathscr{H}^m \,\llcorner\, \operatorname{set}(T_1)], x\big).$$

(ii) For each $\delta > 0$ there is $0 < r(\lambda, \delta) < 1$ and a Borel subset $A(\lambda, \delta)$ of $\mathbf{R}^m \sim \Pi(\lambda)\operatorname{set}(T_1)$ such that

$$\mathscr{L}^m\big(\big[\mathbf{R}^m \sim A(\lambda, \delta)\big] \sim \Pi(\lambda)\operatorname{set}(T_1)\big) < \delta$$

and, for each $x \in A(\lambda, \delta)$ and each $0 < r \leqslant r(\lambda, \delta)$,

$$\Theta^m\big(\Pi(\lambda)_{\#}\|T_1\| + \Pi(\lambda)_{\#}[\mathscr{H}^m \,\llcorner\, \operatorname{set}(T_1)], x, r\big) < \delta.$$

Step 3. We choose and fix $0 < \delta_1 < 1$ sufficiently small so that

$$\delta_1 < (\varepsilon/2)\sup\{\mathscr{L}^m[\Pi(\lambda)\operatorname{set}(T)] \cdot \lambda \in \Lambda(m + n, m)\}$$

and

$$\Gamma_{1.15}\delta_1\alpha(m)\big[3(m + n)^{1/2}\big]^m < \varepsilon_1/2$$

and set

$$r_1 = \inf\{r(\lambda, \delta_1): \lambda \in \Lambda(m + n, m)\}$$

in accordance with Step 2. We further choose and fix $N \in \mathbf{Z}$ sufficiently large so that $3(m + n)^{1/2}2^{-N} < r_1$ and $2^{-N}\Gamma_{1.15}\mathbf{N}S(S + T) < \varepsilon$. For each $K \in \mathbf{K}(N)$ we slice $S_1 + T_1$ by the function $\mathbf{EM}(\cdot, \bigcup \operatorname{nbs}(K)): \mathbf{R}^m \to \mathbf{R}^+$ and choose and fix $0 < s < 2^{-N}$ such that, with $U(K) = \{x: \mathbf{EM}(x, \bigcup \operatorname{nbs}(K)) < s\}$, we have

$$V(K) \,\llcorner\, U(K) = (S_1 + T_1) \,\llcorner\, U(K) \in \mathbf{S}_m(\mathbf{R}^m).$$

We now apply Theorem 1.15 with $T, \mathscr{F}, V(K), W(K), X(K), Y(K), U(K)$ there ($K \in \mathscr{F}$ there) replaced by $S + T = S_1 + T_1$, $\mathbf{K}(N)$, $V(K)$, $V(K)$, 0, 0, $U(K)$ respectively, in the present context ($K \in \mathbf{K}(N)$ here) to obtain $P, R, Q, V_{*}(K)$ ($K \in \mathbf{K}(N)$). Finally, we set

$$\mathscr{L} = \mathbf{K}(N) \cap \{K: \Pi(\lambda)K \cap A(\lambda, \delta_1) \neq \varnothing \text{ for some } \lambda \in \Lambda(m + n, m)\},$$
$$S_{*} = P \,\llcorner\, \bigcup \mathscr{L}, \qquad T_{*} = P - S_{*}.$$

Step 4. For each $K \in \mathbf{K}(N) \sim \mathscr{L}$ we note

$$\Pi(\lambda)K \cap A(\lambda, \delta_1) = \varnothing \quad \text{for each } \lambda \in \Lambda(m + n, m)$$

so that

$$\Pi(\lambda)\left[\bigcup(\mathbf{K}(N) \sim \mathscr{L})\right] \subset \mathbf{R}^m \sim A(\lambda, \delta_1) \quad \text{for each } \lambda \in \Lambda(m + n, m)$$

which implies, as a consequence of our choices of ε_1 and δ_1, that

$$\begin{aligned}
\mathscr{L}^m&\left(\Pi(\lambda)\left[\bigcup(\mathbf{K}(N) \sim \mathscr{L})\right]\right) \\
&\leqslant \mathscr{L}^m\left(\left[\mathbf{R}^m \sim A(\lambda, \delta_1)\right] \sim \Pi(\lambda)\mathrm{set}(T_1)\right) + \mathscr{L}^m\left[\Pi(\lambda)\mathrm{set}(T_1)\right] \\
&< \delta_1 + \mathscr{L}^m\left[\Pi(\lambda)\mathrm{set}(T)\right] + \mathscr{L}^m\left[\Pi(\lambda)(\mathrm{set}(T) \sim \mathrm{set}(T_1))\right] \\
&< (1 + \varepsilon)\sup\left\{\mathscr{L}^m\left[\Pi(\mu)\mathrm{set}(T)\right] : \mu \in \Lambda(m + n, m)\right\}
\end{aligned}$$

for each $\lambda \in \Lambda(m + n)$.

Step 5. For each $K \in \mathscr{L}$ we use 1.15(2.7) to check $S_* \llcorner K = P \llcorner K = V_*(K) \llcorner K$ and use 1.15(2.5) to estimate

$$\begin{aligned}
\mathbf{S}(S_* \llcorner K) &= \mathbf{S}(V_*(K) \llcorner K) \\
&\leqslant \mathbf{S}(V_*(K)) \leqslant \Gamma_{1.15}\mathbf{S}(V(K)) = \Gamma_{1.15}\mathbf{S}\left[(S_1 + T_1) \llcorner U(K)\right] \\
&\leqslant \Gamma_{1.15}\mathbf{S}(S_1 \llcorner U(K)) + \Gamma_{1.15}\mathbf{S}(T_1 \llcorner U(K)) \\
&< \Gamma_{1.15}\mathbf{S}(S_1 \llcorner U(K)) + \Gamma_{1.15}\delta_1\alpha(m)\left[3(m + n)^{1/2}2^{-N}\right]^m \\
&\qquad\qquad\qquad\qquad\qquad\qquad \text{(by Step 2(ii) since } K \in \mathscr{L}) \\
&< \Gamma_{1.15}\mathbf{S}(S_1 \llcorner U(K)) + (1/2)(2^{-N})^m \quad \text{(by the choice of } \delta_1).
\end{aligned}$$

Since $S_* \in \mathbf{S}_m(\mathbf{K}(N))$ we conclude either $S_* \llcorner K = 0$ or $\Gamma_{1.15}\mathbf{S}(S_1 \llcorner U(K)) > (1/2)(2^{-N})^m$. In either case, we estimate

$$\mathbf{S}(S_* \llcorner K) \leqslant 2\Gamma_{1.15}\mathbf{S}(S_1 \llcorner U(K)) \leqslant 2\Gamma_{1.15}\mathbf{S}(S \llcorner U(K)).$$

In case $S_* \llcorner K \neq 0$, we use Step 1 to infer, additionally, that

$$\Gamma_{1.15}\mathbf{M}(S_1 \llcorner U(K)) > (\varepsilon_1/2)(2^{-N})^m.$$

In a similar manner, we use 1.15(2.5) and Step 2(ii) to estimate

$$\begin{aligned}
\mathbf{M}(S_* \llcorner K) &< \Gamma_{1.15}\mathbf{M}(S_1 \llcorner U(K)) + \Gamma_{1.15}\delta_1\alpha(m)\left[3(m + n)^{1/2}2^{-N}\right]^m \\
&< \Gamma_{1.15}\mathbf{M}(S_1 \llcorner U(K)) + (\varepsilon_1/2)(2^{-N})^m
\end{aligned}$$

and hence,

$$\mathbf{M}(S_* \llcorner K) \leqslant 2\Gamma_{1.15}\mathbf{M}(S_1 \llcorner U(K)) \leqslant 2\Gamma_{1.15}\mathbf{M}(S \llcorner U(K)).$$

Step 6. We infer the conclusions of the present theorem from the definitions of

Step 3, the estimates of Step 4, the estimates of Step 5, the cardinality definitions of 1.5, and conclusion (1) of 1.15.

CHAPTER TWO
APPLICATIONS OF DEFORMATIONS

2.1. Cubical Approximation Theorem. *Corresponding to each* $T \in \mathbf{S}_m(\mathbf{R}^{m+n})$ *[resp.* $\in \mathbf{I}_m(\mathbf{R}^{m+n})$*] and each* $N \in \mathbf{Z}$ *there exist* $P \in \mathbf{S}_m(\mathbf{K}(N))$ *[resp.* $\in \mathbf{I}_m(\mathbf{K}(N))$*], $R \in \mathbf{S}_m(\mathbf{R}^{m+n})$ [resp. $\in \mathbf{I}_m(\mathbf{R}^{m+n})$], and $Q \in \mathbf{S}_{m+1}(\mathbf{R}^{m+n})$ [resp. $\in \mathbf{I}_{m+1}(\mathbf{R}^{m+n})$] with the following properties*:

(1) $P - T = \partial Q - R$,

(2) $\operatorname{spt}(P) \cup \operatorname{spt}(Q) \subset \{x: \mathbf{EM}(x, \operatorname{spt}(T)) \leqslant 2^{-N}\}$ *and* $\operatorname{spt}(\partial P) \cup \operatorname{spt}(R) \subset \{x: \mathbf{EM}(x, \operatorname{spt}(\partial T))\} \leqslant 2^{-N}$,

(3) $\mathbf{M}(P) \leqslant \Gamma_{1.15}\mathbf{M}(T)$ *and* $\mathbf{S}(P) \leqslant \Gamma_{1.15}\mathbf{M}(T)$,

(4) $\mathbf{M}(\partial P) \leqslant \Gamma_{1.15}\mathbf{M}(\partial T)$ *and* $\mathbf{S}(\partial P) \leqslant \Gamma_{1.15}\mathbf{S}(\partial T)$,

(5) $\mathbf{M}(Q) \leqslant 2^{-N}\Gamma_{1.15}\mathbf{M}(T)$ *and* $\mathbf{S}(Q) \leqslant 2^{-N}\Gamma_{1.15}\mathbf{S}(T)$,

(6) $\mathbf{M}(R) \leqslant 2^{-N}\Gamma_{1.15}\mathbf{M}(\partial T)$ *and* $\mathbf{S}(R) \leqslant 2^{-N}\Gamma_{1.15}\mathbf{S}(\partial T)$,

(7) $\mathbf{F}(P, T) \leqslant 2^{1-N}\Gamma_{1.15}\mathbf{N}(T)$ *[resp.* $\mathscr{F}(P, T) \leqslant 2^{1-N}\Gamma_{1.15}\mathbf{N}(T)$*] and* $\mathbf{FS}(P, T) \leqslant 2^{1-N}\Gamma_{1.15}\mathbf{NS}(T)$,

(8) *in case* $T \in \mathbf{P}_m(\mathbf{R}^{m+n})$ *[resp.* $\in \mathbf{IP}_m(\mathbf{R}^{m+n})$*] then* $Q \in \mathbf{P}_{m+1}(\mathbf{R}^{m+n})$ *[resp.* $\in \mathbf{IP}_{m+1}(\mathbf{R}^{m+n})$*] and* $R \in \mathbf{P}_m(\mathbf{R}^{m+n})$ *[resp.* $\subset \mathbf{IP}_m(\mathbf{R}^{m+n})$*],*

(9) *in case* $\partial T \in \mathbf{P}_{m-1}(\mathbf{R}^{m+n})$ *[resp.* $\in \mathbf{IP}_{m-1}(\mathbf{R}^{m+n})$*] then* $R \in \mathbf{P}_m(\mathbf{R}^{m+n})$ *[resp.* $\in \mathbf{IP}_m(\mathbf{R}^{m+n})$*],*

(10) *in case* $T \in \mathbf{S}_m(\mathbf{R}^{m+n})$ *with* $\partial T \in \mathbf{I}_{m-1}(\mathbf{R}^{m+n})$ *then* $\partial P \in \mathbf{I}_{m-1}(\mathbf{K}(N))$ *and* $R \in \mathbf{I}_m(\mathbf{R}^{m+n})$.

PROOF. Use 1.15 in conjunction with 1.12, 1.13 and 1.14.

REMARK. Note that in 2.1(3) the estimate on $\mathbf{M}(P)$ does not depend on $\mathbf{M}(\partial T)$. This is in contrast to the corresponding estimate of [**FH**, 4.2.9(2), Deformation Theorem].

REMARK. Following [**FH**, 4.4], we infer, as a consequence of the Cubical Approximation Theorem, that the homology groups of the chain complex

$$\bigoplus_{k=0}^{m+n} \mathbf{S}_{k,0}(A, B) \quad \left[\text{resp. } \bigoplus_{k=1}^{m+n} \mathbf{I}_{k,0}(A, B)\right]$$

are naturally isomorphic with the singular homology groups of the pair (A, B) with coefficients in $\mathbf{R}$ [resp. in $\mathbf{Z}$] whenever $B \subset A$ are closed subsets of $\mathbf{R}^{m+n}$ which are local Lipschitz neighborhood retracts.

2.2. Isoperimetric Inequality (Size concentration form).

Hypotheses. (a) $T \in \mathbf{S}_m(\mathbf{R}^{m+n})$ *[resp.* $\in \mathbf{I}_m(\mathbf{R}^{m+n})$*] with* $\partial T = 0$,

(b) $N \in \mathbf{Z}$ *such that, for each* $p \in \mathbf{R}^{m+n}$,

$$\Gamma_{1.15}\mathbf{S}\left[T \, \llcorner \, \mathbf{B}^{m+n}\left(p, \tfrac{3}{2}(m + n)^{1/2}2^{-N}\right)\right] < 2^{-mN}.$$

Conclusion. There is $Q \in \mathbf{S}_{m+1}(\mathbf{R}^{m+n})$ *[resp.* $\in \mathbf{I}_{m+1}(\mathbf{R}^{m+n})$*] such that*

(1) $\partial Q = T$,

(2) $\mathbf{M}(Q) \leqslant 2^{-N}\Gamma_{1.15}\mathbf{M}(T)$,

(3) $\mathbf{S}(Q) \leqslant 2^{-N}\Gamma_{1.15}\mathbf{S}(T)$.

PROOF. We apply 1.15, with T, $\mathcal{F}$, Q there corresponding to T, $\mathbf{K}(N)$, $-Q$ here, so that, by 1.15(1.1), $P - T = \partial Q + R$ (in the terminology there). 1.15(1.8) implies $R = 0$ since $\partial T = 0$. Since $P \in \mathbf{S}_m(\mathbf{K}(N))$, either $P = 0$ or $\mathbf{S}(P \llcorner K) \geqslant 2^{-mN}$ for some $K \in \mathbf{K}(N)$. We check for each $K \in \mathbf{K}(N)$ that

$$\bigcup \mathrm{nbs}(K) \subset \mathbf{B}^m\left(\mathrm{center}(K), \tfrac{3}{2}(m + n)^{1/2}2^{-N}\right)$$

and use Hypothesis (b) together with 1.15(1.5) to infer $P = 0$. Hence $-T = \partial Q$ in the terminology of 1.15. Conclusions (2) and (3) above follow from 1.15(1.7).

2.3. Isoperimetric Inequality (Size form). *Corresponding to each $T \in \mathbf{S}_m(\mathbf{R}^{m+n})$ [resp. $\in \mathbf{I}_m(\mathbf{R}^{m+n})$] with $\partial T = 0$ there is $Q \in \mathbf{S}_{m+1}(\mathbf{R}^{m+n})$ [resp. $\in \mathbf{I}_{m+1}(\mathbf{R}^{m+n})$] with*

(1) $\partial Q = T$,

(2) $\mathbf{M}(Q) \leqslant \Gamma_{2.3}\mathbf{M}(T)\mathbf{S}(T)^{1/m}$ *[resp.* $\mathbf{M}(Q) \leqslant \Gamma_{2.3}\mathbf{M}(T)\mathbf{S}(T)^{1/m} \leqslant \Gamma_{2.3}\mathbf{M}(T)^{(m+1)/m}$]*,

(3) $\mathbf{S}(Q) \leqslant \Gamma_{2.3}\mathbf{S}(T)^{(m+1)/m}$.

Here, $\Gamma_{2.3} = \Gamma_{1.15}(2\Gamma_{1.15})^{1/m}$.

PROOF. We note that our assertions are invariant under homothetic transformations. Assuming $T \neq 0$ and adjusting $\mathbf{S}(T)$ by a homothetic transformation if necessary, we can choose $N \in \mathbf{Z}$ by requiring $\Gamma_{1.15}\mathbf{S}(T) = \tfrac{1}{2} 2^{-mN}$ so that $2^{-N}\Gamma_{1.15} = [2\Gamma_{1.15}\mathbf{S}(T)]^{1/m}\Gamma_{1.15}$. Our conclusions now follow readily from 2.2.

REMARK. The isoperimetric inequalities of 2.3 are invariant under similarity transformations and, for $T \in \mathbf{S}_m(\mathbf{R}^{m+n})$, under scalar multiplication. The size term in 2.3(2) for integral T is a refinement of the corresponding isoperimetric inequality of [**FH**, 4.2.10]. The optimal (i.e., smallest) value of a constant $\Gamma_{2.3}$ for which the inequalities (2) and (3) of 2.3 hold is

$$\gamma(m + 1) = (m + 1)^{(m+1)/m}\alpha(m + 1)^{1/m}$$

as shown in [**AF2**].

2.4. Theorem (Homotopy groups of the real rectifiable cycle groups). *Suppose A is a compact Lipschitz neighborhood retract in $\mathbf{R}^{m+n}$ and j, $k \in \{0, 1, \ldots, m + n\}$ with $j + k \leqslant m + n$.*

(1) *If*

$$\mathbf{SZ}_{k,0}(A) = \mathbf{S}_{k,0}(A) \cap \{T : \partial T = 0\}$$

*is topologized as the inductive limit of its (**MS**) bounded subsets with the **GS** metric topology, then*

$$\Pi_j(\mathbf{Z}_{k,0}(A); 0) \cong \mathbf{H}_{j+k}(A; \mathbf{R});$$

here Π_j refers to homotopy in dimension j and $\mathbf{H}_{j+k}$ refers to singular homology in dimension $j + k$.

(2) *If* $\mathbf{SZ}_{k,0}(A)$ *is topologized as the inductive limit of its* (**MS**) *bounded subsets with the* **G** *metric topology, then* $\mathbf{SZ}_{k,0}(A)$ *is homotopically trivial.*

PROOF. In establishing (1) one argues as in the integer coefficient case [**AF1**] using the Isoperimetric Inequality 2.3. In (2) scalar multiplication by $t \in [0,1]$ generates a homotopy between various mappings and the 0 map.

2.5. Isoperimetric Inequality (Projected area form). *Corresponding to each* $T \in \mathbf{S}_m(\mathbf{R}^{m+n})$ *[resp.* $\in \mathbf{I}_m(\mathbf{R}^{m+n})$*] with* $\partial T = 0$ *there is* $Q \in \mathbf{S}_m(\mathbf{R}^{m+n})$ *with*

(1) $\partial Q = T$,

(2) $\mathbf{M}(Q) \leqslant \Gamma_{2.5}\mathbf{M}(T)\sup\{\mathscr{L}^m[\Pi(\lambda)\mathrm{set}(T)]\colon \lambda \in \Lambda(m+n,m)\}^{1/m}$,

(3) $\mathbf{S}(Q) \leqslant \Gamma_{2.5}\mathbf{S}_m(T)\sup\{\mathscr{L}^m[\Pi(\lambda)\mathrm{set}(T)]\colon \lambda \in \Lambda(m+n,m)\}^{1/m}$.

Here, $\Gamma_{2.5} = (1 + \Gamma_{1.15}\Gamma_{1.16})[4\binom{m+n}{m}]^{1/m}$.

PROOF. We assume $T \neq 0$ and, noting that our assertion is invariant under homothetic transformations, apply a homothetic transformation if necessary and further assume without loss of generality that

$$2\sup\{\mathscr{L}^m[\Pi(\lambda)\mathrm{set}(T)]\colon \lambda \in \Lambda(m+n,m)\} = 1\big/4\binom{m+n}{m}.$$

We now apply Theorem 1.18 with S, T, ε there replaced by $0, T, \inf\{\mathbf{M}(T), \mathbf{S}(T)\}$, respectively, to obtain $N \in \mathbf{Z}$, $T_* \in \mathbf{S}_m(\mathbf{K}(N))$, and $Q_1 \in \mathbf{S}_{m+1}(\mathbf{R}^{m+n})$ with $T_* - T = \partial Q_1$, $\mathbf{M}(T_*) \leqslant \Gamma_{1.15}\mathbf{M}(T)$, $\mathbf{S}(T_*) \leqslant \Gamma_{1.15}\mathbf{S}(T)$, $\mathbf{MS}(Q_1) < \inf\{\mathbf{M}(T), \mathbf{S}(T)\}$, and

$$\sup\{\mathscr{L}^m[\Pi(\lambda)\mathrm{spt}(T_*)]\colon \lambda \in \Lambda(m+n,m)\}$$

$$\leqslant 2\sup\{\mathscr{L}^m[\Pi(\lambda)\mathrm{set}(T)]\colon \lambda \in \Lambda(m+n,m)\} = 1\big/4\binom{m+n}{m}.$$

We further apply Theorem 1.17 with M, N, ε, S, T there replaced by $0, N, 1, 0, T_*$, respectively, to obtain $a \in [0,1]^{m+n}$ and $Q_2 \in \mathbf{S}_{m+1}(\mathbf{R}^{m+n})$ with $-\tau(a)_{\#}T_* = \partial Q_2$, $\mathbf{M}(Q_2) \leqslant \Gamma_{1.16}\mathbf{M}(T_*)$, and $\mathbf{S}(Q_2) \leqslant \Gamma_{1.16}\mathbf{S}(T_*)$.

Finally, we set $Q = -Q_1 - \tau(-a)_{\#}Q_2$ so that $\partial Q = (T - T_*) + T_* = T$ and

$$\mathbf{M}(Q) < (1 + \Gamma_{1.16}\Gamma_{1.15})\mathbf{M}(T)$$

$$= (1 + \Gamma_{1.15}\Gamma_{1.16})\mathbf{M}(T)\left[4\binom{m+n}{m}\right]^{1/m}$$

$$\cdot \sup\{\mathscr{L}^m[\Pi(\lambda)\mathrm{set}(T)]\colon \lambda \in \Lambda(m+n,m)\}^{1/m}$$

$$= \Gamma_{2.5}\mathbf{M}(T)\sup\{\mathscr{L}^m[\Pi(\lambda)\mathrm{set}(T)]\colon \lambda \in \Lambda(m+n,m)\}^{1/m}$$

with a corresponding estimate for $\mathbf{S}(Q)$.

2.6. Isoperimetric Inequality (Comparison of real and integral coefficients). *Corresponding to each* $T \in \mathbf{I}_m(\mathbf{R}^{m+n})$ *and* $Y \in \mathbf{S}_{m+1}(\mathbf{R}^{m+n})$ *with* $\partial Y = T$ *there is* $Q \in \mathbf{I}_{m+1}(\mathbf{R}^{m+n})$ *with*

(1) $\partial Q = T$,

(2) $\mathbf{M}(Q) \leqslant \Gamma_{2.6}\mathbf{M}(T)\inf\{\mathbf{M}(Y), \mathbf{S}(Y)\}^{1/(m+1)}$,

(3) $\mathbf{S}(Q) \leqslant \Gamma_{2.6}\mathbf{S}(T)\inf\{\mathbf{M}(Y), \mathbf{S}(Y)\}^{1/(m+1)}$.

Here, $\Gamma_{2.6} = \Gamma_{1.15}(2n\Gamma_{1.15})^{1/(m+1)}$.

PROOF. We assume $T \neq 0$ and, noting that our assertion is invariant under homothetic transformations, apply a homothetic transformation if necessary and further assume without loss of generality that $\inf\{\mathbf{M}(Y), \mathbf{S}(Y)\} = 1/[2n\Gamma_{1.15}]$. We then apply 1.15 in an obvious way with $\mathscr{F} = \mathbf{K}(0)$ (note the remark about alternatives there) to get $P \in \mathbf{I}_m(\mathbf{K}(0))$, $Q \in \mathbf{I}_{m+1}(\mathbf{R}^{m+n})$, $Y_* \in \mathbf{S}_{m+1}(\mathbf{K}(0))$ with $P - T = \partial Q$, $P = \partial Y_*$, $\mathbf{M}(Q) \leqslant \Gamma_{1.15}\mathbf{M}(T)$, $\mathbf{S}(Q) \leqslant \Gamma_{1.15}\mathbf{S}(T)$, and

$$\inf\{\mathbf{M}(Y_*), \mathbf{S}(Y_*)\} \leqslant \Gamma_{1.15}\inf\{\mathbf{M}(Y), \mathbf{S}(Y)\} = \Gamma_{1.15}/[2n\Gamma_{1.15}] = 1/2n.$$

We infer from 1.11(4) that $Y_* = 0$; hence, $P = 0$ and $\partial(-Q) = T$ with

$$\mathbf{M}(Q) \leqslant \Gamma_{1.15}\mathbf{M}(T) = \Gamma_{1.15}\mathbf{M}(T)[2n\Gamma_{1.15}]^{1/(m+1)}\inf\{\mathbf{M}(Y), \mathbf{S}(Y)\}^{1/(m+1)}$$

$$= \Gamma_{2.6}\mathbf{M}(T)\inf\{\mathbf{M}(Y), \mathbf{S}(Y)\}^{1/(m+1)}$$

with a corresponding estimate for $\mathbf{S}(Q)$.

REMARK. The Isoperimetric Inequality 2.6 is a generalization of similar inequalities appearing in [S1, Isoperimetric Theorem, Conclusion (2)] and [YL] based on the same essential idea for the proof.

2.7. Special linearly independent families of orthogonal projections. (1) We say that $\Sigma(1), \ldots, \Sigma(N) \in \mathbf{O}(m + n, m)$ are **linearly independent** orthogonal projections if and only if $\sigma(1), \ldots, \sigma(N) \in \wedge_m\mathbf{R}^{m+n}$ are linearly independent vectors (in the vector space $\wedge_m\mathbf{R}^{m+n}$) whenever, for each $i \in \{1, \ldots, N\}$, $\sigma(i)$ is a simple unit m-vector with

$$\ker[\Sigma(i)]^\perp = \mathbf{R}^{m+n} \cap \{v: 0 = \sigma(i) \wedge v \in \wedge_m\mathbf{R}^{m+n}\};$$

the linear independence of such $\sigma(1), \ldots, \sigma(N)$ is equivalent to the condition

$$0 \neq \sigma(1) \wedge \cdots \wedge \sigma(N) \in \wedge_N[\wedge_m\mathbf{R}^{m+n}].$$

If $\Sigma(1), \ldots, \Sigma(\binom{m+n}{m}) \in \mathbf{O}(m + n, m)$ are linearly independent, we readily infer the existence of a constant $1 < C < \infty$ such that

$$|\mu| \leqslant C\sup\left\{|\langle\mu, \wedge_m\Sigma(i)\rangle|: i = 1, \ldots, \binom{m+n}{m}\right\}$$

for each $\mu \in \wedge_m\mathbf{R}^{m+n}$.

(2) Since $\{\mathbf{e}(\lambda): \lambda \in \Lambda(m + n, m)\}$ is an orthonormal basis for $\wedge_m\mathbf{R}^{m+n}$, $|\mu|^2 = \Sigma_\lambda(\mu \cdot \mathbf{e}(\lambda))^2$ for each $\mu \in \wedge_m\mathbf{R}^{m+n}$. We infer the existence of and fix a neighborhood Ω of $\mathbf{1}_{\mathbf{R}^{m+n}}$ in $\mathbf{O}(m + n)$ such that

$$|\mu|^2 \leqslant 2\binom{m+n}{m}\sup\left\{\left[\mu \cdot \langle\mathbf{e}(\lambda), \wedge_m\theta(\lambda)^{-1}\rangle\right]^2: \lambda \in \Lambda(m + n, m)\right\}$$

$$= 2\binom{m+n}{m}\sup\left\{|\langle\mu, \wedge_m[\Pi(\lambda) \circ \theta(\lambda)]\rangle|^2: \lambda \in \Lambda(m + n, m)\right\}$$

for each $\mu \in \wedge_m\mathbf{R}^{m+n}$ and each function $\theta: \Lambda(m + n, m) \to \Omega$.

(3) The Grassmann manifold $\mathbf{G}_0(m + n, m)$ of oriented m-dimensional linear subspaces of $\mathbf{R}^{m+n}$ is naturally (and has been) identified with the algebraic submanifold of $\wedge_m\mathbf{R}^{m+n} \cap \{\mu: |\mu| = 1\}$ consisting of all simple unit m-vectors. We infer that $\mathbf{G}_0(m + n, m)$ lies in no proper linear subspace of $\wedge_m\mathbf{R}^{m+n}$ from the fact that $\{\mathbf{e}(\lambda)\}_\lambda \subset \mathbf{G}_0(m + n, m)$ is a basis for $\wedge_m\mathbf{R}^{m+n}$. As a consequence,

suppose $N \in \{1, \ldots, \binom{m+n}{m}) - 1\}$ and $\mu \colon \{1, \ldots, N\} \to \mathbf{G}_0(m + n, m)$ with $0 \neq \mu(1) \wedge \cdots \wedge \mu(N) \in \wedge_N[\wedge_m \mathbf{R}^{m+n}]$. Then there is at least one $\mu \in \mathbf{G}_0(m + n, m)$ with $0 \neq \mu(1) \wedge \cdots \wedge \mu(N) \wedge \mu$. Furthermore, since $|\mu(1) \wedge \cdots \wedge \mu(n) \wedge (\cdot)|^2$ is real analytic, we infer

$$0 \neq \mu(1) \wedge \cdots \wedge \mu(N) \wedge \mu \quad \text{for almost every } \mu \in \mathbf{G}_0(m + n, m).$$

Finally, since $\mathbf{G}_0(m + n, m)$ is a homogeneous space of $\mathbf{O}(m + n)$, we infer that, for almost every $\theta \in \mathbf{O}(m + n)$,

$$0 \neq \mu(1) \wedge \cdots \wedge \mu(N) \wedge \langle \mathbf{e}(\lambda), \wedge_m \theta^{-1} \rangle$$

for each $\lambda \in \wedge_m \mathbf{R}^{m+n}$.

(4) Suppose $N \in \{1, 2, 3, \ldots\}$ and $\theta(1), \ldots, \theta(N) \in \Omega \subset \mathbf{O}(m + n)$ have been chosen so that

$$0 \neq \langle \mathbf{e}(\lambda(1)), \wedge_m \theta(i(1))^{-1} \rangle \wedge \cdots \wedge \langle \mathbf{e}(\lambda(M)), \wedge_m \theta(i(M))^{-1} \rangle$$
$$\in \wedge_M[\wedge_m \mathbf{R}^{m+n}]$$

whenever $M = \inf\{N, \binom{m+n}{m}\}$, $\lambda(1), \ldots, \lambda(M) \in \Lambda(m + n, m)$ (not necessarily distinct), and $1 \leqslant i(1) < i(2) < \cdots < i(M) \leqslant N$. In view of the finiteness of possible choices of $\lambda(j)$'s we infer the existence of $\theta(N + 1) \in \Omega \subset \mathbf{O}(m + n)$ such that

$$0 \neq \langle \mathbf{e}(\lambda(1)), \wedge_m \theta(i(1))^{-1} \rangle \wedge \cdots \wedge \langle \mathbf{e}(\lambda(L)), \wedge_m \theta(i(L))^{-1} \rangle \in \wedge_L[\wedge_m \mathbf{R}^{m+n}]$$

whenever $L = \inf\{N + 1, \binom{m+n}{m}\}$, $\lambda(1), \ldots, \lambda(L) \in \Lambda(m + n, m)$, and $1 \leqslant i(1) < i(2) < \cdots < i(L) \leqslant N + 1$; indeed, almost every member of $\Omega \subset \mathbf{G}(m + n)$ is a possible choice for $\theta(N + 1)$.

(5) We set $N_{2.7} = 1 + N_{1.5}[1 + \binom{m+n}{m}] \in \mathbf{Z}^+$. Additionally, we use (1), (2), (4) above to choose and fix

$$\omega(1) = \mathbf{1}_{\mathbf{R}^{m+n}}, \omega(2), \ldots, \omega(N_{2.7}) \in \Omega \subset \mathbf{O}(m + n)$$

which satisfy the following two conditions:

(5.1) For each $\mu \in \wedge_m \mathbf{R}^{m+n}$ and each $j \in \{1, \ldots, N_{2.7}\}$,

$$|\mu|^2 \leqslant 2\binom{m + n}{m} \sup\left\{ |\langle \mu, \wedge_m \Pi(\lambda(i)) \circ \omega(j) \rangle|^2 : i = 1, \ldots, \binom{m + n}{m} \right\}$$

whenever $\{\lambda(1), \ldots, \lambda(\binom{m+n}{m})\} = \Lambda(m + n, m)$.

(5.2) For each sequence $\lambda(1), \ldots, \lambda(\binom{m+n}{m}) \in \Lambda(m + n, m)$ and each increasing sequence $1 \leqslant i(1) < i(2) < \cdots < i(\binom{m+n}{m}) \leqslant N_{2.7}$,

$$0 \neq \langle \mathbf{e}(\lambda(1)), \omega(i(1))^{-1} \rangle \wedge \cdots \wedge \left\langle \mathbf{e}\left(\lambda\left(\binom{m + n}{m}\right)\right), \wedge_m \omega\left(i\left(\binom{m + n}{m}\right)\right)^{-1} \right\rangle.$$

In particular, for each such sequence, the orthogonal projections

$$\Pi(\lambda(1)) \circ \omega(i(1)), \ldots, \Pi\left(\lambda\left(\binom{m + n}{m}\right)\right) \circ \omega\left(i\left(\binom{m + n}{m}\right)\right) \in \mathbf{O}(m + n, m)$$

are linearly independent.

(6) If $\theta \colon \Lambda(m + n, m) \to \mathbf{O}(m + n)$ is defined by requiring

$$\theta(\lambda)(x) = (x_{\lambda(1)}, \ldots, x_{\lambda(m)}, x_{*\lambda(1)}, \ldots, x_{*\lambda(n)})$$

for each $\lambda \in \Lambda(m + n, m)$ and each $x \in \mathbf{R}^{m+n}$, then

$$\Pi(\lambda) = \{\mathbf{R}^m \times \{0\}\}_\natural \circ \theta(\lambda)$$

58   F. ALMGREN

for each λ so that

$$\{\Pi(\lambda) \circ \omega(i)\}_{\lambda, i} = \left\{\{\mathbf{R}^m \times \{0\}\}_\natural \circ [\theta(\lambda) \circ \omega(i)]\right\}_{\lambda, i}.$$

Additionally, arguments similar to those in (3) above show the existence of (a single) $\sigma \in \mathbf{O}(m + n)$ and $1 < v < \infty$ with the following property: Corresponding to each λ, $\mu \in \Lambda(m + n, m)$ and each $i \in \{1, \ldots, N_{2.7}\}$, there is a linear function $g(\lambda, i, \mu)$: $\mathbf{R}^m \to \mathbf{R}^n$ with $\|g(\lambda, i, \mu)\| \leq v$ and

$$\operatorname{graph}[g(\lambda, i, \mu)] = [\theta(\mu) \circ \omega(i) \circ \sigma]^{-1}\left[\mathbf{R}^{m+n} \cap \{x : x_{*\lambda(1)} = \cdots = x_{*\lambda(n)} = 0\}\right].$$

2.8. Proposition (Estimating the size of a piece of a rectifiable set by the size of its projections). *Suppose S is an $(\mathcal{H}^m, m)$ rectifiable and $\mathcal{H}^m$ measurable subset of $\mathbf{R}^{m+n}$ and θ: $S \to \mathbf{R}^+$ is $\mathcal{H}^m \mathbin{\llcorner} S$ summable.*

(1) Suppose $\Sigma(1), \ldots, \Sigma\binom{m+n}{m} \in \mathbf{O}(m + n, m)$ are linearly independent.

(1.1) Corresponding to each $\varepsilon > 0$ there is $1 < C < \infty$ such that

$$\mathcal{H}^m(T) \leq \varepsilon + C \sup\left\{\mathcal{L}^m[\Sigma(i)T] : i = 1, \ldots, \binom{m+n}{m}\right\}$$

for each $\mathcal{H}^m$ measurable subset T of S.

(1.2) There is a continuous function σ_1: $\mathbf{R}^+ \to \mathbf{R}^+$ with $\sigma_1(0) = 0$ such that $\mathcal{H}^m(T) \leq \sigma_1(\sup\{\mathcal{L}^m[\Sigma(i)T] : i = 1, \ldots, \binom{m+n}{m}\})$ for each $\mathcal{H}^m$ measurable T of S.

(1.3) There is a continuous function σ_2: $\mathbf{R}^+ \to \mathbf{R}^+$ with $\sigma_2(0) = 0$ such that

$$\mathcal{H}^m + \int_T \theta \, d\mathcal{H}^m \leq \sigma_2\left(\sup\left\{\mathcal{L}^m[\Sigma(i)T] : i = 1, \ldots, \binom{m+n}{m}\right\}\right)$$

for each $\mathcal{H}^m$ measurable subset T of S.

(2) There is a continuous nondecreasing function σ: $\mathbf{R}^+ \to \mathbf{R}^+$ with $\sigma(0) = 0$ such that

$$\mathcal{H}^m(T) + \int_T \theta \, d\mathcal{H}^m$$

$$\leq \sigma\left[\inf\left\{\sup\left\{\mathcal{L}^m[\Pi(\lambda(j)) \circ \omega(i(j))T] : j = 1, \ldots, \binom{m+n}{m}\right\} : \right.\right.$$

$$\lambda(1), \ldots, \lambda\binom{m+n}{m} \in \Lambda(m + n, m) \text{ and}$$

$$\left.\left. 1 \leq i(1) < i(2) < \cdots < i\binom{m+n}{m} \leq N_{2.7}\right\}\right].$$

PROOF. Conclusion (1.1) follows in a straightforward way from 2.7(1), the Hausdorff area formula [**FH**, 3.2.20], and the fact that S can be changed on a set of arbitrarily small measure to become a continuously differentiable m-dimensional submanifold of $\mathbf{R}^{m+n}$. Conclusion (1.1) readily implies Conclusion (1.2), and Conclusion (1.2), together with [**FH**, 2.4.11], readily implies Conclusion (1.3). Conclusion (1.3), together with 2.7(5.2), readily implies Conclusion (2).

2.9. Theorem (Deforming out parts of currents with small projected areas).
Hypotheses. (a) $S, T \in \mathscr{S}_m(\mathbf{R}^{m+n})$ [*resp.* $\in \mathscr{I}_m(\mathbf{R}^{m+n})$] *with* $\partial(S + T) = 0$,
(b) $0 < \varepsilon < 1$ *and* $M \in \mathbf{Z}$,
(c) $\mathcal{H}^m[\operatorname{set}(S) \cup \operatorname{set}(T)] \leq \varepsilon^{-1} 2^{-mM}(3^m(m + n)^{m/2}\alpha(m)M_{1.5})$,
(d) *for each* $\lambda \in \Lambda(m + n, m)$ *and each* $i \in \{1, \ldots, N_{2.7}\}$,

$$\mathscr{L}^m\big[\Pi(\lambda)\circ\omega(i)(\mathrm{set}(T))\big]\leqslant\varepsilon^m 2^{-mM}\big(4^m(m+n)^{1/2}\alpha(m)\beta(m)/\Gamma_{2.9}^{*}\big);$$

here $\omega(1),\ldots,\omega(N_{2.7})$ are as in 2.7(5) and

$$\Gamma_{2.9}^{*}=\beta(m)\alpha(m)4^m(m+n)^{m/2}2^{4m+3}\binom{m+n}{m}M_{1.5}N_{1.5}(6\Gamma_{1.16}N_{1.5})^m$$

$$\cdot\big(3^m(m+n)^{m/2}\alpha(m)M_{1.5}\big)^m.$$

Conclusion. There exist $S_{*}\in\mathbf{S}_m(\mathbf{R}^{m+n})$ [resp. $\in\mathbf{I}_m(\mathbf{R}^{m+n})$] with $\partial S_{*}=0$, $Q\in\mathbf{S}_{m+1}(\mathbf{R}^{m+n})$ [resp. $\in\mathbf{I}_{m+1}(\mathbf{R}^{m+n})$], an open subset A of $\mathbf{R}^{m+n}$, and Borel subsets $A(\lambda(1),\ldots,\lambda(\binom{m+n}{m});\,i(1),\ldots,i(\binom{m+n}{m}))$ of A corresponding to each sequence $\lambda(1),\ldots,\lambda(\binom{m+n}{m})\in\Lambda(m+n,m)$ and each increasing sequence $1\leqslant i(1)<\cdots<i(\binom{m+n}{m})\leqslant N_{2.7}$ with the following properties:

(1) $S_{*}-(S+T)=\partial Q$,

(2) $S_{*}\,\llcorner\,[\mathbf{R}^{m+n}\sim A]=S\,\llcorner\,[\mathbf{R}^{m+n}\sim A]$,

(3) $\mathbf{M}(S_{*}\,\llcorner\,A)\leqslant\Gamma_{2.9}\mathbf{M}(S\,\llcorner\,A)$ *and* $\mathbf{S}(S_{*}\,\llcorner\,A)\leqslant\Gamma_{2.9}\mathbf{S}(S\,\llcorner\,A)$,

(4) $\mathbf{M}(Q)\leqslant 2^{-M}\Gamma_{2.9}\mathbf{M}(S+T)$ *and* $\mathbf{S}(Q)\leqslant 2^{-M}\Gamma_{2.9}\mathbf{S}(S+T)$.

Here, $\Gamma_{2.9}=8\binom{m+n}{m}N_{1.5}\Gamma_{1.15}\Gamma_{1.18}$.

(5) $A=\bigcup\{A(\lambda(1),\ldots,\lambda(\binom{m+n}{m});\ i(1),\ldots,i(\binom{m+n}{m})):\ \lambda(1),\ldots,\lambda(\binom{m+n}{m})\in\Lambda(m+n,m)$ *and* $1\leqslant i(1)<\cdots<i(\binom{m+n}{m})\leqslant N_{2.7}\}$,

(6) *for each sequence $\lambda(1),\ldots,\lambda(\binom{m+n}{m})\in\Lambda(m+n,m)$ and each increasing sequence $1\leqslant i(1)<\cdots<i(\binom{m+n}{m})\leqslant N_{2.7}$ the orthogonal projections*

$$\Pi(\lambda(1))\circ\omega(i(1)),\ldots,\Pi\Big(\lambda\Big(\binom{m+n}{m}\Big)\Big)\circ\omega\Big(i\Big(\binom{m+n}{m}\Big)\Big)\in\mathbf{O}(m+n,m)$$

are linearly independent and

$$\mathscr{L}^m\Big[\Pi(\lambda(j))\circ\omega(i(j))A\Big(\lambda(1),\ldots,\lambda\Big(\binom{m+n}{m}\Big);\,i(1),\ldots,i\Big(\binom{m+n}{m}\Big)\Big)\Big]$$

$$\leqslant\varepsilon\mathscr{H}^m\big[\mathrm{set}(S)\cup\mathrm{set}(T)\big]+\varepsilon^{-m}\Gamma_{2.9}^{*}\sup\big\{\mathscr{L}^m\big[\Pi(\lambda)\circ\omega(i)(\mathrm{set}(T))\big]:$$

$$\lambda\in\Lambda(m+n,m)\ and\ i\in\{1,\ldots,N_{2.7}\}\big\}$$

for each $j=1,\ldots,\binom{m+n}{m}$.

PROOF. We will utilize the deformation procedures of 1.15, 1.17, 1.18 in our construction. The proof is in eight steps.

Step 1. We define $1<\mu<\infty$, $L\in\{1,\,2,\,3,\ldots\}$, and $0<\varepsilon_0<1$ by the requirements

(1) $\mu\varepsilon=3^m(m+n)^{m/2}\alpha(m)M_{1.5}$,

(2) L is that positive integer for which $2^{L-1}<3\Gamma_{1.16}N_{1.5}\mu\leqslant 2^L$,

(3) $2^{4m+3}\binom{m+n}{m}M_{1.5}N_{1.5}2^{mL}\varepsilon_0=1$.

Suppose $N\in\mathbf{Z}$ and $K\in\mathbf{K}(N)$. We say that K is **manageable** if and only if both of the following two conditions hold:

(i) $\mathscr{H}^m[(\mathrm{sct}(S)\cup\mathrm{sct}(T))\cap K]\leqslant\mu2^{-mN}$,

(ii) there do not exist $\lambda(1),\ldots,\lambda(\binom{m+n}{m})\in\Lambda(m+n,m)$ and $1\leqslant i(1)<\cdots<i(\binom{m+n}{m})\leqslant N_{2.7}$ with

$$\mathscr{L}^m\big[\Pi(\lambda(j))\circ\omega(i(j))(\mathrm{set}(T)\cap K)\big]>\varepsilon_0 2^{-mN}$$

for each $j\in\{1,\ldots,\binom{m+n}{m}\}$.

We infer from Hypotheses (c) and (d) that each $K\in\bigcup_{N\leqslant M}\mathbf{K}(N)$ is manageable.

We define $\mathscr{F}$ to be that admissible family of cubes $K \in \mathbf{K}$ for which the following three conditions are satisfied:

(a) level(K) $\geqslant M$;

(b) Int(K) $\cap$ Int(J) $= \varnothing$ for each $J \in \mathscr{F}$ with level(J) $\leqslant$ level(K) $- 1$;

(c) one or both of the following two conditions hold:

(c.1) there is $J \in \mathscr{F}$ with level(J) $=$ level(K) $- 1$ and $J \cap K \neq \varnothing$,

(c.2) there is some $J \in \mathbf{K}(\text{level}(K) + 1)$ with $J \subset K$ such that J is not manageable.

Additionally, we set $A = \cup \mathscr{F}$, recalling from 1.5 that A is open.

We assert that $\mathscr{H}^m[\text{set}(T) \sim A] = 0$ which is required for Conclusion (2). To see this, we suppose initially that $\Sigma \in \mathbf{G}(m + n, m)$, $N \in \mathbf{Z}$, $z \in [0, 2^{-N}]^{m+n} \in \mathbf{K}(N)$, $D = \tau(z)[\Sigma \cap \mathbf{B}^{m+n}(0, 2^{-N})] \subset [-2^{-N}, 2 \cdot 2^{-N}]^{m+n}$, and $\lambda(1), \ldots, \lambda(\binom{m+n}{m})$ $\in \Lambda(m + n, m)$ with $\{\lambda(1), \ldots, \lambda(\binom{m+n}{m})\} = \Lambda(m + n, m)$. We use 2.7(5.2) to infer the existence of $i \in \{1, \ldots, \binom{m+n}{m}\}$ such that

$$\mathscr{L}^m\big[\Pi(\lambda(i)) \circ \omega(i) D\big] \geqslant \Big[2\binom{m+n}{m}\Big]^{-1/2} \mathscr{H}^m(D) = \Big[2\binom{m+n}{m}\Big]^{-1/2} \alpha(m) 2^{-mN}$$

and hence, for some $K \in \mathbf{K}(N)$ with $K \subset [-2^{-N}, 2 \cdot 2^{-N}]^{m+n}$,

$$\mathscr{L}^m\big[\Pi(\lambda(i)) \circ \omega(i)(D \cap K)\big]$$
$$\geqslant 3^{-(m+n)}\Big[2\binom{m+n}{m}\Big]^{-1/2} \alpha(m) 2^{-mN} >> \varepsilon_0 2^{-mN}.$$

Our assertion now follows in a straightforward way from (ii) and (c.2) above, the Hausdorff area formula [**FH**, 3.2.20], and the fact that set(T) can be changed on a set of arbitrarily small measure to become a continuously differentiable m-dimensional submanifold of $\mathbf{R}^{m+n}$; in regard to this note [**FH**, 3.2.29] or [**FH**, 3.1.6, 3.2.18, 3.2.19]. Since $\mathscr{H}^m[\text{set}(T) \sim A] = 0$, $S \llcorner (\mathbf{R}^{m+n} \sim A) = (S + T) \llcorner (\mathbf{R}^{m+n} \sim A)$.

Step 2. We consider various subsets of $\mathscr{F}$ as follows:

(a) We denote by $\mathscr{F}_1$ the collection of those cubes K in $\mathscr{F}$ for which

$$\mathbf{S}\big[(S + T) \llcorner K\big] > \mu 2^{-m(1 + \text{level}(K))}.$$

(b) We denote by $\mathscr{F}_2$ the collection of those cubes $J \in \mathscr{F}$ for which there exist $K \in \mathscr{F}_1$ and a sequence $J = J(\text{level}(J))$, $J(\text{level}(J) - 1)$, $J(\text{level}(J) - 2), \ldots, J(\text{level}(K)) = K$ of members of $\mathscr{F}$ with level($J(i)$) $= i$ for each $i = \text{level}(K), \ldots, \text{level}(J)$ and $J(j) \cap J(j + 1) \neq \varnothing$ for each $j = \text{level}(K) - 1, \ldots, \text{level}(J)$. We note that $\mathscr{F}_1 \subset \mathscr{F}_2$ and that

$$\bigcup \mathscr{F}_2 \subset \bigcup\big\{\{x: \mathbf{EM}(x, K) < 2^{-\text{level}(K)}\}: K \in \mathscr{F}_1\big\}.$$

(c) Corresponding to each sequence $\lambda(1), \ldots, \lambda(\binom{m+n}{m}) \in \Lambda(m + n, m)$ and each increasing sequence $1 \leqslant i(1) < \cdots < i(\binom{m+n}{m}) \leqslant N_{2.5}$, we denote by

$$\mathscr{F}\Big(\lambda(1), \ldots, \lambda\binom{m+n}{m}; i(1), \ldots, i\binom{m+n}{m}\Big)$$

the collection of all those cubes $K \in \mathscr{F}$ for which

$$\mathscr{L}^m\big[\Pi(\lambda(j)) \circ \omega(i(j))(\text{set}(T) \cap K)\big] > \varepsilon_0 2^{-m(1 + \text{level}(K))}$$

for each $j = 1, \ldots, \binom{m+n}{m}$.

(d) For the same sequences as in (c), we denote by $\mathscr{F}_2(\lambda(1),\ldots;i(1),\ldots)$ the collection of all those cubes $J \in \mathscr{F}$ for which there exist $K \in \mathscr{F}_1(\lambda(1),\ldots;i(1),\ldots)$ and a sequence of members of $\mathscr{F}$ as in (b) above. We note

$$\mathscr{F}_1(\lambda(1),\ldots;i(1),\ldots) \subset \mathscr{F}_2(\lambda(1),\ldots;i(1),\ldots)$$

and

$$\bigcup \mathscr{F}_2(\lambda(1),\ldots;i(1),\ldots)$$

$$\subset \bigcup \left\{ \left\{ x: \mathbf{EM}(x,K) < 2^{-\operatorname{level}(K)} \right\} : K \in \mathscr{F}_1(\lambda(1),\ldots;i(1),\ldots) \right\}.$$

It is entirely straightforward to check

$$\mathscr{F} = \mathscr{F}_2 \cup \bigcup \left\{ \mathscr{F}_2\left(\lambda(1),\ldots,\lambda\left(\genfrac{}{}{0pt}{}{m+n}{m}\right); i(1),\ldots,i\left(\genfrac{}{}{0pt}{}{m+n}{m}\right)\right): \right.$$

$$\lambda(1),\ldots,\lambda\left(\genfrac{}{}{0pt}{}{m+n}{m}\right) \in \Lambda(m+n,m) \text{ and}$$

$$\left. 1 \leqslant i(1) < \cdots < i\left(\genfrac{}{}{0pt}{}{m+n}{m}\right) \leqslant N_{2.7} \right\}.$$

For each sequence $\lambda(1),\ldots,\lambda(\genfrac{}{}{0pt}{}{m+n}{m}) \in \Lambda(m+n,m)$ and each increasing sequence $1 \leqslant i(1) < \cdots < i(\genfrac{}{}{0pt}{}{m+n}{m}) \leqslant N_{2.7}$ we set

$$A\left(\lambda(1),\ldots,\lambda\left(\genfrac{}{}{0pt}{}{m+n}{m}\right); i(1),\ldots,i\left(\genfrac{}{}{0pt}{}{m+n}{m}\right)\right)$$

$$= \bigcup \mathscr{F}_2 \cup \bigcup \mathscr{F}_2\left(\lambda(1),\ldots,\lambda\left(\genfrac{}{}{0pt}{}{m+n}{m}\right); i(1),\ldots,i\left(\genfrac{}{}{0pt}{}{m+n}{m}\right)\right),$$

which is clearly a Borel subset of A.

Step 3. We fix a sequence $\lambda(1),\ldots,\lambda(\genfrac{}{}{0pt}{}{m+n}{m}) \in \Lambda(m+n,m)$ and an increasing sequence $1 \leqslant i(1) < \cdots < i(\genfrac{}{}{0pt}{}{m+n}{m}) \leqslant N_{2.5}$ and $j \in \{1,\ldots,(\genfrac{}{}{0pt}{}{m+n}{m})\}$ and will estimate

$$\mathscr{L}^m\left[\Pi(\lambda(j)) \circ \omega(i(j)) A\left(\lambda(1),\ldots,\lambda\left(\genfrac{}{}{0pt}{}{m+n}{m}\right); i(1),\ldots,i\left(\genfrac{}{}{0pt}{}{m+n}{m}\right)\right)\right].$$

(a) We recall Step 2 and estimate

$$\mathscr{L}^m\left[\Pi(\lambda(j)) \circ \omega(i(j)) \bigcup \mathscr{F}_2\right]$$

$$\leqslant \mathscr{L}^m\left[\Pi(\lambda(j)) \circ \omega(i(j)) \bigcup \left\{\mathbf{B}^{m+n}\left(\operatorname{center}(K), \left(\tfrac{3}{2}\right)(m+n)^{1/2}2^{-\operatorname{level}(K)}\right):\right.\right.$$

$$\left.\left. K \in \mathscr{F}_1\right\}\right]$$

$$\leqslant \sum\left\{\alpha(m)\left(\tfrac{3}{2}\right)^m(m+n)^{m/2}2^{-m\operatorname{level}(K)}: K \in \mathscr{F}_1\right\}$$

$$\leqslant \alpha(m)\left(\tfrac{3}{2}\right)^m(m+n)^{m/2}\sum\left\{2^m\mu^{-1}\mathbf{S}[(S+T) \llcorner K]: K \in \mathscr{F}_1\right\}$$

$$\leqslant \alpha(m)3^m(m+n)^{m/2}\mu^{-1}M_{1.5}\mathbf{S}(S \mid T) = e\mathbf{S}(S \mid T)$$

$$\leqslant \mathscr{H}^m[\operatorname{set}(S) \cup \operatorname{set}(T)].$$

(b) We set

$$\mathscr{E}_1 = \Big\{ \mathbf{B}^{m+n}\big(p, 2(m+n)^{1/2}2^{-\operatorname{level}(K)}\big) \colon K \in \mathscr{F}_1(\lambda(1),\ldots;i(1),\ldots)$$

$$\text{and } p \in \mathbf{R}^{m+n} \text{ with } \mathbf{EM}(p, K) < 2^{-\operatorname{level}(K)}\Big\}$$

and

$$\mathscr{E}_2 = \Big\{ \Pi(\lambda(j)) \circ \omega(i(j))\mathbf{B}^{m+n}(p, r) \colon \mathbf{B}^{m+n}(p, r) \in \mathscr{E}_1 \Big\}$$

$$= \Big\{ \mathbf{B}^m\big(q, 2(m+n)^{1/2}2^{-\operatorname{level}(K)}\big) \colon \text{there is } K \in \mathscr{F}_1(\lambda(1),\ldots;i(1),\ldots)$$

$$\text{and } p \in \mathbf{R}^{m+n} \text{ with } \mathbf{EM}(p, K) < 2^{-\operatorname{level}(K)}$$

$$\text{such that } q = \Pi(\lambda(j)) \circ \omega(i(j))p \Big\}.$$

We use the Besicovitch–Federer Covering Theorem applied to the family $\mathscr{E}_2$ of closed balls in $\mathbf{R}^m$ to choose $K(1), K(2), K(3),\ldots \in \mathscr{F}_1(\lambda(1),\ldots;i(1),\ldots)$ and $p(1), p(2), p(3),\ldots \in \mathbf{R}^{m+n}$ with $\mathbf{EM}(p(k),\ K(k)) < 2^{-\operatorname{level}(K)}$ for each $k = 1, 2, 3,\ldots$ such that

$$\Pi(\lambda(j)) \circ \omega(i(j))\bigcup \mathscr{F}_2(\lambda(1),\ldots;i(1),\ldots)$$

$$\subset \Pi(\lambda(j)) \circ \omega(i(j))\big[\mathbf{R}^{m+n} \cap \{ p \colon \mathbf{EM}(p, K) < 2^{-\operatorname{level}(K)}$$

$$\text{for some } K \in \mathscr{F}_1(\lambda(1),\ldots;i(1),\ldots)\}\big]$$

$$\subset \bigcup \Big\{ \mathbf{B}^m\big(\Pi(\lambda(j)) \circ \omega(i(j))p(k), 2(m+n)^{1/2}2^{-\operatorname{level}(K(k))}\big) \colon$$

$$k = 1, 2, 3,\ldots \Big\}$$

and, for each $x \in \mathbf{R}^m$,

$$\beta(m) \geqslant \operatorname{card}\Big\{ k \colon x \in \mathbf{B}^m\big(\Pi(\lambda(j)) \circ \omega(i(j))p(k), 2(m+n)^{1/2}2^{-\operatorname{level}(K(k))}\big\}.$$

We estimate

$$\mathscr{L}^m\big[\Pi(\lambda(j)) \circ \omega(i(j))\bigcup F_2(\lambda(1),\ldots;i(1),\ldots)\big]$$

$$\leqslant \sum \Big\{ \alpha(m)2^m(m+n)^{m/2}2^{-m\operatorname{level}(K(k))} \colon k = 1, 2, 3,\ldots \Big\}$$

$$\leqslant \alpha(m)2^m(m+n)^{m/2}\sum \Big\{ 2^m\varepsilon_0^{-1}\mathscr{L}^m\big[\Pi(\lambda(j)) \circ \omega(i(j))(\operatorname{set}(T) \cap K(k))\big] \colon$$

$$k = 1, 2, 3,\ldots \Big\}$$

$$(\text{since } K(k) \in \mathscr{F}_1(\lambda(1),\ldots:i(1),\ldots) \text{ for each } k)$$

$$\leqslant \beta(m)\alpha(m)4^m(m+n)^{m/2}\varepsilon_0^{-1}\mathscr{L}^m\big[\Pi(\lambda(j)) \circ \omega(i(j))(\operatorname{set}(T))\big]$$

by our $\beta(m)$ multiplicity of the covering estimate above since

$$\Pi(\lambda(j)) \circ \omega(i(j))K(k)$$

$$\subset \mathbf{B}^m\big(\Pi(\lambda(j)) \circ \omega(i(j))p(k), 2(m+n)^{1/2}2^{-\operatorname{level}(K(k))}\big)$$

for each k.

The final estimate of Conclusion (6) follows.

Step 4. For each $K \in \mathscr{F}$ we infer from Step 1(ii)(a)(b)(c) that

$$\binom{m+n}{m} - 1 \geqslant \operatorname{card}\{ i \colon \mathscr{L}^m\big[\Pi(\lambda) \circ \omega(i)(\operatorname{set}(T) \cap K)\big]$$

$$> \varepsilon_0 2^{-m\operatorname{level}(K)} \text{ for some } \lambda \in \Lambda(m+n, m)\}.$$

We then use 1.5 to infer that, for each $K \in \mathscr{F}$,

$$N_{2.7} - 1 = N_{1.5}\left(\binom{m+n}{m} - 1\right)$$

$$\geqslant \operatorname{card}\left\{i: \mathscr{L}^{m+n}\left[\Pi(\lambda) \circ \omega(i)(\operatorname{set}(T) \cap J)\right] > \varepsilon_0 2^{-m\,\operatorname{level}(J)}\right.$$

$$\left.\text{for some } J \in NBS(K) \text{ and some } \lambda \in \Lambda(m+n, m)\right\}.$$

Hence, we are able to pick $\omega(K) \in \{\omega(1), \ldots, \omega(N_{2.7})\}$ corresponding to each $K \in \mathscr{F}$ such that

$$\mathscr{L}^{m}\left[\Pi(\lambda) \circ \omega(K)(\operatorname{set}(T) \cap J)\right] \leqslant \varepsilon_0 2^{-m\,\operatorname{level}(J)}$$

for each $J \in \mathrm{NBS}(K)$ and each $\lambda \in \Lambda(m+n, m)$. It then follows from 1.5 that

$$\mathscr{L}^{m}\left[\Pi(\lambda) \circ \omega(K)\left(\operatorname{set}(T) \cap \bigcup \mathrm{NBS}(K)\right)\right] \leqslant M_{1.5} N_{1.5} 2^{4m} \varepsilon_0 2^{-m\,\operatorname{level}(K)}$$

for each $\lambda \in \Lambda(m+n, m)$.

Step 5. For fixed $K \in \mathscr{F}$ we recall the terminology of 1.5, note that

$$\left\{x: \mathbf{EM}\left(x, \bigcup \mathrm{nbs}(K)\right) < 2^{-(2+\operatorname{level}(K))}\right\} \subset \bigcup \mathrm{NBS}(K),$$

and use the slicing theory [**FH**, 4.3.4, 4.3.8] to choose

$$2^{-(4+\operatorname{level}(K))} < r(K) < 2^{-(3+\operatorname{level}(K))}$$

such that, with

$$U(K) = \left\{x: \mathbf{EM}\left(x, \bigcup \mathrm{nbs}(K)\right) < r(K)\right\},$$

$$S(K) - S \sqsubset U(K), \qquad T(K) - T \sqsubset U(K),$$

we have $S(K) + T(K) \in \mathbf{S}_m(\mathbf{R}^{m+n})$ [resp. $\in \mathbf{I}_m(\mathbf{R}^{m+n})$] with

$$\operatorname{spt}\left[\partial(S(K) + T(K))\right] \subset \left\{x: \mathbf{EM}\left(x, \bigcup \mathrm{nbs}(K)\right) = r(K)\right\}$$

and

$$\mathscr{L}^{m}\left[\Pi(\lambda) \circ \omega(K)(\operatorname{set}(T) \cap U(K))\right] \leqslant M_{1.5} N_{1.5} 2^{4m} \varepsilon_0 2^{-m\,\operatorname{level}(K)}$$

for each $\lambda \in \Lambda(m+n, m)$ (by Step 4).

Step 6. For fixed $K \in \mathscr{F}$ we set

$$\varepsilon_1(K) = \inf\left\{1, 1/3\Gamma_{1.15} 2^{(m+1)(2+\operatorname{level}(K))}, 1/(m+n)^{1/2} 2^{(6+\operatorname{level}(K))}\right\}$$

and apply Theorem 1.18, with $S, T, \varepsilon, N, S_*, T_*, R, Q$ there corresponding to $\omega(K)_\# S(K)$, $\omega(K)_\# T(K)$, $\varepsilon_1(K)$, $N_1(K)$, $S_1(K)$, $T_1(K)$, $R_1(K)$, $Q_1(K)$ respectively, in the present context.

Step 7. For fixed $K \in \mathscr{F}$ we recall Step 1, set $N_2(K) = \operatorname{level}(K) + L$, and apply Theorem 1.17, with $M, N, S, T, a, \tau(-a)_\# P, \tau(-a)_\# R, \tau(-a)_\# Q$ there corresponding to $N_2(K)$, $\sup\{N_1(K), N_2(K)\}$ (recall 1.11(2)), $S_1(K), T_1(K)$, a, $P_2(K), R_2(K), Q_2(K)$ respectively in the present context.

64 F. ALMGREN

To check that Hypothesis 1.17(b) is satisfied, we estimate, for fixed $\lambda \in \Lambda(m+n, m)$,

$$\mathcal{L}^m \left[\Pi(\lambda) \operatorname{spt}(T_1(K)) \right]$$

$$\leqslant 2 \sup \left\{ \mathcal{L}^m \left[\Pi(\gamma) \circ \omega(K)(\operatorname{set}(T) \llcorner U(K)) \right] : \gamma \in \Lambda(m+n, m) \right\}$$

$$\text{(by Steps 5 and 6 and 1.18(6))}$$

$$\leqslant 2^{4m+1} \varepsilon_0 M_{1.5} N_{1.5} 2^{-m \operatorname{level}(K)} \quad \text{(by Step 5)}$$

$$= 2^{-mN_2(K)} \Big/ 4 \binom{m+n}{m} \quad \text{(by Step 1(3) and Step 7)}.$$

Step 8. We apply Theorem 1.15, with T, $\mathcal{F}$, $V(K)$, $W(K)$, $X(K)$, $Y(K)$, $U(K)$, P, R, Q, $V_*(K)$, $W_*(K)$, $X_*(K)$, $Y_*(K)$ there ($K \in \mathcal{F}$) corresponding to $(S + T)$, $\mathcal{F}$, $\omega(K)_{\#}^{-1} P_2(K)$, $(S(K) + T(K))$, $\omega(K)_{\#}^{-1}(R_1(K) + R_2(K))$, $\omega(K)_{\#}^{-1}(Q_1(K) + Q_2(K))$, $U(K)$, S_*, 0, Q, $V_*(K)$, $W_*(K)$, $X_*(K)$, $Y_*(K)$ respectively, in the present context ($K \in \mathcal{F}$).

With regard to Hypothesis 1.15(c), we note for $K \in \mathcal{F}$,

$$\omega(K)_{\#}^{-1} P_2(K) - (S(K) + T(K))$$

$$= \omega(K)_{\#}^{-1} \left[P_2(K) - (S_1(K) + T_1(K)) \right]$$

$$\qquad + \omega(K)_{\#}^{-1} \left[(S_1(K) + T_1(K)) - \omega(K)_{\#}(S(K) + T(K)) \right]$$

$$= \omega(K)_{\#}^{-1} \left[\partial Q_2(K) + R_2(K) \right] + \omega(K)_{\#}^{-1} \left[\partial Q_1(K) + R_1(K) \right]$$

$$\text{(by Step 7 and 1.17(2) and by Step 6 and 1.18(2))}$$

$$= \partial \left(\omega(K)^{-1} \left[Q_1(K) + Q_2(K) \right] \right) + \omega(K)_{\#}^{-1} \left[R_1(K) + R_2(K) \right].$$

We note that 1.15(1.1) implies our present Conclusion (1), that 1.15(1.2) and Step 1 imply Conclusion (2), that 1.15(1.7) together with Step 1(a) imply Conclusion (4), that Step 1 and Step 2 imply Conclusion (5), that 2.7(5) implies the assertion about linear independence in Conclusion (6), and that Step 3(a)(b) together with Step 1(1)(2)(3) imply the $\mathcal{L}^m$ estimates of Conclusion (6).

With regard to Conclusion (3), we estimate initially for $K \in \mathcal{F}$,

$$\mathbf{S}(Y(K)) \leqslant \Gamma_{1.15} \mathbf{S}(Q_1(K) + Q_2(K)) \quad \text{(by 1.15(2.6) since } \omega(K) \text{ is an isometry)}$$

$$\leqslant \varepsilon_1(K) + 2^{-N_2(K)} \Gamma_{1.16} \mathbf{S}(S_1(K) + T_1(K))$$

$$\text{(by Step 6 and 1.18(4) and by Step 7 and 1.17(4))}$$

$$\leqslant \tfrac{1}{3} p 2^{-(m+1)(2+\operatorname{level}(K))} + 2^{-\operatorname{level}(K)-L} \Gamma_{1.16} N_{1.5} \mu 2^{-m(\operatorname{level}(K)-2)}$$

$$\text{(by Step 6 and by Step 1(i) and 1.5)}$$

$$\leqslant \tfrac{1}{3} 2^{-(m+1)(2+\operatorname{level}(K))} + \tfrac{1}{3} 2^{-(m+1)(2+\operatorname{level}(K))} \quad \text{(by Step 1(2)(3))}$$

$$< 2^{-(m+1)(2+\operatorname{level}(K))}.$$

The condition of 1.15(2.7) is thus satisfied. Therefore, we use 1.15(2.7) to estimate

$$\mathbf{M}(S_* \llcorner A) \leqslant \Gamma_{1.15}\sum\{\mathbf{M}(P_2(K)): K \in \mathscr{F}\}$$

$$\text{(by 1.15(2.7) since each } \omega(K) \text{ is an isometry)}$$

$$\leqslant \Gamma_{1.15}8\binom{m+n}{m}\sum\{\mathbf{M}(S_1(K)): K \in \mathscr{F}\} \quad \text{(by Step 7 and 1.17(3))}$$

$$\leqslant \Gamma_{1.15}8\binom{m+n}{m}\Gamma_{1.18}\sum\{\mathbf{M}(S(K)): K \in \mathscr{F}\}$$

$$\text{(by Step 6 and 1.18(5) since each } \omega(K) \text{ is an isometry)}$$

$$= \Gamma_{1.15}8\binom{m+n}{m}\Gamma_{1.18}\sum\{\mathbf{M}(S \llcorner U(K)): K \in \mathscr{F}\} \quad \text{(by Step 5)}$$

$$\leqslant \Gamma_{1.15}8\binom{m+n}{m}N_{1.15}\mathbf{M}(S \llcorner A) \quad \text{(by 1.5)}$$

$$= \Gamma_{2.9}\mathbf{M}(S \llcorner A).$$

A virtually identical estimate gives $\mathbf{S}(S_* \llcorner A) \leqslant \Gamma_{2.9}\mathbf{S}(S \llcorner A)$. Conclusion (3) follows.

2.10. Proposition (Lower semicontinuity of mass, size, and projected areas). *Suppose* $T, T_1, T_2, T_3,\ldots \in \mathbf{S}_{m,0}(\mathbf{R}^{m+n})$ [*resp.* $\in \mathbf{I}_{m,0}(\mathbf{R}^{m+n})$] *with* $\lim_{i\to\infty}\mathbf{F}(T, T_i)$ $= 0$ [*resp.* $\lim_{i\to\infty}\mathscr{F}(T, T_i) = 0$] *and* $\Pi \in \mathbf{O}(m+n, m)$. *Then*
(1) $\mathbf{M}(T) \leqslant \liminf_{i\to\infty}\mathbf{M}(T_i)$,
(2) $\mathbf{S}(Q) \leqslant \liminf_{i\to\infty}\mathbf{S}(T_i)$,
(3) $\lim_{i\to\infty}\mathscr{L}^m[\Pi(\mathrm{set}(T)) \sim \Pi(\mathrm{set}(T_i))] = 0$ *so that* $\mathscr{L}^m[\Pi(\mathrm{set}(T))] \leqslant$ $\liminf_{i\to\infty}\mathscr{L}^m[\Pi(\mathrm{set}(T_i))]$.

PROOF. Conclusion (1) is well known and follows readily from the definitions. For notational simplicity in the proof of Conclusions (2) and (3) we will treat only the case $T, T_i \in \mathbf{S}_{m,0}(\mathbf{R}^{m+n})$. Corresponding to each $V \in \mathbf{P}_0(\mathbf{R}^{m+n})$, we check the existence of $\varepsilon(V) > 0$ with the following property: Whenever $W, X \in \mathbf{P}_0(\mathbf{R}^{m+n})$ and $Y \in \mathbf{S}_{m+1,0}(\mathbf{R}^{m+n})$ with $V - W = X - \partial Y$ and $\mathbf{M}(X) + \mathbf{M}(Y) < \varepsilon(V)$, then $\mathbf{S}(W) \geqslant \mathbf{S}(V)$. Our hypotheses imply for very large i the existence of $R \in \mathbf{S}_{m,0}(\mathbf{R}^{m+n})$ and $Q \in \mathbf{S}_{m+1,0}(\mathbf{R}^{m+n})$ with $T - T_i = R + \partial Q$ and $\mathbf{M}(R) + \mathbf{M}(Q)$ very small. According to [**FH**, 4.3.1, 4.3.2, 4.3.8], we can, for $\mathscr{L}^m$ almost every $y \in \mathbf{R}^m$, slice T, T_i, R, Q by Π at y to obtain $V(y) = \langle T, \Pi, y\rangle$, $W(y) = \langle T_i, \Pi, y\rangle$, $X(y) = \langle R, \Pi, y\rangle \in \mathbf{P}_0(\mathbf{R}^{m+n})$, and $Y(y) = \langle Q, \Pi, y\rangle \in \mathbf{S}_{1,0}(\mathbf{R}^{m+n})$ with $V(y) - W(y) = X(y) + \partial Y(y)$ and

$$\int[\mathbf{M}(X(y)) + \mathbf{M}(Y(y))]\,d\mathscr{L}^m y \leqslant \mathbf{M}(R) + \mathbf{M}(Q);$$

in particular the $\mathscr{L}^m$ measure of y's with $\mathbf{M}(X(y)) + \mathbf{M}(Y(y)) \geqslant \varepsilon(V(y))$ will be small. Conclusion (3) follows readily. Conclusion (2) follows with use of the integral geometric formula of [**FH**, 2.10.15] applied to $\mathrm{set}(T)$ and $\mathrm{set}(T_i)$.

2.11. Strong Approximation Theorem. *Corresponding to each size bounded rectifiable current* $T = \mathbf{t}(S, \theta, \xi) \in \mathbf{S}_m(\mathbf{R}^{m+n})$ [*resp. each integral current* $T \in \mathbf{I}_m(\mathbf{R}^{m+n})$] *and each* $\varepsilon > 0$, *there exists a bilipschitz diffeomorphism* $f: \mathbf{R}^{m+n} \to \mathbf{R}^{m+n}$

and a polyhedral chain $P \in \mathbf{P}_m(\mathbf{R}^{m+n})$ [*resp.* $\in \mathbf{IP}_m(\mathbf{R}^{m+n})$] *with the following properties*:

(1) $\mathrm{Lip}(f) < 1 + \varepsilon$, $\mathrm{Lip}(f^{-1}) < 1 + \varepsilon$,

(2) $\mathrm{spt}(f) \subset \{x, \mathrm{dist}(x, \mathrm{spt}(T)) < \varepsilon\}$ *and* $|f(x) - x| < \varepsilon$ *for each* $x \in \mathbf{R}^{m+n}$,

(3) $\mathbf{NS}(T - f_\# P) < \varepsilon$.

PROOF. A straightforward adaptation of the devices used in [**FH**, 4.2.19, 4.2.20] to prove the Strong Approximation Theorem for integral currents together with the almost everywhere approximate continuity of the density function θ at points of its Lebesgue set [**FH**, 2.9.9, 2.9.10] enables us to reduce our proof to construction of a polyhedral cycle $P_0 \in \mathbf{P}_m(\mathbf{R}^{m+n})$ [resp. $\in \mathbf{IP}_m(\mathbf{R}^{m+n})$] which is a strong mass and size approximation to $T_0 \in \mathbf{S}_m(\mathbf{R}^{m+n})$ [resp. $\in \mathbf{I}_m(\mathbf{R}^{m+n})$] with $\partial T_0 = 0$ and for which the following is also true: There exist $v \in \{1, 2, 3, \ldots\}$, points $p_1, \ldots, p_v \in \mathbf{R}^{m+n}$, positive numbers $r_1, \ldots, r_v$, and orthogonal mappings $\omega_1, \ldots, \omega_v \in \mathbf{O}(m + n)$ such that

(i) $|p_i - p_j| > r_i + r_j$ for each $i, j \in \{1, \ldots, v\}$ with $i \neq j$,

(ii) $\varepsilon >> \mathbf{MS}(T_0 \vdash \{\mathbf{R}^{m+n} \sim \bigcup_i [(\tau(p_i) \circ \omega_i)(\mathbf{U}^m(0, r_i) \times \{0\})]\})$,

(iii) $\varepsilon r_i^m >> \mathbf{MS}[T_0 \vdash \mathbf{B}^{m+n}(p_i, r_i) \sim (\tau(p_i) \circ \omega_i)(\mathbf{U}^m(0, r_i) \times \{0\})]$ for each $i = 1, \ldots v$.

Caution. We are not able to assume here the existence of $s_1, \ldots, s_v \in \mathbf{R}$ such that T_0 is close in the size metric to

$$\sum_i s_i (\tau(p_i) \circ \omega_i)_\# [(\mathbf{E}^m \vdash \mathbf{U}^m(0, r_i)) \times \llbracket 0, 1 \rrbracket].$$

Hereafter, we adapt the terminology of sections 1.1–1.15 to have the obvious meanings in the context of the Extended Deformation Theorem. We set

$$\mathscr{F} = \{(\tau(p_i) \circ \omega_i)K : i \in \{1, \ldots, v\} \text{ and } K \in \mathbf{WF}(\mathbf{U}^{m+n}(0, r_i))\}$$

and apply the Extended Deformation Theorem of the Remark following the General Deformation Theorem 1.15 to T_0 and the family $\mathscr{F}$ of $m + n$ cubes to obtain $P_1 \in \mathbf{S}_m(\mathbf{R}^{m+n})$ [resp. $\in \mathbf{I}_m(\mathbf{R}^{m+n})$] corresponding to P of 1.15. In particular, $\partial P_1 = 0$, $\mathrm{spt}(P_1) \cap \bigcup \mathscr{F} \subset \bigcup \mathbf{CX}_m(\mathscr{F})$, and $P_1 \vdash \mathbf{R}^{m+n} \sim \bigcup \mathscr{F} = T_0 \vdash \mathbf{R}^{m+n} \sim \bigcup \mathscr{F}$.

For $i \in \{1, \ldots, v\}$, $k \in \{m + 1, \ldots, m + n\}$, and $K \in \mathbf{CX}_k(\mathscr{F})$ with $K \subset (\tau(p_i) \circ \omega_i) \cup \mathbf{WF}(\mathbf{U}^{m+n}(0, r_i))$ we review 1.14 to check that the central projection $\sigma(K, q(K))$ leaves all points in $\bigcup \mathbf{CX}_{k-1}(\mathscr{F})$ fixed, *a fortiori* all points in $(\tau(p_i) \circ \omega_i)(\mathbf{U}^m(0, r_i) \times \{0\})$ fixed. For $K \in \mathbf{CX}_m(\mathscr{F})$, we also check in 1.14 that $\sigma(K, q(K))_\#$ leaves $T(m)$ unchanged, i.e., according to 1.14(2.4),

$$P_1 = T(m - 1) = T(m) + R(m - 1) = T(m)$$

since the condition $\partial T(m) = 0$ (true because $\partial T_0 = 0$) implies $R(m - 1) = 0$. We leave it to the reader to check that these conditions imply $\mathbf{MS}(P_1 - T_0) << \varepsilon$.

We recall the Cubical Theorem 1.10 and choose $M \in \{1, 2, 3, \ldots\}$ sufficiently large so that

$$P_2 = P_1 \vdash \bigcup \{K : K \in \mathbf{CX}_m(\mathscr{F}) \text{ with level}(K) \leqslant M\} \in \mathbf{P}_m(\mathbf{R}^{m+n})$$

$$[\text{resp.} \in \mathbf{IP}_m(\mathbf{R}^{m+n})]$$

with $\mathbf{MS}(P_1 - P_2) < < \varepsilon$ and with

$$\partial(P_1 - P_2) = \partial(-P_2) \in \mathbf{P}_{m-1}(\mathbf{R}^{m+n}) \quad \left[\text{resp. } \in \mathbf{IP}_{m-1}(\mathbf{R}^{m+n})\right]$$

—possibly with huge mass and size.

We now apply the Cubical Approximation Theorem 2.1 with T replaced by $P_1 - P_2$ there and huge $N \in \{1, 2, 3, \ldots\}$ to obtain P_3, $R \in \mathbf{P}_m(\mathbf{R}^{m+n})$ [resp. $\in \mathbf{IP}_m(\mathbf{R}^{m+n})$] (by 2.1(9)) and $Q \in \mathbf{S}_{m+1}(\mathbf{R}^{m+n})$ [resp. $\in \mathbf{I}_{m+1}(\mathbf{R}^{m+n})$] with $P_3 - (P_1 - P_2) = \partial Q + R$. We set $P_0 = P_2 + P_3 - R$ and check that $\partial P_0 = \partial P_1 + \partial \circ \partial Q = 0$ and

$$\begin{aligned}
\mathbf{MS}(P_0 - T_0) &\leqslant \mathbf{MS}(P_2 + P_3 - R - T_0) \\
&\leqslant \mathbf{MS}(P_2 - P_1) + \mathbf{MS}(P_1 - T_0) + \mathbf{MS}(P_3) + \mathbf{MS}(R) \\
&\leqslant (1 + \Gamma_{1.15})\mathbf{MS}(P_1 - P_2) + \mathbf{MS}(P_1 - P_0) \\
&\quad + 2^{-N}\Gamma_{1.15}\mathbf{MS}(\partial(P_1 - P_2)) \quad \text{(by 2.1(3)(6))} \\
&< < \varepsilon \quad \text{(since } N \text{ is huge)}
\end{aligned}$$

CHAPTER THREE
MULTIPLE - VALUED FUNCTIONS

3.1. Examples. It seems useful to give several examples illustrating the terminology and constructions to be used in the present chapter.

First example. Suppose $N \in \{1, 2, 3, \ldots\}$ and set

$$\begin{aligned}
q(j) &= \exp\left[\mathbf{i}(2\pi j/N)\right] \in \mathbf{C} = \mathbf{R}^2 \quad \text{for } j = 1, \ldots, N, \\
p(1) &= (0, 0) \in \mathbf{R}^2, \\
B &= \sum \{\llbracket q(j) \rrbracket; j = 1, \ldots, N\} - N\llbracket p(1) \rrbracket \in \mathbf{IP}_0(\mathbf{R}^2), \\
K &= (1 + N) - 1 = N, \\
a(k) &= p(1) \quad \text{for } k = 1, \ldots, K, \\
b(k) &= q(k) \quad \text{for } k = 1, \ldots, K, \\
P &= \sum \{\llbracket a(k), b(k) \rrbracket : k = 1, \ldots, K\} \in \mathbf{IP}_1(\mathbf{R}^2).
\end{aligned}$$

Then $\partial P = B$ and

$$\mathbf{M}(P) = N = K = \inf\{\mathbf{M}(Q) : Q \in \mathbf{I}_1(\mathbf{R}^2) \text{ with } \partial Q = B\}.$$

Second example. We set

$$\begin{aligned}
&p(1) = (1, \ 1), p(2) = (-1, 1), q(1) = (1, 1), q(2) = (-1, -1) \in \mathbf{R}^2, \\
&R = 2\llbracket q(1) \rrbracket + 2\llbracket q(2) \rrbracket - 2\llbracket p(1) \rrbracket - 2\llbracket p(2) \rrbracket \in \mathbf{IP}_0(\mathbf{R}^2), \\
&Q = \llbracket p(1), q(1) \rrbracket + \llbracket p(2), q(1) \rrbracket + \llbracket p(1), q(2) \rrbracket + \llbracket p(2), q(2) \rrbracket \in \mathbf{IP}_1(\mathbf{R}^2), \\
&C = \llbracket q(1), p(2) \rrbracket + \llbracket p(2), q(2) \rrbracket + \llbracket q(2), p(1) \rrbracket + \llbracket p(1), q(1) \rrbracket \in \mathbf{IP}_1(\mathbf{R}^2), \\
&P = 2\llbracket p(2), q(1) \rrbracket + 2\llbracket p(1), q(2) \rrbracket \in \mathbf{IP}_1(\mathbf{R}^2).
\end{aligned}$$

The following conditions hold:

(1) $\partial Q = B$, $\partial C = 0$, $P = Q + C$, $\partial P = B$,

(2) $\mathbf{M}(Q) = \mathbf{M}(P) = 8 = \inf\{\mathbf{M}(W) : W \in \mathbf{I}_1(\mathbf{R}^2) \text{ with } \partial W = B\}$,

(3) spt(Q) is topologically a circle which is not simply connected,

(4) spt(P) consists of two line segments, which is a simply connected space.

Third example. Suppose $0 \leqslant \varepsilon < \infty$ and $0 < \theta < \Pi/2$ is defined by requiring $1 + 2\varepsilon = 2(1 + \varepsilon)\cos\theta$. We set

$$p(1) = (-1,0),\, q(1) = (\cos\theta, \sin\theta),\, q(2) = (\cos\theta, -\sin\theta) \in \mathbf{R}^2,$$

$$B = [\![q(1)]\!] + [\![q(2)]\!] - 2[\![p(1)]\!] \in \mathbf{IP}_0(\mathbf{R}^2),$$

$$K = 3 \leqslant 3 \cdot 1 + 3 \cdot 2 - 3 = 6,$$

$$a(1) = a(2) = b(3) = (0,0),\, b(1) = q(1),\, b(2) = q(2),\, a(3) = p(1) \in \mathbf{R}^2,$$

$$P = [\![a(1), b(1)]\!] + [\![a(2), b(2)]\!] + 2[\![a(3), b(3)]\!] \in \mathbf{IP}_1(\mathbf{R}^2).$$

Then the following conditions hold:

(1) $\partial P = B$,

(2) $\mathbf{M}(P) = 4$, $\mathbf{S}(P) = 3$,

(3) $\varepsilon\mathbf{M}(P) + \mathbf{S}(P) = 4\varepsilon + 3 = \inf\{\varepsilon\mathbf{M}(Q) + \mathbf{S}(Q): Q \in \mathbf{I}_1(\mathbf{R}^2) \text{ with } \partial Q = B\}$.

Fourth example. Suppose $T_v = (1/v)[\![1]\!] - (1/v)[\![0]\!] \in \mathbf{P}_0(\mathbf{R})$ for each $v = 1, 2, 3, \ldots$. Then $Q_v = (1/v)[\![0,1]\!] \in \mathbf{P}_1(\mathbf{R})$ is the unique member of $\mathbf{S}_1(\mathbf{R})$ having T_v as boundary for each v. We note

$$\lim_v \mathbf{G}(T_v, 0) = \lim_v \mathbf{M}(T_v) = 0,$$

but

$$\lim_v \mathbf{GS}(T_v) = \lim_v \mathbf{MS}(Q_v) = 1.$$

Fifth example. Suppose $N \in \{1, 2, 3, \ldots\}$, $p_1, \ldots, p_N$ are distinct points in $\mathbf{R}^n$ and set

$$a = \inf\{|p_i - p_j|: i, j \in \{1, \ldots, N\} \text{ with } i \neq j\},$$

$$b = \sup\{|p_i - q_i|: i \in \{1, \ldots, N\}\}.$$

We note any path in $\mathbf{R}^n$ connecting distinct points among the collection $\{p_1, \ldots, p_N,\, q_1, \ldots, q_N\}$ other than a path connecting p_i to q_i for some $i \in \{1, \ldots, N\}$ must have length (hence size) at least $a - 2b$.

Suppose also $\sigma_1, \ldots, \sigma_N,\, \tau_1, \ldots, \tau_N \in \mathbf{R}$ with $\sigma_1 + \cdots + \sigma_N = \tau_1 + \cdots + \tau_N$ and set

$$P = \sum\{\sigma_i[\![p_i]\!]: i = 1, \ldots, N\} \in \mathbf{P}_0(\mathbf{R}^n),$$

$$Q = \sum\{\tau_i[\![q_i]\!]: i = 1, \ldots, N\} \in \mathbf{P}_0(\mathbf{R}^n).$$

(1) In case $\mathbf{GS}(P, Q) < a - 2b$ then $\sigma_i = \tau_i$ for each $i = 1, \ldots, N$ and $W = \Sigma\{\sigma_i[\![p_i, q_i]\!]: i = 1, \ldots, N\}$ is the unique member of $\mathbf{S}_1(\mathbf{R}^n)$ for which $\partial W = Q - P$ and $\mathbf{MS}(W) = \mathbf{GS}(P, Q)$.

(2) In case $\sigma_i = \tau_i$ for each $i = 1, \ldots, N$ then

$$\mathbf{GS}(P, Q) \leqslant \sum_i (1 + |\sigma_i|)|p_i - q_i| \leqslant \sum_i (1 + |\sigma_i|)b.$$

Sixth example. Suppose $N \in \{1, 2, 3, \ldots\}$ and $L_1, L_2, \ldots, L_N$: $\mathbf{R}^m \to \mathbf{R}^n$ are distinct linear functions with $\{0\} \times \mathbf{R}^{m-1} \subset \ker(L_i)$ for each $i = 1, \ldots, N$. Suppose also f: $\mathbf{B}^m(0, 1) \to \mathbf{P}_0(\mathbf{R}^n)$ is **GS** continuous and that $\mathrm{spt}[f(x)] \subset \{L_1(x), \ldots, L_N(x)\}$ for each $x \in \mathbf{B}^m(0, 1)$. Then there exist $\sigma_1, \ldots, \sigma_N, \tau_1, \ldots, \tau_N \in \mathbf{R}$ with $\sigma_1 + \cdots + \sigma_N = \tau_1 + \cdots + \tau_N$ such that, for each $x \in \mathbf{B}^m(0, 1)$,

$$f(x) = \begin{cases} \sum_i \sigma_i [\![L_i(x)]\!] & \text{in case } x_1 \leqslant 0, \\ \sum_i \tau_i [\![L_i(x)]\!] & \text{in case } x_i \geqslant 0 \end{cases}$$

(recall the fifth example). In particular, f is **GS** Lipschitz with constant not exceeding

$$\sup\left\{ \sum \left\{ \left(1 + |\sigma_i|\right) \|L_i\| : i = 1, \ldots, N \right\}, \sum \left\{ \left(1 + |\tau_i|\right) \|L_i\| : i = 1, \ldots, N \right\} \right\},$$

and, additionally,

$$\partial\left[\sum_i \sigma_i ([\![\mathbf{1}_{\mathbf{R}^m}]\!] \bowtie L_i)_\sharp (\mathbf{E}^m \llcorner \mathbf{B}^m(0, 1) \cap \{x\colon x_1 \leqslant 0\}) \right] \llcorner \mathbf{U}^m(0, 1) \times \mathbf{R}^n$$

$$= \left(\sum_i \sigma_i \right) [\partial (\mathbf{E}^m \llcorner \mathbf{B}^m(0, 1) \cap \{x\colon x_1 \leqslant 0\}) \times [\![0]\!]] \llcorner \mathbf{U}^m(0, 1) \times \mathbf{R}^n,$$

$$\partial\left[\sum_i \tau_i ([\![\mathbf{1}_{\mathbf{R}^m}]\!] \bowtie L_i)_\sharp (\mathbf{E}^m \llcorner \mathbf{B}^m(0, 1) \cap \{x\colon x_1 \geqslant 0\}) \right] \llcorner \mathbf{U}^m(0, 1) \times \mathbf{R}^n$$

$$= \left(\sum_i \tau_i \right) [\partial (\mathbf{E}^m \llcorner \mathbf{B}^m(0, 1) \cap \{x\colon x_1 \geqslant 0\}) \times [\![0]\!]] \llcorner \mathbf{U}^m(0, 1) \times \mathbf{R}^n,$$

so that (since $\sigma_1 + \cdots + \sigma_N = \tau_1 + \cdots + \tau_N$)

$$0 = \partial\left[\sum_i ([\![\mathbf{1}_{\mathbf{R}^m}]\!] \bowtie L_i)_\sharp (\mathbf{E}^m \llcorner \mathbf{B}^m(0, 1) \cap \{x\colon x_1 \leqslant 0\}) \right.$$

$$\left. + \sum_i \tau_i ([\![\mathbf{1}_{\mathbf{R}^m}]\!] \bowtie L_i)_\sharp (\mathbf{E}^m \llcorner \mathbf{B}^m(0, 1) \cap \{x\colon x_1 \geqslant 0\}) \right] \llcorner \mathbf{U}^m(0, 1) \times \mathbf{R}^n.$$

Seventh example. Suppose $N \in \{1, 2, 3, \ldots\}$ and $J_1, \ldots, J_N$: $\mathbf{R}^m \to \mathbf{R}^n$ are distinct affine functions. Suppose $\varnothing \neq U \subset \mathbf{R}^m$ is open and convex and $A_1, \ldots, A_M$ are the distinct pathwise connected components of

$$U \sim \bigcup \left\{ \{x\colon J_i(x) = J_j(x)\} : i, j \in \{1, \ldots, N\} \text{ with } i \neq j \right\}.$$

Finally, suppose f: $U \to \mathbf{P}_0(\mathbf{R}^n)$ is **GS** continuous with $\mathrm{spt}[f(x)] \subset \{J_1(x), \ldots, J_N(x)\}$ for each $x \in U$. Then there exist $\sigma(i, j) \in \mathbf{R}$ for each $i = 1, \ldots, M$ and $j = 1, \ldots, N$ with $\sum_j \sigma(i, j) = \sum_j \sigma(k, j)$ for each i and k so that, for each $i \in \{1, \ldots, M\}$ and each $x \in A_i$,

$$f(x) = \sum_i \sigma(i, j) [\![J_j(x)]\!]$$

(recall the fifth example). Furthermore, f is **GS** Lipschitz with Lipschitz constant not exceeding

$$\sup_i \sum_j \left(1 + |\sigma(i, j)|\right) \|J_j - J_j(0)\|.$$

Finally,

$$0 = \partial\left[\sum_i \sum_j \sigma(i, j)(\llbracket\mathbf{1}_{\mathbf{R}^m}\rrbracket \bowtie J_j)_{\#}(\mathbf{E}^m \llcorner A_i)\right] \llcorner U \times \mathbf{R}^n$$

(recall the sixth example and adapt the argument there).

Eighth example. Suppose M, N, $J_1,\ldots,J_N$, U, $A_1,\ldots,A_M$, f, and $\{\sigma(i, j)\}_{i,j}$ are as in the seventh example, and define

$$(\llbracket\mathbf{1}_{\mathbf{R}^m}\rrbracket \bowtie f)_{\#}\colon \mathbf{P}_m(\mathbf{R}^m) \cap \{T\colon \mathrm{spt}(T) \subset U\} \to \mathbf{P}_m(\mathbf{R}^{m+n})$$

by setting

$$(\llbracket\mathbf{1}_{\mathbf{R}^m}\rrbracket \bowtie f)_{\#}T = \sum_{i,j}\sigma(i, j)(\llbracket\mathbf{1}_{\mathbf{R}^m}\rrbracket \bowtie J_j)_{\#}(T \llcorner A_i)$$

for each $T \in \mathbf{P}_m(\mathbf{R}^m)$ with $\mathrm{spt}(T) \subset U$. The function $(\llbracket\mathbf{1}_{\mathbf{R}^m}\rrbracket \bowtie f)_{\#}$ is readily checked to be linear and **MS** continuous.

Corresponding to each $k \in \{1, 2,\ldots,m - 1\}$ and each isometric injection $g\colon \mathbf{R}^k \to \mathbf{R}^m$ with $g(\mathbf{R}^k) \cap U \neq \varnothing$, we define

$$(\llbracket\mathbf{1}_U\rrbracket \bowtie f)_{\#}\colon \mathbf{P}_k(\mathbf{R}^m) \cap \{T\colon \mathrm{spt}(T) \subset g(\mathbf{R}^k) \cap U\} \to \mathbf{P}_k(\mathbf{R}^{m+n})$$

by setting $V = g^{-1}(U)$ and

$$(\llbracket\mathbf{1}_U\rrbracket \bowtie f)_{\#}T = (g \times \mathbf{1}_{\mathbf{R}^n})_{\#} \circ (\llbracket\mathbf{1}_V\rrbracket \bowtie (f \circ g))_{\#}\big((g^{-1})_{\#}T\big)$$

for each $T \in \mathbf{P}_k(\mathbf{R}^m)$ with $\mathrm{spt}(T) \subset g(\mathbf{R}^k) \cap U$ (defined in the obvious way). We extend the definition by linearity to obtain a linear mapping

$$(\llbracket\mathbf{1}_U\rrbracket \bowtie f)_{\#}\colon \mathbf{P}_k(\mathbf{R}^m) \cap \{T\colon \mathrm{spt}(T) \subset U\} \to \mathbf{P}_k(\mathbf{R}^{m+n}).$$

More generally, we obtain

$$(\llbracket\mathbf{1}_U\rrbracket \bowtie f)_{\#}\colon \bigoplus_{k=0}^m \mathbf{P}_k(\mathbf{R}^m) \cap \{T\colon \mathrm{spt}(T) \subset U\} \to \bigoplus_{k=0}^m \mathbf{P}_k(\mathbf{R}^{m+n})$$

which can be checked to be a chain mapping (of degree 0), e.g.,

$$(\llbracket\mathbf{1}_U\rrbracket \bowtie f)_{\#} \circ \partial = \partial \circ (\llbracket\mathbf{1}_U\rrbracket \bowtie f)_{\#}$$

(note the estimate of the seventh example).

Ninth example. Suppose

(a) A_1, A_2, $A_3,\ldots$ are open subintervals of $[0, 1]$ with $A_i \cap A_j = \varnothing$ for each i, $j \in \{1, 2, 3,\ldots\}$ with $i \neq j$,

(b) $B = [0, 1] \sim \bigcup_i A_i$ with $B \subset \mathrm{Clos}[\bigcup_i A_i]$; such Cantor sets B can have $\mathscr{L}^1$ measure nearly equal to 1,

(c) $q_1, q_2, q_3,\ldots \in \mathbf{R}$ is a listing of the rational numbers.

We define $f\colon [0, 1] \to \mathbf{IP}_0(\mathbf{R})$ by setting $f(z) = 0$ for each $z \in B$ and

$$f(x) = \llbracket q_i + \mathrm{dist}(x, \partial A_i)\rrbracket - \llbracket q_i - \mathrm{dist}(x, \partial A_i)\rrbracket$$

for each $i \in \{1, 2, 3,\ldots\}$ and each $x \in A_i$. We check

$$\mathbf{GS}(f(u), f(v)) \leqslant 4|u - v| \quad \text{for each } u, v \in [0, 1]$$

and

$$B \times \mathbf{R} \subset \mathrm{Clos}\{(x, y): x \in [0,1] \text{ and } y \in \mathrm{spt}[f(x)]\};$$

a similar example was noted in [**S1**].

3.2. Proposition (Modification of a 1-dimensional polyhedral chain spanning a given boundary).

Hypotheses. (a) $M, N \in \{1, 2, 3, \ldots\}$,

(b) $p(1), \ldots, p(M), q(1), \ldots, q(N) \in \mathbf{R}^n$ *are distinct,*

(c) $r(1), \ldots, r(M), s(1), \ldots, s(N)$ *are positive numbers with* $\Sigma_i r(i) = \Sigma_j s(j)$,

(d) $B = \Sigma_j s(j)[\![q(j)]\!] - \Sigma_i r(i)[\![p(i)]\!] \in \mathbf{P}_0(\mathbf{R}^n)$,

(e) $Q \in \mathbf{P}_1(\mathbf{R}^n)$ *with* $\partial Q = B$.

Conclusions. There exist

(i) $K \in \{1, 2, \ldots, 3M + 3N - 3\}$,

(ii) $a(1), \ldots, a(K), b(1), \ldots, b(K) \in \mathbf{R}^n$ *with* $a(k) \neq b(k)$ *for each* $k = 1, \ldots, K$,

(iii) $t(1), \ldots, t(K) \in (0, r(1) + \cdots + r(M)\,]$,

(iv) $P = \Sigma\{t(k)[\![a(k), b(k)]\!]: k = 1, \ldots, K\} \in \mathbf{P}_1(\mathbf{R}^n)$,

such that

(1) $B = \partial P = \Sigma_k t(k)([\![b(k)]\!] - [\![a(k)]\!])$,

(2) $\mathbf{M}(P) \leqslant \Sigma_k t(k)|a(k) - b(k)|$,

(3) $\mathbf{S}(P) \leqslant \Sigma_k |a(k) - b(k)| \leqslant \mathbf{S}(Q)$,

(4) *for each* $z \in \{a(1), \ldots, a(K), b(1), \ldots, b(K)\} \sim \{p(1), \ldots, p(M), q(1), \ldots, q(N)\}$,

$$\mathrm{card}\{k: k \in \{1, \ldots, K\} \text{ with } z \in \{a(k), b(k)\}\} \geqslant 3,$$

(5) *in case* $\{a(1), \ldots, a(K), b(1), \ldots, b(K)\} = \{p(1), \ldots, p(M), q(1), \ldots, q(N)\}$, *then* $K \leqslant M + N - 1$,

(6) *in case* $Q \in \mathbf{I}_1(\mathbf{R}^n)$, *then* $t(1), \ldots, t(K)$ *are positive integers.*

PROOF. The proof is in five steps.

Step 1. Suppose $J \in \{1, 2, 3, \ldots\}$, $t_1, \ldots, t_J \in \mathbf{R}$, $L_1, \ldots, L_J$ are positive numbers, and $f: \mathbf{R} \to \mathbf{R}$ is defined by setting

$$f(t) = \Sigma\{L_i|t_i - t|: i = 1, \ldots, J\}$$

for each $t \in \mathbf{R}$. Then f is continuous and is affine on each connected component of $\mathbf{R} \sim \{t_1, \ldots, t_J\}$. It is easily checked that f assumes its minimum value at (at least) one of the points $t_1, \ldots, t_J$, i.e., there is $j_0 \in \{1, \ldots, J\}$ such that

$$f(t_{j_0}) = \inf\{f(t): t \in \mathbf{R}\} = \Sigma\{L_i|t_i - t_{j_0}|: j_0 \neq i \in \{1, \ldots, J\}\}$$

$$\leqslant f(0).$$

Step 2. We check the existence of and fix $L \in \{1, 2, 3, \ldots\}$ together with $c(1,0), c(1,1), c(2,0), c(2,1), c(3,0), c(3,1), \ldots, c(L,0), c(L,1) \in \mathbf{R}^n$ and positive numbers $\tau(1), \ldots, \tau(L)$ such that

$$Q = \Sigma\{\tau(k)[\![c(k,0), c(k,1)]\!]: k = 1, \ldots, L\}$$

and, also, $c(i,0) \neq c(i,1)$ for each $i = 1, \ldots, L$ and

$$\mathrm{spt}[\![c(j,0), c(j,1)]\!] \cap \mathrm{spt}[\![c(k,0), c(k,1)]\!]$$

$$\subset \{c(j,0), c(j,1), c(k,0), c(k,1)\}$$

whenever $j, k \in \{1,\dots,L\}$ with $j \neq k$. We note

$$\mathbf{M}(Q) = \sum_k \tau(k)|c(k,0) - c(k,1)| \quad \text{and} \quad \mathbf{S}(Q) = \sum_k |c(k,0) - c(k,1)|.$$

Step 3 (*elimination of closed loops*). Suppose $J \in \{3,\dots,L\}$ and there exist distinct $i(1),\dots,i(J) \in \{1,\dots,L\}$ and a function $\varepsilon\colon \{1,\dots,J\} \to \{0,1\}$ such that

$$c(i(1), 1 - \varepsilon(1)) = c(i(2), \varepsilon(2)), \quad c(i(2), 1 - \varepsilon(2)) = c(i(3), \varepsilon(3)),\dots,$$

$$c(i(J-1), 1 - \varepsilon(J-1)) = c(i(J), \varepsilon(J)), \quad c(i(J), 1 - \varepsilon(J)) = c(1, \varepsilon(1)).$$

If this supposition is valid, we say that $\mathrm{spt}(Q)$ contains a closed loop. We then set

$$L_j = |c(i(j), 0) - c(i(j), 1)| \quad \text{and} \quad t_j = (-1)^{\varepsilon(j)} \tau(i(j))$$

for each $j = 1,\dots,J$ and we use Step 1 to choose $j_0 \in \{1,\dots,J\}$ so that

$$\sum \left\{ L_j |t_j - t_{j_0}| : j_0 \neq j \in \{1,\dots,J\} \right\} \leqslant \sum \left\{ L_j |t_j| : j = 1,\dots,J \right\}.$$

We further set

$$C_1 = \sum \left\{ t_{j_0} [\![c(i(j), \varepsilon(j)), c(i(j), 1 - \varepsilon(j))]\!] : j = 1,\dots,J \right\} \in \mathbf{P}_1(\mathbf{R}^n)$$

checking that $\partial C_1 = 0$. We then set

$$Q_1 = Q - C_1$$

$$= \sum \left\{ (t_j - t_{j_0}) [\![c(i(j), \varepsilon(j)), c(i(j), 1 - \varepsilon(j))]\!] : j = 1,\dots,J \right\}$$

$$\quad + \sum \left\{ \tau(k) [\![c(k,0), c(k,1)]\!] : k \in \{1,\dots,L\} \sim \{i(1),\dots,i(J)\} \right\}$$

so that $\partial Q_1 = \partial Q = B$, $\mathbf{M}(Q_1) \leqslant \mathbf{M}(Q)$, and $\mathbf{S}(Q_1) \leqslant \mathbf{S}(Q) - L_{j_0} < \mathbf{S}(Q)$. We note further that our right-hand expression for Q_1 above contains at most $L - 1$ nonzero summands in contrast with the L nonzero summands in our expression for Q in Step 2.

In case $\mathrm{spt}(Q)$ does contain a closed loop, we carry out the construction described to obtain a cycle C_1 and new $Q_1 \in \mathbf{P}_1(\mathbf{R}^n)$ with $\partial Q_1 = B$. If $\mathrm{spt}(Q_1)$ should also contain a closed loop (with the obvious adaptation of terminology), we again carry out the construction described (with Q_1 replacing Q, etc.) to obtain $C_2 \in \mathbf{P}_1(\mathbf{R}^n)$ and $Q_2 = Q_1 - C_2 \in \mathbf{P}_1(\mathbf{R}^n)$ with $\mathbf{M}(Q_2) \leqslant \mathbf{M}(Q_1) \leqslant \mathbf{M}(Q)$, $\mathbf{S}(Q_2) < \mathbf{S}(Q_1) < \mathbf{S}(Q)$, and the obvious expression for Q_2 containing not more than $L - 2$ summands. Continuing in this manner if necessary, we obtain $C_3, Q_3,\dots,C_\mu, Q_\mu$ (with the obvious meanings) so that $\partial Q_\mu = B$, $\mathbf{M}(Q_\mu) \leqslant \mathbf{M}(Q)$, $\mathbf{S}(Q_\mu) < \mathbf{S}(Q)$, and $\mathrm{spt}(Q_\mu)$ contains no closed loops. An examination of the hypotheses and conclusion of our Proposition shows (by virtue of our possible constructions described above) that it is sufficient to prove the Proposition under the additional assumption in Hypothesis (e) that $\mathrm{spt}(Q)$ contains no closed loops, which assumption we now make (in order to avoid excessive terminology).

Step 4 (*elimination of bendings and excess vertices*). Since the construction of this step can introduce new crossings of line segments, we introduce an (abstract) oriented graph with densities Γ whose vertices are the (distinct) points in the set $V = \{c(1,0), c(1,1),\dots,c(L,1)\} \subset \mathbf{R}$ and whose edges are the ordered pairs $(c(k,0), c(k,1))$ with density $\tau(k)$ corresponding to $k \in \{1,\dots,L\}$. Associated with any such oriented graph Γ with densities is

$$\sum_k \tau(k) [\![c(k,0), c(k,1)]\!] \in \mathbf{P}_1(\mathbf{R}^n)$$

(which for the present graph Γ equals Q). We set

$$V(1) = \{ p(1),\dots,p(M), q(1),\dots,q(N) \} \subset V,$$

$$V(2) = (V \sim V(1)) \cap \{ v\colon \mathrm{card}\{ k\colon k \in \{1,\dots,L\}$$

$$\text{with } v \in \{ c(k,0), c(k,1) \} \} \geqslant 3\}$$

$$V(3) = V \sim (V(1) \cup V(2)) = (V \sim V(1)) \cap \{ v\colon \mathrm{card}\{ k\colon k \in \{1,\dots,L\}$$

$$\text{with } v \in \{ c(k,0), c(k,1) \} \} = 2\}$$

(this last equality is readily checked).

In case $V(3) \neq \varnothing$ there will exist (as is readily checked) distinct k_0, $k_1 \in \{1,\dots,L\}$ such that $c(k_0,1) = c(k_1,0) \in V(3)$ and $\tau(k_0) = \tau(k_1)$. We then denote by Γ_1 the (abstract) oriented graph with densities whose vertices are the set $V \sim \{ c(k_0,1) \}$ and whose edges are the pair $(c(k_0,0), c(k_1,1))$ with density $\tau(k_0)$ together with the pairs $(c(k,0), c(k,1))$ with density $\tau(k)$ associated to $k \in \{1,\dots,L\} \sim \{ k_0, k_1 \}$. For the associated

$$Q_1 = \tau(k_0) [\![c(k_0,0), c(k_1,1)]\!]$$

$$+ \sum \{ \tau(k) [\![c(k,0), c(k,1)]\!] \colon k \in \{1,\dots,L\} \sim \{ k_0, k_1 \} \} \in \mathbf{P}_1(\mathbf{R}^n)$$

we note that $\mathbf{M}(Q_1) \leqslant \mathbf{M}(Q)$ and $\mathbf{S}(Q_1) \leqslant \mathbf{S}(Q)$.

We set $v = \mathrm{card}[V(3)]$ and (in case $v \geqslant 2$) proceed in a similar way to modify Γ_1, etc., to obtain graphs $\Gamma_2,\dots,\Gamma_v$ and polyhedral currents $Q_2,\dots,Q_v$ such that the vertices of Γ_v consist only of members of $V(1) \cup V(2)$ and each member of $V(2)$ meets at least 3 edges of Γ_v. Additionally, as a consequence of Step 3 and the nature of our construction above, we infer that any (abstract) geometric realization of Γ_v (most likely **not** $\mathrm{spt}[Q_v]$) will be a finite one-dimensional simplicial complex containing no closed loops. Letting E_v denote the collection of edges Γ_v, we use Euler's formula to infer that

$$\mathrm{card}[V(1) \cup V(2)] - \mathrm{card}(E_v) = (\text{the number of components of } \Gamma_v) \geqslant 1.$$

We infer from the definition of $V(2)$ that

$$\mathrm{card}(E_v) \geqslant \tfrac{3}{2}\,\mathrm{card}[V(2)]$$

so that

$$M + N + \mathrm{card}[V(2)] = \mathrm{card}[V(1) \cup V(2)]$$

$$\geqslant \mathrm{card}(E_v) + 1 \geqslant \tfrac{3}{2}\,\mathrm{card}[V(2)] + 1,$$

which gives

$$M + N \geqslant \tfrac{1}{2}\,\mathrm{card}[V(2)] + 1, \qquad \mathrm{card}[V(2)] \leqslant 2M + 2N - 2,$$

$$\mathrm{card}(E_v) \leqslant 3M + 3N - 3.$$

We set $P = Q_v$ in Conclusion (iii), $K = \mathrm{card}(E_v)$ in Conclusion (i), and select $a(k)$, $b(k) \in V(1) \cup V(2)$ for $k = 1,\dots,K$ so that $E_v = \{(a(i), b(i))\colon i = 1,\dots,K\}$ for Conclusion (ii); the numbers $t(1),\dots,t(K)$ of Conclusion (iii) are the corresponding densities associated with members of E_v. Conclusions

(1), (2), (3), (4) are readily checked. Conclusion (5) follows from Euler's formula above with $\mathrm{card}[V(2)] = 0$. Conclusion (6) follows from inspection of the constructions in Steps 1, 2, 3 and 4. It remains to check the bounds on $t(1), \ldots, t(K)$ asserted with Conclusion (iii).

Step 5. We retain the terminology of Step 4. As noted there, any (abstract) geometric realization of Γ_v contains no (closed) loops. From consideration of the ends of maximal paths in a geometric realization of Γ_v, we infer the existence of $k_0 \in \{1, \ldots, K\}$ and $z_0 \in \{a(k_0),\ b(k_0)\}$ such that $z_0 \in \{p(1), \ldots, p(M),\ q(1), \ldots, q(N)\}$ and $z_0 \notin \{a(k),\ b(k)\}$ for each $k \in \{1, \ldots, K\} \sim \{k_0\}$. Since our various constructions have preserved combinatorial boundaries we can estimate

$$t(k_0) = r(i_0) \quad \text{in case } z_0 = p(i_0)$$
$$= s_{j_0} \quad \text{in case } z_0 = q(j_0).$$

We define w_0 by requiring $\{z_0,\ w_0\} = \{a(k_0),\ b(k_0)\}$ and let Γ_{v+1} be the subgraph of Γ_v obtained by deleting the vertex z_0 and the edge $(a(k_0),\ b(k_0))$. In order to continue this procedure for estimating $t(k)$'s in the graph Γ_{v+1}, we replace our boundary B by B_{v+1} defined by setting

$$B_{v+1} = \sum \{s(j)[\![q(j)]\!] : j \in \{1, \ldots, N\} \text{ with } q(j) \notin \{z_0, w_0\}\}$$
$$- \sum \{r(i)[\![p(i)]\!] : i \in \{1, \ldots, M\} \text{ with } p(i) \notin \{z_0, w_0\}\} + s_0[\![w_0]\!],$$

where

$$s_0 = -r(i_0) \quad \text{if } z_0 = p(i_0) \text{ and } w_0 \notin \{p(1), \ldots, p(M), q(1), \ldots, q(N)\},$$
$$= s(j_0) \quad \text{if } z_0 = q(j_0) \text{ and } w_0 \in \{p(1), \ldots, p(M), q(1), \ldots, q(N)\},$$
$$= -r(i_0) - r(i_1) \quad \text{if } z_0 = p(i_0) \text{ and } w_0 = p(i_1),$$
$$= -r(i_0) + s(j_1) \quad \text{if } z_0 = p(i_0) \text{ and } w_0 = q(j_1),$$
$$= s(j_0) - r(i_1) \quad \text{if } z_0 = q(j_0) \text{ and } w_0 = p(i_1),$$
$$= s(j_0) + s(j_1) \quad \text{if } z_0 = q(j_0) \text{ and } w_0 = q(j_1).$$

Since any (abstract) geometric realization of Γ_{v+1} contains no (closed) loops we similarly infer the existence of $k_1 \in \{1, \ldots, K\} \sim \{k_0\}$ with

$$t(k_1) \in \{|s_0|\} \cup \{s(j) : j \in \{1, \ldots, N\} \text{ with } q(j) \notin \{z_0, w_0\}\}$$
$$\cup \{r(i) : i \in \{1, \ldots, M\} \text{ with } p(i) \notin \{z_0, w_0\}\},$$

etc. The bound asserted in Conclusion (iii) follows in a straightforward manner.

3.3. Proposition (Existence of special 1-dimensional mass minimizing polyhedral chains). *Suppose*

$$B = \sum \{s(j)[\![q(j)]\!] : j = 1, \ldots, N\} - \sum \{r(i)[\![p(i)]\!] : i = 1, \ldots, M\} \in \mathbf{P}_0(\mathbf{R}^n)$$
$$[resp. \in \mathbf{IP}_0(\mathbf{R}^n)]$$

as in Proposition 3.2. Then there exist

(i) $K \in \{1, \ldots, M + N - 1\}$,
(ii) $a(1), \ldots, a(K), b(1), \ldots, b(K) \in \{p(1), \ldots, p(M), q(1), \ldots, q(N)\}$,
(iii) *positive numbers* [*resp. positive integers*] $t(1), \ldots t(K) \leqslant r(1) + \cdots + r(M)$,

(iv) $P = \Sigma\{t(k)[\![a(k), b(k)]\!]: k = 1,\ldots,K\} \in \mathbf{P}_1(\mathbf{R}^n)$ [*resp.* $\in \mathbf{IP}_1(\mathbf{R}^n)$] *such that*

(1) $B = \partial P$,

(2) $\mathbf{M}(P) = \Sigma\{t(k)|a(k) - b(k)|: k = 1,\ldots,K\}$,

(3) $\mathbf{M}(P) = \inf\{\mathbf{M}(Q): Q \in \mathbf{S}_1(\mathbf{R}^n)$ [*resp.* $\mathbf{I}_1(\mathbf{R}^n)$] *with* $\partial Q = P\}$,

(4) $\mathrm{spt}[\![a(j), b(j)]\!] \cap \mathrm{spt}[\![a(k), b(k)]\!] \subset \{a(j), b(j)\}$ *whenever* j, $k \in \{1,\ldots,K\}$ *with* $j \neq k$,

(5) $\mathrm{spt}(P)$ *contains no closed loops*; *i.e., the components of* $\mathrm{spt}(P)$ *are simply connected.*

PROOF. We use the Strong Approximation Theorem 2.11 together with Proposition 3.2 and the obvious compactness properties of expressions of the form 3.2(iv) to conclude the existence of $Q \in \mathbf{P}_1(\mathbf{R}^n)$ [resp. $\in \mathbf{IP}_1(\mathbf{R}^n)$] such that $\partial P = B$ and

$$\mathbf{M}(Q) = \inf\{\mathbf{M}(T): T \in \mathbf{R}_1(\mathbf{R}^n) \text{ [resp. } \in \mathbf{I}_1(\mathbf{R}^n)] \text{ with } \partial T = B\}.$$

We now apply 3.2 with such minimizing Q as in Hypothesis 3.2(d). We modify Q if necessary by the procedure described in Step 3 of the proof of 3.2 (elimination of closed loops) to obtain Q_μ, as there, for which $\partial Q_\mu = B$, $\mathbf{M}(Q_\mu) = \mathbf{M}(Q)$ (since Q was $\mathbf{M}$ minimizing), and $\mathrm{spt}(Q_\mu)$ contains no closed loops. We then modify Q_μ if necessary by the procedure described in Step 4 of the proof of 3.2 (elimination of bendings and excess vertices) to obtain an oriented graph Γ_ν, as there, and associated K, $a(1),\ldots,a(K)$, $b(1),\ldots,b(K)$, $t(1),\ldots,t(K)$, and P. Using obvious comparison curves together with the triangle inequality for length, we conclude that if k_0, k_1 are as in Step 4 of the proof of 3.2 then the line segment $c(k_0, 0)—c(k_0, 1)$ is colinear with the line segment $c(k_1, 0)—c(k_1, 1)$ and that

$$[\![c(k_0, 0), c(k_0, 1)]\!] + [\![c(k_1, 0), c(k_1, 1)]\!] = [\![c(k_0, 0), c(k_1, 1)]\!] \in \mathbf{P}_1(\mathbf{R}^n)$$

there. We readily conclude that $\mathrm{spt}(P) = \mathrm{spt}(Q_\nu)$ so that $\mathrm{spt}(P)$ is free of closed loops as required in 3.3(5).

Obvious comparison curves together with estimates based on the triangle inequality for length enable us to conclude from 3.3(3) that $\{a(1),\ldots,a(K),$ $b(1),\ldots,b(K)\} \sim \{p(1),\ldots,p(M), q(1),\ldots,q(N)\}$ is empty. Conclusion 3.2(5) then implies that $K \leqslant M + N - 1$ as required for 3.3(i). Similar comparison curves enable us to conclude that $\mathrm{spt}(P) \sim \{p(1),\ldots,p(M), q(1),\ldots,q(N)\}$ has no nontrivial crossings, i.e., is a one-dimensional submanifold of $\mathbf{R}^n$ as required in 3.3(4). Conclusion 3.3(2) is inferred from the minimality of Q in the context of our constructions. The inequality of 3.3(iii) follows from 3.2(iii) as proved in Step 5 of the proof of 3.2.

3.4. Proposition (Existence of special 1-dimensional mass and size minimizing polyhedral chains). *Suppose* $0 \leqslant \varepsilon < \infty$ *and*

$$B = \Sigma\{s(j)[\![q(j)]\!] . j = 1,\ldots,N\} - \Sigma\{r(i)[\![p(i)]\!]: i = 1,\ldots,M\}$$

is a member of $\mathbf{P}_0(\mathbf{R}^n)$ [*resp. of* $\mathbf{IP}_0(\mathbf{R}^n)$] *as in Proposition 3.2. Then there exist*

(i) $K \subset \{1,\ldots,3M + 3N - 3\}$,

(ii) $a(1),\ldots,a(K), b(1),\ldots,b(K) \in \mathbf{R}^n$ *with* $a(k) \neq b(k)$ *for each* $k = 1,\ldots,K$,

(iii) *positive numbers [resp. positive integers]* $t(1),\ldots,t(K) \leqslant r(1) + \cdots + r(M)$,

(iv) $P = \Sigma\{t(k)[\![a(k), b(k)]\!]: k = 1,\ldots,K\} \in \mathbf{P}_1(\mathbf{R}^n)$ *[resp.* $\in \mathbf{IP}_1(\mathbf{R}^n)]$ *such that*

(1) $\partial P = B$,

(2) $\mathbf{M}(P) = \Sigma_k t(k)|a(k) - b(k)|$,

(3) $\mathbf{S}(P) = \Sigma_k |a(k) - b(k)|$,

(4) $\varepsilon\mathbf{M}(P) + \mathbf{S}(P) = \inf\{\varepsilon\mathbf{M}(Q) + \mathbf{S}(Q): Q \in \mathbf{S}_1(\mathbf{R}^n)$ *[resp.* $\in \mathbf{I}_1(\mathbf{R}^n)]$ *with* $\partial Q = B\}$,

(5) $\operatorname{spt}[\![a(j), b(j)]\!] \cap \operatorname{spt}[\![a(k), b(k)]\!] \subset \{a(j), b(j)\}$ *whenever* $j,\ k \in \{1,\ldots,K\}$ *with* $j \neq k$,

(6) $\operatorname{spt}(P)$ *contains no closed loops, i.e., the components of* $\operatorname{spt}(P)$ *are simply connected,*

(7) *for each* $z \in \{a(1),\ldots,a(K),\ b(1),\ldots,b(K)\} \sim \{p(1),\ldots,p(M), q(1),\ldots,q(N)\}$,

$$\operatorname{card}\{k: K \in \{1,\ldots,K\} \text{ with } z \in \{a(k), b(k)\}\} \geqslant 3,$$

(8) $\{a(1),\ldots,a(K),\ b(1),\ldots,b(K)\}$ *lies within the convex hull of* $\{p(1),\ldots,p(M), q(1),\ldots,q(N)\}$.

PROOF. We largely adapt the proof of Proposition 3.3 in an entirely straightforward way. With regard to 3.4(5), we note that, in the context of Step 2 of the proof of Proposition 3.2,

$$\operatorname{spt}(Q) \sim \{c(1,0), c(1,1), c(2,0),\ldots,c(L,1)\}$$

is a one-dimensional submanifold of $\mathbf{R}^n$ and that the constructions of Step 3 of the proof of 3.2 (as applied in the present context as discussed in the proof of 3.3) do not introduce new crossings in $\operatorname{spt}(Q)$. The minimality property of 3.4(4) readily implies 3.4(8).

3.5. Proposition (Properties of the G metric on $\mathbf{P}_0(\mathbf{R}^n)$).

Hypotheses. (a) $L \in \mathbf{R}, M \in \mathbf{R}^+, N \in \mathbf{Z}^+,$ *and* $0 < R \in \mathbf{R}^+$,

(b)

$$\mathbf{Q} = \mathbf{P}_0(\mathbf{R}^n) \cap \{T: \langle T, 1 \rangle = L, \mathbf{M}(T) \leqslant M, \mathbf{S}(T) \leqslant N, \operatorname{spt}(T) \subset \mathbf{B}^n(0, R)\}$$

$$= \mathbf{P}_0(\mathbf{R}^n) \cap \{r_1[\![p_1]\!] + \cdots + r_N[\![p_N]\!]: r_1,\ldots,r_N \in \mathbf{R}$$

$$\text{with } L = r_1 + \cdots + r_N \leqslant |r_1| + \cdots + |r_N| \leqslant M$$

$$\text{and } p_1,\ldots,p_N \in \mathbf{B}^n(0, R)\}.$$

Conclusions. (1) *Suppose for each* $i \in \{1,\ldots,N\}$ *we have*

$$r(i), r_1(i), r_2(i), r_3(i),\ldots \in \mathbf{R} \quad \text{with } \lim_{v \to \infty} r_v(i) = r(i)$$

and also have

$$p(i), p_1(i), p_2(i), p_3(i),\ldots \in \mathbf{B}^n(0, R) \quad \text{with } \lim_{v \to \infty} p_v(i) = p(i).$$

Suppose, additionally, for each $v \in \{1, 2, 3, \ldots\}$, $r_v(1) + r_v(2) + \cdots + r_v(N) = L$. Then, for each $v \in \{1, 2, 3, \ldots\}$,

$$\sum_i \left(r(i) \llbracket p(i) \rrbracket - r_v(i) \llbracket p_v(i) \rrbracket \right)$$

$$= \sum_i r(i) \left(\llbracket p(i) \rrbracket - \llbracket p_v(i) \rrbracket \right) + \sum_i \left(r(i) - r_v(i) \right) \llbracket p_v(i) \rrbracket$$

$$= \partial \left[\sum_i r(i) \llbracket p_v(i), p(i) \rrbracket \right] + \partial \left(\llbracket 0 \rrbracket \times \sum_i \left(r(i) - r_v(i) \right) \llbracket p_v(i) \rrbracket \right)$$

(since $\sum_i (r(i) - r_v(i)) = L - L = 0$) so that

$$\mathbf{G}\left(\sum_i r(i) \llbracket p(i) \rrbracket, \sum_i r_v(i) \llbracket p_v(i) \rrbracket \right)$$

$$\leqslant \sum_i |r(i)| \cdot |p_v(i) - p(i)| + \sum_i |r(i) - r_v(i)| R.$$

In particular,

$$0 = \lim_{v \to \infty} \mathbf{G}\left(\sum_i r(i) \llbracket p(i) \rrbracket, \sum_i r_v(i) \llbracket p_v(i) \rrbracket \right).$$

We note additionally for each $v \in \{1, 2, 3, \ldots\}$ that if

$$S_v = \sum_i r_v \llbracket p_v(i) \rrbracket \quad \text{and} \quad P_v = \sum_i \left(r(i) - r_v(i) \right) \llbracket p_v(i) \rrbracket,$$

then $S_v + P_v = \sum_i r(i) \llbracket p_v(i) \rrbracket$ so that $\mathbf{M}(S_v + P_v) \leqslant M$, $\mathbf{S}(S_v + P_v) \leqslant N$, and $\mathrm{spt}(S_v + P_v) \subset \mathbf{B}^n(0, R)$.

(2) $\mathbf{Q}$ *is sequentially compact and the mass function $\mathbf{M}$ and size function $\mathbf{S}$ are lower semicontinuous in the $\mathbf{G}$ metric topology on $\mathbf{Q}$.*

(3) *Corresponding to each $T \in \mathbf{Q}$ and each $0 < \varepsilon < \infty$ there is $0 < \delta < 1$ with the following property. Whenever $S \in \mathbf{Q}$ with $\mathbf{G}(T, S) < \delta$ then there will exist $P \in \mathbf{P}_0(\mathbf{R}^n)$ and $Q \in \mathbf{P}_1(\mathbf{R}^n)$ such that*

(3.1) $T - S = \partial Q + \partial(\llbracket 0 \rrbracket \times P) = \partial Q + P$,

(3.2) $\mathbf{M}(P) < \varepsilon$, $\mathbf{M}(\llbracket 0 \rrbracket \times P) < \varepsilon$, $\mathbf{S}(P) \leqslant N$, *and* $\mathrm{spt}(P) \subset \mathbf{B}^n(0, R)$,

(3.3) $\mathbf{MS}(Q) < \varepsilon$ *and* $S + P \in \mathbf{Q}$.

(4) *Suppose $S, T, S_1, T_1, S_2, T_2, S_3, T_3, \ldots \in \mathbf{Q}$ with $\lim_{i \to \infty} \mathbf{G}(S, S_i) = 0$ and $\lim_{i \to \infty} \mathbf{G}(T, T_i) = 0$. Then*

$$\inf\{ \mathbf{MS}(Q) \colon Q \in S_1(\mathbf{R}^n) \text{ with } \partial Q = S - T \}$$

$$\leqslant \liminf_{i \to \infty} \inf\{ \mathbf{MS}(Q) \colon Q \in \mathbf{S}_1(\mathbf{R}^n) \text{ with } \partial Q = S_i - T_i \},$$

i.e., $\mathbf{GS}(S, T) \leqslant \liminf_{i \to \infty} \mathbf{GS}(S_i, T_i)$.

(5) *Suppose $f(1), f(2), f(3), \ldots \colon \mathbf{R}^m \to \mathbf{Q}$ such that*

(5.1) $\bigcup \{ \mathrm{spt}[f(i)(x)] \colon x \in \mathbf{R}^m, i = 1, 2, 3, \ldots \}$ *is bounded,*

(5.2) $\infty > K = \sup\{ |x - y|^{-1} \mathbf{G}(f(i)(x), f(i)(y)) \colon x, y \in \mathbf{R}^m \text{ with } x \neq y \text{ and } i = 1, 2, 3, \ldots \}$.

Then there exists a subsequence $i(1), i(2), i(3), \ldots$ of $1, 2, 3, \ldots$ and $f \colon \mathbf{R}^m \to \mathbf{Q}$ such that

(5.3) $0 = \lim_{j \to \infty} \sup\{ \mathbf{G}(f(x), f(i(j))(x)) \colon x \in \mathbf{R}^m \}$ *(uniform convergence),*

(5.4) $\mathbf{G}(f(x), f(y)) \leqslant K|x - y|$ *for each $x, y \in \mathbf{R}^m$,*

(5.5) *for each* $x, y \in \mathbf{R}^m$,

$$\mathbf{GS}(f(x), f(y)) \leqslant \liminf_{j \to \infty} \mathbf{GS}(f(i(j))(x), f(i(j))(y)).$$

PROOF. Conclusion (1) is clear. Conclusion (2) follows readily from Conclusion (1) (note 3.3(2)). An obvious argument shows that the negation of Conclusion (3) implies the negation of Conclusion (1). Conclusion (4) follows from obvious compactness and lower semicontinuity properties of expressions of the form of 3.4(iv). Conclusion (5) follows from Conclusion (4) and well-known properties of equicontinuous families of functions.

3.6. A basic construction to be used in extending multiple-valued mappings from 0 skeletons to 1 skeletons.

Assumptions. (a) $K, M_0, M_1, N_0, N_1 \in \{0, 1, 2, \ldots\}$ with $M_0 + M_1 + N_0 + N_1 > 0$.

(b) $a(1), \ldots, a(K), b(1), \ldots, b(K) \in \mathbf{R}^n$ with $a(k) \neq b(k)$ for each $k = 1, \ldots, K$.

(c) $p(1), \ldots, p(M_0), q(1), \ldots, q(N_0), P(1), \ldots, P(M_1), Q(1), \ldots, Q(N_1) \in \mathbf{R}^n$ are distinct.

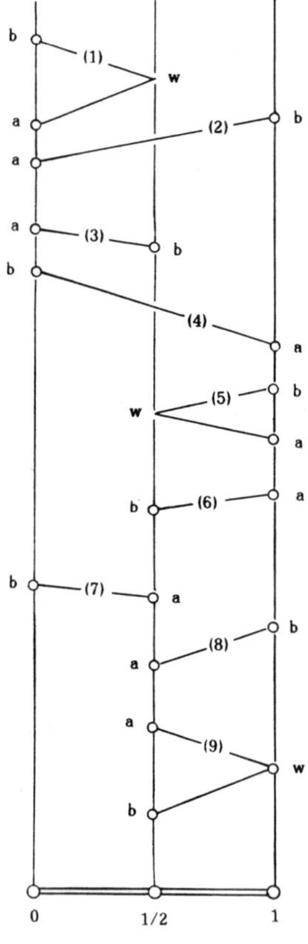

FIGURE 3.6(A). Multiple-valued extension from $\partial[0, 1]$ to $[0, 1]$.

(d) $t(1),\ldots,t(K)$, $r(1),\ldots,r(M_0)$, $s(1),\ldots,s(N_0)$, $R(1),\ldots,R(M_1)$, $S(1),\ldots,S(N_1)$ are positive numbers with $r(1) + \cdots + r(M_0) + S(1) + \cdots + S(N_1) = R(1) + \cdots + R(M_1) + s(1) + \cdots + s(N_0)$.

(e)
$$\partial \sum \{ t(k)[\![a(k), b(k)]\!] : k = 1,\ldots,K \}$$
$$= \sum \{ t(k)([\![b(k)]\!] - [\![a(k)]\!]) : k = 1,\ldots,K \}$$
$$= \left(\sum \{ S(j)[\![Q(j)]\!] : j = 1,\ldots,N_1 \} - \sum \{ R(i)[\![P(i)]\!] : i = 1,\ldots,M_1 \} \right)$$
$$- \left(\sum \{ s(j)[\![q(j)]\!] : j = 1,\ldots,N_0 \} - \sum \{ r(i)[\![p(i)]\!] : i = 1,\ldots,M_0 \} \right);$$

we note that Assumption (e) combinatorially implies the equality of Assumption (d).

Terminology. As useful terminology we set

(i) $w(k) = [a(k) + b(k)]/2$ for each $k = 1,\ldots,K$,

(ii) $\Pi = \{ a(1),\ldots,a(K), b(1),\ldots,b(K) \}$,

(iii) $\Pi(0) = \{ p(1),\ldots,p(M_0), q(1),\ldots,q(N_0) \} \subset \Pi$,

(iv) $\Pi(1) = \{ P(1),\ldots,P(M_1), Q(1),\ldots,Q(N_1) \} \subset \Pi \sim \Pi(0)$,

(v) $\Pi(\tfrac{1}{2}) = \Pi \sim (\Pi(0) \cup \Pi(1))$.

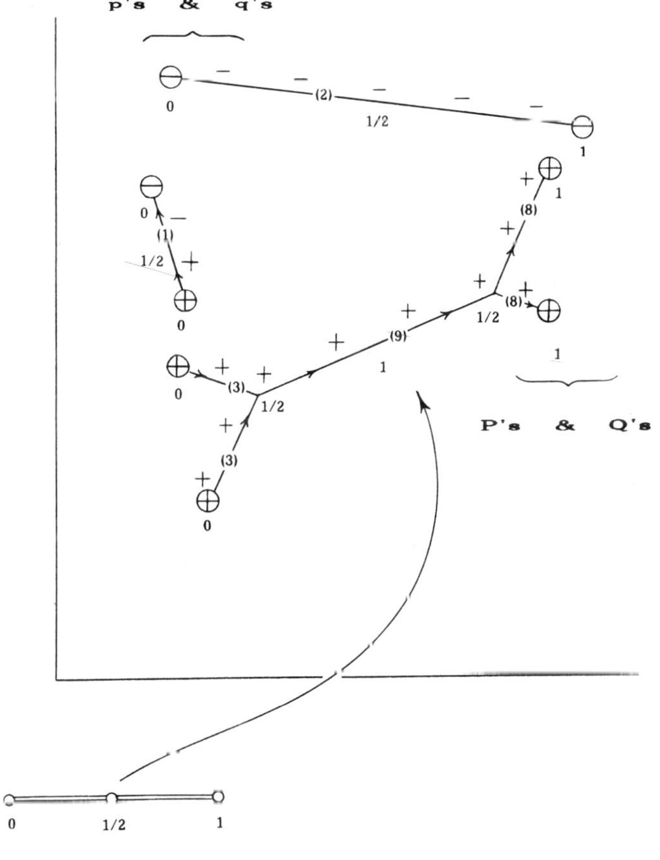

FIGURE 3.6(B). Multiple-valued extension from $\partial[0,1]$ to $[0,1]$.

Construction. We will construct mappings $f(1),\ldots,f(K)\colon [0,1] \to \mathbf{P}_0(\mathbf{R}^n)$. For $k \in \{1,\ldots,K\}$, the definition of $f(k)$ depends on which of the sets $\Pi(0)$, $\Pi(1)$, $\Pi(\frac{1}{2})$ contains $a(k)$ and which contains $b(k)$; there are nine possibilities which we list; see Figures 3.6(A) and 3.6(B).

(1) $a(k) \in \Pi(0)$, $b(k) \in \Pi(0)$:

$$f(k)(\tau) = \begin{cases} t(k)[\![(1 - 2\tau)a(k) + 2\tau w(k)]\!] \\ \qquad\qquad -t(k)[\![(1 - 2\tau)b(k) + 2\tau w(k)]\!] & \text{if } 0 \leqslant \tau \leqslant \tfrac{1}{2}, \\ 0 \quad \text{if } \tfrac{1}{2} \leqslant \tau \leqslant 1. \end{cases}$$

(2) $a(k) \in \Pi(0)$, $b(k) \in \Pi(1)$:

$$f(k)(\tau) = t(k)[\![(1 - \tau)a(k) + \tau b(k)]\!] \quad \text{for } 0 \leqslant \tau \leqslant 1.$$

(3) $a(k) \in \Pi(0)$, $b(k) \in \Pi(\tfrac{1}{2})$:

$$f(k)(\tau) = \begin{cases} t(k)[\![(1 - 2\tau)a(k) + 2\tau b(k)]\!] & \text{if } 0 \leqslant \tau \leqslant \tfrac{1}{2}, \\ 0 & \text{if } \tfrac{1}{2} \leqslant \tau \leqslant 1. \end{cases}$$

(4) $a(k) \in \Pi(1)$, $b(k) \in \Pi(0)$:

$$f(k)(\tau) = -t(k)[\![(1 - \tau)b(k) + \tau a(k)]\!] \quad \text{for } 0 \leqslant \tau \leqslant 1.$$

(5) $a(k) \in \Pi(1)$, $b(k) \in \Pi(1)$:

$$f(k)(\tau) = \begin{cases} 0 \quad \text{if } 0 \leqslant \tau \leqslant \tfrac{1}{2}, \\ t(k)[\![(2 - 2\tau)w(k) + (2\tau - 1)b(k)]\!] \\ -t(k)[\![(2 - 2\tau)w(k) + (2\tau - 1)a(k)]\!] & \text{if } \tfrac{1}{2} \leqslant \tau \leqslant 1. \end{cases}$$

(6) $a(k) \in \Pi(1)$, $b(k) \in \Pi(\tfrac{1}{2})$:

$$f(k)(\tau) = \begin{cases} 0 & \text{if } 0 \leqslant \tau \leqslant \tfrac{1}{2}, \\ -t(k)[\![(2 - 2\tau)b(k) + (2\tau - 1)a(k)]\!] & \text{if } \tfrac{1}{2} \leqslant \tau \leqslant 1. \end{cases}$$

(7) $a(k) \in \Pi(\tfrac{1}{2})$, $b(k) \in \Pi(0)$:

$$f(k)(\tau) = \begin{cases} -t(k)[\![(1 - 2\tau)b(k) + 2\tau a(k)]\!] & \text{if } 0 \leqslant \tau \leqslant \tfrac{1}{2}, \\ 0 & \text{if } \tfrac{1}{2} \leqslant \tau \leqslant 1. \end{cases}$$

(8) $a(k) \in \Pi(\tfrac{1}{2})$, $b(k) \in \Pi(1)$:

$$f(k)(\tau) = \begin{cases} 0 & \text{if } 0 \leqslant \tau \leqslant \tfrac{1}{2}, \\ t(k)[\![(2 - 2\tau)a(k) + (2\tau - 1)b(k)]\!] & \text{if } \tfrac{1}{2} \leqslant \tau \leqslant 1. \end{cases}$$

(9) $a(k) \in \Pi(\tfrac{1}{2})$, $b(k) \in \Pi(\tfrac{1}{2})$:

$$f(k)(z) = \begin{cases} 0 \quad \text{if } 0 \leqslant \tau \leqslant \tfrac{1}{2}, \\ t(k)[\![(2 - 2\tau)b(k) + (2\tau - 1)w(k)]\!] \\ -t(k)[\![(2 - 2\tau)a(k) + (2\tau - 1)w(k)]\!] & \text{if } \tfrac{1}{2} \leqslant \tau \leqslant 1. \end{cases}$$

Finally, we set

$$f = f(1) + \cdots + f(K)\colon [0,1] \to \mathbf{P}_0(\mathbf{R}^n).$$

Properties of f. (1) For each $x \in [0,1]$, $\mathbf{S}(f(x)) \leqslant 2K$ and $\mathbf{M}(f(x)) \leqslant 2\Sigma_k t(k)$.

(2) $f(0) = \Sigma\{s(j)[\![q(j)]\!]: j = 1,\ldots,N_0\} - \Sigma\{r(i)[\![p(i)]\!]: i = 1,\ldots,M_0\}$.

(3) $f(1) = \Sigma\{S(j)[\![Q(j)]\!]: j = 1,\ldots,N_1\} - \Sigma\{R(i)[\![P(i)]\!]: i = 1,\ldots,M_1\}$.

(4) f is **GS** continuous.

(5) For each $k \in \{1,\ldots,K\}$ and each $x, y \in [\,0,\frac{1}{2})$ [resp. $\in (\frac{1}{2},1\,]\,]$ there is $Q \in \mathbf{P}_1(\mathbf{R}^n)$ such that

$$\partial Q = f(k)(x) - f(k)(y), \qquad \mathbf{S}(Q) \leqslant 2|x - y| \cdot |a(k) - b(k)|,$$

$$\mathbf{M}(Q) \leqslant 2|x - y| \cdot t(k) \cdot |a(k) - b(k)|.$$

(6) For each $x, y \in [0,1]$ there is $Q \in \mathbf{P}_1(\mathbf{R}^n)$ such that

$$\partial Q = f(x) - f(y),$$

$$\mathbf{S}(Q) \leqslant 2|x - y| \sum \{|a(k) - b(k)|: k = 1,\ldots,K\},$$

$$\mathbf{M}(Q) \leqslant 2|x - y| \sum \{t(k)|a(k) - b(k)|: k = 1,\ldots,K\},$$

$$\langle f(x),1\rangle = \langle f(y) + \partial Q, 1\rangle = \langle f(y),1\rangle + \langle Q, d(1)\rangle = \langle f(y),1\rangle.$$

(7) For each $x \in [0,1]$,

$$\langle f(x),1\rangle = S(1) + \cdots + S(N_1) - R(1) - \cdots - R(M_1)$$

$$= s(1) + \cdots + s(n_0) - r(1) - \cdots - r(M_0).$$

(8) $\mathrm{Clos}(\mathbf{R}^n \cap \{y: y \in \mathrm{spt}[f(x)] \text{ for some } x \in [0,1]\})$

$$= \bigcup\{\tau a(k) + (1 - \tau)b(k): 0 \leqslant \tau \leqslant 1 \text{ and } k = 1,\ldots,K\}.$$

(9) There are affine functions $g(1),\ldots,g(2K)$: $\mathbf{R} \to \mathbf{R}^n$ with $2|a(k) - b(k)| \geqslant \sup\{\|Dg(2k - 1)\|, \|Dg(2k)\|\}$ for each $k = 1,\ldots,K$ and

$$\sum \{\|Dg(i)\|: i = 1,\ldots,2K\} \leqslant 4\sum \{|a(k) - b(k)|: k = 1,\ldots,K\}$$

such that

$$\mathrm{Clos}(\mathbf{R} \times \mathbf{R}^n \cap \{(\tau, y): 0 \leqslant \tau \leqslant 1 \text{ and } y \in \mathrm{spt}[f(x)]\})$$

$$\subset \mathbf{R} \times \mathbf{R}^n \cap \{(x, y): y = g(j)(x) \text{ for some } j = 1,\ldots,2K\}.$$

3.7. Corollary (Extension of a multiple-valued function from $\partial[0,1]$ to $[0,1]$).

Hypotheses. (a) $L, M, N, R, \mathbf{Q}$ *are as in* 3.5(a)(b),

(b) $V, W \in \mathbf{Q}$ [*resp.* $\in \mathbf{Q} \cap \mathbf{I}_0(\mathbf{R}^n)$],

(c) $\lambda = \inf\{\mathbf{M}(Q) + \mathbf{S}(Q): Q \in \mathbf{P}_1(\mathbf{R}^n) \text{ with } \partial Q = W - V\}$ [*resp.* $\lambda = \inf\{\mathbf{M}(Q): Q \in \mathbf{IP}_1(\mathbf{R}^n) \text{ with } \partial Q = W - V\}$].

Conclusion. There is a Lipschitz function f: $[0,1] \to \mathbf{P}_0(\mathbf{R}^n)$ [resp. $\to \mathbf{IP}_0(\mathbf{R}^n)$] together with affine functions $g(1),\ldots,g(12N - 6)$: $\mathbf{R} \to \mathbf{R}^n$ [resp. $g(1),\ldots, g(4N - 2)$: $\mathbf{R} \to \mathbf{R}^n$] with the following properties:

(1) *for each $0 \leqslant x \leqslant 1$,*

$$\mathbf{S}(f(x)) \leqslant 12N - 6 \quad [resp. \mathbf{S}(f(x)) \leqslant 4N - 2],$$

$$\Theta^0(f(x), y) \leqslant M \quad for\ each\ y \in \mathbf{R}^n,$$

$$\mathbf{M}(f(x)) \leqslant M(12N - 6) \quad [resp. \mathbf{M}(f(x)) \leqslant M(4N - 2)],$$

and $\langle f(x),1\rangle = L$,

(2) $f(0) = V$ *and* $f(1) = W$,

(3) *for each* $x, y \in [0, 1]$,

$$\mathbf{GS}(f(x), f(y)) = \inf\{\mathbf{M}(Q) + \mathbf{S}(Q): Q \in \mathbf{S}_1(\mathbf{R}^n) \text{ with } \partial Q = f(x) - f(y)\}$$
$$\leqslant 2\lambda|x - y|$$
$$[resp. \ \mathscr{G}(f(x), f(y)) = \inf\{\mathbf{M}(Q): Q \in \mathbf{I}_1(\mathbf{R}^n) \text{ with } \partial Q = f(x) - f(y)\}$$
$$\leqslant 2\lambda|x - y|],$$

(4) $\Sigma\{\|Dg(j)\|: j = 1, 2, \ldots, 12N - 6\} \leqslant 4\lambda$ *and*

$$\mathrm{Clos}(\mathbf{R} \times \mathbf{R}^n \cap \{(x, y): 0 \leqslant x \leqslant 1 \text{ and } y \in \mathrm{spt}[f(x)]\})$$
$$\subset \mathbf{R} \times \mathbf{R}^n \cap \{(x, y): y = g(j)(x) \text{ for some } j = 1, \ldots, 12N - 6\},$$

(5) $\mathscr{H}^1[\mathrm{Clos}(\mathbf{R}^n \cap \{y: y \in \mathrm{spt}[f(x)] \text{ for some } x \in [0, 1]\})] \leqslant \lambda$,

(6) *the construction of f adapted from* 3.6 *is symmetric with respect to the symmetry of* $[0, 1]$, *i.e.,* $\tau \to 1 - \tau$.

PROOF. In case $V = W = 0$ we set $f(x) = 0$ for each $x \in [0, 1]$ and properties (1), (2) and (3) are immediate. For the remainder of the proof we assume V and W are not both 0. Therefore, we can choose M_0, M_1, N_0, $N_1 \in \{0, 1, 2, \ldots\}$ with $M_0 + M_1 + N_0 + N_1 > 0$ together with distinct points $p(1), \ldots, p(M_0)$, $q(1), \ldots, q(N_0) \in \mathbf{R}^n$, distinct points $P(1), \ldots, P(M_1)$, $Q(1), \ldots, Q(N_1) \in \mathbf{R}^n$, and positive numbers [resp. positive integers] $r(1), \ldots, r(M_0)$, $R(1), \ldots, R(M_1)$, $s(1), \ldots, s(N_0)$, $S(1), \ldots, S(N_1)$ such that

$$V = \sum\{s(j)[\![q(j)]\!]: j = 1, \ldots, N_0\} - \sum\{r(i)[\![p(i)]\!]: i = 1, \ldots, M_1\},$$
$$W = \sum\{S(j)[\![Q(j)]\!]: j = 1, \ldots, N_1\} - \sum\{R(i)[\![P(i)]\!]: i = 1, \ldots, M_1\}.$$

It may not be true that all the points $p(1), \ldots, p(M_0)$, $q(1), \ldots, q(N_0)$, $P(1), \ldots, P(M_1)$, $Q(1), \ldots, Q(N_1)$ are distinct. If this situation occurs it seems easiest (in terms of terminology) to construct our mapping f as the **G** limit of mappings associated with slightly perturbed collections of such points which are distinct. The details of such a limit construction are entirely straightforward and are left to the reader. For the remainder of our proof we will assume that the points $p(1), \ldots, q(1), \ldots, P(1), \ldots, Q(1), \ldots$ are distinct. Since V, $W \in \mathbf{Q}$ we infer $M_0 + N_0 \leqslant N$, $M_1 + N_1 \leqslant N$, $\Sigma_j s(j) + \Sigma_i r(i) \leqslant M$, $\Sigma_j S(j) + \Sigma_i R(i) \leqslant M$, $\Sigma_j s(j) - \Sigma_i r(i) = L$, $\Sigma_j S(j) - \Sigma_i R(i) = L$. It follows that

$$2\sum_j s(j) \leqslant M + L, \qquad 2\sum_i r(i) \leqslant M - L,$$
$$\sum_j s(j) + \sum_i R(i) = \sum_j S(j) + \sum_i r(i) \leqslant M.$$

We now apply Proposition 3.4 with $\varepsilon = 1$ [resp. Proposition 3.3] and $B = W - V$ to obtain

$$K \in \{1, 2, \ldots, 3(M_1 + N_0) + (M_0 + N_1) - 1\} \subset \{1, 2, \ldots, 6N - 3\}$$
$$[\text{resp. } K \in \{1, 2, \ldots, (M_1 + N_0) + (M_0 + N_1) - 1\} \subset \{1, 2, \ldots, 2N - 1\}]$$

together with $a(1),\dots,a(k)$, $b(1),\dots,b(K) \in \mathbf{R}^n$ with $a(k) \neq b(k)$ for each $k \in \{1,\dots,K\}$ and positive numbers [resp. positive integers]

$$t(1),\dots,t(L) \leqslant \sum_j s(j) + \sum_i R(i) \leqslant M$$

and

$$P = \sum \{ t(k)[\![a(k), b(k)]\!] : k \in 1,\dots,K \} \in \mathbf{P}_1(\mathbf{R}^n) \quad [\text{resp. } \in \mathbf{IP}_1(\mathbf{R}^n)]$$

such that, in particular, $\mathbf{MS}(P) \leqslant \lambda$ [resp. $\mathbf{M}(P) \leqslant \lambda$]. With regard to the density inequalities of Conclusion (1) we note 3.3(4), 3.4(5).

3.8. How to extend Lipschitz multiple-valued functions defined on the 0 skeleton of an admissible cubical complex.

Assumptions. (a) We replace $m + n$ by m in the obvious way in the terminology of 1.5 and 1.8 and here assume that $\mathcal{F}$ is an admissible family of m-dimensional cubes in $\mathbf{R}^m$ with associated cubical complex $\mathbf{CX}(\mathcal{F})$.

(b) $\lambda \in \mathbf{R}^+$, $L \in \mathbf{R}$, $M \in \mathbf{R}^+$, $N \in \{1, 2, 3,\dots\}$.

(c) $f_0: \bigcup\mathbf{CX}_0(\mathcal{F}) \to \mathbf{P}_0(\mathbf{R}^n)$ [resp. $\mathbf{IP}_0(\mathbf{R}^n)$].

(d) $\langle f(v), 1 \rangle = L$, $\mathbf{M}(f(v)) \leqslant M$, and $\mathbf{S}(f(v)) \leqslant N$ for each $\{v\} \in \mathbf{CX}_0(\mathcal{F})$.

(e) $\mathbf{GS}(f(v), f(w)) \leqslant \lambda|v - w|$ [resp. $\mathcal{G}(f(v), f(w)) \leqslant \lambda|v - w|]$ whenever $J \in \mathbf{CX}_1(\mathcal{F})$ with $\{v, w\} = \partial J$.

(f) Corresponding to each bounded subset A of $\mathbf{R}^m$ there is $0 < R < \infty$ such that $\mathrm{spt}[f(v)] \subset \mathbf{B}^n(0, R)$ for each $v \subset A \cap \bigcup\mathbf{CX}_0(\mathcal{F})$.

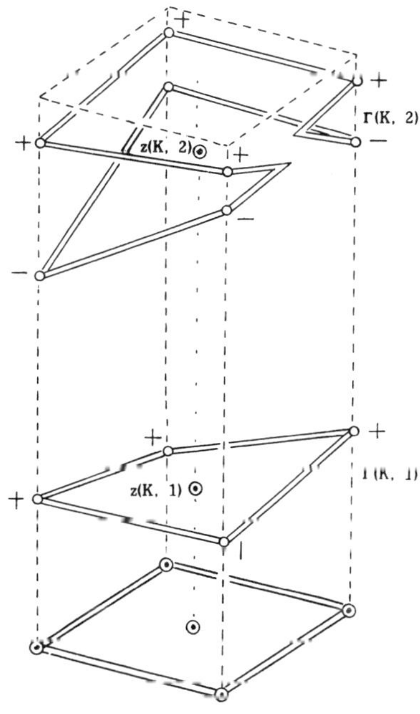

FIGURE 3.8(A). Multiple-valued mapping on the 1 skeleton of ∂K.

Extension to the 1 *skeleton* (*see Figure* 3.8(A)). *We construct*

$$f_1\colon \bigcup \mathbf{CX}_1(\mathscr{F}) \to \mathbf{P}_0(\mathbf{R}^n) \quad [\text{resp. } \to \mathbf{IP}_0(\mathbf{R}^n)]$$

by requiring $f_1((1 - t)v + tw) = f(t)$ whenever $J \in \mathbf{CX}_1(\mathscr{F})$ with $\{v, w\} = \partial J$; here $f\colon [0, 1] \to \mathbf{P}_0(\mathbf{R}^n)$ [resp. $\to \mathbf{IP}_0(\mathbf{R}^n)$] is constructed in accordance with Corollary 3.7 with L, M, N, V, W, λ there corresponding to $L, M, N, f_0(v), f_0(w)$, $\lambda|v - w|$, respectively, in the present context (and R having a suitable value). In particular, for each $x \in \bigcup \mathbf{CX}_0(\mathscr{F})$, $\mathbf{S}(f_1(x)) \leqslant 12N - 6$ [resp. $\leqslant 4N - 2$], $\Theta^0(f_1(x), y) \leqslant M$ for each $y \in \mathbf{R}^n$, and $\mathbf{M}(f_1(x)) \leqslant M(12N - 6)$ [resp. $\leqslant M(4N - 1)$]. Additionally, for each $J \in \mathbf{CX}_1(\mathscr{F})$, there exist affine functions $g(1),\ldots,g(12N - 6)\colon \mathbf{R}^m \to \mathbf{R}^n$ with $\Sigma_i\|Dg(i)\| \leqslant 4\lambda$ such that $\mathrm{spt}[f_1(x)] \subset \{g(1)(x),\ldots,g(12N - 6)(x)\}$ for each $x \in J$.

Some terminology and estimates. Whenever $K \in \{2, 3,\ldots,m\}$, $v \in \{0, 1, 2\ldots\}$, and $K \in \mathbf{CX}_k(\mathscr{F})$, we can check that

$$\mathrm{card}\big(\mathbf{CX}_1(\mathscr{F}) \cap \{J\colon J \subset \partial K\}\big) \leqslant 2k(3^{k-1} - 1)$$

and

$$\mathscr{H}^1\Big(\bigcup\{J\colon J \in \mathbf{CX}_1(\mathscr{F}) \text{ with } J \subset \partial K\}\Big) \leqslant k(3^{k-1} - 1)2^{-v}.$$

For such k, v, K we get

$$\Gamma(K) = \mathrm{Clos}\big(\mathbf{R}^n \cap \{y\colon y \in \mathrm{spt}[f(x)] \text{ for some } x \in J \in \mathbf{CX}_1(\mathscr{F}) \text{ with } J \subset \partial K\}\big)$$

and use 3.7(5) to infer

$$\mathscr{H}^1(\Gamma(K)) \leqslant k(3^{k-1} - 1)2^{-v}\lambda.$$

Additionally, for such k, v, K we let $\gamma(K)$ denote the number of pathwise connected components of $\Gamma(K)$, let $\Gamma(K, 1),\ldots,\Gamma(K, \gamma(K))$ be a listing of these components and choose $z(K, 1),\ldots,z(K, \gamma(K)) \in \mathbf{R}^n$ with $z(K, i) \in \Gamma(K, i)$ for each $i = 1,\ldots,\gamma(K)$. We conclude from 3.7 and our estimates above that

$$\gamma(K) < (12N - 6)k(3^{k-1} - 1) \quad \big[\text{resp. } < (4N - 2)k(3^{k-1} - 1)\big]$$

and

$$\mathrm{diam}[\Gamma(K, i)] \leqslant \mathscr{H}^1(\Gamma(K)) \leqslant k(3^{k-1} - 1)2^{-v}\lambda$$

for each $i = 1,\ldots,\gamma(K)$.

We also note that whenever $k \in \{3, 4,\ldots,m\}$, $K \in \mathbf{CX}_k(\mathscr{F})$, $J \in \mathbf{CX}_{k-1}(\mathscr{F})$ with $J \subset \partial K$, and $\alpha \in \{1,\ldots,\gamma(K)\}$, then there is a unique $\beta \in \{1,\ldots,\gamma(K)\}$ with $\Gamma(J, \alpha) \subset \Gamma(K, \beta)$; for such β we denote $\beta = \beta(K, J, \alpha)$.

Sequential extension of mappings (*see Figure* 3.8 (B)). (1) Suppose $K \in \mathbf{CX}_2(\mathscr{F})$. According to 1.8, $\partial K \subset \mathrm{dmn}(f_1)$ and we define

$$f_1(K, 1),\ldots,f_1(K, \gamma(K))\colon \partial K \to \mathbf{P}_0(\mathbf{R}^n) \quad [\text{resp. } \to \mathbf{IP}_0(\mathbf{R}^n)]$$

by setting $f_1(K, i)(x) = f_1(x) \lfloor \Gamma(K, i)$ for each $i = 1,\ldots,\gamma(K)$ and each $x \in \partial K$.

We further define

$$f_2(K,1),\ldots,f_2(K,\gamma(K)),f_2(K)\colon K \to \mathbf{P}_0(\mathbf{R}^n) \quad [\text{resp.} \to \mathbf{IP}_0(\mathbf{R}^n)]$$

by setting

$$f_2(K,i)((1-t)\mathrm{center}(K)+tx)$$
$$= \tau(z(K,i))_\# \circ \mu(t)_\# \circ \tau(-z(K,i))_\# f_1(K,i)(x)$$

for each $i \in \{1,\ldots,\gamma(K)\}$, each $x \in \partial K$, and each $0 \leqslant t \leqslant 1$, and

$$f_2(K) = f_2(K,1) + \cdots + f_2(K,\gamma(K)).$$

Finally, we set

$$f_2 = \bigcup\{f_2(K)\colon K \in \mathbf{CX}_2(\mathscr{F})\}\colon \bigcup \mathbf{CX}_2(\mathscr{F}) \to \mathbf{P}_0(\mathbf{R}^n) \quad [\text{resp.} \to \mathbf{IP}_0(\mathbf{R}^n)].$$

Clearly, for each $x \in \bigcup\mathbf{CX}_2(\mathscr{F})$, $\mathbf{S}(f_2(x)) \leqslant 12N-6$ [resp. $\leqslant 4N-1$], $\Theta^0(f_2(x),y) \leqslant M$ for each $y \in \mathbf{R}^n$, and $\mathbf{M}(f_2(x)) \leqslant M(12N-6)$ [resp. $\leqslant M(4N-1)$].

Additionally, we readily infer the existence of $1 < C_2 < \infty$ (depending only on the dimensions 2 and n) such that, whenever $J \in \mathbf{CX}_1(\mathscr{F})$ and $K \in \mathbf{CX}_2(\mathscr{F})$ with $J \subset K$ and

$$L_2 = \{(1-t)\mathrm{center}(K)+tx\colon 0 \leqslant t \leqslant 1 \text{ and } x \in J\},$$

there will exist affine functions $g(1),\ldots,g(12N-6)\colon \mathbf{R}^m \to \mathbf{R}^n$ with $\Sigma_j\|Dg(j)\| \leqslant C_2\lambda$ such that

$$\mathrm{spt}[f_2(z)] \subset \{g(1)(z),\ldots,g(12N-6)(z)\}$$

for each $z \in L_2$.

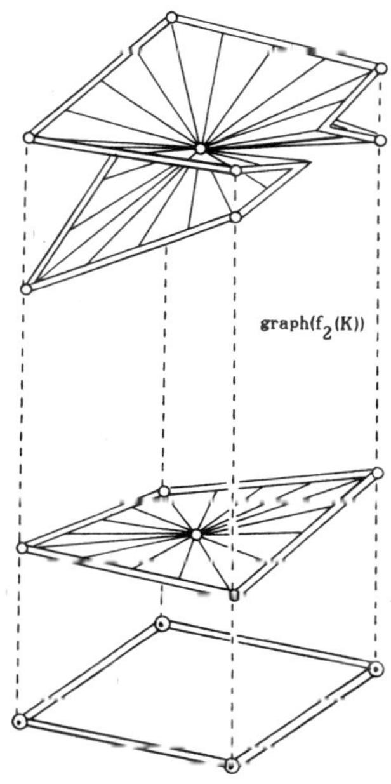

FIGURE 3.8(B). Multiple-valued extension from ∂K to K.

(2) We now assume inductively that $k \in \{3, 4, \ldots, m\}$ and that we have defined mappings

$$f_{k-1}(J, 1), \ldots, f_{k-1}(J, \gamma(J)), f_{k-1}(J): J \to \mathbf{P}_0(\mathbf{R}^n) \ [\text{resp.} \to \mathbf{IP}_0(\mathbf{R}^n)]$$

for each $J \in \mathbf{CX}_{k-1}(\mathscr{F})$ with $f_{k-1}(J) = f_{k-1}(J, 1) + \cdots + f_{k-1}(J, \gamma(J))$ and also

$$f_{k-1} = \bigcup \{ f_{k-1}(J): J \in \mathbf{CX}_{k-1}(\mathscr{F}) \}: \bigcup \mathbf{CX}_{k-1}(\mathscr{F}) \to \mathbf{P}_0(\mathbf{R}^n)$$

$$[\text{resp.} \to \mathbf{IP}_0(\mathbf{R}^n)].$$

Suppose $K \in \mathbf{CX}_k(\mathscr{F})$. According to 1.8, $\partial L \subset \mathrm{dmn}(f_{k-1})$, and we define

$$f_{k-1}(K, 1), \ldots, f_{k-1}(K, \gamma(K)): \partial K \to \mathbf{P}_0(\mathbf{R}^n) \quad [\text{resp.} \to \mathbf{IP}_0(\mathbf{R}^n)]$$

by setting

$$f_{k-1}(K, \beta) = \sum \{ f_{k-1}(J, \alpha): J \in \mathbf{CX}_{k-1}(\mathscr{F}) \text{ with } J \subset \partial K \text{ and}$$

$$\alpha \in \{1, \ldots, \gamma(J)\} \text{ with } \beta = \beta(K, J, \alpha) \}$$

for each $\beta \in \{1, \ldots, \gamma(K)\}$.

We further define

$$f_k(K, 1), \ldots, f_k(K, \gamma(K)), f_k(K): K \to \mathbf{P}_0(\mathbf{R}^n) \quad [\text{resp.} \to \mathbf{IP}_0(\mathbf{R}^n)]$$

by setting

$$f_k(K, \beta)((1 - t)\mathrm{center}(K) + tx)$$
$$= \tau(z(K, i))_{\#} \circ \mu(t)_{\#} \circ \tau(-z(K, i))_{\#} f_k(K, \beta)(x)$$

for each $\beta \in \{1, \ldots, \gamma(\mathrm{K})\}$, each $x \in \partial K$, and each $0 \leqslant t \leqslant 1$.

Finally, we set

$$f_k = \bigcup \{ f_k(K): K \in \mathbf{CX}_k(\mathscr{F}) \}: \bigcup \mathbf{CX}_k(\mathscr{F}) \to \mathbf{P}_0(\mathbf{R}^n) \quad [\text{resp.} \to \mathbf{IP}_0(\mathbf{R}^n)].$$

Clearly, for each $x \in \mathbf{UCX}_k(\mathscr{F})$, $\mathbf{S}(f_k(x)) \leqslant 12N - 6$ [resp. $\leqslant 4N - 2$], $\Theta^0(f_k(x), y) \leqslant M$ for each $y \in \mathbf{R}^n$, and $\mathbf{M}(f_k(x)) \leqslant M(12N - 6)$ [resp. $\leqslant M(4N - 1)$].

Additionally, we check the existence of $1 < C_k < \infty$ (depending only on the dimensions k and n) with the following property: Corresponding to any sequence $K_1 \in \mathbf{CX}_1(\mathscr{F})$, $K_2 \in \mathbf{CX}_2(\mathscr{F}), \ldots, K_k \in \mathbf{CX}_k(\mathscr{F})$ with $K_1 \subset \partial K_2$, $K_2 \subset \partial K_3, \ldots, K_{k-1} \subset \partial K_k$ and associated $L_2, \ldots, L_k$ defined inductively by the requirements

$$L_j = \{ (1 - t)\mathrm{center}(K_j) + tx: 0 \leqslant t \leqslant 1 \text{ and } x \in L_{j-1} \}$$

for $j = 2, \ldots, k$, there will exist affine functions $g(1), \ldots, g(12N - 6): \mathbf{R}^m \to \mathbf{R}^n$ with $\sum_j \|Dg(j)\| \leqslant C_k \lambda$ such that $\mathrm{spt}[f(z)] \subset \{ g(1)(z), \ldots, g(12N - 6)(z) \}$ for each $z \in L_k$.

(3) As our final function we set

$$f = f_m: \bigcup \mathscr{F} \to \mathbf{P}_0(\mathbf{R}^n) \quad [\text{resp.} \to \mathbf{IP}_0(\mathbf{R}^n)]$$

and summarize some of its properties in the following corollary.

3.9. Corollary (Extension of Lipschitz multiple-valued functions within a cubical complex). *There is a constant $\Gamma_{3.9}$ depending only on m and n with the following properties*:

Suppose. (a) $\mathscr{F}$ *is an admissible family of m-dimensional cubes in* $\mathbf{R}^m$ *with associated cubical complex* $\mathbf{CX}(\mathscr{F})$ *(with the obvious meanings adapted from* 1.5, 1.8 *with m replacing m + n there),*

(b) $0 \leqslant \lambda < \infty$, $L \in \mathbf{R}$, $M \in \mathbf{R}^+$, $N \in \{1, 2, 3, \ldots\}$,

(c) $f_0: \bigcup \mathbf{CX}_0(\mathscr{F}) \to \mathbf{P}_0(\mathbf{R}^n)$ *[resp.* $\to \mathbf{IP}_0(\mathbf{R}^n)$*]*,

(d) $\langle f_0(v), 1 \rangle = L$, $\mathbf{M}(f_0(v)) \leqslant M$, $\mathbf{S}(f_0(v)) \leqslant N$ *whenever* $\{v\} \in \mathbf{CX}_0(\mathscr{F})$,

(e) $\mathbf{GS}(f_0(v), f_0(w)) \leqslant \lambda|v - w|$ *[resp.* $\mathscr{G}(f_0(v), f_0(w)) \leqslant \lambda|v - w|$*] whenever* $J \in \mathbf{CX}_1(\mathscr{F})$ *with* $\{v, w\} = \partial J$,

(f) *corresponding to each bounded subset A of* $\mathbf{R}^m$ *there is* $0 < R < \infty$ *such that* $\mathrm{spt}[f_0(v)] \subset \mathbf{B}^n(0, R)$ *for each* $v \in A \cap \bigcup \mathbf{CX}_0(\mathscr{F})$.

Then there is a function $f: \bigcup \mathscr{F} \to \mathbf{P}_0(\mathbf{R}^n)$ *[resp.* $\to \mathbf{IP}_0(\mathbf{R}^n)$*] with the following properties*:

(1) $f \mid \bigcup \mathbf{CX}_0(\mathscr{F}) = f_0$;

(2) *for each* $x \in \bigcup \mathscr{F}$, $\mathbf{S}(f(x)) \leqslant 12N - 6$ *[resp.* $\leqslant 4N - 2$*]*, $\Theta^0(f(x), y) \leqslant M$ *for each* $y \in \mathbf{R}^n$, $\mathbf{M}(f(x)) \leqslant M(12N - 6)$ *[resp.* $\leqslant M(4N - 2)$*], and* $\langle f(x), 1 \rangle = L$;

(3) $\mathbf{GS}(f(x), f(y)) \leqslant \Gamma_{3.9}(1 + M)N\lambda|x - y|$ *[resp.* $\mathscr{G}(f(x), f(y)) \leqslant \Gamma_{3.9}MN\lambda|x - y|$*] whenever* $x, y \in \bigcup \mathscr{F}$ *with* $(1 - t)x + ty \subset \bigcup \mathscr{F}$ *for each* $0 \leqslant t \leqslant 1$;

(4) *corresponding to each compact subset A of* $\bigcup \mathscr{F}$ *there exist* $v \in \{1, 2, 3, \ldots\}$ *and affine functions* $g(1), \ldots, g(v): \mathbf{R}^m \to \mathbf{R}^n$ *with* $\|Dg(i)\| \leqslant \Gamma_{3.9}\lambda$ *for each* $i = 1, \ldots, v$ *so that* $\mathrm{spt}[f(x)] \subset \{g(1)(x), \ldots, g(v)(x)\}$ *for each* $x \in A$.

3.10. Lipschitz Extension Theorem for multiple-valued functions. *There is a constant* $1 < \Gamma_{3.10} < \infty$ *depending only on m and n with the following properties.*

Suppose. (a) $0 \leqslant \lambda < \infty$, $L \in \mathbf{R}$, $M \in \mathbf{R}^+$, $N \in \{1, 2, 3, \ldots\}$,

(b) $\varnothing \neq A \subset \mathbf{R}^m$,

(c) $f: A \to \mathbf{P}_0(\mathbf{R}^n)$ *[resp.* $\to \mathbf{IP}_0(\mathbf{R}^n)$*]*,

(d) $\langle f(a), 1 \rangle = L$, $\mathbf{M}(f(a)) \leqslant M$, $\mathbf{S}(f(a)) \leqslant N$ *for each* $a \in A$,

(e) $\mathbf{GS}(f(a), f(b)) \leqslant \lambda|a - b|$ *[resp.* $\mathscr{G}(f(a), f(b)) \leqslant \lambda|a - b|$*] for each* $a, b \in A$,

(f) *corresponding to each bounded subset B of* $\mathbf{R}^m$ *there is* $0 < R < \infty$ *such that* $\mathrm{spt}[f(a)] \subset \mathbf{B}^n(0, R)$ *for each* $a \in A \cap B$.

Then there is a function $g: \mathbf{R}^m \to \mathbf{P}_0(\mathbf{R}^n)$ *[resp.* $\to \mathbf{IP}_0(\mathbf{R}^n)$*] with the following properties.*

(1) $g \mid A = f$;

(2) $\langle g(x), 1 \rangle = L$, $\mathbf{M}[g(x)] \leqslant M(12N - 6)$ *[resp.* $\leqslant M(4N - 2)$*], and* $\mathbf{S}[g(x)] \leqslant 12N - 6$ *[resp.* $\leqslant 4N - 2$*] for each* $x \in \mathbf{R}^m$;

(3) $\mathbf{GS}(g(x), g(y)) \leqslant \Gamma_{3.10}(1 + M)N\lambda|x - y|$ *[resp.* $\mathscr{G}(g(x), g(y)) \leqslant \Gamma_{3.10}MN\lambda|x - y|$*] for each* $x, y \in \mathbf{R}^m$;

(4) *Corresponding to each compact subset B of $\mathbf{R}^m \sim \mathrm{Clos}(A)$ there exist $v \in \{1, 2, 3, \ldots\}$ and affine functions $g(1), \ldots, g(v)$: $\mathbf{R}^m \to \mathbf{R}^n$ with $\|Dg(i)\| \leqslant \Gamma_{3.10}\lambda$ for each $i = 1, \ldots, v$ such that $\mathrm{spt}[g(x)] \subset \{g(1)(x), \ldots, g(v)(x)\}$ for each $x \in B$.*

(5) *Suppose $p \in A$ and there exist a neighborhood U of p in $\mathbf{R}^m$, $v \in \{1, 2, \ldots, N\}$, Lipschitz functions $f(1), \ldots, f(v)$: $A \cap U \to \mathbf{R}^n$, and $\theta(1), \ldots, \theta(v) \in \mathbf{R} \sim \{0\}$ such that*

$$0 < \inf\left\{ |f(i)(x) - f(j)(x)| : i, j \in \{1, \ldots, v\} \text{ with } i \neq j \text{ and } x \in A \cap U \right\}$$

and

$$f(z) = \sum \left\{ \theta(i)[\![f(i)(z)]\!] : i = 1, \ldots, v \right\}$$

for each $z \in A \cap U$. Then there is a neighborhood V of p in $\mathbf{R}^m$ and Lipschitz functions $h(1), \ldots, h(v)$: $V \to \mathbf{R}^n$ with

$$0 < \inf\left\{ |h(i)(x) - h(j)(x)| : i, j \in \{1, \ldots, v\} \text{ with } i \neq j \text{ and } x \in V \right\}$$

such that

$$g(x) = \sum \left\{ \theta(i)[\![h(i)(x)]\!] : i = 1, \ldots, v \right\}$$

for each $x \in V$.

PROOF. We use 3.5(2) to extend the domain of f to $\mathrm{Clos}(A)$ as a $\mathbf{G}$ Lipschitz function with supposition (d) still holding for the extended domain. We infer from 3.5(4) that supposition (e) still holds for the extended domain. It is easy to check that supposition (f) also holds for the extended domain. Henceforth, we assume that A is closed. We replace $m + n$ by m in the terminology of 1.5, 1.6, 1.8 and let $\mathscr{F}$ denote the Whitney family $\mathbf{WF}(\mathbf{R}^m \sim \mathrm{Clos}(A))$ of cubes in $\mathbf{R}^m$ with associated cubical complex $\mathbf{CX}(\mathscr{F})$. We select a function f_0: $\mathbf{UCX}_0(\mathscr{F}) \to \mathbf{P}_0(\mathbf{R}^n)$ [resp. $\to \mathbf{IP}_0(\mathbf{R}^n)$] subject to the requirement that for each $x \in \mathbf{UCX}_0(\mathscr{F})$ there is $y \in \mathrm{Clos}(A)$ with $|x - y| = \mathrm{dist}(x, A)$ so that $f_0(x) = f(y)$. We let F: $\bigcup \mathscr{F} \to \mathbf{P}_0(\mathbf{R}^n)$ [resp. $\to \mathbf{IP}_0(\mathbf{R}^n)$] be the extension of f_0 according to Corollary 3.9 (called "f" there) and set $g = f \cup F$ for the present theorem. The asserted properties are readily checked.

3.11. A procedure to be used in extended multiple-valued mappings from 0 skeletons to 1 skeletons when constructing G continuous homotopies.

Assumptions. (a) $0 < \varepsilon < 1, 0 \leqslant \lambda < \infty$, and $L, M, N, R, \mathbf{Q}$ are as in 3.5(a)(b).

(b) $V_0, V_1, W_0, W_1 \in \mathbf{Q}$ and $X_0, X_1 \in \mathbf{P}_1(\mathbf{R}^n)$ such that $\partial X_0 = W_0 - V_0$, $\partial X_1 = W_1 - V_1$, $\mathbf{GS}(V_0, W_0) = \mathbf{MS}(X_0) \leqslant \lambda$, and $\mathbf{GS}(V_1, W_1) = \mathbf{MS}(X_1) \leqslant \lambda$.

(c) $P_0, P_1 \in \mathbf{P}_0(\mathbf{R}^n)$ and $Q_0, Q_1 \in \mathbf{P}_1(\mathbf{R}^n)$ such that the following conditions hold:

$$V_1 - V_0 = \partial Q_0 + \partial([\![0]\!] \times P_0) = \partial Q_0 + P_0,$$

$$W_1 - W_0 = \partial Q_1 + \partial([\![0]\!] \times P_1) = \partial Q_1 = P_1,$$

$$\mathbf{S}(P_0) \leqslant N, \qquad \mathbf{S}(P_1) \leqslant N, \qquad \mathrm{spt}(P_0) \subset \mathbf{B}^n(0, R), \qquad \mathrm{spt}(P_1) \subset \mathbf{B}^n(0, R),$$

and each of the following quantities is less than ε:

$$\mathbf{MS}(Q_0), \quad \mathbf{MS}(Q_1), \quad \mathbf{M}(P_0), \quad \mathbf{M}(P_1), \quad \mathbf{M}([\![0]\!] \times P_0), \quad \mathbf{M}([\![0]\!] \times P_1).$$

(d) $V_* = V_1 - P_0, W_* = W_1 - P_1 \in \mathbf{Q}$.

Observations and mappings based on Corollary 3.7 (*see Figure* 3.11). (i) f_0: $[0,1] \to \mathbf{P}_0(\mathbf{R}^n)$ *is constructed as in* 3.7 *with* V, W, λ *there replaced by* V_0, W_0, $\mathbf{GS}(V_0, W_0)$, *respectively.*

(ii) g_0: $[0,1] \to \{0\} \in \mathbf{P}_0(\mathbf{R}^n)$.

(iii) $W_* - V_* = (W_1 - P_1) - (V_1 - P_0) = (\partial Q_1 + W_0) - (\partial Q_0 + V_0) = \partial Q_1 + \partial X_0 - \partial Q_0$ so that $\mathbf{GS}(V_*, W_*) < 2\varepsilon + \lambda$.

(iv) f_1: $[0,1] \to \mathbf{P}_0(\mathbf{R}^n)$ is constructed as in 3.7 with V, W, λ there replaced by $V_*, W_*, \mathbf{GS}(V_*, W_*)$.

(v) $P_1 - P_0 = (W_1 - W_0 - \partial Q_1) - (V_1 - V_0 - \partial Q_0) = \partial X_1 - \partial X_0 - \partial Q_1 + \partial Q_0$ so that $\mathbf{GS}(P_0, P_1) < 2\varepsilon + 2\lambda$.

(vi) g_1: $[0,1] \to \mathbf{P}_0(\mathbf{R}^n)$ is constructed as in 3.7 with $L, M, N, R, V, W, \lambda$ there replaced by $0, \varepsilon, N, R, P_0, P_1, \mathbf{GS}(P_0, P_1)$, respectively.

(vii) $V_* - V_0 = V_1 - P_0 - V_0 = \partial Q_0$ so that $\mathbf{GS}(V_0, V_*) < \varepsilon$.

(viii) ϕ_0: $[0,1] \to \mathbf{P}_0(\mathbf{R}^n)$ is constructed as in 3.7 with V, W, λ there replaced by $V_0, V_*, \mathbf{GS}(v_0, V_*)$, respectively.

(ix) ψ_0: $[0,1] \to \mathbf{P}_0(\mathbf{R}^n)$ is defined by setting $\psi_0(t) = \mu(t)_\# P_0$ for each $0 \leqslant t \leqslant 1$.

(x) $W_* - W_0 = W_1 - P_1 - W_0 = \partial Q_1$ so that $\mathbf{GS}(W_0, W_*) < \varepsilon$.

(xi) ϕ_1: $[0,1] \to \mathbf{P}_0(\mathbf{R}^n)$ is constructed as in 3.7 with V, W, λ there replaced by $W_0, W_*, \mathbf{GS}(W_0, W_*)$, respectively.

(xii) ψ_1: $[0,1] \to \mathbf{P}_0(\mathbf{R}^n)$ is defined by setting $\psi_1(t) = \mu(t)_\# P_1$ for each $0 \leqslant t \leqslant 1$.

The extended mappings F, G, H. We define F, G, H: $\partial[0,1]^2 \to \mathbf{P}_0(\mathbf{R}^n)$ by setting $H = F + G$ and, additionally, setting for each $0 \leqslant t \leqslant 1$,

$$F(t,0) = f_0(t), \qquad G(t,0) = g_0(t),$$
$$F(t,1) = f_1(t), \qquad G(t,1) = g_1(t),$$
$$F(0,t) = \phi_0(t), \qquad G(0,t) = \psi_0(t),$$
$$F(1,t) = \phi_1(t), \qquad G(1,t) = \psi_1(t).$$

Properties of F, G, H. (1) $H(0,0) = V_0$, $H(1,0) = W_0$, $H(0,1) = V_1$, $H(1,1) = W_1$.

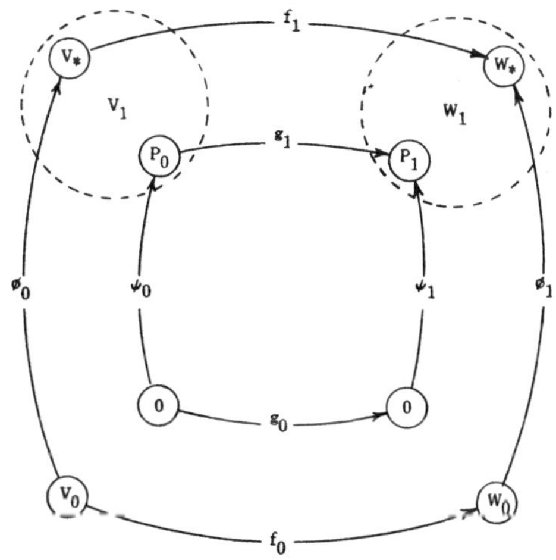

FIGURE 3.11. Mappings used in assembling homotopies.

(2) For each $x \in \partial[0,1]^2$, $\mathbf{S}(F(x)) \leqslant 12N - 6$, $\Theta^0(F(x), y) \leqslant M$ for each $y \in \mathbf{R}^n$, $\mathbf{M}(F(x)) \leqslant M(12N - 6)$, and $\langle F(x), 1 \rangle = 0$.

(3) For each $x \in \partial[0,1]^2$, $\mathbf{S}(G(x)) \leqslant 12N - 6$, $\Theta^0(G(x), y) < \varepsilon$ for each $y \in \mathbf{R}^n$, $\mathbf{M}(G(x)) \leqslant \varepsilon(12N - 6)$, and $\langle G(x), 1 \rangle = 0$.

(4) Corresponding to each $J \in \{[0,1] \times \{0\}, [0,1] \times \{1\}, \{0\} \times [0,1], \{1\} \times [0,1]\}$ there exist affine functions $g(1), \ldots, g(12N - 6)$, $h(1), \ldots, h(12N - 6)$: $\mathbf{R}^m \to \mathbf{R}^n$ with the following properties:

(4.1) for each $x \in J$, $\mathrm{spt}[F(x)] \subset \{g(1)(x), \ldots, g(12N - 6)(x)\}$ and $\mathrm{spt}[G(x)] \subset \{h(1)(x), \ldots, h(12N - 6)(x)\}$,

(4.2) in case $J = [0,1] \times \{1\}$ then $\Sigma_i \|Dg(i)\| < 8\varepsilon + 4\lambda$ and $\Sigma_i \|Dh(i)\| < 8\varepsilon + 8\lambda$,

(4.3) in case $J = [0,1] \times \{0\}$ then $\Sigma_i \|Dg(i)\| \leqslant 4\lambda$ and $\Sigma_i \|h(i)\| = 0$,

(4.4) in case $J \in \{\{0\} \times [0,1], \{1\} \times [0,1]\}$ then $\Sigma_i \|Dg(i)\| < 4\varepsilon$ and $\|Dh(j)\| \leqslant R$ for each $j = 1, \ldots, 12N - 6$.

(5) The mappings ϕ_0, ψ_0 [resp. ϕ_1, ψ_1] are determined by V_0, P_0, V_1, Q_0, V_* [resp. by W_0, P_1, W_1, Q_1, W_*] and, in particular, are independent of W_0, P_1, W_1, Q_1, W_* [resp. of V_0, P_0, V_1, Q_0, V_*]. Corresponding dependence and independence hold for F, G, H.

(6) The construction procedure used in obtaining the mappings f_0, g_0, f_1, g_1 is symmetric with respect to reversal of the interval (note 3.7(6)). Corresponding symmetry holds for F, G, H.

3.12. How to construct homotopies h between GS Lipschitz multiple-valued mappings which are G close.

Assumptions. (a) $0 < \varepsilon < 1$, $v \in \{1, 2, 3, \ldots\}$, $0 < \lambda < \infty$, and $L, M, N, R, \mathbf{Q}$ are as in 3.5(a)(b).

(b) $f, g \colon \mathbf{R}^m \to \mathbf{Q}$.

(c) $\mathbf{GS}(f(x), f(y)) \leqslant \lambda|x - y|$ and $\mathbf{GS}(g(x), g(y)) \leqslant \lambda|x - y|$ for each $x, y \in \mathbf{R}^m$.

(d) There is $0 < R_1 < \infty$ such that $f(x) = f(y) = g(x)g(y)$ for each $x, y \in \mathbf{R}^m \sim \mathbf{U}^m(0, R_1)$.

(e) We recall 1.2 and for each $\mu \in \{v, v + 1, v + 2, \ldots\}$, each $j \in \{0, 1, \ldots, m\}$, and each $k \in \{0, 1, \ldots, m + 1\}$, set

$$\mathbf{K}_j^m(\mu) = \{ K \colon K \times \{0\} \subset \mathbf{R}^m \times \mathbf{R}^n \text{ with } K \times \{0\} \in \mathbf{K}_j(\mu) \}$$

and

$$\mathbf{K}_k^{m+1}(\mu) = \{ K \colon K \times \{0\} \subset \mathbf{R}^{m+1} \times \mathbf{R}^{n-1} \text{ with } K \times \{0\} \in \mathbf{K}_k(\mu) \}.$$

(f) We assume that f and g are sufficiently close in the $\mathbf{G}$ metric so that (by virtue of 3.5(3)) there will exist functions

$$P \colon \bigcup \mathbf{K}_0^m(v) \to \mathbf{P}_0(\mathbf{R}^n) \quad \text{and} \quad Q \colon \bigcup \mathbf{K}_0^m(v) \to \mathbf{P}_1(\mathbf{R}^n)$$

such that for each $a \in \bigcup \mathbf{K}_0^m(v)$ the following is true:

$$g(a) - f(a) = \partial Q(a) + \partial(\llbracket 0 \rrbracket \bowtie P(a)) = \partial Q(a) + P(a),$$

$$g(a) - P(a) \in \mathbf{Q}, \quad \mathbf{MS}(Q) < \varepsilon, \quad \mathbf{S}(P(a)) \leqslant N,$$

$$\mathbf{M}(P(a)) < \varepsilon, \quad \mathbf{M}(\llbracket 0 \rrbracket \bowtie P(a)) < \varepsilon, \quad \mathrm{spt}[P(a)] \subset \mathbf{B}^n(0, R).$$

Some terminology. See Figures 3.12(A) and 3.12(B).

(i) For each $k \in \{0, 1, 2, \ldots\}$ we set $\lambda(k) = 2^{-v} + 2^{-(v+1)} + 2^{-(v+2)} + \cdots + 2^{-(v+k)}$ so that $\lambda(\infty) = \lim_{k \to \infty} \lambda(k) = 2^{1-v}$.

(ii) We set $I(0) = [0, 2^{-v}]$, $I[0, +] = [2^{-v}, 2^{-v} + \lambda(0)]$, $I(0, -) = [-\lambda(0), 0]$, $I(1, +) = [2^{-v} + \lambda(0), 2^{-v} + \lambda(1)]$, $I(1, -) = [-\lambda(1), -\lambda(0)]$, and, for each $k \in \{2, 3, 4, \ldots\}$, set

$$I(k, +) = \left[2^{-v} + \lambda(k - 1), 2^{v} + \lambda(k) \right], \qquad I(k, -) = \left[-\lambda(k), -\lambda(k - 1) \right].$$

(iii) We set

$$\mathscr{F} = \left\{ I(0) \times K : K \in \mathbf{K}^m_m(v) \right\}$$

$$\cup \left\{ I(k, +) \times K : k \in \{0, 1, 2, \ldots\}, K \in \mathbf{K}^m_m(v + k) \right\}$$

$$\cup \left\{ I(k, -) \times K : k \in \{0, 1, 2, \ldots\}, K \in \mathbf{K}^m_m(v + k) \right\}$$

$$\subset \mathbf{K}^{m+1}_{m+1}(v) \cup \mathbf{K}^{m+1}_{m+1}(v + 1) \cup \mathbf{K}^{m+1}_{m+1}(v + 2) \cup \cdots.$$

(iv) We replace $m + n$ by $m + 1$ in the obvious way in the terminology of 1.5, 1.6, 1.7, 1.8 and check that $\mathscr{F}$ is an admissible family of $m + 1$ cubes in $\mathbf{R}^{m+1}$ and let $\mathbf{CX}(\mathscr{F})$ denote the associated cubical complex.

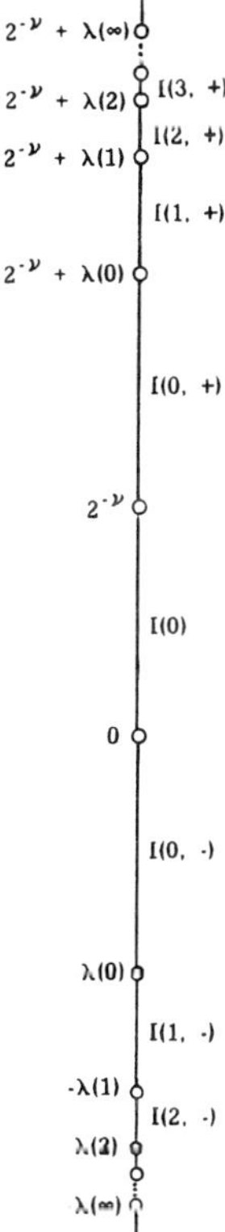

FIGURE 3.12(A). The various homotopy time intervals.

(v) Suppose $a, b \in \mathbf{R}^m$ so that $J = \{(1 - s)a + sb\colon 0 \leqslant s \leqslant 1\} \in \mathbf{K}_1^m(v)$. We note 3.12($c$) and use 3.4 to choose $X_0(a, b)$, $X_1(a, b) \in \mathbf{P}_1(\mathbf{R}^n)$ so that $\partial X_0(a, b) = f(b) - f(a)$, $\partial X_1(a, b) = g(b) - g(a)$,

$$\mathbf{GS}(f(a), f(b)) = \mathbf{MS}[X_0(a, b)] \leqslant 2^{-v}\lambda,$$

$$\mathbf{GS}(g(a), g(b)) = \mathbf{MS}[X_1(a, b)] \leqslant 2^{-v}\lambda.$$

We then apply 3.11 with ε, λ, L, M, N, R, $\mathbf{Q}$, V_0, V_1, W_0, W_1, X_0, X_1, P_0, P_1, Q_0, Q_1, V_*, W_* there replaced by ε, $2^{-v}\lambda$, L, M, N, R, $\mathbf{Q}$, $f(a)$, $g(a)$, $f(b)$, $g(b)$, $X_0(a, b)$, $X_1(a, b)$, $P(a)$, $P(b)$, $Q(a)$, $Q(b)$, $g(a) - P(a)$, $g(b) - P(b)$ respectively in the present context to obtain mappings f_0, g_0, f_1, g_1, $\phi_0, \psi_0, \phi_1, \psi_1$ (in the terminology of 3.11). We set

$F_0(J), G_0(J)\colon \{0\} \times J \to \mathbf{P}_0(\mathbf{R}^n)$,

$$F_0(J)(0, (1 - s)a + sb) = f_0(s),$$
$$G_0(0, (1 - s)a + sb) = g_0(s) = 0;$$

$F_1(J), G_1(J)\colon \{2^v\} \times J \to \mathbf{P}_0(\mathbf{R}^n)$,

$$F_1(J)(2^{-v}, (1 - s)a + sb) = f_1(s),$$
$$G_1(J)(2^{-v}, (1 - s)a + sb) = g_1(s)$$

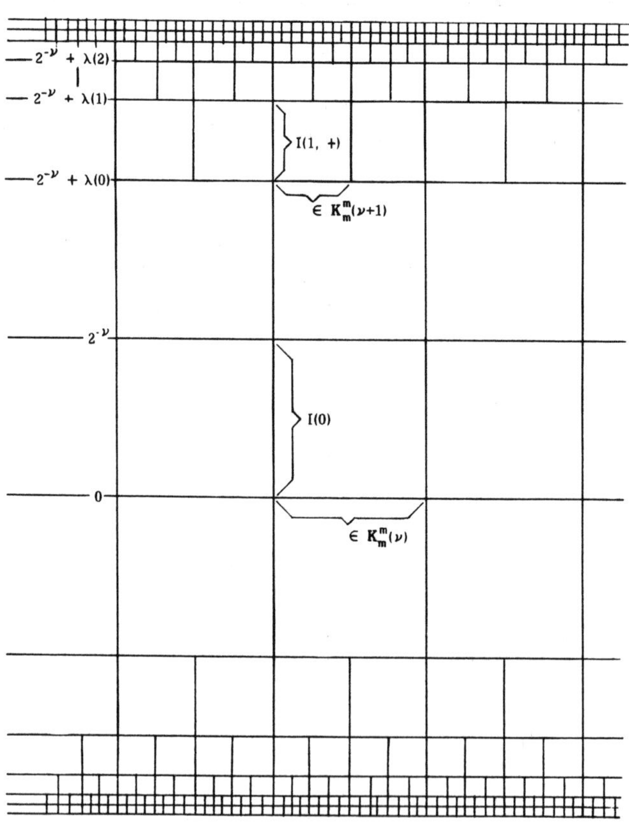

FIGURE 3.12(B). The cubical complex of the homotopy.

for each $0 \leqslant s \leqslant 1$;

$$\phi(a), \psi(a): I(0) \times \{a\} \to \mathbf{P}_0(\mathbf{R}^n),$$
$$\phi(a)(2^{-\upsilon}t, a) = \phi_0(t),$$
$$\psi(a)(2^{-\upsilon}t, a) = \psi_0(t);$$
$$\phi(b), \psi(b): I(0) \times \{b\} \to \mathbf{P}_0(\mathbf{R}^n),$$
$$\phi(b)(2^{-\upsilon}t, b) = \phi_1(t),$$
$$\psi(b)(2^{-\upsilon}t, b) = \psi_1(t)$$

for each $0 \leqslant t \leqslant 1$.

We note items (5) and (6) among the properties of F, G, H in 3.11 and select our various $\phi(a), \psi(a)$ to depend only on a (and not on the particular interval having a as endpoint), etc., and select $F_0(J), G_0(J), F_1(J), G_1(J)$ to depend only on J (and not, for example, on higher-dimensional cubes containing J).

Definition of h on 0 skeleton of $\mathscr{F}$. We define $h_0: \mathrm{U}\mathbf{CX}_0(\mathscr{F}) \to \mathbf{Q}$ by setting, for each $(t, x) \in \mathbf{R}^m$ with $\{(t, x)\} \in \mathbf{CX}_0(\mathscr{F})$,

$$h_0(t, x) = \begin{cases} g(x) & \text{in case } t \geqslant 2^{-\upsilon}, \\ f(x) & \text{in case } t \leqslant 0. \end{cases}$$

Extension of h to 1 skeleton of $\mathscr{F}$. The definition of $h_1: \mathrm{U}\mathbf{CX}_1(\mathscr{F}) \to \mathbf{Q}$ is given in various cases:

Case A. Whenever $k \in \{0, 1, 2,\dots\}$, $a \in \mathbf{R}^m$, and $(t, a) \in I(k, +) \times \{a\} \in \mathbf{CX}_1(\mathscr{F})$, we set $h_1(t, a) = g(a)$.

Case B. Whenever $k \in \{0, 1, 2,\dots\}$, $a \in \mathbf{R}^m$, and $(t, a) \in I(k, -) \times \{a\} \in \mathbf{CX}_1(\mathscr{F})$, we set $h_1(t, a) = f(a)$.

Case C. Suppose $k \in \{0, 1, 2,\dots\}$ and $a, b \in \mathbf{R}^m$ such that $J = \{(1 - s)a + sb: 0 \leqslant s \leqslant 1\} \subset \mathbf{K}_1^m(\upsilon \mid k \mid 1)$ so that $\{2^{-\upsilon} \mid \lambda(k)\} \times J$, $\{\,\lambda(k)\} \times J \subset \mathbf{CX}_1(\mathscr{F})$. For each $0 \leqslant s \leqslant 1$, we define

$$h_1(2^{-\upsilon} + \lambda(k), (1 - s)a + sb) = f(s)$$
$$\big[\text{resp. } h_1(-\lambda(k), (1 - s)a + sb) = f(s)\big],$$

where $f: [0, 1] \to \mathbf{Q}$ is constructed in accordance with Corollary 3.7 with $L, M, N, R, \mathbf{Q}, V, W, \lambda$ there corresponding to $L, M, N, R, \mathbf{Q}, g(a)$ [resp. $f(a)$], $g(b)$ [resp. $f(b)$], $\mathbf{GS}(g(a), g(b))$ [resp. $\mathbf{GS}(f(a), f(b))$] in the present context.

Case D. Whenever $J \in \mathbf{K}_1^m(\upsilon)$ so that $\{0\} \times J, \{2^{-\upsilon}\} \times J \in \mathbf{CX}_1(\mathscr{F})$, we set

$$h_1|\{0\} \times J = F_0(J) + G_0(J) = F_0(J)$$

and

$$h_1|\{2^{-\upsilon}\} \times J = F_1(J) + G_1(J).$$

Case E. Whenever $\{a\} \in \mathbf{K}_0^m(\upsilon)$ so that $I(0) \times \{a\} \in \mathbf{CX}_1(\mathscr{F})$, we set

$$h_1|I(0) \times \{a\} = \phi(a) + \psi(a).$$

Extension of h to higher-dimensional skeletons. The extension of h_1 to mappings to higher-dimensional skeletons of $(-\lambda(\infty), 2^{-\upsilon} + \lambda(\infty)) \times \mathbf{R}^m$ requires several steps. For each $k \in \{2,\dots,m + 1\}$ and each $K \in \mathbf{CX}_k(\mathscr{F})$, $h_k|K$ will be constructed from mappings defined on ∂K (as was the case, for example, in the construction of 3.7). In general, $h_k|K$ will not be determined uniquely by mappings defined on ∂K (recall our choices of the $z(K, i)$ in 3.8). We understand throughout our construction, however, that some particular choice of $h_k|K$ is made and is fixed at the appropriate stage of our construction. In particular, if $k < m + 1$ so that K is part of the boundary of at least two distinct $k + 1$ cubes in $\mathbf{CX}_{k+1}(\mathscr{F})$, then the values of h or of other relevant mappings on K do not depend on which $k + 1$ cubes might be under later consideration.

Step 1. Suppose $k \in \{1, 2, 3, \ldots\}$ and $K \in \mathbf{K}_m^{m+1}(v + k)$ so that $L = I(k, +) \times K \in \mathbf{CX}_{m+1}(\mathscr{F})$. We replace m by $m + 1$ in the construction of 3.8 to obtain $h_{m+1}(L)\colon L \to \mathbf{P}_0(\mathbf{R}^n)$ so that, in particular,

$$h_{m+1}(L)\big|\big[\partial L \cap \bigcup \mathbf{CX}_1(\mathscr{F})\big] = h_1\big|\big[\partial L \cap \bigcup \mathbf{CX}_1(\mathscr{F})\big].$$

Step 2. Suppose $k \in \{0, 1, 2, \ldots\}$ and $K \in \mathbf{K}_m^{m+1}(v + k)$ so that $L = I(k, -) \times K \in \mathbf{CX}_{m+1}(\mathscr{F})$. Since each $G_0(J) = 0$ in Case D above, we again can replace m by $m + 1$ in the construction of 3.8 to obtain $h_{m+1}(L)\colon L \to \mathbf{P}_0(\mathbf{R}^n)$ so that, in particular,

$$h_{m+1}(L)\big|\big[\partial L \cap \bigcup \mathbf{CX}_1(\mathscr{F})\big] = h_1\big|\big[\partial L \cap \bigcup \mathbf{CX}_1(\mathscr{F})\big].$$

Step 3. Suppose $K \in \mathbf{K}_m^m(v)$ so that $L = \{2^{-v}\} \times K \in \mathbf{CX}_m(\mathscr{F})$. We recall from Case D above that

$$h_1\big|\big[\partial L \cap \bigcup \mathbf{CX}_1(\mathscr{F})\big] \quad = \bigcup \{ F_1(J)\colon J \in \mathbf{K}_1^m(v) \text{ with } J \subset \partial K \}$$

$$\cup \bigcup \{ G_1(J)\colon J \in \mathbf{K}_1^m(v) \text{ with } J \subset \partial K \}.$$

Since each $F_1(J)$ and each $G_1(J)$ is constructed from functions f_1 and g_1 above which are obtained in 3.11(iv)(vi) from the construction device of 3.7, we again are able to use the construction of 3.8 in an obvious way to obtain functions $F_1(L), G_1(L)\colon L \to \mathbf{P}_0(\mathbf{R}^n)$ such that, in particular,

$$F_1(L)\big|\big[\partial L \cap \bigcup \mathbf{CX}_1(\mathscr{F})\big] = \bigcup \{ F_1(J)\colon J \in \mathbf{K}_1^m(v) \text{ with } J \subset \partial K \}$$

and

$$G_1(L)\big|\big[\partial L \cap \bigcup \mathbf{CX}_1(\mathscr{F})\big] = \bigcup \{ G_1(J)\colon J \in \mathbf{K}_1^m(v) \text{ with } J \subset \partial K \}.$$

We set $h_m(L) = F_1(L) + G_1(L)$.

For use in Step 4, we adapt the terminology of 3.8 in the obvious way (as for use in constructing $F_1(L), G_1(L)$) so that for relevant $J \subset H \subset \{2^{-v}\} \times L$ we have sets $\Gamma_F(J), \Gamma_G(J)$, integers $\gamma_F(J), \gamma_G(J)$, partitionings $\{\Gamma_F(J, i)\}_i$, $\{\Gamma_G(J, i)\}_i$, points $\{z_F(J, i)\}_i$, $\{z_G(J, i)\}_i$, indexing functions $\beta_F(H, J, \cdot)$, $\beta_G(H, J, \cdot)$, and mappings $\{F_1(H, \beta)\}_\beta$, $\{G_1(H, \beta)\}_\beta$; note that $F_1(J) = F_1(\{2^{-v}\} \times J)$, $G_1(J) = G_1(\{2^{-v}\} \times J)$, etc., for $J \in \mathbf{K}_1^m(v)$ for consistency of notation.

Step 4. Suppose $H \in \mathbf{K}_m^m(v)$ so that $L = I(0, +) \times H \in \mathbf{CX}_{m+1}(\mathscr{F})$. For each $k \in \{2, \ldots, m + 1\}$ and each $K \in \mathbf{CX}_k(\mathscr{F})$ with $K \subset L$, we follow the general procedure of 3.8 and set

$$\Gamma(K) = \mathrm{Clos}\big(\mathbf{R}^n \cap \big\{ y\colon y \in \mathrm{spt}\big[h_1(x)\big]$$

$$\text{for some } x \in J \in \mathbf{CX}_1(\mathscr{F}) \text{ with } J \subset \partial K \big\}\big)$$

(disregarding, at this point, the fact that for $x \in J \subset \{2^{-v}\} \times H$ one can write $h_1(x) = F_1(x) + G_1(x)$) and give the obvious meanings to $\gamma(K)$, $\{\Gamma(K, i)\}_i$, $\{z(K, i)\}_i$, $\beta(K', K, \cdot)$ (by our understanding these choices are made compatible with those of Step 1). We recall the terminology introduced in Step 3 and, in case $K = \{2^{-v}\} \times K_0$, $K_* = I(0, +) \times K_0$, define

$$\beta_F(K_*, K, \cdot): \{1,\ldots,\gamma_F(K)\} \to \{1,\ldots,\gamma(K_*)\}$$

and

$$\beta_G(K_*, K, \cdot): \{1,\ldots,\gamma_G(K)\} \to \{1,\ldots,\gamma(K_*)\}$$

by requiring

$$\Gamma_F(K, \alpha) \subset \Gamma(K_*, \beta_F(K_*, K, \alpha)), \qquad \Gamma_G(K, \alpha) \subset \Gamma(K_*, \beta_G(K_*, K, \alpha))$$

for each relevant α. We then define

$$h_{k+1}[K_*, 1),\ldots,h_{k+1}(K_*, \gamma(K_*)), h_{k+1}(K_*): K_* \to \mathbf{P}_0(\mathbf{R}^n)$$

by setting

$$h_{k+1}(K_*, \beta)((1 - t)\mathrm{center}(K_*) + tx)$$
$$= \tau(z(K_*, \beta))_\# \circ \mu(t)_\# \circ \tau(-z(K_*, \beta))_\# \circ h_k(K_*, \beta)(x)$$

for each $\beta \in \{1,\ldots,\gamma(K_*)\}$, each $x \in \partial K_*$, and each $0 \leqslant t \leqslant 1$ and

$$h_{k+1}(K_*) - h_{k+1}(K_*, 1) + \cdots + h_{k+1}(K_*, \gamma(K_*));$$

here, for each $\beta \in \{1,\ldots,\gamma(K_*)\}$ and $x \in J \in \mathbf{CX}_k(\mathscr{F})$ with $J \subset \partial K_*$,

$$h_k(K_*, \beta)(x) = \begin{cases} \sum\{h_k(J, \alpha)(x): \beta = \beta(K_*, J, \alpha)\} & \text{if } J \not\subset \{2^{-v}\} \times H \\ \sum\{F_1(J, \alpha)(x): \beta = \beta_F(K_*, J, \alpha)\} \\ \quad + \sum\{G_1(J, \alpha)(x): \beta = \beta_G(K_*, J, \alpha)\} \text{if } J \subset \{2^{-v}\} \times H. \end{cases}$$

As in 3.8 we are led to our definition of $h_{m+1}(L)$.

Step 5. Suppose $K \in \mathbf{K}_m^m(v)$ so that $L = I(0) \times K \in \mathbf{CX}_{m+1}(\mathscr{F})$. We recall 3.11(i)(ii)(iv)(viii) and (v) above and adapt the procedure of 3.8 in a straightforward way to define $F(L): L \to \mathbf{P}_0(\mathbf{R}^n)$ so that

$$F(L)|[\{2^{-v}\} \times K] = F_1(\{2^{-v}\} \times K) \quad \text{(as defined in Step 3)},$$

$$F(L)|[\{0\} \times K] = h_{m+1}(I(0, -) \times K)|[\{0\} \times K] \quad \text{(as constructed in Step 2)},$$

$$F(L)|[I(0) \times \{a\}] = \phi(a) \quad \text{whenever } a \in \partial K \text{ with } \{a\} \in \mathbf{K}_0^m(v).$$

Step 6. Suppose $K \in \mathbf{K}_m^m(v)$ so that $L = I(0) \times K \in \mathbf{CX}_{m+1}(\mathscr{F})$. We will use an inductive procedure to construct $G(L): L \to \mathbf{P}_0(\mathbf{R}^n)$ so that

$$G(L)|[\{2^{-v}\} \times K] - G_1(K) \quad \text{(as defined in Step 2)},$$

$$G(L)|[\{0\} \times K] = G_0(K) = 0 \quad \text{(note Step 2)},$$

$$G(L)|[I(0) \times \{a\}] - \psi(a) \quad \text{whenever } a \subset \partial K \text{ with } \{a\} \in \mathbf{K}_0^m(v).$$

In this case we have no reasonable method for choosing centers of retractions (i.e., $z(J, i)$'s) and we simply require

$$G(L)((1 - t)\text{center}(H) + tx) = \mu(t)_{\#}G(L)(x)$$

whenever $k \in \{2, \ldots, m + 1\}$, $H \in \mathbf{CX}_k(\mathscr{F})$ with $H \subset L$, $0 \leqslant t \leqslant 1$, and $x \in \partial H$; this, together with our requirements above, characterizes $G(L)$.

Step 7. Our final map h: $[-\lambda(\infty), 2^{-v} + \lambda(\infty)] \times \mathbf{R}^m \to \mathbf{P}_0(\mathbf{R}^n)$ is defined by the requirements

(7.1) $h(-\lambda(\infty), x) = f(x)$ for each $x \in \mathbf{R}^m$,

(7.2) $h(2^{-v} + \lambda(\infty), x) = g(x)$ for each $x \in \mathbf{R}^m$,

(7.3) $h|([(-\lambda(\infty), 2^{-v} + \lambda(\infty)) \sim (0, 2^{-v})] \times \mathbf{R}^m)$

$$= \bigcup \{ h_{m+1}(K) : K \in \mathbf{CX}_{m+1}(\mathscr{F}), K \cap (0, 2^{-v}) \times \mathbf{R}^m = \varnothing \},$$

(7.4) $h|([0, 2^{-v}] \times \mathbf{R}^m) = F + G$, where

$$F = \bigcup \{ F(I(0) \times K) : K \in \mathbf{K}_m^m(v) \}, \qquad G = \bigcup \{ G(I(0) \times K) : K \in \mathbf{K}_m^m(v) \}.$$

Properties of h, F, G. (1) H is continuous with $h(-\lambda(\infty), x) = f(x)$ and $h(2^{-v} + \lambda(\infty), x) = g(x)$ for each $x \in \mathbf{R}^m$; e.g., h is a homotopy between f and g.

(2) For each $(t, x) \in [-\lambda(\infty), 2^{-v} + \lambda(\infty)] \times \mathbf{R}^m$, $\mathbf{S}(h(t, x)) \leqslant 24N + 12$, $\Theta^0(h(t, x), y) \leqslant M + \varepsilon$ for each $y \in \mathbf{R}^n$, $\mathbf{M}(h(t, x)) \leqslant (M + \varepsilon)(12N - 6)$, $\langle h(t, x), 1 \rangle = L$, and $\text{spt}[h(t, x)] \subset \mathbf{B}^n(0, R)$.

(3) $h(t, x) = f(x) = g(x)$ for each $t \in [-\lambda(\infty), 2^{-v} + \lambda(\infty)]$ and each $x \in \mathbf{R}^m \sim \mathbf{B}^m(0, R + m^{1/2}2^{-v})$.

There is also a constant $1 < \Gamma_{3.12} < \infty$ depending only on m and n for which the following additional estimates hold:

(4) $\mathbf{GS}(h(s, x), h(t, y)) \leqslant \Gamma_{3.12}(1 + M + \varepsilon)N(2^v\varepsilon + \lambda)|(s, x) - (t, y)|$ for each $(s, x), (t, y) \in [2^{-v}, 2^{-v} + \lambda(\infty)] \times \mathbf{R}^m$.

(5) Corresponding to each compact subset A of $[2^{-v}, 2^{-v} + \lambda(\infty)) \times \mathbf{R}^m$, there exist $\mu \in \{1, 2, 3, \ldots\}$ and affine functions $g(1), \ldots, g(\mu)$: $\mathbf{R}^{m+1} \to \mathbf{R}^n$ with $\|Dg(i)\| \leqslant \Gamma_{3.12}(2^v\varepsilon + \lambda)$ for each $i = 1, \ldots, \mu$ such that

$$\text{spt}[h(t, x)] \subset \{ g(1)(t, x), \ldots, g(\mu)(t, x) \}$$

for each $(t, x) \in A$.

(6) $\mathbf{GS}(h(s, x), h(t, y)) \leqslant \Gamma_{3.12}(1 + M)N\lambda|(s, x) - (t, y)|$ for each $(s, x), (t, y) \in [-\lambda(\infty), 0] \times \mathbf{R}^m$.

(7) Corresponding to each compact subset A of $(-\lambda(\infty), 0] \times \mathbf{R}^m$, there exist $\mu \in \{1, 2, 3, \ldots\}$ and affine functions $g(1), \ldots, g(\mu)$: $\mathbf{R}^{m+1} \to \mathbf{R}^n$ with $\|Dg(i)\| \leqslant \Gamma_{3.12}\lambda$ for each $i = 1, \ldots, \mu$ such that

$$\text{spt}[h(t, x)] \subset \{ g(1)(t, x), \ldots, g(\mu)(t, x) \}$$

for each $(t, x) \in A$.

(8) For each $(t, x) \in [0, 2^{-v}] \times \mathbf{R}^m$, $\mathbf{S}(F(t, x)) \leqslant 12N - 6$, $\mathbf{S}(G(t, x)) \leqslant 12N - 6$, $\Theta^0(F(t, x), y) \leqslant M + \varepsilon$ and $\Theta^0(G(t, x), y) \leqslant \varepsilon$ for each $y \in \mathbf{R}^n$, $\mathbf{M}(F(t, x)) \leqslant (M + \varepsilon)(12N - 6)$, $\mathbf{M}(G(t, x)) \leqslant \varepsilon(12N - 6)$, $\langle F(t, x), 1 \rangle = L$, $\langle G(t, x), 1 \rangle = 0$, $\text{spt}[F(t, x)] \subset \mathbf{B}^n(0, R)$, and $\text{spt}[G(t, x)] \subset \mathbf{B}^n(0, R)$.

(9) $\mathbf{GS}(F(s, x), F(t, y)) \leqslant \Gamma_{3.12}(1 + M + \varepsilon)N(2^v\varepsilon + \lambda)|(s, x) - (t, y)|$ and

$\mathbf{GS}(G(s, x), G(t, y)) \leqslant \Gamma_{3.12}(1 + \varepsilon)N(2^vR + 2^v\varepsilon + \lambda)|(s, x) - (t, y)|$

for each $(s, x), (t, y) \in [0, 2^{-v}] \times \mathbf{R}^m$.

(10) Corresponding to each bounded subset A of $\mathbf{R}^m$ there exist $\mu \in \{1, 2, 3,\ldots\}$ and affine functions $g(1),\ldots,g(\mu)$, $h(1),\ldots,h(\mu)$: $\mathbf{R}^{m+1} \to \mathbf{R}^n$ with $\|Dg(i)\| \leqslant \Gamma_{3.12}(2^v\varepsilon + \lambda)$ and $\|Dh(i)\| \leqslant (2^vR + 2^v\varepsilon + \lambda)$ for each $i = 1,\ldots,\mu$ such that

$$\mathrm{spt}[F(t, x)] \subset \{ g(1)(t, x),\ldots,g(\mu)(t, x)\}$$

and

$$\mathrm{spt}[G(t, x)] \subset \{ h(1)(t, x),\ldots,h(\mu)(t, x)\}$$

for each $(t, x) \subset [0, 2^{-\mu}] \times A$.

Some mass and size estimates. Suppose $k \in \{0, 1,\ldots,m\}$, $T \in \mathscr{S}_k(\mathbf{R}^m)$, and $I = [-\lambda(\infty), 2^{-v} + \lambda(\infty)]$. In anticipation of the definitions and estimates of 3.13 and 3.14 we record here some useful information about our function h which is a consequence of those definitions and estimates together with the usual mapping estimates for rectifiable currents derived from the Hausdorff area formula.

(1) $\{\{0\} \times \mathbf{R}^m \times \mathbf{R}^n\}_{\#} \circ (\llbracket \mathbf{1}_{I \times \mathbf{R}^m} \rrbracket \bowtie h)_{\#}(\llbracket 2^{-v} + \lambda(\infty) \rrbracket \times T) = (\llbracket \mathbf{1}_{\mathbf{R}^m} \rrbracket \bowtie g)_{\#}T.$

(2) $\{\{0\} \times \mathbf{R}^m \times \mathbf{R}^n\}_{\#} \circ (\llbracket \mathbf{1}_{I \times \mathbf{R}^m} \rrbracket \bowtie h)_{\#}(\llbracket -\lambda(\infty) \rrbracket \times T) = (\llbracket \mathbf{1}_{\mathbf{R}^m} \rrbracket \bowtie f)_{\#}T.$

(3) In case $k > 0$ and $T \in \mathbf{S}_k(\mathbf{R}^m)$ then

$(\llbracket \mathbf{1}_{\mathbf{R}^m} \rrbracket \bowtie g)_{\#}T - (\llbracket \mathbf{1}_{\mathbf{R}^m} \rrbracket \bowtie f)_{\#}T$

$\qquad = \partial \{\{0\} \times \mathbf{R}^m \times \mathbf{R}^n\}_{\#} \circ (\llbracket \mathbf{1}_{I \times \mathbf{R}^m} \rrbracket \bowtie h)_{\#}(\llbracket -\lambda(\infty), 2^{-v} + \lambda(\infty) \rrbracket \times T)$

$\qquad\quad + \{\{0\} \times \mathbf{R}^m \times \mathbf{R}^n\}_{\#} \circ (\llbracket \mathbf{1}_{I \times \mathbf{R}^m} \rrbracket \bowtie h)_{\#}(\llbracket -\lambda(\infty), 2^{-v} + \lambda(\infty) \rrbracket \times \partial T).$

(4) There is a constant $1 < \Gamma_{3.12}^* < \infty$ depending only on m and n such that

$\mathbf{M}[(\llbracket \mathbf{1}_{I \times \mathbf{R}^m} \rrbracket \bowtie h)_{\#}(\llbracket -\lambda(\infty), 2^{-v} + \lambda(\infty) \rrbracket \times T)]$

$\qquad \leqslant \Gamma_{3.12}^* 2^{-v}(M + \varepsilon)(1 + 2^v\varepsilon + \lambda)^{k+1}\mathbf{M}(T) + \Gamma_{3.12}^* 2^{-v}\varepsilon N(1 + 2^vR)^{k+1}\mathbf{M}(T)$

and

$\mathbf{S}[(\llbracket \mathbf{1}_{I \times \mathbf{R}^m} \rrbracket \bowtie h)_{\#}(\llbracket -\lambda(\infty), 2^{-v} + \lambda(\infty) \rrbracket \times T)]$

$\qquad \leqslant \Gamma_{3.12}^* 2^{-v}N(1 + 2^v\varepsilon + \lambda)^{k+1}\mathbf{S}(T) + \Gamma_{3.12}^* 2^{-v}N(1 + 2^vR)^{k+1}\mathbf{S}(T).$

To check these estimates, we write

$\llbracket -\lambda(\infty), 2^{-v} + \lambda(\infty) \rrbracket \times T$

$\qquad\qquad = \llbracket -\lambda(\infty), 0 \rrbracket \times T + \llbracket 0, 2^{-v} \rrbracket \times T + \llbracket 2^{-v}, 2^{-v} + \lambda(\infty) \rrbracket \times T$

and use (2), (7), (8) of the "Properties of h, F, G" above together with (7) among the "Additional properties of $(\llbracket \mathbf{1}_{\mathbf{R}^m} \rrbracket \bowtie f)_{\#}$ and $f_{\#}$" in 3.13.

3.13. How to map rectifiable currents by Lipschitz multiple-valued maps.

Assumptions. (a) $\lambda \in \mathbf{R}^+$, $L \in \mathbf{R}$, $M \in \mathbf{R}^+$, $N \in \{1, 2, 3,\ldots\}$, $0 < R < \infty$,

(b) $f: \mathbf{R}^m \to \mathbf{P}_0(\mathbf{R}^n)$,

(c) for each $x \in \mathbf{R}^m$, $\mathbf{S}(f(x)) \leqslant N$, $\mathbf{M}(f(x)) \leqslant M$, $\langle f(x), 1 \rangle = L$, and $\mathrm{spt}[f(x)] \subset \mathbf{B}^n(0, R)$,

(d) $\mathbf{GS}(f(x), f(y)) \leqslant \lambda |x - y|$ for each $x, y \in \mathbf{R}^m$,

(e) $k \in \{0, 1, \ldots, m\}$.

It is our object to define mappings

$$([\![\mathbf{1}_{\mathbf{R}^m}]\!] \bowtie f)_{\#} \colon \mathscr{S}_k(\mathbf{R}^m) \to \mathscr{S}_k(\mathbf{R}^{m+n})$$

and

$$f_{\#} = \{\{0\} \times \mathbf{R}^n\}_{\#} \circ ([\![\mathbf{1}_{\mathbf{R}^m}]\!] \bowtie f)_{\#} \colon \mathscr{S}_k(\mathbf{R}^m) \to \mathscr{S}_k(\mathbf{R}^n).$$

Construction. (A) For each $v \in \{0, 1, \ldots, N\}$ we set $A_v = \mathbf{R}^m \cap \{x \colon \mathbf{S}(f(x)) = v\}$. We check that A_N is an open subset of $\mathbf{R}^m$ and that A_v is a (relatively) open subset of $\mathbf{R}^m \sim (A_{v+1} \cup \cdots \cup A_N)$ for each $v = 0, \ldots, N - 1$. In particular, $A_0, A_1, \ldots, A_N$ are all Borel subsets of $\mathbf{R}^m$.

(B) Suppose $v \in \{1, 2, \ldots, N\}$ and $p \in A_v$. We use the Fifth Example of 3.1 together with Kirszbraun's theorem [**FH**, 2.10.43] to conclude the existence of $\sigma_1, \ldots, \sigma_v \in \mathbf{R}$ (with $|\sigma_1| + \cdots + |\sigma_v| \leqslant M$) together with an open neighborhood U of p in $\mathbf{R}^m$ and Lipschitz functions $g_1, \ldots, g_v \colon U \to \mathbf{R}^n$ (each with Lipschitz constant not exceeding λ) such that

$$0 < \inf\left\{ |g_i(x) - g_j(x)| \colon i, j \in \{1, \ldots, v\} \text{ with } i \neq j \text{ and } x \in U \right\}$$

and

$$f(x) = \sum \{\sigma_i [\![g_i(x)]\!] \colon i = 1, \ldots, v\} \quad \text{for each } x \in A_v \cap U.$$

(C) Suppose $T \in \mathscr{S}_k(\mathbf{R}^m)$. We set $T_v = T \llcorner A_v$ for each $v \in \{1, 2, \ldots, N\}$ and define

$$([\![\mathbf{1}_{\mathbf{R}^m}]\!] \bowtie f)_{\#} T = ([\![\mathbf{1}_{\mathbf{R}^m}]\!] \bowtie f)_{\#} T_1 + \cdots + ([\![\mathbf{1}_{\mathbf{R}^m}]\!] \bowtie f)_{\#} T_N.$$

Here, for each $v \in \{1, \ldots, N\}$, $([\![\mathbf{1}_{\mathbf{R}^m}]\!] \bowtie f)_{\#} T_v$ is characterized by the requirement

$$\left[([\![\mathbf{1}_{\mathbf{R}^m}]\!] \bowtie f)_{\#} T_v \right] \llcorner (U \times \mathbf{R}^n) = \sum \{\sigma_i (\mathbf{1}_U \bowtie g_i)_{\#}(T_v \llcorner U) \colon i = 1, \ldots, v\}$$

whenever p, U, $\sigma_1, \ldots, \sigma_v$, $g_1, \ldots, g_v$ are as in (B) above.

This completes the definition of $([\![\mathbf{1}_{\mathbf{R}^m}]\!] \bowtie f)_{\#}$.

Properties of $([\![\mathbf{1}_{\mathbf{R}^m}]\!] \bowtie f)_{\#}$ *and* $f_{\#}$.

(1) $([\![\mathbf{1}_{\mathbf{R}^m}]\!] \bowtie f)_{\#}$ and $f_{\#}$ are linear.

(2) $\mathbf{M}[([\![\mathbf{1}_{\mathbf{R}^m}]\!] \bowtie f)_{\#} T] \leqslant M(1 + \lambda^2)^{k/2} \mathbf{M}(T)$ and

$$\mathbf{S}\left[([\![\mathbf{1}_{\mathbf{R}^m}]\!] \bowtie f)_{\#} T \right] \leqslant N(1 + \lambda^2)^{k/2} \mathbf{S}(T)$$

for each $T \in \mathscr{S}_k(\mathbf{R}^m)$.

(3) $\mathbf{M}(f_{\#} T) \leqslant M \lambda^k \mathbf{M}(T)$ and $\mathbf{S}(f_{\#} T) \leqslant N \lambda^k \mathbf{S}(T)$ for each $T \in \mathscr{S}_k(\mathbf{R}^m)$; to check this we note for each $0 < r < \infty$ that $\mathrm{Lip}(f \circ \mu(1/r)) = (1/r)\mathrm{Lip}(f)$ and $[f \circ \mu(1/r)]_{\#}[\mu(r)_{\#} T] = f_{\#} T$, etc.

(4) $([\![\mathbf{1}_{\mathbf{R}^m}]\!] \bowtie f)_{\#}$ and $f_{\#}$ are Lipschitz with respect to the **MS** metrics on $\mathscr{S}_k(\mathbf{R}^m)$, $\mathscr{S}_k(\mathbf{R}^{m+n})$, and $\mathscr{S}_k(\mathbf{R}^n)$ with Lipschitz constants $(M + N)(1 + \lambda^2)^{k/2}$ and $(M + N)\lambda^k$, respectively.

(5) Whenever $h \colon \mathbf{R}^m \to \mathbf{R}^m$ is a bilipschitz homeomorphism then

$$([\![\mathbf{1}_{\mathbf{R}^m}]\!] \bowtie f)_{\#} \circ h_{\#} = (h \times \mathbf{1}_{\mathbf{R}^n})_{\#} \circ ([\![\mathbf{1}_{\mathbf{R}^m}]\!] \bowtie f)_{\#}$$

so that

$$([\![\mathbf{1}_{\mathbf{R}^m}]\!] \bowtie f)_{\#} = \left(h^{-1} \times \mathbf{1}_{\mathbf{R}^n} \right)_{\#} \circ ([\![\mathbf{1}_{\mathbf{R}^m}]\!] \bowtie f)_{\#} \circ h_{\#}.$$

(6) In case $f(x) \in \mathbf{IP}_0(\mathbf{R}^n)$ for each $x \in \mathbf{R}^m$ then

$$([\![\mathbf{1}_{\mathbf{R}^m}]\!] \bowtie f)_{\#}\mathscr{I}_k(\mathbf{R}^m) \subset \mathscr{I}_k(\mathbf{R}^{m+n}) \quad \text{and} \quad f_{\#}\mathscr{I}_k(\mathbf{R}^m) \subset \mathscr{I}_k(\mathbf{R}^n).$$

Additional properties of $([\![\mathbf{1}_{\mathbf{R}^m}]\!] \bowtie f)_{\#}$ *and* $f_{\#}$. Suppose $0 \leqslant \lambda_1 < \infty$, $N_1 \in \{1, 2, 3,\ldots\}$, and $g(1),\ldots,g(N_1)\colon \mathbf{R}^m \to \mathbf{R}^n$ are affine functions with $\|Dg(i)\| \leqslant \lambda_1$ for each $i = 1,\ldots,N_1$. Suppose also U is an open subset of $\mathbf{R}^m$ such that

$$\mathrm{spt}[\,f(u)\,] \subset \{\, g(1)(u),\ldots,g(N_1)(u)\,\}$$

for each $u \in U$. Then, whenever $T_1 \in \mathscr{S}_k(\mathbf{R}^m)$ with $\mathrm{spt}(T_1) \subset U$ we readily check the alternative estimates

$$(7) \qquad \mathbf{M}\big[([\![\mathbf{1}_{\mathbf{R}^m}]\!] \bowtie f)_{\#}T_1\big] \leqslant M\big(1 + \lambda_1^2\big)^{k/2}\mathbf{M}(T_1),$$

$$\mathbf{S}\big[([\![\mathbf{1}_{\mathbf{R}^m}]\!] \bowtie f)_{\#}T_1\big] \leqslant N\big(1 + \lambda_1^2\big)^{k/2}\mathbf{S}(T_1),$$

$$\mathbf{M}(f_{\#}T_1) \leqslant M\lambda_1^k\mathbf{M}(T_1), \qquad \mathbf{S}(f_{\#}T_1) \leqslant N\lambda_1^k\mathbf{S}(T_1).$$

3.14. Theorem (Multiple-valued mappings of currents) (*see Figure* 3.14).
Hypotheses. (a) $\lambda, \lambda_g, \lambda_h \in \mathbf{R}^+$; $L, L_g, L_h \in \mathbf{R}$; $M, M_g, M_h \in \mathbf{R}^+$; N, N_g, N_h, $p \in \{1, 2, 3,\ldots\}$, *and* $0 < R < \infty$, $0 < R_g < \infty$, $0 < R_h < \infty$;
(b) $f, g\colon \mathbf{R}^m \to \mathbf{P}_0(\mathbf{R}^n)$ [*resp.* $\to \mathbf{IP}_0(\mathbf{R}^n)$] *and* $h\colon \mathbf{R}^n \to \mathbf{P}_0(\mathbf{R}^p)$ [*resp.* $\to \mathbf{IP}_0(\mathbf{R}^p)$];
(c) *for each* $x \in \mathbf{R}^m$ *and* $y \in \mathbf{R}^n$,

$$\mathbf{S}(f(x)) \leqslant N, \qquad \mathbf{S}(g(x)) \leqslant N_g, \qquad \mathbf{S}(h(x)) \leqslant N_h,$$

$$\mathbf{M}(f(x)) \leqslant M, \qquad \mathbf{M}(g(x)) \leqslant M_g, \qquad \mathbf{M}(h(x)) \leqslant N_h,$$

$$\langle f(x),1\rangle = L, \qquad \langle g(x),1\rangle = L_g, \qquad \langle h(x),1\rangle = L_h,$$

$$\mathrm{spt}[\,f(x)\,] \subset \mathbf{B}^n(0, R), \quad \mathrm{spt}[\,g(x)\,] \subset \mathbf{B}^n(0, R_g), \quad \mathrm{spt}[\,h(x)\,] \subset \mathbf{B}^p(0, R_h);$$

(d) *for each* $w, x \in \mathbf{R}^m$ *and* $y, z \in \mathbf{R}^n$,

$$\mathbf{GS}(f(w), f(x)) \leqslant \lambda|w - x| \quad \big[\textit{resp. } \mathscr{G}(f(w), f(x)) \leqslant \lambda|w - x|\big],$$

$$\mathbf{GS}(g(w), g(x)) \leqslant \lambda_g|w - x| \quad \big[\textit{resp. } \mathscr{G}(g(w), g(x)) \leqslant \lambda_g|w - x|\big],$$

$$\mathbf{GS}(h(y), h(z)) \leqslant \lambda_h|y - z| \quad \big[\textit{resp. } \mathscr{G}(h(y), h(z)) \leqslant \lambda_h|y - z|\big];$$

(e) *for each* $k \in \{0, 1,\ldots,m\}$ *and each* $j \in \{0, 1,\ldots,n\}$,

$$([\![\mathbf{1}_{\mathbf{R}^m}]\!] \bowtie f)_{\#}\colon \mathscr{S}_k(\mathbf{R}^m) \to \mathscr{S}_k(\mathbf{R}^{m+n}), \qquad f_{\#}\colon \mathscr{S}_k(\mathbf{R}^m) \to \mathscr{S}_k(\mathbf{R}^n),$$

$$([\![\mathbf{1}_{\mathbf{R}^m}]\!] \bowtie g)_{\#}\colon \mathscr{S}_k(\mathbf{R}^m) \in \mathscr{S}_k(\mathbf{R}^{m+n}), \qquad g_{\#}\colon \mathscr{S}_k(\mathbf{R}^m) \to \mathscr{S}_k(\mathbf{R}^n),$$

$$([\![\mathbf{1}_{\mathbf{R}^n}]\!] \bowtie h)_{\#}\colon \mathscr{S}_j(\mathbf{R}^n) \to \mathscr{S}_j(\mathbf{R}^{n+p}), \qquad h_{\#}\colon \mathscr{S}_j(\mathbf{R}^n) \to \mathscr{S}_j(\mathbf{R}^p)$$

are the mappings constructed in 3.13;
(f) $H - h_{\#} \circ f\colon \mathbf{R}^m \to \mathbf{P}_0(\mathbf{R}^p)$.
Conclusions. (1) *For each* $x \in \mathbf{R}^m$, $f(x) = f_{\#}([\![x]\!])$,

$$([\![\mathbf{1}_{\mathbf{R}^m}]\!] \bowtie f)(x) = [\![x]\!] \times f(x) = ([\![\mathbf{1}_{\mathbf{R}^m}]\!] \bowtie f)_{\#}[\![x]\!],$$

F. ALMGREN

and

$$\{\{0\} \times \{0\} \times \mathbf{R}^{p}\}_{\#} \circ \left([\![\mathbf{1}_{\mathbf{R}^{m} \times \mathbf{R}^{n}}]\!] \bowtie \left[h \circ \{\{0\} \times \mathbf{R}^{n}\}_{\natural}\right]\right)_{\#} \circ \left([\![\mathbf{1}_{\mathbf{R}^{m}}]\!] \bowtie f\right)(x)$$
$$= H(x).$$

(2) *For each* $k \in \{1, 2, \ldots, m\}$,
$$\left([\![\mathbf{1}_{\mathbf{R}^{m}}]\!] \bowtie f\right)_{\#} \mathbf{S}_{k}(\mathbf{R}^{m}) \subset \mathbf{S}_{k}(\mathbf{R}^{m+n}), \qquad f_{\#} \mathbf{S}_{k}(\mathbf{R}^{m}) \subset \mathbf{S}_{k}(\mathbf{R}^{n})$$
$$\left[resp. \left([\![\mathbf{1}_{\mathbf{R}^{m}}]\!] \bowtie f\right)_{\#} \mathbf{I}_{k}(\mathbf{R}^{m}) \subset \mathbf{I}_{k}(\mathbf{R}^{m+n}), f_{\#} \mathbf{I}_{k}(\mathbf{R}^{m}) \subset \mathbf{I}_{k}(\mathbf{R}^{n})\right].$$

(3) *The mappings*
$$\left([\![\mathbf{1}_{\mathbf{R}^{m}}]\!] \bowtie f\right)_{\#} \colon \bigoplus_{k=0}^{m} \mathbf{S}_{k}(\mathbf{R}^{m}) \to \bigoplus_{k=0}^{m} \mathbf{S}_{k}(\mathbf{R}^{m+n}), f_{\#} \colon \bigoplus_{k=0}^{m} \mathbf{S}_{k}(\mathbf{R}^{m}) \to \bigoplus_{k=0}^{m} \mathbf{S}_{k}(\mathbf{R}^{n})$$
$$\left[resp. \left([\![\mathbf{1}_{\mathbf{R}^{m}}]\!] \bowtie f\right)_{\#} \colon \bigoplus_{k=0}^{m} \mathbf{I}_{k}(\mathbf{R}^{m}) \to \bigoplus_{k=0}^{m} \mathbf{I}_{k}(\mathbf{R}^{m+n}),\right.$$
$$\left. f_{\#} \colon \bigoplus_{k=0}^{m} \mathbf{I}_{k}(\mathbf{R}^{m}) \to \bigoplus_{k=0}^{m} \mathbf{I}_{k}(\mathbf{R}^{n})\right]$$

are chain mappings (of degree 0); in particular, they are linear [resp. group homomorphisms] and commute with the boundary operator ∂.

(4) *For each* $k \in \{0, 1, \ldots, m\}$ *and each* $T \in \mathscr{S}_{k}(\mathbf{R}^{m})$,
$$\mathbf{M}\left[\left([\![\mathbf{1}_{\mathbf{R}^{m}}]\!] \bowtie f\right)_{\#} T\right] \leqslant M(1 + \lambda^{2})^{k/2} \mathbf{M}(T),$$
$$\mathbf{S}\left[\left([\![\mathbf{1}_{\mathbf{R}^{m}}]\!] \bowtie f\right)_{\#} T\right] \leqslant N(1 + \lambda^{2})^{k/2} \mathbf{S}(T),$$
$$\mathbf{M}(f_{\#} T) \leqslant M\lambda^{k} \mathbf{M}(T), \qquad \mathbf{S}(T) \leqslant N\lambda^{k} \mathbf{S}(T).$$

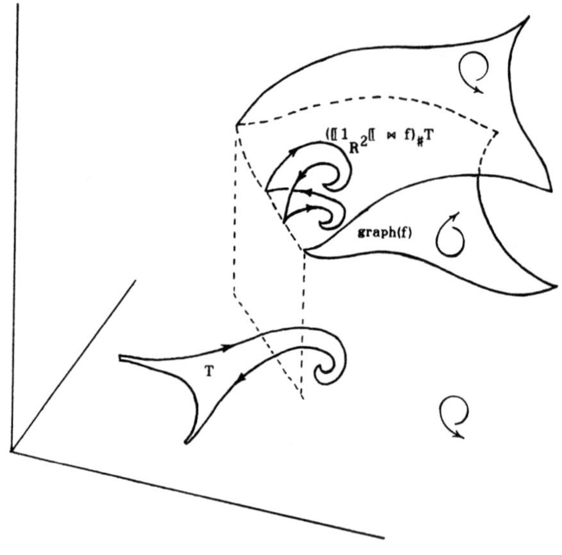

FIGURE 3.14. Multiple-valued mapping of currents.

(5) *Suppose* $0 \leqslant \lambda_1 < \infty$, $N_1 \in \{1, 2, 3, \ldots\}$, *and* $g(1), \ldots, g(N_1) \colon \mathbf{R}^m \to \mathbf{R}^n$ *are affine functions with* $\|Dg(i)\| \leqslant \lambda_1$ *for each* $i = 1, \ldots, N_1$. *Suppose also* U_1 *is an open subset of* $\mathbf{R}^m$ *such that*

$$\mathrm{spt}[f(u)] \subset \{g(1)(u), \ldots, g(N_1)(u)\}$$

for each $u \in U_1$. *Finally, suppose* $k \in \{0, 1, \ldots, m\}$ *and* $T_1 \in \mathscr{S}_k(\mathbf{R}^m)$ *with* $\mathrm{spt}(T_1)$ $\subset U_1$. *Then*

$$\mathbf{M}\big[(\llbracket \mathbf{1}_{\mathbf{R}^m}\rrbracket \bowtie f)_\sharp T_1\big] \leqslant M\big(1 + \lambda_1^2\big)^{k/2}\mathbf{M}(T_1),$$

$$\mathbf{S}\big[(\llbracket \mathbf{1}_{\mathbf{R}^m}\rrbracket \bowtie f)_\sharp T_1\big] \leqslant N\big(1 + \lambda_1^2\big)^{k/2}\mathbf{S}(T_1),$$

$$\mathbf{M}(f_\sharp T_1) \leqslant M\lambda_1^k \mathbf{M}(T_1), \qquad \mathbf{S}(f_\sharp T_1) \leqslant N\lambda_1^k \mathbf{S}(T_1).$$

(6) *For each bounded open subset* U *of* $\mathbf{R}^m$,

$$(\llbracket \mathbf{1}_{\mathbf{R}^m}\rrbracket \bowtie f)_\sharp(\mathbf{E}^m \llcorner U) = (\llbracket \mathbf{1}_{\mathbf{R}^m}\rrbracket \bowtie g)_\sharp(\mathbf{E}^m \llcorner U)$$

if and only if $f(u) = g(u)$ *for each* $u \in U$.

(7) *For each* w, $x \in \mathbf{R}^m$; $\mathbf{S}[(f + g)(x)] \leqslant N + N_g$; $\mathbf{S}(H(x)) \leqslant N\,N_h$; $\mathbf{M}[(f + g)(x)] \leqslant M + M_g$; $\mathbf{M}(H(x)) \leqslant M\,M_h$; $\langle(f + g)(x), 1\rangle = L + L_g$; $\langle H(x), 1\rangle = L\,L_h$;

$$\mathbf{GS}((f + g)(w), (f + g)(x)) \leqslant (\lambda + \lambda_g)|w - x|$$

$$\big[resp. \ \mathscr{G}((f + g)(w), (f + g)(x)) \leqslant (\lambda + \lambda_g)|w - x|\big];$$

and $\mathbf{GS}(H(w), H(x)) \leqslant \lambda\,\lambda_h|w - x|$ *[resp.* $\mathscr{G}(H(w), H(x)) \leqslant \lambda\,\lambda_h|w - x|$]*.

(8) $(\llbracket \mathbf{1}_{\mathbf{R}^m}\rrbracket \bowtie (f + g))_\sharp = (\llbracket \mathbf{1}_{\mathbf{R}^m}\rrbracket \bowtie f)_\sharp + (\llbracket \mathbf{1}_{\mathbf{R}^m}\rrbracket \bowtie g)_\sharp$ *and* $(f + g)_\sharp = f_\sharp + g_\sharp$.

(9) $H_\sharp = h_\sharp \circ f_\sharp$.

PROOF. In view of 3.13 the essential new thing to show here is that the mapping $(\llbracket \mathbf{1}_{\mathbf{R}^m}\rrbracket \bowtie f)_\sharp$ commutes with the boundary operator. Suppose that $k \in \{1, 2, \ldots, m\}$ and $T \in \mathbf{S}_k(\mathbf{R}^m)$. We will check that

$$(*) \qquad (\llbracket \mathbf{1}_{\mathbf{R}^m}\rrbracket \bowtie f)_\sharp \circ \partial T = \partial \circ (\llbracket \mathbf{1}_{\mathbf{R}^m}\rrbracket \bowtie f)_\sharp T.$$

In view of the linearity of $(\llbracket \mathbf{1}_{\mathbf{R}^m}\rrbracket \bowtie f)_\sharp$ and the continuity of $(\llbracket \mathbf{1}_{\mathbf{R}^m}\rrbracket \bowtie f)_\sharp$ in the **MS** metric topology noted in 3.13(1)(4), the remark concerning bilipschitz homeomorphisms in 3.13(5), and the Strong Approximation Theorem 2.11, we conclude that it is sufficient to establish $(*)$ in case $T = \phi_\sharp(\mathbf{E}^k \llcorner W)$ corresponding to some bounded convex polyhedron $W \subset \mathbf{U}^k(0, 1)$ and some isometric injection $\phi \colon \mathbf{R}^k \to \mathbf{R}^m$. From inspection of the definitions in 3.13 we infer

$$(\phi \times \mathbf{1}_{\mathbf{R}^n})_\sharp \circ (\llbracket \mathbf{1}_{\mathbf{R}^k}\rrbracket \bowtie (f \circ \phi))_\sharp(\mathbf{E}^k \llcorner W) = (\llbracket \mathbf{1}_{\mathbf{R}^m}\rrbracket \bowtie f)_\sharp \circ \phi_\sharp(\mathbf{E}^k \llcorner W)$$

and

$$(\phi \times \mathbf{1}_{\mathbf{R}^n})_\sharp \circ (\llbracket \mathbf{1}_{\mathbf{R}^k}\rrbracket \bowtie (f \circ \phi))_\sharp \circ \partial(\mathbf{E}^k \llcorner W)$$

$$= (\llbracket \mathbf{1}_{\mathbf{R}^m}\rrbracket \bowtie f)_\sharp \circ \phi_\sharp \circ \partial(\mathbf{E}^k \llcorner W).$$

Since nothing special has been assumed about ϕ, m, or f, it is sufficient to establish (∗) in the case when $k = m$ and $\phi = \mathbf{1}_{\mathbf{R}^m}$, which we now assume.

Suppose then $0 < \varepsilon < \alpha(m)$ and $A_0, A_1,\ldots,A_N$ are the Borel sets of 3.13(A) associated with f. We fix compact sets $C_i \subset \mathbf{B}^m(0,2) \cap A_i$ for each $i = 1,\ldots,N$ such that $\Sigma_i \mathscr{L}^m[\mathbf{B}^m(0,2) \cap A_i \sim C_i] < \varepsilon$ and $\Sigma_i \mathscr{H}^{m-1}[\partial W \cap A_i \sim C_i] < \varepsilon$ [**FH**, 2.2.2] and set $C = C_1 \cup \cdots \cup C_N$. We further let $G\colon \mathbf{R}^m \to \mathbf{P}_0(\mathbf{R}^n)$ be a function obtained as in the conclusion of 3.10 when $L, M, N, R, \lambda, A, f, g$ there are replaced by $L, M, N, R, \lambda, C, f|C, G$ in the present context. We use 3.13(3) and 3.10(1)(3) to estimate

$$\mathbf{M}\big[(\llbracket\mathbf{1}_{\mathbf{R}^m}\rrbracket \bowtie G)_{\#}(\mathbf{E}^m \llcorner W) - (\llbracket\mathbf{1}_{\mathbf{R}^m}\rrbracket \bowtie f)_{\#}(\mathbf{E}^m \llcorner W)\big]$$
$$\leqslant \varepsilon\Big[M(1 + \lambda^2)^{m/2} + M(12N - 6)\big(1 + \Gamma_{3.10}^2\lambda^2\big)^{m/2}\Big],$$

$$\mathbf{S}\big[(\llbracket\mathbf{1}_{\mathbf{R}^m}\rrbracket \bowtie G)_{\#}(\mathbf{E}^m \llcorner W) - (\llbracket\mathbf{1}_{\mathbf{R}^m}\rrbracket \bowtie f)_{\#}(\mathbf{E}^m \llcorner W)\big]$$
$$\leqslant \varepsilon\Big[N(1 + \lambda^2)^{m/2} + (12N - 6)\big(1 + \Gamma_{3.10}^2\lambda^2\big)^{m/2}\Big],$$

$$\mathbf{M}\big[(\llbracket\mathbf{1}_{\mathbf{R}^m}\rrbracket \bowtie G)_{\#}\circ\partial(\mathbf{E}^m \llcorner W) - (\llbracket\mathbf{1}_{\mathbf{R}^m}\rrbracket \bowtie f)_{\#}\circ\partial(\mathbf{E}^m \llcorner W)\big]$$
$$\leqslant \varepsilon\Big[M(1 + \lambda^2)^{m/2} + M(12N - 6)\big(1 + \Gamma_{3.10}^2\lambda^2\big)^{m/2}\Big],$$

$$\mathbf{S}\big[(\llbracket\mathbf{1}_{\mathbf{R}^m}\rrbracket \bowtie G)_{\#}\circ\partial(\mathbf{E}^m \llcorner W) - (\llbracket\mathbf{1}_{\mathbf{R}^m}\rrbracket \bowtie f)_{\#}\circ\partial(\mathbf{E}^m \llcorner W)\big]$$
$$\leqslant \varepsilon\Big[N(1 + \lambda^2)^{m/2} + (12N - 6)\big(1 + \Gamma_{3.10}^2\lambda^2\big)^{m/2}\Big].$$

Since ε is arbitrary we conclude that in order to verify (∗) it is sufficient to show

$$(\#) \qquad \partial\circ(\llbracket\mathbf{1}_{\mathbf{R}^m}\rrbracket \bowtie G)_{\#}(\mathbf{E}^m \llcorner W) = (\llbracket\mathbf{1}_{\mathbf{R}^m}\rrbracket \bowtie G)_{\#}\circ\partial(\mathbf{E}^m \llcorner c_i W).$$

Since the boundary operator ∂ is local it is sufficient to suppose $q \in \mathbf{R}^m$ and show the existence of an open neighborhood U of $\mathbf{Q}$ in which (#) holds (with an obvious meaning). We consider two cases:

Case 1. Suppose $q \in C_v$ for some $v \in \{0,\ldots,N\}$. We note since $C_i \cap C_j = \varnothing$ for each $i \neq j$ and the C_k's are compact that $\mathrm{dist}(C_i, C_j) > 0$ for each $i \neq j$. We utilize the continuity of f, the definitions of $C_i \subset A_i$, and the observations of the Fifth Example in 3.1 to conclude the applicability of 3.10 with C replacing A there in obtaining a neighborhood V of q and Lipschitz functions $g(1),\ldots,g(v)\colon V \to \mathbf{R}^n$ as in 3.10(5) and $\theta(1),\ldots,\theta(v) \in \mathbf{R}$ such that $G(x) = \Sigma\{\theta(i)\llbracket g(i)(x)\rrbracket\colon i = 1,\ldots,v\}$ for each $x \in V$; if $v = 0$ we understand that $G(x) = 0$ for each $x \in V$. It is immediate that (#) holds near q.

Case 2. Suppose $\mathrm{dist}(q, C) > 0$. In this case, according to 3.10(4) there will exist a neighborhood V of q, $v \in \{1, 2, 3,\ldots\}$, and affine functions $h(1),\ldots,h(v)\colon \mathbf{R}^m \to \mathbf{R}^n$ such that $\mathrm{spt}[G(x)] \subset \{h(1)(x),\ldots,h(v)(x)\}$ for each $x \in V$. We are essentially in the case of the Eighth Example of 3.1 and we conclude as there that (#) holds near q. This finishes the proof.

3.15. Theorem (Sequences of multiple-valued mapped currents).

Hypotheses. (a) $0 \leqslant \lambda < \infty$ *and* $L, M, N, R, \mathbf{Q}$ *are as in* 3.5$(a)(b)$ *with* $L = 0$,
  (b) $f(1), f(2), f(3),\ldots\colon \mathbf{R}^m \to \mathbf{Q}$ [*resp.* $\to \mathbf{Q} \cap \mathbf{IP}_0(\mathbf{R}^n)$],

(c) $\mathbf{GS}(f(i)(x), f(i)(y)) \leqslant \lambda|x - y|$ [resp. $\mathscr{G}(f(i)(x), f(i)(y)) \leqslant \lambda|x - y|$] for each $i \in \{1, 2, 3, \ldots\}$ and each $x, y \in \mathbf{R}^m$,

(d) $\bigcup\{\mathrm{spt}[f(i)]: i = 1, 2, 3, \cdots \} \subset \mathbf{R}^m$ is bounded.

Conclusion. There exists a subsequence $i(1), i(2), i(3), \ldots$ of $1, 2, 3, \ldots$ and a function

$$f: \mathbf{R}^m \to \mathbf{Q} \quad [resp. \to \mathbf{Q} \cap \mathbf{IP}_0(\mathbf{R}^n)]$$

with the following properties:

(1) $\lim_{j \to \infty} \sup\{\mathbf{G}(f(x), f(i(j))(x)): x \in \mathbf{R}^m\} = 0$ (*not* $\mathbf{GS}$) [*resp.* $\lim_{j \to \infty} \sup\{\mathscr{G}(f(x), f(i(j))(x)): x \in \mathbf{R}^m\} = 0$].

(2) $\mathbf{GS}(f(x), f(y)) \leqslant \lambda|x - y|$ [*resp.* $\mathscr{G}(f(x), f(y)) \leqslant \lambda|x - y|$] *for each x, $y \in \mathbf{R}^m$.*

(3) $\mathrm{spt}(f) \subset \mathbf{R}^m$ *is bounded.*

(4) *Corresponding to each $0 < \delta < \infty$, there exists $j_0 \in \{1, 2, 3, \ldots\}$ with the following property: Whenever $j \in \{j_0, j_0 + 1, j_0 + 2, \ldots\}$, $k \in \{1, \ldots, m\}$, and $T \in \mathbf{S}_k(\mathbf{R}^m)$ [resp. $T \in \mathbf{I}_k(\mathbf{R}^m)$], then (with the obvious meanings in view of Hypothesis* (d))

$$\mathbf{F}\big([\![\mathbf{1}_{\mathbf{R}^m}]\!] \bowtie f]_\sharp T, [\![\mathbf{1}_{\mathbf{R}^m}]\!] \bowtie f(i(j))]_\sharp T\big) \leqslant \delta \mathbf{N}(T)$$
$$\big[resp. \mathscr{F}\big([\![\mathbf{1}_{\mathbf{R}^m}]\!] \bowtie f]_\sharp T, [\![\mathbf{1}_{\mathbf{R}^m}]\!] \bowtie f(i(j))]_\sharp T\big) \leqslant \delta \mathbf{N}(T)\big];$$

in case $\partial T = 0$, then

$$\mathbf{G}\big([\![\mathbf{1}_{\mathbf{R}^m}]\!] \bowtie f]_\sharp T, [\![\mathbf{1}_{\mathbf{R}^m}]\!] \bowtie f(i(j))]_\sharp T\big) \leqslant \delta \mathbf{M}(T) \quad (not \ \mathbf{GS})$$
$$\big[resp. \mathscr{G}\big([\![\mathbf{1}_{\mathbf{R}^m}]\!] \bowtie f]_\sharp T, [\![\mathbf{1}_{\mathbf{R}^m}]\!] \bowtie f(i(j))]_\sharp T\big) \leqslant \delta \mathbf{M}(T)\big];$$

so that, in particular,

$$\mathbf{F}\big(f_\sharp T, f(i(j))_\sharp T\big) \leqslant \delta \mathbf{N}(T) \quad \big[resp. \mathscr{F}\big(f_\sharp T, f(i(j))_\sharp T\big) \leqslant \mathbf{N}(T)\big]$$

and also, in case $\partial T = 0$,

$$\mathbf{G}\big(f_\sharp T, f(i(j))_\sharp T\big) \leqslant \delta \mathbf{M}(T) \quad \big[resp. \mathscr{G}\big(f_\sharp T, f(i(j))_\sharp T\big) \leqslant \delta \mathbf{M}(T)\big].$$

(5) $\lim_{j \to \infty} \mathbf{G}(f_\sharp \mathbf{E}^m, f(i(j))_\sharp \mathbf{E}^m) = 0$ [*resp.* $\lim_{j \to \infty} \mathscr{G}(f_\sharp \mathbf{E}^m, f(i(j))_\sharp \mathbf{E}^m) = 0$].

PROOF. Conclusions $(1), (2), (3)$ follow readily from 3.5(5). Conclusion (4) readily implies Conclusion (5). We now consider Conclusion (4). Suppose then $0 < \delta < \infty$.

With the construction of 3.12 in mind, we choose and fix $v \in \{1, 2, 3, \ldots\}$ sufficiently large so that

$$\delta/2 \geqslant \Gamma^*_{3.12} 2^{-v}(M + 1)(1 + 1 + \lambda)^{m+1}.$$

With the choice of v fixed, we now choose and fix $0 < \varepsilon < 1$ sufficiently small so that $2^v \varepsilon \leqslant 1$ and $\delta/2 > \Gamma^*_{3.12} 2^{-v} \varepsilon N(1 + 2^v)^{m+1}$. As in 3.12, we set

$$\mathbf{K}_0^m(v) = \{\{v\}: \{(v, 0)\} \in \mathbf{K}_0(v)\}$$

and choose $j_0 \in \{1, 2, 3, \ldots\}$ sufficiently large so that whenever $j \in \{j_0, j_0 + 1, j_0 + 2, \ldots\}$ and $a \in \bigcup \mathbf{K}_0^m(v)$ there will exist $P_j(a) \in \mathbf{P}_0(\mathbf{R}^n)$ [resp. $P_j(a) = 0$] and $Q_j(a) \in \mathbf{P}_1(\mathbf{R}^n)$ [resp. $\in \mathbf{IP}_0(\mathbf{R}^n)$] such that the following is true:

$$f(i(j))(a) - f(a) = \partial Q_j(a) + \partial(\llbracket 0 \rrbracket \times\!\!\!\times P_j(a)) = \partial Q_j(a) + P_j(a),$$

$$f(i(j))(a) - P_j(a) \in Q, \qquad \mathbf{MS}(Q_j(a)) < \varepsilon, \qquad \mathbf{S}(P_j(a)) \leqslant N,$$

$$\mathbf{M}(P_j(a)) < \varepsilon, \qquad \mathbf{M}(\llbracket 0 \rrbracket \times\!\!\!\times P_j(a)) < \varepsilon, \qquad \mathrm{spt}[P_j(a)] \subset \mathbf{B}^n(0, R).$$

The ability to select such $P_j(a)$ and $Q_j(a)$ is a consequence of Conclusions (1) and (3) above together with 3.5(3) with T, S, P, Q there corresponding to $f(a)$, $f(i(j))(a), -P_j(a), -Q_j(a)$, respectively, for each j and a.

For each $j \in \{j_0, j_0 + 1, j_0 + 2, \ldots\}$ we utilize the construction of 3.12 with $\varepsilon, v, \lambda, \mathbf{Q}, f, g, P, Q, h$ there corresponding to $\varepsilon, v, \lambda, \mathbf{Q}, f, f(i(j)), P_j, Q_j, h$ respectively in the present context. The assertions of Conclusion (4) now follow readily from (1), (2), (3), (4) of "Some mass and size estimates" in 3.12.

3.16. How to construct a Lipschitz multiple-valued approximation of a rectifiable cycle with error of small projected area.

Assumptions. (a) $0 < \mu < \infty$ and $\theta \in \mathbf{O}(m + n)$,

(b) corresponding to each $\lambda \in \Lambda(m + n, m)$, there is a linear function $g(\lambda)$: $\mathbf{R}^m \to \mathbf{R}^n$ with $\|Dg(\lambda)\| \leqslant \mu$ such that

$$\mathrm{graph}[g(\lambda)] = \theta\big[\mathbf{R}^{m+n} \cap \{x : x_{*\lambda(1)} = x_{*\lambda(2)} = \cdots = x_{*\lambda(n)} = 0\}\big],$$

(c) $0 \neq T \in \mathbf{S}_{m,0}(\mathbf{R}^{m+n})$ [resp. $\in \mathbf{I}_{m,0}(\mathbf{R}^{m+n})$] with $\partial T = 0$,

(d) $0 < R < \infty$ with $\mathrm{spt}[T] \subset \mathbf{B}^{m+n}(0, R)$,

(e) $0 < N < \infty$.

The construction. (A) We set $\Pi = \{\mathbf{R}^m \times \{0\}\}\natural: \mathbf{R}^{m+n} \to \mathbf{R}^m$ and $S = \theta_\# T$.

(B) We let $A \subset \mathbf{R}^m$ denote the set of points $a \in \mathbf{R}^m$ for which

$$\mathbf{MS}[S \llcorner \Pi^{-1}\mathbf{B}^m(a, r)] \leqslant N\alpha(m)r^m$$

for *each* $0 < r < \infty$.

(C) We set $B = \mathbf{R}^m \sim A$ so that $b \in B$ implies the existence of $0 < r(b) < \infty$ (we pick and fix one such $r(b)$) with $\mathbf{MS}[S \llcorner \Pi^{-1}\mathbf{B}^m(b, r(b))] \geqslant N\alpha(m)r(b)^m$. The set B (which will contain the projection of our error set) has the following properties:

(C.1) B is open (obvious).

(C.2) $B \subset \mathbf{U}^m(0, R + [\mathbf{MS}(T)/N\alpha(m)]^{1/m})$; otherwise we could use Assumption (d) to infer the existence of $b \in B$ with $r(b)^m \geqslant \mathbf{MS}(T)/N\alpha(m)$ which is not possible.

(C.3) $\mathscr{L}^m(B) \leqslant N^{-1}\beta(m)\mathbf{MS}(T)$. To see this estimate we use the Besicovitch–Federer covering theorem to infer the existence of $b_1, b_2, b_3, \ldots \in B$ with $B \subset \bigcup_i \mathbf{B}^m(b_i, r(b_i))$ and

$$\beta(m) \geqslant \mathrm{card}\{i : i \in \{1, 2, 3, \ldots\} \text{ with } z \in \mathbf{B}^m(b_i, r(b_i))\}$$

for each $z \in \mathbf{R}^m$. Hence

$$\mathscr{L}^m(B) \leqslant \mathscr{L}^m \left[\bigcup_i \mathbf{B}^m(b_i, r(b_i)) \right] \leqslant \sum_i \alpha(m) r(b_i)^m$$

$$< N^{-1} \sum_i \mathbf{MS}\left[S \llcorner \Pi^{-1}\mathbf{B}^m(b_i, r(b_i)) \right] \leqslant N^{-1}\beta(m)\mathbf{MS}(T).$$

Property (C.1) guarantees for each $b \in B$ the existence of $a(b)$ (which we now pick and fix) with $|b - a(b)| = \mathrm{dist}(b, A)$ and for which the following conditions hold:

(C.4) For each $b \in B$,

$$|b - a(b)| \leqslant \left[N^{-1}\alpha(m)^{-1}\beta(m)\mathbf{MS}(T) \right]^{1/m}$$

as a consequence of (C.3) and the observation that $\mathbf{U}^m(b, |b - a(b)|) \subset B$ so that $\alpha(m)|b - a(b)|^m \leqslant \mathscr{L}^m(B)$.

(C.5) For each $b \in B$ and each $0 < r < \infty$,

$$\mathbf{B}^m(b, r) \subset \mathbf{B}^m\big(a(b), r + |b - a(b)|\big)$$

so that

$$\mathbf{MS}\big(S \llcorner \Pi^{-1}\mathbf{B}^m(b, r)\big) \leqslant N\alpha(m)\big(r + |b - a(b)|\big)^m$$

(since $a(b) \in A$).

(D) For each $k \in \{1, 2, 3, \ldots\}$ we let $\mathscr{F}(k) = \mathbf{WF}((\Pi \circ \theta)^{-1}B; k)$ and also $\mathscr{F} = \mathbf{WF}((\Pi \circ \theta)^{-1}B)$ be the admissible families of $m + n$ cubes constructed in 1.6. We note the following properties of $\mathscr{F}$ and $\{\mathscr{F}(k)\}_k$:

(D.1) $\bigcup \mathscr{F} = (\Pi \circ \theta)^{-1}B$ and $\bigcup \mathscr{F}(k) = \mathbf{R}^{m+n}$ for each $k = 1, 2, 3, \ldots$ (by 1.6).

(D.2) For each $k \in \{1, 2, 3, \ldots\}$, each $J \in \mathscr{F}(k)$, and each $K \in \mathscr{F}$ we infer from (C.3) and the fact that $\mathbf{B}^{m+n}(\mathrm{center}(K), 2^{-(1+\mathrm{level}(K))}) \subset K \subset \bigcup \mathscr{F}$ that

$$\alpha(m)2^{-m(1+\mathrm{level}(K))} \leqslant \beta(m)\mathbf{MS}(T)/N,$$

$$2^{-\mathrm{level}(K)} \leqslant 2\left[\beta(m)\mathbf{MS}(T)/N\alpha(m)\right]^{1/m},$$

$$2^{-\mathrm{level}(J)} \leqslant \sup\left\{2^{-k}, 2\left[\beta(m)\mathbf{MS}(T)/N\alpha(m)\right]^{1/m}\right\}.$$

(D.3) Suppose $a \in A$, $k \in \{1, 2, 3, \ldots\}$ and $J, K, L \in \mathscr{F}(k)$ with $a \in (\Pi \circ \theta)J$, $a \in (\Pi \circ \theta)K$, and $J \cap L \neq \varnothing$. Using the definitions in 1.5 and 1.6, we check that $k = \mathrm{level}(J) = \mathrm{level}(K)$ and $k - 1 \leqslant \mathrm{level}(L)$ so that

$$\theta(J \cup K \cup L) \subset \Pi^{-1}\mathbf{B}^m\big(a, 3(m + n)^{1/2}2^{-\mathrm{level}(K)}\big).$$

(D.4) Suppose $b \in B$ and $J, K, L \in \mathscr{F}$ with $b \in (\Pi \circ \theta)J$, $b \in (\Pi \circ \theta)K$, and $J \cap L \neq \varnothing$; then

$$\theta(J \cup K \cup L) \subset \Pi^{-1}\mathbf{B}^m\big(a(b), 15(m + n)2^{-\mathrm{level}(K)}\big).$$

We note

(i)
$$2^{-\mathrm{level}(J)} < 3(m + n)^{1/2}2^{-\mathrm{level}(K)}$$

by 1.6(∗), and check

$$|b - a(b)| + \sup\{\operatorname{diam}(K), \operatorname{diam}(J) + \operatorname{diam}(L)\}$$

$$\leqslant 6(m + n)^{1/2}2^{-\operatorname{level}(K)}$$

$$+ (m + n)^{1/2}\sup\{2^{-\operatorname{level}(K)}, 2^{-\operatorname{level}(J)} + 2^{-\operatorname{level}(L)}\}$$

$$\leqslant 6(m + n)^{1/2}2^{-\operatorname{level}(K)} + (m + n)^{1/2}\sup\{2^{-\operatorname{level}(K)}, 3 \cdot 2^{-\operatorname{level}(J)}\}$$

$$< 15(m + n)2^{-\operatorname{level}(K)}$$

by 1.6(∗), our estimate above, and the fact that $\mathscr{F}$ is admissible; we further infer

$$(\Pi \circ \theta)(J \cup K \cup L)$$

$$\subset \mathbf{B}^m(b, \sup\{\operatorname{diam}(K), \operatorname{diam}(J) + \operatorname{diam}(L)\})$$

$$\subset \mathbf{B}^m(a(b), |b - a(b)| + \sup\{\operatorname{diam}(K), \operatorname{diam}(J) + \operatorname{diam}(L)\})$$

from which our initially asserted inclusion follows.

(D.5) Suppose $b \in B$, $k \in \{1, 2, 3, \dots\}$, and $J, K, L \in \mathscr{F}(k)$ with $b \in (\Pi \circ \theta)J$, $b \in (\Pi \circ \theta)K$, and $J \cap L \neq \varnothing$; then

$$\theta(J \cup K \cup L) \subset \Pi^{-1}\mathbf{B}^m(a(b), 15(m + n)2^{-\operatorname{level}(K)}).$$

To see this we let J', $K' \in \mathscr{F}$ with $J' \subset J$, $K' \subset K$ so that $\operatorname{level}(J) = \inf\{k, \operatorname{level}(J')\}$, $\operatorname{level}(K) = \inf\{k, \operatorname{level}(K')\}$ in accordance with our definition in 1.6; we use (D.4)(∗) and 1.6(∗) to check $2^{-\operatorname{level}(J)} < 3(m + n)^{1/2}2^{-\operatorname{level}(K)}$, $|b - a(b)| \leqslant 6(m + n)^{1/2}2^{-\operatorname{level}(K)}$; the inclusion asserted above now follows in a manner similar to that of (D.4).

(E) We apply the deformation construction of 1.14 and the General Deformation Theorem 1.15 with $\mathscr{F}, T, P, Q, R$ there corresponding to $\mathscr{F}, T, P, Q, 0$ (since $\partial T = 0$), respectively, in the present context and $V(m + n), W(m + n)$, $X(m + n), Y(m + n), V, W, X, Y$ there all equal to 0; see Figure 3.16. From a review of the procedure of 1.14(2) we see that the various deformed currents and homotopy currents are effectively obtained from a sequence of (singular) deformation mappings $\sigma(K, q(K))$ and (singular) homotopies $\tau(K, q(K))$ associated with various cubes $K \in \mathbf{CX}(\mathscr{F})$; in particular, $q(K) \in \mathbf{PC}(K)$ is chosen to be suitable with respect to the various criteria of 1.12 and 1.13, $\sigma(K, q(K))$ corresponds to a projection of K onto ∂K from center $q(K)$, and $\tau(K, q(K))$ is a homotopy between $\sigma(K, q(K))$ and the identity mapping.

(F) For each $k \in \{1, 2, 3, \dots\}$ we again apply the deformation construction of 1.14 and the General Deformation Theorem 1.15 with $\mathscr{F}, T, P, Q, R$ there corresponding to $\mathscr{F}(k), T, P(k), Q(k), 0$ (since $\partial T = 0$), respectively, in the present context and $V(m + n), W(m + n), X(m + n), Y(m + n), V, W, X, Y$ all equal to 0. As in (E), various centers of projection $q(K) \in \mathbf{PC}(K)$ are chosen according to the relevant criteria of 1.12 and 1.13, and these choices alone determine the deformations $\sigma(K, q(K))$ and the homotopies $\tau(K, q(K))$. As an additional condition of our present construction, we require that whenever $k \in \{1, 2, 3, \dots\}, j \in \{m, m + 1, \dots, m + n\}, K \in \mathbf{CX}_j(\mathscr{F}) \cap \mathbf{CX}_j(\mathscr{F}(k))$, and the point $q(K)$ (already fixed in (E) above) is a suitable choice for a center of projection $q(K, k)$ (being selected) then we indeed choose $q(K, k) = q(K)$. As a consequence of this additional condition we have

(F.1) $\infty = \lim_{k \to \infty} \inf\{\text{level}(K): K \in \mathscr{F} \cup \mathscr{F}(k)$ with either $P \llcorner K \neq P(k) \llcorner K$ or $Q \llcorner K \neq Q(k) \llcorner K\}$.

(F.2) $0 = \lim_{k \to \infty} \mathbf{GS}(P, P(k))$. To see this we use 1.15(1) to write, for each k,

$$C(k) = \bigcup \{K: K \in \mathscr{F} \cup \mathscr{F}(k) \text{ with } Q \llcorner K \neq Q(k) \llcorner K\},$$

$$P - P(k) = (P - T) - (P(k) - T) = \partial Q - \partial Q(k) = \partial[Q - Q(k)]$$
$$= \partial[(Q \llcorner C(k)) - (Q(k) \llcorner C(k))],$$

and then use 1.15(1.7) and (F.1) in a straightforward way. We further conclude for each $k \in \{1, 2, 3, \dots\}$

(F.3) $P(k) \in \mathbf{S}_m(\mathscr{F}(k)) \subset \mathbf{P}_m(\mathbf{R}^{m+n})$ [resp. $\in \mathbf{I}_m(\mathscr{F}(k)) \subset \mathbf{IP}_m(\mathbf{R}^{m+n})$] with $\partial P(k) = 0$ so that, in particular, there is a function $r(k): \mathbf{CX}_m(\mathscr{F}(k)) \to \mathbf{R}$ [resp. $\to \mathbf{Z}$] so that (recall 1.11)

$$P(k) = \sum \{r(k)(K)\mathbf{t}(K, 1, \text{direction}(K)): K \in \mathscr{F}(k)\}.$$

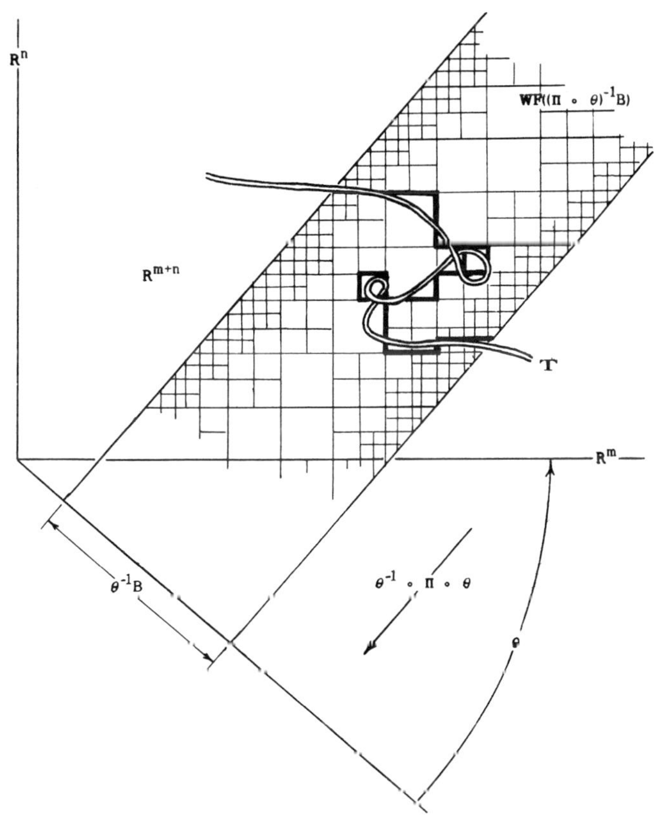

FIGURE 3.16. Deforming a current into a nonparametric multiple-valued graph.

(F.4) For each $a \in A$,

$$\mathrm{card}\{\mathbf{CX}_m(\mathscr{F}(k)) \cap \{K \colon a \in (\Pi \circ \theta)K \text{ with } r(k)(K) \neq 0\}\}$$
$$\leqslant N\left[\alpha(m)3^m(m+n)^{m/2}\Gamma_{1.15}\right]$$

and

$$\sum\{|r(k)(K)| \colon K \in \mathbf{CX}_m(\mathscr{F}(k)) \text{ with } a \in (\Pi \circ \theta)K\}$$
$$\leqslant N\left[\alpha(m)3^m(m+n)^{m/2}\Gamma_{1.15}\right].$$

To check these inequalities, we set

$$\mathbf{L} = \mathbf{CX}_m(\mathscr{F}(k)) \cap \{K \colon a \in (\Pi \circ \theta)K \text{ with } r(k)(K) \neq 0\},$$
$$\mathbf{NL} = \bigcup \{\mathrm{nbs}(L) \colon K \subset L \in \mathscr{F}(k) \text{ for some } K \in \mathbf{L}\},$$
$$v = \sup\{\mathrm{level}(K) \colon K \in \mathbf{L}\} \qquad (v = k \text{ here}).$$

Clearly,

$$\mathbf{S}\!\left(P(k) \llcorner \bigcup \mathbf{L}\right) \geqslant 2^{-mv}\,\mathrm{card}(\mathbf{L})$$

and

$$\mathbf{M}\!\left(P(k) \llcorner \bigcup \mathbf{L}\right) \geqslant 2^{-mv}\sum\{|r(k)(K)| \colon K \in \mathbf{L}\}.$$

We infer from 1.15(1.5) that

$$\mathbf{S}\!\left(P(k) \llcorner \bigcup \mathbf{L}\right) \leqslant \Gamma_{1.15}\mathbf{S}(T \llcorner \mathbf{NL})$$

and

$$\mathbf{M}\!\left(P(k) \llcorner \bigcup \mathbf{L}\right) \leqslant \Gamma_{1.15}\mathbf{M}(T \llcorner \mathbf{NL}).$$

We then use (D.3) to infer

$$\theta(\mathbf{NL}) \subset \Pi^{-1}\mathbf{B}^m\!\left(a, 3(m+n)^{1/2}2^{-v}\right)$$

and, finally, use (B) to conclude

$$\mathbf{MS}\!\left[T \llcorner (\Pi \circ \theta)^{-1}\mathbf{B}^m\!\left(a, 3(m+n)^{1/2}2^{-v}\right)\right] \leqslant N\alpha(m)3^m(m+n)^{m/2}2^{-mv}.$$

The asserted inequalities can now be readily checked.

(F.5) For each $b \in B$,

$$\mathrm{card}(\mathbf{CX}_m(\mathscr{F}(k)) \cap \{K \colon b \in (\Pi \circ \theta)K \text{ with } r(k)(K) \neq 0\})$$
$$\leqslant N\left[\alpha(m)15^m(m+n)^m\Gamma_{1.15}\right]$$

and

$$\sum\{|r(k)(K)| \colon K \in \mathbf{CX}_m(\mathscr{F}(k)) \text{ with } b \in (\Pi \circ \theta)K\}$$
$$\leqslant N\left[\alpha(m)15^m(m+n)^m\Gamma_{1.15}\right].$$

An argument similar to that of (F.4) but with (D.5) replacing (D.3), establishes the asserted inequalities.

(F.6) For each

$$c \in \mathbf{R}^m \sim \mathbf{U}^m\Big(0, R + (m + n)^{1/2}\sup\big\{2^{-k}, 2[\beta(m)\mathbf{MS}(T)/N\alpha(m)]^{1/m}\big\}\Big)$$

we infer from Hypothesis (d) and (D.2) that

$$0 = \mathrm{card}\big(\mathbf{CX}_m(\mathscr{F}(k)) \cap \{K: c \in (\Pi \circ \theta)K \text{ with } r(k)(K) \ne 0\}\big).$$

(G) For each $k \in \{1, 2, 3, \dots\}$ we define

$$U(k) = \mathbf{R}^m \sim \bigcup\{(\Pi \circ \theta)(\partial K): K \in \mathbf{CX}_m(\mathscr{F}(k)) \text{ with } r(k)(K) \ne 0\},$$

$$g(k) = \langle P(k), \Pi \circ \theta, \cdot \rangle: U(k) \to \mathbf{P}_0(\mathbf{R}^{m+n}) \quad [\text{resp. } \to \mathbf{IP}_0(\mathbf{R}^{m+n})]$$

(i.e., the 0-dimensional slices of $P(k)$ by the map $\Pi \circ \theta$), and

$$\mathrm{sign}: \mathbf{CX}_m(\mathscr{F}(k)) \to \{-1, +1\}$$

by requiring for each $K \in \mathbf{CX}_m(\mathscr{F}(k))$ that $\mathrm{sign}(K) = +1$ if and only if

$$0 < (e_1 \wedge \cdots \wedge e_m) \cdot \langle e_{\mathrm{direction}(K)}, \wedge_m \theta \rangle.$$

We then have the following:

(G.1) For each $k \in \{1, 2, 3, \dots\}$, $\mathbf{R}^m \sim U(k)$ is compact with

$$\mathscr{H}^{m-1}[\mathbf{R}^m \sim U(k)] < \infty.$$

(G.2) For each $k \in \{1, 2, 3, \dots\}$ and each $u \in U(k)$,

$$g(k)(u) = \sum \big\{\mathrm{sign}(K) \cdot r(k)(K)[\![p(K)]\!]: K \in \mathbf{CX}_m(\mathscr{F}(k)) \text{ with }$$

$$r(k)(K) \ne 0 \text{ and } \varnothing \ne (K \sim \partial K) \cap (\Pi \circ \theta)^{-1}\{u\} = \{p(K)\}\big\}$$

so that, in particular, (F.4) and (F.5) imply

$$\mathbf{S}(g(k)(u)) \le N\big[\alpha(m)15^m(m + n)^m\Gamma_{1.15}\big]$$

and

$$\mathbf{M}(g(k)(u)) \le N\big[\alpha(m)15^m(m + n)^m\Gamma_{1.15}\big];$$

additionally, Hypothesis (d), together with (D.2), implies

$$\mathrm{spt}[g(k)(u)]$$

$$\subset \mathbf{B}^{m+n}\Big(0, R + (m + n)^{1/2}\sup\big\{2^{-k}, 2[\beta(m)\mathbf{MS}(T)/N\alpha(m)]^{1/m}\big\}\Big).$$

(G.3) For each $k \in \{1, 2, 3, \dots\}$ and each $u, v \in U(k)$,

$$\mathbf{GS}(g(k)(u), g(k)(v))$$

$$\le N \cdot |u - v| \cdot (1 + \mu^2)^{1/2} \cdot \big[2\alpha(m)15^m(m + n)^m\Gamma_{1.15}\big]$$

$$[\text{resp. } \mathscr{G}(g(k)(u), g(k)(v))$$

$$\le N \cdot |u - v| \cdot (1 + \mu^2)^{1/2} \cdot \big[2\alpha(m)15^m(m + n)^m\Gamma_{1.15}\big]].$$

To see this, we assume $u \ne v$, set $\Sigma = \mathbf{R}^m \cap \{x: x \cdot (u - v) = 0\}$, slice $P(k)$ by the function $\{\Sigma\}_\natural \circ \Pi \circ \theta$, and use the estimates of the slicing theory given in [**FH**, 4.3.4, 4.3.8] (which are elementary in the present case) together with straightforward estimates based on Assumption (b), (G.2), and 3.14(5).

(G.4) For each $k \in \{1, 2, 3, \ldots\}$ there is a unique **GS** continuous function $f(k)$: $\mathbf{R}^m \to \mathbf{P}_0(\mathbf{R}^{m+n})$ [resp. $\to \mathbf{IP}_0(\mathbf{R}^{m+n})$] such that $f(k)|U(k) = g(k)$. Furthermore, the size, mass, support, and Lipschitz constant estimates of (G.2) and (G.3) hold with $f(k)$ replacing $g(k)$.

(G.5) As a consequence of (F.4), $f(k)(x) = 0$ for each $k \in \{1, 2, 3, \ldots\}$ and each

$$x \in \mathbf{R}^m \sim \mathbf{U}^m\big(0, R + (m + n)^{1/2} \sup\{2^{-k}, 2[\beta(m)\mathbf{MS}(T)/N\alpha(m)]^{1/m}\}\big).$$

(G.6) For each $k \in \{1, 2, 3, \ldots\}$ we review our construction (everything is polyhedral) to check

$$P(k) = f(k)_{\#}\mathbf{E}^m = f(k)_{\#}[\mathbf{E}^m \llcorner U(k)]$$
$$= f(k)_{\#}\Big[\mathbf{E}^m \llcorner \mathbf{U}^m\big(0, R + (m + n)^{1/2} \sup\{2^{-k}, 2[\beta(m)\mathbf{MS}(T)/N\alpha(m)]^{1/m}\}\big)\Big].$$

(H) We use 3.15 to construct a subsequence $i(1), i(2), i(3), \ldots$ of $1, 2, 3, \ldots$ and a function f: $\mathbf{R}^m \to \mathbf{P}_0(\mathbf{R}^{m+n})$ [resp. $\to \mathbf{IP}_0(\mathbf{R}^{m+n})$] such that

$$0 = \lim_{j \to \infty} \sup\{\mathbf{G}(f(x), f(i(j))(x)): x \in \mathbf{R}^m\}$$
$$\Big[\text{resp. } 0 = \lim_{j \to \infty} \sup\{\mathscr{G}(f(x), f(i(j))(x)): x \in \mathbf{R}^m\}\Big].$$

We then use 3.15(5) together with (F.2) and (G.6) to conclude

(H.1) $f_{\#}\mathbf{E}^m = P$.

Since the limit current P is independent of the subsequence, we use 3.14(6) to infer

(H.2) $0 = \lim_{k \to \infty} \sup\{\mathbf{G}(f(x), f(k)(x)): x \in \mathbf{R}^m\}$ [resp. $0 = \lim_{k \to \infty} \sup\{\mathscr{G}(f(x), f(k)(x)): x \in \mathbf{R}^m\}$].

Additionally, we infer from 3.15 and (G.4), (G.5) that

(H.3) For each $x, y \in \mathbf{R}^m$,

$$\mathbf{S}(f(x)) \leqslant N\big[\alpha(m)15^m(m + n)^m\Gamma_{1.15}\big],$$
$$\mathbf{M}(f(x)) \leqslant N\big[\alpha(m)15^m(m + n)^m\Gamma_{1.15}\big],$$
$$\mathrm{spt}[f(x)] \subset \mathbf{B}^{m+n}\big(0, R + 2[\beta(m)\mathbf{MS}(T)/N\alpha(m)]^{1/m}\big),$$
$$\mathbf{GS}(f(x), f(y)) \leqslant N \cdot |x - y| \cdot (1 + \mu^2)^{1/2}\big[2\alpha(m)15^m(m + n)^m\Gamma_{1.15}\big]$$
$$\Big[\text{resp. } \mathscr{G}(f(x), f(y)) \leqslant N \cdot |x - y| \cdot (1 + \mu^2)^{1/2}\big[2\alpha(m)15^m(m + n)^m\Gamma_{1.15}\big]\Big].$$

(H.4) $f(x) = 0$ for each

$$x \in \mathbf{R}^m \sim \mathbf{U}^m\big(0, R + 2[\beta(m)\mathbf{MS}(T)/N\alpha(m)]^{1/m}\big).$$

Since $T - f_{\#}\mathbf{E}^m = T - P = (T - f_{\#}\mathbf{E}^m) \llcorner \cup \mathscr{F}$ by 1.15(1.2) we conclude

(H.5) $T - f_{\#}\mathbf{E}^m = (T - f_{\#}\mathbf{E}^m) \llcorner (\Pi \circ \theta)^{-1}B$.

We can write

(H.6) $T = f_{\#}\mathbf{E}^m + (T - f_{\#}\mathbf{E}^m)$ where the "error current" $T - f_{\#}\mathbf{E}^m$ under the orthogonal projection $\Pi \circ \theta$ projects within the set B with

$$\mathscr{L}^m(B) \leqslant N^{-1}\beta(m)\mathbf{MS}(T),$$

by (C.3), which is small if N is large, e.g.,

$$\left(T - f_{\#}\mathbf{E}^m\right) \mathbin{\llcorner} \left[\mathbf{R}^{m+n} \sim \left(\Pi \circ \theta\right)^{-1}B\right] = 0.$$

(H.7) Furthermore,

$$\mathbf{MS}\left(f_{\#}\mathbf{E}^m\right) = \mathbf{MS}(P)$$

$$\leqslant \mathbf{MS}\left(T \mathbin{\llcorner} \left[\mathbf{R}^{m+n} \sim \left(\Pi \circ \theta\right)^{-1}B\right]\right) + \Gamma_{1.15}\mathbf{MS}\left(T \mathbin{\llcorner} \left(\Pi \circ \theta\right)^{-1}B\right)$$

$$\leqslant \Gamma_{1.15}\mathbf{MS}(T)$$

by 1.15(1.2)(1.5), and

(H.8) $\mathbf{MS}(T - f_{\#}\mathbf{E}^m) \leqslant (1 + \Gamma_{1.15})\mathbf{MS}(T \mathbin{\llcorner} (\Pi \circ \theta)^{-1}B)$.

As a special case, we check

(H.9) In case $B = \varnothing$ then $T = f_{\#}\mathbf{E}^m$ and, for each $x, y \in \mathbf{R}^m$,

$$\mathbf{S}(f(x)) \leqslant N, \qquad \mathbf{M}(f(x)) \leqslant N, \qquad \mathbf{GS}(f(x), f(y)) \leqslant 2N|x - y|.$$

3.17. Corollary (Condition for exact representation). *Suppose* $S \in \mathbf{S}_{m,0}(\mathbf{R}^{m+n})$ *[resp.* $\in \mathbf{I}_{m,0}(\mathbf{R}^{m+n})$*] with* $\partial S = 0$. *Then the following two conditions are equivalent:*

(1) $\infty > N = \sup\{(\alpha(m)r^m)^{-1}\mathbf{MS}(S \mathbin{\llcorner} \mathbf{B}^m(x, r) \times \mathbf{R}^n): x \in \mathbf{R}^m, 0 < r < \infty\}$.

(2) $S = (\llbracket \mathbf{1}_{\mathbf{R}^m} \rrbracket \bowtie g)_{\#}\mathbf{E}^m$ *for some* **MS** *bounded and* **GS** *Lipschitz function* g: $\mathbf{R}^m \to \mathbf{P}_0(\mathbf{R}^n) \cap \{P: \langle P, 1 \rangle = 0\}$ *[resp.* $\to \mathbf{IP}_0(\mathbf{R}^n) \cap \{P: \langle P, 1 \rangle = 0\}$*] with compact support and with* $\bigcup\{\mathrm{spt}[g(x)]: x \in \mathbf{R}^m\}$ *bounded.*

PROOF. In case $S \neq 0$ and Condition (1) holds we choose some $0 < \mu < \infty$ and $\theta \in \mathbf{O}(m + n)$ so that Assumption (b) of 3.16 holds for suitable $g(\lambda)$'s and apply 3.16 with $T = \theta_{\#}^{-1}S$ in Assumption (c) there and the present N as in Assumption (e) there to obtain $f: \mathbf{R}^m \to \mathbf{P}_m(\mathbf{R}^{m+n})$ as in (H) there with $f_{\#}\mathbf{E}^m = T$. We set $g = \{\{0\} \times \mathbf{R}^n\}_{\#} \circ \theta_{\#} \circ f$.

CHAPTER FOUR
APPLICATIONS OF MULTIPLE-VALUED FUNCTIONS

4.1. Theorem (Lipschitz multiple-valued approximation of a rectifiable cycle with error of small projected areas).

Hypotheses. (a) $1 \leqslant \mu < \infty$ *and* $v \in \{1, 2, 3, \dots\}$,

(b) $\theta(1), \dots, \theta(v) \in \mathbf{O}(m + n)$ *and* $\Pi = \{\mathbf{R}^m \times \{0\}\}_{\#} \in \mathbf{O}(m + n, m)$,

(c) *corresponding to each* $\lambda \in \Lambda(m + n, m)$ *and each* $j \in \{1, 2, \dots, v\}$ *there is a linear function* $g(\lambda, j): \mathbf{R}^m \to \mathbf{R}^n$ *with* $\|g(\lambda, j)\| \leqslant \mu$ *such that*

$$\mathrm{graph}[g(\lambda, j)] = \theta(j)^{-1}\left[\mathbf{R}^{m+n} \cap \{x: x_{*\lambda(1)} = x_{*\lambda(2)} = \cdots = x_{*\lambda(n)} = 0\}\right],$$

(d) $T \in \mathbf{S}_{m,0}(\mathbf{R}^{m+n})$ *[resp.* $\in \mathbf{I}_{m,0}(\mathbf{R}^{m+n})$*] with* $\partial T = 0$,

(e) $0 < R < \infty$ *with* $\mathrm{spt}(T) \subset \mathbf{B}^{m+n}(0, R)$,

(f) $0 < N < \infty$,

(g) $\Gamma_{4.1} = 4 \cdot 60^m(m + n)^m\alpha(m)\Gamma_{1.15}$.

Conclusion. There exist a **GS** [*resp.* $\mathcal{G}$] *Lipschitz function* $f\colon \mathbf{R}^m \to \mathbf{P}_0(\mathbf{R}^{m+n})$ [*resp.* $\to \mathbf{IP}_0(\mathbf{R}^{m+n})$] *together with an* (*error*) *current* $S \in \mathbf{S}_{m,0}(\mathbf{R}^{m+n})$ [*resp.* $\in \mathbf{I}_{m,0}(\mathbf{R}^{m+n})$] *with* $\partial S = 0$ *and an open subset* U *of* $\mathbf{R}^{m+n}$ *with the following properties*:

(1) $T = f_\# \mathbf{E}^m + S$.

(2) $S = S \llcorner U$ *and*

$$\mathscr{L}^m\big[(\Pi \circ \theta(j))U\big] \leqslant N^{-1}(1 + \Gamma_{1.15})^v \mathbf{MS}(T)$$

for each $j = 1,\ldots,v$.

(3) *For each* $x, y \in \mathbf{R}^m$,

$$\mathbf{S}(f(x)) \leqslant \Gamma_{4.1} v \mu^m N, \qquad \mathbf{M}(f(x)) \leqslant \Gamma_{4.1} v \mu^m N, \qquad \langle f(x), 1 \rangle = 0,$$

$$\operatorname{spt}[f(x)] \subset \mathbf{B}^{m+n}\big(0, R + v(m + n)^{1/2}2\big[\beta(m)(1 + \Gamma_{1.15})^v \mathbf{MS}(T)/N\alpha(m)\big]^{1/m}\big),$$

$$\mathbf{GS}(f(x), f(y)) \leqslant \Gamma_{4.1} v \mu^m N|x - y|$$

$$\big[resp.\ \mathcal{G}(f(x), f(y)) \leqslant \Gamma_{4.1} v \mu^m N|x - y|\big].$$

(4) $f(x) = 0$ *for each*

$$x \in \mathbf{R}^m \sim \mathbf{U}^m\big(0, R + v(m + n)^{1/2}2\big[\beta(m)(1 + \Gamma_{1.15})^v \mathbf{MS}(T)/N\alpha(m)\big]^{1/m}\big).$$

(5) $\mathbf{MS}(f_\# \mathbf{E}^m) \leqslant (1 + \Gamma_{4.1})^v \mathbf{MS}(T)$ *and* $\mathbf{MS}(S) \leqslant (1 + \Gamma_{4.1})^v \mathbf{MS}(T)$.

PROOF. Among the ingredients of our proof are

(a) Bounded open sets $B(k) \subset \mathbf{R}^m$ for $k = 1, 2,\ldots,v$,

(b) Whitney families of $m + n$ cubes

$$\mathscr{F}(k) = \mathbf{WF}\big((\Pi \circ \theta(1))^{-1} B(1) \cap (\Pi \circ \theta(2))^{-1} B(2)$$

$$\cap \cdots \cap (\Pi \circ \theta(k))^{-1} B(k)\big) \subset \mathbf{R}^{m+n}$$

for $k = 1, 2,\ldots,v$ defined in accordance with 1.6—these are not the same $\mathscr{F}(k)$'s as in 3.16(D),

(c) for each $k = 1, 2,\ldots,v$ a singular mapping $\sigma(k)\colon \mathbf{R}^{m+n} \to \mathbf{R}^{m+n}$ associated with the construction of 1.14 based on $\mathscr{F}(k)$ so that, in particular,

$$\sigma(k)\big[\bigcup \mathscr{F}(k) \cap \operatorname{dmn}(\sigma(k))\big] \subset \bigcup \mathbf{CX}_m(\mathscr{F}(k))$$

and

$$\sigma(k)(x) = x \quad \text{for each } x \in \bigcup \mathbf{CX}_m(\mathscr{F}(k)) \cup \big[\mathbf{R}^{m+n} \sim \bigcup \mathscr{F}(k)\big],$$

(d) **GS** [resp. $\mathcal{G}$] Lipschitz maps $f(k)\colon \mathbf{R}^m \to \mathbf{P}_0(\mathbf{R}^{m+n})$ [resp. $\to \mathbf{IP}_0(\mathbf{R}^{m+n})$] with $f = f(1) + f(2) + \cdots + f(v)$ in the conclusion of our theorem,

(e) currents $T(0) = T$ and, for each $k \in \{1, 2,\ldots,v\}$, $P(k), T(k) \in \mathbf{S}_{m,0}(\mathbf{R}^{m+n})$ [resp. $\in \mathbf{I}_{m,0}(\mathbf{R}^{m+n})$] with $\partial P(k) = \partial T(k) = 0$ such that

(e.1) $P(k) = \sigma(k)_\# T(k - 1) = f(k)_\# \mathbf{E}^m$ is an approximation of $T(k - 1)$, and

(e.2) $T(k) = T(k - 1) - P(k)$ is the error in the approximation of $T(k - 1)$ by $P(k)$.

We check

$$T = P(1) + [T(0) - P(1)] = P(1) + T(1)$$
$$= P(1) + P(2) + [T(1) - P(2)] = P(1) + P(2) + T(2)$$
$$= \cdots = P(1) + P(2) + \cdots + P(v) + T(v)$$
$$= f(1)_\sharp \mathbf{E}^m + f(2)_\sharp \mathbf{E}^m + \cdots + f(v)_\sharp \mathbf{E}^m + T(v) = f_\sharp \mathbf{E}^m + S$$

where we set $S = T(v)$ as the error current in the conclusion of our theorem.

The sequential construction of $B(1)$, $\mathcal{F}(1)$, $\sigma(1)$, $P(1)$, $T(1)$, $B(2)$, $\mathcal{F}(2)$, $\sigma(2)$, $P(2)$, $T(2)$, $B(3),\ldots,T(v)$ in the order indicated is specified by the following two conditions in addition to those given above. See Figure 4.1.

(f) For each $k \in \{1, 2,\ldots,v\}$, $B(k)$ consists of those points $b \in \mathbf{R}^{m+n}$ for which

$$\mathbf{MS}\big(T(k-1) \llcorner (\Pi \circ \theta(k))^{-1} \mathbf{B}^m(b, r)\big) > N\alpha(m)r^m$$

for some $0 < r < \infty$; compare 3.16(C).

(g) For each $k \in \{1, 2,\ldots,v\}$, $P(k)$ is the deformed image of $T(k-1)$ obtained by the procedure of 1.14 and 1.15; in particular, we apply 1.15 with T, $\mathcal{F}$, P there corresponding to $T(k-1), \mathcal{F}(k), P(k)$ in the present context and V, W, X, Y there all equal to 0.

The construction of and estimates on our mappings $f(1), f(2),\ldots,f(v)$ require that we adapt and record information about the particular deformations producing $P(1),\ldots,P(v)$. From a review of the procedure of 1.14 we see that the various

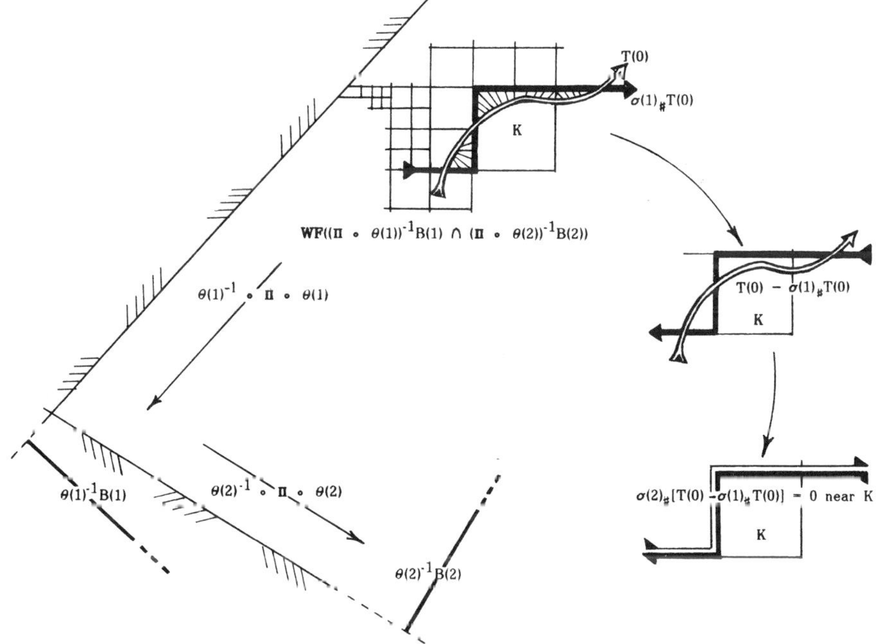

FIGURE 4.1. Multiple-valued representation of a current with error of small projected areas.

intermediate deformed currents are associated with a series of singular deformation mappings $\sigma(K, q(K, k))$ corresponding to various cubes $K \in \mathbf{CX}(\mathscr{F})$ for each $k \in \{1, \ldots, v\}$; in particular, $q(K, k) \in \mathbf{PC}(K) \subset K$ is chosen to be suitable with respect to the various criteria of 1.12 and 1.13, and $\sigma(K, q(K, k))$ corresponds to projection of K onto ∂K from center $q(K, k)$.

As an adaptation of 1.14 and 1.15 for our present construction of $P(1), \ldots, P(v)$ we require for each $k \in \{1, \ldots, v\}$ that the only deformation mappings $\sigma(K, q(K, k))$ which are utilized are those associated with cubes $K \in \mathbf{CX}(\mathscr{F})$ with $\dim(K) \in \{m + 1, m + 2, \ldots, m + n\}$. Since $\partial T(k - 1) = 0$, deformations $\sigma(K, q(K, k))$ with $\dim(K) = m$ would not affect the construction of $P(k)$ in any event; in this regard note that, in the terminology of 1.14(2), $T(m - 1) = T(m)$ because $R(m - 1) = 0$ and $Q(m - 1) = 0$.

As another condition on our deformations, we require that whenever $k \in \{2, 3, \ldots, v\}$, $j \in \{m + 1, m + 2, \ldots, m + n\}$, $K \in \mathbf{CX}_j(\mathscr{F}(k - 1)) \cap \mathbf{CX}_j(\mathscr{F}(k))$, and the point $q(K, k - 1)$ (already fixed) is a suitable choice for the point $q(K, k)$ (being selected), then we indeed take $q(K, k) = q(K, k - 1)$.

For each $k \in \{1, \ldots, v\}$, we compose the various singular deformations $\{\sigma(K, q(K, k))\}_K$ in an order consistent with that used in 1.14 to obtain a single singular mapping $\sigma(k): \mathbf{R}^{m+n} \to \mathbf{R}^{m+n}$ as in (c) above and write, in a symbolic way, $P(k) = \sigma(k)_{\#} T(k - 1)$, etc. We are using such $\sigma(k)$'s as a condensation of the fairly cumbersome notation of 1.14 in order to make the essential ideas of the present proof more transparent. We record several observations.

(i) Since we are using $1.14, 1.15$ with $V(m + n), W(m + n), X(m + n),$ $Y(m + n), V, W, X, Y$ all identically 0 and since $\partial T = 0$, the suitability requirements for choosing the various $q(K, k)$'s are determined solely by the behavior of integrals corresponding to those of 1.12(a) with $\dim(K) > m$ associated with various intermediate deformed images of $T(k - 1)$.

(ii) Suppose $k \in \{1, \ldots, v\}$, $j \in \{m + 1, \ldots, m + n\}$, and $K \in \mathbf{CX}_j(\mathscr{F}(k))$. Then that part of $T(k - 1)$ which has any influence on the conditions for suitability of a choice of $q(K) \in \mathbf{PC}(K)$ is not more than

$$T(k - 1) \, \llcorner \, \left[\bigcup \{ L : L \in \mathbf{CX}_{m+n}(\mathscr{F}(k)) \text{ with } K \subset L \} \sim \bigcup \mathbf{CX}_{j-1}(\mathscr{F}(k)) \right]$$

in accordance with the observations of 1.14(2).

(iii) Suppose $k \in \{1, \ldots, v\}$ and $J \in \mathbf{CX}_m(\mathscr{F}(k))$. Then, according to 1.8, there is $K \in \mathscr{F}(k)$ such that J is an m-dimensional face of K, and, according to 1.7, one (or both) of the following two possibilities occurs: **Either** there is $L \in \mathbf{WF}((\Pi \circ \theta)^{-1} B(k))$ with $K \subset L$ (possibly $K = L$) and $\mathrm{level}(K) \leqslant \mathrm{level}(L) + 1$, **or** $K \in \mathscr{F}(k - 1)$ and $\mathrm{level}(K) \geqslant \mathrm{level}(L) + 1$ for that unique $L \in \mathbf{WF}((\Pi \circ \theta)^{-1} B(k))$ with $K \subset L$; furthermore, in this case, $\mathscr{F}(k - 1) \cap \mathrm{nbs}(K) \subset \mathscr{F}(k)$ so that, in particular, $J \in \mathbf{CX}_m(\mathscr{F}(k - 1))$ and

$$\bigcup \{ H : H \in \mathscr{F}(k - 1) \text{ with } H \cap J \neq \varnothing \}$$
$$= \bigcup \{ H : H \in \mathscr{F}(k) \text{ with } H \cap J \neq \varnothing \}.$$

(iv) For each $k \in \{1,\dots,v\}$,

$$\bigcup \mathbf{CX}_m(\mathscr{F}(k)) \cap \bigcup \mathscr{F}(k+1) \subset \bigcup \mathbf{CX}_m(\mathscr{F}(k+1))$$

by the remarks in 1.8 and (b) above.

(v) For each $k \in \{1, 2,\dots,v-1\}$,

$$\sigma(k+1)_\# \left\{ \sigma(k)_\# \left[T(0) \, \llcorner \, \bigcup \mathscr{F}(k) \right] \right\} = \sigma(k)_\# \left[T(0) \, \llcorner \, \bigcup \mathscr{F}(k) \right],$$

with the obvious meanings, provided $\sigma(k)_\#[T(0) \, \llcorner \, \bigcup \mathscr{F}(k)]$ is well defined. This is because

$$0 = \left\{ \sigma(k)_\# \left[T(0) \, \llcorner \, \bigcup \mathscr{F}(k) \right] \right\} \, \llcorner \, \left[\mathbf{R}^{m+n} \sim \bigcup \mathbf{CX}_m(\mathscr{F}(k)) \right] \quad \text{(by } 1.15(1.2)(1.3)),$$

$$\bigcup \mathbf{CX}_m(\mathscr{F}(k)) \cap \bigcup \mathscr{F}(k+1) \subset \mathbf{CX}_m(\mathscr{F}(k+1)) \qquad \text{(by (iv))},$$

and

$$\sigma(k+1)(x) = x \quad \text{for each } x \in \bigcup \mathbf{CX}_m(\mathscr{F}(k+1)) \cup \left[\mathbf{R}^{m+n} \sim \bigcup \mathscr{F}(k+1) \right]$$

(in accordance with (c) above).

(vi) The conditions for suitability of the various $q(K,1)$'s clearly depend only on $T(0) \, \llcorner \, \bigcup \mathscr{F}(1)$. According to (e.1) and (g), $P(1) = \sigma(1)_\# T(0)$ so that

$$T(1) = T(0) - P(1) = T(0) - \sigma(1)_\# T(0)$$

$$= \{ T(0) - \sigma(1)_\# \} \, \llcorner \, \bigcup \mathscr{F}(1) \quad \text{(by (c); recall } 1.15(1.2)(1.3))$$

$$= T(0) \, \llcorner \, \bigcup \mathscr{F}(1) - \sigma(1)_\# \left[T(0) \, \llcorner \, \bigcup \mathscr{F}(1) \right] \quad \text{(by (c) again)}.$$

As a consequence of (iv) and (v),

$$\left[T(1) \, \llcorner \, \bigcup \mathscr{F}(2) \right] \, \llcorner \, \left[\mathbf{R}^{m+n} \sim \bigcup \mathbf{CX}_m(\mathscr{F}(2)) \right]$$

$$= \left[T(0) \, \llcorner \, \bigcup \mathscr{F}(2) \right] \, \llcorner \, \left[\mathbf{R}^{m+n} \sim \bigcup \mathbf{CX}_m(\mathscr{F}(2)) \right].$$

As a further consequence of (ii), we infer that the conditions for suitability of the various $q(K,2)$'s depend only on $T(0) \, \llcorner \, \bigcup \mathscr{F}(2)$; in particular, $\sigma(2)_\#[T(0) \, \llcorner \, \bigcup \mathscr{F}(2)]$ is well defined.

(vii)

$$P(2) = \sigma(2)_\# T(1) \quad \text{(by (e.1) and (g))}$$

$$= \sigma(2)_\# \left[T(0) \, \llcorner \, \bigcup \mathscr{F}(1) \right] - \sigma(2)_\# \circ \sigma(1)_\# \left[T(0) \, \llcorner \, \bigcup \mathscr{F}(1) \right]$$

$$\text{(with existence and equality following from (vi))}$$

$$= \sigma(2)_\# \left[T(0) \, \llcorner \, \bigcup \mathscr{F}(1) \right] - \sigma(1)_\# \left[T(0) \, \llcorner \, \bigcup \mathscr{F}(1) \right] \quad \text{(by (v) and (vi))}.$$

(viii) Suppose $J \in \mathbf{CX}_m(\mathscr{F}(2))$. We combine (ii), (iii), (vi), (vii) to infer that one of the following two possibilities occurs: **Either** there is $L \in \mathbf{WF}((\Pi \circ \theta(2))^{-1} B(2))$ with $J \subset L$ and $\mathrm{level}(J) \leqslant \mathrm{level}(L) + 1$, or there is $K \in \mathscr{F}(1) \cap \mathscr{F}(2)$ such that J is an m-dimensional face of K, $\mathscr{F}(1) \cap \mathrm{nbs}(K) \subset \mathscr{F}(2)$, and

$$\bigcup \{ H \colon H \in \mathscr{F}(1) \text{ with } H \cap J \neq \varnothing \} = \bigcup \{ H \colon H \in \mathscr{F}(2) \text{ with } H \cap J \neq \varnothing \},$$

so that, in particular,

$$\left\{ \sigma(2)_\# \left[T(0) \, \llcorner \, \bigcup \mathscr{F}(1) \right] \right\} \, \llcorner \, J = \left\{ \sigma(1)_\# \left[T(0) \, \llcorner \, \bigcup \mathscr{F}(1) \right] \right\} \, \llcorner \, J$$

(by the suitability observations in (vi) and the compatibility requirement in (g) relating our choices of $q(K, 1)$'s and $q(K, 2)$'s, which implies $P(2) \mathop{\llcorner} J = 0$ (by (vii)).

(ix)

$$T(2) = T(1) - P(2) \quad \text{(by (e.2))}$$

$$= T(0) \mathop{\llcorner} \bigcup \mathscr{F}(1) - \sigma(1)_{\#}\Big[T(0) \mathop{\llcorner} \bigcup \mathscr{F}(1)\Big]$$

$$-\sigma(2)_{\#}\Big[T(0) \mathop{\llcorner} \bigcup \mathscr{F}(1)\Big] + \sigma(1)_{\#}\Big[T(0) \mathop{\llcorner} \bigcup \mathscr{F}(1)\Big] \quad \text{(by (vi) and (vii))}$$

$$= T(0) \mathop{\llcorner} \bigcup \mathscr{F}(1) - \sigma(2)_{\#}\Big[T(0) \mathop{\llcorner} \bigcup \mathscr{F}(1)\Big]$$

$$= \Big\{ T(0) \mathop{\llcorner} \bigcup \mathscr{F}(1) - \sigma(2)_{\#}\Big[T(0) \mathop{\llcorner} \bigcup \mathscr{F}(1)\Big]\Big\} \mathop{\llcorner} \bigcup \mathscr{F}(2) \quad \text{(by (c))}$$

$$= T(0) \mathop{\llcorner} \bigcup \mathscr{F}(2) - \sigma(2)_{\#}\Big[T(0) \mathop{\llcorner} \bigcup \mathscr{F}(2)\Big] \quad \text{(by (c) again)}.$$

As a consequence of (iv) and (v),

$$\Big[T(2) \mathop{\llcorner} \bigcup \mathscr{F}(3)\Big] \mathop{\llcorner} \Big[\mathbf{R}^{m+n} \sim \bigcup \mathbf{CX}_m(\mathscr{F}(3))\Big]$$

$$= \Big[T(0) \mathop{\llcorner} \bigcup \mathscr{F}(3)\Big] \mathop{\llcorner} \Big[\mathbf{R}^{m+n} \sim \bigcup \mathbf{CX}_m(\mathscr{F}(3))\Big].$$

As a further consequence of (ii), we infer that the conditions for suitability of the various $q(k, 3)$'s depend only on $T(0) \mathop{\llcorner} \bigcup \mathscr{F}(3)$ so that, in particular, $\sigma(3)_{\#}[T(0) \mathop{\llcorner} \bigcup \mathscr{F}(3)]$ and $\sigma(3)_{\#}[T(0) \mathop{\llcorner} \bigcup \mathscr{F}(2)]$ are well defined.

(x)

$$P(3) = \sigma(3)_{\#}T(2) \quad \text{(by (e.1) and (g))}$$

$$= \sigma(3)_{\#}\Big[T(0) \mathop{\llcorner} \bigcup \mathscr{F}(2)\Big] - \sigma(3)_{\#} \circ \sigma(2)_{\#}\Big[T(0) \mathop{\llcorner} \bigcup \mathscr{F}(2)\Big] \quad \text{(by (ix))}$$

$$= \sigma(3)_{\#}\Big[T(0) \mathop{\llcorner} \bigcup \mathscr{F}(2)\Big] - \sigma(2)_{\#}\Big[T(0) \mathop{\llcorner} \bigcup \mathscr{F}(2)\Big] \quad \text{(by (v) and (ix))}.$$

We assert the following:

(xi) Suppose $k \in \{1, 2, \ldots, v\}$ and $J \in \mathbf{CX}_m(\mathscr{F}(k))$. Then: **Either** there is $L \in \mathbf{WF}((\Pi \circ \theta(k))^{-1}B(k))$ with $J \subset L$ and $\operatorname{level}(J) \leqslant \operatorname{level}(K) + 1$, **or** $P(k) \mathop{\llcorner} J = 0$. Additionally, $S = T(v) = T(0) \mathop{\llcorner} \bigcup \mathscr{F}(v) - \sigma(v)_{\#}[T(0) \mathop{\llcorner} \bigcup \mathscr{F}(v)]$.

Proof of Assertion (xi). There is nothing to prove for $k = 1$ since $\mathscr{F}(1) = \mathbf{WF}((\Pi \circ \theta(1))^{-1}B(1))$. For $k = 2$ our assertion was checked in (viii). For $k = 3, 4, \ldots, v$ we will prove our assertion by an inductive argument.

For $k \in \{2, 3, \ldots, v\}$ we denote by $\mathbf{A}_k, \mathbf{B}_k, \mathbf{C}_k$ the following statements.

$\mathbf{A}_k$: The conditions for suitability of the various $q(K, k - 1)$'s depend only on $T(0) \mathop{\llcorner} \bigcup \mathscr{F}(k - 1)$ and the conditions for suitability of the various $q(K, k)$'s depend only on $T(0) \mathop{\llcorner} \bigcup \mathscr{F}(k)$. As a particular consequence,

$$\sigma(k - 1)_{\#}\Big[T(0) \mathop{\llcorner} \bigcup \mathscr{F}(k - 2)\Big], \qquad \sigma(k - 1)_{\#}\Big[T(0) \mathop{\llcorner} \bigcup \mathscr{F}(k - 1)\Big],$$

$$\sigma(k)_{\#}\Big[T(0) \mathop{\llcorner} \mathscr{F}(k - 1)\Big], \quad \text{and} \quad \sigma(k)_{\#}\Big[T(0) \mathop{\llcorner} \bigcup \mathscr{F}(k)\Big]$$

are all well defined.

$\mathbf{B}_k$: $T(k) = T(0) \mathop{\llcorner} \bigcup \mathscr{F}(k) - \sigma(k)_{\#}[T(0) \mathop{\llcorner} \bigcup \mathscr{F}(k)]$.

$\mathbf{C}_k$: $P(k) = \sigma(k)_{\#}[T(0) \mathop{\llcorner} \bigcup \mathscr{F}(k - 1)] - \sigma(k - 1)_{\#}[T(0) \mathop{\llcorner} \bigcup \mathscr{F}(k - 1)]$.

Assuming $k \in \{3, 4, \ldots, v\}$ and statements A_k, B_{k-1}, and C_k hold we conclude the following:

(1)

$$T(k) = T(k-1) - P(k) \quad (\text{by (e.1)})$$

$$= T(0) \, \llcorner \, \bigcup \mathscr{F}(k-1) - \sigma(k-1)_\# \big[T(0) \, \llcorner \, \bigcup \mathscr{F}(k-1)\big]$$

$$-\sigma(k)_\# \big[T(0) \, \llcorner \, \bigcup \mathscr{F}(k-1)\big] + \sigma(k-1)_\# \big[T(0) \, \llcorner \, \bigcup \mathscr{F}(k-1)\big]$$

$$(\text{by } B_{k-1} \text{ and } C_k)$$

$$= T(0) \, \llcorner \, \bigcup \mathscr{F}(k-1) - \sigma(k)_\# \big[T(0) \, \llcorner \, \bigcup \mathscr{F}(k-1)\big]$$

$$= \big\{ T(0) \, \llcorner \, \bigcup \mathscr{F}(k-1) - \sigma(k)_\# \big[T(0) \, \llcorner \, \bigcup \mathscr{F}(k-1)\big] \big\} \, \llcorner \, \bigcup \mathscr{F}(k)$$

$$(\text{by (c); recall } 1.15(1.2)(1.3))$$

$$= T(0) \, \llcorner \, \bigcup \mathscr{F}(k) - \sigma(k)_\# \big[T(0) \, \llcorner \, \bigcup \mathscr{F}(k)\big] \quad (\text{by (c) again}).$$

This is statement B_k.

In case $k \leqslant v - 1$, we then use (iv) and (v) above to conclude

$$\big[T(k) \, \llcorner \, \bigcup \mathscr{F}(k-1)\big] \, \llcorner \, \big[\mathbf{R}^{m+n} \sim \bigcup \mathbf{CX}_m(\mathscr{F}(k+1))\big]$$

$$= \big[T(0) \, \llcorner \, \bigcup \mathscr{F}(k+1)\big] \, \llcorner \, \big[\mathbf{R}^{m+n} \sim \bigcup \mathbf{CX}_m(\mathscr{F}(k+1))\big].$$

Additionally, in case $k \leqslant v - 1$, we use (ii) to infer that the conditions for suitability of the various $q(K, k+1)$'s depend only on $T(0) \, \llcorner \, \bigcup \mathscr{F}(k+1)$ so that, in particular, $\sigma(k+1)_\# [T(0) \, \llcorner \, \bigcup \mathscr{F}(k)]$ and $\sigma(k+1)_\# [T(0) \, \llcorner \, \bigcup \mathscr{F}(k+1)]$ are well defined. In particular, in case $k \leqslant v - 1$, then statement A_{k+1} holds.

(2) In case $k \leqslant v - 1$, we estimate

$$P(k+1) - \sigma(k+1)_\# T(k) \quad (\text{by (e.1) and (g)})$$

$$= \sigma(k+1)_\# \big[T(0) \, \llcorner \, \bigcup \mathscr{F}(k)\big] - \sigma(k+1)_\# \circ \sigma(k)_\# \big[T(0) \, \llcorner \, \bigcup \mathscr{F}(k)\big]$$

$$(\text{by } B_k \text{ established in (1) above})$$

$$= \sigma(k+1)_\# \big[T(0) \, \llcorner \, \bigcup \mathscr{F}(k)\big] - \sigma(k)_\# [T(0) \, \llcorner \, \mathscr{F}(k)]$$

$$(\text{by (v) and } A_{k+1} \text{ established in (1) above}).$$

In particular, if $k \leqslant v - 1$, then statement C_{k+1} holds.

(3) Suppose $k \in \{2, \ldots, v\}$ and $J \in \mathbf{CX}_m(\mathscr{F}(k))$. We combine (ii) and (iii) to infer that one of the following two possibilities occurs: **Either** there is $L \in \mathbf{WF}((\Pi \circ \theta(k))^{-1} B(k))$ with $J \subset L$ and $\text{level}(J) \leqslant \text{level}(L) + 1$, **or** there is $K \in \mathscr{F}(k-1) \cap \mathscr{F}(k)$ such that J is an m-dimensional face of K, $\mathscr{F}(k-1) \cap \text{nbs}(K) \subset \mathscr{F}(k)$, and

$$\bigcup \{ H \colon H \in \mathscr{F}(k-1) \text{ with } H \cap J \neq \varnothing \}$$

$$= \bigcup \{ H \colon H \in \mathscr{F}(k) \text{ with } H \cap J \neq \varnothing \}$$

so that, in particular,

$$\big\{ \sigma(k)_\# \big[T(0) \, \llcorner \, \bigcup \mathscr{F}(k-1)\big] \big\} \, \llcorner \, J - \big\{ \sigma(k-1)_\# \big[T(0) \, \llcorner \, \bigcup \mathscr{F}(k-1)\big] \big\} \, \llcorner \, J$$

(by A_k and the compatibility requirement in (g) relating our choices of $q(K, k-1)$'s and $q(K, k)$'s) which implies $P(k) \mathbin{\llcorner} J = 0$.

We note that (vi) and (ix) above imply A_3, that (ix) above implies B_2, and that (x) above implies C_3. We then use (1), (2), (3) above inductively to conclude that A_k, B_k, C_k all hold for $k = 3, 4, \ldots, v$. Assertion (xi) now follows from (3) for $k = 3, 4, \ldots, v$ together with B_v. This finishes the proof of assertion (xi).

We record some further observations and estimates.

(xii) We use (e.1), (e.2), and 1.15(1.2)(1.5) to estimate for each $k \in \{1, 2, \ldots, v\}$,

$$\mathbf{MS}(P(k)) \leqslant \Gamma_{1.15}\mathbf{MS}(T(k-1)),$$

$$\mathbf{MS}(T(k)) \leqslant \mathbf{MS}(T(k-1)) + \mathbf{MS}(P(k)) \leqslant (1 + \Gamma_{1.15})^v\mathbf{MS}(T),$$

$$\mathbf{MS}(P(k)) \leqslant (1 + \Gamma_{1.15})^v\mathbf{MS}(T).$$

We then estimate as in 3.16(C.3) to conclude

$$\mathscr{L}^m(B(k)) \leqslant N^{-1}(1 + \Gamma_{1.15})^v\mathbf{MS}(T)$$

for each $k \in \{1, 2, \ldots, v\}$.

(xiii) Suppose $k \in \{1, \ldots, v\}$, $b \in B(k)$, and $a(b) \in \mathbf{R}^m \sim B(k)$ with $|b - a(b)| = \operatorname{dist}(b, \mathbf{R}^m \sim B(k))$ (compare 3.16(C)). We combine (xi) with 3.16(D.4) applied in the obvious way to conclude the following: Let $J, K \in \mathbf{CX}_m(\mathscr{F}(k))$ with $P(k) \mathbin{\llcorner} J \neq 0$, $P(k) \mathbin{\llcorner} K \neq 0$, and with $b \in (\Pi \circ \theta(k))J$, $b \in (\Pi \circ \theta(k))K$, and let $L \in \mathscr{F}(k)$ with $J \cap L \neq \varnothing$. Then

$$\theta(k)(J \cup K \cup L) \subset \Pi^{-1}\mathbf{B}^m\bigl(a(b), 30(m + n)2^{-\operatorname{level}(K)}\bigr).$$

(xiv) Suppose $k \in \{1, \ldots, v\}$. Since $P(k) \mathbin{\llcorner} \bigcup\mathscr{F}(k) \in \mathscr{S}_m(\mathscr{F}(k))$ [resp. $\in \mathscr{I}_m(\mathscr{F}(k))$] (by 1.15(1.3)) there is a function $r(k)$: $\mathbf{CX}_m(\mathscr{F}(k)) \to \mathbf{R}$ [resp. $\to \mathbf{Z}$] so that (recall 1.11)

$$P(k) \mathbin{\llcorner} \bigcup\mathscr{F}(k) = \sum \{r(k)(K) \cdot \mathbf{t}(K, 1, \operatorname{direction}(K)): K \in \mathbf{CX}_m(\mathscr{F}(k))\}.$$

We adapt the estimate of 3.16(F.5) in the obvious way utilizing (xiii) above in place of 3.16(D.4) to conclude for each $b \in B(k)$ that

$$\operatorname{card}(\mathbf{CX}_m(\mathscr{F}(k)) \cap \{K: b \in (\Pi \circ \theta(k))K \text{ with } r(k)(K) \neq 0\})$$

$$\leqslant N\bigl[\alpha(m)30^m(m + n)^m\Gamma_{1.15}\bigr]$$

and

$$\sum \{|r(k)(K)|: K \in \mathbf{CX}_m(\mathscr{F}(k)) \text{ with } b \in (\Pi \circ \theta(k))K\}$$

$$\leqslant N\bigl[\alpha(m)30^m(m + n)^m\Gamma_{1.15}\bigr].$$

(xv) Suppose $k \in \{1, \ldots, v\}$. We combine the estimates of (xiv) together with Hypothesis (c) to conclude

$$\mathbf{MS}\bigl[P(k) \mathbin{\llcorner} \theta(k)^{-1}(\mathbf{B}^m(b, r) \times \mathbf{R}^n)\bigr]$$

$$\leqslant N(1 + \mu^2)^{m/2}\bigl[2\alpha(m)30^m(m + n)^m\Gamma_{1.15}\bigr]\alpha(m)r^m$$

whenever $b \in B(k)$, $0 < r < \infty$, and $\mathbf{B}^m(b, r) \subset B(k)$ (recall $B(k)$ is open). We then use the Besicovitch–Federer covering theorem in a straightforward way together with the definition of $B(k)$ (note the definition of A in 3.16(B)) to conclude that

$$\mathbf{MS}\left[P(k) \llcorner (\Pi \circ \theta(k))^{-1}\mathbf{B}^m(x, r)\right]$$
$$\leqslant N\left\{ 1 + (1 + \mu^2)^{m/2}\left[2\alpha(m)30^m(m + n)^m\Gamma_{1.15}\right]\right\}\alpha(m)r^m$$

for each $x \in \mathbf{R}^m$ and each $0 < r < \infty$.

(xvi) We combine the final estimate of (xv) with the special case (H.9) of 3.16 (with N replaced by N times $\Gamma = 1 + (1 + \mu^2)^{m/2}[2\alpha(m)30^m(m + n)^m\Gamma_{1.15}]$) to conclude the existence of $f(k)\colon \mathbf{R}^m \to \mathbf{P}_0(\mathbf{R}^{m+n})$ [resp. $\to \mathbf{IP}_0(\mathbf{R}^{m+n})$] with $P(k) = f(k)_\# \mathbf{E}^m$ and with the property that, for each $x, y \in \mathbf{R}^m$,

$$\mathbf{S}(f(k)(x)) \leqslant \Gamma N, \qquad \mathbf{M}(f(k)(x)) \leqslant \Gamma N,$$
$$\mathbf{GS}(f(k)(x), f(k)(y)) \leqslant 2\Gamma N|x - y|$$
$$\left[\text{resp. } \mathscr{G}(f(k)(x), f(k)(y)) \leqslant 2\Gamma N|x - y|\right].$$

Conclusion (1) follows by construction as noted in (e). Conclusion (2) follows from (b), (e), (xi), (xii) and 1.15(1.2)(1.3). Conclusion (5) follows from (xii). The support estimates of Conclusions (3) and (4) follow from Conclusion (5) combined with the observation that

$$\text{spt}\left[\sigma(k)_\# T(k - 1)\right]$$
$$\subset \left\{ x\colon \mathbf{EM}(x, \text{spt}[T(k - 1)]) < \sup\left\{2^{-\text{level}(K)}\colon K \in \mathscr{F}(k)\right\}\right\};$$

recall 3.16(D.2). The estimate $\langle f(x), 1\rangle = 0$ in Conclusion (3) follows from the definitions in 3.13, 3.14(3)(6), and the fact that each $P(k)$ is a cycle (which implies that $(\Pi \circ \theta(k))_\# P(k) = 0$ and hence $(\Pi \circ \theta(k))_\# \circ f(k)_\# \mathbf{E}^m = 0$).

4.2. Compactness Theorem for size bounded rectifiable cycles with bounded supports.

Hypotheses. (a) $0 < M < \infty$,

(b) $T_1, T_2, T_3, \ldots \in \mathbf{S}_{m,0}(\mathbf{R}^{m+n})$ [resp. $\in \mathbf{I}_{m,0}(\mathbf{R}^{m+n})$] *with* $\partial T_i = 0$ *for each* $i = 1, 2, 3, \ldots$,

(c) $\text{spt}[T_i] \subset \mathbf{B}^{m+n}(0, M)$ *for each* $i = 1, 2, 3, \ldots$,

(d) $\mathbf{MS}(T_i) \leqslant M$ *for each* $i = 1, 2, 3, \ldots$.

Conclusion. There exists a subsequence $i(1), i(2), i(3), \ldots$ *of* $1, 2, 3, \ldots$ *and* $T \in \mathbf{S}_{m,0}(\mathbf{R}^{m+n})$ [*resp.* $\in \mathbf{I}_{m,0}(\mathbf{R}^{m+n})$] *with the following properties:*

(1) $\lim_{j \to \infty}\mathbf{G}(T, T_{i(j)}) = 0$ [*resp.* $\lim_{j \to \infty}\mathscr{G}(T, T_{i(j)}) = 0$],

(2) $\text{spt}(T) \subset \mathbf{R}^{m+n}(0, M)$,

(3) $M(T) \leqslant \liminf_{j \to \infty}\mathbf{M}(T_{i(j)})$ *and* $\mathbf{S}(T) \leqslant \liminf_{j \to \infty}\mathbf{S}(T_{i(j)})$.

PROOF. The proof is in five parts. For notational simplicity we will indicate only the case in which $T_1, T_2, T_3, \ldots \in \mathbf{S}_{m,0}(\mathbf{R}^{m+n})$ (not $\in \mathbf{I}_{m,0}(\mathbf{R}^{m+n})$).

Part 1. We let $\omega(1), \omega(2), \ldots, \omega(N_{2.7}) \in \mathbf{O}(m + n)$ be as in 2.7(5) and assert the existence of a constant $1 < \Gamma < \infty$ (which we now fix) with the following property: Corresponding to each $0 < N < \infty$ and each $i \in \{1, 2, 3, \ldots\}$ there is a

GS Lipschitz function $f\colon \mathbf{R}^m \to \mathbf{P}_0(\mathbf{R}^{m+n})$ together with a current $S \in \mathbf{S}_{m,0}(\mathbf{R}^{m+n})$ with $\partial S = 0$ and an open subset U of $\mathbf{R}^{m+n}$ with the following properties:

(1.1) $T_i = f_{\#}\mathbf{E}^m + S$,

(1.2) $U \subset \mathbf{U}^{m+n}(0, \Gamma)$, $S = S \,\llcorner\, U$, and $\mathscr{L}^m[(\Pi(\lambda) \circ \omega(j))U] \leqslant N^{-1}\Gamma$ for each $\lambda \in \Lambda(m + n, m)$ and each $j \in \{1, 2, \dots, N_{2.7}\}$,

(1.3) for each x, $y \in \mathbf{R}^m$, $\mathbf{S}(f(x)) \leqslant N\Gamma$, $\mathbf{M}(f(x)) \leqslant N\Gamma$, $\langle f(x), 1 \rangle = 0$, $\mathrm{spt}[f(x)] \subset \mathbf{B}^{m+n}(0, \Gamma)$, and $\mathbf{GS}(f(x), f(y)) \leqslant N|x - y|\Gamma$,

(1.4) $f(x) = 0$ for each $x \in \mathbf{R}^m \sim \mathbf{U}^m(0, \Gamma)$,

(1.5) $\mathbf{MS}(f_{\#}\mathbf{E}^m) \leqslant \Gamma$ and $\mathbf{MS}(S) \leqslant \Gamma$.

To check this assertion, we let $\theta(\lambda)$, $\sigma \in \mathbf{O}(m + n)$, and $1 < v < \infty$ be as in 2.7(6) and then apply 4.1 with μ, v, $\{\theta(j)\}_j$, T, N, f, S, U there corresponding to v, $\binom{m+n}{m}N_{2.7}$, $\{\theta(\lambda) \circ \omega(j) \circ \sigma\}_{\lambda, j}$, $\sigma_{\#}^{-1}T_i$, N, f_1, S_1, U_1, respectively, in the present context. Finally, we set $f = \sigma_{\#} \circ f_1$, $S = \sigma_{\#}S_1$, $U = \sigma(U_1)$. Γ, of course, depends on $N_{2.7}$ and v from 2.7; these numbers, however, are fixed in the present theorem.

Part 2. For each i, $v \in \{1, 2, 3, \dots\}$ we use Part 1 with T_i and N there corresponding to T_i and 2^v here to obtain $g_i(v)\colon \mathbf{R}^m \to \mathbf{P}_0(\mathbf{R}^{m+n})$, $S_i(v)$, $U_i(v)$ (corresponding to f, S, U there) such that

(2.1) $T_i = g_i(v)_{\#}\mathbf{E}^m + S_i$,

(2.2) $S_i(v) = S_i(v) \,\llcorner\, U_i(v)$ and $\mathscr{L}^m[(\Pi(\lambda) \circ \omega(j))U_i(v)] \leqslant 2^{-v}\Gamma$ for each $\lambda \in \Lambda(m + n, m)$ and each $j \in \{1, \dots, N_{2.7}\}$,

(2.3) $\mathbf{MS}(g_i(v)_{\#}\mathbf{E}^m) \leqslant \Gamma$ and $\mathbf{MS}(S_i(v)) \leqslant \Gamma$,

and conditions (1.3) and (1.4) also hold with the obvious replacements. We combine (2.1), (2.2), (2.3) above with the Isoperimetric Inequality 2.5 to estimate

(2.4) $\mathbf{GS}(T_i, g_i(v)_{\#}\mathbf{E}^m) = \mathbf{GS}(S_i(v), 0) \leqslant \Gamma_{2.5}\Gamma^{(m+1)/m}2^{-v/m}$.

In particular,

(2.5) $\lim_{v \to \infty}\mathbf{GS}(T_i, g_i(v)_{\#}\mathbf{E}^m) = 0$ for each (fixed) $i \in \{1, 2, 3, \dots\}$.

We now use 3.15(5) (for each fixed v) together with Cantor's diagonal process to obtain a single subsequence $i(1)$, $i(2)$, $i(3)$, $\dots$ of $1, 2, 3, \dots$ and functions $g(v)\colon \mathbf{R}^m \to \mathbf{P}_0(\mathbf{R}^{m+n})$ for each v such that

$$0 = \lim_{j \to \infty} \sup\{\mathbf{G}(g(v)(x), g_{i(j)}(v)(x))\colon x \in \mathbf{R}^m\}$$

and

(2.6) $0 = \lim_{j \to \infty}\mathbf{G}(g(v)_{\#}\mathbf{E}^m, g_{i(j)}(v)_{\#}\mathbf{E}^m)$ with

(2.7) $\mathbf{MS}(g(v)_{\#}\mathbf{E}^m) \leqslant \Gamma$ for each v.

For each i and v we infer from (1.1) that

$$g_i(v)_{\#}\mathbf{E}^m - g_i(v + 1)_{\#}\mathbf{E}^m = -S_i(v) + S_i(v + 1)$$

and further infer from (1.2) that

$$\mathscr{L}^m[(\Pi(\lambda) \circ \omega(j))\mathrm{set}(-S_i(v) + S_i(v + 1))] \leqslant \tfrac{3}{2}\Gamma 2^{-v}$$

for each $\lambda \in \Lambda(m + n, m)$ and each $j \in \{1, \dots, N_{2.7}\}$.

For each v we set $C(v) = \text{set}[g(v)_\#\mathbf{E}^m - g(v+1)_\#\mathbf{E}^m]$ and infer from Proposition 2.10 on the lower semicontinuity of projected areas that

(2.8) $\mathscr{L}^m[(\Pi(\lambda)\circ\omega(j))C(v)] \leqslant \frac{3}{2}2^{-v}$ for each $\lambda \in \Lambda(m+n, m)$ and each $j \in \{1, \ldots, N_{2.7}\}$; we note here, of course, that

$$\lim_i\left(-S_i(v) + S_i(v+1)\right) = g(v)_\#\mathbf{E}^m - g(v+1)_\#\mathbf{E}^m \in \mathbf{S}_{m,0}(\mathbf{R}^{m+n}).$$

Part 3. For each $v \in \{1, 2, 3, \ldots\}$ we set

$$T(v) = g(v)_\#\mathbf{E}^m \, \llcorner \, \bigcap\{[\mathbf{R}^{m+n} \sim C(\mu)] : \mu = v, v+1, v+2, \ldots\}$$

and estimate

$$g(v)_\#\mathbf{E}^m + [g(v+1)_\#\mathbf{E}^m - g(v)_\#\mathbf{E}^m] = g(v+1)_\#\mathbf{E}^m,$$

$$[g(v)_\#\mathbf{E}^m] \llcorner [\mathbf{R}^{m+n} \sim C(v)] + [g(v+1)_\#\mathbf{E}^m - g(v)_\#\mathbf{E}^m] \llcorner [\mathbf{R}^{m+n} \sim C(v)]$$
$$= [g(v+1)_\#\mathbf{E}^m] \llcorner [\mathbf{R}^{m+n} \sim C(v)],$$

$$[g(v)_\#\mathbf{E}^m] \llcorner [\mathbf{R}^{m+n} \sim C(v)] = [g(v+1)_\#\mathbf{E}^m] \llcorner [\mathbf{R}^{m+n} \sim C(v)],$$

$$[g(v)_\#\mathbf{E}^m] \llcorner \bigcap\{[\mathbf{R}^{m+n} \sim C(\mu)] : \mu = v, v+1, v+2, \ldots\}$$
$$= \left([g(v+1)_\#\mathbf{E}^m] \llcorner \bigcap\{[\mathbf{R}^{m+n} \sim C(\mu)] : \mu = v, v+1, v+2, \ldots\}\right)$$
$$\llcorner [\mathbf{R}^{m+n} \sim C(v)]$$

so that, finally

(3.1) $T(v) = T(v+1) \llcorner [\mathbf{R}^{m+n} \sim C(v)]$.

Since $\sup_v \mathbf{MS}(T(v)) \leqslant \Gamma$ (by (2.3)) we conclude the existence of $T \in \mathscr{S}_{m,0}(\mathbf{R}^{m+n})$ with

(3.2) $0 = \lim_{v \to \infty} \mathbf{MS}(T - T(v))$.

The existence (as noted) and rectifiability of such T are clear. We have not yet shown that $T \in \mathbf{S}_{m,0}(\mathbf{R}^{m+n})$ with $\partial T = 0$ or that $0 = \lim_i \mathbf{G}(T, T_i)$. It is also clear, however, that

(3.3) $T(v) = T \llcorner \bigcap\{[\mathbf{R}^{m+n} \sim C(\mu)] : \mu = v, v+1, v+2, \ldots\}$.

Part 4. In order to show $\partial T = 0$ we will apply Theorem 2.9 for each v with

S replaced by $T(v) = g(v)_\#\mathbf{E}^m \llcorner \bigcap\{[\mathbf{R}^{m+n} \sim C(\mu)] : \mu = v, v+1, \ldots\}$,

T replaced by $g(v)_\#\mathbf{E}^m - T(v) = g(v)_\#\mathbf{E}^m \llcorner \bigcup\{C(\mu) : \mu = v, v+1, \ldots\}$,

ε replaced by $\varepsilon(v)$,

M replaced by $M(v)$,

where

(4.1) $\varepsilon(v)$ equals the quantity

$$2^{-v/(m+1)}\left\{3\left[3^m(m+n)^{m/2}\alpha(m)M_{1.5}\right]\Big/\left[4^m(m+n)^{1/2}\alpha(m)\beta(m)/\Gamma_{2.9}^*\right]\right\}^{1/(m+1)},$$

and

(4.2) $M(v) \in \mathbf{Z}$ is defined by the requirement

$$2^{mM(v)} \leqslant \varepsilon(v)^{-1}\left[3^m(m+n)^{m/2}\alpha(m)M_{1.5}\right]/\Gamma < 2^{m(M(v)+1)}$$

so that

(4.3) $\lim_{v \to \infty} \varepsilon(v) = 0$, $\lim_{v \to \infty} M(v) = \infty$,

(4.4) $\Gamma \leqslant \varepsilon(v)^{-1} 2^{-mM(v)} [3^m (m + n)^{m/2} \alpha(m) M_{1.5}]$,

(4.5) $3 \cdot 2^{-v} \Gamma \leqslant \varepsilon(v)^m 2^{-mM(v)} [4^m (m + n)^{1/2} \alpha(m) \beta(m)/\Gamma_{2.9}^*]$.

In view of (2.7) and (2.8) we conclude that the hypotheses of 2.9 are satisfied for each v and we denote by $S_*(v)$ and $Q(v)$ the currents of the conclusion of 2.9 (called S_* and Q respectively there) and by A_v and $\{A_v(\lambda(1),\ldots;i(1),\ldots)\}_{\lambda(1),\ldots;i(1),\ldots}$, respectively, the open set and Borel subsets of the conclusion of 2.9. We estimate

(4.6) $\mathbf{GS}(S_*(v), g(v)_\# \mathbf{E}^m) \leqslant 2 \cdot 2^{-M(v)} \Gamma_{1.15} \Gamma$ (by 2.9(1)(4) and (2.7)) so that

(4.7) $0 = \lim_{v \to \infty} \mathbf{GS}(S_*(v), g(v)_\# \mathbf{E}^m)$, and

(4.8) for each sequence $\lambda(1),\ldots,\lambda(\binom{m+n}{m}) \in \Lambda(m + n, m)$ and each increasing sequence $1 \leqslant i(1) < \cdots < i(\binom{m+n}{m}) \leqslant N_{2.7}$ the orthogonal projections

$$\Pi(\lambda(1)) \circ \omega(i(1)),\ldots,\Pi\left(\lambda\left(\frac{m + n}{m}\right)\right) \circ \omega\left(i\left(\frac{m + n}{m}\right)\right) \in \mathbf{O}(m + n, m)$$

are linearly independent and

$$\mathscr{L}^m\left[\left(\Pi(\lambda(j)) \circ \omega(i(j))\right) A_v\left(\lambda(1),\ldots,\lambda\left(\frac{m + n}{m}\right); i(1),\ldots,i\left(\frac{m + n}{m}\right)\right)\right]$$
$$\leqslant \varepsilon(v)\Gamma + \varepsilon(v)^{-m} \cdot 3 \cdot 2^{-v}\Gamma$$

for each $j = 1,\ldots,(\binom{m+n}{m})$ (by 2.9(6), (2.7), and (2.8)) so that, by virtue of (4.1) and (4.3),

$$(4.9) \qquad 0 = \lim_{v \to \infty} \varepsilon(v)\Gamma + \varepsilon(v)^{-m}\Gamma_{2.9}^* 3 \cdot 2^{-v}\Gamma$$
$$= \lim_{v \to \infty} \mathscr{L}^m\left[\left(\Pi(\lambda(j)) \circ \omega(i(j))\right) A_v(\lambda(1),\ldots;i(1),\ldots)\right].$$

We use 2.8 with S and θ there associated with our present T, i.e., $T = \mathbf{t}(S, \theta, \xi)$ for some ξ, to obtain a continuous nondecreasing function $\sigma: \mathbf{R}^+ \to \mathbf{R}^+$ having the properties given in 2.8(2); in particular, $\lim_{t \to 0} \sigma(t) = 0$.

In order to estimate $\mathbf{MS}(T - S_*(v))$, we write

$$T - S_*(v) = [T - T(v)] + [T(v) - S_*(v)]$$
$$= [T - T(v)] + [T(v) \llcorner A_v - S_* \llcorner A_v]$$
$$+ [T(v) \llcorner (\mathbf{R}^{m+n} \sim A_v) - S_*(v) \llcorner (\mathbf{R}^{m+n} \sim A_v)]$$
$$= [T - T(v)] + [T(v) \llcorner A_v] - [S_* \llcorner A_v]$$

(by 2.9(2)) so that

$$\mathbf{MS}(T - S_*(v)) \leqslant \mathbf{MS}(T - T(v)) + \mathbf{MS}(T(v) \llcorner A_v) + \mathbf{MS}(S_*(v) \llcorner A_v)$$
$$\leqslant \mathbf{MS}(T - T(v)) + (1 + \Gamma_{1.15})\mathbf{MS}(T(v) \llcorner A_v)   \text{ (by 2.9(3))}$$
$$\leqslant \mathbf{MS}(T - T(v)) + (1 + \Gamma_{1.15})\mathbf{MS}(T \llcorner A_v)   \text{ (by (3.3))}$$
$$\leqslant \mathbf{MS}(T - T(v)) + (1 + \Gamma_{1.15})\sigma\left(\varepsilon(v)\Gamma + \varepsilon(v)^{-m}\Gamma_{2.9}^* 3 \cdot 2^{-v}\Gamma\right)$$
$$\text{(by 2.8(2) and (4.8))}$$

and hence

(4.10) $0 = \lim_{v \to \infty} \mathbf{MS}(T - S_*(v))$ (by (3.2) and (4.9)); this with (2.8) implies

(4.11) $T \in \mathbf{S}_{m,0}(\mathbf{R}^{m+n})$ with $\partial T = 0$ and $\mathbf{MS}(T) \leqslant \Gamma$. Furthermore,

(4.12) $\mathbf{GS}(T, S_*(v)) \leqslant 2\Gamma_{2.3}[\mathbf{MS}(T - S_*(v))]^{(m+1)/m}$ (by the Isoperimetric Inequality 2.3) so that as a consequence of (4.10),

(4.13) $\lim_{v \to \infty} \mathbf{GS}(T, S_*(v)) = 0$.

Part 5. We assert $\lim_{j \to \infty} \mathbf{G}(T, T_{i(j)}) = 0$ (which will finish the proof of the theorem). To see this, we suppose $0 < \varepsilon < 1$ and use (4.13) and (4.7) to choose and fix $\mu \in \{1, 2, 3, \ldots\}$ such that $\Gamma < 2^\mu$ and

$$\mathbf{G}(T, S_*(\mu)) + \mathbf{G}(S_*(\mu), g(\mu)_\# \mathbf{E}^m) + \Gamma_{2.5}\Gamma^{(m+1)/m}2^{-\mu/m} < \varepsilon/2.$$

For such fixed μ, we use (2.6) to choose and fix $k \in \{1, 2, 3, \ldots\}$ such that

$$\mathbf{G}(g(\mu)_\# \mathbf{E}^m, g_{i(j)}(\mu)_\# \mathbf{E}^m) < \varepsilon/2$$

for each $j \in \{k, k+1, k+2, \ldots\}$.

For each $j \in \{k, k+1, k+2, \ldots\}$, we combine these conditions with (2.4) to conclude

$$\begin{aligned}
\mathbf{G}(T, T_{i(j)}) &\leqslant \mathbf{G}(T, S_*(\mu)) + \mathbf{G}(S_*(\mu), g(\mu)_\# \mathbf{E}^m) \\
&\quad + \mathbf{G}(g(\mu)_\# \mathbf{E}^m, g_{i(j)}(\mu)_\# \mathbf{E}^m) + \mathbf{G}(g_{i(j)}(\mu)_\# \mathbf{E}^m, T_{i(j)}) \\
&< \mathbf{G}(T, S_*(\mu)) + \mathbf{G}(S_*(\mu), g(\mu)_\# \mathbf{E}^m) + \varepsilon/2 + \Gamma_{2.5}\Gamma^{(m+1)/m}2^{-\mu/m} \\
&< \varepsilon
\end{aligned}$$

which finishes the proof of the theorem.

4.3. Compactness Theorem for size bounded rectifiable currents with bounded supports.

Hypotheses. (a) $0 < M < \infty$,

(b) $T_1, T_2, T_3, \ldots \in \mathbf{S}_{m,0}(\mathbf{R}^{m+n})$ [*resp.* $\in \mathbf{I}_{m,0}(\mathbf{R}^{m+n})$],

(c) $\mathrm{spt}[T_i] \subset \mathbf{B}^{m+n}(0, M)$ *for each* $i = 1, 2, 3, \ldots$,

(d) $\mathbf{NS}(T_i) \leqslant M$ *for each* $i = 1, 2, 3, \ldots$

Conclusion. There exists a subsequence $i(1), i(2), i(3), \ldots$ *of* $1, 2, 3, \ldots$ *and* $T \in \mathbf{S}_{m,0}(\mathbf{R}^{m+n})$ [*resp.* $\in \mathbf{I}_{m,0}(\mathbf{R}^{m+n})$] *with the following properties:*

(1) $\lim_{j \to \infty} \mathbf{F}(T, T_{i(j)}) = 0$ [*resp.* $\lim_{j \to \infty} \mathscr{F}(T, T_{i(j)}) = 0$],

(2) $\mathrm{spt}(T) \subset \mathbf{B}^{m+n}(0, M)$,

(3) $\mathbf{M}(T) \leqslant \liminf_{j \to \infty} \mathbf{M}(T_{i(j)})$, $\mathbf{S}(T) \leqslant \liminf_{j \to \infty} \mathbf{S}(T_{i(j)})$, $\mathbf{M}(\partial T) \leqslant \liminf_{j \to \infty} \mathbf{M}(\partial T_{i(j)})$, *and* $\mathbf{S}(\partial T) \leqslant \liminf_{j \to \infty} \mathbf{S}(T_{i(j)})$.

PROOF. We first apply 4.2 to the cycles $\partial T_1, \partial T_2, \partial T_3, \ldots$ to obtain a subsequence $k(1), k(2), k(3), \ldots$ of $1, 2, 3, \ldots$ and $C \in \mathbf{S}_{m-1,0}(\mathbf{R}^{m+n})$ [resp. $\in \mathbf{I}_{m,0}(\mathbf{R}^{m+n})$] with $\partial C = 0$ and $\lim_{j \to \infty} \mathbf{G}(C, \partial T_{k(j)}) = 0$ [resp. $\lim_{j \to \infty} \mathscr{G}(C, \partial T_{k(j)}) = 0$]. We check that

$$\lim_{j \to \infty} \mathbf{F}(\llbracket 0 \rrbracket \times C, \llbracket 0 \rrbracket \times \partial T_{k(j)}) = 0$$

$$\left[\text{resp. } \lim_{j \to \infty} \mathscr{F}(\llbracket 0 \rrbracket \times C, \llbracket 0 \rrbracket \times T_{k(j)}) = 0\right].$$

Then we apply 4.2 to the cycles $T_{k(j)} - [\![0]\!] \times \partial T_{k(j)}$, $j = 1, 2, 3, \ldots$, to obtain a subsequence $i(1), i(2), i(3), \ldots$ of $k(1), k(2), k(3), \ldots$ and $S \in \mathbf{S}_{m,0}(\mathbf{R}^{m+n})$ [resp. $\in \mathbf{I}_{m,0}(\mathbf{R}^{m+n})$] with $\partial S = 0$ and

$$\lim_{j \to \infty} \mathbf{G}\big(S, T_{i(j)} - [\![0]\!] \times \partial T_{i(k)}\big) = 0$$

$$\Big[\text{resp. } \mathscr{G}\big(S, T_{i(j)} - [\![0]\!] \times T_{i(j)}\big) = 0\Big].$$

We set $T = S + [\![0]\!] \times C \in \mathbf{S}_{m,0}(\mathbf{R}^{m+n})$ [resp. $\in \mathbf{I}_{m,0}(\mathbf{R}^{m+n})$] and check that Conclusion (1) holds. Conclusion (2) is immediate. Conclusion (3) is a consequence of Proposition 2.10.

4.4. Compactness Theorem for currents which are locally size bounded and rectifiable. *Suppose $T_1, T_2, T_3, \ldots \in \mathbf{S}_{m,\mathrm{loc}}(\mathbf{R}^{m+n})$ [resp. $\in \mathbf{I}_{m,\mathrm{loc}}(\mathbf{R}^{m+n})$] with*

$$\infty > \sup\{\mathbf{MS}(T_i \, \llcorner \, U) + \mathbf{MS}(\partial T_i \, \llcorner \, U) : i = 1, 2, 3, \ldots\}$$

for each bounded open set $U \subset \mathbf{R}^{m+n}$. Then there is a subsequence $i(1), i(2), i(3), \ldots$ of $1, 2, 3, \ldots$ and $T \in \mathbf{S}_{m,\mathrm{loc}}(\mathbf{R}^{m+n})$ [resp. $\in \mathbf{I}_{m,\mathrm{loc}}(\mathbf{R}^{m+n})$] with

$$0 = \lim_{j \to \infty} \mathbf{F}(T, T_i; U) \quad \Big[resp. \, 0 = \lim_{j \to \infty} \mathscr{F}(T, T_i; U)\Big]$$

for each bounded open set $U \subset \mathbf{R}^{m+n}$. In case $\partial T_i = 0$ for each i then $\partial T = 0$ and one can replace $\mathbf{F}$ [resp. replace $\mathscr{F}$] by $\mathbf{G}$ [resp. by $\mathscr{G}$].

PROOF. Localize by slicing [**FH**, 4.3.4, 4.3.8] and use 4.3, 2.3, and 1.13.

APPENDIX

Much of the terminology of this paper will be that of pages 669–671 of H. Federer's treatise, *Geometric Measure Theory* [**FH**]. Exceptions to such terminology will be indicated specifically in Appendix A or in Appendix B; in particular, we will follow the newer, standardized terminology of the 1984 AMS Summer Research Institute on Geometric Measure Theory and the Calculus of Variations.

Appendix A. General terminology. We specify the following general terminology:

(1) $\mathbf{Z}$ denotes the integers. $\mathbf{Z}^+ = \{1, 2, 3, \ldots\}$. The integers m, $n \in \mathbf{Z}^+$ are fixed throughout this paper. $\mathbf{R}$ denotes the real numbers. $\mathbf{R}^+ = \mathbf{R} \cap \{r : 0 \leqslant r < \infty\}$.

(2) For each $k \in \mathbf{Z}^+$, $\mathbf{R}^k$ denotes k-dimensional Euclidean space, and for each $p \in \mathbf{R}^k$ and $0 < r < \infty$, we set

$$\mathbf{U}^k(p, r) = \mathbf{R}^k \cap \{x : |x - p| < r\}, \qquad \mathbf{B}^k(p, r) = \mathbf{R}^k \cap \{x : |x - p| \leqslant r\},$$

$$\partial \mathbf{B}^k(p, r) = \mathbf{R}^k \cap \{x : |x - p| = r\},$$

and we also define

$$\mu(r), \tau(p) : \mathbf{R}^k \to \mathbf{R}^k$$

by setting $\mu(r)(x) = rx$ and $\tau(p)(x) = p + x$ for each $x \in \mathbf{R}^k$.

Additionally, $\mathscr{L}^k$ denotes k-dimensional Lebesgue measure in $\mathbf{R}^k$,

$$\alpha(k) = \mathscr{L}^k\big[\mathbf{U}^k(0, 1)\big],$$

and

$$\gamma(k+1) = 1/(k+1)^{(k+1)/k}\alpha(k+1)^{1/k}$$

is the **optimal isoperimetric constant** of [A2]. We also set $\alpha(0) = 1$.

(3) For each $k \in \mathbf{Z}^+$, $\beta(k)$ denotes the optimal constant for the **Besicovitch–Federer Covering Theorem** [FH, 2.8.14] when covering by Euclidean balls in $\mathbf{R}^k$. The theorem states the following:

Suppose $0 < R < \infty$ and $\varnothing \neq A \subset \mathbf{R}^k \times (0, R)$. Then there exist (p_1, r_1), $(p_2, r_2), (p_3, r_3), \ldots \in A$ such that

$$\{\, p\colon (p, r) \in A \text{ for some } r \,\} \subset \bigcup_i \mathbf{B}^k(p_i, r_i),$$

and

$$\beta(k) \geqslant \mathrm{card}\{\, i\colon z \in \mathbf{B}^k(p_i, r_i)\} \quad \text{for each } z \in \mathbf{R}^k.$$

(4) For each $k \in \mathbf{Z}^+$ and each $j \in \{0, 1, \ldots, k\}$, $\mathscr{H}^j$ denotes j-dimensional Hausdorff measure in $\mathbf{R}^k$. In particular,

$$\mathscr{H}^{k-1}\big[\partial\mathbf{B}^k(0,1)\big] = k\alpha(k).$$

Also, whenever μ measures $\mathbf{R}^k$ and $p \in \mathbf{R}^k$ we have the j-**dimensional density ratios of** μ **at** p,

$$\Theta^j(\mu, p, r) = \mu\big[\mathbf{B}^k(p, r)\big]/\alpha(j)r^j, \qquad 0 < r < \infty,$$

and the j-**dimensional density of** μ **at** p,

$$\Theta^j(\mu, p) = \lim_{r \to 0+} \Theta^j(\mu, p, r),$$

provided the limit exists.

(5) $\mathbf{O}(m+n)$ denotes the orthogonal group of $\mathbf{R}^{m+n}$, and $\mathbf{O}(m+n, m)$ denotes the space of all orthogonal projections $\mathbf{R}^{m+n} \to \mathbf{R}^m$.

(6) $\mathbf{G}(m+n, m)$ denotes the Grassmann manifold of all m-dimensional linear subspaces of $\mathbf{R}^{m+n}$, and $\mathbf{G}_0(m+n, m)$ denotes the Grassmann manifold of all oriented m-dimensional linear subspaces of $\mathbf{R}^{m+n}$. $\Lambda_m\mathbf{R}^{m+n}$ denotes the Grassmann vector space of all m vectors in $\mathbf{R}^{m+n}$, and we naturally identify $\mathbf{G}_0(m+n, m)$ with the algebraic submanifold of $\Lambda_m\mathbf{R}^{m+n}$ consisting of all simple unit m vectors. We also say that $\xi \in \mathbf{G}_0(m+n, m) \subset \Lambda_m\mathbf{R}^{m+n}$ is associated with $\Pi \in \mathbf{G}(m+n, m)$ if and only if $\Pi = \mathbf{R}^{m+n} \cap \{v\colon \xi \wedge v = 0\}$.

(7) For each $\Pi \in \mathbf{G}(m+n, m)$,

$$\Pi_{\natural}\colon \mathbf{R}^{m+n} \to \mathbf{R}^{m+n}$$

denotes orthogonal projection onto Π. Also,

$$\{\mathbf{R}^m \times \{0\}\}_{\natural}\colon \mathbf{R}^m \times \mathbf{R}^n \to \mathbf{R}^m \quad \big[\text{resp. } \{\{0\} \times \mathbf{R}^n\}_{\natural}\colon \mathbf{R}^m \times \mathbf{R}^n \to \mathbf{R}^n\big]$$

denotes projection onto the first [resp. second] factor; we abbreviate $(\Pi_{\natural})_{\#} = \Pi_{\#}$, etc.

(8) Whenever $U \subset \mathbf{R}^m$ and $f\colon U \to \mathbf{R}^n$ we set

$$\mathrm{graph}(f) = \{(x, f(x))\colon x \in U\} \subset \mathbf{R}^{m+n}.$$

It is, of course, formally true that f equals graph(f). In this paper we will sometimes deal with a function whose graph lies in a rotated and translated image of the graph of another function. It seems useful in such cases to emphasize graph(f) as a subset of $\mathbf{R}^{m+n}$.

(9) Whenever $f: A \to B$ and $g: A \to C$ we write

$$f \bowtie g: A \to B \times C, \qquad (f \bowtie g)(a) = (f(a), g(a)) \quad \text{for } a \in A.$$

Appendix B. Terminology for currents. The definitions and terminology below have the obvious meanings when m and n are replaced by other relevant integers.

(1) For each $k \in \{0, 1, \ldots, m + n\}$ we denote by $\mathscr{D}^k(\mathbf{R}^{m+n})$ the vector space of all infinitely differentiable functions $\mathbf{R}^{m+n} \to \wedge^k \mathbf{R}^{m+n}$ having compact support (smooth differential k-forms with compact support). Members of the dual space $\mathscr{D}_k(\mathbf{R}^{m+n})$ are called k-**dimensional currents in** $\mathbf{R}^{m+n}$.

(2) We denote by $\mathscr{M}_{m,\text{loc}}(\mathbf{R}^{m+n})$ the linear subspace of $\mathscr{D}_m(\mathbf{R}^{m+n})$ consisting of currents T which are representable by integration; i.e., $\|T\|$ is a Radon measure [**FH**, 4.1.7]. For such T we, as usual, define the **mass of** T [**FH**, 4.1.7] as

$$\mathbf{M}(T) = \|T\|(\mathbf{R}^{m+n}) = \sup\{T(\phi): \phi \in \mathscr{D}^m(\mathbf{R}^{m+n}) \text{ with } \mathbf{M}(\phi) \leqslant 1\}.$$

We also set

$$\mathscr{M}_m(\mathbf{R}^{m+n}) = \mathscr{M}_{m,\text{loc}}(\mathbf{R}^{m+n}) \cap \{T: \mathbf{M}(T) < \infty\},$$

$$\mathscr{M}_{m,0}(\mathbf{R}^{m+n}) = \mathscr{M}_m(\mathbf{R}^{m+n}) \cap \{T: \text{spt}(T) = \text{spt}\|T\| \text{ is compact}\}.$$

For $T, T' \in \mathscr{M}_m(\mathbf{R}^{m+n})$ we set $\mathbf{M}(T, T') = \mathbf{M}(T - T')$. $\mathbf{M}(\cdot)$ is thus a norm with associated metric $\mathbf{M}(\cdot, \cdot)$ on $\mathscr{M}_m(\mathbf{R}^{m+n})$. Corresponding definitions hold, as noted above, with m replaced by $k \in \{0, 1, \ldots, m + n\}$.

We further set

$$\mathbf{N}_{m,\text{loc}}(\mathbf{R}^{m+n}) = \mathscr{M}_{m,\text{loc}}(\mathbf{R}^{m+n}) \cap \{T: \partial T \in \mathscr{M}_{m-1,\text{loc}}(\mathbf{R}^{m+n})\},$$

$$\mathbf{N}_m(\mathbf{R}^{m+n}) = \mathscr{M}_m(\mathbf{R}^{m+n}) \cap \{T: \partial T \in \mathscr{M}_{m-1}(\mathbf{R}^{m+n})\},$$

$$\mathbf{N}_{m,0}(\mathbf{R}^{m+n}) = \mathbf{N}_m(\mathbf{R}^{m+n}) \cap \mathscr{M}_{m,0}(\mathbf{R}^{m+n});$$

$\mathbf{N}_{m,0}(\mathbf{R}^{m+n})$ is called $\mathbf{N}_m(\mathbf{R}^{m+n})$ in [**FH**, 4.1.7], while our $\mathbf{N}_m(\mathbf{R}^{m+n})$ has no special name there.

For $T, T' \in \mathbf{N}_m(\mathbf{R}^{m+n})$ we set $\mathbf{N}(T) = \mathbf{M}(T) + \mathbf{M}(\partial T)$ and $\mathbf{N}(T, T') = \mathbf{N}(T - T')$. $\mathbf{N}(\cdot)$ is thus a norm with associated metric $\mathbf{N}(\cdot, \cdot)$ on $\mathbf{N}_m(\mathbf{R}^{m+n})$.

(3) A subset S of $\mathbf{R}^{m+n}$ is called $(\mathscr{H}^m, m)$ **rectifiable** and $\mathscr{H}^m$ **measurable** if and only if for each $\varepsilon > 0$ there exists a compact m-dimensional submanifold M of $\mathbf{R}^{m+n}$ with boundary and of class 1 such that

$$\mathscr{H}^m(M \sim S) + \mathscr{H}^m(S \sim M) < \varepsilon.$$

Equivalent conditions are the following:

(3.1) S is $\mathscr{H}^m$ measurable, and for each $\varepsilon > 0$ there is a Lipschitz function f: $\mathbf{U}^m(0, 1) \to \mathbf{R}^{m+n}$ such that $\mathscr{H}^m(S \sim f[\mathbf{U}^m(0, 1)]) < \varepsilon$.

(3.2) S is $\mathscr{H}^m$ measurable with $\mathscr{H}^m(S) < \infty$ and $\mathscr{H}^m$ almost all of S is contained in the union of a countable family of nonparametric Lipschitz m-dimensional surfaces in $\mathbf{R}^{m+n}$.

(3.3) $\mathscr{H}^m(S) < \infty$ and for each $1 < \lambda < \infty$ there exist compact subsets $K_1, K_2, K_3, \ldots$ of $\mathbf{R}^m$ and Lipschitz maps $f_1, f_2, f_3, \ldots : \mathbf{R}^m \to \mathbf{R}^{m+n}$ such that $f_1(K_1), f_2(K_2), f_3(K_3), \ldots$ are disjoint subsets of S with

$$\mathscr{H}^m\left[S \sim \bigcup_{i=1}^{\infty} f_i(K_i) \right] = 0,$$

and for each $i = 1, 2, 3, \ldots,$

$$\mathrm{Lip}(f_i) \leqslant \lambda, \qquad f_i|K_i \text{ is univalent}, \qquad \mathrm{Lip}\left[\left(f_i|K_i \right)^{-1} \right] \leqslant \lambda,$$

$$\lambda^{-1}|v| \leqslant |\langle v, Df_i(a) \rangle| \leqslant \lambda |v| \quad \text{for } a \in K_i, v \in \mathbf{R}^m.$$

With regard to the equivalence of these conditions note [**FH**, 2.10.19(5), 3.1.6, 3.1.15, 3.2.14, 3.2.18, 3.2.19, 3.2.29]. Finite subsets of $\mathbf{R}^{m+n}$ are called 0 **rectifiable** or, alternatively, $(\mathscr{H}^0, 0)$ **rectifiable and** $\mathscr{H}^0$ **measurable** when convenient for exposition.

(4) When we write $\mathbf{t}(S, \theta, \xi)$ (as a member of $\mathscr{S}_m(\mathbf{R}^{m+n})$) we mean, in particular, that the **set** S is an $(\mathscr{H}^m, m)$ rectifiable and $\mathscr{H}^m$ measurable subset of $\mathbf{R}^{m+n}$, the **density function** $\theta \colon S \to \mathbf{R}^+$ is $\mathscr{H}^m \llcorner S$ almost everywhere positive and is $\mathscr{H}^m \llcorner S$ summable, and the **orientation function** $\xi \colon S \to \mathbf{G}_0(m+n, m) \subset \Lambda_m \mathbf{R}^{m+n}$ is $\mathscr{H}^m$ measurable, and, additionally, for $\mathscr{H}^m$ almost every $x \in S$, $\xi(x)$ is associated with $\mathrm{Tan}^m(\mathscr{H}^m \llcorner S, x)$. We further mean that $\mathbf{t}(S, \theta, \xi) \in \mathscr{M}_m(\mathbf{R}^{m+n})$ with

$$\langle \mathbf{t}(S, \theta, \xi), \phi \rangle = \int_{x \in S} \langle \xi(x), \phi(x) \rangle \theta(x) \, d\mathscr{H}^m x$$

for each $\phi \in \mathscr{D}^m(\mathbf{R}^{m+n})$. The space $\mathscr{S}_m(\mathbf{R}^{m+n})$ of m-**dimensional size bounded rectifiable currents in** $\mathbf{R}^{m+n}$ is the linear subspace of $\mathscr{M}_m(\mathbf{R}^{m+n})$ consisting of currents which can be so written. If $T = \mathbf{t}(S, \theta, \xi) \in \mathbf{S}_m(\mathbf{R}^{m+n})$ then the **mass of** T equals the number [**FH**, 4.1.7]

$$\mathbf{M}(T) = \int_S \theta \, d\mathscr{H}^m = \|T\| \mathbf{R}^{m+n} = \sup\{ T(\phi) : \mathbf{M}(\phi) \leqslant 1 \},$$

the **set of** T is the set

$$\mathrm{set}(T) = \mathbf{R}^{m+n} \cap \{ x : \Theta^m(\|T\|, x) > 0 \},$$

which is $\mathscr{H}^m$ almost equal to S, and the **size of** T is the number

$$\mathbf{S}(T) = \mathscr{H}^m(\mathrm{set}(T)) = \mathscr{H}^m(S).$$

T, of course, uniquely determines $\mathrm{set}(T)$, while possible S's are determined only up to sets of $\mathscr{H}^m$ measure 0.

Caution. $\mathrm{set}(T)$ usually is not closed, and $\mathrm{Clos}[\mathrm{set}(T)]$ can be much larger than $\mathrm{set}(T)$.

We check $\theta(x) = \Theta^m(\|T\|, x)$ for $\mathscr{H}^m$ almost every $x \in S$.

The size function $\mathbf{S}$ is invariant under nonzero scalar multiplication in $\mathscr{S}_m(\mathbf{R}^{m+n})$ and thus is not a norm. We do set $\mathbf{S}(T, T') = \mathbf{S}(T - T')$ for $T, T' \in \mathscr{S}_m(\mathbf{R}^{m+n})$ so that $\mathbf{S}(\cdot, \cdot)$ is a metric on $\mathbf{S}_m(\mathbf{R}^{m+n})$.

Corresponding definitions apply with m above replaced by $k \in \{0, 1,\ldots,m + n\}$, and we set

$$\mathscr{S}_{m,0}(\mathbf{R}^{m+n}) = \mathscr{S}_m(\mathbf{R}^{m+n}) \cap \mathscr{M}_{m,0}(\mathbf{R}^{m+n}),$$

$$\mathbf{S}_m(\mathbf{R}^{m+n}) = \mathscr{S}_m(\mathbf{R}^{m+n}) \cap \{T: \partial T \in \mathscr{S}_{m-1}(\mathbf{R}^{m+n})\},$$

$$\mathbf{S}_{m,0}(\mathbf{R}^{m+n}) = \mathbf{S}_m(\mathbf{R}^{m+n}) \cap \mathscr{S}_{m,0}(\mathbf{R}^{m+n}).$$

$\mathscr{S}_{m,\mathrm{loc}}(\mathbf{R}^{m+n})$ and $\mathbf{S}_{m,\mathrm{loc}}(\mathbf{R}^{m+n})$ have the obvious meanings.

For $T, T' \in \mathscr{S}_m(\mathbf{R}^{m+n})$ [resp. $\in \mathbf{S}_m(\mathbf{R}^{m+n})$] it is sometimes convenient to write $\mathbf{MS}(T) = \mathbf{M}(T) + \mathbf{S}(T)$ [resp. $\mathbf{NS}(T) = \mathbf{MS}(T) + \mathbf{MS}(\partial T)$], $\mathbf{MS}(T, T') = \mathbf{MS}(T - T')$ [resp. $\mathbf{NS}(T, T') = \mathbf{NS}(T - T')$]. $\mathbf{MS}(\cdot, \cdot)$ and $\mathbf{MS}(\cdot, \cdot)$ are both metrics.

Integer multiplicity currents have been of special geometric and analytic significance, and we set

$$\mathscr{I}_{m,\mathrm{loc}}(\mathbf{R}^{m+n}) = \mathscr{S}_{m,\mathrm{loc}}(\mathbf{R}^{m+n}) \cap \{\mathbf{t}(S, \theta, \xi): \mathrm{im}(\theta) \subset \mathbf{Z}^+\},$$

$$\mathbf{I}_{m,\mathrm{loc}}(\mathbf{R}^{m+n}) = \mathscr{I}_{m,\mathrm{loc}}(\mathbf{R}^{m+n}) \cap \{T: \partial T \in \mathscr{I}_{m-1,\mathrm{loc}}(\mathbf{R}^{m+n})\}$$

with corresponding definitions for $\mathscr{I}_m(\mathbf{R}^{m+n})$, $\mathbf{I}_m(\mathbf{R}^{m+m})$ and $\mathscr{I}_{m,0}(\mathbf{R}^{m+n})$, $\mathbf{I}_{m,0}(\mathbf{R}^{m+n})$.

The m**-dimensional Euclidean current**

$$\mathbf{E}^m = \mathbf{t}(\mathbf{R}^{m+n}, 1, e_1 \wedge \cdots \wedge e_m) \in \mathbf{I}_{m,\mathrm{loc}}(\mathbf{R}^{m+n})$$

(with the obvious meanings) is used many times in this paper.

Whenever $B \subset A \subset \mathbf{R}^{m+n}$ are closed we will write

$$X_m(A, B) = X_m(\mathbf{R}^{m+n}) \cap \{T: \mathrm{spt}(T) \subset A \text{ and } \mathrm{spt}(\partial T) \subset B\},$$

$$X_{m,0}(A, B) = X_{m,0}(\mathbf{R}^{m+n}) \cap X_m(A, B),$$

$$X_m(A) = X_m(A, A), \qquad X_{m,0}(A) = X_{m,0}(A, A),$$

etc. for $X = \mathbf{M}, \mathbf{N}, \mathscr{S}, \mathbf{S}, \mathscr{I}, \mathbf{I}$.

(5) For $p, p_0, p_1,\ldots,p_m \in \mathbf{R}^{m+n}$ we set

$$[\![p]\!] \in \mathbf{I}_0(\mathbf{R}^{m+n}), \qquad \langle[\![p]\!], \phi\rangle = \phi(p) \quad \text{for } \phi \in \mathscr{D}^0(\mathbf{R}^{m+n}),$$

and let

$$[\![p_0, p_1,\ldots,p_m]\!] \in \mathbf{I}_{m,0}(\mathbf{R}^{m+n})$$

denote the **oriented m simplex** of [FH, 4.1.11]. We denote by $\mathbf{P}_0(\mathbf{R}^{m+n})$ [resp. by $\mathbf{IP}_0(\mathbf{R}^{m+n})$] the real vector space [resp. abelian group] of 0**-dimensional polyhedral chains** [resp. 0**-dimensional integral polyhedral chains**] generated by such $[\![p]\!]$'s. We also denote by $\mathbf{P}_m(\mathbf{R}^{m+n})$ [resp. by $\mathbf{IP}_m(\mathbf{R}^{m+n})$] the real vector space [resp. abelian group] of m**-dimensional polyhedral chains** [resp. m**-dimensional integral polyhedral chains**] generated by such $[\![p_0, p_1,\ldots,p_m]\!]$'s. We note that

$$\mathbf{P}_0(\mathbf{R}^{m+n}) = \mathscr{S}_0(\mathbf{R}^{m+n}) = \mathscr{S}_{0,0}(\mathbf{R}^{m+n}) = \mathbf{S}_0(\mathbf{R}^{m+n}) = \mathbf{S}_{0,0}(\mathbf{R}^{m+n})$$

with similar equalities for integer coefficients. We use polyhedral notation in dimension 0 especially when such currents are being used in constructions of higher-dimensional polyhedral currents.

If $f: \mathbf{R}^m \to \mathbf{P}_0(\mathbf{R}^n)$ [resp. $\to \mathbf{IP}_0(\mathbf{R}^n)$], then

$$(\llbracket \mathbf{1}_{\mathbf{R}^m} \rrbracket \bowtie f): \mathbf{R}^m \to \mathbf{P}_0(\mathbf{R}^{m+n}) \quad \left[\text{resp.} \to \mathbf{IP}_0(\mathbf{R}^{m+n})\right]$$

maps $x \in \mathbf{R}^m$ to $\llbracket x \rrbracket \times f(x) \in \mathbf{P}_0(\mathbf{R}^{m+n})$ [resp. $\in \mathbf{IP}_0(\mathbf{R}^{m+n})$].

(6) Several functions on spaces of currents are important in this paper.

Whenever $S, T \in \mathscr{S}_{m,\mathrm{loc}}(\mathbf{R}^{m+n})$ and U is a **bounded** open subset of $\mathbf{R}^{m+n}$, we set

$$\mathbf{F}(S, T; U) = \inf\{\mathbf{M}(X) + \mathbf{M}(Y): X \in \mathbf{S}_{m,0}(\mathbf{R}^{m+n}) \text{ and } Y \in \mathbf{S}_{m+1,0}(\mathbf{R}^{m+n})$$
$$\text{with spt}[(S - T) - (X + \partial Y)] \subset [\mathbf{R}^{m+n} \sim U]\},$$

$$\mathbf{FS}(S, T; U) = \inf\{\mathbf{MS}(X) + \mathbf{MS}(Y): X \in \mathbf{S}_{m,0}(\mathbf{R}^{m+n}) \text{ and }$$
$$Y \in \mathbf{S}_{m+1,0}(\mathbf{R}^{m+n}) \text{with spt}[(S - T) - (X + \partial Y)] \subset [\mathbf{R}^{m+n} \sim U]\}.$$

Whenever $S, T \in \mathscr{S}_m(\mathbf{R}^{m+n})$, we set

$$\mathbf{F}(S, T) = \inf\{\mathbf{M}(X) + \mathbf{M}(Y): X \in \mathbf{S}_m(\mathbf{R}^{m+n}) \text{ and } Y \in \mathbf{S}_{m+1}(\mathbf{R}^{m+n})$$
$$\text{with } S - T = X + \partial Y\},$$

$$\mathbf{F}(S) = \mathbf{F}(S, 0),$$

$$\mathbf{FS}(S, T) = \inf\{\mathbf{MS}(X) + \mathbf{MS}(Y): X \in \mathbf{S}_m(\mathbf{R}^{m+n}) \text{ and } Y \in \mathbf{S}_{m+1}(\mathbf{R}^{m+n})$$
$$\text{with } S - T = X + \partial Y\},$$

$$\mathbf{FS}(S) = \mathbf{FS}(S, 0);$$

in case S and T have compact support here, it is no further restriction to require that X and Y also have compact support. $\mathbf{F}(\cdot, \cdot)$ and $\mathbf{FS}(\cdot, \cdot)$ are both metrics. $\mathbf{F}(\cdot)$ is a norm.

Whenever $S, T \in \mathscr{S}_{m,\mathrm{loc}}(\mathbf{R}^{m+n})$ with $\partial(S - T) = 0$ and U is a **bounded** open subset of $\mathbf{R}^{m+n}$, we set

$$\mathbf{G}(S, T; U) = \inf\{\mathbf{M}(Y): Y \in \mathbf{S}_{m+1,0}(\mathbf{R}^{m+n})$$
$$\text{with spt}[(S - T) - \partial Y] \subset [\mathbf{R}^{m+n} \sim U]\}.$$

Whenever $S, T \in \mathbf{S}_m(\mathbf{R}^{m+n})$ with $\partial(S - T) = 0$, we set

$$\mathbf{G}(S, T) = \inf\{\mathbf{M}(Y): Y \in \mathbf{S}_{m+1}(\mathbf{R}^{m+n}) \text{ with } S - T = \partial Y\},$$

$$\mathbf{G}(S) = \mathbf{G}(S, 0) \quad (\text{in case } \partial S = 0),$$

$$\mathbf{GS}(S, T) = \inf\{\mathbf{MS}(Y): Y \in \mathbf{S}_{m+1}(\mathbf{R}^{m+n}) \text{ with } S - T = \partial Y\},$$

$$\mathbf{GS}(S) = \mathbf{GS}(S, 0) \quad (\text{in case } \partial S = 0);$$

in case $S, T \in \mathbf{S}_{m,0}(\mathbf{R}^{m+n})$ it is no further restriction to require that $Y \in \mathbf{S}_{m+1,0}(\mathbf{R}^{m+n})$. $\mathbf{G}(\cdot, \cdot)$ and $\mathbf{GS}(\cdot, \cdot)$ are metrics. $\mathbf{G}(\cdot)$ is a norm.

Caution. The $\mathbf{G}$ metric on real cycles does not admit a natural restriction to subsets of $\mathbf{R}^{m+n}$ since, in particular, cycles no longer necessarily bound. The $\mathbf{F}$ and $\mathbf{FS}$ metrics are well behaved under restrictions to, say, compact Lipschitz neighborhood retracts: e.g., slice the X's and Y's by the function dist$(A, \cdot)$ and retract. One can additionally use the Isoperimetric Inequality 2.3 to show that the **GS** metric is well behaved under restriction to compact Lipschitz neighborhood retracts.

When S, T, X, Y above are restricted to being integer multiplicity currents, corresponding definitions apply with **F** replaced by $\mathscr{F}$ and **G** replaced by $\mathscr{G}$; there is no need for an "**FS**" or a "**GS**" since mass dominates size for integer multiplicity currents.

REFERENCES

[**A1**] F. Almgren, *The homotopy groups of the integral cycle groups*, Topology **1** (1962), 257–299.

[**A2**] ______ , *Optimal isoperimetric inequalities* (preprint).

[**AS**] F. Almgren and B. Super, *Multiple valued functions in the geometric calculus of variations*, Astérisque **118** (1984), 13–22.

[**FH**] H. Federer, *Geometric measure theory*, Springer-Verlag, 1969.

[**FF**] H. Federer and W. H. Fleming, *Normal and integral currents*, Ann. of Math. **72** (1960), 458–520.

[**FW**] W. H. Fleming, *Flat chains over a finite coefficient group*, Trans. Amer. Math. Soc. **121** (1966), 160–186.

[**S1**] B. Solomon, *Lipschitz spaces of multiple valued functions and the closure theorem*, Ph.D. Thesis, Princeton University, June 1982.

[**S2**] ______ , *A new proof of the closure theorem for integral currents*, Indiana Univ. Math. J. **33** (1984), no. 3, 393–418.

[**YL**] L. C. Young, *Some extremal questions for simplicial complexes*, V., Rend. Circ. Mat. Palermo Ser. II **2** (1963), 257–274.

PRINCETON UNIVERSITY

Proceedings of Symposia in Pure Mathematics
Volume **44** (1986)

Local Estimates for Minimal Submanifolds
in Dimensions Greater than Two

MICHAEL T. ANDERSON[1]

1. Introduction. This paper deals with local curvature estimates for minimal submanifolds of Euclidean space $\mathbf{E}^N$ and more general Riemannian manifolds N. An interesting feature of our estimates is that they hold only in dimensions greater than two, and for arbitrary codimension. Curvature estimates for minimal surfaces in 3-manifolds have been obtained by a number of authors under certain conditions: for surfaces in $\mathbf{E}^3$ with assumptions on the Gauss map [9], for stable surfaces [14, 11], for embedded discs [13, 2], and for embedded surfaces in 3-manifolds of positive Ricci curvature [6]. In higher dimensions, there are curvature estimates for stable minimal hypersurfaces [14, 12], up to dimension $N \leqslant 7$.

Our main results are as follows: The proofs appear in §2.

THEOREM A (INTERIOR CURVATURE ESTIMATE). *Let Ω be a domain in $\mathbf{E}^N$ and let $\psi\colon B^n \to \Omega$ be a proper minimal immersion of the n-disc in Ω. If $n \geqslant 3$, then*

$$(1.1) \qquad \sup_{x \in \Omega_s} |A|^2(x) \leqslant C\left(\frac{1}{s}, \int_{B^n} |A|^n \, dV \right),$$

where $s(x) = \mathrm{dist}(x, \partial\Omega)$, $\Omega_s = \{ p \in B^n \colon \mathrm{dist}(\psi(p), \partial\Omega) \geqslant s \}$ and $|A|$ is the norm of the second fundamental form of ψ.

THEOREM B (BOUNDARY CURVATURE ESTIMATE). *Let S be a C^1 embedded $(n-1)$-sphere in $\mathbf{E}^N$ and let $\psi\colon B^n \to \mathbf{E}^N$ be a C^1 minimal immersion of the closed n-disc with $\psi(\partial B^n) = S$. If $n \geqslant 3$, then*

$$(1.2) \qquad \sup_{x \in B^n} |A|^2(x) \leqslant C\left(\int_{B^n} |A|^n \, dV \right).$$

1980 *Mathematics Subject Classification.* Primary 53A10, 53C42, 49F15.
[1]N.S.F. Mathematical Sciences Postdoctoral Fellow.

© 1986 American Mathematical Society
0082-0717/86 $1.00 + $.25 per page

More generally, if S belongs to a compact subset $\mathscr{D}$ of the space of C^1 embeddings $\mathrm{Emb}(S^{n-1}(1), \mathbf{E}^N)$, *then (1.2) holds, with constant C depending in addition only on* $\mathscr{D}$.

It is worthwhile to point out that there are no stability or embeddedness assumptions in these theorems. Roughly speaking, the theorems indicate that an L^n bound on the curvature of a minimal n-disc implies the existence of an L^∞ bound. We note that it is rather well known that if $\int_{B^n} |A|^n \, dV$ is sufficiently small, then one may obtain an L^∞ estimate using the de Giorgi–Nash–Moser method, cf. §2. The same results hold if one assumes a bound on $\int_{B^n} |A|^p \, dV$, $p > n$, but are much easier to prove in this case.

We note that $\mathscr{A} = \int_{B^n} |A|^n \, dV$ is an intrinsic isometric invariant of the Riemannian manifold B^n; in fact, $|A|^2$ is the negative of the scalar curvature of a minimal immersion, suitably normalized. Actually, one may show that $\mathscr{A}$ is a projective invariant. It would be interesting to determine if the estimates above hold with bounds on other curvature integrals replacing the bound on $\mathscr{A}$. We show however in §2 below, that a bound on the total absolute curvature, i.e. the volume of the image of the Gauss map, does not suffice.

Theorems A and B lead immediately to certain compactness theorems for spaces of minimal immersions of the n-disc in $\mathbf{E}^N$; these are presented at the end of §2. In §3, we show that these theorems also hold for "sufficiently small" domains in arbitrary Riemannian manifolds. We also show that one may relax somewhat the assumption on the minimality of the immersion.

2. Proof of the estimates. In this section, unless otherwise stated, Ω will denote an arbitrary domain, i.e. a connected open set, in Euclidean space $\mathbf{E}^N$. We consider proper minimal immersions ψ of an n-manifold M^n in Ω, i.e. immersions whose mean curvature vector is identically zero. We will let $\Sigma^n = \psi(M^n)$ denote the immersed submanifold of $\mathbf{E}^N$.

We begin with the following well-known result.

THEOREM 2.1 (COMPACTNESS THEOREM). *Let $\{\Sigma_i\}$ be a sequence of properly immersed minimal submanifolds in a ball $U \subset \mathbf{E}^N$. Suppose there is a constant C such that $|A_{\Sigma_i}|^2(x) \leqslant C, \ \forall x \in \Sigma_i$. Then there is a subsequence $\{\Sigma_{i'}\}$ which converges in the C^k topology, for all $k \geqslant 2$, to a properly immersed minimal submanifold $\Sigma_\infty \subset U$.*

By C^k convergence, we mean the following. For every $p \in \Sigma_\infty$, there is a neighborhood V in $\mathbf{E}^N$ such that $\Sigma_\infty \cap V$ and $\Sigma_{i'} \cap V$, for i' sufficiently large, may be graphed over the tangent plane $T_p \Sigma_\infty$ by the normal exponential map to Σ_∞ in $\mathbf{E}^N$: we require these graphing functions to converge, in the usual C^k topology, to that of Σ_∞.

The proof of the theorem is not difficult; we refer to [10] for details.

The next lemma is also well known; we include a proof to incorporate the generalizations in §3.

LEMMA 2.2. *Let Σ be a proper minimal immersion of a manifold M^n in Ω. Suppose $H^c_{n-1}(M^n, \mathbf{Z}) = 0$. Then, for almost all balls $B \Subset \Omega$, $\Sigma \cap \bar{B}$ is a collection of minimal submanifolds each having exactly one boundary component.*

PROOF. Let $\psi\colon M^n \to \Omega$ denote the immersion and let r_x denote the distance function in $\mathbf{E}^N$ from x. Then $\Sigma \cap \bar{B} = \psi[(r_x \circ \psi)^{-1}[0, s]]$, for some s. For almost all s, this is a collection of smooth manifolds with boundary. The set $(r_x \circ \psi)^{-1}[0, s]$ is a collection of disjoint domains in M^n; suppose one of them, denoted by W, has more than one boundary component. Since $H^c_{n-1}(M^n, \mathbf{Z}) = 0$, each boundary component bounds a domain $\Delta \subset M$ and we have $\psi(\Delta) \not\subset B$. However, $\psi(\partial\Delta) \subset \partial B$. Since $r_x \circ \psi$ is a subharmonic function on Δ and $r_x \circ \psi|_{\partial\Delta} \equiv s$, we obtain a contradiction. $\square$

We now begin the proofs of Theorems A and B.

PROOF OF THEOREM A. If (1.1) were false, there must exist a sequence of minimal immersions Σ_i of the disc B^n, $n \geqslant 3$, such that

$$(2.1) \qquad \int_{\Sigma_i} |A_i|^n \, dV < M, \qquad \sup_{x \in \Omega_s} |A_i|^2 \to \infty \quad \text{as } i \to \infty,$$

for some $s \in (0, \operatorname{diam}\Omega)$, constant M; A_i denotes the second fundamental form of Σ_i.

First, let $\varepsilon_0 = 1/4\sqrt{n} \cdot 1/C_s$, where C_s is the Sobolev constant of M, cf. [8]. Divide the interval $[0, s/2]$ into N equal subintervals, where $N = M/\varepsilon_0$. Then for a subsequence of $\{\Sigma_i\}$, denoted also $\{\Sigma_i\}$, there is a fixed interval $I_0 \subset [0, s/2]$, such that

$$(2.2) \qquad \int_{\Sigma_i \cap A(I_0)} |A_i|^n \, dV < \varepsilon_0,$$

where $A(I_0) = \{x \in \Omega\colon \operatorname{dist}(x, \partial\Omega) \in I_0\}$. It follows from Simons' equation [15] that $|A|$ satisfies the elliptic inequality $\Delta|A| + 2|A|^3 \geqslant 0$. From the de Giorgi–Nash–Moser method in elliptic P.D.E., one may deduce from (2.2) that

$$(2.3) \qquad \sup_{x \in A(I_1)} |A_i|^2(x) \leqslant \frac{C}{L_0^2}\left[\int_{\Sigma_i \cap A(I_0)} |A_i|^n \, dV\right]^{2/n},$$

where I_1 is the middle interval of I_0, $L_0 = MR/2\varepsilon_0$ (see e.g. Proposition 2.1 of [3] for further details).

Next, let $\alpha_i^2 = \sup_{x \in \Omega_{L_1}} |A_i|^2$, where L_1 is the smaller of the two endpoints of I_1. Let $x_i \in \Sigma_i \cap \bar{\Omega}_{L_1}$ be a point realizing the supremum. By (2.1), $\alpha_i^2 \to \infty$ and by (2.3), $x_i \notin A(I_1)$, for i sufficiently large. Define minimal submanifolds $\tilde{\Sigma}_i$ by $\tilde{\Sigma}_i = \delta_i(\Sigma_i)$, where δ_i is the dilation of $\mathbf{E}^N$ by the factor α_i, centered at x_i. Let $\tilde{A}_i$ denote the second fundamental form of $\tilde{\Sigma}_i$. Clearly, $\sup|\tilde{A}_i|^2 \leqslant 1$ and $|\tilde{A}_i|^2(x_i) = 1$. Further

$$(2.4) \qquad \operatorname{dist}(x_i, \partial\tilde{\Omega}) = \alpha_i \cdot \operatorname{dist}(x_i, \partial\Omega) \geqslant \alpha_i \cdot L_0 \overset{i \to \infty}{\to} \infty.$$

Theorem 2.1 implies that a subsequence of $\{\tilde{\Sigma}_i\}$, again denoted by $\{\tilde{\Sigma}_i\}$, converges in the C^k topology on compact sets in $\mathbf{E}^N$, $k \geqslant 2$, to a smoothly immersed minimal submanifold $\tilde{\Sigma}_\infty$ in $\mathbf{E}^N$. Note that for $x_\infty = \lim x_i$,

$$(2.5) \qquad |\tilde{A}_\infty|^2(x_\infty) = 1.$$

Equation (2.4) also implies that $\tilde{\Sigma}_\infty$ is a complete minimal immersion. If $\tilde{\Sigma}_\infty$ had more than one end, then for R sufficiently large, $\tilde{\Sigma}_\infty \cap B^{N-1}(R)$ must have components with more than one boundary component in $S^{N-1}(R)$. Since $\tilde{\Sigma}_i$ converges to $\tilde{\Sigma}_\infty$ smoothly, this contradicts Lemma 2.2. Further, by the dominated convergence theorem

$$\int_{\tilde{\Sigma}_\infty} |\tilde{A}_\infty|^n \, dV = \lim_{i \to \infty} \int_{\tilde{\Sigma}_i} |\tilde{A}_i|^n \, dV \leqslant M,$$

so that Σ_∞ has finite total scalar curvature in the sense of [3]. However, Theorem 5.2 of [3] implies that any complete minimal immersion of finite total curvature in $\mathbf{E}^N$, having one end and of dimension > 2, must be a plane. This contradicts (2.5). $\square$

It is appropriate at this point to discuss various aspects of the hypotheses in Theorem A.

REMARKS. (1) Theorem A is of course false for $n = 2$. For example, let E denote Enneper's minimal surface in $\mathbf{E}^3$ and consider the family of surfaces $E_r = r^{-1}(E \cap B(r))$. However, an estimate of the form (1.1) does hold for embedded minimal discs in $\mathbf{E}^3$; cf. [13, 2]. There are no known analogous estimates for surfaces of higher codimension.

(2) A brief examination of the proof shows that the topological assumption in Theorem A may be relaxed. In fact, (1.1) holds for a proper minimal immersion $\psi\colon M^n \to \Omega$, provided $H^c_{n-1}(M^n, \mathbf{Z}) = 0$. (The proof is identical.) We note that some topological assumption is necessary. Consider for instance the higher dimensional catenoid $C \subset \mathbf{E}^{n+1}$, diffeomorphic to $S^{n-1} \times \mathbf{R}$. The family of surfaces $C_r = r^{-1}(C \cap B(r))$ contradict the estimate (1.1), for r large.

(3) It would be very interesting to obtain estimates of the form (1.1) with other curvature integrals in place of the total scalar curvature $\mathscr{A}$. For instance, one might consider $\int |K| \, dV$, the total absolute curvature in the sense of Chern–Lashof [5], i.e. the volume of the image of the Gauss map. However, in [4], Bombieri has shown the existence of an infinite family of minimal submanifolds, all diffeomorphic to $B^3 \times S^2$, with boundary the "Clifford torus" $S^2 \times S^2 \subset S^5 \subset \mathbf{E}^6$, converging to the cone $C(S^2 \times S^2)$. In particular, the curvature is not bounded in a neighborhood of the origin. However, one may compute, although we do not include the computation, that the total absolute curvature remains uniformly bounded.

We now turn to the proof of the boundary estimate, Theorem B. Given an embedding $F\colon S \to \mathbf{E}^N$ of a closed manifold in $\mathbf{E}^N$, define the C^1 norm, $\|S\|_1$, as follows: first, take the usual C^1 norm $\|F\|_1$, where F is considered as a vector-valued function on S and S is given the induced metric: set $\|S\|_1 = \inf \|F \circ \psi\|_1$, where ψ runs over the diffeomorphism group of S.

PROOF OF THEOREM B. The proof resembles that of Theorem A. If (1.2) were false, there would exist a sequence of minimal discs Σ_i and C^1 boundaries $S_i = \partial\Sigma_i \in \mathcal{D}$ such that $\int_{\Sigma_i} |A_i|^n \, dV < M$, for some constant M, and $\alpha_i^2 = \sup_{\Sigma_i} |A_i|^2 \to \infty$ as $i \to \infty$. Let $x_i \in \Sigma_i$ be points realizing the supremum and let $\tilde{\Sigma}_i$ be the dilation of Σ_i by the factor d_i, centered at x_i. As in Theorem A, $\tilde{\Sigma}_i$ subconverges smoothly to a minimal submanifold $\tilde{\Sigma}_\infty$ having exactly one end. Further $\int_{\tilde{\Sigma}_\infty} |\tilde{A}_\infty|^n \leqslant M$ and, for $x_\infty = \lim x_i$, $|\tilde{A}_\infty|(x_\infty) = 1$. If $\tilde{\Sigma}_\infty$ is complete, one obtains a contradiction as before. If $\tilde{\Sigma}_\infty$ is not complete, then $\partial\tilde{\Sigma}_\infty$ is the limit of the boundaries $\partial\tilde{\Sigma}_i$. Since $\partial\Sigma_i \in \mathcal{D}$, so that $\|\partial\Sigma_i\|_1$ is uniformly bounded, it follows that $\|\partial\tilde{\Sigma}_\infty\|_1 = 0$, i.e. $\partial\tilde{\Sigma}_\infty$ is an affine $(n-1)$-plane. By the reflection principle for minimal submanifolds in $\mathbf{E}^N$ (cf. [1]), one may reflect $\tilde{\Sigma}_\infty$ through $\partial\tilde{\Sigma}_\infty$ to obtain a complete minimally immersed submanifold of $\mathbf{E}^N$. Clearly, this complete submanifold also has exactly one end and has finite total scalar curvature. Thus one obtains once more a contradiction to Theorem 5.2 of [3]. $\square$

The estimates (1.1) and (1.2), in conjunction with the compactness theorem, Theorem 2.1, yield immediately the compactness of certain spaces of minimal immersions.

COROLLARY 2.3. *Let Ω be a domain in $\mathbf{E}^N$ containing the origin 0. Then the space of proper minimal immersions $\Sigma = \psi(B^n)$ of the n-disc B^n in Ω satisfying*

(i) $0 \in \Sigma$ *and* $n \geqslant 3$,

(ii) $\int_{B^n} |A|^n \, dV \leqslant C$,

is compact in the C^k topology, for all $k \geqslant 2$. $\square$

COROLLARY 2.4. *Let $\mathcal{D}$ be a compact subset of the space of C^r embeddings of the sphere S^{n-1} in $\mathbf{E}^N$, $r \geqslant 1$. Then the space of C^r minimal immersions $\Sigma = \psi(\overline{B^n})$ of the closed n-disc in $\mathbf{E}^N$ with $\partial\Sigma \in \mathcal{D}$, satisfying*

(i) $n \geqslant 3$,

(ii) $\int_{B^n} |A|^n \, dv \leqslant C$,

is compact in the C^r topology. $\square$

3. Generalizations. We conclude with several remarks concerning generalizations of Theorems A and B.

I. *Riemannian manifolds.* Let N be a Riemannian manifold and $\Omega \subset N$ a domain with the following property: for every $x \in \Omega$, there is a nonnegative, convex function $f_x \colon \Omega \to \mathbf{R}$ such that $f_x^{-1}(0) = x$. We note that every sufficiently small domain in a given Riemannian manifold will satisfy this condition. If N is simply connected and has nonpositive sectional curvature, then Ω may be an arbitrary domain in N.

All the results of §2 hold for minimal discs immersed in $\Omega \subset N$ as above; for the boundary estimate, we require that the boundary lie within $\overline{\Omega}$. To see this, one may prove the compactness theorem and Lemma 2.2 as before. For Theorem A, one may establish the estimate (2.3) in a similar manner. One must use Simons' equation for a general Riemannian manifold, cf [14] as well as the Sobolev inequality, cf. [7]. The dilations in this case are dilations of the metric on Ω. More precisely, one may isometrically embed N in a Euclidean space $\mathbf{E}^L$. The dilation of the metric ds^2 at $p \in \Omega$ by the factor α is the metric induced on Ω by dilating the submanifold $N \subset \mathbf{E}^L$ by the factor α at p. As $\alpha \to \infty$, these dilations converge to a flat metric on $\mathbf{E}^N$, identified as $T_p N$. With these modifications, the proof of Theorem A, as well as Theorem B, proceeds as before.

II. *Nonminimal discs.* The results of §2 did not make use of minimality in a particular strong way. As an example of estimates for more generally immersed discs, we mention the following result. A submanifold S of $\mathbf{E}^N$ is said to have the convex-hull property if, for all balls B in $\mathbf{E}^N$, $S \cap B \subset \mathscr{C}(\partial(S \cap B))$, where $\mathscr{C}$ denotes convex hull.

PROPOSITION 3.1. *Let* $\psi\colon \overline{B}^n \to \mathbf{E}^N$ *be a* C^2 *immersion of the closed n-disc with* $\psi|_{S^{n-1}}\colon S^{n-1} \to \mathbf{E}^N$ *a* C^2 *embedding of the sphere. Suppose* ψ *satisfies the convex-hull property. Then, if* $n \geqslant 3$ *and* $p > n$,

$$\sup_{x \in B^n} |A|^2(x) \leqslant C\left(\int_{B^n} |A|^n \, dV, \int_{B^n} |H|^p \, dV \right),$$

where H is the mean curvature vector of the immersion.

PROOF. The proof is the same as that of Theorem B. The convex-hull property ensures the validity of Lemma 2.2. Note that under a dilation by the factor α, the integral $\int_{B^n} |H|^p \, dV$ transforms by the factor α^{n-p}: in particular, as $\alpha \to \infty$, the integral converges to 0. Thus as in Theorem B, the dilated surfaces $\tilde{\Sigma}_i$ will converge, in the C^2 topology, to a minimal surface. The rest of the argument proceeds as before. $\square$

REFERENCES

1. W. K. Allard, *On the first variation of a varifold: boundary behavior*, Ann. of Math. (2) **101** (1975), 418–446.

2. M. T. Anderson, *Curvature estimates for minimal surfaces in 3-manifolds*, Ann. Sci. École Norm. Sup. **18** (1) (1985).

3. ______, *The compactification of a minimal submanifold in Euclidean space by the Gauss map* (submitted to Acta Math.)

4. E. Bombieri, *Recent progress in the theory of minimal surfaces*, Enseign. Math. (2) **25** (1979), 1–8.

5. S.-S. Chern and R. Lashof, *On the total curvature of immersed manifolds*. I, Amer. J. Math. **79** (1957), 306–318.

6. H. I. Choi and R. Schoen, *The space of minimal embeddings of a surface into a three-dimensional manifold of positive Ricci curvature* (preprint).

7. D. Hoffman and J. Spruck, *Sobolev and isoperimetric inequalities for Riemannian manifolds*, Comm. Pure Appl. Math. **27** (1974), 715–727.

8. J. H. Michael and L. Simon, *Sobolev and mean-value inequalities on generalized submanifolds of $\mathbf{R}^n$*, Comm. Pure Appl. Math. **26** (1973), 361–379.

9. R. Osserman, *On the Gauss curvature of minimal surfaces*, Trans. Amer. Math. Soc. **96** (1960), 115–128.

10. J. Pitts, *Existence and regularity of minimal submanifolds in Riemannian manifolds*, Math. Notes, Princeton Univ. Press, Princeton, N. J., 1981.

11. R. Schoen, *Estimates for stable minimal surfaces in three-dimensional manifolds*, Ann. of Math. Studies vol. 103, Princeton Univ. Press, Princeton, N. J., 1983.

12. R. Schoen and L. Simon, *Regularity of stable minimal hypersurfaces*, Comm. Pure Appl. Math. **34** (1981), 741–797.

13. ______, *Regularity of simply connected surfaces with quasi-conformal Gauss map*, Ann. of Math. Studies vol. 103, Princeton Univ. Press, Princeton N. J., 1983.

14. R. Schoen, L. Simon and S.-T. Yau, *Curvature estimates for minimal hypersurfaces*, Acta Math. **134** (1975), 270–288.

15. J. Simons, *Minimal varieties in Riemannian manifolds*, Ann. of Math. (2) **88** (1968), 62–105.

INSTITUT DES HAUTES ÉTUDES SCIENTIFIQUES, FRANCE

Current address: California Institute of Technology

Proceedings of Symposia in Pure Mathematics
Volume **44** (1986)

Second Variation Estimates for Minimal Orbits

JOHN E. BROTHERS

1. Introduction. Consider a compact Lie group G of isometries of a riemannian manifold M. It is well known that the minimal principal orbits of G are precisely those on which the volume function $\mathbf{v}$, which corresponds to $p \in M$, the volume of the orbit of p, is critical. In [**B**] we began an investigation of the stability of minimal orbits, obtaining results which, in the degenerate case where the hessian $H_\mathbf{v}$ of $\mathbf{v}$ vanishes on the minimal orbit, were sharp enough to enable us to verify stability of several naturally occuring examples.

However, the methods used in [**B**] did not involve *a priori* estimates of second variation, resulting in a theory which did not relate the size of $H_\mathbf{v}$ with stability. In the present paper we derive a lower estimate for second variation of a minimal orbit N, which involves the relationship between $H_\mathbf{v}$ and a measure of the degree to which the distribution of normal planes to the orbits is not involutive. This implies that if $H_\mathbf{v}$ is sufficiently large then N is stable. We also use eigenfunctions of the laplacian on N to show that if $H_\mathbf{v}$ is sufficiently small relative to this degree of noninvolutivity then N is not stable.

2. Second variation. Our notation and terminology will be consistent with [**B**]. For future reference we recall the formula for the second variation $\delta^{(2)}$ of a minimal principal orbit N which was derived in [**B**]. In order to simplify notation we assume $\mathbf{v}|N = 1$.

Denote $\dim M = m$, $\dim N = n$, and choose smooth invariant orthonormal vector fields $W_1, \ldots, W_{m-n}$ which are orthogonal to the orbits of G in a neighborhood of N. Then the vector fields $W_{ik} = [W_i, W_k]$ are invariant and tangent to N. For $i, k = 1, \ldots, m - n$, denote

$$\nu_i = \operatorname{card}\{ k: W_{ik} \neq 0\};$$

assume $\nu_i \neq 0$ for $i \leqslant i_1$, $\nu_i = 0$ for $i > i_1$.

1980 *Mathematics Subject Classification.* Primary 53C42; Secondary 49F22.

© 1986 American Mathematical Society
0082-0717/86 $1.00 + $.25 per page

Recalling that for V a smooth vector field on N, $\delta^{(2)}(V)$ depends only on the normal component W of V, we denote $W = \sum_{i=1}^{m-n} a^i W_i$. Then

$$\delta^{(2)}(W) = \int_N H_\nu(W, W) + \sum_{i=1}^{m-n} \|da^i\|^2 - \sum_{i<k} \langle W_{ik}, a^i da^k - a^k da^i \rangle \, d\mathcal{H}^n.$$

For the remainder of the paper we assume the existence of a unit G-invariant vector field X on N such that $W_{ik} = c_{ik}(\nu_i \nu_k)^{-1/2} X$ for $i, k = 1, \ldots, i_1$. We also assume the trajectories of X to be circles; these have constant length L since X is invariant.

3. Stability.

3.1. LEMMA. *Denote by C_L the circle of length L. Fix $\alpha \in \mathbf{R}$, $0 \leqslant \lambda < 1/4$, $0 < r \in C^\infty(C_L)$, and denote by $\mu_\lambda(r, \alpha)$ the infimum of the numbers*

$$L^{-1} \int_{C_L} r^2 (\phi^2 - \phi + \lambda) \, d\mathcal{H}^1$$

corresponding to $\phi \in C^\infty(C_L)$ with $L^{-1} \int_{C_L} \phi \, d\mathcal{H}^1 = \alpha$. Denote $\beta = [(2 - 4\lambda - (1 - 4\lambda)^{1/2} - 2\alpha)/(1 - 4\lambda)]^2$.

(i) *If $\beta < 1$ then $\lim_{c \to \infty} \mu_\lambda(c + f, \alpha) = -\infty$ whenever $f \in C^\infty(C_L)$.*
(ii) *If $\beta \geqslant 1$ then*

$$\mu_\lambda(r, \alpha) \geqslant 1.52(\lambda - 1/4)\pi^{-2} L \int_{C_L} |r'|^2 \, d\mathcal{H}^1 .$$

PROOF. We can assume $L = 1$; denote $C = C_L$. We first consider the case $\lambda = 0$, hence $\beta = (1 - 2\alpha)^2$. Consider the functional I on $\mathcal{L}^2(C, r^2 \mathcal{H}^1)$ defined by

$$I(\phi) = \int_C r^2 (\phi^2 - \phi) \, d\mathcal{H}^1 .$$

I is continuous and strictly convex, hence the restriction of I to $\{\phi: \int_C \phi \, d\mathcal{H}^1 = \alpha\}$ achieves its absolute minimum uniquely at some ϕ_0.

Applying the method of Lagrange multipliers we consider the functional J on $\mathcal{L}^2(C, r^2 \mathcal{H}^1)$ defined by $J(\phi) = I(\phi) + \lambda \int_C \phi \, d\mathcal{H}^1$ and obtain $r^2(2\phi_0 - 1) + \lambda = 0$; hence

$$\lambda = (1 - 2\alpha)/\int_C r^{-2} \, d\mathcal{H}^1 ,$$

$$\phi_0 = \frac{1}{2}\left[1 - (1 - 2\alpha)r^{-2}/\int_C r^{-2} \, d\mathcal{H}^1\right],$$

$$I(\phi_0) = \frac{1}{4}\left[-\int_C r^2 \, d\mathcal{H}^1 + \beta/\int_C r^{-2} \, d\mathcal{H}^1\right].$$

Next write $r = c + f$ with $c = \int_C r \, d\mathcal{H}^1$, so that $\int_C f \, d\mathcal{H}^1 = 0$, and denote $F(c) = -c^2 + \beta/\int_C r^{-2} \, d\mathcal{H}^1$. Thus

$$I(\phi_0) = \frac{1}{4}\left(F(c) - \int_C f^2 \, d\mathcal{H}^1\right).$$

We estimate $F(c)$ as follows:

Denote $c_0 = -\inf f \geqslant 0$. $c_0 = 0$ implies $f = 0$, hence $I(\phi_0) = c^2(\beta - 1)/4$; this is consistent with our assertions because $0 < \alpha < 1$ if and only if $\beta < 1$. Assume $c_0 > 0$; thus $c > c_0$. We can assume $c_0 = 1$. Denote $x = 1/c \in (0, 1)$ and

$$G(x) = c^{-2}F(c) = -1 + \beta \left[\int_C (1 + xf)^{-2} \, d\mathscr{H}^1 \right]^{-1}.$$

For the proof of (i) we observe that $G(x) \to \beta - 1$ as $x \to 0^+$; consequently, if $\beta < 1$, $F(c) \to -\infty$ as $c \to \infty$. Assuming $\beta \geqslant 1$ we next apply Taylor's theorem to the function $g: [\,0, 1) \to \mathbf{R}$ defined by

$$g(t) = \int_C (1 + tf)^{-2} \, d\mathscr{H}^1$$

to obtain

$$g(x) = 1 + 6 \int_0^x (x - t) \int_C f^2 (1 + tf)^{-4} \, d\mathscr{H}^1 \, dt,$$

because $\int_C f \, d\mathscr{H}^1 = 0$ implies $g'(0) = 0$. Now $f > -1$ implies $1 + tf > 1 - t > 0$ for $t \in [0, 1\,)$; hence denoting $\varepsilon = \int_C f^2 \, d\mathscr{H}^1$ we have

$$g(x) < 1 + \varepsilon Q(x), \qquad Q(x) = \varepsilon x^2(3 - 2x)/(1 - x)^2,$$

$$F(c) = x^{-2}G(x) > -\varepsilon \left[x^2 \left(Q(x)^{-1} + \varepsilon \right) \right]^{-1} = \Phi(x, \varepsilon).$$

For each $\varepsilon > 0$, $\Phi(x, \varepsilon)$ achieves its absolute minimum $m(\varepsilon)$ on $[0, 1]$ at a unique $x_\varepsilon \in (0, 1)$; moreover, m is nonincreasing. One uses the implicit function theorem to show that x_ε is a smooth function of $\varepsilon > 0$, whence it follows that $m(\varepsilon)/\varepsilon$ is an increasing function of ε because

$$\frac{d}{d\varepsilon} \left[x_\varepsilon^2 \left(Q(x_\varepsilon)^{-1} + \varepsilon \right) \right] = x_\varepsilon^2 > 0.$$

We conclude that for each $\varepsilon_0 > 0$,

$$F(c) > \begin{cases} m(\varepsilon_0), & \varepsilon \leqslant \varepsilon_0, \\ \left(\dfrac{m(\varepsilon_0)}{\varepsilon_0} \right) \varepsilon, & \varepsilon > \varepsilon_0, \end{cases}$$

hence

$$\mu(r, \alpha) - I(\phi_0)$$

$$> -\tfrac{1}{4} \sup\left\{ -m(\varepsilon_0) + \varepsilon, \left(1 - m(\varepsilon_0)/\varepsilon_0\right)\varepsilon \right\}$$

$$(*) \qquad \geqslant -\tfrac{1}{4} \sup\left\{ (2\pi)^2 \varepsilon_0 \left(c_{\varepsilon_0}^2 - (2\pi)^{-2} \right) \right.$$

$$\left. + \int_C |f|^2 \, d\mathscr{H}^1 \,,\, c_{\varepsilon_0}^2 \int_C |f'|^2 \, d\mathscr{H}^1 \right\}$$

by Poincaré's inequality, where $c_{\varepsilon_0} = (2\pi)^{-1}(1 - m(\varepsilon_0)/\varepsilon_0)^{1/2}$.

On the other hand, for $c \geqslant (2\pi)^{-1}$,

$$\int_C c^2|f'|^2 - |f|^2 \, d\mathcal{H}^1 \geqslant \left(c^2 - (2\pi)^{-2}\right)\int_C |f'|^2 \, d\mathcal{H}^1 .$$

Minimizing $\int_0^1 h'(x)^2 \, dx$ over the family of smooth $h\colon [0,1] \to \mathbf{R}$ with $h(0) = -1$, $h(1) = -1 + 2k\pi$, k an integer, and $\int_0^1 h(x)\, dx = 0$, one finds that an extremal satisfies $h'' = $ constant and so the minimum is achieved by $h(x) = 6(x - x^2) - 1$ with $\int_0^1 h'(x)^2 \, dx = 12$. We conclude that

$$\int_C c^2|f'|^2 - |f|^2 \, d\mathcal{H}^1 \geqslant 12\left(c^2 - (2\pi)^{-2}\right);$$

setting $\varepsilon_0 = 3/\pi^2$, $c = c_{\varepsilon_0}$, and referring to $(*)$ we thus obtain

$$\mu(r, \alpha) > -\frac{1}{4} c_{3/\pi^2}^2 \int_C |f'|^2 \, d\mathcal{H}^1 .$$

Finally, x_{3/π^2} is the unique root in $(0,1)$ of

$$12x^3 - (36 + \pi^2)x^2 + 3(9 + \pi^2)x - 2\pi^2 = 0$$

and one approximates $x_{3/\pi^2} \in (0.5796962, 0.58)$, $c_{3/\pi^2}^2 \leqslant 1.52/\pi^2$.

Turning to the case where $0 < \lambda < 1/4$ we apply the preceding with ϕ replaced by

$$\psi = (1 - 4\lambda)^{-1}(\phi - \lambda_0), \qquad \lambda_0 = \tfrac{1}{2}\left(1 - (1 - 4\lambda)^{1/2}\right),$$

and observe that

$$L^{-1}\int_{C_L} \psi \, d\mathcal{H}^1 = (1 - 4\lambda)^{-1}(\alpha - \lambda_0) \in (0,1)$$

if and only if $\beta < 1$. Furthermore,

$$\int_{C_L} r^2(\phi^2 - \phi + \lambda) \, d\mathcal{H}^1 = (1 - 4\lambda)\int_{C_L} r^2(\psi^2 - \psi) \, d\mathcal{H}^1 .$$

3.2. Denote by λ the least eigenvalue of H_v; assume $\lambda \geqslant 0$ (for otherwise N clearly can not be stable).

Let $W = \sum_{i=1}^{m-n} a^i W_i$ be smooth on N and define

$$r^{ik} = \left[v_i^{-1}(a^i)^2 + v_k^{-1}(a^k)^2 \right]^{1/2}, \qquad i, k = 1,\ldots,i_1.$$

Note that r^{ik} is lipschitzian, hence dr^{ik} exists $\mathcal{H}^n$ almost everywhere on N.

Lemma. *Assume that for each $i < k$ it is true that either $c_{ik}^2 \leqslant 4\lambda$ or $c_{ik}^2 > 4\lambda$ and the interval*

$$\left(|c_{ik}|L/2\pi\right)\left(1 - \mu_{ik},\, 1 - \mu_{ik} + 2\mu_{ik}^2\right)$$

contains no even integer, where $\mu_{ik} = (1 - 4c_{ik}^{-2}\lambda)^{1/2}$. If $c_{ik}^2 \leqslant 4\lambda$ for each $i < k$ then $\delta^{(2)}(W) \geqslant 0$. Otherwise,

$$\delta^{(2)}(W) \geqslant \sum_{i<k}^{i_1} \int_N \|dr^{ik}\|^2 + 1.52(4\lambda - c_{ik}^2)(L/2\pi)^2\langle X, dr^{ik}\rangle^2 \, d\mathcal{H}^n,$$

where the sum is extended over those $i < k$ with $c_{ik}^2 > 4\lambda$.

PROOF. Following [**B**, 3.3] we set $A_{ik} = \{ p: r^{ik}(p) \neq 0 \}$, define $b_{ik} = 1$ if $c_{ik} \neq 0$ and $b_{ik} = 0$ otherwise, and for $i < k$ with $b_{ik} = 1$ define $\theta^{ik}: A_{ik} \to \mathbf{S}^1$ so that

$$\nu_i^{-1/2} a^i = r^{ik} \cos \theta^{ik}, \qquad \nu_k^{-1/2} a^k = r^{ik} \sin \theta^{ik}.$$

Denote $d\theta^{ik} = \theta^{ik*}\Omega$, where Ω is the positively oriented unit 1-form on $\mathbf{S}^1$. Then

$$\delta^{(2)}(W) \geq \sum_{i<k} b_{ik} \int_{A_{ik}} \nu_i^{-1}\|da^i\|^2 + \nu_k^{-1}\|da^k\|^2 + \lambda\left(\nu_i^{-1}(a^i)^2 + \nu_k^{-1}(a^k)^2\right)$$

$$-c_{ik}\langle X, (\nu_i \nu_k)^{-1/2}(a^i da^k - a^k da^i)\rangle\, d\mathscr{H}^n,$$

and each of these integrals I_{ik} is not less than

$$\int_{A_{ik}} \|dr^{ik}\|^2 + (r^{ik})^2\left(\langle X, d\theta^{ik}\rangle^2 - c_{ik}\langle X, d\theta^{ik}\rangle + \lambda\right) d\mathscr{H}^n.$$

Fix $i < k$ with $b_{ik} = 1$ and denote $r = r^{ik}$, etc; since the polynomial $x^2 - cx + \lambda \geq 0$ for $c^2 \leq 4\lambda$, we can assume $c^2 > 4\lambda$.

Inasmuch as the action of G commutes with the flow of X, the set P of orbits of X has a smooth manifold structure and the map Φ which associates the orbit of p with each $p \in N$ is a submersion. Fix a riemannian metric on P. It is shown in [**B**, 4.5] that, since $\operatorname{div} X = 0$, the Jacobian $J\Phi$ is constant on the orbits of X. Denoting $(J\Phi)^{-1}(q) = J\Phi(p)^{-1}$ for $\Phi(p) = q$, we use the co-area formula [**F**, 3.2.12] to write

$$\int_{A^*} r^2\left(\langle X, d\theta\rangle^2 - c\langle X, d\theta\rangle + \lambda\right) d\mathscr{H}^n = c^2 \int_{A_*} I(q)(J\Phi)^{-1}(q)\, d\mathscr{H}^{n-1}q,$$

where $A_* = P \sim \Phi(N \sim A_{ik})$, $A^* = \Phi^{-1}(A_*) \subset A_{ik}$, and for $q \in A_*$,

$$I(q) = \int_{\Phi^{-1}\{q\}} r^2(\phi^2 - \phi + \lambda/c^2)\, d\mathscr{H}^1, \qquad \phi = c^{-1}\langle X, d\theta\rangle.$$

Fixing $q \in A_*$ we apply 3.1 with $C_L = \Phi^{-1}\{q\}$ and λ replaced by λ/c^2. Since the curve $\theta|C_L$ represents an element of the fundamental group of $\mathbf{S}^1$, $\theta|C_L$ is homotopic to an integral multiple γ of the identity map of $\mathbf{S}^1$, and so

$$\alpha = L^{-1}\int_{C_L} \phi\, d\mathscr{H}^1 = (cL)^{-1}\int_{C_L} d\theta = (cL)^{-1}\int_{\theta|C_L} \Omega = 2\pi\gamma(cL)^{-1}.$$

Now $\beta = [(1 + \mu^2 - \mu - 2\alpha)/\mu^2]^2 \geq 1$ for each integer γ if and only if

$$\left|\frac{|c|L}{4\pi}(1 - \mu + \mu^2) - \gamma\right| \geq \frac{|c|L}{4\pi}\mu^2,$$

which is equivalent to there being no even integer in the interval $(|c|L/2\pi)(1 - \mu, 1 - \mu + 2\mu^2)$. We conclude that

$$I(q) \geq 1.52(c^{-2}\lambda - 1/4)(L/\pi)^2\int_{C_L} \langle X, dr\rangle^2\, d\mathscr{H}^1,$$

hence

$$I_{ik} \geq \int_{A^*} \|dr\|^2 + 1.52(4\lambda - c^2)(L/2\pi)^2\langle X, dr\rangle^2\, d\mathscr{H}^n.$$

We next write $(N \sim A_{ik}) \times \{0\} = \Pi^{-1}\{0\} \cap \Gamma$ where $\Pi: N \times \mathbf{R}^2 \to \mathbf{R}^2$ is the projection and

$$\Gamma = N \times \mathbf{R}^2 \cap \{(p, z): z = (a^i(p), a^k(p))\}.$$

Using the Morse–Sard theorem we choose a sequence $z_l = (\varepsilon_l, \delta_l) \in \mathbf{R}^2$, $l = 1, 2, \ldots$, of regular values of $\Pi|\Gamma$ such that $z_l \to 0$ as $l \to \infty$, and denote

$$a_l^i = a^i - \varepsilon_l, \qquad a_l^k = a^k - \delta_l,$$

$$A_l = N \sim \{p: a_l^i(p) = 0 = a_l^k(p)\}.$$

Then $N \sim A_l$ is the projection on N of $\Pi^{-1}\{z_l\} \cap \Gamma$, hence is a (possibly empty) smooth submanifold of N of codimension 2; we conclude that $\mathscr{H}^{n-1}(\Phi(N \sim A_l)) = 0$, hence $\mathscr{H}^n(N \sim A_l^*) = 0$, $A_l^* = \Phi^{-1}(P \sim \Phi(N \sim A_l)) \subset A_l$. From the preceding discussion it follows that the integrals $I_{ik,l}$ corresponding to a_l^i, a_l^k satisfy

$$I_{ik,l} \geq \int_N \|dr_l\|^2 + 1.52(4\lambda - c^2)(L/2\pi)^2 \langle X, dr_l \rangle^2 \, d\mathscr{H}^n.$$

Finally, it is clear that $I_{ik,l} \to I_{ik}$ as $l \to \infty$. Further, r_l is lipschitzian with $\|dr_l\| \leq n^{1/2}(\|da^i\| + \|da^k\|)$; we will show that $dr_l(p) \to dr(p)$ for $\mathscr{H}^n$ almost all $p \in N$, hence conclude using Lebesgue's dominated convergence theorem that the right side of the above inequality converges as $l \to \infty$ to

$$\int_N \|dr\|^2 + 1.52(4\lambda - c^2)(L/2\pi)^2 \langle X, dr \rangle^2 \, d\mathscr{H}^n.$$

For this purpose we first note that $\mathscr{H}^n(N \sim \cap_{l=1}^\infty A_l) = 0$ and fix $p \in \cap_{l=1}^\infty A_l$ such that $dr(p)$ exists. If $p \in A_{ik}$ then clearly $dr_l(p) \to dr(p)$ as $l \to \infty$. If $p \notin A_{ik}$ and p is a point of Lebesgue density 1 of $N \sim A_{ik}$ then $dr(p) = 0$ and

$$dr_l(p) = (v_i^{-1}\varepsilon_l^2 + v_k^{-1}\delta_l^2)^{-1/2}(-v_i^{-1}\varepsilon_l \, da^i(p) - v_k^{-1}\delta_l \, da^k(p)) = 0.$$

Inasmuch as $\mathscr{H}^n$ almost every $p \in N \sim A_{ik}$ is a point of density 1, our assertion follows.

3.3 THEOREM. *Assume that for each $i < k$ either $4c_{ik}^{-2}\lambda \geq 1$ or $4c_{ik}^{-2}\lambda < 1$ and*

$$|c_{ik}|L/2\pi \leq \max\{1, (0.8/\mu_{ik})(1 - \mu_{ik})/(1 - \mu_{ik} + 2\mu_{ik}^2)\},$$

where $\mu_{ik} = (1 - 4c_{ik}^{-2}\lambda)^{1/2}$. Then N is stable.

PROOF. Assume $4c_{ik}^{-2}\lambda < 1$. If $|c_{ik}|L/2\pi \leq 1$ then the corresponding integral in $\delta^{(2)}$ will be nonnegative by [**B**, 4.5]. For the remainder of the proof we can assume $\mu_{ik} \leq 1/2$ because $(0.8/x)(1 - x)/(1 - x + 2x^2) < 1$ for $1/2 \leq x \leq 1$.

We first verify stability for the case where

$$|c_{ik}|L/2\pi = 0.8/\mu_{ik} \quad \text{if } [\![(0.4/\mu_{ik})(1 - \mu_{ik} + 2\mu_{ik}^2)]\!] \leq (0.4/\mu_{ik})(1 - \mu_{ik}),$$

where $[\![\rho]\!]$ is the greatest integer in ρ, and otherwise

$$|c_{ik}|L/2\pi = (0.8/\mu_{ik})(1 - \mu_{ik})/(1 - \mu_{ik} + 2\mu_{ik}^2).$$

In the former case there is no even integer in the interval

$$(0.8/\mu_{ik})(1 - \mu_{ik}, 1 - \mu_{ik} + 2\mu_{ik}^2).$$

In the latter case there is no even integer in the interval

$$(0.8/\mu_{ik})(1 - \mu_{ik})/(1 - \mu_{ik} + 2\mu_{ik}^2)(1 - \mu_{ik}, 1 - \mu_{ik} + 2\mu_{ik}^2)$$

because

$$(0.8/\mu_{ik})\left[(1 - \mu_{ik} + 2\mu_{ik}^2) - (1 - \mu_{ik})^2/(1 - \mu_{ik} + 2\mu_{ik}^2)\right]$$
$$\leqslant 3.2\mu_{ik} < 2.$$

We conclude that 3.2 applies, hence N is stable because

$$\|dr^{ik}\|^2 + 1.52(4\lambda - c_{ik}^2)(L/2\pi)^2\langle X, dr^{ik}\rangle^2$$
$$\geqslant \left(1 - 1.52(c_{ik}\mu_{ik}L/2\pi)^2\right)\langle X, dr^{ik}\rangle^2.$$

Next assume $|c_{ik}|L/2\pi < 0.8/\mu_{ik}$ with

$$[\![(0.4/\mu_{ik})(1 - \mu_{ik} + 2\mu_{ik}^2)]\!] \leqslant (0.4/\mu_{ik})(1 - \mu_{ik})$$

and set $\tau_0 = (0.8/\mu_{ik})(|c_{ik}|L/2\pi)^{-1}$. Denote

$$\mu_{ik}(\tau) = \left(1 - 4\lambda/(\tau c_{ik}^2)\right)^{1/2}, \qquad 1 \leqslant \tau \leqslant \tau_0,$$

and choose $\tau_1 \in (1, \tau_0)$ such that

$$|\tau_1 c_{ik}|L/2\pi = 0.8/\mu_{ik}(\tau_1).$$

The preceding argument implies that the corresponding term of $\delta^{(2)}$ with c_{ik} replaced by $\tau_1 c_{ik}$ and λ replaced by $\tau_1\lambda$ is nonnegative, whence we conclude that, since $\tau_1 \geqslant 1$, this term of $\delta^{(2)}$ is nonnegative for c_{ik} and λ.

Finally, in case $|c_{ik}|L/2\pi < (0.8/\mu_{ik})(1 - \mu_{ik})/(1 - \mu_{ik} + 2\mu_{ik}^2)$ with

$$[\![(0.4/\mu_{ik})(1 - \mu_{ik} + 2\mu_{ik}^2)]\!] > (0.4/\mu_{ik})(1 - \mu_{ik}),$$

a similar argument using the fact that $(1 - x)/(1 - x + 2x^2)x$ is a decreasing function of $x \in (0, 1]$ shows that the corresponding term of $\delta^{(2)}$ is nonnegative.

3.4. REMARKS. The assumption concerning the vector fields W_{ik} can be relaxed to obtain stronger versions of [**B**, 4.5 and 4.6]. The assumption that the orbits of X are compact can also be eliminated to obtain a stronger version of [**B**, 4.8].

We now discuss the possibility of using spectral properties of N to strengthen the inequality

$$\int_N \|dr\|^2 \, d\mathcal{H}^n > \int_N \langle X, dr\rangle^2 \, d\mathcal{H}^n$$

used in the proof of 3.3. Inasmuch as r is lipschitzian we can assume r to be smooth. Integration by parts yields

$$\int_N \|dr\|^2 \, d\mathcal{H}^n = \int_N (\Delta r)r \, d\mathcal{H}^n,$$

$$\int_N X(r)^2 \, d\mathcal{H}^n = \int_N (-X^2 r)r \, d\mathcal{H}^n,$$

where in the latter case we have used the fact that X is divergence free and so is a skew-symmetric transformation of $C^{\infty}(N; \mathbf{C})$ with the standard inner product.

Assume X to be Killing, which, for example, would be the case if N were equivariantly isomorphic with G and the metric on N were bi-invariant. Then X commutes with Δ, and there therefore exists a complete orthonormal basis of $\mathscr{L}^2(N)$ consisting of simultaneous eigenfunctions $\phi_0 = 1, \phi_1, \phi_2, \ldots,$ of Δ, and of X^2. Denote $\Delta\phi_i = \Lambda_i\phi_i$, $-X^2\phi_i = K_i\phi_i$. Representing $r = \sum_{i=1}^{\infty} a_i\phi_i$ we therefore have

$$(*) \qquad \int_N \|dr\|^2 - \langle X, dr \rangle^2 \, d\mathscr{H}^n = \sum_{i=1}^{\infty}(\Lambda_i - K_i)a_i^2 > 0.$$

In case N is isometric to a product $P \times C_L$, where C_L is the circle of length L and X is tangent to the circles $\{p\} \times C_L$, each Λ_i is also an eigenvalue of $-X^2$ and so $(*)$ is sharp; i.e., whenever $c > 1$ there exists r with

$$\int_N \|dr\|^2 - c\langle X, dr \rangle^2 \, d\mathscr{H}^n < 0.$$

More generally it is shown in [**BD**] that Δ and $-X^2$ have coinciding eigenvalues if and only if the distribution of normal planes to X is involutive. Thus assume this distribution to be not involutive. Clearly, $(*)$ is sharp if and only if $\liminf_{i \to \infty}\Lambda_i/K_i = 1$. An example is $\mathbf{S}^{2n-1} \subset \mathbf{C}^n$ for $n > 1$ where X is induced by the action of $\mathbf{S}^1$ on $\mathbf{C}^n$ through scalar multiplication. It is not difficult to verify in this case that $\liminf_{i \to \infty}\Lambda_i/K_i = 1$. According to Jiri Dadok this appears in general to be the case for N a symmetric space, hence for all practical purposes $(*)$ should be regarded as being sharp.

4. Instability.

4.1. Assume $c_{ik} \geqslant c > 0$ for $1 \leqslant i < k \leqslant i_2$. Fix $p \in N$ and denote

$$\lambda_{i_2} = \sup\left\{ H_v(p)(w, w): w \in \text{span}\{W_1(p), \ldots, W_{i_2}(p)\}, \|w\| = 1\right\};$$

clearly λ_{i_2} does not depend on p.

Assume X to be Killing, and let $\Lambda \geqslant K > 0$ be simultaneous eigenvalues of Δ and of $-X^2$, respectively (see 3.4).

THEOREM. *There exists a variation W of N such that $\int_N\|W\|^2 \, d\mathscr{H}^n = 1$ and*

$$\delta^{(2)}(W) \leqslant \lambda_{i_2} + \Lambda - K^{1/2}c(i_2 - 1)^{-1}\cot(\pi/2i_2).$$

PROOF. Choose $\phi \in C^{\infty}(N)$ such that $\Delta\phi = \Lambda\phi$ and $-X^2\phi = K\phi$. Denote $K^{1/2} = 2\pi\omega/L$ where ω is a positive integer, set $\delta = L/(2i_2\omega)$, denote by $\alpha: \mathbf{R} \times N \to N$ the one parameter group of transformations induced by X, and define for $p \in N$,

$$a^i(p) = \phi(\alpha(-(i-1)\delta, p)), \qquad i = 1, \ldots, i_2.$$

Since these transformations are isometries we have

$$\Delta a^i = \Lambda a^i, \qquad -X^2 a^i = K a^i;$$

in particular, if ϕ achieves its maximum value on the orbit of p at p then

$$\phi(\alpha(\theta, p)) = \phi(p)\cos(K^{1/2}\theta)$$

and

$$a^i(\alpha(\theta, p)) = \phi(p)\cos(K^{1/2}(\theta - (i - 1)\delta)).$$

Consequently,

$$\langle X, a^i da^k - a^k da^i \rangle = \phi(p)^2 K^{1/2}\sin((k - i)\pi/i_2).$$

It follows that, since $i_2 \geqslant 2$,

$$\sum_{i < k} \langle X, a^i da^k - a^k da^i \rangle = \phi(p)^2 K^{1/2} \sum_{k=1}^{i_2} (i_2 - k)\sin(k\pi/i_2)$$

$$= \frac{1}{2}\phi(p)^2 K^{1/2} i_2 \cot(\pi/2i_2).$$

On the other hand, if we define $\Phi\colon N \to P$ as in 3.2 then

$$\int_{\Phi^{-1}\{q\}} \phi^2 \, d\mathscr{H}^1 = \phi(p)^2 L/2, \qquad \Phi(p) = q,$$

hence application of the co-area formula [**F**, 3.2.12] yields

$$\int_N \sum_{i < k} \langle X, a^i da^k - a^k da^i \rangle \, d\mathscr{H}^n = K^{1/2} i_2 \cot(\pi/2i_2)\int_N \phi^2 \, d\mathscr{H}^n .$$

Further,

$$\int_N \|da^i\|^2 \, d\mathscr{H}^n = \int_N \|d\phi\|^2 \, d\mathscr{H}^n = \int_N (\Delta\phi)\phi \, d\mathscr{H}^n = \Lambda \int_N \phi^2 \, d\mathscr{H}^n .$$

Consequently, defining

$$W = \left(i_2 \int_N \phi^2 \, d\mathscr{H}^n\right)^{-1/2} \sum_{i=1}^{i_2} a^i W_i$$

we have $\int_N \|W\|^2 \, d\mathscr{H}^n = 1$ and thus

$$\delta^2(W) \leqslant \lambda_{i_2} + \Lambda - K^{1/2} c(i_2 - 1)^{-1} \cos(\pi/2i_2)$$

because $\|W_{ik}\| \geqslant c(i_2 - 1)^{-1}$.

4.2. REMARKS. Note that $(i_2 - 1)^{-1} \cot(\pi/2i_2) \in (2/\pi, 1\,]$.

In case $W_1, \ldots, W_{i_2}$ are eigenvectors of $H_\mathbf{v}$, λ_{i_2} can be replaced with $i_2^{-1}\operatorname{trace} H_\mathbf{v}|\operatorname{span}_\mathbf{R}\{W_1, \ldots, W_{i_2}\}$.

In case $i_2 \leqslant 3$ the vector fields $W_1, \ldots, W_{i_2}$ can be reindexed so that $c_{ij} \geqslant 0$ for $1 \leqslant i < j \leqslant i_2$. However, this may not be possible for $i_2 \geqslant 4$.

In case $m = n + 2$, $H_\mathbf{v} = 0$, and $\Lambda = K = (2\pi\omega/L)^2$, 4.1 implies that N is not stable provided $cL/2\pi\omega > 1$. On the other hand, 3.3 implies that N is stable if $cL/2\pi\omega \leqslant 1$. It is shown in [**BD**] that if Λ is the first eigenvalue of Δ which is also an eigenvalue of $-X^2$ then the normal distribution of X is involutive and, in the notation of the proof of 3.2, there is an ω-to-1 covering isometry $\Phi_\omega\colon N \to P \times C_{L/\omega}$ such that $\Pi \circ \Phi_\omega = \Phi$, where Π is the projection onto P. Thus the case

$\omega = 1$ where the above criteria for stability meet corresponds to the case where N is isometric to $P \times C_L$. For $\omega > 1$ the variation which gives instability for $CL/2\pi\omega > 1$ is the pullback by Φ_ω of the variation of $P \times C_{L/\omega}$ giving instability.

4.3. EXAMPLE. We consider the case where the orbits of G are isometric to spheres.

In case the dimension of the orbits is even, the nonexistence of nonvanishing vector fields on spheres of even dimension implies that a minimal orbit N is stable if and only if $H_v \geqslant 0$ on N.

The situation with regard to orbits of odd dimension is of course much more complex. We consider the case where G is the unitary group $\mathbf{U}(n)$ and the minimal orbit N is equivariantly isometric to $\mathbf{S}^{2n-1}$, $n \geqslant 1$, with $\mathbf{U}(n)$ acting on $\mathbf{S}^{2n-1} \subset \mathbf{C}^n$ in the standard way. The flow of an invariant vector field X such as we have been considering must commute with the action of $\mathbf{U}(n)$ hence must be the action of $\mathbf{S}^1$ through scalar multiplication on $\mathbf{C}^n$. Note that the trajectories of X are geodesics, hence $L = 2\pi$. For simplicity we assume the orbits of G to be of codimension 2. $\|W_{12}\| = c$ is independent of choice of W_1, W_2; we choose W_1, W_2 so that $W_{12} = cX$ and W_1, W_2 are eigenvectors of $\mathscr{H}^{2n-1}(N)^{-1}H_v$. As before denote by λ the smallest eigenvalue and assume $\lambda \geqslant 0$.

Using 3.3 we infer that N is stable provided either $4c^{-2}\lambda \geqslant 1$ or $4c^{-2}\lambda < 1$ and

$$c \leqslant \max\left\{1, (0.8/\mu)(1 - \mu)/(1 - \mu + 2\mu^2)\right\},$$

where $\mu = (1 - 4c^{-2}\lambda)^{1/2}$.

On the other hand, the l-th eigenvalue of Δ on N is $\Lambda_l = l(l + 2n - 2)$. The largest simultaneous eigenvalue of $-X^2$ is l^2. (Eigenfunctions are real and imaginary parts of restrictions to $\mathbf{S}^{2n-1}$ of complex monomials of degree l.) Setting $l = 1$ we conclude from 4.1 that N is not stable if

$$c > 2n - 1 + \tfrac{1}{2}\mathscr{H}^{2n-1}(N)^{-1}\,\text{trace}\,H_v.$$

REFERENCES

[**BD**] D. Bleecker, *The spectrum of a Riemannian manifold with a unit Killing vector field*, Trans. Amer. Math. Soc. **275** (1983), 409–416.

[**B**] J. Brothers, *Stability of minimal orbits*, Trans. Amer. Math. Soc. (to appear).

[**F**] H. Federer, *Geometric measure theory*, Springer-Verlag, 1969.

INDIANA UNIVERSITY

Proceedings of Symposia in Pure Mathematics
Volume **44** (1986)

Index Theory for Operator Ranges
and Geometric Measure Theory

RICHARD W. CAREY AND JOEL D. PINCUS[1]

0. Introduction. We have made some investigations of commutative algebras of operators on Hilbert spaces which are associated with holomorphic chains. The local degree of these chains (which lie in $\mathbf{C}^n$) is an "index" for the maximal ideals of the operator algebra; and the specification of this local degree corresponds to the construction of certain varieties carrying weights (specified by the lengths of certain composition series associated with the algebra's module action), which span distinguished cycles concentrated on the joint essential spectrum of the generators. For these considerations, the reader is referred to [3] and to a forthcoming exposition. We remark also that this kind of construction of a spanning surface has also been carried forward by us in certain cases of operator algebras with noncommuting generators. In [15], for example, an application is made to the theory of symmetric or selfadjoint Toeplitz operators on multiply-connected plane domains where a flat current in R^3 is considered.

But the present paper returns to the foundations of our subject, the case of a single operator; and gives complete proofs of new results which bear upon the meaning of our "index" for an operator $T \in \mathscr{B}(\mathscr{H})$ for which the ranges $\mathscr{R}(T)$, $\mathscr{R}(T^*)$, of the operator T and its adjoint may not be closed. Our presentation should make it clear that geometric measure theory and the index theory of linear operators on infinite-dimensional spaces are nicely intertwined.

To begin, suppose that U and V are bounded selfadjoint operators on Hilbert space and that $T \equiv U + iV$ with $T^*T - TT^* \in \mathscr{I}_1$, the trace ideal in $\mathscr{B}(\mathscr{H})$. We seek, as a natural generalization of Fritz Noether's original construction of index, a 1-cycle $\mathbf{L}$ such that: Index $f(T) = \mathbf{L}(df/f)$ for operators $f(T)$ defined by some

1980 *Mathematics Subject Classification.* Primary 47A53.
[1] This work is supported by grants from the National Science Foundation.

© 1986 American Mathematical Society
0082-0717/80 $1.00 + $.25 per page

smooth functional calculus on the generators $\{U, V\}$ (see [2]), which correspond to symbols f not vanishing on $\sigma_{\text{ess}}(T)$, the essential spectrum of T.

The solution of this problem is obtained through the construction of a spanning 2-current $\mathbf{T}$ for $\mathbf{L}$ which has fundamental properties. This so-called principal current will have $\partial\mathbf{T} = \mathbf{L}$, and is the object, whose study in the case of operator algebras with more than one generator, leads to the multidimensional theory alluded to above.

The current $\mathbf{T}$ has the form $(1/2\pi i)E^2 {\llcorner} g_T$, where g_T is a real compactly supported L^1 function in the plane. It is fundamental to the theory of this current that

(i) $$\partial\mathbf{T}(df/f) = \text{Index } f(U, V) \quad \text{if } f \neq 0 \text{ on } \sigma_{\text{ess}}(T),$$

(ii) $$\partial\mathbf{T}(R\,dS) = \text{trace}[S(U, V), R(U, V)],$$

(iii) $$\text{Mass}(\mathbf{T}) \leqslant \tfrac{1}{2}\|[T; T]\|_{\mathscr{I}_1},$$

and that there is a transformation law of the principal current which has operator theoretic meaning.

Set $A = \int\!\int \alpha(x, y)\, d\mathscr{E}_x\, d\mathscr{F}_y$ and $B = \int\!\int \beta(x, y)\, d\mathscr{E}_x\, d\mathscr{F}_y$, where $\mathscr{E}_x$ and $\mathscr{F}_y$ are respectively the spectral resolutions of U and V. Form the operator $\Omega(U, V)$ with symbol $\Omega(x, y) = \alpha(x, y) + i\beta(x, y)$. Thus $\Omega = A + iB$. It is easily seen, if α and β are sufficiently smooth, that $[\Omega, \Omega^*] \in \mathscr{I}_1$.

Thus there is a principal current $\mathbf{T}$ associated to Ω. It is fundamental that $\mathbf{T}_\Omega = \Omega_\#(\mathbf{T})$. Furthermore, $\mathbf{T}$ is invariant under trace class perturbations T.

We now outline the characterization of the principal function g_T and the problem which is solved in the present investigation.

Let $T_z = T - z$, $T_z^* = (T - z)^*$. Form the selfadjoint operators $T_z^*T_z$ and $T_z T_z^*$. Let the corresponding spectral resolutions for these operators be E_λ^z and F_λ^z, respectively.

If $z \notin \sigma_{\text{ess}}(T)$, it is clear that E_λ^z and F_λ^z are constant finite rank operators for sufficiently small positive λ. Thus, if $p(T)$ is a polynomial in T, we have

$$J(p)(z) \equiv \text{trace}\{p(T)(E_\lambda^z - F_\lambda^z)\} = p(z)\,\text{Index}(T - z).$$

In order to probe the meaning of index when $z \in \sigma_{\text{ess}}(T)$, we consider another functional. Suppose that ψ is a smooth, real, compactly supported function normalized so that $\pi\int_0^\infty s\psi^2(s)\, ds = 1$.

Let

$$J_\varepsilon(p)(z) \equiv \frac{\pi}{2\varepsilon^2}\, \text{trace}\left\{ p(T)\int_0^\infty (E_\lambda^z - F_\lambda^z)\psi\left(\frac{\sqrt{\lambda}}{\varepsilon}\right) d\lambda \right\}.$$

It is clear that $J_\varepsilon(p)(z) = J(p)(z)$ for $z \notin \sigma_{\text{ess}}(T)$, provided that ε is sufficiently small. But there exists a compactly supported real summable function g_T, the principal function of T, so that

$$J_\varepsilon(p)(z) = \frac{1}{\varepsilon^2}\int p(\zeta)\psi\left(\frac{|\zeta - z|}{\varepsilon}\right) g_T(\zeta)\, d\mathscr{L}_\zeta^2.$$

This fact implies that

$$\lim_{\varepsilon \downarrow 0} J_\varepsilon(p)(z) = g_T(z)p(z)$$

at Lebesgue points of g_T and that $g_T(z) = \mathrm{Index}(T - z)$ for $z \notin \sigma_{\mathrm{ess}}(T)$.

The construction just presented leaves one point unclear. We have used the families of projections E_λ^z and F_λ^z, and it is not clear if $g_T(z)$ depends only upon $\mathscr{R}(T_z)$, $\mathscr{R}(T_z^*)$, or upon other information associated with T.

We will prove theorems in the present paper which show that we are studying only a relationship between these ranges.

In this connection, we point out that there is a substantial literature about operators with nonclosed ranges. For example, Prössdorf [16] considers singular integral operators on smooth contours in the plane with symbols that vanish. He shows that these operators do not have closed range and calculates the difference between the dimensions of the null spaces of the operator and its adjoint. This difference is shown by example to be altered even by finite rank perturbations. It is not the winding number of the symbol, yet the theorems developed here apply.

Geometric measure theory enters into our theory in a particularly clear way if g_T is integer valued and in BV, and we have proved [4] that if U and V have summable spectral multiplicity functions, then this is the case.

In such a situation $g_T = \Sigma_m m \psi_m(z)$ with ψ_m the characteristic function of $g_T^{-1}(m)$. Each of those sets, Γ_m, has finite perimeter, and

$$g_T(z) = \sum m \cdot \frac{1}{2\pi} \int_{\partial \Gamma_m} \frac{n_1^{(m)}(\zeta) + i n_2^{(m)}(\zeta)}{\zeta - z} \, d\mathscr{H}_\zeta^1 \,,$$

where $(n_1^{(m)}, n_2^{(m)})$ denotes the Federer exterior normal to Γ_m and $\mathscr{H}^1$ is one-dimensional Hausdorff measure in R^2. Thus, as with ordinary index, the principal function is a winding number. But it counts the winding only of a Federer–De Giorgi measure boundary $1 n \sigma_{\mathrm{ess}}(T)$ about z.

In our study of the relationship of $g_T(z)$ with $\{\mathscr{R}(T_z), \mathscr{R}(T_z^*)\}$ we will make use of a connection between the principal function and the phase shift δ_z for the selfadjoint perturbation problem $T_z^*T_z \to T_z T_z^*$.

It is known from the work of M. G. Krein [13] that there is an L^1 function $\delta_z(\lambda)$ such that

$$\mathrm{trace}\big[f(T_z^*T_z) - f(T_z T_z^*) \big] = \int f'(\lambda) \delta_z(\lambda) \, d\lambda$$

for smooth functions $f(\cdot)$.

But we [2] have shown that

$$\delta_z(\lambda) - \frac{1}{2\pi} \int_0^{2\pi} g_{T_z}(\sqrt{\lambda}\, e^{i\theta}) \, d\theta.$$

Thus, to study the local behavior of g_{T_z} near zero, it is reasonable to study the local behavior of $\delta_z(\cdot)$ near zero. For, by a change of variables,

$$\lim_{r \downarrow 0} \frac{1}{r} \int_0^r \delta_z(\lambda) \, d\lambda = \lim_{r \downarrow 0} \frac{1}{\pi r} \int_0^{\sqrt{r}} \int_0^{2\pi} g_{T_z}(t e^{i\theta}) t \, dt \, d\theta.$$

Hence, if g_T has a Lebesgue point at z, then

$$\lim_{r\downarrow 0} \frac{1}{r} \int_0^r \delta_z(\lambda)\, d\lambda = g_{T_z}(0).$$

When g_T is in BV we can redefine it to be $\frac{1}{2}(\lambda + \mu)$, the average of the upper and lower $\mathfrak{L}^2$-approximate limits. By Federer [9, Property VIII, p. 306], this can be done for $\mathcal{H}^1$ almost all points in R^2, and then

$$\lim_{r\downarrow 0} \frac{1}{r} \int_0^r \delta_z(\lambda)\, d\lambda = \frac{1}{2}(\lambda + \mu).$$

Recall that these approximate limits have the property that the set E, where the upper limit λ is less than the lower limit μ, is a countable union of $(\mathcal{H}^1, 1)$ rectifiable sets, and that for $\mathcal{H}^1$ almost all x in $R^2 \sim E$ there exists a $u \in S^1$ so that

$$\lim_{r\downarrow 0} r^{-2} \int_{B(x,\,r)\cap N} \big|g_T(z) - \lambda(z)\big|\, d\mathfrak{L}^2 = 0,$$

$$\lim_{r\downarrow 0} r^{-2} \int_{B(x,\,r)\cap P} \big|g_T(z) - \mu(z)\big|\, d\mathfrak{L}^2 = 0,$$

where $N = \{z | (z - x) \cdot u < 0\}$, and $P = \{z | (z - x) \cdot u > 0\}$.

Thus, in this case, it is certainly of clear geometric significance to try and understand the meaning of $\delta_z(0)$ as a new index which measures the relationship of two possibly dense nonclosed subspaces $\mathcal{R}(T)$ and $\mathcal{R}(T^*)$.

As a simple example, we note that if T is the unilateral shift operator and w is a complex number with modulus 1, then $\mathcal{R}((T - w)^*) \supset \mathcal{R}(T - w)$ and both ranges are dense subspaces of $\mathcal{H}$. The quotient space $\mathcal{R}((T - \omega)^*)/\mathcal{R}(T - w)$ is a one-dimensional vector space, yet the correct assignment of their *relative dimension* is $\frac{1}{2}$.

The construction which follows introduces a measure-theoretic index group to pairs of operator ranges in a way that goes essentially beyond the context just outlined.

We conclude this introduction by giving some notation and terminology. An operator on a Hilbert space $\mathcal{H}$ will always be understood to be a bounded linear transformation from $\mathcal{H}$ to itself. The algebra of all bounded operators on $\mathcal{H}$ will be denoted by $\mathcal{B}(\mathcal{H})$. If $T \in \mathcal{B}(\mathcal{H})$, then $\mathcal{R}(T)$ or $T\mathcal{H}$ denote the generally nonclosed set of values of T. An *operator range* in $\mathcal{H}$ is a linear subspace of $\mathcal{H}$, that is the range of some operator on $\mathcal{H}$. The basic sources of material on operator ranges are the treatises of Dixmier [5, 6, 7] and Fillmore–Williams [10].

The collection $\mathcal{L}$ of all operator ranges in a Hilbert space is the lattice generated by the closed subspaces. An ideal in $\mathcal{L}$ is a subset $\mathcal{I}$ of $\mathcal{L}$ such that if $\mathcal{D}$ and $\mathcal{D}'$ are in $\mathcal{I}$, so is $\mathcal{D} + \mathcal{D}'$, and if $\mathcal{D} \in \mathcal{I}$, $\mathcal{D}' \in \mathcal{I}$, and $\mathcal{D}' \subset \mathcal{D}$, then $\mathcal{D}' \in \mathcal{I}$. The ideal $\mathcal{I}$ is invariant if whenever $\mathcal{D} \in \mathcal{I}$, then every $\mathcal{D}'$ unitary equivalent to $\mathcal{D}$ is also in $\mathcal{I}$. If I is an arbitrary two-sided ideal in $\mathcal{B}(\mathcal{H})$, then the ranges of the elements in I form an invariant ideal.

If A and C are selfadjoint operators with C trace class, then $A \to^{\xi} A + C$ designates that ξ is the phase shift for the pair $\{A, C\}$.

1. The extended Lebesgue domain for elements of L^∞. Suppose f is an element of $L^\infty \equiv L^\infty((-\infty, \infty); \mathfrak{L}^1)$. Recall that v is an approximate limit of the function f at x provided for every $\varepsilon > 0$, $\{ y\colon |v - f(y)| < \varepsilon \}$ has positive density at x. The concepts of right- and left-approximate limits are analogously defined by requiring positive right and left density respectively at x. The sets $\omega^+(x, f)$ and $\omega^-(x, f)$ are the (closed) sets of right and left approximate limits of f at x. With $\omega(x, f) = \omega^+(x, f) \cup \omega^-(x, f)$, f is said to be approximately continuous at x when $\{ f(x) \} = \omega(x, f)$. It is of interest to note that for a measurable function g, a theorem of G. C. Young states that for at most countably many points x can $\sup \omega^\pm(x, g) < \inf \omega^\mp(x, g)$. However, our interest is in the behavior of the Hilbert transform at a single point.

DEFINITION 1.1. Let f, g, h belong to L^∞, $h \geqslant 0$, with f, g, h having compact support in the closed interval $[0, \infty)$. The function f is said to be h-related to g at the origin, provided

$$\left| \int \frac{f(x) - g(x)}{x + l} \, dx \right| \leqslant \int \frac{h(x)}{x + l} \, dx \quad \text{for } l > 0.$$

Denoted by $f \sim_h g$, the condition f is h-related to g at the origin. If $\omega^+(0, h) = \{0\}$, we shall write $f \simeq g$. Note that $\simeq$ is an equivalence relation and the family of equivalence classes $[f]$ forms an abelian group, $K(\mathscr{L})$, under the natural identification $[f \pm f'] = [f] \pm [f']$. This group will serve as an index group for the lattice of operator ranges. Note that if $f \circ a(x) \equiv f(ax)$, for $a > 0$, then $[f] = [f \circ a]$.

The next lemma appears in [1].

LEMMA 1.2. *If g is a nonnegative L^∞-function with compact support and $\sup \omega^\pm(0, g) < \inf \omega^\mp(0, g)$, then*

$$\lim_{\varepsilon \downarrow 0} \int_{|x-y|>\varepsilon} \frac{g(x)}{x - y} \, dx = \pm \infty.$$

PROPOSITION 1.3. *Suppose g and g' are representatives of $[f]$ with $\{g(0)\} = \omega^+(0, g)$ and $\{g'(0)\} = \omega^+(0, g')$. Then $g(0) = g'(0)$. In other words, there is a natural injective homomorphism of the additive group of real numbers into $K(\mathscr{L})$.*

PROOF. By hypothesis $g \sim_h g'$ for some h in $L^\infty[0, \infty)$ with $\{0\} = \omega^+(0, h)$. Consider $(g - g') \mp h$. If $g(0) \neq g'(0)$, Lemma 1.2 states that

$$\lim_{\varepsilon \downarrow 0} \int_\varepsilon^1 \frac{(g - g')(x) \mp h(x)}{x} \, dx = \pm \infty,$$

depending upon whether $g(0) - g'(0)$ is positive or negative. On the other hand, if $k \in L^\infty$, the difference

$$\left| \int_l^1 \frac{k(x)}{x} \, dx - \int_0^1 \frac{k(x)}{x + l} \, dx \right| \quad \text{is bounded in } l.$$

Hence we obtain a contradiction that $g \sim_h g'$ which says that h dominates the asymptotic behavior as $l \to 0^+$.

This proves the proposition.

DEFINITION 1.4. Let $f \in L^\infty([\,0, \infty)\,)$. The origin is said to be in the extended Lebesgue domain of $[f]$ if there is a $g \in [f]$ which is right-approximately continuous at the origin.

By Proposition 1.3, the value $g(0)$ does not depend on the choices of g and h. We shall denote this value by $f^*(0)$. It is clear that when f itself is approximately continuous at $x = 0$, then $\{f^*(0)\} = \omega^+(0, f)$. On the other hand, if $f(x) = \sin(1/x)$ on $(0, 1)$, then

$$ f^*(0) = 0 = \frac{\sup \omega^+(0, f) + \inf \omega^+(0, f)}{2} . $$

In general, $f^*(0)$ is not an average of the upper- and lower-approximate limits.

2. Definition of the index: The finite rank case. First we describe the concept of congruence with respect to an ideal.

DEFINITION 2.1. Suppose $\mathcal{R}$ and $\mathcal{S}$ are elements of the lattice $\mathcal{L}$ and suppose $\mathcal{I}$ is an ideal in $\mathcal{L}$. Then $\mathcal{R}$ and $\mathcal{S}$ are said to be congruent modulo $\mathcal{I}$, provided $\mathcal{R} + d = \mathcal{S} + d'$ for some elements d, d' in $\mathcal{I}$.

Congruence is of course an equivalence relation; we denote it by $\mathcal{R} \equiv \mathcal{S}(\mathcal{I})$.

Suppose $\mathcal{F}$ is taken as the invariant ideal of $\mathcal{L}$ consisting of the finite-dimensional subspaces. If $\mathcal{R} \equiv \mathcal{S}(\mathcal{F})$, then there are positive operators $\{A, B, C, D\}$ with $\mathcal{R}(\sqrt{A}) = \mathcal{R}$, $\mathcal{R}(\sqrt{B}) = \mathcal{S}$, $\mathcal{R}(\sqrt{A + C}) = \mathcal{R}(\sqrt{B + D})$, and $\mathcal{R}(\sqrt{C}) + \mathcal{R}(\sqrt{D}) \in \mathcal{F}$. Let $A \to^\xi A + C$ and $B \to^\tau B + D$ denote the associated phase shifts.

THEOREM 2.2. *The equivalence class $[\xi\text{-}\tau] = [\xi] - [\tau]$ is independent of the choice of implementing operators $\{A, B, C, D\}$ and therefore only depends on the subspaces $\mathcal{R}$ and $\mathcal{S}$. In other words, if $\{V, W, X, Y\}$ is another choice of positive operators with $\mathcal{R}(\sqrt{V}) = \mathcal{R}, \mathcal{R}(\sqrt{W}) = \mathcal{S}, \mathcal{R}(\sqrt{V + X}) = \mathcal{R}(\sqrt{W + Y}), V \to^\eta V + X$, and $W \to^\gamma W + Y$, then $\xi - \tau \simeq \eta - \gamma$.*

DEFINITION 2.3. *The subspaces $\mathcal{R}$ and $\mathcal{S}$ in $\mathcal{L}$ are said to be $\mathcal{F}$-comparable if $\mathcal{R} \equiv \mathcal{S}(\mathcal{F})$ and the origin belongs to the extended Lebesgue domain of the class $[\xi - \tau]$.*

If $\mathcal{R} \equiv \mathcal{S}(\mathcal{F})$, the class $[\xi - \tau]$ is called the index of the pair $(\mathcal{R}, \mathcal{S})$ and is denoted by $\mathrm{ind}(\mathcal{R}, \mathcal{S})$. Further, if $\mathcal{R}$ and $\mathcal{S}$ are $\mathcal{F}$-comparable, the value $(\xi - \tau)^*(0)$ is also to be regarded as the index.

Note that when $\mathcal{R}$ and $\mathcal{S}$ are $\mathcal{F}$-comparable, $\mathrm{ind}(\mathcal{R}, \mathcal{S}) = -\mathrm{ind}(\mathcal{S}, \mathcal{R})$. It is easy to check that for any Fredholm operator T, the subspaces $T\mathcal{H}$ and $T^*\mathcal{H}$ are $\mathcal{F}$-comparable and $\mathrm{ind}(T\mathcal{H}, T^*\mathcal{H})$ is the ordinary Fredholm index of T. Moreover, whenever $T\mathcal{H}$ and $T^*\mathcal{H}$ are $\mathcal{F}$-comparable and K is a finite rank operator, then $(T + K)\mathcal{H}$ and $(T + K)^*$ are $\mathcal{F}$-comparable and

$$ \mathrm{ind}\big((T + K)\mathcal{H}, (T + K)^*\mathcal{H}\big) = \mathrm{ind}(T\mathcal{H}, T^*\mathcal{H}). $$

The proof of Theorem 2.2 is split into a series of propositions. We begin with a result which appears in Douglas [8].

PROPOSITION 2.4. *Let A and B be bounded operators on $\mathcal{H}$. The following conditions are equivalent:*
(1) $\mathscr{R}(A) \subset \mathscr{R}(B)$;
(2) $AA^* \leqslant aBB^*$ *for some* $a > 0$;
(3) $A = BC$ *for some bounded operator* C *on* $\mathcal{H}$.

PROPOSITION 2.5. *If V and W are positive operators with $\mathscr{R}(\sqrt{V}) = \mathscr{R}(\sqrt{W})$ and ϕ is a vector in $\mathcal{H}$, then $\xi \simeq \tau$ where $V \to^{\xi} V + \phi \otimes \phi$ and $W \to^{\tau} W + \phi \otimes \phi$. Further, if $\phi \in \mathscr{R}(\sqrt{V})$, then $\xi \simeq 0$.*

PROOF. Since $\mathscr{R}(\sqrt{V}) = \mathscr{R}(\sqrt{W})$ by Proposition 2.4 there are $0 < a < 1 < b$ such that $aV \leqslant W \leqslant bV$. Therefore, if $l > 0$,

$$1 + \left([bV + l]^{-1}\phi, \phi\right) \leqslant 1 + \left([W + l]^{-1}\phi, \phi\right) \leqslant 1 + \left([aV + l]^{-1}\phi, \phi\right).$$

Since $\det\{[A - z]^{-1}[B - z]\} = \det\{1 + [A - z]^{-1}(B - A)\}$, it follows that for $l > 0$ and $tV \to^{\xi_t} tV + C$,

$$\int \frac{\xi_b(x)}{x + l}\,dx \leqslant \int \frac{\tau(x)}{x + l}\,dx \leqslant \frac{\xi_a(x)}{x + l}\,dx.$$

But for any $0 < \alpha < \beta$,

$$1 \leqslant \frac{1 + \left([\alpha V + l]^{-1}\phi, \phi\right)}{1 + \left([\beta V + l]^{-1}\phi, \phi\right)} \leqslant \frac{\beta}{\alpha},$$

and therefore with $\alpha = a$, $\beta = b$,

$$(2.6) \qquad 0 \leqslant \int \frac{\tau(x) - \xi_b(x)}{x + l}\,dx \leqslant \int \frac{\xi_a(x) - \xi_b(x)}{x + l}\,dx \leqslant \mathrm{Log}\,\frac{b}{a}.$$

Also for $\alpha = 1$ and $\beta = b$,

$$(2.7) \qquad 0 \leqslant \int \frac{\xi(x) - \xi_b(x)}{x + l}\,dx \leqslant \mathrm{Log}\,b.$$

Combining (2.6) and (2.7) we have

$$\left| \int \frac{\tau(x) - \xi(x)}{x + l}\,dx \right| \leqslant \mathrm{Log}\,\frac{b^2}{a},$$

and so $\xi \simeq \tau$. Lastly, if $\phi \in \mathscr{R}(\sqrt{V})$, then $\xi(x)/x$ is summable so that $\xi \simeq 0$.
This completes the proof of Proposition 2.5.

COROLLARY 2.8. *If V, W, and C are positive operators with $\mathscr{R}(\sqrt{V}) = \mathscr{R}(\sqrt{W})$ and $\mathscr{R}(C) \in \mathscr{F}$, then $\xi \simeq \eta$ where $V \to^{\xi} V + C$ and $W \to^{\eta} W + C$. Moreover, if $\mathscr{R}(\sqrt{C}) \subseteq \mathscr{R}(\sqrt{V})$, then $\xi \simeq 0$.*

PROOF. Suppose $C = \Sigma_1^n \phi_\alpha \otimes \phi_\alpha$. Consider the transition diagram

$$V \xrightarrow{\xi_1} V + \phi_1 \otimes \phi_1 \xrightarrow{\xi_2} V + \phi_1 \otimes \phi_1 + \phi_2 \otimes \phi_2 \xrightarrow{\xi_3} \cdots$$

$$\xrightarrow{\xi_{n-1}} \left[V + \sum_1^{n-1} \phi_\alpha \otimes \phi_\alpha \right] \xrightarrow{\xi_n} \left[V + \sum_1^{n} \phi_\alpha \otimes \phi_\alpha \right].$$

Note that $\xi = \Sigma_1^n \xi_\alpha$. Replacing V by W gives a similar relation $\eta = \Sigma_1^n \eta_\alpha$. As

$$\mathscr{R}\left(\sqrt{V + \Sigma_1^k \phi_\alpha \otimes \phi_\alpha} \right) = \mathscr{R}\left(\sqrt{W + \Sigma_1^k \phi_\alpha \otimes \phi_\alpha} \right),$$

we conclude by Proposition 2.5 that for each α, $\xi_\alpha \simeq \eta_\alpha$. Consequently $\xi \simeq \eta$. Finally, if $\mathscr{R}(\sqrt{C}) \subset \mathscr{R}(\sqrt{V})$, we have $\xi \simeq 0$ since $\xi_\alpha \simeq 0$ for each α.

PROPOSITION 2.9. *Suppose V and W are positive operators with $\mathscr{R}(\sqrt{V}) = \mathscr{R}(\sqrt{W})$ and $V - W$ of finite rank. Then $\xi \simeq 0$ where $V \xrightarrow{\xi} W$.*

PROOF. Let $V - W = L_+ - L_-$ where $L_\pm \geqslant 0$ is the spectral decomposition. Since $\mathscr{R}(V) \subset \mathscr{R}(\sqrt{V})$ and $\mathscr{R}(W) \subset \mathscr{R}(\sqrt{W}) = \mathscr{R}(\sqrt{V})$, we have $\mathscr{R}(V - W) = \mathscr{R}(L_+) + \mathscr{R}(L_-) = \mathscr{R}(\sqrt{L_+}) + \mathscr{R}(\sqrt{L_-}) \subset \mathscr{R}(\sqrt{V})$. As the diagram commutes

$$
\begin{array}{ccc}
V & \xrightarrow{\xi_-} & V + L_- \\[4pt]
\xi \downarrow & & \downarrow 0 \\[4pt]
W & \xleftarrow[\xi_+]{} & W + L_+
\end{array}
$$

we have $\xi = \xi_- + \xi_+$. By Corollary 2.8, $\xi_\pm \simeq 0$, hence $\xi \simeq 0$.

PROPOSITION 2.10. *Suppose A, B, V, and W are positive operators such that $\mathscr{R}(\sqrt{A}) = \mathscr{R}(\sqrt{V}), \mathscr{R}(\sqrt{B}) = \mathscr{R}(\sqrt{W}), \mathscr{R}(A - B) \in \mathscr{F}, $ and $\mathscr{R}(V - W) \in \mathscr{F}$. Then $\xi \simeq \tau$ where $A \xrightarrow{\xi} B$ and $V \xrightarrow{\tau} W$.*

PROOF. Suppose $A - B = C_+ - C_-$. Then $A \xrightarrow{\xi} B$ splits up as

$$A \xrightarrow{\xi_-} A + C_- \xrightarrow{\xi_+} A + C_- - C_+ = B.$$

On the other hand, $\xi_- \simeq \tau_-$ and $\xi_+ \simeq \tau_+$, where $V \xrightarrow{\tau_-} V + C_-$ and $W + C_+ \xrightarrow{\tau_+} W$, since $A + C_- = B + C_+$, $\mathscr{R}(\sqrt{V}) = \mathscr{R}(\sqrt{A})$, and $\mathscr{R}(\sqrt{W}) = \mathscr{R}(\sqrt{B})$. Thus, $\xi \simeq \tau_+ + \tau_-$. Furthermore, as

$$V \xrightarrow{\tau_-} V + C_- \xrightarrow{\delta} W + C_+ \xrightarrow{\tau_+} W,$$

we have $\tau = \tau_- + \delta + \tau_+$. But $\delta \simeq 0$ since $\mathscr{R}(\sqrt{V + C_-}) = \mathscr{R}(\sqrt{W + C_+})$, so that $\xi \simeq \tau$.

This completes the proof of Proposition 2.10.

PROOF OF THEOREM 2.2. Suppose $\{A, B, C, D\}$ and $\{V, W, X, Y\}$ are positive operators which implement the congruence relation $\mathscr{R} \equiv \mathscr{S}(\mathscr{F})$. We must check that $(\xi - \tau) \simeq (\eta - \gamma)$ where

$$A \xrightarrow{\xi} A + C, \quad B \xrightarrow{\tau} B + D, \quad V \xrightarrow{\eta} V + X, \quad W \xrightarrow{\gamma} W + Y,$$

and

$$\mathcal{R} = \mathcal{R}(\sqrt{A}) = \mathcal{R}(\sqrt{V}), \qquad \mathcal{S} = \mathcal{R}(\sqrt{B}) = \mathcal{R}(\sqrt{W}),$$

$$\mathcal{R}(\sqrt{A + C}) = \mathcal{R}(\sqrt{B + D}), \qquad \mathcal{R}(\sqrt{V + X}) = \mathcal{R}(\sqrt{W + Y}).$$

Introduce the transitions

$$A \overset{\xi}{\to} A + C \overset{\theta}{\to} A + X \overset{\rho}{\to} A, \qquad B \overset{\tau}{\to} B + D \overset{\chi}{\to} B + Y \overset{\delta}{\to} B,$$

so that $\xi + \theta + \rho = \tau + \chi + \delta = 0$.

As $\mathcal{R}(\sqrt{A + C}) = \mathcal{R}(\sqrt{B + D})$ and $\mathcal{R}(\sqrt{A + X}) = \mathcal{R}(\sqrt{B + Y})$, it follows from Proposition 2.10 that $\theta \simeq \chi$. Similarly, $\rho \simeq -\eta$ and $\delta \simeq -\gamma$, and we conclude that $\xi - \tau \simeq \eta - \gamma$.

This finishes the proof of Theorem 2.2.

PROPOSITION 2.11. *Suppose T and K are (bounded) operators on the Hilbert space $\mathcal{H}$ with $\mathcal{R}(K) \in \mathcal{F}$. Suppose that $T\mathcal{H}$ and $T^*\mathcal{H}$ are $\mathcal{F}$-comparable. Then $[T + K]\mathcal{H}$ and $[T + K]^*\mathcal{H}$ are $\mathcal{F}$-comparable and*

$$\operatorname{ind}\big([T + K]\mathcal{H}, [T + K]^*\mathcal{H}\big) = \operatorname{ind}(T\mathcal{H}, T^*\mathcal{H}).$$

PROOF. By hypothesis there are finite subspaces d, d', such that $\mathcal{R}(T) + d = \mathcal{R}(T)^* + d'$. Since

$$\mathcal{R}(T + K) + \mathcal{R}(K) = \mathcal{R}(T) + \mathcal{R}(K)$$

and

$$\mathcal{R}\big([T + K]^*\big) + \mathcal{R}(K^*) = \mathcal{R}(T^*) + \mathcal{R}(K^*),$$

it follows that

$$\mathcal{R}(T + K) + \mathcal{R}(K) + \mathcal{R}(K^*) + d = \mathcal{R}\big([T + K]^*\big) + \mathcal{R}(K) + \mathcal{R}(K^*) + d'.$$

Consequently, $\mathcal{R}(T + K) \equiv \mathcal{R}([T + K]^*)(\mathcal{F})$. To show that the perturbed operators are $\mathcal{F}$-comparable and the equality of the respective indices, choose positive operators C and D having finite rank so that $TT^* \overset{\xi}{\to} T^*T + C$, $T^*T \overset{\eta}{\to} T^*T + D$, and $\mathcal{R}(T) + \mathcal{R}(\sqrt{C}) = \mathcal{R}(T^*) + \mathcal{R}(\sqrt{D})$.

Let X and Y have finite rank so that $\mathcal{R}(T + K) + \mathcal{R}(\sqrt{X}) = \mathcal{R}([T + K]^*) + \mathcal{R}(\sqrt{Y})$, and introduce phase shifts

$$[T + K][T + K]^* \overset{\rho}{\to} [T + K][T + K]^* + X \overset{\gamma}{\to} TT^* + C \overset{-\xi}{\to} TT^*,$$

$$[T + K]^*[T + K] \overset{\tau}{\to} [T + K]^*[T + K] + Y \overset{\theta}{\to} T^*T + D \overset{-\eta}{\to} T^*T.$$

Since $\mathcal{R}(T + K) + \mathcal{R}(\sqrt{X}) = \mathcal{R}([T + K]^*) + \mathcal{R}(\sqrt{Y})$ and $\mathcal{R}(T) + \mathcal{R}(\sqrt{C}) = \mathcal{R}(\sqrt{T^*}) + \mathcal{R}(\sqrt{D})$, it follows that $\gamma \sim \theta$. Moreover, $\rho + \gamma - \xi = \tau + \theta - \eta$ as elements in L^{∞}. To see this, note that

$$(n+1)\left\{\int x^n[\tau + \theta - \eta](x)\,dx - \int x^n[\rho + \gamma - \xi](x)\,dx\right\}$$

$$= \text{trace}\left\{\left((T+K)^*(T+K)\right)^{n+1} - (T^*T)^{n+1} - \left((T+K)(T+K)^*\right)^{n+1}\right.$$

$$\left. + (TT^*)^{n+1}\right\}$$

$$= \text{trace}\left\{\left[(T+K)^*, (T+K)\left((T+K)^*(T+K)\right)^n\right] - \left[T^*, T(T^*T)^n\right]\right\}$$

$$= 0$$

for $n = 0, 1, \ldots$, since $\text{trace}(KN) = \text{trace}(NK)$ for any bounded operator N when K has finite rank.

Consequently, $\xi - \eta \simeq \rho - \tau$. Therefore, $(T+K)\mathcal{H}$ and $(T+K)^*\mathcal{H}$ are $\mathcal{F}$-comparable and

$$\text{ind}\left((T+K)\mathcal{H}, (T+K)^*\mathcal{H}\right) = \text{ind}(T\mathcal{H}, T^*\mathcal{H}).$$

PROPOSITION 2.12. *Suppose A and B are selfadjoint operators such that $A \to^\xi B$ where $A - B$ has finite rank. For any polynomial p,*

$$\mathcal{R}(p(A)) \equiv \mathcal{R}(p(B))(\mathcal{F}) \quad \text{and} \quad \text{ind}(\mathcal{R}(p(A)), \mathcal{R}(p(B))) = \left[\xi_{|p|^2}\right]$$

where $\xi_{|p|^2}(x) = \Sigma_{t:\,|p(t)|^2 = x}\xi(t)\text{sgn}(|p(t)|^2)'$. In particular, for $\mathcal{L}^1$-almost all x, $\mathcal{R}(A - x)$ and $\mathcal{R}(B - x)$ are $\mathcal{F}$-comparable and $\text{ind}(\mathcal{R}(A - x), (B - x)) = \omega^+(x, \xi) - \omega^-(x, \xi) = 0$.

PROOF. The proof is just an application of the standard transformation law for the phase shift formula $\text{trace}\{\phi(B) - \phi(A)\} = \int\xi(x)\,d\phi$.

3. Definition of the index relative to the Hilbert–Schmidt operators. We shall extend the results of §2 to a larger ideal and give somewhat alternative proofs. First we note that for general trace class perturbations $A \to^\xi A + C$, we must consider L^1-functions and not just L^∞. Thus we introduce a new group, $K^{(2)}(\mathcal{L})$, of equivalence classes $[f]^{(2)}$, $f \in L^1[0, \infty)$, support f compact, by the relation $f \sim_h g$ where f, g, h are L^1-functions of compact support with the origin in the Lebesgue set of h and $\omega^+(0, h) = \{0\}$. Note that the natural map $K(\mathcal{L}) \to K^{(2)}(\mathcal{L})$ given by $[f] \to [f]^{(2)}$ is injective on the range of **R**.

DEFINITION 3.1. Elements $\mathcal{R}$ and $\mathcal{S}$ in $\mathcal{L}$ are said to be $\mathcal{I}_2$-consistent if there exist positive operators $\{A, B, C, D\}$ with $\sqrt{C}$ and $\sqrt{D}$ Hilbert–Schmidt such that:

(1) $\mathcal{R}(\sqrt{A}) = \mathcal{R}, \mathcal{R}(\sqrt{B}) = \mathcal{S}, \mathcal{R}(\sqrt{A+C}) = \mathcal{R}(\sqrt{B+D})$;

(2) The classes $[\xi^{(\mu)}]^{(2)}$ and $[\tau^{(\nu)}]^{(2)}$ are constant for $\mu > 0, \nu > 0$, where $A \to^{\xi^{(\mu)}} A + \mu C$ and $B \to^{\tau^{(\nu)}} B + \nu D$.

The pair $\{A, C\}$ will also be called consistent if condition (2) holds.

Note that consistency implies stability in the sense that $[\xi^{(\mu)}]^{(2)} = [\xi]^{(2)}$ for $\mu > 0$ implies $[\xi \circ a]^{(2)} = [\xi]^{(2)}$ for $a > 0$. Consistency is necessary if we are to obtain an invariant of the operator ranges and not on the implementing operators. If C has finite rank, then $\{A, C\}$ is always consistent. However, it is quite easy to construct trace class operators C with $C \to^\xi C + C$ and $\omega^+(0, \xi) = \{1\}$ despite the fact that $\mathscr{R}(\sqrt{C}) = \mathscr{R}(\sqrt{C + C})$.

THEOREM 3.2. *Suppose $\mathscr{R}$ and $\mathscr{S}$ are $\mathscr{I}_2$-consistent. The class $[\xi]^{(2)} - [\tau]^{(2)}$ is independent of the choice of implementing operators $\{A, B, C, D\}$.*

The proof is split into a series of lemmas.

LEMMA 3.3. *Suppose $\{A, C\}$ is consistent and $\mathscr{R}(\sqrt{A}) \subset \mathscr{R}(\sqrt{V})$ for some positive operator V. Then $\{V, C\}$ is consistent. If $\mathscr{R}(\sqrt{A}) = \mathscr{R}(\sqrt{V})$, then also $[\xi]^{(2)} = [\eta]^{(2)}$ where $A \to^\xi A + C$ and $V \to^\eta V + C$. Furthermore, if $\mathscr{R}(\sqrt{C}) \subset \mathscr{R}(\sqrt{A})$, then $\xi \simeq 0$.*

PROOF. Since $\mathscr{R}(\sqrt{A}) \subset \mathscr{R}(\sqrt{V})$, we have $aA \leqslant V$ for some $a > 0$. Let $\mu > 1$ and consider $V + C \to^{\theta^{(\mu)}} V + \mu C$. Then

$$1 + (\mu - 1)\sqrt{C}\,[(V + C) + l]^{-1}\sqrt{C} \leqslant 1 + \left(\frac{\mu - 1}{a}\right)\sqrt{C}\left[A + \frac{C}{a} + \frac{l}{a}\right]^{-1}\sqrt{C}.$$

Therefore with $A + C/a \to^{\rho^{(\mu/a)}} A + \mu C/a$, we have

$$0 \leqslant \int \frac{\theta^{(\mu)}(x)}{x + l}\,dx \leqslant \int \frac{\rho^{(\mu/a)}(x/a)}{x + l}\,dx$$

for $l > 0$. But $\theta^{(\mu)} = \eta^{(\mu)} - \eta \geqslant 0$ and $\rho^{(\mu/a)} = \xi^{(\mu/a)} - \xi^{(1/a)} \geqslant 0$. Consequently, $\eta^{(\mu)} \sim \eta$ since $\xi^{(\mu/a)} \circ (1/a) \simeq \xi^{(1/a)} \circ (1/a)$. Similarly, $\eta^{(\mu)} \simeq \eta$ for $0 < \mu < 1$. Next, suppose that $\mathscr{R}(\sqrt{A}) = \mathscr{R}(\sqrt{V})$. Then $aA \leqslant V \leqslant bA$ for some $b > 0$. Consequently, if $l > 0$,

$$\int \frac{\xi^{(1/b)}(x/b)}{x + l}\,dx \leqslant \int \frac{\eta(x)}{x + l}\,dx \leqslant \int \frac{\xi^{(1/a)}(x/a)}{x + l}\,dx,$$

so that

$$0 \leqslant \int \frac{\eta(x) - \xi^{(1/b)}(x/b)}{x + l}\,dx \leqslant \int \frac{\xi^{(1/a)}(x/a) - \xi^{(1/b)}(x/b)}{x + l}\,dx,$$

and therefore, $[\eta]^{(2)} = [\xi^{(1/b)} \circ (1/b)]^{(2)} = [\xi^{(1/b)}]^{(2)} = [\xi]^{(2)}$.

Lastly, suppose $\mathscr{R}(\sqrt{C}) \subset \mathscr{R}(\sqrt{A})$. Consider the two-fold perturbation $A \to^\xi A + C \to^\theta A + 2C$. Since $\{A, C\}$ is consistent, $\theta \simeq 0$. On the other hand, $\mathscr{R}(\sqrt{A}) = \mathscr{R}(\sqrt{A + C})$ and therefore, by what has already been proved with $V = A + C$, we have $\theta \simeq \xi$.

LEMMA 3.4. *Suppose $\{A, C\}$ and $\{A, X\}$ are consistent and $\mathscr{R}(\sqrt{A + C}) = \mathscr{R}(\sqrt{A + X})$. Then $[\xi]^{(2)} = [\eta]^{(2)}$ where $A \to^\xi A + C$ and $A \to^\eta A + X$.*

PROOF. Introduce the phase shifts

$$A \overset{\xi}{\to} A + C \overset{\theta}{\to} A + C + X \overset{\rho}{\to} A + X \overset{-\eta}{\to} A$$

so that $\xi + \theta + \rho - \eta = 0$. Observe that $\theta \simeq 0$ and $\rho \simeq 0$ by Lemma 3.3, since $\{A, C\}$ and $\{A, X\}$ are consistent and $\mathscr{R}(\sqrt{A + C}) = \mathscr{R}(\sqrt{A + X})$. Consequently, $\xi - \eta \simeq 0$ or $[\xi]^{(2)} = [\eta]^{(2)}$.

COROLLARY 3.5. *If $\{A, C\}$ and $\{V, X\}$ are consistent, $\mathscr{R}(\sqrt{A}) = \mathscr{R}(\sqrt{V})$, and $\mathscr{R}(\sqrt{A + C}) = \mathscr{R}(\sqrt{V + X})$, then $\xi \simeq \eta$ where $A \to^{\xi} A + C$ and $V \to^{\eta} V + X$.*

PROOF. Since $\mathscr{R}(\sqrt{A}) = \mathscr{R}(\sqrt{V})$, $\xi \simeq \gamma$ by Lemma 3.3, where $V \to^{\gamma} V + C$. Also by Lemma 3.4, $\gamma \simeq \eta$.

PROOF OF THEOREM 3.2. Suppose $\{A, B, C, D\}$ and $\{V, W, X, Y\}$ are positive operators with $\{A, C\}$, $\{B, D\}$, $\{V, X\}$, and $\{W, Y\}$ consistent pairs, and

$$\mathscr{R} = \mathscr{R}(\sqrt{A}) = \mathscr{R}(\sqrt{V}), \qquad \mathscr{S} = \mathscr{R}(\sqrt{B}) = \mathscr{R}(\sqrt{W}),$$
$$\mathscr{R}(\sqrt{A + C}) = \mathscr{R}(\sqrt{B + D}), \qquad \mathscr{R}(\sqrt{V + X}) = \mathscr{R}(\sqrt{W + Y}).$$

Introduce the transitions

$$A \xrightarrow{\xi} A + C \xrightarrow{\theta} A + X \xrightarrow{\rho} A, \qquad B \xrightarrow{\tau} B + D \xrightarrow{\chi} B + Y \xrightarrow{\delta} B,$$

so that $\xi + \theta + \rho = \tau + \chi + \delta = 0$. It suffices to show that $\theta \simeq \chi$ since $-\rho \simeq \eta$ and $\delta \simeq -\gamma$ where $V \to^{\eta} V + X$ and $W \to^{\gamma} W + Y$. With

$$A + C \xrightarrow{\lambda} A + C + X \xrightarrow{\kappa} A + X, \qquad B + D \xrightarrow{\alpha} B + D + Y \xrightarrow{\beta} B + Y,$$

we have $\theta = \lambda + \kappa$ and $\chi = \alpha + \beta$. Now $\lambda \simeq \alpha$ since $\mathscr{R}(\sqrt{A + C + X}) = \mathscr{R}(\sqrt{B + D + Y})$ and $\{A + C, X\}$ and $\{B + D, Y\}$ are consistent.

Finally, $\kappa \simeq \beta$ since $\mathscr{R}(\sqrt{A + X}) = \mathscr{R}(\sqrt{B + Y})$, $\mathscr{R}(\sqrt{B + D + Y}) = \mathscr{R}(\sqrt{A + C + X})$; and $\{A + X, C\}$ and $\{B + Y, D\}$ are consistent. Therefore, $\theta \simeq \chi$ and the proof of Theorem 3.2 is completed.

REFERENCES

1. R. W. Carey and W. D. Pepe, *The phase shift and singular measures*, Indiana Univ. Math. J. **22** (1973), 1049–1064.

2. R. W. Carey and J. D. Pincus, *An invariant for certain operator algebras*, Proc. Nat. Acad. Sci. USA. **71** (1974), 1952–1956.

3. ______, *Operator theory and boundaries of complex curves*, preprint.

4. ______, *Principal functions, index theory, geometric measure theory and function algebras*, Integral Equations and Operator Theory **2** (1979), 441–483.

5. J. Dixmier, *Etude sur les variétés et les operateurs de Julia*, Bull. Soc. Math. France **77** (1949), 11–101.

6. ______, *Sur les variétés J d'un espace de Hilbert*, J. Math. Pures Appl. **28** (1949), 321–358.

7. ______, *Les idéaux dans l'ensemble des variétés d'un espace hilbertien*, Ann. Fac. Sci. Toulouse Math. **10** (1949), 91–114.

8. R. G. Douglas, *On majorization, factorization and range inclusion of operators in Hilbert space*, Proc. Amer. Math. Soc. **17** (1966), 413–416.

9. H. Federer, *Colloquium on geometric measure theory*, Bull. Amer. Math. Soc. **84** (1978), 291–338.

10. P. A. Fillmore and J. P. Williams, *On operator ranges*, Adv. in Math. **7** (1971), 254–281.

11. C. Foiaş, *Invariant para-closed subspaces*, Indiana Univ. Math. J. **20** (1971), 897–900.

12. ______, *Invariant para-closed subspaces*, Indiana Univ. Math. J. **21** (1972), 887–906.

13. M. G. Krein, *Perturbation determinants and a formula for the traces of unitary and self-adjoint operators*, Dokl. Akad. Nauk. SSSR. **144** (1962), 268–271; English transl. in Soviet Math. Dokl. **3** (1962), 707–710.

14. G. W. Mackey, *On the domains of closed linear transformations in Hilbert space*, Bull. Amer. Math. Soc. **52** (1946), 1009.

15. J. D. Pincus and J. Xia, *Symmetric and self-adjoint Toeplitz operators on multi-connected plane domains*, J. Funct. Anal. **59** (1984), 111–159.

16. S. Prössdorf, *Stability of the index of a one-dimensional singular integral operator with a vanishing symbol*, Probl. Math. Anal. Boundary Value Problems Integr. Equations, Izdat. Leningrad Univ., Leningrad (1966), 70–79.

UNIVERSITY OF KENTUCKY

STATE UNIVERSITY OF NEW YORK, STONY BROOK

Proceedings of Symposia in Pure Mathematics
Volume **44** (1986)

MACSYMA and Minimal Surfaces

PAUL CONCUS AND MARIO MIRANDA[1]

The first proof of the existence of minimal singular surfaces of codimension one was given in 1969 by E. Bombieri, E. De Giorgi, and E. Giusti [1]. They proved that the cone

$$\{(x, y)|x \in R^{k+1}, y \in R^{k+1}, |x|^2 = |y|^2\},$$

is a $(2k + 1)$-dimensional minimal surface if $k \geqslant 3$. In 1972, B. H. Lawson proved that all cones

$$\{(x, y)|x \in R^{k+1}, y \in R^{h+1}, h|x|^2 = k|y|^2\}$$

are $(h + k + 1)$-dimensional minimal surfaces if $h + k \geqslant 7$ or $h = k = 3$ [6]. In 1983, U. Massari and M. Miranda presented an elementary proof of the Bombieri–De Giorgi–Giusti result [8].

The aim of this note is to apply Massari and Miranda's method to the cones considered by Lawson. Such an application turns out to be successful with the additional assumptions $h < 5k$, $k < 5h$.

In §1 we sketch briefly the mathematical framework of the problem. In §2 we show how the use of symbolic algebraic manipulation, as provided to us by VAXIMA, the Berkeley VAX/UNIX version of the MACSYMA computer programming system [7], makes possible the application of Massari and Miranda's method to the nonsymmetric cones.

1. Minimal surfaces of codimension one. Throughout this paper we use the following definition of a codimension one surface in $(n + 1)$-Euclidean space.

DEFINITION 1. An *n-dimensional surface* is the boundary of any measurable set of R^{n+1}.

---

[1] This work was supported in part by the Applied Mathematics Subprogram of the Office of Energy Research, U.S. Department of Energy under contract DE-AC03-76SF00098.

1980 *Mathematics Subject Classification.* Primary 49F10.

© 1986 American Mathematical Society
0082-0717/80 $1.00 + $.25 per page

Definition 1 attaches the label *n-dimensional surface* to closed subsets of R^{n+1}, which may be singular and have nonzero $(n + 1)$-dimensional measure. Nevertheless, Definition 1 together with a convenient definition of n-dimensional measure has been successfully used for the study of codimension one surfaces since 1954, when it was first introduced by E. De Giorgi [2, 3]. The appropriate definition of n-measure for our surfaces is possible through use of the divergence theorem.

DEFINITION 2. If Ω is a measurable subset of R^{n+1} and A is an open subset of R^{n+1}, we take as *n-dimensional measure of the portion of $\partial\Omega$ contained in A*, the quantity

$$\sup\left\{ \int_\Omega \operatorname{div}\phi(x)\, dx \,\middle|\, \phi \in \left[C_0^1(A)\right]^{n+1}, |\phi(x)| \leqslant 1 \text{ for all } x \right\},$$

which we denote by $M_n(\partial\Omega \cap A)$.

The following remark indicates the basis for Definition 2.

REMARK 1. If $\partial\Omega$ is sufficiently smooth, then the divergence theorem

$$(1) \qquad \int_\Omega \operatorname{div}\phi(x)\, dx = \int_{\partial\Omega} \phi(x) \cdot \nu(x)\, dH_n$$

holds for any vector function $\phi \in [C_0^1(R^{n+1})]^{n+1}$. In (1) $\nu(x)$ denotes the outward unit vector normal to $\partial\Omega$ at x and H_n is the Hausdorff n-dimensional measure in R^{n+1}. The validity of (1) obviously implies

$$(2) \qquad M_n(\partial\Omega \cap A) = H_n(\partial\Omega \cap A), \text{ for all open } A.$$

We give finally the following definition of a minimal surface.

DEFINITION 3. $\partial\Omega$ is a *minimal surface* in R^{n+1} if the following inequalities hold

$$(3) \qquad M_n(\partial\Omega \cap A) < +\infty, \qquad M_n(\partial\Omega \cap A) \leqslant M_n(\partial\Omega^* \cap A),$$

for all bounded open sets A and any measurable set Ω^* satisfying

$$(4) \qquad (\Omega - \Omega^*) \cup (\Omega^* - \Omega) \Subset A.$$

A theory based on Definitions 1–3 and rich in remarkable results was developed between 1960 and 1970. The cones considered by Bombieri–De Giorgi–Giusti and Lawson belong to this theory, providing examples of singular minimal surfaces. Their lowest dimension is 7. That no lower dimension is possible follows from Theorem 1.

THEOREM 1 (W. H. FLEMING [5], E. DE GIORGI, et al. [4], J. SIMONS [10]). *The only minimal six-dimensional surfaces in R^7 are planes.*

Two more basic definitions of this paper are the following.

DEFINITION 4. A measurable set F of R^{n+1} is *M-concave* if its boundary is minimal with respect to modifications contained in F itself.

In other words, in order to be called *M*-concave the set F must satisfy the inequalities

$$(5) \qquad M_n(\partial F \cap A) < +\infty, \qquad M_n(\partial F \cap A) \leqslant M_n(\partial F^* \cap A),$$

for all bounded open sets A and all open sets $F^* \subset F$ such that

$$(6) \qquad\qquad (F - F^*) \Subset A.$$

DEFINITION 5. *The measurable set F of R^{n+1} is said to be M-convex if the measurable set $R^{n+1} - F$ is M-concave.*

The interest of Definitions 4 and 5 is underlined by the following obvious remark.

REMARK 2. *The measurable set F has minimal boundary if and only if it is M-concave and M-convex.*

A sufficient condition for M-concavity is provided by the following theorem.

THEOREM 2. *If $\{f_j\}$ is a sequence of subsolutions to the minimal surface equation (m.s.e.) in the open set F of R^{n+1}, i.e.*

$$(7) \qquad\qquad \operatorname{div}\left(\frac{\operatorname{grad} f_j(x)}{\sqrt{1 + |\operatorname{grad} f_j(x)|^2}} \right) \geqslant 0, \quad \text{for } x \in F, \text{for all } j,$$

if $M_n(\partial F \cap A) < +\infty$ for all open bounded sets A, if

$$(8) \qquad\qquad f_j(x) = 0, \quad \text{for } x \in \partial F, \text{for all } j,$$

and if

$$(9) \qquad\qquad \lim_{j \to \infty} f_j(x) = +\infty, \quad \text{for } x \in F,$$

then F is M-concave.

PROOF. Since f_j is a subsolution to the m.s.e., the open set

$$\{(x, z) | x \in F, z \in R, z < f_j(x)\},$$

is M-concave with respect to compact modifications contained in $F \times R$. Conditions (8) and (9) imply that the cylinder $F \times R$ itself is M-concave with respect to compact modifications contained in $F \times R_+$, where $R_+ = \{z \in R | z > 0\}$. Therefore F must be M-concave. ∎

Theorem 2 will be particularly useful in the special case of cones, as shown by the following corollaries.

COROLLARY 1. *If F is an open cone of R^{n+1}, i.e., if F satisfies*

$$(10) \qquad\qquad x \in F, \qquad \rho > 0 \Rightarrow \rho x \in F,$$

and if $M_n(\partial F \cap A) < +\infty$ for all open bounded sets A, then for F to be M-concave it is sufficient that there exist one subsolution f to the m.s.e. satisfying

$$(11) \qquad\qquad f(x) = 0, \quad \text{for } x \in \partial F,$$

$$(12) \qquad\qquad f(x) > 0, \quad \text{for } x \in F,$$

and

$$(13) \qquad\qquad f(\rho x) = \rho^\alpha f(x), \quad \text{for } x \in F, \text{for all } \rho > 0,$$

with $\alpha \neq 1$.

PROOF. For any $\rho > 0$, the function

$$(14) \qquad \rho^{-1}f(\rho x) = \rho^{\alpha-1}f(x),$$

is a subsolution to the m.s.e. in F. Since $\alpha \neq 1$ and $\rho > 0$ is arbitrary, (14) implies that any multiple of f is a subsolution. Therefore the sequence $\{f_j\} = \{jf\}$ satisfies the hypotheses of Theorem 2, which implies that F is M-concave. $\blacksquare$

COROLLARY 2. *If* P: $R^{n+1} \to R$ *is a homogeneous nonconstant polynomial, satisfying*

$$(15) \qquad P(x)\mathrm{div}\left(\frac{\mathrm{grad}\, P(x)}{\sqrt{1 + |\mathrm{grad}\, P(x)|^2}} \right) \geqslant 0, \quad \text{for } x \in R^{n+1},$$

then the cone $\{ x \in R^{n+1} | P(x) > 0\}$ *has minimal boundary.*

PROOF. If P is homogeneous of degree 1 the cone (16) is a plane; therefore, it is obviously minimal. We can thus assume that P is homogeneous of degree different from one.

The open set $F = \{ x \in R^{n+1} | P(x) > 0\}$ and the function $f(x) = P(x)$ satisfy the hypotheses of Corollary 1; therefore, F is M-concave. Again, the open set $\tilde{F} = \{ x \in R^{n+1} | P(x) < 0\}$ and the function $\tilde{f} = -P(x)$ satisfy the hypotheses of Corollary 1; therefore, $\tilde{F}$ is M-concave. Since $\tilde{F} = R^{n+1} - (F \cup \partial F)$ is equivalent to $R^{n+1} - F$, we get that F is M-convex, thus F has minimal boundary.

2. MACSYMA. MACSYMA is a symbolic algebraic manipulation computer programming system that has been developed by the Mathlab Group at the Massachusetts Institute of Technology Laboratory for Computer Science [7]. With it, a computer user can interactively carry out many operations on symbolic expressions, including those particularly useful for our study, such as algebraic manipulation, differentiation, factorization, and regrouping of polynomials, as well as solving polynomial equations. A system like MACSYMA permits lengthy algebraic manipulation to be carried out easily and efficiently that would normally be almost impossible by hand, and without the inherent probability of human error to which lengthy hand calculations are prone.

In order to prove (with the aid of MACSYMA) that Lawson's cones are minimal, we consider the following polynomial

$$P(x, y) = (hx^2 - ky^2)(ax^2 + \beta y^2),$$

where $x \in R^{k+1}, y \in R^{h+1}$, α, β, are real numbers, and

$$x^2 = \sum_{i=1}^{k+1} x_i^2, \qquad y^2 = \sum_{j=1}^{h+1} y_j^2.$$

Let us denote by $\mathcal{M}P$ the polynomial

$$\mathcal{M}P \equiv (1 + |\mathrm{grad}\, P|)^{3/2}\mathrm{div}\left(\frac{\mathrm{grad}\, P}{\sqrt{1 + |\mathrm{grad}\, P|^2}} \right).$$

We obtain that the remainder after dividing $\mathcal{M}P$ by $hx^2 - ky^2$ is

$$(2/k)x^2\{\alpha(5h - k)k - \beta(5k - h)h\}.$$

This implies that the polynomial

$$P^*(x, y) = (hx^2 - ky^2)\{(5k - h)hx^2 + (5h - k)ky^2\}$$

divides $\mathcal{M}P^*$ exactly.

We have then the relationship

$$P^*\mathcal{M}P^* = (hx^2 - ky^2)^2\{(5k - h)hx^2 + (5h - k)ky^2\}Q,$$

where Q is the polynomial $Q = \mathcal{M}P^*/(hx^2 - ky^2)$. If we assume

$$(17) \qquad\qquad k < 5h, \qquad h < 5k,$$

we have

$$\{(x, y)|P^*(x, y) = 0\} = \{(x, y)|hx^2 = ky^2\},$$

and for $P^* \neq 0$

$$P^*\mathcal{M}P^* \geqslant 0 \Leftrightarrow Q \geqslant 0.$$

Our problem is thereby reduced to the study of the sign of Q. First we compute

$$Q_0 = Q(0,0) = (8h - 12)k^2 + (8k - 12)h^2 + 72hk,$$

which is easily seen to be positive for all integers h, k, satisfying (17). Let

$$q = (Q - Q_0)/(64hk),$$

and consider q as a function of $t = hx^2$ and $s = ky^2$. The computation of q at $s = t$ gives

$$q|_{s=t} = (h + k - 6)(h + k)^3(2t)^3,$$

which is nonnegative for $h + k \geqslant 6$.

Assuming

$$(18) \qquad\qquad h + k \geqslant 6,$$

we have only to study the behavior of q as a function of s for $s \geqslant t$, and the behavior of q as a function of t for $t \geqslant s$. The computation of q as a function of s and t gives a homogeneous polynomial of degree 3

$$q(s, t) = as^3 + 3bs^2t + 3cst^2 + dt^3,$$

where a, b, c, d depend on h, k. Computation of a and d gives

$$a = (5h - k)^2(2k^2 + 2hk + 3k - 3h), \quad d = (5k - h)^2(2h^2 + 2hk + 3h - 3k).$$

Assumptions (17) imply

$$(19) \qquad\qquad a > 0, \qquad d > 0.$$

We have $d^2q/ds^2 = 6(as + bt)$. Therefore

$$\left.\frac{d^2q}{ds^2}\right|_{s \geqslant t} \geqslant 0 \Leftrightarrow a + b \geqslant 0.$$

Computation of $a + b$ yields

$$a + b = -2(k + h)(k^3 - 5hk^2 + 2k^2 - 9h^2k - 16hk - 3h^3 + 30h^2),$$

which shows that $a + b \geqslant 0$ if

$$(20) \qquad\qquad\qquad\qquad h \leqslant k \leqslant 5h.$$

Assuming (20), we get that q is a convex function of s for $s \geqslant t$; therefore, it will also be nondecreasing if

$$\left.\frac{dq}{ds}\right|_{s=t} = 3t^2(a + 2b + c) \geqslant 0.$$

Computing $a + 2b + c$, we obtain

$$a + 2b + c = 4(h + k)^2(2hk + k + 2h^2 - 13h).$$

Therefore $a + 2b + c \geqslant 0$ if (20) holds. Thus, our first conclusion is $q \geqslant 0$ for $s \geqslant t$ if (18) and (20) hold.

The proof that $q \geqslant 0$ for all $t \geqslant s$ proceeds similarly. We obtain

$$\left.\frac{d^2q}{dt^2}\right|_{t=s} = 6(d + c)s, \qquad \left.\frac{dq}{dt}\right|_{t=s} = 3(d + 2c + b)s^2$$

and the expressions

$$d + c = 2(k + h)(3k^3 + 9hk^2 - 30k^2 + 5h^2k + 16hk - h^3 - 2h^2),$$

$$d + 2c + b = 4(k + h)^2(2k^2 + 2hk - 13k + h).$$

Assuming (18), (20), and $(h, k) \neq (1, 5)$, $(h, k) \neq (2, 4)$, there follows

$$d + c \geqslant 0, \qquad d + 2c + b \geqslant 0.$$

We conclude that $q \geqslant 0$ for $t \geqslant s$.

Therefore, $q \geqslant 0$ for $s \geqslant 0$, $t \geqslant 0$, if (18), (20) hold and $(h, k) \neq (1, 5)$, $(h, k) \neq (2, 4)$. In other words, Lawson's cones corresponding to (h, k) satisfying (18), (20) and different from $(1, 5)$, $(2, 4)$ are minimal. Since the interchange of x and y does not change the cone, we can conclude that our method confirms Lawson's results under the additional restriction (17).

References

[1] E. Bombieri, E. De Giorgi and E. Giusti, *Minimal cones and the Bernstein problem*, Invent. Math. **7** (1969), 243–268.

[2] E. De Giorgi, *Su una teoria generale della misura $(r - 1)$-dimensionale in uno spazio euclideo ad r dimensioni*, Ann. Mat. Pura Appl. (4) **36** (1954), 191–213.

[3] E. De Giorgi, *Nuovi teoremi relativi alle misure $(r - 1)$-dimensionali in uno spazio ad r dimensioni*, Ricerche Mat. **4** (1955), 95–113.

[4] E. De Giorgi, F. Colombini and L. C. Piccinini, *Frontiere orientate di misura minima e questioni collegate*, Scuola Norm. Sup. Pisa, (1972).

[5] W. H. Fleming, *On the oriented Plateau problem*, Rend. Circ. Mat. Palermo (2) **11** (1962), 69–90.

[6] H. B. Lawson, Jr., *The equivariant Plateau problem and interior regularity*, Trans. Amer. Math. Soc. **173** (1972), 231–249.

[7] MACSYMA Reference Manual, Version Ten, Mathlab Group, Laboratory for Computer Science, Massachusetts Institute of Technology, Cambridge, Mass., 1983.

[8] U. Massari and M. Miranda, *A remark on minimal cones*, Boll. Un. Mat. Ital. D(6) **2-A** (1983), 123–125.

[9] ______, *Minimal surfaces of codimension one*, Notas de Matemática, North-Holland, Amsterdam, 1984.

[10] J. Simons, *Minimal varieties in Riemannian manifolds*, Ann. of Math. (2) **88** (1968), 62–105.

UNIVERSITY OF CALIFORNIA AT BERKELEY AND LAWRENCE BERKELEY LABORATORY

Proceedings of Symposia in Pure Mathematics
Volume **44** (1986)

Sur les Chaînes Maximalement Complexes
de Bord Donné

PIERRE DOLBEAULT

Introduction. On appelle p-chaîne holomorphe d'une variété analytique complexe Z, de dimension complexe n, tout courant $T = \Sigma \, n_j [V_j]$ où V_j désigne un ensemble analytique complexe irréductible de Z, de dimension complexe p, $[V_j]$ le courant d'intégration sur V_j et n_j un entier.

Soit M une sous-variété réelle, compacte, de classe C^1, orientée, de dimension $(2p - 1)$ d'une variété analytique complexe X, de dimension complexe n. S'il existe une p-chaîne holomorphe T de $Z = X \setminus M$, à support relativement compact dans X, ayant une extension simple à X, notée encore T et telle que $dT = [M]$, courant d'intégration sur M, on dira que M est le bord de T. Plus généralement, on permettra à M des singularités contenues dans un ensemble fermé, de mesure de Hausdorff $(2p - 1)$-dimensionnelle nulle.

Les théorèmes de prolongement à un ouvert de $\mathbf{C}^n$, d'une fonction holomorphe définie au voisinage du bord de l'ouvert (Hartogs, Bochner, Martinelli [5]) conduisent au problème suivant:

Problème 1. X et M étant données comme ci-dessus, trouver une p-chaîne holomorphe de bord M.

Une condition nécessaire est que M soit maximalement complexe. Harvey et Lawson ont montré que cette condition est suffisante si $X = \mathbf{C}^n$ (ou un espace de Stein) pour $p \geqslant 2$ [6, 5] et aussi, si $X = \mathbf{P}^n(\mathbf{C}) \setminus \mathbf{P}^{n-q}(\mathbf{C})$ (ou un sous-espace analytique de $\mathbf{P}^n(\mathbf{C}) \setminus \mathbf{P}^{n-q}(\mathbf{C})$), pour $p \geqslant q + 1$ [7].

On appellera *chaîne maximalement complexe*, de dimension $(2p - 1)$, dans une variété analytique complexe Z, tout courant $M = \Sigma_j \, n_j [W_j]$ où $[W_j]$ désigne le courant d'intégration sur une sous-variété orientée W_j, de classe C^1, de dimension $(2n - 1)$, maximalement complexe de Z, à singularités $(2p - 1)$-négligeables.

1980 *Mathematics Subject Classification*. Primary 32C25.

© 1986 American Mathematical Society
0082-0717/86 $1.00 + $.25 per page

En adaptant la méthode de Harvey–Lawson dans $\mathbf{C}^n$ [5], on cherche à résoudre:

Problème 2. Soit E un hyperplan réel de $\mathbf{C}^n$. Soit N une sous-variété compacte orientée de E, de dimension $(2p - 2)$, de dimension holomorphe $(p - 2)$ avec, éventuellement, des singularités contenues dans un fermé de mesure de Hausdorff $(2p - 2)$-dimensionnelle nulle. On désigne par $[N]$ le courant d'intégration sur N dans $\mathbf{C}^n$; on suppose $[N]$ de masse finie et tel que $d[N] = 0$. Trouver une $(2p - 1)$-chaîne maximalement complexe M de $\mathbf{C}^n \setminus N$, à support relativement compact dans E, de masse finie, ayant une extension simple à $\mathbf{C}^n$, notée encore M et vérifiant $dM = [N]$; N sera dite le bord de M.

On se borne au cas $p = n - 1$, $n \geqslant 3$, auquel se ramène le cas général par la méthode des projections de [6] si $p \geqslant 3$; le cas $p = 1$ n'est pas étudié ici. Un programme pour une résolution du Problème 2 a été exposé dans [1]; l'application à une solution du Problème 1 pour $X = \mathbf{P}^n(\mathbf{C})$ a été indiquée schématiquement dans [1] et, avec une condition très particulière, dans [3].

Voici un résumé des résultats obtenus. Le $n° 1$ donne une solution, non unique, du Problème 2, pour des courants d'une variété de Stein de dimension $\geqslant 4$, sans condition sur les supports autre que la compacité et pour $p = n - 1$ (problème affaibli). Désignant par j l'inclusion: $E \to \mathbf{C}^n$ et par M un courant plat de $\mathbf{C}^n$, d-fermé, maximalement complexe, à support dans E, on étudie le courant S de E tel que $j_* S = M$ et on l'explicite à l'aide de courants de degré 0 ($n° 2$). Puis, en supposant M rectifiable, on précise l'un des courants ci-dessus ($n° 3$). Supposant, ensuite, que M est, sauf sur un fermé négligeable, une variété analytique réelle (C^ω), on précise le résultat du $n° 2$ et on obtient une formule proche de celle de Poincaré–Lelong ($n° 4$).

Soit N une sous-variété C^ω de $\mathbf{C}^n$, à singularités négligeables, de dimension $(2n - 4)$, de dimension holomorphe $(n - 3)$, contenue dans E; on suppose, en outre, qu'en tout point z de N en dehors d'un fermé de mesure nulle, N est transversale au sous-espace affine complexe maximal de E passant par z; alors on calcule les coefficients d'une fonction CR ($n° 5$) qui définira, après prolongement ($n° 6$) une chaîne maximalement complexe C^ω, à support relativement compact contenu dans E, de bord N; la construction n'est cependant achevée que dans les cas suivants pour lesquels, après complexification locale, des procédés d'extension de Harvey–Lawson sont utilisés:

(a) N est une sous-variété C^ω lisse.

(b) N est l'intersection de E et d'une chaîne C^ω maximalement complexe de $\mathbf{C}^n$.

Les nn. 2 à 4 donnent des résultats sur les chaînes maximalement complexes qui seront peut-être utiles pour établir des théorèmes de structure; une partie seulement de ces résultats est utilisée pour la solution du Problème 2. Les définitions relatives aux chaînes à singularités négligeables se trouvent au début du $n° 4$. On a appelé type d'une forme différentielle homogène par rapport aux différentielles des coordonnées complexes et de leurs imaginaires conjuguées le bidegré en ces différentielles et on a noté d', d'' les différentiations extérieures par rapport aux coordonnées complexes et à leurs imaginaires conjuguées respectivement.

Jacques Besnault a participé à la phase initiale de la préparation de cet article, en particulier des nn. 1, 2 et 3.

Table des matières.

1. Solution du problème affaibli sur une variété de Stein.

1.1. Sur une variété de Stein X, on donne un courant N réel, à support compact, de degré 4, sans terme de type $(0,4)$, d-fermé. On cherche un courant M réel, à support compact, maximalement complexe, tel que $dM = N$.

1.2. PROPOSITION. *Si X est une variété de Stein de dimension $n \geqslant 4$, le problème 1.1 a une infinité de solutions.*

PREUVE. Le courant N étant réel, sans terme de type $(0,4)$ s'écrit

$$(1.1) \qquad N = N^{1,3} + N^{2,2} + N^{3,1}$$

avec

$$(1.2) \qquad N^{2,2} = \overline{N^{2,2}}, \qquad N^{3,1} = \overline{N^{1,3}}.$$

La condition $dN = 0$, compte tenu de (1.2) équivaut à

$$(1.3) \qquad d''N^{1,3} = 0, \qquad d'N^{1,3} + d''N^{2,2} = 0.$$

Le courant cherché M étant maximalement complexe et réel satisfait à

$$(1.4) \qquad M = M^{1,2} + M^{2,1},$$

$$(1.5) \qquad M^{2,1} = \overline{M^{1,2}}.$$

La condition $dM = N$, compte tenu de (1.2) et (1.5) équivaut à

$$(1.6) \qquad d''M^{1,2} = N^{1,3},$$

$$(1.7) \qquad d'M^{1,2} + \overline{d'M^{1,2}} = N^{2,2}.$$

D'après le théorème de dualité de Serre $H_c^{p,q}(X, \mathbf{C}) = H^{n-p,n-q}(X, \mathbf{C})$, ce dernier espace est nul pour $n - q \geqslant 1$ d'après le Théorème B; comme $n \geqslant 4$, l'espace est nul pour $q \leqslant 3$; donc l'équation (1.6) a une solution $M^{1,2}$ à support compact.

Soit $M_0^{1,2}$ une solution particulière de (1.6). Alors, pour toute solution $M^{1,2}$ de (1.6) on a

$$d''\left(M^{1,2} - M_0^{1,2} \right) = 0.$$

Comme $H_c^{1,2}(X, \mathbf{C}) = 0$, il existe un courant $\mu^{1,1}$ à support compact tel que $d''\mu^{1,1} = M^{1,2} - M_0^{1,2}$. Reste à déterminer $\mu^{1,1}$ pour que $M^{1,2} = M_0^{1,2} + d''\mu^{1,1}$ satisfasse à (1.7).

$$(1.8) \qquad d'd''\mu^{1,1} + \overline{d'd''\mu^{1,1}} = N^{2,2} - d'M_0^{1,2} - \overline{d'M_0^{1,2}} = \nu^{2,2},$$

courant réel à support compact. Le courant $\mu^{1,1}$ satisfait à

$$(1.9) \qquad d'd''\left(\mu^{1,1} - \overline{\mu^{1,1}}\right) = \nu^{2,2}.$$

On a $d'\nu^{2,2} = d'N^{2,2} - d'(\overline{d'M_0^{1,2}})$, mais $d'\left(\overline{d'M_0^{1,2}}\right) = -\overline{d'd''M_0^{1,2}} = -\overline{d'N^{1,3}}$ $= \overline{d''N^{2,2}}$, d'après (1.3); en outre $\overline{d''N^{2,2}} = \overline{d'N^{2,2}} = d'N^{2,2}$, d'après (1.2); donc $d'\nu^{2,2} = 0$.

On a $d''\nu^{2,2} = d''N^{2,2} - d''d'M_0^{1,2}$, mais $-d''d'M_0^{1,2} = d'd''M_0^{1,2} = -d''N^{2,2}$, d'après (1.6) et (1.3), d'où $d''\nu^{2,2} = 0$.

En désignant par $\Lambda_c^{p,q}$ le quotient

$$\mathrm{Ker}\left(\mathscr{D}^{p,q}(X) \overset{d}{\to} \mathscr{D}^{p+1,q}(X) \oplus \mathscr{D}^{p,q+1}(X)\right)\Big/ d'd''\mathscr{D}^{p-1,q-1}(X),$$

on a [3, Théorème (2.4)B]: *si X est une variété de Stein de dimension n et si $p \geq 1$, $q \geq 1$, $p + q \leq n$ et p ou $q < n$, alors $\Lambda_c^{p,q}(X) = 0$.* Le courant $\nu^{2,2}$ étant d'- et d''-fermé, il existe un courant $\lambda^{1,1}$ à support compact tel que

$$(1.10) \qquad d'd''\lambda^{1,1} = \nu^{2,2}.$$

D'après (1.8), le courant $\nu^{2,2}$ est réel. Soit $l^{1,1} = \mathrm{Re}\,\lambda^{1,1}$; l'opérateur $d'd''$ étant imaginaire pur, on a $d'd''l^{1,1} = 0$; donc $\lambda^{1,1} - l^{1,1}$ est imaginaire pur et solution de (1.10), i.e., il existe une solution imaginaire pure $\lambda^{1,1}$ de (1.10). Alors $\mu^{1,1} = \frac{1}{2}\lambda^{1,1}$ est solution de (1.9) et

$$M^{1,2} = M_0^{1,2} + d''\mu^{1,1}$$

satisfait à (1.6) car $d''M^{1,2} = d''M_0^{1,2} = N^{1,3}$ et à (1.7) car

$$d'M^{1,2} + \overline{d'M^{1,2}} = d'M_0^{1,2} + \overline{d'M_0^{1,2}} + d'd''\mu^{1,1} + \overline{d'd''\mu^{1,1}}$$

$$= d'M_0^{1,2} + \overline{d'M_0^{1,2}} + d'd''\lambda^{1,1} = d'M_0^{1,2} + \overline{d'M_0^{1,2}} + \nu^{2,2} = N^{2,2},$$

d'après (1.10) et (1.8).

On a donc obtenu un courant M réel, à support compact, maximalement complexe tel que $dM = N$.

La solution imaginaire pure $\lambda^{1,1}$ de (1.10) n'est pas unique; la différence de deux telles solutions est un $(1,1)$-courant imaginaire pur, à support compact, $d'd''$-fermé; en particulier, pour toute fonction f à valeurs réelles, C^2, à support compact, $d'd''f$ est un tel courant. $\square$

2. Courants maximalement complexes, d-fermés, localement plats, à support dans E.

2.1. *Notations.* Dans $\mathbf{C}^n(z_1, \ldots, z_n)$ on considère l'hyperplan réel $E = \{ z \in \mathbf{C}^n; \mathrm{Im}\, z_1(z) = 0\}$; soit $z_1 = x_1 + iy_1$ $(x_1, y_1 \in \mathbf{R})$; alors $(x_1, z_2, \ldots, z_n)$ est un système de coordonnées de E.

Dans E, tout courant T s'écrit $T = T' + dx_1 \wedge T''$ où T' et T'' sont des formes différentielles en dz_j, $d\bar{z}_j$ ($j = 2,\ldots,n$) à coefficients courants de degré 0. Pour un courant de E, on indiquera, éventuellement, par un couple d'exposants, le type par rapport à dz_j, $d\bar{z}_j$ ($j = 2,\ldots,n$).

On considère la projection $\pi\colon \mathbf{C}^n \to \mathbf{C}^{n-1}$ et on note encore π

$$z = (z_1,\ldots,z_n) \mapsto z' = (z_1,\ldots,z_{n-1})$$

sa restriction à E.

Un *courant M* de $\mathbf{C}^n$, de dimension $(2n - 3)$, *maximalement complexe*, s'écrit $M = M^{1,2} + M^{2,1}$. Dans ce numéro, on considère un courant M de $\mathbf{C}^n$ possédant les propriétés suivantes:

(A) $M = M^{1,2} + M^{2,1}$, *M est réel, d-fermé, à support dans E, localement plat.*

Soit $j\colon E \to \mathbf{C}^n$ l'injection canonique; d'après le théorème du support ([**8, 4**]), il existe un courant S de E localement plat, de degré 2, tel que

$$(2.1) \qquad\qquad j_*S = M.$$

2.2. *Expression de S.*

2.2.1. Soit $\mathbf{R}_1$ (resp. $\mathbf{R}_1'$) l'espace $\mathbf{R}$ muni de la coordonnée x_1 (resp. y_1). On a $E = \mathbf{R}_1 \times \mathbf{C}^{n-1}$, $\mathbf{C}^n = \mathbf{R}_1 \times \mathbf{R}_1' \times \mathbf{C}^{n-1}$, $\mathbf{C}^{n-1}$ étant décrit par $(z_2,\ldots,z_n)$. Soient T un courant de degré p de E et W un courant de degré 1 de $\mathbf{R}_1'$. On a $W = w\,dy_1$ où w est un courant de degré 0. On pose $T = T' + dx_1 \wedge T''$ où T' et T'' sont indépendants de dx_1. Toute forme différentielle appartenant à $\mathscr{D}^{2n-p-1}(\mathbf{C}^n)$ s'écrit $\varphi = \varphi_1 + dy_1 \wedge \varphi_2$ où φ_1 et $\varphi_2 \in \mathscr{D}^{\cdot}(\mathbf{C}^n)$ et sont indépendantes de dy_1.

2.2.2. *Produit tensoriel de courants.* A partir du produit tensoriel des distributions, on va définir le produit tensoriel d'un courant T de E et d'un courant de $\mathbf{R}_1'$

Soit d'abord $W = w\,dy_1$ le courant considéré sur $\mathbf{R}_1'$; on va définir $T \otimes W$; ce courant contiendra dy_1 en facteur, donc sera nul sur les formes $dy_1 \wedge d\varphi_2 \in \mathscr{D}^{\cdot}(\mathbf{C}^n)$; il suffit de le définir sur les formes $\varphi_1 \in \mathscr{D}^{2n-p-1}(\mathbf{C}^n)$ indépendantes de dy_1.

$$\varphi_1 = \sum_{I,\,I'|I+I'|=2n-p-1} \alpha_{II'}\,dz_I \wedge d\bar{z}_{I'} + \sum_{J,\,J'|J+J'|=2n-p-2} dx_1 \wedge \beta_{JJ'} \wedge dz_J \wedge d\bar{z}_{J'}.$$

Il suffit de définir $T \otimes W$ sur les formes monômes.

$$T \otimes W(dx_1 \wedge \beta\,dz_J \wedge d\bar{z}_{J'}) = T' \otimes W(dx_1 \wedge \beta\,dz_J \wedge d\bar{z}_{J'})$$
$$= (-1)^{2n-p-1} T' \wedge dx_1 \wedge dz_J \wedge d\bar{z}_J \otimes W(\beta)$$

où, dans la dernière expression $\otimes$ désigne le produit tensoriel des distributions $T' \wedge dx_1 \wedge dz_J \wedge d\bar{z}_{J'}$ de E et W de $\mathbf{R}_1'$. De la même façon, par définition, on pose

$$T \otimes W(\alpha\,dz_I \wedge d\bar{z}_{I'}) = dx_1 \wedge T'' \otimes W(\alpha\,dz_I \wedge d\bar{z}_{I'})$$
$$= (-1)^{2n-p-1} dx_1 \wedge T'' \wedge dz_I \wedge d\bar{z}_I \otimes W(\alpha),$$

où le produit tensoriel est calculé au sens des distributions dans le dernier membre.

Si w est un courant de degré 0 de $\mathbf{R}'_1$, on définit $T \otimes w$ par

$$(2.2) \qquad dy_1 \wedge T \otimes w = (-1)^p T \otimes w \, dy_1 = (-1)^p T \otimes W.$$

2.2.3. PROPOSITION. *Dans les hypothèses de 2.1, le courant S est de degré 2 et a pour expression*

$$(2.3) \qquad S = S^{1,1} + dx_1 \wedge (S^{1,0} + S^{0,1}).$$

DÉMONSTRATION. Considérons le courant M du n. 2.1. et $\varphi \in \mathscr{D}^{2n-3}(\mathbf{C}^n)$; comme ci-dessus, $\varphi = \varphi_1 + dy_1 \wedge \varphi_2$. On a $j^*\varphi = \varphi_1(x_1, 0, \underline{z})$ avec $\underline{z} = (z_2, \ldots, z_n)$ et

$$M(\varphi) = j_* S(\varphi) = S(j^*\varphi) = S(\varphi_1(x_1, 0, \underline{z})).$$

Soit $\delta(y_1) = \varepsilon(y_1)\, dy_1$ la mesure de Dirac de $\mathbf{R}'_1$ en 0; alors $M(\varphi) = S \otimes \delta(y_1)(\varphi_1) = dy_1 \wedge S \otimes \varepsilon(y_1)(\varphi)$ d'après (2.2), i.e., $M = dy_1 \wedge S \otimes \varepsilon(y_1)$. Le courant S étant de degré 2 dans E s'écrit

$$(2.4) \qquad S = S^{2,0} + S^{1,1} + S^{0,2} + dx_1 \wedge (S^{1,0} + S^{0,1}).$$

Compte tenu de la condition (A), de (2.4) et de $dx_1 = \tfrac{1}{2}(dz_1 + d\bar{z}_1)$, $dy_1 = (i/2)(d\bar{z}_1 - dz_1)$, on a

$$M^{1,2} + M^{2,1} = (i/2)(d\bar{z}_1 - dz_1)$$
$$\wedge \left(S^{2,0} + S^{1,1} + S^{0,2} + \tfrac{1}{2}(dz_1 + d\bar{z}_1) \wedge (S^{1,0} + S^{0,1}) \right) \otimes \varepsilon(y_1)$$
$$= (i/2)\left[-dz_1 \wedge S^{2,0} + d\bar{z}_1 \wedge S^{0,2} + d\bar{z}_1 \wedge S^{2,0} - dz_1 \wedge dS^{1,1} \right.$$
$$\left. - dz_1 \wedge d\bar{z}_1 \wedge S^{1,0} + d\bar{z}_1 \wedge S^{1,1} - dz_1 \wedge S^{0,2} - dz_1 \wedge d\bar{z}_1 \wedge S^{0,1} \right] \otimes \varepsilon(y_1).$$

En comparant les types, dans $\mathbf{C}^n$, des membres extrêmes des égalités ci-dessus, on obtient $dz_1 \wedge S^{2,0} = 0$ et $d\bar{z}_1 \wedge S^{0,2} = 0$; comme $S^{2,0}$ et $S^{0,2}$ sont indépendants de dz_1, $d\bar{z}_1$, on a $S^{2,0} = S^{0,2} = 0$, i.e., l'expression (2.3). $\square$

2.3. *Conditions de réalité.* M étant réel, S est aussi réel, i.e., $\bar{S} = S$, d'où $\overline{S^{1,1}} = S^{1,1}$ et $S^{0,1} = \overline{S^{1,0}}$.

2.4. *Condition de d-fermeture.* On a $dM = 0$; donc $dj_* S = 0$; pour toute forme $\varphi \in \mathscr{D}^\cdot(\mathbf{C}^n)$, on a

$$dj_* S(\varphi) = -j_* S(d\varphi) = -S(j^* d\varphi) = -S(dj^*\varphi) = dS(j^*\varphi) = 0;$$

comme toute forme de $\mathscr{D}^\cdot(E)$ est égale à la restriction à E d'un élément de $\mathscr{D}^\cdot(\mathbf{C}^n)$, la relation précédente entraîne

$$(2.5) \qquad dS = 0.$$

On pose

$$d'_E = \sum_{j=2}^{n} \frac{\partial}{\partial z_j} dz_j, \qquad d''_E = \sum_{j=2}^{n} \frac{\partial}{\partial \bar{z}_j} d\bar{z}_j.$$

Alors, d'après (2.4),

$$dS = d'_E S^{1,1} + d''_E S^{1,1} + dx_1 \wedge \left(\partial S^{1,1}/\partial x_1 \right)$$
$$- dx_1 \wedge \left(d'_E S^{1,0} + d''_E S^{0,1} + d'_E S^{0,1} + d''_E S^{1,0} \right).$$

Compte tenu des types et du facteur dx_1 éventuel, (2.5) entraîne cinq équations qui, à cause de 2.3 se réduisent aux trois équations

$$(2.6) \qquad\qquad d''_E S^{1,1} = 0$$

et

$$(2.7) \qquad d'_E S^{1,0} = 0, \qquad d''_E S^{1,0} + \overline{d''_E S^{1,0}} - \partial S^{1,1}/\partial x_1 = 0.$$

2.5. *Une expression de $S^{1,1}$.*

2.5.1. Dans les notations définies en 2.1, on pose $z'' = (z_2,\ldots,z_{n-1})$ et $\zeta' = (x_1; z'')$. Soient B'' une boule de $\mathbf{C}^{n-2}$ (variable z'') et I un intervalle de $\mathbf{R}$ (variable x_1). On pose

$$(2.8) \qquad\qquad S^{1,1} = \left(\frac{i}{2} \right) \sum_{j,k=2}^{n} S_{j\bar{k}} \, dz_j \wedge d\bar{z}_k$$

de sorte que les coefficients $S_{j\bar{k}}$ sont réels.

2.5.2. LEMME. *Dans les notations ci-dessus, on pose $A = (2/\pi)\delta_0(\zeta') \otimes \log|z_n|$ et on suppose que $\pi|\mathrm{Spt}\, S$ est propre dans $I \times B'' \times \mathbf{C}$; alors le courant*

$$(2.9) \qquad\qquad \varphi = A * S_{n\bar{n}}$$

satisfait à

$$(2.10) \qquad\qquad (i/2) d'_E d''_E \varphi = S^{1,1}.$$

DÉMONSTRATION. (On reprend la démonstration du Lemme 2.5 de [5]). On a

$$d''_E S^{1,1} = \left(\frac{i}{2} \right) \sum_{l,j,k=2}^{n} \left(\partial S_{j\bar{k}}/\partial \bar{z}_l \right) d\bar{z}_l \wedge dz_j \wedge d\bar{z}_k$$
$$= -\left(\frac{i}{2} \right) \sum_{j=2}^{n} \sum_{2 \leqslant l < k \leqslant n} \left(\partial S_{j\bar{k}}/\partial \bar{z}_l - \partial S_{j\bar{l}}/\partial \bar{z}_k \right) dz_j \wedge d\bar{z}_l \wedge d\bar{z}_k.$$

D'après (2.6), on a

$$(2.11) \qquad \partial S_{j\bar{k}}/\partial \bar{z}_l - \partial S_{j\bar{l}}/\partial \bar{z}_k = 0, \qquad j,k,l = 2,\ldots,n.$$

De même

$$0 = d'_E S^{1,1} = \left(\frac{i}{2} \right) \sum_{k=2}^{n} \sum_{2 \leqslant l < j \leqslant n} \left(\partial S_{j\bar{k}}/\partial z_l - \partial S_{l\bar{k}}/\partial z_j \right) dz_l \wedge dz_j \wedge \partial \bar{z}_k,$$

d'où

$$(2.11') \qquad \partial S_{j\bar{k}}/\partial z_l - \partial S_{l\bar{k}}/\partial z_j = 0, \qquad j,k,l = 2,\ldots,n.$$

D'autre part, pour une variable complexe w, on a

$$(2/\pi)(\partial^2/\partial w\, \partial \bar{w}) \log|w| = \delta_0(w);$$

alors

$$\partial^2 A/\partial z_n\,\partial\bar z_n = \delta_0(\zeta') \otimes \delta_0(z_n) = \delta_0(\zeta', z_n),$$

en identifiant les mesures de Dirac à des courants de degré 0.

Le support de A est le produit d'un compact (point de l'espace des ζ') et de $\mathbf{C}$ (variable z_n); par hypothèse, le support de $S_{n\bar n}$ a une intersection compacte avec $\zeta' = \zeta_0'$ fixe; alors le produit de convolution $A * S_{n\bar n}$ est bien défini.

$$\partial^2\varphi/\partial z_n\,\partial\bar z_n = \big(\partial^2 A/\partial z_n\,\partial\bar z_n\big) * S_{n\bar n},$$

mais

$$\partial\varphi/\partial\bar z_j = \partial/\partial\bar z_j(A * S_{n\bar n}) = A * \big(\partial S_{n\bar n}/\partial\bar z_j\big) = A * \big(\partial S_{nj}/\partial\bar z_n\big)$$

d'après (2.11); donc $\partial\varphi/\partial\bar z_j = (\partial A/\partial\bar z_n) * S_{nj}$;

$$\partial^2\varphi/\partial z_k\,\partial\bar z_j = \big(\partial A/\partial\bar z_n\big) * \big(\partial S_{nj}/\partial z_k\big) = \big(\partial A/\partial\bar z_n\big) * \big(\partial S_{kj}/\partial z_n\big)$$

d'après (2.11'), d'où

$$\partial^2\varphi/\partial z_k\,\partial\bar z_j = \big(\partial^2 A/\partial z_n\,\partial\bar z_n\big) * S_{kj} = \delta_0(\zeta'; z_n) * S_{kj} = S_{kj}.$$

Finalement,

$$(i/2)\,d_E'd_E''\varphi = (i/2) \sum_{j,k=2}^{n} \big(\partial^2\varphi/\partial z_k\,d\bar z_j\big)\,dz_k \wedge d\bar z_j$$

$$= (i/2) \sum_{j,k=2}^{n} S_{kj}\,dz_k \wedge d\bar z_j = S^{1,1}. \quad \square$$

2.6. *Evaluation de $S^{1,0} + S^{0,1}$.*

2.6.1. On a les opérateurs réels $d_E = d_E' + d_E''$, $d_E^c = i(d_E'' - d_E')$, d'où $d_E' = \frac12(d_E + id_E^c)$, $d_E'' = \frac12(d_E - id_E^c)$,

$$(2.12) \qquad\qquad id_E'd_E'' = \tfrac12 d_E d_E^c.$$

Compte tenu des conditions de réalité, d'après (2.7), on a

$$d_E\big(S^{1,0} + S^{0,1}\big) = \partial S^{1,1}/\partial x_1,$$

d'où, d'après (2.10) et (2.12),

$$(2.13) \qquad\qquad d_E\big(S^{1,0} + S^{0,1}\big) = \tfrac14 d_E d_E^c(\partial\varphi/\partial x_1).$$

D'après (2.13) $S^{1,0} + S^{0,1}$ et $\tfrac14 d_E^c(\partial\varphi/\partial x_1)$ diffèrent d'une d_E-différentielle.

2.6.2. Calculons $d_E^c\,\partial\varphi/\partial x_1$. On pose

$$(2.14) \qquad S^{1,0} = (-i/2)\sum_{j=2}^{n} S_{j1}\,dz_j, \qquad S^{0,1} = (i/2)\sum_{j=2}^{n} S_{1j}\,d\bar z_j;$$

alors $S_{1j} = \bar S_{j1}$. D'autre part

$$d_E''S^{1,0} = \cdots - (i/2)(\partial S_{n1}/\partial\bar z_n)\,d\bar z_n \wedge dz_n = \cdots + (i/2)(\partial S_{n1}/\partial\bar z_n)\,dz_n \wedge d\bar z_n,$$

$$d_E'S^{0,1} = \cdots + (i/2)(\partial S_{1\bar n}/\partial z_n)\,dz_n \wedge d\bar z_n;$$

en explicitant les termes en $dz_n \wedge d\bar{z}_n$ dans la seconde équation (2.7), on trouve

$$(2.15) \qquad \partial S_{n1}/\partial \bar{z}_n + \partial S_{1\bar{n}}/\partial z_n = \partial S_{n\bar{n}}/\partial x_1.$$

Pour simplifier l'écriture, on pose

$$(2.16) \qquad \varphi_1 = \partial \varphi/\partial x_1 = A * (\partial S_{n\bar{n}}/\partial x_1);$$

alors $\partial \varphi_1/\partial z_j = (\partial A/\partial z_j) * (\partial S_{n\bar{n}}/\partial x_1) = (\partial A/\partial z_j) * (\partial S_{n1}/\partial \bar{z}_n + \partial S_{1\bar{n}}/\partial z_n)$, d'après (2.15), d'où

$$(2.17) \qquad \partial \varphi_1/\partial z_j = (\partial A/\partial \bar{z}_n) * (\partial S_{n1}/\partial z_j) + (\partial A/\partial z_n) * (\partial S_{1\bar{n}}/\partial z_j).$$

Mais, d'après la première équation (2.7), on a

$$0 = 2id_E' S^{1,0} = \sum_{k=2}^{n} \sum_{j=2}^{n} (\partial S_{j1}/\partial z_k)\, dz_k \wedge dz_j$$

$$= \sum_{2 \leqslant k < j \leqslant n} (\partial S_{j1}/\partial z_k - \partial S_{k1}/\partial z_j)\, dz_k \wedge dz_j,$$

i.e.,

$$(2.18) \qquad \partial S_{j1}/\partial z_k = \partial S_{k1}/\partial z_j \quad \text{pour } j, k \geqslant 2;$$

en particulier

$$(2.19) \qquad \partial S_{n1}/\partial z_k = \partial S_{k1}/\partial z_n \quad \text{pour } k \geqslant 2.$$

Alors $(\partial A/\partial \bar{z}_n) * (\partial S_{n1}/\partial z_j) = (\partial A/\partial \bar{z}_n) * (\partial S_{j1}/\partial z_n) = (\partial^2 A/\partial z_n \partial \bar{z}_n) * S_{j1} = S_{j1}$ car $d^2A/\partial z_n\, \partial \bar{z}_n = \delta_0(\zeta'; z_n)$ d'après la démonstration de 2.5.2.; d'après (2.17), on a

$$\partial \varphi_1/\partial z_j = (\partial A/\partial z_n) * (\partial S_{1\bar{n}}/\partial z_j) + S_{j1},$$

d'où

$$d_E' \varphi_1 = \sum_{j=2}^{n} (\partial A/\partial z_n) * (\partial S_{1\bar{n}}/\partial z_j)\, dz_j + 2iS^{1,0}.$$

On pose

$$(2.20) \qquad v = (\partial A/\partial z_n) * S_{1\bar{n}} = A * (\partial S_{1\bar{n}}/\partial z_n);$$

alors

$$d_E' \varphi_1 = d_E' v + 2iS^{1,0};$$

par passage à l'imaginaire conjugué, on obtient

$$d_E'' \varphi_1 = d_E'' \bar{v} - 2iS^{0,1},$$

d'où

$$(2.21) \qquad d_E^c \varphi_1 = i(d_E'' - d_E')\varphi_1 = 2(S^{1,0} + S^{0,1}) - i(d_E' v - d_E'' \bar{v}).$$

D'autre part $v + \bar{v} = A * (\partial S_{1\bar{n}}/\partial z_n + \partial S_{n1}/\partial \bar{z}_n) = A * (\partial S_{n\bar{n}}/\partial x_1) = \varphi_1$ d'après (2.9), (2.15) et (2.16), soit

$$(2.22) \qquad i\psi = A * (\partial S_{1\bar{n}}/\partial z_n - \partial S_{n1}/\partial \bar{z}_n),$$

alors $v - \bar{v} = i\psi$, $v = \frac{1}{2}(\varphi_1 + i\psi)$ et, d'après (2.21)

$$d_E^c\varphi_1 = 2(S^{1,0} + S^{0,1}) + \tfrac{1}{2}d_E^c\varphi_1 + \tfrac{1}{2}d_E\psi,$$

i.e.,

$$(2.23) \qquad\qquad S^{1,0} + S^{0,1} = \tfrac{1}{4}d_E^c\varphi_1 - \tfrac{1}{4}d_E\psi.$$

Posons $\underline{d}_1 = \partial/\partial x_1\, dx_1$; alors $dx_1 \wedge d_E^c\varphi_1 = \underline{d}_1 d_E^c\varphi$, d'où

$$(2.24) \qquad\qquad dx_1 \wedge (S^{1,0} + S^{0,1}) = \tfrac{1}{4}\underline{d}_1 d_E^c\varphi - \tfrac{1}{4}dx_1 \wedge d_E\psi.$$

Finalement, compte tenu de (2.10), (2.12) et (2.24), du Lemme 2.5.2 et de (2.3), on obtient

2.6.3. PROPOSITION. *Le courant S, défini en 2.1 et satisfaisant à la condition*: $\pi|\mathrm{Spt}\,S$ *est propre dans $I \times B'' \times \mathbf{C}$ (voir 2.5.1), s'exprime sous la forme*

$$(2.25) \qquad\qquad S = \tfrac{1}{4}d_E d_E^c\varphi = \tfrac{1}{4}\underline{d}_1 d_E^c\varphi - \tfrac{1}{4}dx_1 \wedge d_E\psi$$

dans laquelle (2.9) $\varphi = A * S_{n\bar{n}}$, (2.22) $\psi = iA *(\partial S_{n1}/\partial \bar{z}_n - \partial S_{1\bar{n}}/\partial z_n)$; $S_{n\bar{n}}$, $S_{1\bar{n}}$ *et* S_{n1} *étant les coefficients de S définis dans* (2.8) *et* (2.14) *respectivement.*

2.6.4. REMARQUE. Dans les hypothèses de 2.1, la Proposition 2.6.3 montre que le courant S est déterminé par la connaissance des coefficients $S_{n\bar{n}}$ et S_{n1} de son expression comme forme différentielle à coefficients courants de degré 0.

3. Courants localement rectifiables.

3.1. Dans le $n°$ 3, on remplace (A) du $n°$ 2 par une condition plus forte: soit M un courant d'un ouvert U de $\mathbf{C}^n$ satisfaisant à

(B) $M = M^{1,2} + M^{2,1}$, M est *réel, d-fermé, à support dans E, localement rectifiable.*

Alors le courant S associé à M par (2.1) est *localement rectifiable.*

3.1.1. Dans les notations de 2.5.2, on considère la restriction de S à l'ouvert $I \times B'' \times \mathbf{C} \cap U$ qu'on note encore S; on suppose $\pi|S$ propre. On considère la tranche $\langle S, \pi, \zeta' \rangle$ de S relativement à π en ζ' aux points de $I \times B''$ où elle est définie. Rappelons la construction de la tranche: pour toute fonction $\alpha \in \mathscr{D}(I \times B'' \times \Delta)$, αS est plat, de dimension $(2n - 3)$; il en est de même de $\pi_*(\alpha S)$ qui est donc de degré 0; c'est une fonction L^1 qui a des valeurs sur une partie F_S de $I \times B''$ de complémentaire Lebesgue-négligeable. Pour tout $\zeta' \in F_S$, on a, par définition $\langle S, \pi, \zeta' \rangle(\alpha) = $ valeur de $\pi_*(\alpha S)$ en ζ'.

3.1.2. LEMME. *On a* $\langle S, \pi, \zeta' \rangle = \langle S^{1,1}, \pi, \zeta' \rangle = (i/2)\langle S_{n\bar{n}}\, dz_n \wedge d\bar{z}_n, \zeta' \rangle$.

DÉMONSTRATION. Le courant $\pi_*(\alpha S)$ étant de degré 0, opère sur les formes β de $I \times B''$ de degré $(2n - 3)$ comme suit: $\pi_*(\alpha S)(\beta) = \alpha S(\pi^*\beta)$; de plus, $\beta = \beta_0\, dx_1 \wedge dz_2 \wedge d\bar{z}_2 \wedge \cdots \wedge dz_{n-1} \wedge d\bar{z}_{n-1}$, où β_0 est une fonction; donc $\pi^*(\beta) = (\pi^*\beta_0)\, dx_1 \wedge dz_2 \wedge \cdots \wedge d\bar{z}_{n-1}$; alors $\alpha S(\pi^*\beta) = (i/2)\alpha S_{n\bar{n}}\, dz_n \wedge d\bar{z}_n(\pi^*\beta)$. $\square$

3.1.3. LEMME (FEDERER [4]). *Dans les notations ci-dessus, si $\pi|\mathrm{spt}\, S$ est propre, alors il existe un disque Δ de $\mathbf{C}$ (avec la variable z_n) centré en 0, tel que, pour tout $\zeta' \in F_S$, la tranche $\langle S, \pi, \zeta' \rangle$ soit un courant rectifiable, de dimension 0, i.e., une chaîne entière finie, de dimension 0,*

$$\langle S, \pi, \zeta' \rangle = \sum_{j=1}^{P} \left[\left(\zeta', r_j^+(\zeta') \right) \right] - \sum_{l=1}^{Q} \left[\left(\zeta', r_l^-(\zeta') \right) \right]$$

où $r_j^{\pm}(\zeta') \in \Delta$ et où [ ] désigne le courant d'intégration sur le point entre crochets, i.e., la mesure de Dirac en ce point. $\square$

3.2. *La fonction R_S.*

3.2.1. Dans les notations de 3.1, on pose

$$(3.1) \qquad R_S(\zeta'; z_n) = \prod_{j=1}^{P} \left(z_n - r_j^+(\zeta') \right) \prod_{l=1}^{Q} \left(z_n - r_l^-(\zeta') \right)^{-1}.$$

3.2.2. PROPOSITION. *Pour I, B'', Δ comme dans 3.1, $\pi|\mathrm{spt}\, S$ étant propre, à S est associée la fonction $R_S(\zeta'; z_n)$ définie presque partout par (3.1); le courant φ étant défini dans 2.2.5, on a*

$$\varphi(\zeta'; z_n) = \log\left| R_S(\zeta'; z_n) \right|$$

presque partout sur $I \times B'' \times \Delta$ et pour ζ' fixé, en posant $\varphi_{\zeta'}(z_n) = \varphi(\zeta'; z_n)$, on a

$$\tfrac{1}{2} d'_{z_n} d''_{z_n} \varphi_{\zeta'} = \sum_{j=1}^{P} \left[r_j^+(\zeta') \right] - \sum_{l=1}^{Q} \left[r_l^-(\zeta') \right].$$

DÉMONSTRATION. (Selon [5]). On rappelle $\varphi(\zeta'; z_n) = A * S_{n\bar{n}}$ et on fait la démonstration pour $S_{n\bar{n}}\, C^{\infty}$; pour $S_{n\bar{n}}$ général, on approche $S_{n\bar{n}}$ par des fonctions C^{∞}. $\square$

3.2.3. PROPOSITION (cf. [5, Lemma 2.7, (2) et (3)]). *Pour $w \notin \Delta$ et $\zeta' \in F_S$, on a*

$$(3.2) \qquad \log R_S(\zeta'; w) = (P - Q)\log w + \sum_{m=1}^{\infty} (1/m) C_m(\zeta') w^{-m}$$

avec $C_m(\zeta') = \sum_{j=1}^{P}(r_j^+(\zeta'))^m - \sum_{l=1}^{Q}(r_l^-(\zeta'))^m$; de plus $C_m(\zeta') = \pi_*(z_n^m S) = \pi_*(z_n^m S^{1,1})$; enfin (3.2) converge faiblement sur $I \times B'' \times (\mathbf{C} \setminus \overline{\Delta})$.

DÉMONSTRATION. Soit $\log(1 - u) = \sum_{m}^{\infty} m^{-1} u^m$ la détermination principale du logarithme pour $|u| < 1$. On a

$$\log R_S(\zeta'; w) = \sum_{j=1}^{P} \log\left[w\left(1 - w^{-1} r_j^+(\zeta')\right) \right] - \sum_{l=1}^{Q} \log\left[w\left(1 - w^{-1} r_l^-(\zeta')\right) \right];$$

pour $w \in \mathbf{C} \setminus \Delta$, on a $|r_j^{\pm}(\zeta')w^{-1}| < 1$; alors

$$(3.3) \qquad \log\left[w\left(1 - w^{-1} r_j^{\pm}(\zeta')\right) \right] = \log w + \sum_{m=1}^{\infty} m^{-1}\left[w^{-1} r_j^{\pm}(\zeta') \right]^m.$$

Les séries (3.3) qui figurent dans l'expression de $\log R_S(\zeta'; w)$ sont absolument convergentes en w pour ζ' fixé; par regroupement des termes, on obtient

$$\log R_S(\zeta'; w) = (P - Q) \log w$$
$$- \sum_{m=1}^{\infty} m^{-1} w^{-m} \left[\sum_{j=1}^{P} \left(r_j^+(\zeta') \right)^m - \sum_{l=1}^{Q} \left(r_l^-(\zeta') \right)^m \right].$$

S étant localement rectifiable, S est localement plat; alors $\pi_*(z_n^m S)$ est localement plat et de degré 0, donc localement intégrable; pour $\zeta' \in F_S$, on a

$$\pi_*\left(z_n^m S \right)(\zeta') = (i/2) \pi_*\left(z_n^m S_{n\bar{n}} \, dz_n \wedge d\bar{z}_n \right)(\zeta') = \left\langle S, \pi, \zeta' \right\rangle \left(z_n^m \right)$$
$$= \sum_{j=1}^{P} \left(r_j^+(\zeta') \right)^m - \sum_{l=1}^{Q} \left(r_l^-(\zeta') \right)^m,$$

d'après la définition du tranchage et 3.1.3, d'où la relation (3.2) et les expressions de $C_m(\zeta')$.

Pour toute fonction ψC^∞ à support compact K dans $I \times B'' \times (\mathbf{C} \setminus \overline{\Delta})$, on a

$$(3.4) \quad \log w \left(1 - w^{-1} r_j^{\pm}(\zeta') \right)(\psi) = \int_K \left[\log w + \sum_m m^{-1} \left[w^{-1} r_j^{\pm}(\zeta') \right]^m \right] \psi \, dv$$

où dv est la mesure de Lebesgue. Sur K, il existe une constante $k \in [0, 1[$ telle que $|w^{-1} r_j^{\pm}(\zeta')| \leqslant k$, de sorte que

$$m^{-1} \left| \int_K \left[w^{-1} r_j^{\pm}(\zeta') \right]^m \psi \, dv \right| \leqslant m^{-1} k^m \int_K |\psi| dv;$$

donc la série (3.3) converge faiblement sur $I \times B'' \times (\mathbf{C} \setminus \overline{\Delta})$, d'où la dernière assertion de la proposition. $\square$

3.3. *Une propriété des coefficients $C_m(\zeta')$.*

3.3.1. On pose $\mathscr{E} = \{ z \in E; \ z_n(z) = 0 \}$. On dira qu'une fonction f localement intégrable sur $I \times B'' \subset \mathscr{E}$ est CR si, au sens des courants $d_{\mathscr{E}}'' f = 0$.

3.3.2. LEMME. *Soit U un courant sur E; alors on a $d_E'' \pi_* U = \pi_* d_E'' U$, $(\partial/\partial x_1) \pi_* U = \pi_*(\partial U/\partial x_1)$.*

DÉMONSTRATION. Pour toute $\alpha \in \mathscr{D}'(\mathscr{E})$, on a $d_E'' \pi_* U(\alpha) = \pm U(\pi^* d_E'' \alpha)$; α et $\pi^* \alpha$ ayant la même expression indépendante de z_n et de dz_n, $d\bar{z}_n$, on a $\pi^* d_E'' \alpha = d_E'' \pi^* \alpha$; donc

$$d_E'' \pi_* U(\alpha) = \pm U \left(d_E'' \pi^* \alpha \right) = \pi_* d_E'' U(\alpha).$$

De même

$$(\partial/\partial x_1) \pi_* U(\alpha) = -U \left(\pi^* (\partial/\partial x_1) \alpha \right) = -U \left((\partial/\partial x_1) \pi^* \alpha \right) = \pi_* (\partial/\partial x_1) U(\alpha)$$

parce que α et $\pi^* \alpha$ ont la même expression.

3.3.3. PROPOSITION. *Les fonctions $C_m(\zeta')$ sont CR sur $I \times B'' \subset \mathscr{E}$.*

On a vu en 3.2.3 que $C_m(\zeta')$ est un courant L^1_{loc}; on a

$$d_{\mathscr{E}}'' C_m(\zeta') = d_E'' C_m(\zeta') = d_E'' \pi_* \left(z_n^m S^{1,1} \right) = \pi_* d_E'' \left(z_n^m S^{1,1} \right)$$

d'après 3.3.2; donc $d_E'' C_m(\zeta') = \pi_*(z_n^m d_E'' S^{1,1}) = 0$ d'après (2.6).

4. Cas où M possède des propriétés de régularité.

4.1. *q-chaînes maximalement complexes de classe C^ω.*

4.1.1. Soit Z une variété analytique complexe. Pour $k \in \mathbf{N} \cup \{\infty\} \cup \{\omega\}$, un fermé W de Z est appelé une *C^k-sous-variété à singularités négligeables* de dimension q, s'il existe un fermé σ de W, de mesure de Hausdorff q-dimensionnelle nulle ($\mathscr{H}_q(\sigma) = 0$) tel que $W \setminus \sigma$ soit une sous-variété fermée, de classe C^k, orientée, de $Z \setminus \sigma$, de dimension q, de volume q-dimensionnel localement fini; σ est appelé *l'ensemble singulier de W.*

Le courant d'intégration sur $W \setminus \sigma$ est alors défini et désigné par $[W]$; il a pour support W.

On appelle *chaîne de Z*, de dimension q, toute combinaison linéaire localement finie, à coefficients entiers, de courants d'intégration $[W_j]$ définis par des sous-variétés W_j à singularités négligeables de Z, de dimension q, $V = \sum_j n_j [W_j]$; le support $|V|$ de V est $\bigcup_j W_j$; il est de volume q-dimensionnel localement fini; alors si $|V|$ est compact, il est de volume fini.

On dira que la chaîne est *fermée* ou est un *cycle* si $dV = 0$.

4.1.2. Soit W une sous-variété à singularités négligeables, de dimension q, d'ensemble singulier σ. On dit que $W \setminus \sigma$ est de dimension holomorphe r en $x \in W \setminus \sigma$ si l'espace tangent $T_x(W)$ a un sous-espace vectoriel complexe maximal de dimension complexe r; on pose $dh_x(W \setminus \sigma) = r$. S'il existe un entier $r \geqslant 0$ et un fermé $\sigma' \supset \sigma$ de W avec $\mathscr{H}_q(\sigma') = 0$, tels que, pour tout point $x \in W \setminus \sigma'$, $dh_x(W \setminus \sigma') = r$, on dit que W est de *dimension holomorphe r* et on pose $r = dh(W)$. Soit $V = \sum_j n_j [W_j]$ une chaîne; si pour tout j, $dh(W_j) = r$, on dit que V est de dimension holomorphe r et on pose $r = dh(V)$.

4.1.3. On utilisera, dans la suite, les notations de 2.1, 2.2.1 et $\mathbf{C}_{(n)}^{n-1} = \mathbf{C}^{n-1}$ avec la variable $z' = (z_1, \ldots, z_{n-1})$.

Considérons maintenant une sous-variété W contenue dans E, de dimension $2n - 3$, à singularités négligeables, d'ensemble singulier σ, de classe C^ω, maximalement complexe, i.e., telle que $dh(W) = n - 2$. $\mathbf{C}$ muni de la variable z_j sera noté $\mathbf{C}_j$.

On a $E = \mathbf{R}_1 \times \mathbf{C}^{n-1}$, toute hypersurface C^ω H de $\{x_0\} \times \mathbf{C}^{n-1}$ où $x_0 \in \mathbf{R}_1$ satisfait aux conditions ci-dessus. Dans la suite on suppose qu'*aucune composante connexe de W n'est une telle sous-variété H.* Alors, pour un système de coordonnées $(z_2, \ldots, z_n)$ convenable, d'après le théorème des fonctions implicites, il existe un fermé $\sigma' \supset \sigma$ de W, de mesure de Hausdorff $(2n - 3)$-dimensionnelle nulle tel que, pour tout $x_0 - (x_1^0, z_0'', z_n^0) \in W \setminus \sigma'$, il existe un voisinage U de x_0 dans E possédant la propriété ci-dessous: pour $u = \pi(U)$ assez petit, il existe une fonction γ, CR et C^ω sur l'ouvert u de $\mathscr{E}$, telle que $W \cap U$ soit le graphe de γ [5]; pour z'' fixé et u assez petit, $\gamma(\zeta') = \gamma(x_1; z'')$ est une série entière convergente en $(x_1 - x_1^0)$ sur $\mathbf{R}_1 \cap u$; elle se prolonge donc à $\mathbf{C}_1$, au voisinage de $\mathbf{R}_1 \cap u$, en une fonction $\Gamma(z_1; z'')$ holomorphe en z_1. La fonction γ étant CR, on a $d''_\mathscr{E} \gamma = 0$; donc γ et Γ sont holomorphes en z''; alors Γ, étant séparément holomorphe en z_1 et z'' et localement bornée, est holomorphe en $z' = (z_1, \ldots, z_{n-1})$ au voisinage de

$\pi(x_0)$ dans $\mathbf{C}^{n-1}_{(n)}$; $F = z_n - \Gamma$ est une fonction holomorphe au voisinage de x_0 dans $\mathbf{C}^n$, qui prolonge la fonction CR $f = z_n - \gamma$ définie au voisinage de x_0 dans E; en outre, $W \cap U$ est l'ensemble des zéros de f dans U.

4.1.4. Plus généralement, soit M une $(2n - 3)$-chaîne maximalement complexe de $\mathbf{C}^n$, de classe C^ω dont le support $|M|$ est contenu dans E. Alors, pour un système de coordonnées $(z_2,\ldots,z_n)$ convenable, il existe un fermé σ' de $|M|$, tel que, pour tout $x_0 \in |M| \setminus \sigma'$, il existe un voisinage U de x_0 dans E possédant les propriétés suivantes:

(i) $M \cap U = \sum_j m_j [Z_j] - \sum_l n_l [P_l]$ où Z_j, P_l sont des sous-variétés à singularités négligeables, maximalement complexes, de dimension $(2n - 3)$ de U; $m_j, n_l \in \mathbf{N}^*$;

(ii) il existe des fonctions CR, de classe C^ω, g_j, h_l, restrictions à U de fonctions G_j, H_l holomorphes dans un voisinage V de x_0 dans $\mathbf{C}^n$ telles que: $V \cap E = U$; $Z_j = \{ x \in U;\ g_j(x) = 0 \}$, $P_l = \{ x \in U;\ h_l(x) = 0 \}$, où les fonctions g_j et h_j sont définies comme la fonction f du $n°$ 4.1.3.

La fonction $f = \prod_j g_j^{m_j} \prod_l h_l^{-n_l}$, quotient de deux fonctions CR sur U, *sera appelée aussi fonction* CR. C'est la restriction à U de la fonction méromorphe $F = \prod_j G_j^{m_j} \prod_l H_l^{-n_l}$.

4.2. *Expression de M.*

4.2.1. La construction qui suit est locale, au voisinage d'un point de $|M| \setminus \sigma'$. On considère la chaîne holomorphe définie par le diviseur de F, i.e., d'après le théorème de Poincaré–Lelong, $T = (i/\pi) d' d'' \log |F|$. On pose $T = (i/2) \sum_{j,k=1}^n T_{j\bar{k}}\, dz_j \wedge d\bar{z}_k$ avec $dT = 0$. Soit h la projection $\mathbf{C}^n \to \mathbf{R}'_1$. Alors, localement, au voisinage de x_0, on a $M = \langle T, h, 0 \rangle$ d'après la construction de T. Soit $\beta_{0,\varepsilon}$ une unité approchée C^∞ de support $[-\varepsilon, +\varepsilon]$ dans $\mathbf{R}'_1$. Par définition

$$M = \langle T, h, 0 \rangle = \lim_{\varepsilon \to 0} T \wedge h^*(\beta_{0,\varepsilon}\, dy_1)$$

$$= (i/2) \lim_{\varepsilon \to 0} \left[\sum_{j,k=2}^n T_{j\bar{k}}\, dz_j \wedge d\bar{z}_k + \sum_{k=2}^n T_{1\bar{k}}\, dz_1 \wedge d\bar{z}_k \right.$$

$$\left. + \sum_{j=2}^n T_{j\bar{1}}\, dz_j \wedge d\bar{z}_1 - 2i T_{1\bar{1}}\, dx_1 \wedge dy_1 \right] \wedge \left(h^* \beta_{0,\varepsilon} \right) dy_1$$

$$= (i/2) \lim_{\varepsilon \to 0} \left[\sum_{j,k=2}^n T_{j\bar{k}}\, dz_j \wedge d\bar{z}_k + dx_1 \wedge \sum_{k=2}^n T_{1\bar{k}}\, d\bar{z}_k \right.$$

$$\left. + \sum_{j=2}^n T_{j\bar{1}}\, dz_j \wedge dx_1 \right] \wedge \left(h^* \beta_{0,\varepsilon} \right) dy_1.$$

4.2.2. F est définie dans un voisinage de x_0 qu'après rétrécissement on peut supposer de la forme $B \times \Delta_n$ où B est la boule ouverte de centre $\pi(x_0)$ et de rayon r'_0 et où Δ_n est un disque ouvert de $\mathbf{C}_n$ centré en z_n^0; d'après la construction de F, en diminuant au besoin r'_0, on peut supposer que $\pi|\mathrm{spt}\, T$ est propre.

Soit $C = (2/\pi)\delta(z') \otimes \log|z_n|$; on pose $c = C(\partial/\partial z_n)(\partial/\partial \bar{z}_n)$. Soit U un courant de degré 2 de $B \times \Delta_n$; alors $U = u_{n\bar{n}}\, dz_n \wedge d\bar{z}_n + \dots$ où les termes non écrits sont indépendants de $dz_n \wedge d\bar{z}_n$. On définit le produit de convolution-contraction $\sharp$ comme dans [5]: $c\sharp U = c * u_{n\bar{n}}$. D'après [5], pour T de 4.2.1, on a $T = id'd''\phi$ avec $\phi = c\sharp(T_{n\bar{n}}\, dz_n \wedge d\bar{z}_n) = (2/i)c\sharp T$.

On considère la projection h (4.2.1) et le tranchage relativement à h en 0.

4.2.3. LEMME. *Pour tout courant U localement plat (de degré 2), $c\sharp U$ est localement plat et on a $\langle c\sharp U, h, 0 \rangle = c\sharp \langle U, h, 0 \rangle$ quand les deux membres sont définis.*

DÉMONSTRATION. U étant localement plat, il existe deux courants U' et U'' à coefficients L^1_{loc} tels que $U = U' + dU''$; alors $c\sharp U = c\sharp U' + c\sharp dU''$. D'après une propriété du produit de convolution, $c\sharp U'$ est L^1_{loc}.

Considérons $c\sharp dU''$; on a $U'' = u''_n\, dz_n + u''_{\bar{n}}\, d\bar{z}_n + \dots$, les termes non écrits étant indépendants de dz_n et de $d\bar{z}_n$; $dU'' = ((-\partial/\partial \bar{z}_n)u''_n + (\partial/\partial z_n)u''_{\bar{n}})\, dz_n \wedge d\bar{z}_n + \dots$. $c\sharp dU'' = C * (-(\partial/\partial \bar{z}_n)u''_n + (\partial/\partial z_n)u''_{\bar{n}}) = -(\partial C/\partial \bar{z}_n) * u''_n + (\partial C/\partial z_n) * u''_{\bar{n}}$ est L^1_{loc}; donc $c\sharp U$ est localement plat.

Dans les notations de 4.2.1 et de 4.2.2, on a

$$\Theta = \langle c\sharp U, h, 0 \rangle = (2/\pi) \lim_{\varepsilon \to 0} \left[\delta(z') \otimes \log|z_n| * u_{n\bar{n}}h^*\beta_{0,\varepsilon}\, dy_1\right].$$

Pour toute $\alpha \in \mathscr{D}^2(\mathbf{C}^n)$, on a

$$\Theta(\alpha) = (2/\pi) \lim_{\varepsilon \to 0} \delta_\xi(z') \otimes \log|z_n|_\xi$$
$$\otimes (u_{n\bar{n}})_\eta (\beta_{0,\varepsilon}(y_1(\xi + \eta))(dy_1(\xi) + dy_1(\eta)) \wedge \alpha(\xi + \eta))$$

où ξ et η sont les variables dans deux exemplaires de $\mathbf{C}^n$; en indices des symboles des courants, elles signifient que ces courants sont définis par rapport aux variables ξ et η respectivement; alors

$$\Theta(\alpha) = (2/\pi) \lim_{\varepsilon \to 0} (u_{n\bar{n}})_\eta (\beta_{0,\varepsilon}(y_1(\eta))\, dy_1(\eta))$$
$$\cdot \int_{z_n(\xi)} \alpha(z'(\eta); z_n(\xi) + z_n(\eta)) \log|z_n(\xi)|dx_n(\xi) \wedge dy_n(\xi);$$

d'autre part,

$$c\sharp\langle U, h, 0 \rangle(\alpha) = (2/\pi)\delta(z') \otimes \log|z_n| * \lim_{\varepsilon \to 0}(u_{n\bar{n}})h^*\beta_{0,\varepsilon}\, dy_1(\alpha)$$
$$= (2/\pi) \lim_{\varepsilon \to 0}(u_{n\bar{n}})\beta_{0,\varepsilon}(y_1(\eta))\, dy_1(\eta)$$
$$\cdot \int_{z_n(\xi)} \alpha(z'(\eta); z_n(\xi) + z_n(\eta))\log|z_n(\xi)|dx_n(\xi) \wedge dy_n(\xi). \qquad \square$$

4.2.4. Remarquons que $\langle T, h, 0 \rangle$, étant égal à M, est bien défini. On a $M = j_* S = S \otimes \delta(y_1) = S \otimes \varepsilon_1(y_1)\, dy_1$; en particulier, d'après 4.2.1, on a

$$(4.1) \qquad j_* S_{n\bar{n}} = \langle T_{n\bar{n}}, h, 0 \rangle = \lim_{\varepsilon \to 0} T_{n\bar{n}}h^*\beta_{0,\varepsilon}\, dy_1.$$

De plus (4.2.2), on a $\phi = (2/i)c\#T$; alors, d'après 4.2.3,

$$\langle \phi, h, 0 \rangle = (2/i)c\#\langle T, h, 0 \rangle = (2/i)\#j_*S = (2/i)c\#S \otimes \delta(y_1)$$
$$= (A \otimes \delta(y_1)) * S_{n\bar{n}} \otimes \delta(y_1) = (A * S_{n\bar{n}}) \otimes \delta(y_1) = j_*\varphi.$$

4.2.5. *Expression de S.*

De 4.2.1 résultent: $j_*S_{1\bar{n}} = \langle T_{1\bar{n}}, h, 0 \rangle = \lim_{\varepsilon \to 0} T_{1\bar{n}} h^* \beta_{0,\varepsilon} dy_1$; de même: $j_*S_{n1} = \lim_{\varepsilon \to 0} T_{n\bar{1}} h^* \beta_{0,\varepsilon} dy_1$; alors

$$(4.2) \qquad j_*(\partial S_{1\bar{n}}/\partial z_n) = (\partial/\partial z_n)(j_*S_{1\bar{n}}) = \lim_{\varepsilon \to 0} (\partial T_{1\bar{n}}/\partial z_n) h^* \beta_{0,\varepsilon} dy_1,$$

$$(4.3) \qquad j_*(\partial S_{n1}/\partial \bar{z}_n) = (\partial/\partial \bar{z}_n)(j_*S_{n1}) = \lim_{\varepsilon \to 0} (\partial T_{n\bar{1}}/\partial \bar{z}_n) h^* \beta_{0,\varepsilon} dy_1,$$

parce que $h^*\beta_{0,\varepsilon}$ est indépendant de z_n; cela implique que $\langle \partial T_{1\bar{n}}/\partial z_n, h, 0 \rangle$ et $\langle \partial T_{n\bar{1}}/\partial \bar{z}_n, h, 0 \rangle$ sont définis.

La d-fermeture de T implique

$$(4.4) \qquad \partial T_{n\bar{n}}/\partial z_1 = \partial T_{1\bar{n}}/\partial z_n; \qquad \partial T_{n\bar{n}}/\partial \bar{z}_1 = \partial T_{n\bar{1}}/\partial \bar{z}_n,$$

car

$$dT = d(T_{n\bar{n}} dz_n \wedge d\bar{z}_n + T_{1\bar{n}} dz_1 \wedge d\bar{z}_n + \cdots)$$
$$= (\partial T_{n\bar{n}}/\partial z_1 - \partial T_{1\bar{n}}/\partial z_n) dz_1 \wedge dz_n \wedge d\bar{z}_n + \cdots.$$

On a

$$(4.5) \quad j_*(\partial S_{1\bar{n}}/\partial z_n - \partial S_{n1}/\partial \bar{z}_n) = \lim_{\varepsilon \to 0} (\partial T_{1\bar{n}}/\partial z_n - \partial T_{n\bar{1}}/\partial \bar{z}_n) h^* \beta_{0,\varepsilon} dy_1$$
$$= \lim_{\varepsilon \to 0} (\partial T_{n\bar{n}}/\partial z_1 - \partial T_{n\bar{n}}/\partial \bar{z}_1) h^* \beta_{0,\varepsilon} dy_1$$

d'après (4.4). D'après 4.2.3 et 4.2.4, on a

$$\lim_{\varepsilon \to 0} (C * (\partial T_{1\bar{n}}/\partial z_n)) h^* \beta_{0,\varepsilon} dy_1 = C * \left(\lim_{\varepsilon \to 0} (\partial T_{1\bar{n}}/\partial z_n) h^* \beta_{0,\varepsilon} dy_1 \right)$$
$$= C * j_*(\partial S_{1\bar{n}}/\partial z_n) = (\delta(y_1) \otimes A) * (\partial S_{1\bar{n}}/\partial z_n) \otimes \delta(y_1)$$
$$= \delta(y_1) \otimes (A * (\partial S_{1\bar{n}}/\partial z_n)) = j_*(A * (\partial S_{1\bar{n}}/\partial z_n));$$

de même: $\lim_{\varepsilon \to 0}(C * (\partial T_{n\bar{1}}/\partial \bar{z}_n)) h^* \beta_{0,\varepsilon} dy_1 = j_*(A * (\partial S_{n1}/\partial \bar{z}_n))$; donc dans la notation (2.25)

$$j_*\psi = ij_*(A * (\partial S_{n1}/\partial \bar{z}_n - \partial S_{1\bar{n}}/\partial z_n))$$
$$= i \lim_{\varepsilon \to 0} [C * (\partial T_{n\bar{n}}/\partial \bar{z}_1 - \partial T_{n\bar{n}}/\partial z_1)] h^* \beta_{0,\varepsilon} dy_1;$$

d'après (4.5)

$$(4.6) \quad j_*\psi = i \lim_{\varepsilon \to 0} [\partial \phi/\partial \bar{z}_1 - \partial \phi/\partial z_1] h^* \beta_{0,\varepsilon} dy_1 = - \lim_{\varepsilon \to 0} (\partial \phi/\partial y_1) h^* \beta_{0,\varepsilon} dy_1.$$

4.2.6. PROPOSITION. *Si M est analytique réelle, pour $\phi = C * T_{n\bar{n}}$, dans l'expression (2.25) de la Proposition 2.6.3, au voisinage d'un point de $M \setminus \sigma$, on a*

$$-j_* \tfrac{1}{4} dx_1 \wedge d_E \psi = \tfrac{1}{4} dx_1 \wedge \langle d_E(\partial \phi/\partial y_1), h, 0 \rangle$$

DÉMONSTRATION.

$$-j_* \, dx_1 \wedge d_E \psi = -dx_1 \wedge d_E j_* \psi$$

$$= dx_1 \wedge d_E \lim_{\varepsilon \to 0} (\partial \phi / \partial y_1) h^* \beta_{0,\varepsilon} \, dy_1$$

$$(\text{d'après } (4.6)) = dx_1 \wedge \lim_{\varepsilon \to 0} \left(d_E (\partial \phi / \partial y_1) h^* \beta_{0,\varepsilon} \, dy_1 \right)$$

$$= dx_1 \wedge \left\langle d_E (\partial \phi / \partial y_1), h, 0 \right\rangle$$

parce que $\beta_{0,\varepsilon}$ est indépendant de (z'', z_n) et parce que, d'après les calculs ci-dessus, le tranchage est bien défini. $\square$

4.2.7. REMARQUE. Le Proposition 4.2.6 montre que, dans le cas analytique réel, l'information contenue dans le coefficient $S_{n\bar{n}}$ est suffisante pour reconstituer le courant $[M] = j_* S$, cela n'est pas connu dans le cas différentiable, d'après 2.6.4.

4.3. *Une formule de Poincaré–Lelong.*

4.3.1. *L'opérateur D.* On a $d' = d'_E + (\partial / \partial z_1) \, dz_1$; $d'' = d''_E + (\partial / \partial \bar{z}_1) \, d\bar{z}_1$; d'autre part, $(\partial / \partial z_1) = \frac{1}{2}(\partial / \partial x_1 - i\partial / \partial y_1)$; $(\partial / \partial \bar{z}_1) = \frac{1}{2}(\partial / \partial x_1 + i\partial / \partial y_1)$; d'où

$$d'd'' = d'_E d''_E + d'_E \wedge \tfrac{1}{2}(\partial / \partial x_1 + i\partial / \partial y_1) \, d\bar{z}_1$$

$$+ \tfrac{1}{2}(\partial / \partial x_1 - i\partial / \partial y_1) \, dz_1 \wedge d''_E + (\partial / \partial z_1) \, dz_1 \wedge (\partial / \partial \bar{z}_1) \, d\bar{z}_1.$$

On pose

$$D = d'd'' |_E = d'_E d''_E + \tfrac{1}{2} dx_1 \wedge \left[(\partial / \partial x_1)(d''_E - d'_E) - i(\partial / \partial y_1)(d'_E + d''_E) \right]$$

$$= d'_E d''_E - (i/2) \, dx_1 \wedge \left[(\partial / \partial x_1) d^c_E + (\partial / \partial y_1) \, d_E \right].$$

4.3.2. Dans les notations de 4.1.4, on définit, localement, une fonction méromorphe F; d'après la formule de Poincaré–Lelong classique, la chaîne holomorphe T de F satisfait à $(i/\pi) d'd'' \log|F| = T$; alors, localement, on a $j_* S = M = \langle T, h, 0 \rangle = (i/\pi) \langle d'd'' \log|F|, h, 0 \rangle = (i/\pi) d'd'' \langle \log|F|, h, 0 \rangle$. Localement, pour $\alpha \in \mathscr{D}\cdot(\mathbf{C}^n)$, on a

$$M(\alpha) = (i/\pi) \langle \log|F|, h, 0 \rangle (d'd''\alpha) = (i/\pi) \lim_{\varepsilon \to 0} \left(\log|F| h^* \beta_{0,\varepsilon} \, dy_1 \right)(d'd''\alpha)$$

$$= (i/\pi) \lim_{\varepsilon \to 0} \log|F| \left(h^* \beta_{0,\varepsilon} \, dy_1 \wedge d'd''\alpha \right);$$

$\log|F|$ étant localement intégrable, on a

$$M(\alpha) = (i/\pi) \lim_{\varepsilon \to 0} \int \log|F| h^* \beta_{0,\varepsilon} \, dy_1 \wedge d'd''\alpha = (i/\pi) \int_E \log|F| D\alpha$$

$$= (i/\pi) \int_E \log|f| D\alpha = (i/\pi) \log|f| [E](D\alpha) = (i/\pi) D(\log|f| [E])(\alpha),$$

d'où

4.3.3. PROPOSITION. *Dans les hypothèses et notations de 4.3.2., on a*

$$M(\alpha) = (i/\pi) D(\log|f| [E])(\alpha)$$

où f est la fonction CR définissant localement M dans E.

5. Construction d'une chaîne maximalement complexe, contenue dans E, de bord donné: calcul des coefficients C_m.

On utilise les notions de sous-variété à singularités négligeables, chaîne, et cycle définies en 4.1.

5.1. $(2n - 4)$-*cycle de dimension holomorphe* $(n - 3)$ *de* E.

5.1.1. Soit N *une sous-variété à singularités négligeables, compacte*, d'ensemble singulier τ, contenue dans E. On désigne aussi par N le courant d'intégration $[N]$ et on suppose $dN = 0$, i.e., que N est *un cycle*. On suppose, en outre, que N satisfait aux conditions suivantes: N est de classe C^k $(k \geqslant 1)$, de dimension $(2n - 4)$, de dimension holomorphe $(n - 3)$.

(H) *Il existe un fermé* $\sigma \supset \tau$ *tel que* $\mathscr{H}_{2n-4}(\sigma) = 0$ *et que, en tout point* $z^0 \in N \setminus \sigma$, *la sous-variété* $N \setminus \sigma$ *soit transverse au sous-espace affine complexe maximal* $H_{z^0}(E)$ *de* E *passant par* z^0.

Y étant un sous-espace affine de $\mathbf{R}^{2n}$, on note $\tilde{Y}$ la direction de Y, i.e., le sous-espace vectoriel de $\mathbf{R}^{2n}$ défini par Y. Les directions complexes γ de dimension 1 de E sont les éléments de l'espace projectif complexe $\mathbf{P}^{n-2}(\mathbf{C})$. Alors pour γ hors d'un ensemble de volume nul de $\mathbf{P}^{n-2}(\mathbf{C})$, on a

(H′) *l'ensemble* σ' *des* $z \in N \setminus \tau$ *tels que* $\overline{T_z(N)}$ *contienne une direction réelle de* γ *est de mesure de Hausdorff* $(2n - 4)$-*dimensionnelle nulle; en outre*, σ' *est fermé*.

Dans la suite du $n°$ 5, on suppose les coordonnées $(z_2, \ldots, z_n)$ choisies de sorte que, pour la direction de l'axe des z_n, la condition (H′) soit satisfaite.

De (H′) résulte: pour tout point de $N \setminus (\tau \cup \sigma')$, le rang de $\pi | N$ est $(2n - 4)$; alors, on a la propriété:

(c) $\mathscr{N} = \pi(N)$ *est une sous-variété à singularités négligeables, de codimension* 1 *de* $\mathscr{E} = \pi(E)$ *et*, d'après le théorème des fonctions implicites, *pour tout point* $z^0 \in N \setminus (\tau \cup \sigma')$, *il existe un voisinage de* z^0 *dans* N *dans lequel* N *est le graphe d'une fonction* l *sur un voisinage de* $\pi(z^0)$ *dans* $\mathscr{N}$.

5.1.2. Soit $k: E \to \mathbf{R}_1$ telle que $k((x_1; z''; z_n)) = x_1$; alors, d'après (H), on a

(e) $k(N)$ *est d'intérieur non vide dans* $\mathbf{R}_1$.

5.1.3. N étant de dimension holomorphe $(n - 3)$, le courant N, dans $\mathbf{C}^n$, s'écrit: $N = N^{3,1} + N^{2,2} + N^{1,3}$. Comme pour M (2.1 et 2.2), N étant localement plat, il existe un courant P de E, de degré 3, tel que $j: E \to \mathbf{C}^n$ étant l'inclusion, on ait $j_* P = N$; alors

$$P = P^{2,1} + P^{1,2} + dx_1 \wedge (P^{2,0} + P^{1,1} + P^{0,2}),$$

où les types sont relatifs aux dz_j, $d\bar{z}_j$ $(j = 2, \ldots, n)$. N est un courant réel, ce qui équivaut à

(5.1) $\qquad P^{1,2} = \overline{P^{2,1}}; \qquad P^{0,2} = \overline{P^{2,0}}; \qquad \overline{P^{1,1}} = P^{1,1};$

N est d-fermé, ce qui équivaut à

(5.2) $\qquad\qquad\qquad dP = 0.$

Compte tenu des conditions (5.1), la condition (5.2) s'écrit

(5.3) $\qquad \overline{d'_E P^{1,2}} + d'_E P^{1,2} = 0, \qquad d''_E P^{1,2} = 0;$

(5.4) $\qquad d''_E P^{1,1} + d'_E P^{0,2} = \partial P^{1,2}/\partial x_1, \qquad d''_E P^{0,2} = 0.$

5.2. *L'équation $dM = N$ dans E.*

5.2.1. Dans $E \setminus N$, on cherche à construire un courant S de E, défini par une chaîne maximalement complexe de $\mathbf{C}^n \setminus N$ à support dans E, admettant une extension simple de $E \setminus N$ à E notée encore S telle que $dS = P$, i.e., satisfaisant à

$$(5.5) \qquad \begin{cases} d_E'' S^{1,1} = P^{1,2}; \\ d_E'' S^{1,0} + d_E' S^{0,1} = \partial S^{1,1}/\partial x_1 - P^{1,1}; \\ d_E'' S^{0,1} = -P^{0,2}. \end{cases}$$

5.2.2. Compte tenu de (H), pour presque tout $x_1^0 \in k(N)$, on a

$$\mathscr{H}_{2n-5}\big(\tau \cap \{x_1 = x_1^0\}\big) = 0 \ [\mathbf{9}, \text{Corollaire 4}].$$

D'après (H$'$) l'ensemble K des points critiques de $\pi|N \setminus \tau$ est négligeable; alors, pour presque tout $x_1^0 \in k(N)$, on a

$$\mathscr{H}_{2n-5}\big(K \cap \{x_1 = x_1^0\}\big) = 0 \ [\mathbf{9}, \text{Corollaire 4}];$$

donc, *pour presque tout $x_1^0 \in k(N)$, $N \cap \{x = x_1^0\}$ et $\mathscr{N} \cap \{x_1 = x_1^0\}$ sont des sous-variétés de dimension $(2n - 5)$ à singularités négligeables.*

5.3. *Cas où N est de classe C^ω.*

5.3.1. On suppose que N satisfait aux hypothèses de 5.1 et est de classe C^ω. Alors, d'après 5.1.1(c) et (5.2.2), $\mathscr{N}$ et $\mathscr{N} \cap \{x_1 = x_1^0\}$ pour presque tout $x_1^0 \in k(N)$, sont des sous-variétés à singularités négligeables, de classe C^ω; de plus, pour tout $z^0 \in N \setminus (\tau \cup \sigma')$, il existe un voisinage de z^0 dans N dans lequel N est le graphe d'une fonction l C^ω au voisinage de $\pi(z^0)$ dans $\mathscr{N}$. Pour calculer le courant P, il suffit de calculer le courant d'intégration sur N, dans E, au voisinage des points z^0 ci-dessus, i.e., sur les formes différentielles à support compact dans tout ouvert V de E dans lequel N est un graphe.

De plus, d'après les conditions (H$'$) et (H), $\mathscr{N}$ est, sauf sur $\pi(\sigma \cup \sigma')$, une hypersurface C^ω de $\mathscr{E}$ sur laquelle x_1 est une coordonnée locale. On suppose qu'il en est ainsi sur un voisinage u_0 de $\pi(z^0)$ dans $\mathscr{E}$.

Supposant u_0 assez petit, $\mathscr{N} \cap u_0$ est une variété et il existe un système de coordonnées $\xi = x_1$, $t = (t_1, \ldots, t_{n-5})$ sur $\mathscr{N} \cap u_0$; par translation dans $\mathbf{R}_1$, on peut supposer $x_1(z^0) = 0$ et on choisit t tel que $t(z^0) = 0$; N est le graphe d'une fonction l C^ω définie sur u_0; en outre, il existe un disque ouvert $\Delta(z_n^0)$ centré en z_n^0 dans $\mathbf{C}_n$ tel que $V = u_0 \times \Delta(z_n^0)$ contienne le graphe de l. $N \cap V$ est défini par

$$(5.6) \qquad x_1 = \xi; \qquad z_k = \alpha_k(\xi; t), \quad \text{où } \alpha_k \text{ est } C^\omega, k = 2, \ldots, n.$$

Alors $N \cap V$ est l'image du plongement $i \colon W \to V$ défini par (5.6) où W est un ouvert de $\mathbf{R}^{2n-4}$ muni des coordonnées (ξ, t).

Pour $\varphi \in \mathscr{D}^{2n-4}(V)$, on a

$$P(\varphi) = \int_N \varphi = \int_{i(W)} \varphi = \int_W \varepsilon i^* \varphi$$

où $\varepsilon = \pm 1$ suivant l'orientation de N.

Soient $(\varphi_A)_{A \in \mathscr{A}}$ les coefficients de φ dans le système de coordonnées $(x_1, z_2, \ldots, z_n)$, $\mathscr{A}$ étant l'ensemble des multiindices $[1, 2, \ldots, n, \bar{2}, \ldots, \bar{n}]^{2n-4}$. Sur $V \cap N$, on a

$$dx_1 = d\xi; \qquad dz_k = (\partial\alpha_k/\partial\xi)\,d\xi + \sum_{j=1}^{2n-5} (\partial\alpha_k/\partial t_j)\,dt_j, k = 2, \ldots, n;$$

alors $\varepsilon i^*\varphi = \sum_A (i^*\varphi_A)\rho_A\,d\xi \wedge dt$ avec $dt = dt_1 \wedge \cdots \wedge dt_{2n-5}$ et ρ_A une fonction C^ω de $(\xi; t)$; on a donc

$$P(\varphi) = \sum_A \int_W i^*\varphi_A \rho_A\,d\xi \wedge dt.$$

En rétrécissant W au besoin, on peut supposer $W = I \times W'$, où I est un intervalle de $\mathbf{R}_1$ centré en 0 et W' un produit d'intervalles centrés en 0 des facteurs de $\mathbf{R}^{2n-5}$ avec les variables t_k $(k = 1, \ldots, 2n - 5)$ et $\rho_A = \sum_{j=0}^\infty \xi^j r_A^{(j)}$ où $r_A^{(j)}$ est une série convergente en t sur W' et ρ_A une série convergente en (ξ, t) sur W, ordonnée par rapport aux puissances de ξ. Alors

$$P[\varphi] = \sum_A \int_I \sum_{j=0}^\infty \xi^j\,d\xi \int_{W'} r_A^{(j)}(t)(\varphi_A \circ i)\,dt.$$

Le courant P est donc, au voisinage de z^0, égal à une somme finie de distributions définies par des fonctions analytiques réelles en (ξ, t), avec $\xi = x_1$. D'où

5.3.2. Lemme et Définition. *Dans les hypothèses et notations ci-dessus, dans un voisinage V de z^0 assez petit, le courant P est égal à une somme finie de distributions (appliquées aux coefficients des formes C^∞ à support compact dans V) définies par des fonctions C^ω en x_1. On dira que le courant P est C^ω en x_1.* $\square$

5.3.3. Corollaire. *Le lemme (et la définition) sont valides pour les courants $dx_1 \wedge P^{2,0}$, $dx_1 \wedge P^{1,1}$ et $dx_1 \wedge P^{0,2}$.*

Démonstration. Pour $\varphi \in \mathscr{D}^{2n-4}(V)$, $dx_1 \wedge P^{i,j}(\varphi)$ $(i + j = 2, i = 0, 1, 2)$, est la valeur de P sur la somme des composantes de φ indépendantes de dx_1, de type $(n - 1 - i, n - 1 - j)$. $\square$

5.4. *Calcul de* $S^{0,1}$.

5.4.1. Lemme. *L'équation* $(5.5)_3$ *$d_E''S^{0,1} = -P^{0,2}$, avec $d_E''P^{0,2} = 0$, a une solution à support compact définie à l'addition près de $d_E''T^{0,0}$ où $T^{0,0}$ est à support compact. Si N est analytique réel, il existe une solution $U^{0,1}$ C^ω en x_1.*

Démonstration. On va résoudre l'équation $(5.5)_3$ à l'aide d'un noyau. Soit A le noyau de Harvey–Polking (noté E dans [10, Theorem 6.31]) dans $\mathbf{C}^{n-1}$ $(z'' = (z_2, \ldots, z_n))$.

On considère le noyau $K = \delta_0(x_1 - \xi_1) \otimes A$ où le premier facteur agit par convolution. On pose

$$(5.7) \qquad\qquad U^{0,1} = -KP^{0,2}.$$

Alors $d_E'' U^{0,1} = -P^{0,2}$. Soit k la projection: $E \to \mathbf{R}_1$, $\zeta \mapsto x_1$; on a $\langle U^{0,1}, k, x_1 \rangle$ $= A\langle P^{0,2}, k, x_1 \rangle$; donc $d''\langle U^{0,1}, k, x_1 \rangle = \langle d'' U^{0,1}, k, x_1 \rangle = \langle P^{0,2}, k, x_1 \rangle$. D'après [10], $\langle U^{0,1}, k, x_1 \rangle$ est à support compact. En faisant varier x_1, on obtient la solution (5.7) qui est identiquement nulle pour x_1 assez grand, car il en est ainsi de $\langle P^{0,2}, k, x_1 \rangle$ pour x_1 assez grand. D'après 5.3.3, $P^{0,2}$ est C^ω en x_1, le produit de convolution conservant l'analyticité en x_1, $U^{0,1}$ est C^ω en x_1.

Soient U et U' deux solutions à support compact de $(5.5)_3$; alors

$$(5.8) \qquad d_E''(U - U') = 0;$$

soit $v^{0,1} = \langle U - U', k, x_1 \rangle$; (5.8) équivaut à $d''v^{0,1} = 0$ dans $\{x_1\} \times \mathbf{C}^{n-1}$ pour tout $x_1 \in \mathbf{R}$; $v^{0,1}$ définit une classe de

$$H_c^{0,1}\left(\{x_1\} \times \mathbf{C}^{n-1}\right) \approx H^{n-1,n-2}\left(\{x_1\} \times \mathbf{C}^{n-1}\right)$$

par la dualité de Serre; le dernier groupe est nul pour $n \geqslant 3$, donc aussi le premier, i.e., il existe $T^{0,0}$ à support compact dans E tel que $U' = U + d_E'' T^{0,0}$. $\square$

5.5. *Calcul de $C_m(\zeta')$.*

5.5.1. D'après 3.2.3, si le courant S est défini par une chaîne maximalement complexe, on a

$$(5.9) \qquad C_m(\zeta') = \pi_*\left(z_n^m S^{1,1}\right).$$

On se propose de calculer $C_m(\zeta')$ à partir de la donnée de P.

5.5.2. LEMME. *On a*

$$(5.10) \qquad \begin{cases} d_E'' C_m(\zeta') = \pi_*\left(z_n^m d_E'' S^{1,1}\right); \\ \partial C_m(\zeta')/\partial x_1 = \pi_*\left(z_n^m \left(\partial S^{1,1}/\partial x_1\right)\right). \end{cases}$$

DÉMONSTRATION. Soit $\varphi \in \mathscr{D}'(E)$; on a

$$\langle d_E'' C_m(\zeta'), \varphi \rangle = -\langle C_m(\zeta'), d_E'' \varphi \rangle = -\langle \pi_*\left(z_n^m S^{1,1}\right), d_E'' \varphi \rangle$$

$$= -\langle S^{1,1}, z_n^m \pi^* d_E'' \varphi \rangle = -\langle S^{1,1}, d_E''\left(z_n^m \pi^* \varphi\right) \rangle = \langle \pi_*\left(z_n^m d_E'' S^{1,1}\right), \varphi \rangle;$$

de même

$$\langle \partial C_m(\zeta')/\partial x_1, \varphi \rangle = -\langle C_m(\zeta'), \partial \varphi/\partial x_1 \rangle = -\langle \pi_*\left(z_n^m S^{1,1}\right), \partial \varphi/\partial x_1 \rangle$$

$$= -\langle z_n^m S^{1,1}, \pi^*\left(\partial \varphi/\partial x_1\right) \rangle = -\langle z_n^m S^{1,1}, \left(\partial/\partial x_1\right)\left(\pi^* \varphi\right) \rangle$$

$$= \langle \pi_*\left(z_n^m \left(\partial S/\partial x_1\right)\right), \varphi \rangle;$$

la seule propriété non classique utilisée est la commutation de d_E'' (resp. $(\partial/\partial x_1)$) et de π^*; elle est valide puisque les expressions de φ et de $\pi^* \varphi$ sont identiques. $\square$

5.5.3. D'après $(5.5)_1$, $(5.10)_1$ s'écrit

$$(5.11) \qquad d_E'' C_m(\zeta') - \pi_*\left(z_n^m P^{1,2}\right),$$

$'U^{0,1} = U^{0,1} + d_E'' T^{0,0}$ étant donné par le lemme 5.4.1; d'après $(5.5)_2$, on a

$$\partial S^{1,1}/\partial x_1 - d_E' \, 'U^{0,1} + \overline{d_E' \, 'U^{0,1}} + P^{1,1}.$$

Alors $C_m(\zeta')$ satisfait à l'équation

(5.12)

$$\left(d_E'' + (\partial/\partial x_1)\right)C_m(\zeta') = \pi_*\left[z_n^m\left(P^{1,2} + \left(P^{1,1} + d_E'\,'U^{0,1} + \overline{d_E'\,'U^{0,1}}\right) \wedge dx_1\right)\right].$$

On notera le courant de degré 1 du second membre de (5.12) $R = R^{0,1} + R^{0,0}\,dx_1$, où $R^{0,1}$ et $R^{0,0}$ sont des courants sur $\mathscr{E}$; $R^{0,1}$ étant indépendant de dx_1 et de type $(0,1)$ en $dz_2,\ldots,dz_{n-1}$.

5.5.4. LEMME. *On a* $(d_E'' + (\partial/\partial x_1)\,dx_1)R = 0$.

DÉMONSTRATION. D'après la démonstration de 5.5.2, on a

$$\left(d_E'' + (\partial/\partial x_1)\right)R = \pi_* z_n^m\left[d_E''P^{1,2} + d_E''\left(P^{1,1} + d_E'\,'U^{0,1} + \overline{d_E'\,'U^{0,1}}\right) \wedge dx_1\right.$$
$$\left. + dx_1 \wedge \left(\partial P^{1,2}/\partial x_1\right)\right].$$

D'après (5.3) le crochet du second membre est égal à

$$\left(d_E''P^{1,1} - d_E'd_E''\,'U^{0,1} - \left(\partial P^{1,2}/\partial x_1\right)\right) \wedge dx_1$$
$$= \left(d_E''P^{1,1} + d_E'P^{0,2} - \left(\partial P^{1,2}/\partial x_1\right)\right) \wedge dx_1 = 0,$$

d'après 5.4.1 et (5.4) successivement. $\square$

5.5.5. $H(x_1)$ étant la fonction de Heaviside sur $\mathbf{R}_1$, on considère le noyau $K_E = \delta_0(z'') \otimes H(x_1)(\partial/\partial x_1)$ sur $\mathscr{E}$; pour tout courant u à support compact dans $\mathscr{E}$, on pose $u = dx_1 \wedge u_1 + u_2$, u_1 et u_2 étant indépendants de dx_1 et $K_E\#u = (\delta_0(z'') \otimes H(x_1)) * u_1$.

5.5.6. LEMME. *Pour tout courant u, à support compact sur $\mathscr{E}\,(= \mathbf{R}_1 \times \mathbf{C}^{n-2})$, avec la variable $(x_1; z'')$, on a*

(5.13) $\left(d_E'' + (\partial/\partial x_1)\,dx_1\right)(K_E\#u) + K_E\#\left(d_E'' + (\partial/\partial x_1)\,dx_1\right)u = u.$

DÉMONSTRATION. On a $(d_E'' + (\partial/\partial x_1)\,dx_1)u = d_E''u_2 - dx_1 \wedge d_E''u_1 + dx_1 \wedge (\partial u_2/\partial x_1)$; le premier membre de (5.13) est

$$\left(d_E'' + (\partial/\partial x_1)\,dx_1\right)\left[\left(\delta_0(z'') \otimes H(x_1)\right) * u_1\right]$$
$$+ \left(\delta_0(z'') \otimes H(x_1)\right) * \left(-d_E''u_1 + (\partial u_2/\partial x_1)\right)$$
$$= \left(\delta_0(z'') \otimes H(x_1)\right) * d_E''u_1 + dx_1 \wedge u_1 - \delta_0(z'') \otimes H(x_1) * d_E''u_1 + u_2 = u;$$

l'avant-dernière égalité résulte des propriétés des dérivations d'un produit de convolution et de $(\partial/\partial x_1)H(x_1) = \delta_0(x_1)$. $\square$

5.5.7. PROPOSITION. *$U^{0,1}$ étant construit, à partir de $P^{0,2}$, dans le lemme 5.4.1, l'équation (5.12) possède la solution à support compact unique*

(5.14) $C_m(\zeta') = K_E\#\pi_*\left[z_n^m\left(P^{1,1} + d_E'U^{0,1}\right) \wedge dx_1\right].$

En outre $C_m(\zeta')$ est CR sur $\mathscr{E}\setminus\mathscr{N}$.

DÉMONSTRATION. D'après 5.5.4 et 5.5.6, $C_m(\zeta')$, définie par $K_E\sharp R$ (cf. 5.5.3), est une solution de (5.12). Pour $|z''(\eta')|$ assez grand, les supports de $K_E(\zeta' - \eta')$ et de $R(\zeta')$ sont disjoints; alors $C_m(\eta') = K_E\sharp R(\eta') = 0$. D'autre part, on a $(d_E'' + (\partial/\partial x_1)\, dx_1)C_m(\zeta') = R(\zeta') = 0$ pour $\zeta' \in \mathscr{E}\backslash$ spt R. Pour $x_1 \in \mathbf{R}_1$, posant $\mathscr{E}_{x_1} = \mathscr{E} \cap \{z \in \mathscr{E},\, x_1(z) = x_1\}$, $k\colon \mathscr{E} \to \mathbf{R}_1$ étant la projection: $\zeta' \mapsto x_1$, considérons la tranche $\langle C_m, k, x_1 \rangle$ de C_m relative à k en x_1; c'est un courant contenant dx_1 en facteur; alors

$$(5.15) \quad \left(d_{\mathscr{E}}'' + (\partial/\partial x_1)\, dx_1\right)\langle C_m, k, x_1 \rangle = d_{\mathscr{E}}''\langle C_m, k, x_1 \rangle = \langle R, k, x_1 \rangle = 0$$

pour $\zeta' \in (\mathscr{E}\backslash$ spt $R) \cap \mathscr{E}_{x_1}$; $i_{x_1}\colon \mathscr{E}_{x_1} \to \mathscr{E}$ étant l'injection canonique, il existe un courant localement plat S_{x_1} de $\mathscr{E}_{x_1}$ tel que $\langle C_m, k, x_1 \rangle = i_{x_1}{}_*S_{x_1}$; $\mathscr{E}_{x_1}$ est isomorphe à $\mathbf{C}^{n-2}$ (avec les variables $z_2,\dots,z_{n-1}$). Pour φ C^∞ à support compact de $\mathscr{E}$, on a

$$d_{\mathscr{E}}''\left(i_{x_1}{}_*S_{x_1}\right)(\varphi) = -S_{x_1}\left(i_{x_1}^* d_{\mathscr{E}}''\varphi\right) = -S_{x_1}\left(d_{\mathscr{E}}'' i_{x_1}^*\varphi\right)$$

puisque d commute avec $i_{x_1}^*$ et que $i_{x_1}^* dx_1 = 0$. Mais $d_{\mathscr{E}}''|\mathscr{E}_{x_1} = d''|\mathscr{E}_{x_1} = d''$ sur l'espace complexe $\mathscr{E}_{x_1}$ et, toute $\psi \in \mathscr{D}(\mathscr{E}_{x_1})$ étant restriction à $\mathscr{E}_{x_1}$, d'un élément de $\mathscr{D}(\mathscr{E})$, la relation (5.15) entraîne $d''S_{x_1} = 0$; donc S_{x_1} est holomorphe dans $(\mathscr{E}\backslash$ spt $R) \cap \mathscr{E}_{x_1}$. $i_{x_1}{}_*S_{x_1}$, donc S_{x_1}, étant nul pour $|z''(\eta')|$ assez grand, la fonction holomorphe S_{x_1} est nulle sur la composante connexe infinie de $(\mathscr{E}\backslash$ spt $R) \cap \mathscr{E}_{x_1}$, i.e., en dehors d'un compact; d'autre part, $i_{x_1}{}_*S_{x_1} = \langle C_m, k, x_1 \rangle$ est nul pour $|x_1|$ assez grand; alors C_m est nul en dehors d'un compact de $\mathscr{E}$.

Si $C_m'(\zeta')$ est une autre solution à support compact de (5.12), alors $d_{\mathscr{E}}''(C_m(\zeta') - C_m'(\zeta')) = 0$ et $(\partial/\partial x_1)(C_m(\zeta') - C_m'(\zeta')) = 0$; donc $c_m = C_m - C_m'$ est holomorphe en z'' et constante en x_1; comme c_m est à support compact, on a $c_m = 0$, i.e., $C_m(\zeta')$ définie par $K_E\sharp R$ est l'unique solution à support compact de (5.12).

Montrons maintenant que $K_E\sharp R$ est indépendant des courants $T^{0,0}$. Considérons

$$(5.16) \qquad K_E\sharp\pi_*\left[z_n^m\left(d_E'd_E''T^{0,0} + d_E''\overline{d_E''T^{0,0}}\right) \wedge dx_1\right];$$

c'est un courant à support compact; de plus, d'après 3.3.2, $d_E''\pi_* = \pi_*d_E''$, et, pour dériver un produit de convolution, on remplace le second facteur par sa dérivée; donc, lorsqu'on applique d_E'' au courant (5.16), on obtient $K_E\sharp\pi_*[z_n^m d_E''d_E'd_E''T^{0,0} \wedge dx_1] = 0$. Le courant (5.16) est donc holomorphe en $(z_2,\dots,z_{n-1})$; son support étant compact, il est nul.

L'unique solution à support compact de (5.12) est donc

$$C_m(\zeta') = K_E\sharp\pi_*\left[z_n^m\left(P^{1,2} + \left(P^{1,1} + d_E'U^{0,1} + \overline{d_E'U^{0,1}}\right) \wedge dx_1\right)\right].$$

$P^{1,2}$ n'ayant pas de terme en dx_1, on a $K_E\sharp\pi_*z_n^m P^{1,2} = 0$.

Soit $\psi = \psi_0\, dz'' \wedge d\bar{z}''$ une $(n-2, n-2)$-forme C^∞ à support compact de $\mathscr{E}$.

$$\pi_*\left[z_n^m\overline{d_E'U^{0,1}} \wedge dx_1\right](\psi) = d_E''\overline{U^{0,1}} \wedge dx_1\left(z_n^m\pi^*\psi\right) = \overline{U^{0,1}} \wedge dx_1\left(d_E''z_n^m\pi^*\psi\right)$$

$$= -\overline{U^{0,1}} \wedge dx_1\left(z_n^m\pi^*d_{\mathscr{E}}''\psi\right);$$

mais ψ étant de degré maximum en $d\bar{z}_j (j = 2,\ldots,n-1)$, on a $d''_{\mathscr{E}}\psi = 0$; donc $\pi_*[z_n^m d'_E U^{0,1} \wedge dx_1] = 0$.

Compte tenu de ces deux réductions, $C_m(\zeta')$ a l'expression (5.14) et est CR sur $\mathscr{E} \setminus \mathscr{N}$ puisque $d''_E C_m(\zeta') = 0$ (Lemme 5.5.2) pour $\zeta' \in \mathscr{E} \setminus \mathscr{N}$. $\square$

5.5.8. PROPOSITION. *Si N est de classe C^ω, alors, dans le complémentaire de $\pi(N)$ dans $\mathscr{E}$, la fonction $C_m(\zeta')$ est C^ω.*

DÉMONSTRATION. $U^{0,1}$ étant C^ω en x_1 d'après 5.4.1, il en est de même de $d'_E U^{0,1}$; par produit de convolution, la propriété est conservée: on la vérifie en complexifiant localement puis en appliquant d''_{z_1} au produit de convolution, il suffit de l'appliquer au facteur holomorphe en z_1, ce qui donne 0; donc le produit de convolution est holomorphe en z_1, i.e., C^ω en x_1. Alors $C_m(\zeta')$ est C^ω en x_1; de plus, comme $d''_E C_m(\zeta') = 0$ dans $\mathscr{E} \setminus \pi(N)$, $C_m(\zeta')$ est holomorphe en z''; donc elle est C^ω sur $\mathscr{E} \setminus \pi(N)$. $\square$

6. Théorème principal.

6.1. Soit N un cycle de dimension $(2n-4)$, à singularités négligeables, C^ω, de dimension holomorphe $(n-3)$, contenu dans E. Pour une direction complexe fixée γ de E, on choisit un système de coordonnées complexes $(z_1,\ldots,z_n)$ dans $\mathbf{C}^n$ tel que $E = \{z \in \mathbf{C}^n;\ \mathrm{Im}\, z_1(z) = 0\}$ et que l'axe des z_n ait la direction γ. On considère la projection $\pi\colon E \to \mathscr{E}$ et on appelle $\Gamma = \Gamma_\pi$ l'ensemble des points $z \in E$ tels que chaque point de $\pi^{-1}(\pi(z)) \cap N$ soit un point régulier de $\pi|N$; un point $x \in N$ non-régulier est dit critique. Soit K l'ensemble des points critiques de $\pi|N$; alors $\pi(\Gamma) \cup \pi(K) = \mathscr{E}$, en particulier, on a $\tau \subset K$. Pour simplifier l'exposé, on suppose que N est une sous-variété à singularités négligeables.

6.1.1. LEMME. *On a $\mathscr{H}_{2n-4}(\pi(K)) = 0$; donc $\pi(\Gamma)$ est un ouvert connexe dense dans $\mathscr{E}$. Pour tout $\zeta' \in \mathscr{N} \cap \pi(\Gamma)$, $N \cap \pi^{-1}(\zeta')$ est un ensemble fini de points et il existe un voisinage Δ' de ζ' tel que $N \cap \pi^{-1}(\Delta') = N_1 \cup \cdots \cup N_r$, où N_j ($j = 1,\ldots,r$) est une sous-variété connexe telle que $\pi|N_j\colon N_j \to \Delta'$ soit un plongement.*

[La démonstration est la même que celle du Lemme 3.4 (a) de [5], l'énoncé ne faisant intervenir que la structure réelle de N].

6.1.2. *Dans la suite du n° 6, dans les notations de 5.1, on suppose que N satisfait à la condition (H) et que γ est choisie telle que (H') soit satisfaite.*

D'après les conditions ci-dessus, la réunion $\mathrm{Sing}\, \mathscr{N}$ des valeurs critiques $\pi(K)$ et des points où $\mathscr{N}$ n'est pas une variété est partout non dense dans $\mathscr{N}$ et $\mathscr{E} \setminus \mathrm{Sing}\, \mathscr{N}$ est un ouvert connexe de $\mathscr{E}$.

On désigne par $V_0, V_1, \ldots, V_p, \ldots$ les composantes connexes de $\mathscr{E} \setminus \mathscr{N}$ où V_0 est la composante non bornée.

6.2. *Calcul de C_0.*

On a $C_0 = \pi_*(S^{1,1}) = \pi_*(S)$; $dC_0 = d\pi_*(S) = \pi_*(dS) = \pi_*(P) = [\mathscr{N}]_{\mathscr{E}}$, courant d'intégration défini par $\mathscr{N}$ dans $\mathscr{E}$; alors, à cause de l'unicité de la solution à support compact, on a $C_0 = \sum_{j=0}^\infty m_j[V_j]$ où $m_j \in \mathbf{Z}$ et $m_0 = 0$.

6.3. *La fonction R.*

6.3.1. Considérons une boule $B(0, \rho)$ de E contenant N; soit π_n la projection de E sur $\mathbf{C}_n$; alors $\pi_n(B(0, \rho)) = \Delta(0, \rho)$, disque de $\mathbf{C}$ centré en 0, de rayon ρ. On considère la fonction

$$(6.1) \qquad \phi_w = C_0 \log w + \sum_{m=1}^{\infty} m^{-1} C_m w^{-m}$$

pour w fixé dans $\mathbf{C} \setminus \overline{\Delta(0, \rho)}$. Les coefficients $C_m(\zeta')$ satisfont à l'équation (5.14), dans laquelle $U^{0,1} = -KP^{0,2}$ (5.7). Alors l'équation (5.14) s'écrit

$$(6.2) \qquad \left(d_E'' + (\partial/\partial x_1) \right) C_m(\zeta') = \pi_* \left[z_n^m \left(P^{1,1} - d_E' KP^{0,2} \right) \wedge dx_1 \right].$$

$C_m(\zeta')$ étant calculé au n° 5, dans $(\mathscr{E} \setminus \mathscr{N}) \times (\mathbf{C} \setminus \overline{\Delta(0, \rho)})$, la série (6.1) converge faiblement d'après 3.2.3, donc uniformément sur tout compact comme fonction C^ω d'après 5.5.7 et, d'après l'unicité de la solution à support compact pour l'opérateur $d_E'' + (\partial/\partial x_1)\, dx_1$, on a, pour $w \in \mathbf{C} \setminus \overline{\Delta(0, \rho)}$ fixé,

$$(6.3) \qquad \phi_w = K_E \sharp \pi_* \left[\log(w - z_n)\left(P^{1,1} - d_E' KP^{0,2} \right) \wedge dx_1 \right].$$

Pour $(\zeta', w) \in (\mathscr{E} \setminus \mathscr{N}) \times (\mathbf{C} \setminus \overline{\Delta(0, \rho)})$, on pose $R(\zeta'; w) = \exp \phi_w(\zeta')$. Ce qui précède entraîne

6.3.2. **LEMME.** $R(\zeta'; w)$ *est une fonction* C^ω *sur* $(\mathscr{E} \setminus \mathscr{N}) \times (\mathbf{C} \setminus \overline{\Delta(0, \rho)})$ *qui possède les propriétés suivantes:*

(a) $$R = 1 \; sur \; V_0 \times \left(\mathbf{C} \setminus \overline{\Delta(0, \rho)} \right);$$

(b) $$R(\zeta'; w) = \sum_{m=-\infty}^{C_0} A_m(\zeta') w^m,$$

la série convergeant uniformément sur les compacts de $(\mathscr{E} \setminus \mathscr{N}) \times (\mathbf{C} \setminus \overline{\Delta(0, \rho)})$;

(c) *chaque coefficient* A_m *dans* (b) *est un polynôme en un nombre fini de* C_m.

[La démonstration est celle du Lemme 3.13 de [5].]

6.3.3. On connait R au-dessus de V_0; on va maintenant prolonger R au-dessus des divers V_j, établir des relations entre ces prolongements quand on franchit $\mathscr{N}$ et en déduire l'extension de $R(\zeta'; w)$ aux $(\overline{V}_j \setminus \operatorname{Sing} \mathscr{N}) \times \mathbf{C}$.

6.4. *Extension de R aux* $(\overline{V}_j \setminus \operatorname{Sing} \mathscr{N}) \times (\mathbf{C} \setminus \overline{\Delta(0, \rho)})$.

Il s'agit d'étendre $C_m(\zeta')$ et $R(\zeta'; w)$ qui sont C^ω sur chaque ouvert V_j de $\mathscr{E}$ à $\overline{V}_j \setminus \operatorname{Sing} \mathscr{N}$ et d'évaluer le saut de C_m quand on passe de V_j à $V_{j'}$ contigu en traversant $\mathscr{N}$.

On rappelle (5.14): $C_m = K_E \sharp \pi_* [z_n^m (P^{1,1} + d_E' U^{0,1}) \wedge dx_1]$.

Le noyau $K_E = \delta(z'') \otimes H(x_1)(\partial/\partial x_1)$ ne fait pas intervenir la variable z_n; donc $K_E \sharp$ et π_* commutent.

6.4.1. *Terme en* $P^{1,1}$.

Pour $\psi = \psi_0\, dx_1 \wedge dz'' \wedge d\bar{z}'' \in \mathscr{D}^{2n-3}(\mathscr{E} \setminus \operatorname{Sing} \mathscr{N})$, on a

$$(6.4) \qquad \pi_* \left[K_E \sharp z_n^m P^{1,1} \wedge dx_1 \right](\psi) = K_E \sharp z_n^m P_{n\bar{n}}\, dz_n \wedge dz_n \wedge dx_1 (\pi^* \psi);$$

H_1 étant la fonction de Heaviside en ξ_1, l'expression (6.4) est égale à

$$H_1 * P_{n\bar{n}} dz_n \wedge d\bar{z}_n\left(z_n^m \pi^*\psi\right) = \int_N \int H(\xi_1 - x_1) z_n^m \pi^* \psi_0\left(\xi_1; z''\right) d\xi_1 \wedge dz'' \wedge d\bar{z}''.$$

Sur N, ζ_n est une fonction à plusieurs déterminations, $\zeta_n = \zeta_n^l(\xi_1, \zeta_2,\ldots,\zeta_{n-1})$ de $(\xi_1, \zeta_2,\ldots,\zeta_{n-1}) \in \mathcal{N}$ (voir Lemma 6.1.1) au voisinage du support de ψ supposé assez petit. Alors (6.4) s'écrit

$$\sum_l \int_{\mathcal{N}} \int_{\xi_1 \geq x_1} \left(\zeta_n^l\right)^m \psi_0\left(\xi_1, z''\right) d\xi_1 \wedge dz'' \wedge d\bar{z}'';$$

donc le terme en $P^{1,1}$ est un courant de degré 0 défini par une fonction continue sur $\mathcal{E}$.

6.4.2. *Evaluation du terme en $d_E' U^{0,1}$.*

(a) *Réduction.*

Pour $\psi = \psi_0 \, dz'' \wedge d\bar{z}'' \in \mathcal{D}^{n-2,n-2}(\mathcal{E} \setminus \operatorname{Sing} \mathcal{N})$, on a

$$\pi_*\left[z_n^m\left(d_E' U^{0,1} \wedge dx_1\right)\right](\psi)$$
$$= d_E'\left(U^{0,1} \wedge dx_1\right)\left(z_n^m \pi^*\psi\right) = -U^{0,1} \wedge dx_1\left(d_E'\left(z_n^m \pi^*\psi\right)\right);$$

de plus $d_E'(z_n^m \pi^*\psi) = (d_E' z_n^m) \wedge \pi^*\psi + z_n^m(d_E' \pi^*\psi)$; mais $\pi^*\psi$ est indépendant de z_n; donc $d_E' \pi^*\psi = d_{\mathcal{E}}' \pi^*\psi = 0$ car ψ est de degré maximum en dz_j, $d\bar{z}_j$ $(j = 2,\ldots,n-1)$; alors $d_E'(z_n^m \pi^*\psi) = m z_n^{m-1} dz_n \wedge \pi^*\psi$, d'où

$$\pi_*\left[z_n^m\left(d_E' U^{0,1} \wedge dx_1\right)\right](\psi)$$
$$= -U^{0,1} \wedge dx_1 \wedge \left(m z_n^{m-1} dz_n \wedge \pi^*\psi\right) = -\pi_*\left(m z_n^{m-1} dz_n \wedge U^{0,1} \wedge dx_1\right)(\psi).$$

On pose, comme pour $S^{0,1}$ précédemment:

$$U^{0,1} = (i/2)\left(\sum_{j=2}^{n-1} U_{1j} \, d\bar{z}_j + U_{1\bar{n}} \, d\bar{z}_n\right).$$

Alors

$$(6.5) \quad \pi_*\left[z_n^m\left(d_E' U^{0,1} \wedge dx_1\right)\right] = -m(i/2)\pi_*\left(z_n^{m-1} U_{1\bar{n}} \, dz_n \wedge d\bar{z}_n \wedge dx_1\right),$$

ce qui élimine l'opérateur d_E' de l'expression de C_m.

(b) *Evaluation de $U^{0,1}$.*

Introduisons le noyau

$$(6.6) \qquad\qquad K_{BM}' = \sum_{j=2}^{n-1} K_j\left(\partial/\partial\bar{z}_j\right)$$

avec

$$K_j = \delta_0(x_1) \otimes \frac{(n-3)!}{\pi^{n-2}|z''|^{2(n-2)}} \otimes \delta_0(z_n)\bar{z}_j.$$

La divergence $\sum_{j=2}^{n-1}(\partial/\partial\bar{z}_j)K_j$ de K_{BM}' est δ_0, d'après la propriété du noyau de Bochner–Martinelli dans $\mathbf{C}^{n-2}(z'')$; donc K_{BM}' agissant par convolution-contraction est un noyau pour d_E'', mais le courant $K_{BM}' \# P^{0,2}$ n'est pas à support compact; on a

$$K'_{BM}\sharp P^{0,2} = -K'_{BM}\sharp d''_E U^{0,1} = -U^{0,1} + d''_E\left(K'_{BM}\sharp U^{0,1}\right),$$

i.e.,

$$U^{0,1} = -K'_{BM}\sharp P^{0,2} + d''_E\left(K'_{BM}\sharp U^{0,1}\right).$$

Le terme en $d'_E U^{0,1}$ de l'expression (5.14) de C_m est égal à $\mu_1 + \mu_2$ avec

$$\mu_1 = -K_E\sharp\pi_*\left[z_n^m\, d'_E K'_{BM}\sharp P^{0,2} \wedge dx_1\right];$$

$$\mu_2 = K_E\sharp\pi_*\left[z_n^m\, d'_E d''_E K'_{BM}\sharp U^{0,1} \wedge dx_1\right].$$

On a $d''_E \mu_2 = 0$; donc μ_2 est holomorphe en z''; d'autre part $U^{0,1}$ étant C^ω en x_1, il en est de même de μ_2; alors μ_2 est C^ω sur $\mathcal{E}$. Compte tenu de 6.4.1, modulo les fonctions C^ω sur $\mathcal{E}$, $C_m(\zeta)$ est égal à μ_1.

(c) *Evaluation de $K'_{BM}\sharp P^{0,2}$.*

Posant

$$P^{0,2} = \sum_{2 \leqslant l < r \leqslant n-1} P_{lr}\, d\bar{z}_l \wedge d\bar{z}_r + \sum_{j=2}^{n-1} P_{jn}\, d\bar{z}_j \wedge d\bar{z}_n,$$

on a

$$K'_{BM}\sharp P^{0,2} = \sum_{j=2}^{n-1} K_j * P_{jn}\, d\bar{z}_n + \text{termes ne contenant pas } d\bar{z}_n;$$

Compte tenu de (6.5), on a
(6.7)

$$-\pi_*\left[z_n^m\, d'_E K'_{BM}\sharp P^{0,2} \wedge dx_1\right] = m\pi_*\left(z_n^{m-1}\sum_{j=2}^{n-1} K_j * P_{jn}\, dz_n \wedge d\bar{z}_n \wedge dx_1\right).$$

Calculons $K_j * P_{jn} = \delta_0(x_1) \otimes \left((n-3)!/\pi^{n-2}|z''|^{2(n-2)}\right) \otimes \delta_0(z_n)\bar{z}_j * P_{jn}$ sur $\varphi \in \mathcal{D}(E)$. En indiquant, en indice des distributions, les variables des fonctions sur lesquelles elles portent, on a

$$K_j * P_{jn}(\varphi) = P_{jn,\xi_1,\zeta'',\zeta_n}\Big[\delta_0(\eta_1 - \xi_1)$$

$$\otimes \frac{(n-3)!}{\pi^{n-2}|v'' - \zeta''|^{2(n-2)}} \otimes \delta_0(v_n - \zeta_n)\left(\bar{v}_j - \bar{\xi}_j\right)\left(\varphi(\eta_1, v'', v_n)\right)\Big]$$

$$= P_{jn,\xi_1,\zeta'',\zeta_n}\left[\int_{\mathbb{C}^{n-2}(v'')} C\frac{\left(\bar{v}_j - \bar{\xi}_j\right)}{|v'' - \zeta''|^{2(n-2)}}\varphi(\xi_1, v'', \zeta_n)\, dv'' \wedge d\bar{v}''\right]$$

avec $C = (-1)^{(n-3)(n-4)/2}(i/2)^{n-2}(n-3)!\pi^{2-n}$.

En explicitant le courant P_{jn}, on obtient

$$K_j * P_{jn}(\varphi) = \int_N\left[\int_{\mathbb{C}^{n-2}(v'')} C\frac{\left(\bar{v}_j - \bar{\xi}_j\right)}{|v'' - \zeta''|^{2(n-2)}}\varphi(\xi_1, v'', \zeta_n)\, dv''\right.$$

$$\left.\wedge d\bar{v}''\right]d\zeta_2 \wedge \cdots \wedge d\zeta_{n-1} \wedge d\zeta_n \wedge d\bar{\zeta}_2 \wedge \cdots \wedge \widehat{d\bar{\zeta}_j} \wedge \cdots \wedge d\bar{\zeta}_{n-1}.$$

(d) *Calcul de* μ_1.

Soit $\psi = \psi_0 \, dx_1 \wedge dz'' \wedge d\bar{z}'' \in \mathcal{D}^{2n-3}(\mathscr{E})$; on a, d'après (6.7),

$$\mu_1(\psi) = mK_E \# z_n^{m-1} \sum_{j=2}^{n-1} K_j * P_{jn} \, dz_n \wedge d\bar{z}_n \wedge dx_1 (\pi^*\psi)$$

$$= m \sum_{j=2}^{n-1} H_1 * K_j * P_{jn} \, dz_n \wedge d\bar{z}_n \big(z_n^{m-1} \pi^* \psi \big)$$

$$= m \sum_{j=2}^{n-1} P_{jn,\xi_1,\zeta'',\zeta_n} \left[\int_{\xi_1} H(\xi_1 - \eta_1) \int_{v'' \in \mathbf{C}^{n-2}} C \frac{\bar{v}_j - \bar{\zeta}_j}{|v'' - \zeta''|^{2(n-2)}} \right.$$

$$\left. \zeta_n^{m-1} \pi^* \psi_0 (\eta_1, v'') \, d\eta_1 \wedge dv'' \wedge d\bar{v}'' \right]$$

$$= m \sum_{j=2}^{n-1} \int_N [\ ] \, d\zeta_2 \wedge \cdots \wedge d\zeta_n \wedge \cdots \wedge \widehat{d\bar{\zeta}_j} \wedge \cdots \wedge d\bar{\zeta}_n.$$

Alors, en explicitant les déterminations locales ζ_n^l de ζ_n sur $\mathcal{N}$,

$$\mu_1(\psi) = m \int_{\mathcal{N}} \left[\int_{\eta_1 \geqslant \xi_1} \int_{v'' \in \mathbf{C}^{n-2}} \sum_l \sum_{j=2}^{n-1} C(\bar{v}_j - \bar{\zeta}_j) |v'' - \zeta''|^{2(2-n)} (\zeta_n^l)^{m-1} \right.$$

$$\times \psi_0(\eta_1, v'') \, d\eta_1 \wedge dv'' \wedge d\bar{v}'' \times d\zeta_2 \wedge \cdots \wedge d\zeta_n \wedge \cdots \wedge \widehat{d\bar{\zeta}_j} \wedge \cdots \wedge d\bar{\zeta}_{n-1}.$$

La fonction sous les signes $\int$ est intégrable sur $\mathcal{N} \times \mathbf{R} \times \mathbf{C}^{n-2}$; alors, d'après le théorème de Fubini,

$$\mu_1(\psi) = \int_{\eta_1, v''} F_m(\eta_1, v'') \psi(\eta_1, v'') \, d\eta_1 \wedge dv'' \wedge d\bar{v}'',$$

avec

$$(6.8) \quad F_m(\eta_1, v'') = m \int_{\mathcal{N}:\, \xi_1 \leqslant \eta_1} \sum_l \sum_{j=2}^{n-1} C(\zeta_n^l)^{m-1} (\bar{v}_j - \bar{\zeta}_j) |v'' - \zeta''|^{2(2-n)}$$

$$d\zeta_2 \wedge \cdots \wedge d\zeta_{n-1} \wedge d\zeta_n^l \wedge \cdots \wedge \widehat{d\bar{\zeta}_j} \wedge \cdots \wedge d\bar{\zeta}_{n-1}$$

où $d\zeta_n^l(\zeta_1, \zeta_2, \ldots, \zeta_{n-1})$ est à remplacer par $(\partial \zeta_n^l / \partial \xi_1) \, d\xi_1 + (\partial \zeta_n^l / \partial \bar{\zeta}_j) \, d\bar{\zeta}_j$.

(e) *Calcul de* $F_m(\eta_1, v'')$.

Pour le calcul de (6.8), on va prendre ξ_1 comme coordonnée locale sur $\mathcal{N}$, ce qui est possible au voisinage de tout point de $\mathcal{N}$ n'appartenant pas au fermé $\pi(\sigma \cup \sigma')$ (5.3.1), de mesure de Hausdorff $(2n - 4)$-dimensionnelle nulle; *dans la suite la réunion de cet ensemble et de* Sing $\mathcal{N}$ (6.1.2) *sera désignée par* Sing $\mathcal{N}$.

On a

$$d\zeta_2 \wedge \cdots \wedge d\zeta_{n-1} \wedge d\zeta_n^l \wedge d\bar{\zeta}_2 \wedge \cdots \wedge \widehat{d\bar{\zeta}_j} \wedge \cdots \wedge d\bar{\zeta}_{n-1}$$

$$= (-1)^n (\partial \zeta_n^l / \partial \xi_1) \, d\xi_1 \wedge d\zeta_2 \wedge \cdots \wedge d\zeta_{n-1} \wedge d\bar{\zeta}_2 \wedge \cdots \wedge \widehat{d\bar{\zeta}_j} \wedge \cdots \wedge d\bar{\zeta}_{n-1}$$

$$+ (-1)^{j-2} (\partial \zeta_n^l / \partial \bar{\zeta}_j) \, d\zeta_2 \wedge \cdots \wedge d\zeta_{n-1} \wedge d\bar{\zeta}_2 \wedge \cdots \wedge d\bar{\zeta}_{n-1}.$$

Alors

$$(6.9) \qquad F_m\big(\eta_1, v''\big) = \sum_l \sum_{j=2}^{n-1} \big(G_m^{j,l} + H_m^{j,l}\big)$$

où

$$G_m^{j,l} = C(-1)^n m \int_{\mathcal{N}\,:\,\xi_1 \leqslant \eta_1} \big(\zeta_n^l\big)^{m-1}\big(\partial\zeta_n^l/\partial\xi_1\big)\, d\xi_1 \wedge \big(\bar{v}_j - \bar{\zeta}_j\big)\big|v'' - \zeta''\big|^{2(2-n)}\, d\zeta_2$$

$$\wedge \cdots \wedge d\zeta_{n-1} \wedge d\bar{\zeta}_2 \wedge \cdots \wedge \widehat{d\bar{\zeta}_j} \wedge \cdots \wedge d\bar{\zeta}_{n-1},$$

$$H_m^{j,l} = C(-1)^j m \int_{\mathcal{N}\,:\,\xi_1 \leqslant \eta_1} \big(\zeta_n^l\big)^{m-1}\big(\bar{v}_j - \bar{\zeta}_j\big)\big|v'' - \zeta''\big|^{2(2-n)}\big(\partial\zeta_n^l/\partial\bar{\zeta}_j\big)\, d\zeta_2 \wedge$$

$$\cdots \wedge d\zeta_{n-1} \wedge d\bar{\zeta}_2 \wedge \cdots \wedge d\bar{\zeta}_{n-1}.$$

$G_m^{j,l}$ est définie par l'intégrale d'une fonction de (ξ_1, ζ'') dont seul le facteur $(\zeta_n^l)^{m-1}(\partial\zeta_n^l/\partial\xi_1)$ dépend de ξ_1. Pour $(\xi_1, \zeta'') \notin \mathcal{N}$, on intègre d'abord pour ξ_1 constant; alors

$$G_m^{j,l} = \int_{\xi_1 \leqslant \eta_1} \left[(-1)^{n+1} C \int_{\mathcal{N} \cap \{x_1 = \xi_1\}} m\big(\zeta_n^l(x_1, \zeta'')\big)^{m-1}\big(\partial\zeta_n^l/\partial x_1\big)\big(x_1, \zeta''\big)\big(\bar{v}_j - \bar{\zeta}_j\big) \right.$$

$$\left. \big|v'' - \zeta''\big|^{2(2-n)} d\zeta_2 \wedge \cdots \wedge d\zeta_{n-1} \wedge d\bar{\zeta}_2 \wedge \cdots \wedge \widehat{d\bar{\zeta}_j} \wedge \cdots \wedge d\bar{\zeta}_{n-1} \right] d\xi_1.$$

On montre, comme dans ([**6**, App. B]), en désignant par $\Gamma_m^{j,l}$ le crochet sous le signe $\int_{\xi_1 \leqslant \eta_1}$ de l'expression de $G_m^{j,l}$, que $\Gamma_m^l = \sum_{j=2}^{n-1} \Gamma_m^{jl}$ est continu sur $\overline{V}_j \cap \{x_1 = \xi_1\}$ et, aux points où $\mathcal{N} \cap \{x_1 = \xi_1\}$ est lisse, le saut, à la traversée de $\mathcal{N}$ est $\pm m(\zeta_n^l(\xi_1, v''))^{m-1}(\partial\zeta_n^l/\partial x_1)(\xi_1, v'')$. La fonction saut de $G_m^l = \sum_{j=2}^{n-1} G_m^{j,l}$ au passage de $\mathcal{N}$ est donc $\pm \int_{\xi_1 \leqslant \eta_1} m(\zeta_n^l(\xi_1, v''))^{m-1}(\partial\zeta_n^l/\partial x_1)(\xi_1, v'')\, d\xi_1$, c'est-à-dire, à cause de la compacité de N dans E, $\pm[\zeta_n^l(\eta_1, v'')]^m$.

D'après la condition (H) (5.1.1), on a $H_m^{j,l} = 0$. On a établi

6.4.3. PROPOSITION. *Dans les hypothèses de 6.1, les coefficients C_m ont les propriétés suivantes:*

1) *sur chaque composante connexe V_j de $\mathcal{E} \setminus \mathcal{N}$, C_m est une fonction C^ω et se prolonge à $\overline{V}_j \setminus \operatorname{Sing} \mathcal{N}$ en une fonction continue notée C_m^j;*

2) *si V_j et V_j' sont deux composantes connexes contiguës de $\mathcal{E} \setminus \mathcal{N}$ et si $\mathcal{N}_0$ est un ouvert lisse assez petit de $\mathcal{N}$ contenu dans $\overline{V}_j \cap \overline{V}_{j'}$, alors, dans les notations de 6.4.2(d), on a*

$$C_m^j - C_m^{j'}\big|\mathcal{N}_0 = \pm \sum_l \big[\zeta_n^l(\eta_1, z'')\big]^m \text{ où } \zeta_n^l(\eta_1, z'') \text{ est } C^\omega \text{ sur } \mathcal{N}_0.$$

6.4.4. PROPOSITION. *Dans les notations de 6.3, on a les résultats suivants:*

1) *$A_m|V_j$ s'étend en une fonction continue sur $\overline{V}_j \setminus \operatorname{Sing} \mathcal{N}$, qu'on notera $\overline{A}_m^j$;*

2) *$R(\zeta'; w)|V_j \times (\mathbf{C} \setminus \overline{\Delta}(0, \rho))$ s'étend en une fonction continue sur $(\overline{V}_j \setminus \operatorname{Sing} \mathcal{N}) \times (\mathbf{C}^n \setminus \overline{\Delta}(0, \rho))$, qu'on notera $R_j(\zeta'; w)$;*

3) *pour chaque $\zeta' \in \overline{V}_j \setminus \operatorname{Sing} \mathcal{N}$ fixé, $R_j(\zeta'; w)$ est holomorphe en w et possède un développement en série de Laurent*

$$R_j(\zeta'; w) = \sum_{m=-\infty}^{C_0} \overline{A}_m^j(\zeta') w^m$$

sur $\mathbf{C} \setminus \overline{\Delta(0, \rho)}$;

4) *soit $\mathcal{N}_0$ une composante connexe de la variété $\mathcal{N} \setminus \operatorname{Sing} \mathcal{N}$, contenue dans $\overline{V}_j \cap \overline{V}_{j'}$, telle que $\pi^{-1}(\mathcal{N}_0) \cap N$ ait r composantes connexes $N_1, \ldots, N_r$ avec $N_l = $ graphe (ζ_n^l), $l = 1, \ldots, r$; on a, sur $\mathcal{N}_0 \times (\mathbf{C} \setminus \Delta(0, \rho))$*

$$(6.9) \qquad R_j(\zeta'; w) = \prod_{l=1}^{r} \left(w - \zeta_n^l(\zeta')\right)^{\pm 1} R_{j'}(\zeta'; w);$$

en outre $\zeta_n^l(\zeta')$ est C^ω sur $\mathcal{N}_0$.

DÉMONSTRATION. La Proposition, à l'exception de 2), résulte immédiatement de 6.4.3. L'assertion 2) se démontre à l'aide d'une estimation de la norme ∞ de $F_m(\eta_1, v'')$ en fonction de la norme C^1 de ζ_n sur les compacts de $\mathcal{N}_0$. $\square$

6.5. *Rationalité de R.*

6.5.1. PROPOSITION. *Pour tout $\zeta' \in \overline{V}_j \setminus \operatorname{Sing} \mathcal{N}$ fixé, la fonction $R_j(\zeta'; w)$ est rationnelle en w et à coefficients* CR *et C^ω sur V_j, continus sur $\overline{V}_j \setminus \operatorname{Sing} \mathcal{N}$.*

DÉMONSTRATION. On sait que $R_0 \equiv 1$ sur V_0; $\mathscr{E} \setminus \operatorname{Sing} \mathcal{N}$ est connexe; donc un point de V_0 peut être joint, par un arc continu, à un point quelconque de $\mathscr{E} \setminus \mathcal{N}$ en traversant un nombre fini de composantes connexes $\mathcal{N}_0$ de $\mathcal{N} \setminus \operatorname{Sing} \mathcal{N}$. Soient V_j, $V_{j'}$ deux composantes connexes contiguës de $\mathscr{E} \setminus \mathcal{N}$ situées de part et d'autre d'une composante $\mathcal{N}_0$ au voisinage d'un point de $\mathcal{N} \setminus \operatorname{Sing} \mathcal{N}$. Supposons que, pour tout $\zeta' \in \mathcal{N}_0$, $R_{j'}(\zeta'; w)$ soit rationnelle en w, à coefficients continus en ζ' et d_E''-fermés; alors, d'après (6.9), il en est de même de $R_j(\zeta'; w)$. Après complexification locale, le théorème de E. E. Levi (cf. [5, 3.8]) donne le résultat. $\square$

6.6. *Construction de la chaîne maximalement complexe dans $\Gamma \setminus N$*

6.6.1. La fonction R, à coefficients CR et rationnelle en $w = z_n$ est CR au sens de 4.1.4. D'autre part, pour $x_1 = x_1^0$ fixé, R restreinte à $(\mathscr{E} \setminus \mathcal{N}) \cap \{x_1 = x_1^0\}$ est une fonction méromorphe possédant un diviseur $M_{x_1^0}$; quand x_1^0 varie, $M_{x_1^0}$ décrit la chaîne maximalement complexe $M = (i/\pi) D(\log|R[\![E]\!]])$ dans $(\mathscr{E} \setminus \mathcal{N}) \times \mathbf{C}$, dans les notations de 4.3 et de 6.1.

On va montrer que M s'étend en une chaîne maximalement complexe de $\Gamma \setminus N$ et que $dM = [N]$ sur Γ.

6.6.2. PROPOSITION. *Dans les notations de 6.6.1, M a une extension en une $(2n - 3)$-chaîne maximalement complexe dans $\Gamma \setminus N$.*

DÉMONSTRATION. Pour tout $z_0 \in (E \setminus N) \cap \Gamma$, on va construire un voisinage B de z_0 et une chaîne maximalement complexe M_B sur B qui prolonge M de $B \setminus \pi^{-1}(\mathcal{N})$ à B. On suppose $\zeta_0' = \pi(z_0) \in \mathcal{N}$.

(a) $\zeta_0' \notin \operatorname{Sing} \mathscr{N}$; on prend $B = \Delta' \times \Delta$ où $\Delta' = I \times \Delta''$ avec I intervalle de $\mathbf{R}_1$ et Δ'' boule de $\mathbf{C}^{n-2}$ (variable z''); Δ boule de $\mathbf{C}_n$ (variable z_n), toutes les boules étant centrées en les projections correspondantes de z_0. On choisit Δ' assez petit pour que $\mathscr{N} \cap \Delta'$ soit une sous-variété connexe et que $\Delta' \setminus \mathscr{N}$ soit réunion de deux composantes connexes $V_j \cap \Delta'$ et $V_{j'} \cap \Delta'$. On a $N \cap \pi^{-1}(\Delta') = N_1 \cup \cdots \cup N_r$ où N_s est une composante connexe et est le graphe d'une fonction $f_s \in C^1(\mathscr{N} \cap \Delta')$ d'après le Lemme 6.1.1. Soit $M_j = (i/\pi)D(\log|R|[E])$ au-dessus de V_j; les coefficients de R_j sont continus sur $\overline{V}_j \cap \Delta'$; donc les points de $\pi^{-1}(\zeta_0')$ appartiennent à $\overline{W}_j = \overline{\operatorname{Spt} M_j}$ et forment un ensemble fini qui est l'ensemble des zéros et des pôles de $R_j(\zeta'; w)$; alors, il existe un disque Δ centré en $z_n^0 = z_n(z_0)$ dans $\mathbf{C}_n$ tel que $\{\zeta_0'\} \times \overline{\Delta}$ ne contienne aucun point de $\overline{W}_j \cup \overline{W}_{j'}$, différent de z_0 et qui soit disjoint de N. On prend pour Δ' un produit de boules assez petit pour que $\Delta' \times b\Delta$ soit disjoint de $\overline{W}_j \cup \overline{W}_{j'}$, et que $\Delta' \times \overline{\Delta}$ soit disjoint de N. Alors $\pi|B \cap \overline{\operatorname{Spt} S}$ est propre.

Soient $S' = S|B \setminus \pi^{-1}(\mathscr{N})$ et $c_m = \pi_*(w^m S')$; $m \geqslant 0$; $\pi|\operatorname{Spt}(S \cap B) \setminus \pi^{-1}(\mathscr{N})$ est propre; alors c_m est une fonction CR, C^ω sur $\Delta' \setminus \mathscr{N}$. Soit a_m le coefficient A_m défini par les c_p; on pose $r(\zeta'; w) = \sum_{m=-\infty}^{c_0} a_m(\zeta')w^m$. On va montrer que les c_m, donc les coefficients de r, et r elle-même se prolongent de $\Delta' \setminus \mathscr{N}$ à Δ' en des fonctions CR, C^ω.

On a $j_* S = M = S \otimes \delta(y_1) = S \otimes \varepsilon_1(y_1)\,dy_1$ (voir 2.2) où j est l'inclusion: $E \to \mathbf{C}^n$. Mais, d'après 4.3.3, on a $M = (i/\pi)D(\log|R|[E])$. D'autre part, pour $\varphi \in \mathscr{D}^{2n-3}(\pi(B) \setminus \mathscr{N})$, on a $c_m(\varphi) = \pi_*(w^m S')(\varphi) = \pi_*(w^m S)(\varphi)$. Pour $\alpha \in \mathscr{D}^{2n-3}(\pi(B) \times \mathbf{R}_1' \setminus \mathscr{N})$, on a $j_* c_m(\alpha) = c_m(j^*\alpha)$; puisque $j^*\,dy_1 = 0$, il suffit de faire agir $j_* c_m$ sur les formes $\alpha = \alpha_0\,dx_1 \wedge dz'' \wedge d\bar{z}''$; alors

$$j_* c_m(\alpha) = j_*(\pi_*(w^m S))(\alpha) = \pi_*(w^m S)(j^*\alpha) = w^m S(\pi_* j^*\alpha)$$

$$= S(w^m(\pi^* j^*\alpha)) = S(w^m j^*\alpha),$$

car $\pi^* j^*\alpha$ et $j^*\alpha$ ont la même expression. Donc

$$j_* c_m(\alpha) = j_* S(w^m\alpha) = M(w^m\alpha) = (i/\pi)D(\log|r|[E])(w^m\alpha)$$

$$= (i/\pi)w^m d'_w d''_w(\log|R|[E])(\alpha)$$

à cause du type de α et du fait que α est indépendant de dw et de $d\bar{w}$. $d'd''\log|R| = (-1/2)d''d'\log R$; alors

$$j_* c_m(\alpha) = -(i/2\pi)w^m d''_w d'_w(\log R[E])(\alpha)$$

$$= (i/2\pi)d''_w(w^m R^{-1}(\partial R/\partial w)[E]\,dw)(\alpha)$$

$$= (i/2\pi)(w^m R^{-1}(\partial R/\partial w)[E]\,dw)\,d((\alpha_0\,dx_1 \wedge dz'' \wedge d\bar{z}''))$$

$$= -(1/2\pi i)\lim_{\varepsilon \to 0}\int_{B\setminus\{|R(z)|\geqslant\varepsilon\}\cap E} d(w^m R^{-1}(\partial R/\partial w)\,dw \wedge \alpha_0\,dx_1 \wedge dz'' \wedge d\bar{z}'')$$

$$= (1/2\pi i)\lim_{\varepsilon \to 0}\int_{\{|R(z)|=\varepsilon\}\cap B} w^m R^{-1}(\partial R/\partial w)\,dw \wedge \alpha_0\,dx_1 \wedge dz'' \wedge d\bar{z}''.$$

Pour ζ' fixe, la fonction sous le signe $\int$ est holomorphe en w dans $B \setminus \{|R(z)| = \varepsilon\}$; donc, d'après le théorème de Cauchy

$$j_* c_m(\alpha) = (1/2\pi i) \int_{I \times \Delta''} \left(\int_{b\Delta} w^m R^{-1}(\partial R/\partial w)\, dw \right) \alpha_0\, dx_1 \wedge dz'' \wedge d\bar{z}'', \text{ i.e.,}$$

$$c_m = (1/2\pi i) \int_{b\Delta} w^m R^{-1}(\partial R/\partial w)\, dw.$$

D'après la formule du saut, dans les notations de 6.4.4, on a, pour $\zeta' \in \mathcal{N}_0 \cap \pi(B)$

$$R_j(\zeta'; w) = \prod_{l=1}^{r} \left(w - \zeta_n^l(\zeta') \right)^{\pm 1} R_{j'}(\zeta'; w)$$

où $\zeta_n^l(\zeta')$ est C^ω sur $\mathcal{N}_0$, d'où:

$$c_m^j(\zeta') - c_m^{j'}(\zeta') = \pm (1/2\pi i) \int_{b\Delta} w^m \sum_l \pm \left(w - \zeta_n^l(\zeta') \right)^{-1} dw.$$

Puisque $\Sigma_l \pm (w - \zeta_n^l(\zeta'))^{-1}$ est holomorphe sur Δ pour ζ' fixé, le deuxième membre s'annule; donc c_m^j se prolonge continûment à Δ'; puis après complexification de x_1, holomorphiquement, donc c_m^j se prolonge sur Δ', en une fonction C^ω.

(b) $\zeta_0' \in \operatorname{Sing} \mathcal{N}$; on remarque que ζ_0' est la projection de points réguliers de $\pi \setminus N$ puisque $\zeta_0' \in \pi(\Gamma)$. On choisit un voisinage (produit d'un intervalle et de disques) Δ' de ζ_0' suffisamment petit pour que, d'après le Lemme 6.1.1, on ait $N = \pi^{-1}(\Delta' \cap \mathcal{N}) = N_1 \cup \cdots \cup N_r$ où N_s est une sous-variété et π un plongement de N_s dans Δ', pour $s = 1,\ldots,r$. Soit $\mathcal{N}_s = \pi(N_s)$ orientée conformément à N_s; contrairement au cas (a), les projections $\mathcal{N}_s$ ne sont pas toutes confondues, mais la démonstration se ramène aisément à celle du cas (a), comme dans [5, Lemma 3.2.2]. $\square$

6.7. *Extension de M à Γ.*

6.7.1. PROPOSITION. *La $(2n - 3)$-chaîne M de $\Gamma \setminus N$ a un volume localement fini au voisinage de $N \cap \Gamma$, donc possède, en tant que courant d'intégration, une extension simple, notée encore M, en un courant rectifiable de Γ, de dimension $(2n - 3)$.*

DÉMONSTRATION. $N \cap \Gamma$ est une sous-variété lisse. Considérons la chaîne maximalement complexe M de $\Gamma \setminus N$ et un point de $(N \cap \overline{|M|}) \cap \Gamma$ qu'on peut (après translation) supposer être 0 de $\mathbf{C}^n$; M étant une $(2n - 3)$-chaîne et N une $(2n - 4)$-chaîne, on a $\mathscr{H}^{2(n-1)}(|M| \cup N) = 0$. Pour tout choix de coordonnées $x_1, z_2,\ldots,z_n$ dans E, pour tout multiindice ordonné $J = (i_1,\ldots,i_{n-2})$, on considère la projection

$$\pi_J \colon E \to \mathbf{R}_1 \times \mathbf{C}^{n-2}\left(= \pi_J' \right),$$
$$\left(x_1, z_2,\ldots,z_n \right) \mapsto \left(x_1, z_{i_1},\ldots,z_{i_{n-2}} \right)$$

où π_J' désigne le $(2n - 3)$-espace vectoriel sur lequel π_J projette E orthogonalement.

D'après un raisonnement de Shiffman (Lemmes 1.5 et 1.3 de [5], Lemme 2 et p. 118 de [9]), on peut choisir des coordonnées $z_2, \ldots, z_n$ et un voisinage produit d'un intervalle de $\mathbf{R}_1$ et de disques $\Delta = \Delta'_J \times \Delta''_J$ (où Δ''_J est un disque de $\mathbf{C}$) tels que

$$\pi_J \colon \left(N \cap \overline{|M|} \right) \cap \Delta \to \Delta'_J \subset \pi'_J$$

soit propre.

Alors, on vérifie, par complexification locale de x_1, que l'application $\pi_J|(N \cap \overline{|M|}) \cap \Delta_J$ est la restriction d'un isomorphisme analytique complexe local en dehors d'un sous-ensemble analytique complexe Σ de la complexification locale de $|M|$. En outre, pour chaque J, $\pi_J \colon N \cap \Delta \to \Delta'_J$ est un plongement. Soit $\mathcal{N}_J$ l'image de N dans Δ'_J et S_J le cylindre projetant N dans Δ'_J. Alors l'ensemble $(|M| \setminus \Sigma) \cap S_J$ est une sous-variété lisse de dimension $(2n - 4)$ et $\mathcal{H}_{2n-5}(\Sigma \cap E)$ est finie. Il en résulte que, pour une boule B de E centrée en 0 et contenue dans Δ, on a

$$\mathrm{vol}\bigl(|M| \cap B\bigr) \leqslant \sideset{}{'}\sum_{|J| = n-2} \int_{\Delta'_J \setminus N_J} m_J \, dv'_j,$$

où Σ' est la somme sur les multiindices ordonnés J et m_J est la borne supérieure du nombre fini de points de la fibre de $\pi||M| \cap B$ comptés aves leurs multiplicités dans la chaîne M. $\square$

6.7.2. PROPOSITION. *L'extension simple de M à Γ étant encore désignée par M, on a $dM = \sum n_j[N_j]$ où les N_j sont les composantes connexes de $N \cap \Gamma$, $[N_j]$ le courant d'intégration sur N_j et $n_j \in \mathbf{Z}$.*

DÉMONSTRATION. M est un courant localement rectifiable dans Γ, donc M et dM sont localement plats; $M|\Gamma \setminus |N|$ est d-fermé, donc dM est porté par $|N|$; d'après le théorème de support $dM = h[N]$ où h est une fonction localement intégrable sur $N \setminus \Gamma$; h est localement constante parce que N est d-fermée et est à valeurs entières parce que dM est rectifiable. $\square$

6.8. *Lemme de complexification locale.*

6.8.1. *Préliminaire: cas où N est une sous-variété lisse C^ω.*

Pour toute projection $\pi \colon \mathbf{C}^n \to \mathbf{C}^{n-1}$ dont la direction est contenue dans E, on désigne par Γ_π la partie Γ de E associée à π; la sous-variété N étant lisse, on a $E = \bigcup_\pi \Gamma_\pi$. Tout point z^0 de N appartient donc à $\Gamma = \Gamma_\pi$ pour π convenable. Au voisinage de z^0, x_1 est une coordonnée locale sur N; donc il existe un voisinage ouvert $u(z^0)$ de z^0 dans E et une représentation paramétrique de $N \cap \Gamma \cap u(z^0)$:

$$D \to u(z^0),$$

$$(\xi, x_1) \mapsto \Bigl|{\begin{array}{l} x_1 = x_1 \\ z_j = \varphi_j\bigl(\xi, (x_1 - x_1^0)\bigr), j = 2, \ldots, n, \end{array}}$$

`204   PIERRE DOLBEAULT

où D est un ouvert de $\mathbf{R}^{2n-5} \times \mathbf{R}_1$ et φ_j ($j = 2,\dots,n$) une série entière convergente. Cette représentation s'étend à un ouvert $\hat{D}$ de $\mathbf{R}^{2n-5} \times \mathbf{C}_1$ en

$$\hat{D} \to \hat{u}(z^0),$$

$$(\xi, z_1) \to \Big|\begin{matrix} z_1 = z_1 \\ z_j = \hat{\varphi}_j(\xi, z_1 - x_1^0), j = 2,\dots,n, \end{matrix}$$

où $\hat{\varphi}_j$ est la série convergente obtenue en remplaçant x_1 par $z_1 = x_1 + iy_1$ dans φ_j, pour $|z_1 - x_1^0|$ assez petit. Cela définit, dans un voisinage ouvert assez petit $\hat{u}(z^0)$ de z^0 dans $\mathbf{C}^n$, une sous-variété $\hat{N}_{z^0}$ transversale à E. En $z \in N \cap \Gamma \cap u(z^0)$, le sous-espace complexe maximal tangent à $\hat{N}_{z^0}$ en z est engendré par $H_z(N)$ (sous-espace complexe maximal tangent à N en z) et $\partial/\partial z_1$, donc $dh_z(\hat{N}_{z^0}) \geq n - 2$; on a donc aussi $dh_z(\hat{N}_{z^0}) \geq n - 2$ pour tout $z \in \hat{u}(z^0)$ supposé assez petit.

N étant compact, il existe un nombre fini de points z_l^0 ($l = 1,\dots,p$) de N et, pour tout l, une projection π_l à direction contenue dans E et un voisinage $u(z_l^0)$ dans E satisfaisant aux conditions décrites ci-dessus tels que $\bigcup_{l=1}^p u(z_l^0) \supset N$. Soient $\hat{u}_l = \hat{u}(z_l^0)$; $\hat{N}_l = \hat{N}_{z_l^0}$; *alors $\bigcup_{l=1}^p \hat{u}_l$ est un voisinage ouvert U de N dans $\mathbf{C}^n$, les sous-variétés $\hat{N}_l$ se recollant en une sous-variété $\hat{N}$ C^ω, maximalement complexe de $\hat{U}$, transversale à E, telle que $\hat{N} \cap E = N$.*

6.8.2. *Préliminaire: cas où N est l'intersection de E et d'un cycle C^ω maximalement complexe $\hat{N}_0$ de $\mathbf{C}^n$ transversal à E.* Pour simplifier, comme ci-dessus, on suppose que $\hat{N}_0$ et N sont des sous-variétés à singularités négligeables.

Alors, trivialement, il existe un intervalle ouvert I' de $\mathbf{R}_1'$ (variable y_1) contenant 0 tel que, pour $k: \mathbf{C}^n \to \mathbf{R}_1'$, $z \mapsto y_1(z)$, $\hat{U} = k^{-1}(I')$ *soit un voisinage ouvert de N dans $\mathbf{C}^n$ contenant la sous-variété $\hat{N} = \hat{N}_0 \cap \hat{U}$ transversale à E avec $\hat{N} \cap E = N$.*

6.8.3. LEMME. *Dans les cas 6.8.1 et 6.8.2, il existe un voisinage V de Γ dans $\mathbf{C}^n$, un prolongement $\hat{N}$ de $N \cap \Gamma$ dans V et un prolongement $\hat{M}$ de M dans V tels que $\hat{N}$ soit une sous-variété maximalement complexe de V transversale à E et $\hat{M}$ une chaîne holomorphe de $V \setminus \hat{N}$, transversale à E, ayant une extension simple, notée encore $\hat{M}$, à V et telle que $|d\hat{M}| = |\hat{N}|$.*

DÉMONSTRATION. Dans les hypothèses de 6.8.1 ou 6.8.2, il existe un intervalle I' de $\mathbf{R}_1'$ contenant 0 tel que, pour tout $y_1 \in I'$, dans les notations de 6.8.2, $N_{y_1} = \langle \hat{N}, k, y_1 \rangle$ soit le courant d'intégration sur une sous-variété C^ω (éventuellement à singularités négligeables), de dimension $(2n - 4)$, de dimension holomorphe $(n - 3)$. La projection π étant choisie, considérons la fonction CR, R qui définit la chaîne maximalement complexe M dans $\Gamma \setminus N$ (6.6); R est, localement, le quotient de deux fonctions C^ω en x_1, holomorphes en $(z_2,\dots,z_n)$; par complexification locale de x_1 et pour I' assez petit, R se prolonge, sur $\hat{U} = \pi^{-1}(I')$ en une fonction méromorphe définissant une $(n - 1)$-chaîne holomorphe $\hat{M}$ de $\hat{U}$ telle que $M_{y_1} = \langle \hat{M}, k, y_1 \rangle$ soit la chaîne maximalement complexe de $\pi^{-1}(y_1)$ associée à N_{y_1} comme M l'est à N dans E. $\square$

6.8.4. COROLLAIRE. *Dans les hypothèses de 6.8.3, on a $dM = N$ dans Γ.*

DÉMONSTRATION. Le couple $(\hat{M}, \hat{N})$ du Lemma 6.8.3 possède les propriétés décrites dans [5, Lemma 3.24], d'où le résultat par tranchage en 0 relativement à k. $\square$

6.9. THÉORÈME. *Soit N un cycle C^ω, compact, de E, de dimension $(2n - 4)$, de dimension holomorphe $(n - 3)$, satisfaisant à la condition (H) et à l'une des conditions suivantes:*

(a) *N est une sous-variété lisse;*

(b) *N est l'intersection de E et d'un cycle maximalement complexe de $\mathbf{C}^n$ transversal à E.*

Alors, il existe une $(2n - 3)$-chaîne unique M C^ω maximalement complexe, feuilletée par des $(n - 2)$-chaînes holomorphes, à support dans $E \setminus N$, ayant une extension simple (notée encore M), à support compact dans E telle que $dM = N$.

DÉMONSTRATION. L'unicité et le feuilletage résultent de 6.6.1. D'après 6.8.4, pour toute projection π telle que (H′) soit satisfaite (voir 6.1.2), on a $dM = N$ dans Γ_π. L'ensemble singulier de N étant désigné par τ, le Lemme 6.8.3 permet de prolonger M à $E \setminus \tau$ par la méthode de [5, Lemma 3.25].

Ensuite, ou bien $\tau = \varnothing$ (cas (a)), ou bien (cas (b)), d'après [5, fin du $n°$ 3.10], $\hat{M}$ a une extension simple à un voisinage de E dans $\mathbf{C}^n$, ce qui implique l'existence de l'extension simple de M à E et $dM = N$ sur E. $\square$

RÉFÉRENCES

1. J. Besnault et P. Dolbeault, *Sur les bords d'ensembles analytiques complexes dans $\mathbf{P}^n(\mathbf{C})$*, Sympos. Math. **24** (1981), 205–213.

2. B. Bigolin, *Gruppi di Aeppli*, Ann. Scuola Norm. Sup. Pisa (3) **23** (1969), 259–287.

3. P. Dolbeault, *On holomorphic chains with given boundary in $\mathbf{P}^n(\mathbf{C})$*, Springer Lecture notes in Math., 1039, 1983, 118–129.

4. H. Federer, *Geometric measure theory*, Grundlehren Math. Wiss., Bd. **153**, Springer-Verlag, New York, 1969.

5. R. Harvey, *Holomorphic chains and their boundaries*, Proc. Sympos. Pure Math. **30**, Part I, Amer. Math. Soc. (1977), 309–382.

6. R. Harvey and B. Lawson, *On boundaries of complex analytic varieties I*, Ann. of Math. (2) **102** (1975), 233–290.

7. R. Harvey and B. Lawson, *On boundaries of complex analytic varieties II*, Ann. of Math. (2) **106** (1977), 213–238.

8. J. King, *The currents defined by analytic varieties*, Acta Math. **127** (1971), 185–220.

9. B. Shiffman, *On the removal of singularities of analytic sets*, Michigan Math. J. **15** (1968), 111–120.

10. R. Harvey and J. Polking, *Fundamental solutions in complex analysis I*, Duke Math. J. **46** (1979), 233–300.

UNIVERSITÉ DE PARIS VI, FRANCE

Proceedings of Symposia in Pure Mathematics
Volume **44** (1986)

Index and Total Curvature
of Complete Minimal Surfaces

ROBERT GULLIVER

Complete minimal surfaces in three-dimensional euclidean space $\mathbf{E}^3$ were extensively studied during the nineteenth century, but it is only in recent years that some of their more remarkable properties have become clearly understood. A particularly interesting property is stability. We say that a minimal surface M is *stable* if the second variation of area is nonnegative for any variation of M with compact support. It should be observed that stability is significantly weaker than the property of having minimum area. It was shown recently that a stable, complete minimal surface in $\mathbf{E}^3$ must be a plane (doCarmo–Peng [2] and, independently, Fischer-Colbrie–Schoen [3]). A formulation equivalent to stability is that

$$\int_M \left(|\nabla u|^2 + 2Ku^2 \right) dA > 0$$

for any function $u\colon M \to \mathbf{R}$ with compact support, where K is the Gauss curvature of M. This is the condition that the Jacobi operator $Lu = -\Delta u + 2Ku$ have only nonnegative eigenvalues.

A more general condition than stability is that M have *finite index* (of stability). M is said to have *index* k if the Jacobi operator for M has at most k negative eigenvalues on any compact subset of M. Thus, stability corresponds to index zero. For example, a minimal surface found by means of a minimax procedure will be likely to have finite index. The question arises: what *geometric* properties characterize complete minimal surfaces of finite index? Doris Fischer-Colbrie has conjectured that these are precisely the surfaces of finite total Gauss curvature. In

1980 *Mathematics Subject Classification.* Primary 53A10, 35J70.

© 1986 American Mathematical Society
0082-0717/86 $1.00 + $.25 per page

fact, we have the following

THEOREM. *A complete immersed minimal surface in* $\mathbf{E}^3$ *has finite index if and only if it has finite total curvature.*

We would like to thank Doris Fischer-Colbrie for posing this fascinating question in a stimulating fashion, and to thank Blaine Lawson for fruitful conversations.

Recall that the nth eigenvalue $\lambda_n(\Omega)$ of L on a compact domain Ω may be characterized as the infimum, among all n-dimensional subspaces V of $C_0^1(\Omega)$, of the maximum value of

$$\int_\Omega uLu\, dA = \int_\Omega \left(|\nabla u|^2 + 2Ku^2\right) dA$$

over all functions u in V satisfying $\int_\Omega u^2\, dA = 1$. Here $C_0^1(\Omega)$ denotes the space of continuously differentiable functions u with $u = 0$ on $\partial\Omega$. In particular, index (Ω) is precisely the maximum number of independent functions $u_1,\ldots,u_k$ of which any linear combination u satisfies $\int_\Omega uLu\, dA < 0$.

Suppose that a complete minimal surface M has finite index. Then for some compact subset F, $M\setminus F$ is stable. In fact, we may let F be a compact subset with index$(F) = $ index$(M) = k$. Suppose that $M\setminus F$ is not stable. This means that there is a function u_0, supported in a compact subset D of $M\setminus F$, with $\int_D u_0 Lu_0\, dA < 0$. Meanwhile, there are independent functions $u_1,\ldots,u_k$ with support in F, so that $\int_F uLu\, dA < 0$ for any linear combination $u = c_1 u_1 + \cdots + c_k u_k$. Now consider a linear combination $v = c_0 u_0 + u$ of $u_0, u_1,\ldots,u_k$: we have $vLv = c_0^2 u_0 Lu_0 + uLu$, since u_0 and u have disjoint supports. Hence $\int_{F\cup D} vLv\, dA < 0$, which implies that index$(F\cup D) \geqslant k + 1$, a contradiction.

Of course, the stable minimal surface $M\setminus F$ is not complete. Nonetheless, it has quadratic area growth, finite topological type, and finite total curvature, according to Corollary 3.17 of [4]. This proves half of the theorem.

The proof of the converse is based on Osserman's theorem that a minimal surface M of finite total curvature is conformally equivalent to a compact Riemann surface $\hat{M}$, minus a finite number of points $p_1,\ldots,p_m$, and that the Gauss map $G: M \to S^2$ extends to a branched conformal mapping from $\hat{M}$ to S^2 (see [6, p. 82]). Let $\hat{M}$ be equipped with the singular riemannian metric $d\hat{s}^2$ induced via the Gauss map from the canonical metric on S^2. Then a solution of the Jacobi equation $Lu = -\Delta u + 2Ku = 0$ on M satisfies $\hat{L}u: = -\hat{\Delta}u - 2u = 0$ locally on M. Here $\hat{\Delta}$ is the Laplace operator for the metric $d\hat{s}^2$. In fact, $-K$ is the Jacobian determinant of the conformal mapping G, and the equation $\hat{L}u = 0$ follows immediately away from the branch points of G. On the other hand, these branch points are isolated, while u remains continuous on M; it follows that $\hat{L}u = 0$ everywhere on M.

In order to show that a minimal surface M of finite total curvature must have finite index, it therefore will suffice to show that the degenerate elliptic operator $\hat{L}$ has finite index on the compact surface $\hat{M}$. Namely, the index of the Jacobi operator L on M equals the index of $\hat{L}$ on M, which is at most equal to the index of $\hat{L}$ on $\hat{M}$, since M is a subset of $\hat{M}$. It is well known that a (nondegenerate) elliptic operator has finite index on any compact manifold; we shall extend this result to the present degenerate case to complete the proof of the Theorem:

LEMMA. *Let $\hat{M}$ be a compact two-dimensional manifold, $G: \hat{M} \to S^2$ a branched conformal covering, and let $d\hat{s}^2$ be the singular riemannian metric on $\hat{M}$ induced from the canonical metric on S^2. Then $\hat{L} = -\hat{\Delta} - 2$ has finite index.*

REMARK. As will be clear from the proof, the Lemma holds in much greater generality. In particular, S^2 may be replaced by an arbitrary compact riemannian 2-manifold Σ, and we need only require that some neighborhood U of the singular set of $d\hat{s}^2$ is mapped by an open mapping $G: U \to \Sigma$, so that for some $\mu > 0$,

$$\mu G^*ds_\Sigma^2 \leqslant d\hat{s}^2 \leqslant \mu^{-1} G^*ds_\Sigma^2.$$

Then the eigenvalues λ_j of $-\hat{\Delta}$ on $\hat{M}$ tend to $+\infty$ as $j \to \infty$.

PROOF. We shall proceed by the method of conjugate boundaries, analogous to results of J. C. F. Sturm and of M. Morse for ordinary differential equations, and as carried out for (nondegenerate) elliptic partial differential equations by Smale in [8].

Let $q_1, \ldots, q_n$ denote the branch points of G. Choose a connected neighborhood $\hat{B}_\varepsilon$ of q_j to be locally the inverse image of the disk B_ε in S^2 of radius ε, centered at $G(q_j)$. If ε is small enough, then q_j is the only singularity of $d\hat{s}^2$ on $\hat{B}_\varepsilon$.

We claim that the first eigenvalue $\hat{\lambda}_1 = \lambda_1(\hat{B}_\varepsilon)$ for $\hat{L} = -\hat{\Delta} - 2$, with Dirichlet boundary conditions, is equal to the first eigenvalue $\lambda_1 = \lambda_1(B_\varepsilon)$ for the operator $-\Delta - 2$ on B_ε. In fact, there is a positive solution $u_1: B_\varepsilon \to \mathbf{R}$ of $\Delta u_1 + (\lambda_1 + 2)u_1 = 0$ with $u_1 = 0$ on ∂B_ε. The composition $w_1 = u_1 \circ G$ satisfies $\hat{\Delta} w_1 + (\lambda_1 + 2)w_1 = 0$ on $\hat{B}_\varepsilon$, and is positive. This positivity implies that λ_1 is the first eigenvalue on $\hat{B}_\varepsilon$, as is well known for an elliptic operator, and which may be demonstrated in the present, degenerate case following an argument of Lewy [5, pp. 39f]. We need to show that

$$\int_{\hat{B}_\varepsilon} \left(u\hat{L}u - \lambda_1 u^2 \right) dA \geqslant 0$$

for every function $u: \hat{B}_\varepsilon \to \mathbf{R}$ which vanishes on $\partial \hat{B}_\varepsilon$. By approximation, we may assume that u has compact support in $\hat{B}_\varepsilon$. Write $u = vw_1$. Then v has compact support, and

$$\mathrm{div}\left(v^2 w_1 \nabla w_1 \right) - 2w_1 v \langle \nabla v, \nabla w_1 \rangle + v^2 |\nabla w_1|^2 - (2 + \lambda_1)w_1^2 v^2$$
$$= |\nabla u|^2 - (2 + \lambda_1)u^2 - w_1^2 |\nabla v|^2.$$

Upon integrating, we find

$$\int_{\hat{B}_\varepsilon} \left(u\hat{L}u - \lambda_1 u^2 \right) dA = \int_{\hat{B}_\varepsilon} w_1^2 |\nabla v|^2 \, dA \geqslant 0,$$

as desired. (This also proves uniqueness of the first eigenfunction.) We have invoked the divergence theorem for the degenerate metric $d\hat{s}^2$, which follows immediately from the divergence theorem for a nondegenerate conformally equivalent metric.

In particular, whenever $\varepsilon < \pi/2$ we have $\lambda_1(\hat{B}_\varepsilon(q_j)) > 0$. We choose $\varepsilon < \pi/2$ small enough that $\hat{B}_\varepsilon(q_1),\ldots,\hat{B}_\varepsilon(q_n)$ are disjoint, and let U_0 be their union. Then $\lambda_1(U_0) > 0$.

Now choose a polyhedral decomposition of $\hat{M}$, with one-skeleton Γ, such that the branch points $q_1,\ldots,q_n$ of G all occur among the vertices, and choose one point P_I from the interior of each two-dimensional face, $1 \leqslant I \leqslant N$. Given $\eta > 0$, let U_1 be the complement in $\hat{M}$ of the closed balls of radius η centered at the points $P_1,\ldots,P_N$. Then $U_0 \cup \Gamma$ is a deformation retract of U_1.

Now for η sufficiently small, U_1 has the same index as does $\hat{M}$. In fact, according to the proof of Proposition 1.9 of [4], for any $\varepsilon > 0$, there is a cutoff function Ψ which vanishes on a neighborhood of $\{P_1,\ldots,P_N\}$, such that for all u with $|u| \leqslant 1$,

$$(1) \qquad \int_{\hat{M}} (\Psi u)\hat{L}(\Psi u)\, dA \leqslant (1 + \varepsilon)\int_{\hat{M}} u\hat{L}u\, dA + \varepsilon.$$

Furthermore,

$$(2) \qquad (1 - \varepsilon)\int_{\hat{M}} u^2\, dA \leqslant \int_{\hat{M}} \Psi^2 u^2\, dA \leqslant (1 + \varepsilon)\int_{\hat{M}} u^2\, dA.$$

Specifically, Ψ is the product of a standard cutoff function and a function of the form $|\log r|^{-1}$, where r is the distance to the nearest point P_I. Now suppose $\hat{M}$ has index $\geqslant k$, that is, there are nontrivial orthogonal functions $u_1,\ldots,u_k$ in $L^2(\hat{M})$ with

$$\int_{\hat{M}} u_j\hat{L}u_j\, dA < 0, \qquad 1 \leqslant j \leqslant k.$$

Normalize u_j so that $|u_j| \leqslant 1$, and write $v_j = \Psi u_j$. If ε is sufficiently small, then $v_1,\ldots,v_k$ remain independent, via inequality (2), while inequality (1) implies that

$$\int_{U_1} v_j\hat{L}v_j\, dA < 0, \qquad 1 \leqslant j \leqslant k.$$

But this implies that U_1 has index $\geqslant k$. On the other hand, since U_1 is a subset of $\hat{M}$, the opposite inequality $\mathrm{index}(U_1) \leqslant \mathrm{index}(\hat{M})$ is immediate. Therefore U_1 and $\hat{M}$ have the same index.

For $0 < \delta \leqslant \delta_0$, let U_δ be a smooth domain whose boundary lies at distances between δ and 2δ from $U_0 \cup \Gamma$, varying smoothly and monotonically with δ, so that $U_0 \cup \Gamma$ is a deformation retract of each U_δ. Then as $\delta \to 0$, $\lambda_1(U_\delta) \to \lambda_1(U_0)$. Namely, if the first eigenfunction on U_δ is normalized to have $L^2(U_\delta)$-norm equal to one, then it converges strongly on compact subsets of U_0 to the first eigenfunction for U_0. In particular, if δ_0 is chosen small enough, then $\lambda_1(U_\delta) > 0$ for all $0 < \delta \leqslant \delta_0$. For $\delta_0 \leqslant t \leqslant 1$, let $\{U_t\}$ be a smooth monotone deformation between

U_{δ_0} and U_1. According to the proof of Lemma 2 of [8], the eigenvalues $\lambda_1(U_t) < \lambda_2(U_t) \leqslant \lambda_3(U_t) \leqslant \cdots$ are continuous, strictly decreasing functions of t. Further, for each t only a finite number of eigenvalues may vanish. On the other hand, $k = \text{index}(\hat{M}) = \text{index}(U_1)$ is the largest integer with $\lambda_k(U_1) < 0$, and is therefore equal to the number of values of t, $\delta_0 < t < 1$, for which one of the eigenvalues of t is zero (counted with multiplicity). Therefore $k = \text{index}(\hat{M})$ is finite.   q.e.d.

REMARK. In principle, the index of a minimal surface having finite total curvature may be explicitly computed. In fact, it coincides with the index of the operator $\hat{L} = -\hat{\Delta} - 2$, where $\hat{\Delta}$ is locally the spherical Laplacian, on a compact branched covering of the sphere. Simple examples show that the precise relative location of branch points on S^2, as well as the total order of branching of the extended Gauss map $G: \hat{M} \to S^2$, have a significant effect on the index. It should be noted that the branch points of G on $\hat{M}$ may be described geometrically as the flat points of the minimal surface M, plus those ends which lie between two planes (see [7]). In particular, it does not appear that any simple, general *upper* bound for the index may be readily derived using, say, the methods of Barbosa–doCarmo [1]. Equally interesting would be a quantitative *lower* bound; for example, is it true that

$$\text{index}(M) \geqslant \frac{1}{2\pi} \int_M |K|\,dA - N - 1,$$

where N is the total order of branching of the extended Gauss map?

REFERENCES

1. J. L. Barbosa and M. doCarmo, *On the size of a stable minimal surface in* $\mathbf{R}^3$, Amer. J. Math. **98** (1976), 515–528.

2. M. doCarmo and C.-K. Peng, *Stable complete minimal surfaces in* R^3 *are planes*, Bull. Amer. Math. Soc. **1** (1979), 903–906.

3. D. Fischer-Colbrie and R. Schoen, *The structure of complete stable minimal surfaces in 3-manifolds of nonnegative scalar curvature*, Comm. Pure Appl. Math. **33** (1980), 199–211.

4. R.Gulliver and B. Lawson, *The structure of stable minimal hypersurfaces near a singularity*, preprint, 1984.

5. H. Lewy, *Aspects of the calculus of variations*, Univ. of California Press, Berkeley, 1939.

6. R. Osserman, *A survey of minimal surfaces*, Van Nostrand Reinhold, New York, 1969.

7. R. Schoen, *Uniqueness, symmetry and embeddedness of minimal surfaces*, J. Differential Geom. **18** (1983), 791–809.

8. S. Smale, *On the Morse index theorem*, J. Math. Mech. **14** (1965), 1049–1055.

UNIVERSITY OF MINNESOTA

Proceedings of Symposia in Pure Mathematics
Volume **44** (1986)

The Structure of Stable Minimal Hypersurfaces
Near a Singularity

ROBERT GULLIVER[1] AND H. BLAINE LAWSON, JR.[2]

0. Introduction. The general aim of this paper is to study the behavior of a minimal variety in the neighborhood of a singular set, and, in part, to understand the extent to which a singularity is removable. The discussion leads to some interesting intrinsic theorems for complete surfaces.

The question of removable singularities for minimal hypersurfaces has a long history. The case of graphs was done by Bers ($n = 2$) and de Giorgi–Stampacchia [**de G-S**]. In [**H-L**], general extension theorems were proved. The minimizing case essentially constitutes the well-known interior regularity theory for mass-minimizing integral currents. Recently, Simon and Pitts have shown independently that an embedded stable minimal hypersurface of dimension $n \leqslant 5$ with an isolated singularity extends smoothly across the singularity (see [**Sim**$_1$ and **P**, pp. 319 ff.]). It is possible that these methods could be extended to the immersed case. We consider here a quite different approach, that of the "conformal blow-up", which has independent interest and applies in all dimensions. We consider the following basic set-up. Let $M^n \to \overline{B}^{n+1} \setminus \{0\}$ be a properly immersed minimal hypersurface in the punctured euclidean ball. From [**H-L**] we know that if $n \geqslant 2$, then M^n has finite volume and extends across 0 as a minimal current without boundary, not necessarily immersed. The interesting question is: How badly behaved can M^n be near the "singular" point 0? We shall study this question under the hypothesis that M^n is *stable*.

It should be remarked that the extension of a stable variety is stable and it is tempting to study M^n by passing to the tangent cone at 0. Unfortunately even

1980 *Mathematics Subject Classification*. Primary 53A10, 58C40.

[1] Research partially supported by N.S.F. Grant MCS 79-02032.

[2] Research partially supported by N.S.F. Grant MCS 83-01365 and by the Guggenheim Foundation.

© 1986 American Mathematical Society
0082-0717/86 $1.00 + $.25 por page

stable tangent cones can be highly singular (the union of stable cones is stable.) Consequently, the deep work of L. Simon [$\mathbf{Sim}_2$] does not apply directly. Simon shows that if the cone is regular outside the origin, then the variety is strongly asymptotic to it. An example of what might go wrong in the singular case is given by the sequence $C(\Sigma_{1,g})$ of (unstable) minimal cones in $\mathbf{R}^4$, where $\Sigma_{1,g} \subset S^3$ is a compact minimal surface of genus g [$\mathbf{L}_1$]. Here we find that

$$\lim_{g \to \infty} C(\Sigma_{1,g}) = \mathbf{R}^2 \cup \mathbf{R}^2 \quad \text{(a stable cone).}$$

The main new idea of this paper is to apply a conformal "blow-up" to M^n, that is, to change the metric ds^2 induced on M^n by the immersion to the *complete*, conformally related metric $d\tilde{s}^2 = ds^2/r^2$, where the function r is distance to the origin in B^{n+1}. This blow-up converts cones to cylinders, i.e., if M^n is the cone on a submanifold $N^{n-1} \subset S^n$, then the blow-up is a riemannian product $N^{n-1} \times [0, \infty)$. In general, we show that if M^n is stable, then its blow-up is morally a complete manifold of uniformly positive scalar curvature. More precisely, the blow-up has the property that the operator $L \equiv -\Delta + \kappa$ satisfies

$$(*) \qquad\qquad L \geqslant \frac{n(n-2)}{4},$$

where

$$\kappa \equiv \frac{1}{2} \sum_{i,j} R_{ijji}$$

is the scalar curvature. (Note the normalization of κ.)

We shall apply techniques from [$\mathbf{G\text{-}L}_2$] to convert the inequality $(*)$ to statements about the geometry and topology of M. Results are obtained in all dimensions.

Consider first the case of dimension 2.

THEOREM 3.3. *Let M^2 be a minimal surface of finite index properly immersed in $B^3 \setminus \{0\} \subseteq \mathbf{R}^3$. Then M^2 has finite area and extends across $\{0\}$ as a branched minimal immersion. If furthermore M^2 is embedded, then $M^2 \cup \{0\}$ is an embedded minimal surface.*

Note that, to begin, M^2 may have infinite topological type. It could look, for example, like the image of a triply periodic Schwartz surface under conformal inversion. Such possibilities remain open in the unstable case.

We obtain as a corollary a slightly different proof of the Fischer-Colbrie–Schoen result that a complete stable minimal surface in $\mathbf{R}^3$ is a plane (see also [$\mathbf{dC\text{-}P}$].) We also obtain the following corollary.

COROLLARY 3.16. *Let S be a discrete subset of $\mathbf{R}^3$. Then any stable minimal surface M^2 properly embedded in $\mathbf{R}^3 \setminus S$ is a plane (i.e., $M = \mathbf{R}^2 \setminus (\mathbf{R}^2 \cap S)$ for some affine $\mathbf{R}^2 \subset \mathbf{R}^3$).*

In the process of proving the finiteness of topological type, we prove the following intrinsic results.

THEOREM 3.4. *Let M be a connected riemannian 2-manifold for which $-\Delta + K \geqslant 0$ (where K is the Gauss curvature function). Let $\Omega_0 \subset M$ be a compact set with smooth boundary and let $\mathscr{E}$ be any connected component of $M \setminus \operatorname{int}(\Omega_0)$. If $\mathscr{E}$ is metrically complete, then it has finite topological type.*

THEOREM 3.12. *Let M be a complete noncompact riemannian 2-manifold with $-\Delta + \beta K \geqslant 0$ for some $\beta > 1/2$. Then each component of M has the topological type of the plane, the cylinder, or the Möbius band. If furthermore M is simply connected and $K \leqslant 0$, then $\int_M |K|\, dA < 2\pi/(2\beta - 1)$.*

COROLLARY 3.17. *Let M^2 be a complete minimal surface in $\mathbf{R}^3$ of finite index. Then M has finite topological type, quadratic area growth and finite total curvature.*

This last result is related to work of Doris Fischer-Colbrie. Fischer-Colbrie has further conjectured that, conversely, the condition of finite total curvature implies finite index. This conjecture has been recently verified by the first author [$\mathbf{G_4}$].

We now consider hypersurfaces in higher dimensions. Our techniques are based on scalar curvature methods in [$\mathbf{G\text{-}L_2}$]. The central result is the following.

THEOREM 4.1. *Let $M^n \to \overline{B^{n+1}} \setminus \{0\}$ be a properly immersed, stable minimal hypersurface, with induced metric ds^2. Consider the (metrically) complete, conformally equivalent metric*

$$(1) \qquad\qquad d\tilde{s}^2 = ds^2/r^2,$$

where r is the radial distance function in $\mathbf{R}^{n+1}$. Then there exists a complete warped-product metric on $M^n \times S^1$ of the form $d\tilde{s}^2 + f^2\, d\theta^2$, with scalar curvature $\kappa \geqslant \frac{1}{2}n(n-2)$.

COROLLARY 4.4. *M^n has no bad ends.*

The notion of a bad end generalizes that of a hyperbolic cusp. Examples include ends which are homotopy equivalent to $X \times (0, \infty)$, where X is a torus or more generally, either a solvmanifold or a manifold which admits a metric of nonpositive sectional curvature. A precise definition is given in §4.

The general impact of Theorem 4.1 is that (M^n, ds^2) has all the global properties known for a complete manifold of uniformly positive scalar curvature. This has particularly strong topological and metric consequences in dimension three (see §5). However, the general results hold uniformly in all dimensions. Consequently, they are strongest in high dimension where the behavior of stable hypersurfaces becomes increasingly erratic.

The main results have some straightforward generalizations. There is a theorem for stable hypersurfaces in $\mathbf{R}^N \setminus \mathbf{R}^p$. There is also the following general result. Let $\Sigma^p \subset \overline{M}^{n+1}$ be a compact submanifold of any riemannian manifold $\overline{M}^{n+1}$, and let U be a compact neighborhood of Σ^p.

THEOREM 6.4. *Let $M^n \to U \setminus \Sigma^p$ be a properly immersed stable minimal hypersurface where $0 \leqslant p < \frac{1}{2}n - 1$. Then $M^n \times S^1$ carries a complete metric with $\kappa \geqslant 1$ at infinity. In particular, M^n has no bad ends.*

We begin the paper by discussing some examples of stable minimal surfaces in $B^3 \setminus \{0\}$.

1. Examples of stable minimal surfaces. Stable minimal surfaces in a riemannian three-manifold $\overline{M}$ may be constructed by means of "bridges". The precise statement, as proven by Meeks and Yau in [**M-Y**], is as follows:

1.1. Bridge Principle. Let M_1 and M_2 be strictly stable compact minimal surfaces with boundary immersed in $\overline{M}^3$, and choose points $x_i \in \partial M_i$ so that x_i lies on the boundary of the *locally convex hull* $\mathscr{H}(M_i)$. Suppose π is a smooth regular curve in $\overline{M}$ from $x_1 = \pi(0)$ to $x_2 = \pi(1)$, so that $\pi'(0)$ points outward from $\mathscr{H}(M_1)$ and $-\pi'(1)$ points outward from $\mathscr{H}(M_2)$. Let a short arc (y_i, z_i) containing x_i be removed from ∂M_i, and choose arcs τ from y_1 to y_2 and ρ from z_1 to z_2, each C^2-close to π. Then the Jordan curve γ formed from $\partial M_1 \setminus (y_1, z_1)$, τ, $\partial M_2 \setminus (y_2, z_2)$, and ρ is the boundary of a strictly stable immersed minimal surface M, which is C^0-close to the union of M_1 and M_2 with a narrow strip between τ and ρ.

1.2. Remark. The convex hull $\mathscr{H}(M_i)$ referred to in the Bridge Principle should be understood as the intersection of closed, locally geodesically convex sets containing M_i. The proof of Meeks and Yau applies equally if $\mathscr{H}(H_i)$ is replaced by the "H-convex hull", that is, the intersection of closed subsets D of $\overline{M}^3$ containing M_i, such that D has a smooth boundary with nonnegative inward mean curvature.

1.3. Example. The proof of Theorem 3.3 below implies an area estimate for a stable, properly immersed minimal surface $M \subset B_1 \setminus \{0\}$, where B_1 is the unit ball of euclidean $\mathbf{R}^3$. This is not an *a priori* bound, but one which depends on M itself. (To be precise, we derive an estimate for *renormalized* area near 0, and the euclidean area estimate is a consequence.) However, there is no *a priori* area estimate, even for embedded stable minimal disks.

Given any integer N, let $M_1, \ldots, M_N$ be disjoint plane disks in B_1 with $\partial M_i \subset \partial B_1$. Each M_i is a strictly stable minimal surface. Let $\pi_1, \ldots, \pi_{N-1}$ be disjoint arcs in ∂B_1 connecting the boundaries of adjacent disks and meeting the disks transversely. Observe that $\pi_i'(0)$ and $-\pi_i'(1)$ point strictly outward from the convex hull of $M_1 \cup \cdots \cup M_N$. By applying the Bridge Principle 1.1 inductively, we may construct a stable minimal surface M of the type of the disk which is close to a boundary connected sum of $M_1, \ldots, M_N$. But the area of M may be arbitrarily large.

1.4. Example. It follows from Theorem 3.3 below that a stable minimal surface M, properly immersed in $B_1 \setminus \{0\}$, has finite topological type. The estimates given, however, depend on the local geometry of M near ∂B_1. In this and the following two examples, we observe that there is no *a priori* estimate on the topological type of any component of M.

In fact, the minimal catenoid given in rectangular coordinates by $x^2 + y^2 = R^2 \cosh^2(z/R)$ is stable in the region $|z| < 1.1996R$. Now choose $M_1, \ldots, M_N$ to be disjoint stable catenoids of this form in B_1, using distinct values of R, $0.46 < R < 1$. Using the Bridge Principle as before, we may construct a stable orientable minimal surface M, properly embedded in $B_1 \setminus \{0\}$, having genus zero and Euler characteristic $1 - N$.

1.5. EXAMPLE. In this example, we observe that there is no *a priori* bound on the genus of a stable minimal surface M, properly embedded in B_1^3. We first construct a curve γ_1 on B_1 such that the oriented surface M_1 of smallest area with boundary γ_1 has positive genus. The curve γ_1 may be constructed in a fashion analogous to an example of Douglas, as follows [$\mathbf{D_2}$, p. 209]. We choose γ_1 to consist of the unit circle in the (x, y)-plane minus the four arcs $|x| < \varepsilon_1$ and $|y| < \varepsilon_2$, plus the intersection with ∂B_1 of the four half-planes $y = \pm\varepsilon_2$, $z \geqslant 0$, and $x = \pm\varepsilon_1$, $z \leqslant 0$. By considering the orthogonal projections on the four planes $z = \pm y$, $z = \pm x$, one may show that any surface $\tilde{M}$ of the type of the disk has area $\geqslant (\pi - 2\varepsilon)\sqrt{2}$, where $\varepsilon = \max\{\varepsilon_1, \varepsilon_2\}$ (compare Lemma 2.1 of [$\mathbf{F}$].) Meanwhile, the surface $\hat{M}$ of genus one, consisting of a disk in the (x, y)-plane plus two narrow strips, has area $\leqslant \pi + 4\pi\varepsilon$. It follows that for ε sufficiently small, γ_1 satisfies Douglas' hypothesis and therefore bounds an area-minimizing surface M_1 of genus 1 [$\mathbf{D_1}$]. Moreover, Meeks and Yau have shown that M_1 is embedded [$\mathbf{M\text{-}Y}$].

Now choose a disjoint collection $\gamma_1, \ldots, \gamma_N$ of such curves on ∂B_1 bounding minimal surfaces $M_1, \ldots, M_N$ of genus 1. The curves may be made disjoint by slight translations of the central plane and appropriate choices of ε_1 and ε_2. We may assume that M_i is strictly stable: otherwise, we replace $M_1, \ldots, M_N$ by their images under the homothety $(x, y, z) \mapsto (\alpha x, \alpha y, \alpha z)$ for α slightly larger than 1. Applying the Bridge Principle 1.1 and Remark 1.2, we may construct a stable minimal surface M embedded in B_1, with ∂M a smooth Jordan curve on ∂B_1, having genus N.

1.6. EXAMPLE. Theorem 3.3 implies that a stable minimal surface M, properly immersed in $B_1 \setminus \{0\}$, has only a finite number of ends near 0. However, there can be no *a priori* estimate of the number of ends. Namely, let $M_1, \ldots, M_N$ be disks through the origin and apply the Bridge Principle as before. The result is a strictly stable immersed minimal surface M of the type of the disk; $M \setminus \{0\}$ has N ends near the origin.

1.7. EXAMPLE: BRANCH POINTS. As is well known, branched minimal surfaces cannot minimize area [$\mathbf{O_2}$]. Nonetheless, if 0 is a branch point of order $m - 1$ of the minimal surface Σ in B_1, then $M \setminus \{0\}$ is a stable immersed minimal surface in $B_r \setminus \{0\}$ for some $r > 0$. For if we choose the (x, y)-plane as the asymptotic tangent plane to Σ at 0, then Σ has an almost nonparametric representation: $x + iy = w^m$, $z = \mathrm{Re}\{aw^{m+1}\} + O(w^{m+2})$ ($|w| < \varepsilon$), where the complex parameter w is C^1-related to a conformal parameter [$\mathbf{G_1}$, p. 279]. Write $r = \varepsilon^m$. Then $(x(w), y(w), z(w))$ lifts to a nonparametric surface $\tilde{M} = \{(x(w), y(w), z(w)): 0 < |w| < \varepsilon\}$ in the m-fold covering space $\overline{M}^3$ of $B_r \setminus \{x = y = 0\}$. Further, the one-parameter family of surfaces obtained by translating $\tilde{M}$ in the $\tilde{z}$-direction is a foliation of $\overline{M}$ by minimal surfaces. Therefore each leaf of the foliation, and $\tilde{M}$ in particular, is absolutely area-minimizing. This implies that $\tilde{M}$ is stable, and therefore $M \setminus \{0\}$ is stable.

1.8. EXAMPLE: ENNEPER'S SURFACE. Example 1.7, viewed naively, tends to suggest that $M \setminus \{0\}$ may be stable while M itself is not. This turns out to be false if M is immersed. The phenomenon is well illustrated by Enneper's surfaces X: $\mathbf{C} \to \mathbf{R}^3$, defined by

$$X(w) = \mathrm{Re}\left\{ \left(w - w^3/3, \, iw + iw^3/3, \, w^2 \right) \right\}.$$

One radially symmetric normal Jacobi field is given by

$$\varphi_0(w) = \frac{1 - |w|^2}{1 + |w|^2},$$

which is the normal component of the parallel vector field $(0, 0, 1)$. This shows that Enneper's surface is stable in the unit disk $\{|w| < 1\}$, but is unstable in $\{|w| < r\}$ for all $r > 1$. The other radially symmetric Jacobi fields are given by

$$\varphi_R(w) = \frac{\left(1 - |w|^2\right)\log(|w|/R) + 2}{1 + |w|^2}.$$

We have $\varphi_R(w) = 0$ for w in either of two circles, $|w| = r_1(R)$ and $|w| = r_2(R)$, where $r_1(R) \sim Re^{-2}$ and $r_2(R) \sim 1 + |\log R|^{-1}$ as $R \to 0$. In particular, the punctured disk $\{0 < |w| < r\}$ on Enneper's surface is unstable for $r > 1$.

This property of Enneper's surface turns out to be quite general.

1.9. PROPOSITION. *Let S^k be a submanifold of the compact riemannian manifold-with-boundary M^n. Let p: $M \to \mathbf{R}$ be a positive weight function, $0 < c \leqslant p \leqslant C$, and q: $M \to \mathbf{R}$ continuous. Write $\lambda_1(M)$ for the smallest eigenvalue of $-\Delta + q$ with Dirichlet boundary conditions:*

$$-\Delta u + qu = \lambda p u \quad in \ M, \qquad u = 0 \quad on \ \partial M.$$

Define $\lambda_1(M \setminus S)$ to be the greatest lower bound of $\lambda_1(\Omega)$ for compact subsets of $M \setminus S$. If S has codimension $n - k \geqslant 2$, then $\lambda_1(M \setminus S) = \lambda_1(M)$.

PROOF. As is well known, $\lambda_1(M \setminus S)$ is the infimum of the Rayleigh quotient

$$\frac{\int_M \left(|\nabla u|^2 + qu^2 \right) d\mathrm{Vol}}{\int_M pu^2 d\mathrm{Vol}}$$

over functions u with compact support in $M \setminus S$. Let r denote the riemannian distance to S: r is a smooth function for $0 < r < r_0$. Choose $0 < \delta < R < \min\{r_0, 1\}$. We cut off S in two stages. Define $\mathcal{N}(x) = \log R / \log|r(x)|$, $0 < r(x) < R$, and $\mathcal{N}(x) = 1$, $r(x) \geqslant R$. Let $\varphi(x) = \varphi(r(x))$ where $\varphi(x) = 0$ for $r(x) \leqslant \delta/2$, $\varphi(x) = 1$ for $r(x) \geqslant \delta$, $0 \leqslant \varphi \leqslant 1$, and $|\nabla \varphi| \leqslant 3/\delta$. If u is any function with compact support in M, then $\mathcal{N}\varphi u$ has compact support in $M \setminus S$. We may compute in local coordinates that

$$\int_M |\nabla \mathcal{N}|^2 \, d\mathrm{Vol} \leqslant C_1 |\log R|^{-1} \quad \text{and} \quad \int_M |\nabla \varphi|^2 \leqslant C_2,$$

where the constants C_i depend on the geometry of M and S. For any $\varepsilon > 0$,

$$|\nabla(\mathcal{N}\varphi u)|^2 \leqslant 2u^2\left(1 + \frac{1}{\varepsilon}\right)\left(\varphi^2|\nabla\mathcal{N}|^2 + \mathcal{N}^2|\nabla\varphi|^2\right) + (1 + \varepsilon)\mathcal{N}^2\varphi^2|\nabla u|^2.$$

Therefore, normalizing u so that $|u| \leqslant 1$, we have

$$\int_M \left(|\nabla(\mathcal{N}\varphi u)|^2 + q\mathcal{N}^2\varphi^2 u^2\right) d\operatorname{Vol} \leqslant (1 + \varepsilon)\int_M \left(|\nabla u|^2 + qu^2\right) d\operatorname{Vol}$$

$$+ C_3\left(1 + \frac{1}{\varepsilon}\right)|\log R|^{-1} + C_4\left(1 + \frac{1}{\varepsilon}\right)(\log R/\log \delta)^2.$$

With appropriate choices of ε, R, and δ, this is arbitrarily close to $\int_M(|\nabla u|^2 + qu^2)\, d\operatorname{Vol}$. Meanwhile, as $R \to 0$, $\int_M \mathcal{N}^2\varphi^2 u^2 p\, d\operatorname{Vol}$ converges to $\int_M u^2 p\, d\operatorname{Vol}$. In particular, the Rayleigh quotient for $\mathcal{N}\varphi u$ is at most arbitrarily close to the Rayleigh quotient for u, which shows $\lambda_1(M \setminus S) \leqslant \lambda_1(M)$. The opposite inequality is obvious.   q.e.d.

1.10. COROLLARY. *Suppose that M^2 is an immersed minimal surface in a riemannian 3-manifold and $x_0 \in M$. If $M \setminus \{x_0\}$ is stable, then M itself is stable.*

1.11. REMARK: BRANCH POINTS. Corollary 1.10 applies also to a branched minimal surface with a branch point at x_0, in the following sense: If $M \setminus \{x_0\}$ is stable, then M itself is stable *with respect to normal variations*. However, a one-parameter family of branched immersions starting from M need *not* be obtained by moving in the normal direction and reparametrizing (compare [$\mathbf{G_3}$]). In fact, Böhme has shown that in a certain sense, a minimal surface has Jacobi fields, tangent to the boundary curve, in every neighborhood of a branch point [$\mathbf{B}$].

2. The conformal renormalization. Simons' 1968 paper investigated the properties, and in particular the existence, of a stable n-dimensional minimal cone $C(T)$ in euclidean space by means of the separation of variables. Specifically, if r is the euclidean distance to the vertex, then the operator $L = -\Delta + \kappa$ was written in the form $L = r^{-2}(L_1 + L_2)$, where L_2 is an operator on the half-line $0 \leqslant r < \infty$ and L_1 is an operator on the base T, which is a minimal $(n-1)$-submanifold of the unit sphere. The stability of $C(T)$ was thereby reduced to an eigenvalue problem on T [$\mathbf{S}$].

We are interested in studying a stable minimal hypersurface M^n of finite volume in the punctured ball $B^{n+1} \setminus \{0\}$. M has a tangent cone at the origin, which is itself stable; however, it is not known to be unique, and it need not have a smooth base. For example, the tangent cone might be a union of hyperplanes, the union of congruent copies of Simons' cone in $\mathbf{R}^8$, or all of $\mathbf{R}^{n+1}$. In any case, the method of separation of variables appears to entail serious difficulties on M itself. Instead, we propose to "renormalize" the ambient space $\mathbf{R}^{n+1} \setminus \{0\}$, by a process which converts the cone $C(T) \setminus \{0\}$ into the riemannian product $\mathbf{R} \times T$, where $\rho \in \mathbf{R}$ is defined by $\rho = \log(1/r)$. This process is effected by introducing the conformal metric $\tilde{g} = g/r^2$. The hypersurface M is *no longer minimal*, but retains an intrinsic vestige of *stability*. Encouraged by the existence of a tangent cone, one may hope for reasonably uniform behavior as $r \to 0$.

In this section, we collect a few basic formulas necessary for the program carried out in §§3 and 4.

2.1. CONFORMAL CHANGE LEMMA. *Let (M^n, g) be a riemannian manifold, and introduce the conformal metric $\tilde{g} := \varphi g$, where $\varphi: M \to \mathbf{R}$ is smooth and positive. Write $\psi = \log \varphi$. Then*

(a) *the scalar curvature $\tilde{\kappa}$ of $\tilde{g}$ is given by*

$$\varphi \tilde{\kappa} = \kappa - \frac{n-1}{2}\Delta\psi - \frac{(n-1)(n-2)}{8}|\nabla\psi|^2; \text{ and}$$

(b) *the Laplace–Beltrami operator $\tilde{\Delta}$ of $\tilde{g}$ satisfies*

$$\varphi\tilde{\Delta}u = \Delta u + \frac{n-2}{2}\langle \nabla u, \nabla\psi \rangle.$$

Here $\Delta u, |\nabla\psi|^2$, etc. refer to the original metric g.

Both parts of Lemma 2.1 follow by direct computation, for example, from formulas on p. 90 of [G-K-M]. Observe that we define scalar curvature κ as the sum of sectional curvatures in the *distinct* plane sections spanned by pairs of vectors in an orthonormal basis; this is one-half the quantity defined in [G-K-M]. In particular, when $n = 2$ the Gauss curvature $K = \kappa$.

2.2. COROLLARY. *Fix a real number β. With the same notation as above, the operator*

$$L_\beta := -\Delta + \beta\kappa$$

for the metric g and the corresponding operator $\tilde{L}_\beta$ for the conformal metric $\tilde{g}$ are related by the formula

$$\varphi\tilde{L}_\beta u = L_\beta u - \tfrac{1}{2}\Big[(n-2)\langle \nabla\psi, \nabla u \rangle + (n-1)\beta u \Delta\psi + \tfrac{1}{4}(n-1)(n-2)\beta u |\nabla\psi|^2\Big].$$

Suppose $M^n \subset \overline{M}^{n+p}$ is a regular immersed submanifold-with-boundary of a riemannian manifold $\overline{M}$, and write $\overline{\nabla}$ for the connection on $\overline{M}$. A vector field W along the immersion may be written in terms of orthogonal components: $W = W^T + W^N$, where W^T is tangent to M. Then for any orthonormal basis $e_1, \ldots, e_n$ of tangent vectors to M, we have the trace formula

$$(2.3) \qquad \langle \overline{\nabla}_{e_i} W, e_i \rangle = \mathrm{div}_M W^T - \langle H, W \rangle$$

directly from the definition of the mean curvature vector $H = (\overline{\nabla}_{e_i} e_i)^N$; where summation over $1 \leqslant i \leqslant n$ is indicated. If M is minimal, then it follows from the divergence theorem that

$$(2.4) \quad \int_M f\langle \overline{\nabla}_{e_i} W, e_i \rangle d\mathrm{Vol}_M = -\int_M W^T(f)\, d\mathrm{Vol}_M + \int_{\partial M} f\langle W, \nu \rangle d\mathrm{Vol}_{\partial M},$$

where ν is the unit exterior normal vector to ∂M. This formula remains valid when $M = S$ has interior singularities:

2.5. INTEGRATION BY PARTS LEMMA. *Let S be an n-dimensional integral current in $\overline{M}^{n+p}$ which is stationary for volume relative to ∂S. Write a given vector field W on $\overline{M}$ as $W = W^T + W^N$ a.e. $(\|S\|)$, where W^T lies in the span of the tangent n-vector $\vec{S}$, and W^N is orthogonal to $\vec{S}$. Assume that S is a C^1 manifold-with-boundary at those points of ∂S where $f \neq 0$. Then for any $\|S\|$-measurable choice $e_1,\ldots,e_n$ of orthonormal basis for $\vec{S}$,*

$$(2.6) \qquad \int f \langle \overline{\nabla}_{e_i} W, e_i \rangle d\|S\| = -\int W^T(f)\, d\|S\| + \int f \langle W, \nu \rangle d\|\partial S\|.$$

PROOF. We may write $f = f_0 + f_1$, where S is a C^1 manifold-with-boundary on $\text{suppt}(f_1)$ and where $\text{suppt}(f_0)$ is disjoint from ∂S. Then formula (2.6) holds with f_1 in place of f, since it reduces to equation (2.4), and since a stationary submanifold has $H \equiv 0$. Recall the first-variation formula for the volume of a rectifiable current:

$$\frac{d}{dt}\bigg|_{t=0} \mathbf{M}(\varphi_t(S)) = \int \langle \overline{\nabla}_{e_i} V, e_i \rangle d\|S\|,$$

where φ_t is the flow generated by the vector field V (see *e.g.* Theorem 3.9 of [$\mathbf{L_2}$]). With $V = f_0 W$, this first variation vanishes by hypothesis. But

$$\langle \overline{\nabla}_{e_i} V, e_i \rangle = f_0 \langle \overline{\nabla}_{e_i} W, e_i \rangle + W^T(f_0),$$

so that formula (2.6) holds with f_0 in place of f. The result now follows by additivity.   q.e.d.

In euclidean space $\mathbf{R}^{n+p}$, there is an especially useful choice $W = X = \frac{1}{2}\overline{\nabla}(r^2)$ for integration by parts, where r denotes the distance from the origin. (Similarly useful formulas use $r = $ distance to a k-plane, $k < p$). For this choice, X is the familiar position vector field, and one may compute $\overline{\nabla}_{e_i} X = e_i$ directly in standard coordinates.

2.7. COROLLARY. *Under the hypotheses of Lemma 2.5, in case $\overline{M} = \mathbf{R}^{n+p}$ and $W = X = \frac{1}{2}\overline{\nabla}(r^2)$, there holds*

$$(2.8) \qquad n\int f\, d\|S\| = -\int X^T(f)\, d\|S\| + \int f \langle X, \nu \rangle d\|\partial S\|.$$

2.9. LEMMA. *Suppose M^n is an immersed minimal submanifold of euclidean $\mathbf{R}^{n+p}$. Then the intrinsic gradient and Laplacian of the restriction of r to M satisfy:*

$$(2.10) \qquad\qquad\qquad \nabla(r^2) = 2X^T;$$

$$(2.11) \qquad\qquad\qquad \Delta(r^2) = 2n;$$

$$(2.12) \qquad\qquad \Delta(\log r) = r^{-2}\left(n - 2|\nabla r|^2\right) \geq (n-2)/r^2.$$

PROOF. Recall that $\nabla = (\overline{\nabla})^T$; (2.10) is immediate. Applying formula (2.3), we have

$$\Delta(r^2) = \text{div}_M(\nabla r^2) = 2\,\text{div}_M X^T = 2\langle \overline{\nabla}_{e_i} X, e_i \rangle = 2n.$$

Observe that $r|\nabla r| = |X^T| \leqslant |X| = r$, and, thus $|\nabla r| \leqslant 1$ (with equality if M is a cone). Finally, $n = \frac{1}{2}\Delta(r^2) = r\Delta r + |\nabla r|^2$. Therefore $r^2\Delta(\log r) = r\Delta r - |\nabla r|^2 = n - 2|\nabla r|^2$.   q.e.d.

2.13. COROLLARY. *Suppose M^n is an immersed minimal submanifold in euclidean* $\mathbf{R}^{n+p}$, *with the induced metric g and the conformal metric $\tilde{g} = g/r^2$. Then for any constant β,*

$$\tilde{L}_\beta = r^2 L_\beta + (n-2)X^T + (n-1)\left(n - \frac{n+2}{2}|\nabla r|^2\right)\beta.$$

3. Isolated singularities of two-dimensional surfaces. Consider an immersed surface M^2 in euclidean $\mathbf{R}^3$, having infinite topological type near the origin and no other boundary near the origin. Intuitively, M would need to have a great deal of negative Gauss curvature, on the average. In particular, this would seem likely to create negative eigenvalues for the stability operator $L_2 = -\Delta + 2K$. However, the geometry of the surface in a small neighborhood of 0 is virtually arbitrary, and it might seem possible for most of the negative Gauss curvature to be swallowed up by the singularity. It seems rather difficult, in general, to control the topology of an *incomplete* manifold by geometrical means.

The structure of M is greatly clarified by conformal *renormalization*, as suggested in §2. Specifically, we consider the conformal metric

(3.1) $$d\tilde{s}^2 = ds^2/r^2,$$

where r is the euclidean distance to 0, and ds^2 is the riemannian metric induced from $\mathbf{R}^3$. Then any curve γ which tends to 0 has infinite length, since $ds^2 \geqslant dr^2$ and thus

$$\int_\gamma d\tilde{s} = \int_\gamma \frac{1}{r}\,ds \geqslant \int_0^\varepsilon \frac{1}{r}\,dr = +\infty.$$

This shows that any end of M which is mapped to 0 is complete in the new metric. Of course the stability operator L_2 is dramatically changed. For a minimal surface, according to Corollary 2.13, we have

$$\tilde{L}_2 = r^2 L_2 + 4\left(1 - |\nabla r|^2\right) \geqslant r^2 L_2.$$

Thus if M is a stable minimal surface, then the intrinsic operator $\tilde{L}_2 = -\tilde{\Delta} + 2\tilde{K}$ is positive semi-definite. In view of future applications with variable ambient geometry, we consider instead the operator $L = L_1$, noting that $L_1 \geqslant L_2$, since $K \leqslant 0$ on a minimal surface. The same computation shows that $\tilde{L} = -\tilde{\Delta} + \tilde{K} \geqslant 0$. The operator $\tilde{L}$ will be our principal tool in proving the following result.

3.3. THEOREM. *Suppose M^2 is a (nonstrictly) stable minimal surface, properly immersed in the punctured euclidean ball $B_1 \setminus \{0\}$ in $\mathbf{R}^3$. Then $M \cup \{0\}$ is a branched minimal surface. If M is embedded, then $M \cup \{0\}$ is a stable embedded minimal surface. In particular, M has finite topological type near 0.*

It is remarkable that the finiteness of the topological type of M is a consequence of properties of its intrinsic geometry in the conformal metric (3.1). The following theorem will be applied with the metric $\tilde{g}$.

3.4. Intrinsic Theorem. *Let M^2 be a connected, real-analytic riemannian 2-manifold with Gaussian curvature K, such that the operator $-\Delta + K$ is positive semidefinite. Choose a compact subset Ω_0 with real-analytic boundary, and let D be the union of those components of $M \setminus \Omega_0$ whose closures are metrically incomplete. Then $M \setminus D$ has finite topological type and quadratic area growth.*

Remark. The conclusions of Theorem 3.4 continue to hold under the weaker hypothesis that $L_\beta = -\Delta + \beta K$ is positive on the complement of a compact subset of M, for some $\beta > 1/2$. The proof is unchanged: *cf.* Theorem 3.12, Corollary 3.17.

3.5. Corollary. *Let M be a connected riemannian 2-manifold such that $-\Delta + K$ is positive semidefinite. Let $\mathscr{E}_0$ be a closed and open subset of the set of ends of M. If $\mathscr{E}_0$ consists entirely of complete ends, then $\mathscr{E}_0$ is finite and each end in $\mathscr{E}_0$ has finite topological type.*

Proof of Intrinsic Theorem 3.4. Let Ω be a compact subset of M, consisting of an open set and its real-analytic boundary. We shall use different choices of Ω appropriate to different parts of the theorem: first, $\Omega = \Omega_0$ in order to bound the number of complete ends, and second, Ω will be a "bracelet" at a particular end, to show that the end has finite type.

Having chosen Ω, we designate M_+ to be the union of certain *complete* connected components of $M \setminus \Omega$, and write $M_- = M \setminus (\Omega \cup M_+)$. For some $d_0 > 0$, the points of M at distance $\leq d_0$ from Ω form a compact set. Write $s(x) = d(x, \Omega)$ and $\Omega(r) = \Omega \cup \{x \in M_+ : s(x) \leq r\}$. We shall exploit the hypothesis $-\Delta + K \geq 0$ with a family of test functions, parameterized by the constant $R > 0$:

$$f_R(x) = \begin{cases} \varphi(s(x)/R), & x \in M_+, \\ \varphi(s(x)/d_0), & x \in M_-, \\ 1, & x \in \Omega, \end{cases}$$

where $\varphi \colon \mathbf{R} \to [0, \infty)$ is a C^2 function with $\varphi(t) = 1$ for all $t \leq 0$, $\varphi(t) = 0$ for all $t \geq 1$, and $\varphi'(t) \leq 0$. The following argument is similar to Chapter 10 of [**C L**$_2$]; we repeat certain parts for the sake of clarity. The hypothesis $L \geq 0$ implies

$$(3.6) \qquad \int_M \left(|\nabla f_R|^2 + K f_R^2 \right) dA \geq 0.$$

Writing this integral in two terms according to $M = M_- \cup (\Omega \cup M_+)$, we observe that the integral over M_- is a constant C_1 independent of R. Since $\overline{M}_+$ is metrically complete, s is a proper Lipschitz function from $\overline{M}_+$ to $[0, \infty)$, with

$|\nabla s| = 1$ almost everywhere. We have

$$\int_{M_+} |\nabla f_R|^2 \, dA = \int_0^R R^{-2}(\varphi'(s/R))^2 L(s) \, ds,$$

where $L(s_0)$ is the length of the curve $\partial\Omega(s_0) \cap M_+ = \{x \in M_+ : s(x) = s_0\}$.

Similarly, we may write

$$\int_{M_+} K f_R^2 \, dA = \int_0^R \overline{K}(s)(\varphi(s/R))^2 \, ds,$$

where $\overline{K}(s_0)$ is the arc-length integral of Gauss curvature along the curve $\partial\Omega(s_0) \cap M_+$. Thus inequality (3.6) becomes

$$(3.7) \quad \int_0^R \left\{ R^{-2}(\varphi'(s/R))^2 L(s) + (\varphi(s/R))^2 \overline{K}(s) \right\} ds \geq -C_1 - \int_\Omega K \, dA.$$

Now $s: M_+ \to \mathbf{R}$ is a subanalytic function. In fact, its graph is the topological boundary of the subanalytic set $\{(\exp_x(t_1\nu(x)), t_2): x \in \partial\Omega \cap \overline{M}_+, t_2 \geq t_1 \geq 0\}$ of $M \times \mathbf{R}$, where $\nu(x)$ is the exterior unit normal to $\partial\Omega$ at x. It follows that each level set, when parameterized by arc length, is a Hölder continuous, piecewise $C^{1,\alpha} \cap W^{2,1+\alpha}$ curve for some $\alpha > 0$, and is real-analytic except on a finite set (see [**Hi** and **Ha**]). Moreover, the curves $\partial\Omega(s)$ are $C^{1,\alpha}$-isotopic to $\partial\Omega(s_0)$ for s in some interval about s_0, except for a discrete set of values of s_0. Finally, it follows in a straightforward fashion that $L(s)$ and $\overline{K}(s)$ are of class $C^\alpha([0, \infty))$.

We need to find a relation between the coefficients $L(s)$ and $\overline{K}(s)$ in inequality (3.7). Writing the Gauss–Bonnet formula for $\Omega(r)$ as an integral in s, we have

$$\int_\Omega K \, dA + \int_0^r \overline{K}(s) \, ds + \Gamma(r) - \Gamma_0 = 2\pi\chi(\Omega(r)),$$

where $\Gamma(r)$ is the integral of geodesic curvature of $\partial\Omega(r) \cap M_+$, plus the sum of exterior angles, and where $-\Gamma_0$ is the integral of geodesic curvature of $\partial\Omega \cap \overline{M}_-$. Differentiation with respect to r yields

$$(3.8) \qquad \overline{K}(s) = \frac{d}{ds}\left[2\pi\chi(\Omega(s)) - \Gamma(s)\right].$$

Observe that the difference $2\pi\chi(\Omega(s)) - \Gamma(s)$ is a $C^{1,\alpha}$ function, despite the discrete set of discontinuities for each term. When $\partial\Omega(s)$ is smooth, there holds $\Gamma(s) = dL(s)/ds$; in general, we have only the upper bound

$$(3.9) \qquad\qquad \Gamma(s) \geq \frac{dL(s)}{ds}$$

for the distributional derivative of L, with strict inequality occurring in the presence of a reentrant corner (negative exterior angle $\theta > 2\tan(\theta/2)$.)

After an integration by parts, using equation (3.8), the stability inequality (3.7) results in

$$-\Gamma_0 - C_1 \leq \int_0^R \left\{ R^{-2}(\varphi')^2 L - 2R^{-1}\varphi\varphi'[2\pi\chi(\Omega(s)) - \Gamma(s)] \right\} ds.$$

Here, certain boundary terms have been cancelled by way of the Gauss–Bonnet formula for Ω. Recalling inequality (3.9) and $\varphi\varphi' \leqslant 0$, we may replace $\Gamma(s)$ by $dL(s)/ds$ in this integral. A second integration by parts leads to

$$(\Gamma_0 + C_1)R^2 \geqslant \int_0^R L(s)\{2\varphi\varphi'' + (\varphi')^2\}\, ds + 4\pi R \int_0^R \chi(\Omega(s))\varphi\varphi'\, ds.$$

Let $\varphi(t)$ be a smooth approximation to $\varphi_0(t) = 1 - t$ $(0 \leqslant t \leqslant 1)$, $\varphi_0 \equiv 1$ on $(-\infty, 0]$, and $\varphi_0 \equiv 0$ on $[1, \infty)$. Then $-\varphi(t)\varphi''(t)\, dt$ is an approximate convolution identity, and it follows that
(3.10)

$$-4\pi R \int_0^R (1 - s/R)\chi(\Omega(s))\, ds \leqslant 2RL(0) - \int_0^R L(s)\, ds + (\Gamma_0 + C_1)R^2$$

$$\leqslant 2RL(0) + (\Gamma_0 + C_1)R^2.$$

We shall need to estimate the Euler number χ in terms of the *reduced Betti number*

$$\tilde{b}_1(\Omega(r)) = \operatorname{rank\,image}[H_1(\Omega(r)) \to H_1(M)] \leqslant b_1(\Omega(r))$$

(homology with integer coefficients). Observe that $\tilde{b}_1(\Omega(r))$ is a *monotone nondecreasing* function of r. Further, the unreduced Betti numbers $b_0(\Omega(r)) \leqslant b_0(\Omega)$, $b_2(\Omega(r)) = 0$, and hence $\chi(\Omega(r)) \leqslant b_0(\Omega) - \tilde{b}_1(\Omega(r))$.

By exploiting the monotonicity of $\tilde{b}_1(\Omega(r))$ in estimate (3.10) and integrating from $R/2$ to R, we obtain

$$\frac{\pi}{2}R^2\tilde{b}_1(\Omega(R/2)) \leqslant 4\pi R \int_0^R (1 - s/R)\tilde{b}_1(\Omega(s))\, ds$$

$$\leqslant (2\pi b_0(\Omega) + \Gamma_0 + C_1)R^2 + 2L(0)R.$$

Therefore, for all $R \geqslant 0$,

$$\tilde{b}_1(\Omega(R/2)) \leqslant 4b_0(\Omega) + 2(\Gamma_0 + C_1)/\pi + 4L(0)/\pi R.$$

Invoking the monotonicity of $\tilde{b}_1(\Omega(r))$ once again, we conclude that
(3.11)
$$\tilde{b}_1(\Omega(r)) \leqslant 4b_0(\Omega) + 2(\Gamma_0 + C_1)/\pi$$

for all $r \geqslant 0$.

We shall now make the specific choice of $\Omega = \Omega_0$, and choose M_- to include all incomplete components of $M \setminus \Omega$, in order to show that M_+ has only a finite number of ends. In fact, for each $r > 0$, write $e(r)$ for the number of unbounded components of $M_+ \setminus \Omega(r)$. Then if r is sufficiently large, $\Omega(r)$ is connected and $\tilde{b}_1(\Omega(r)) \geqslant e(r) - 1$. Namely, to each unbounded component M_i of $M_+ \setminus \Omega(r)$ there corresponds the homology class $\gamma_i \in H_1(M)$ of $\partial\Omega(r) \cap \overline{M_i}$. But these homology classes satisfy at most the relation $\gamma_1 + \cdots + \gamma_{e(r)} = 0$. The estimate (3.11) now yields a constant upper bound on $e(r)$ as $r \to +\infty$, and hence an upper bound on the number of ends of M_+. This shows that $M \setminus D$ has only a finite number of ends.

Next, we consider any end ε of M corresponding to $M \setminus D$. Since ε is isolated, we may choose a connected compact set Ω with analytic boundary and sufficiently large that ε corresponds uniquely to a component M_+ of $M \setminus \Omega$. Write

$M_- = M \setminus (\Omega \cup M_+)$, as usual. Inequality (3.11) now provides a constant upper bound for the reduced Betti number $\tilde{b}_1(\Omega(r))$ as $r \to \infty$, and hence for $b_1(M_+ \cup \Omega)$. Therefore $M_+ \cup \Omega$ has the homotopy type of a finite complex.

Finally, we observe that inequality (3.10) implies

$$\int_0^R L(s)\, ds = \text{Area}(\Omega(R) \setminus \Omega) < 2\pi R L(0) + (\Gamma_0 + C_1)R^2 + 2\pi R^2 b_0(\Omega),$$

since $\chi(\Omega(r)) \leqslant b_0(\Omega)$. Choosing Ω isotopic to Ω_0, this shows the quadratic area growth for $M \setminus D$.   q.e.d.

PROOF OF THEOREM 3.3. Consider the minimal surface $M^2 \subset \mathbf{R}^3$ as a riemannian 2-manifold with the conformal metric (3.1). As we have seen, the operator $\tilde{L} = -\tilde{\Delta} + \tilde{K}$ is positive semidefinite. Further, every end ε of M which is mapped to 0 is complete. We may now apply Theorem 3.4, choosing, for example, $\Omega = M \cap \overline{B}_{1/2} \setminus B_{1/4}$: we conclude that M has finite topological type in some neighborhood of 0. Thus, it remains only to show that M extends as a minimal surface across a single end which has the topological type of the annulus. This extension is well known (see [$\mathbf{O}_1$, pp. 78 ff.]); we sketch a proof in the interest of completeness.

In some neighborhood of the end, M has the conformal type of an annulus $A = \{w \in \mathbf{C}: \rho_1 < |w| < \rho_2\}$, where $0 \leqslant \rho_1 < \rho_2 < \infty$. In terms of the conformal parameterization, M is locally the image of a harmonic immersion $X: A \to \mathbf{R}^3$, which extends continuously to $\overline{A}$ by setting $X \equiv 0$ on the circle $\{|w| = \rho_1\}$. If $\rho_1 > 0$, then X may be extended as a harmonic function across the inner boundary circle. The extended mapping is conformal and defines a vector-valued analytic function $X_u - iX_v$. This implies that $X_u - iX_v$ has only isolated zeroes. But $X_u = X_v = 0$ on $\{|w| = \rho_1\}$, a contradiction. Therefore, $\rho_1 = 0$. Now X is a bounded harmonic function in the punctured disk A, and by Liouville's theorem X may be extended to a harmonic function in the disk $\{w: |w| < \rho_2\}$. The extended mapping is a branched minimal immersion by continuity.

Finally, suppose that M is embedded. Then the conformal parameterization $X: A \to \mathbf{R}^3$ is one-to-one. It follows that the extended mapping $X: A \cup \{0\} \to \mathbf{R}^3$ cannot have a branch point at $w = 0$. For a true branch point would be accompanied by a curve of self-intersection of M, while a false branch point would entail the coincidence of the images of two disjoint open sets (Theorem 3.1 of [$\mathbf{G}_1$]). Therefore, $M \cup \{0\}$ is an embedded minimal surface, and stability follows from Corollary 1.10.   q.e.d.

In light of the intrinsic character of Theorem 3.4, it is of interest to investigate the consequence of the stability inequality $L = -\Delta + K \geqslant 0$ for a *complete* riemannian 2-manifold M, without assuming any minimal immersion. More generally, we consider the operator $L_\beta = -\Delta + \beta K$, with a constant $\beta > \frac{1}{2}$. The proof of Theorem 3.4 shows that if L_β is positive semidefinite, then M has finite topological type. However, much stronger conclusions hold:

3.12. THEOREM. *Let M be a complete, connected, noncompact real-analytic riemannian 2-manifold such that $L_\beta \geqslant 0$ for some $\beta > \frac{1}{2}$. Then M has quadratic* *area growth and the topological type of the plane, the cylinder or the Möbius band. If M is simply connected and $K \leqslant 0$, then the total absolute curvature of M is less than $\pi/(\beta - \frac{1}{2})$. If M is a cylinder or Möbius band and $K \leqslant 0$, then M is flat.*

PROOF. As in the proof of Theorem 3.4, we choose a compact set $\Omega \subset M$ with analytic boundary. Since M is complete, we may choose M_- to be empty. Now suppose that there is an *upper* bound on Euler number: $\chi(\Omega(R)) \leqslant \chi_0$ for all $R \geqslant 0$. Let the test functions f_R be defined as before: $f_R(x) = \varphi(s(x)/R)$, where $\varphi(t)$ is a smooth monotone approximation to $\varphi_0(t) = \min\{1, \max(1 - t, 0)\}$. The hypothesis $L_\beta \geqslant 0$ implies

(3.13)

$$0 \leqslant \int_M \left(|\nabla f_R|^2 + \beta K f_R^2 \right) dA = \beta \int_\Omega K \, dA + \int_0^R \left\{ R^{-2}(\varphi')^2 L(s) + \beta \varphi^2 \overline{K}(s) \right\} ds$$

$$\leqslant 2\pi\beta\chi_0 - R^{-2} \int_0^R L(s)\left\{ (2\beta - 1)(\varphi')^2 + 2\beta\varphi\varphi'' \right\} ds$$

by means of a twofold integration by parts, using equation (3.8), inequality (3.9) and the Gauss–Bonnet formula with $\Gamma_0 = 0$. As a consequence, we obtain the area growth estimate

$$(3.14) \quad \text{Area}(\Omega(R) \setminus \Omega) = \int_0^R L(s) \, ds \leqslant \frac{2\beta}{2\beta - 1}\left(L(0) + \pi\chi_0 R^2 \right).$$

Now suppose that M has $\geqslant e$ ends and genus $\geqslant g$. If we replace Ω by $\Omega(R_0)$ for some sufficiently large R_0, then we may assume that for all R, $\Omega(R)$ has at least e boundary components and genus $\geqslant g$, so that $\chi(\Omega(R)) \leqslant 2 - 2g - e =: \chi_0$. Therefore, if $e \geqslant 3$ then $\chi_0 \leqslant -1$, contradicting inequality (3.14). Similarly, noncompactness implies $e \geqslant 1$ so that $g \geqslant 1$ would lead to a contradiction. If M is orientable, this shows that M is either the plane or the cylinder. Finally, in the nonorientable case we must have $\chi_0 \geqslant 0$, which implies that M is the Möbius band.

We now assume that $K \leqslant 0$ and return to inequality (3.13). Using the same test functions f_R but without integration by parts, we find

$$\beta \int_\Omega |K| dA + \beta \int_0^R |\overline{K}(s)|(1 - s/R)^2 \, ds \leqslant R^{-2} \int_0^R L(s) \, ds$$

$$\leqslant \frac{2\beta}{2\beta - 1}\left[L(0)/R^2 + \pi\chi_0 \right].$$

Letting $R \to \infty$, we find

$$\int_M |K| dA \leqslant \frac{2\pi\chi_0}{2\beta - 1}.$$

If M is simply connected, we may take $\chi_0 = 1$. Otherwise, $\chi_0 = 0$ and the conclusions follow.   q.e.d.

It is instructive to compare Theorem 3.12 with Theorem 3 of [**FC-S**], where the conclusions partially overlap and quite different methods are used. They share the following corollary, which was also proved in [**dC-P**]:

3.15. COROLLARY. *A complete, stable minimal surface immersed in euclidean* $\mathbf{R}^3$ *must be a flat plane.*

PROOF. We apply Theorem 3.12 with $\beta = 2$, and recall that the total curvature of a complete minimal surface in $\mathbf{R}^3$ is either infinite or an integer multiple of 4π [$\mathbf{O}_1$, Theorem 9.2].   q.e.d.

By applying the Removability Theorem 3.3, we may further strengthen Corollary 3.15 as follows:

3.16. COROLLARY. *Let S be a discrete subset of euclidean* $\mathbf{R}^3$. *If M is a stable minimal surface, properly embedded in* $\mathbf{R}^3 \setminus S$, *then M is a flat plane minus a subset of S.*

Doris Fischer-Colbrie has conjectured that a complete minimal surface in $\mathbf{R}^3$ has finite index if and only if it has finite total curvature. Half of this conjecture is contained in the following Corollary and Remark; the converse was recently proved by the first author [$\mathbf{G}_4$].

3.17. COROLLARY. *Let* M^2 *be a complete immersed surface in* $\mathbf{R}^3$, *such that for some compact subset F,* $M \setminus F$ *is minimal and stable. Then M has finite topological type, quadratic area growth and finite total curvature.*

PROOF. The quadratic area growth follows from the Intrinsic Theorem 3.4, applied to $M \setminus F$. Further, inequalities (3.7) and (3.10) yield

$$-\int_\Omega K\, dA - \int_0^R \overline{K}(s)(1 - s/R)^2\, ds \leqslant \frac{2}{R} L(0) + 2\pi b_0(\Omega) + \Gamma_0 + 2C_1.$$

Here Ω is a compact set in $M \setminus F$ which separates F from the ends of M. Since $K \leqslant 0$ on $M \setminus F$, this inequality converges as $R \to \infty$ and implies

$$\int_M |K|\, dA \leqslant \int_{M_-} |K|\, dA + 2\pi b_0(\Omega) + \Gamma_0 + 2C_1. \quad \text{q.e.d.}$$

3.18. REMARK. If a minimal surface M has finite index, then $M \setminus F$ is stable for some compact subset F. In particular, Corollary 3.17 and Removability Theorem 3.3 continue to hold if M is not assumed to be stable, but only to have finite index. This may be seen as follows.

For a noncompact surface M, the *index* of M is the maximum index of its compact subsets. For a compact subset D of M, the *index* k of D is the maximum number of nontrivial, pairwise $L^2(D)$-orthogonal functions $u_1, \ldots, u_k$ with support in D, such that

$$\int_D \left(|\nabla u_j|^2 + 2K u_j^2 \right) dA < 0.$$

We may assume that each u_j is an eigenfunction for the Dirichlet problem $L_2 u = 0$ inside D, with eigenvalue λ_j, and vanishes outside D, where $\lambda_1 < \lambda_2 \leqslant \cdots \leqslant \lambda_k < 0$. Now given a noncompact surface M with finite index k, let F be a compact subset having index k. We claim that $M \setminus F$ is stable. Otherwise, there is a function u_0 whose support is a compact subset D of $M \setminus F$, with

$$\int_D \left(|\nabla u_0|^2 + 2Ku_0^2 \right) \, dA < 0.$$

But $u_0, u_1, \ldots, u_k$ are pairwise orthogonal functions in $L^2(D \cup F)$, implying that M has index $\geq k + 1$, a contradiction.

The same proof is valid for an arbitrary selfadjoint elliptic operator in a manifold of arbitrary dimension. As a consequence, many of the results proved in the following three sections continue to hold if the hypothesis of stability is replaced by finite index; these include Corollary 4.4 and Theorems 5.1, 5.4, 5.5, and 6.4.

4. Isolated singularities on minimal hypersurfaces. In this section we shall study isolated singularities on minimal hypersurfaces. It is known in general that any minimal variety of dimension n extends naturally as a minimal variety across a submanifold of dimension $\leq n - 2$ (see [**H-L**]). Rather straightforward calculations show that if the original variety is stable so is its extension. Specifically, this means the following. Let S be a closed q-dimensional submanifold of a riemannian manifold M. Let T be a minimal variety (i.e., a stationary integral current without boundary) of dimension $n \geq q + 2$ in $M \setminus S$. Then the mass of T is finite in any neighborhood of a point on S, and the extension of T to M is again minimal. Furthermore, if T is stable, so is its extension.

While such extensions exist in the domain of integral currents, the geometric behavior of the variety near such a "bad" point could be quite complicated. The point of this chapter will be to find some restrictions on this behavior. We begin with the basic case of a smooth hypersurface in $\mathbf{R}^n$.

Let $B^N = \{ x \in \mathbf{R}^N \colon \|x\| \leq 1 \}$ be the *closed* unit ball in euclidean space. We want to consider a properly immersed submanifold $f \colon M^p \to B^N \setminus \{0\}$ which is transversal to ∂B^N. Note that M^p is a manifold with compact boundary $\partial M^p = f^{-1}(\partial B^N)$. However M^p itself is *not necessarily compact*. In what follows, we shall neglect the transversality hypothesis, since by Sard's Theorem M^p is transversal to $\partial B^N(r)$ for almost all r.

A riemannian metric on M^p will be called *complete* if the induced distance makes M^p a complete metric space.

A riemannian metric on M^p is said to have *uniformly positive scalar curvature* if the scalar curvature function κ satisfies $\kappa \geq \kappa_0$ on M^p for some $\kappa_0 > 0$. Our main result is the following.

4.1. THEOREM. *Let M^n be a stable minimal hypersurface properly immersed in $B^{n+1} \setminus \{0\}$ for $n \geq 3$. Then $M^n \times S^1$ carries a complete riemannian metric of uniformly positive scalar curvature.*

4.2. NOTE. This metric will be a warped product of S^1 with a metric on M which is conformally equivalent to the one induced by the immersion.

4.3. NOTE. Suppose M^n is a (punctured) cone, i.e., suppose $M^n = C(N^{n-1}) = \{tx \in \mathbf{R}^{n+1}: x \in N^{n-1} \text{ and } 0 < t \leqslant 1\}$ where N^{n-1} is a compact submanifold of S^n. Then the proof shows directly that the base of the cone N^{n-1} carries a metric of positive scalar curvature. This was originally an observation of R. Schoen [**Sc**]. It gives as a corollary, a result of Almgren [**A**] for $n = 3$.

From the wealth of results on complete metrics of positive scalar curvature (cf. [**G-L$_2$**], [**S-Y**]), we can make substantial conclusions about the topology of the singularity.

Let M^n be as above. Then, as in [**G-L$_2$**], the manifold M^n is said to be a *bad end*, if there is a map $M^n \to K^{n-1}$ onto a compact enlargeable manifold so that the composition $\partial M^n \to M^n \to K^{n-1}$ has nonzero degree. The category of enlargeable manifolds is discussed in [**G-L$_1$, G-L$_2$**]. It includes all compact manifolds which carry metrics of non-positive sectional curvature, and all compact solvmanifolds. It is stable under products and under connected sum with arbitrary manifolds. As an example, the manifold $T^{n-1} \times [0, \infty)$ is a bad end, where T^{n-1} is the $(n-1)$-torus. The same is true of $T^{n-1} \times [0, \infty) \sharp X^n$ where X^n is any compact n-manifold.

The manifold M^n is said to *have a bad end* if there is a proper, noncompact n-dimensional submanifold with boundary $M_0^n \subset M^n$ so that M_0 is a bad end. Using [**G-L$_2$, §7**] we shall prove the following.

4.4. COROLLARY. *Suppose M^n admits a stable minimal immersion into $B^{n+1} \setminus \{0\}$. Then M^n has no bad ends.*

We shall explore further consequences of our main techniques in subsequent sections.

PROOF OF THEOREM 4.1. To any riemannian metric on a manifold, we can associate the family of operators

$$(4.5) \qquad\qquad L_\beta := -\Delta + \beta\kappa$$

for $\beta \geqslant 0$. (Recall our convention that $\kappa = -\sum_{i<j} R_{ijij} = \frac{1}{2}\kappa'$ where κ' is the scalar curvature function defined, for example, in [**G-L$_2$**].) These operators have the following elementary property.

4.6. LEMMA. *If $L_\beta \geqslant 0$, then $L_{\beta'} \geqslant 0$ for $0 \leqslant \beta' \leqslant \beta$.*

PROOF. Assume $\beta' > 0$, since $L_0 = -\Delta \geqslant 0$. Then

$$L_{\beta'} = \frac{\beta'}{\beta}\left[-\frac{\beta}{\beta'}\Delta + \beta\kappa \right] = \frac{\beta'}{\beta}\left[-\left(\frac{\beta}{\beta'} - 1\right)\Delta + L_\beta \right] \geqslant 0. \quad \text{q.e.d.}$$

Suppose now that ds^2 is the riemannian metric on M^n induced by the immersion $M^n \to B^{n+1} \setminus \{0\}$, and let L_β be its associated family of operators.

As in §2 we consider the conformally equivalent, complete metric

$$(4.7) \qquad\qquad d\tilde{s}^2 := ds^2/r^2,$$

where $r = \|X\|$ is the radial distance function in B^{n+1}. We let $\tilde{L}_\beta = -\tilde{\Delta} + \beta\tilde{\kappa}$ denote its associated family of operators.

It will suffice to prove that $\tilde{L}_1 \geqslant C$ for some constant $C > 0$. This is for the following reason. If $\tilde{L}_1 \geqslant C$, then there exists a smooth function $f > 0$ on M with $\tilde{L}_1(f) \geqslant Cf$ (cf. [**Ka, FC-S**]). Consider the warped product metric $ds_f^2 = d\tilde{s}^2 + f^2\, d\theta^2$ on $M^n \times S^1$. This metric has scalar curvature $\kappa_f = (1/f)\tilde{L}_1(f) \geqslant C$ (see [**G-L$_2$**, §11] for example.)

The theorem will now follow directly from the next proposition. For each integer $n > 1$, set

$$(4.8) \qquad \beta_n \equiv \frac{1}{2}\left(\frac{n-2}{n-1}\right).$$

4.9. PROPOSITION. *For any β satisfying $\beta_n \leqslant \beta \leqslant 1$, we have*

$$\tilde{L}_\beta \geqslant \frac{(n-2)(n-1)}{2}(\beta - \beta_n).$$

NOTE. It is interesting that $(n-2)(n-1)/2$ is exactly the scalar curvature of S^{n-1} and that β_n is the constant that makes the operator L_{β_n} conformally invariant (in dimension n).

PROOF. Let $\varphi \in C_0^\infty(M)$ have compact support in $M - \partial M$, and set

$$(4.10) \qquad \varphi =: r^{n/2-1}\psi.$$

Then a straightforward calculation using Lemma 2.9 and Corollary 2.13 shows that for $\beta \geqslant \beta_n$,

(4.11)

$$\int_M (\tilde{L}_\beta\varphi)\varphi\, d\tilde{v} = \int_M (L_\beta\psi)\psi\, dv + \frac{(n+2)}{2}\left[\frac{(n-2)}{2} - \beta(n-1)\right]\int_M |\nabla r|^2\varphi^2\, d\tilde{v}$$

$$+ \left[\beta n(n-1) - \frac{1}{2}n(n-2)\right]\int_M \varphi^2\, d\tilde{v}$$

$$\geqslant \int_M (L_\beta\psi)\psi\, dv + \frac{(n-2)}{2}\left[\beta(n-1) - \frac{(n-2)}{2}\right]\int_M \varphi^2\, d\tilde{v}$$

$$= \int_M (L_\beta\psi)\psi\, dv + \frac{(n-2)(n-1)}{2}(\beta - \beta_n)\int_M \varphi^2\, d\tilde{v}.$$

Here dv denotes the riemannian volume element for ds^2, and $d\tilde{v} = r^{-n}\, dv$ denotes the one for $d\tilde{s}^2$.

The stability of M^n in B^{n+1} implies that $L_1 \geqslant 0$. Hence, using Lemma 4.6 we have

$$\int_M (L_\beta\psi)\psi\, dv \geqslant 0$$

for all $\psi \in C_0^\infty(M \setminus \partial M)$ and all β, $0 \leqslant \beta \leqslant 1$. The proposition and the theorem now follow immediately from (4.11).

PROOF OF NOTE 4.3. If $M^n = C(N^{n-1})$ is a cone, then the metric $d\tilde{s}^2$ on $N^{n-1} \times [0, \infty)$ is a product metric $d\tilde{s}^2 = d\tilde{\tilde{s}}^2 + d\rho^2$, where $d\tilde{\tilde{s}}^2$ is the induced metric on $N^{n-1} \subset S^n$ and where $\rho = \log(1/r)$. The associated operators satisfy

$$\tilde{L}_\beta = \tilde{\tilde{L}}_\beta - \frac{\partial^2}{\partial\rho^2}.$$

It follows very easily that: $\tilde{L}_\beta \geq C \Rightarrow \tilde{\tilde{L}}_\beta \geq C$. We now apply the following well known fact (cf. [**K-W**]).

4.12. THEOREM. *Let ds^2 be a riemannian metric on a compact n-dimensional manifold. If $L_{\beta_n} > 0$, then ds^2 is conformally equivalent to a metric of positive scalar curvature.*

4.13. REMARK. It is tempting to try to prove directly that the metric induced on M^n can be conformally deformed to a complete metric of positive scalar curvature by using the fact that $\tilde{L}_1 \geq C > 0$. Using arguments of J. Kazdan [**Ka**] this could be done if $\tilde{\kappa}$ were nonnegative outside a compact set. Nevertheless, all topological and metric consequences of the existence of such a metric on M will also follow from Theorem 4.1.

PROOF OF COROLLARY 4.4. If M^n has a bad end (say M_0^n), then $M^n \times S^1$ has a bad end (namely, $M_0^n \times S^1$). Hence, we want to apply Theorem 7.46 of [**G-L$_2$**] to the complete metric constructed above on $M^n \times S^1$. There are two additional hypotheses in this theorem. The manifold M^n must be spin. Since M^n is immersible in $\mathbf{R}^{n+1}$, it will be spin (and in fact stably parallelizable) provided it is orientable, which we may assume to be so. The second hypothesis is that the metric on M^n have Ric $\geq -C^2$ for some constant C. This is only needed to provide a proper exhaustion function $F\colon M \to \mathbf{R}^+$ with $|\nabla F| + \Delta F \leq$ constant. However, $F \equiv \log(1/r)$ is such a function. So the proof in [**G-L$_2$**] applies directly. q.e.d.

5. The case of 3-dimensional hypersurfaces. We now specialize the discussion of section 4 to the case where $n = 3$. Throughout this section, let $M^3 \to B^4 \setminus \{0\}$ be a properly immersed, stable minimal 3-manifold (as in Theorem 4.1) and let $d\tilde{s}^2 = (1/r^2) ds^2$ be the complete metric, conformal to the induced one, which was considered above. Our main point here is that the riemannian manifold $(M^3, d\tilde{s}^2)$ enjoys nearly all the global properties established in [**G-L$_2$**, §10] for complete 3-manifolds with scalar curvature ≥ 1. For example, we shall prove the following.

5.1. THEOREM. *Let $\Omega \subset M^3$ be a compact domain with smooth boundary in M^3, and let $\gamma \subset \Omega$ be a closed curve which bounds in $(\Omega, \partial\Omega)$. If distance $(\gamma, \partial\Omega)$ (in the conformal metric) $> 8\pi$, then γ bounds in Ω.*

This leads us to the following concept formulated in [**G-L$_2$**].

5.2. DEFINITION. A circle γ embedded in M^3 is said to be *small* if $[\gamma]$ has infinite order in $H_1(M^3; \mathbf{Z})$ and if the normal circle to γ has infinite order in $M \setminus \gamma$. This second condition means that γ does not meet a closed surface in M^3 with nonzero intersection number; for if it did, by removing a small disk from the surface at each intersection point, we could construct a surface bounding a multiple of the normal circle in $M \setminus \gamma$.

5.3. THEOREM. *Neither M^3 nor any covering space of M^3 can carry a small circle placed at a distance $\geq 8\pi$ from the boundary.*

Thus, for example, M^3 cannot have a connected summand of $K(\pi, 1)$-type placed at a distance $\geq 8\pi$ from ∂M^3.

5.4. THEOREM. *If $\pi_1 M^3$ is finitely generated, then M^3 is simply-connected at infinity.*

5.5. THEOREM. *If M^3 is simply-connected, then there exists a distance nonincreasing map $p: M^3 \to \Gamma$, where Γ is a metric graph, and a constant C, such that*

$$\operatorname{diam}\left[p^{-1}(x)\right] \leq C$$

for all $x \in \Gamma$.

NOTE. The map p will collapse ∂M to a single "boundary point" x_0 on Γ. The constant C can be taken to be 48π for all x at a distance $\geq 8\pi$ from x_0.

From this we shall prove the following corollary.

5.6. COROLLARY. *Suppose M^3 is simply connected. Let ds^2 be the metric on M^3 induced by the stable minimal immersion $M^3 \to B^4 \setminus \{0\}$, and let $r = \|X\|$ be the radial distance function in B^4. Then each connected component of $M_t^3 \equiv r^{-1}(t)$ has intrinsic diameter $\leq 48\pi t$ for all t sufficiently small.*

PROOF OF THEOREM 5.1. We recall that by Theorem 4.1 there is a (complete) warped product metric

$$(5.7) \qquad d\tau^2 = d\tilde{s}^2 + f^2\, d\theta^2$$

on $M^3 \times S^1$ with scalar curvature $\geq 3/4$. Observe that rotations of the S^1-factor give a 1-parameter group of isometries $G \cong SO(2)$.

Consider the surface $\gamma \times S^1$ in the domain $\Omega \times S^1 \subset M^3 \times S^1$. We claim that $\gamma \times S^1$ bounds a stable minimal hypersurface of the form $\Sigma \times S^1$ in $\Omega \times S^1$ relative to the boundary $\partial \Omega \times S^1$. This can be proved in one of two ways. The first method is to solve the Plateau Problem for $\gamma \times S^1$ in $(\Omega \times S^1, \partial \Omega \times S^1)$. We choose the solution homologous to $\Sigma_0 \times S^1$, where $\Sigma_0 \subset \Omega$ is a surface with $\partial \Sigma_0 = \gamma \pmod{\partial \Omega}$. Applying results of [**Gao**], we know that the solution H is either G-invariant, or it has the property that the vector field $\partial/\partial\theta$ (which generates the G-action) satisfies $\langle \partial/\partial\theta, v \rangle > 0$ where v is the unit normal field to H. This second alternative is not possible for the following topological reasons. The hypersurface H defines a relative homology class in the manifold $(\Omega \setminus U) \times S^1$, where U is a tubular neighborhood of γ in M^3. Under the second conclusion, the orbits of the S^1-action meet H with positive intersection number. This contradicts the choice of the homology class of H.

Since H is G-invariant, it is of the form $\Sigma \times S^1$ where Σ is a surface in Ω with $\partial\Sigma = \gamma \pmod{\partial \Omega}$. The metric induced on $\Sigma \times S^1$ is a warped product. Now as in [**G-L$_2$**, §11], the stability of $\Sigma \times S^1$ in $M^3 \times S^1$ implies that there is a warped product metric $d\tau_1^2$ on $(\Sigma \times S^1) \times S^1$ with scalar curvature $\geq 3/4$. This metric has the form

$$(5.8) \qquad d\tau_1^2 = ds^2 + f_1^2(x)\, d\theta_1^2 + f_2^2(x, \theta_1)\, d\theta_2^2$$

where ds^2 is the metric on Σ and where f_2 is the first eigenfunction of the stability operator. Since the stability operator is G-invariant, we may assume f_2 to be also G-invariant, i.e., to be independent of θ_1.

We now make the following claim.

5.9. CLAIM. *No point of Σ lies at a distance $\geqslant 4\pi$ from $\partial\Sigma$.*

PROOF OF CLAIM. Fix $p_0 \in (\Sigma \setminus \partial\Sigma)$ and solve the Plateau Problem for the boundary $p_0 \times S^1 \times S^1$ in $(\Sigma \times S^1 \times S^1, \partial\Sigma \times S^1 \times S^1)$. Arguing exactly as before we find that the solution is $SO(2) \times SO(2)$-invariant, and that stability implies that there is a warped product metric of the form

$$d\tau_2^2 = dt^2 + f_1^2(t)\, d\theta_1^2 + f_2^2(t)\, d\theta_2^2 + f_3^2(t)\, d\theta_3^2,$$

with $\kappa \geqslant 3/4$ on the manifold $[0, l] \times S^1 \times S^1 \times S^1$. Note that the length of the "base" l satisfies $l \leqslant \mathrm{dist}(p_0, \partial\Sigma)$. As in [**G-L$_2$**, §12], the condition $\kappa \geqslant 3/4$ gives a differential inequality which implies that $l \leqslant 4\pi$. (The curvature used in [**G-L$_2$**, §12] was normalized differently; the reader should translate $\kappa \geqslant 3/4$ to be $\kappa \geqslant 1/4$ in that context.) This proves the claim, and the claim proves the theorem.

The second method of proof is much like the first except that one directly solves the *equivariant* Plateau Problem in $\Omega \times S^1$ (see [**L$_1$**] for example.) This is equivalent to directly solving the Plateau Problem for γ in $(\Omega, \partial\Omega)$ with the modified metric

$$(5.10) \qquad\qquad\qquad d\hat{s}^2 = f\, d\tilde{s}^2$$

(compare (5.7)). The solution will be a surface Σ which is stable in the metric (5.10). The stability operator $\hat{L}$ on Σ lifts to become the stability operator L on $\Sigma \times S^1$ in $\Omega \times S^1$ (restricted to the G-invariant functions). This is evident from the definition of $\hat{L}$ and L in terms of the second variation. Now since $\hat{L} > 0$, there is a function $\hat{f} > 0$ so that $\hat{L}\hat{f} > 0$. Hence, $Lf > 0$ where f is the lift of $\hat{f}$ to $\Sigma \times S^1$. We can now continue the warping process using the function f ($= f_2$ in (5.8)). The remainder of the proof proceeds as above.   q.e.d.

PROOF OF THEOREM 5.3. This follows from Theorem 5.1 exactly as in [**G-L$_2$**, 10.13**]**.

PROOF OF THEOREM 5.4. Again we follow [**G-L$_2$**, §10]. Let $K \subset M^3$ be a compact set which carries $\pi_1(M^3)$. Let Ω_0 be a neighborhood of infinity and let $\Omega \subset \Omega_0$ be any subneighborhood such that $\mathrm{dist}(\Omega, \partial\Omega_0 \cup K) > 16\pi$. We shall show that the map $\pi_1(\Omega) \to \pi_1(\Omega_0)$ is zero. Consider an embedded curve $\gamma \subset \Omega$. This curve is homotopic to a curve $\gamma' \subset K$. If $[\gamma'] \neq 0$ in $\pi_1 M$ we can solve the Plateau Problem for a surface of the type of the annulus with $\gamma \cup \gamma'$ as boundary (following Morrey [**Mo**].) We do this in the riemannian manifold $(M, f\, d\tilde{s}^2)$, appropriately compactified far away from γ. (This amounts to solving the equivariant Plateau Problem in $M \times S^1$, as observed above.) The basic argument above applies to show that no point on the solution surface can be more than a distance 8π from the boundary. Since $\mathrm{dist}(\gamma, \gamma') > 16\pi$, we have a contradiction, and we conclude that γ bounds in M. Solving now for a minimal disk-like surface with boundary γ, we apply the same estimate to conclude that this disk is contained in Ω_0. This completes the proof.   q.e.d.

PROOF OF THEOREM 5.5. We follow the argument given in [G-L$_2$, 10.11 and Carr] except that d, instead of being distance from a point, is distance to ∂M^3.

PROOF OF COROLLARY 5.6. One repeats the above argument with d replaced by the function r.

6. Generalizations. The results established above can be carried over directly to a more general setting. For example, one can consider extending minimal varieties across linear subspaces, or more generally, submanifolds of $\mathbf{R}^{n+1}$.

6.1. THEOREM. *Let $M^n \to \mathbf{R}^{n+1} \setminus \mathbf{R}^p$ be a properly immersed, stable minimal hypersurface, where $0 \leqslant p < \frac{1}{2}n - 1$. Then $M^n \times S^1$ carries a complete metric of uniformly positive scalar curvature.*

PROOF. With respect to the orthogonal decomposition $\mathbf{R}^{n+1} = \mathbf{R}^{n-p+1} \oplus \mathbf{R}^p$, we write any vector X as $X = (X_0, X_1)$, and we consider the function $r_0^2 = \|X_0\|^2$. (Note that r_0 is just the distance to $\{0\} \times \mathbf{R}^p$ in $\mathbf{R}^{n+1}$.)

We now proceed exactly as before. Let ds^2 be metric induced on M^n by the immersion, and consider the conformally equivalent, complete metric

$$d\tilde{s}^2 = ds^2/r_0^2.$$

Consider the two associated operators L_β and $\tilde{L}_\beta$. Using the easily verified facts that on M,

$$(6.2) \qquad \Delta(r_0^2) \geqslant 2(n - p), \qquad \nabla(r_0^2) = 2X_0^T,$$

we proceed exactly as before to prove that for $\beta_n \leqslant \beta < 1$,

$$(6.3) \qquad \tilde{L}_\beta \geqslant \frac{(n-1)(n-2-2p)}{2}(\beta - \beta_n).$$

The details are straightforward and are left to the reader. It now follows immediately from, say [G-L$_2$, 11.14], that $M \times S^1$ carries a complete metric of uniformly positive scalar curvature.   q.e.d.

We also have the following general result. Let $\Sigma^p \subset \overline{M}^{n+1}$ be an embedded, compact p-dimensional submanifold of a riemannian manifold $\overline{M}^{n+1}$, and let U be any compact neighborhood of Σ^p in $\overline{M}^{n+1}$.

6.4. THEOREM. *Let $M^n \to U \setminus \Sigma^p$ be a properly immersed, stable minimal hypersurface, and suppose $0 \leqslant p < \frac{1}{2}n - 1$. Then $M^n \times S^1$ carries a complete metric with uniformly positive scalar curvature at infinity. In particular, M has no bad ends.*

PROOF. It suffices to consider U to be a small tubular neighborhood of Σ^p where the distance function $d(x) \equiv \text{dist}(x, \Sigma^p)$ is defined and smooth. Direct calculations in Fermi coordinates show that on the submanifold M^n, we have

$$(6.5) \qquad \Delta(d^2) \geqslant 2(n - p) + O(d^2), \qquad \nabla(d^2) = 2dn^T,$$

where $n := \mathrm{grad}(d)$. We then consider the metric $d\tilde{s}^2 = (1/d^2)\,ds^2$ on M^n. The calculation now proceeds formally as before, with (6.2) replaced by (6.5) to show that

$$\tilde{L}_\beta \geqslant \tfrac{1}{2}(n-1)(n-2-2p) + O(d^2).$$

The argument is completed, as always, by [G-L$_2$, 11.14].   q.e.d.

References

[A] F. Almgren, *Some interior regularity theorems for minimal surfaces and an extension of Bernstein's theorem*, Ann. of Math. (2) **84** (1966), 277–292.

[B] R. Böhme, *Die Jacobifelder zu Minimalflächen im $\mathbf{R}^3$*, Manuscripta Math. **16** (1975), 51–73.

[Carr] Carr, *Manifolds of positive scalar curvature, Yang-Mills fields and the Kalusa-Klein model*, Thesis, S.U.N.Y., Stony Brook, 1984.

[dC-P] M. do Carmo and C. K. Peng, *Stable complete minimal surfaces in R^3 are planes*, Bull. Amer. Math. Soc. (N.S.) **1** (1979), 903–906.

[de G-S] E. deGiorgi and G. Stampacchia, *Sulle singolarità eliminabili delle ipersuperficie minimali*, Atti Accad. Naz. Lincei Rend. Cl. Sci. Fis. Mat. Natur. **38** (1965), 352–357.

[D$_1$] J. Douglas, *Minimal surfaces of general topological structure with any finite number of assigned boundaries*, J. Math. Phys. **15** (1936), 105–123.

[D$_2$] ______, *Minimal surfaces of higher topological structure*, Ann. of Math. (2) **40** (1939), 205–298.

[Fi] F. Fiala, *Le problème des isoperimètres sur les surfaces ouvertes à courbure positive*, Comment. Math. Helv. **13** (1941), 293–346.

[FC-S] D. Fischer-Colbrie and R. Schoen, *The structure of complete stable minimal surfaces in 3-manifolds of nonnegative scalar curvature*, Comm. Pure Appl. Math. **33** (1980), 199–211.

[F] W. Fleming, *An example in the problem of least area*, Proc. Amer. Math. Soc. **7** (1956), 1063–1074.

[Gao] Z. Gao, *Applications of minimal surface theory to topology and riemannian geometry: constructions of negatively Ricci-curved manifolds*, Thesis, S.U.N.Y., Stony Brook, 1983.

[G-K-M] D. Gromoll, W. Klingenberg and W. Meyer, *Riemannsche Geometrie im Grossen*, Lecture Notes in Math., vol. 55, Springer-Verlag, 1968.

[G-L$_1$] M. Gromov and H. B. Lawson, *Spin and scalar curvature in the presence of a fundamental group*, Ann. of Math. (2) **111** (1980), 209–230.

[G-L$_2$] ______, *Positive scalar curvature and the Dirac operator on complete Riemannian manifolds*, Publ. Math. **58** (1983), 295–408.

[G$_1$] R. Gulliver, *Regularity of minimizing surfaces of prescribed mean curvature*, Ann. of Math. (2) **97** (1973), 275–305.

[G$_2$] ______, *Removability of singular points on surfaces of bounded mean curvature*, J. Differential Geometry **11** (1976), 345–350.

[G$_3$] ______, *Representation of surfaces near a branched minimal surface*, Minimal Submanifolds and Geodesics, Kaigai Publications, Tokyo, 1978, pp. 31–42.

[G$_4$] ______, *Index and total curvature of complete minimal surfaces*, this volume.

[Ha] R. Hardt, *Topological properties of subanalytic sets*, Trans. Amer. Math. Soc. **211** (1975), 57–70.

[H-L] R. Harvey and B. Lawson, *Extending minimal varieties*, Invent. Math. **28** (1975), 209–226.

[Hi] H. Hironaka, *Subanalytic sets*, Number Theory, Algebraic Geometry, and Commutative Algebra, Kinokuniya, Tokyo, 1973, pp. 453–493.

[Ka] J. Kazdan, *Deforming to positive scalar curvature on complete manifolds*, Math. Ann. **261** (1982), 227–234.

[K-W] J. Kazdan and F. Warner, *Prescribing curvatures*, Proc. Sympos. Pure Math., vol. 27, Amer. Math. Soc., Providence, R.I., 1975, 309–319.

[L$_1$] H. B. Lawson, *The equivariant Plateau problem and interior regularity*, Trans. Amer. Math. Soc. **173** (1972), 231–249.

[**L₂**] ______, *Minimal varieties in real and complex geometry*, Univ. of Montreal Press, Montreal, 1973.

[**M-Y**] W. Meeks and S.-T. Yau, *The existence of embedded minimal surfaces and the problem of uniqueness*, Math. Z. **179** (1982), 151–168.

[**Mo**] C. B. Morrey, *The problem of Plateau on a Riemannian manifold*, Ann. of Math. (2) **49** (1948), 807–851.

[**O₁**] R. Osserman, *A survey of minimal surfaces*, Van Nostrand Reinhold, New York, 1969.

[**O₂**] ______, *A proof of the regularity everywhere of the classical solution to Plateau's problem*, Ann. of Math. (2) **91** (1970), 550–569.

[**P**] J. Pitts, *Existence and regularity of minimal surfaces on Riemannian manifolds*, Math. Notes 27, Princeton Univ. Press, Princeton, N.J., 1981.

[**Sc**] R. Schoen, *A remark on minimal hypercones*, Proc. Nat. Acad. Sci. **79** (1982), 4523–4524.

[**S-Y**] R. Schoen and S.-T. Yau, *Existence of incompressible minimal surfaces and the topology of 3-dimensional manifolds with nonnegative scalar curvature*, Ann. of Math. (2) **110** (1979), 127–142.

[**Sim₁**] L. Simon, *Isolated singularities of minimal surfaces*, Proc. Centre for Math. Anal., Vol. 1, 1982, pp. 70–100.

[**Sim₂**] ______, *Asymptotics for a class of nonlinear evolution equations with application to geometric problems*, Ann. of Math. (2) **118** (1983), 525–571.

[**S**] J. Simons, *Minimal varieties in riemannian manifolds*, Ann. of Math. (2) **88** (1968), 62–105.

UNIVERSITY OF MINNESOTA, MINNEAPOLIS

STATE UNIVERSITY OF NEW YORK AT STONY BROOK

Proceedings of Symposia in Pure Mathematics
Volume **44** (1986)

Some Regularity Results in Plasticity

ROBERT M. HARDT[1] AND DAVID KINDERLEHRER[1]

The equilibrium configuration of an elastic-plastic body has been the subject of several recent works [1–3, 6, 8–13]. In such works one uses the C^1 functional φ: defined by

$$\varphi(p) = \begin{cases} |p|^2/2 & \text{for } |p| \leqslant 1, \\ |p| - 1/2 & \text{for } |p| \geqslant 1 \end{cases}$$

(in contrast to *elasticity* where one would use a functional with growth like $|p|^2$ near ∞).

Before giving the various technical definitions involved in the vector problem, we consider a simplified scalar problem—called anti-planar shear traction. Here one is interested in the infinitesimal displacement of points in an infinite vertical cylinder subjected to a balanced vertically-invariant, vertical boundary force.

Precisely, given a smooth bounded domain $\Omega \subset \mathbf{R}^n$ and a function $f: \partial\Omega \to \mathbf{R}$ with $\int_{\partial\Omega} f = 0$, one tries to find a function u. $\Omega \to \mathbf{R}$ with $\int_\Omega u = 0$ which minimizes $I(u) = \int_\Omega \varphi(Du) - \int_{\partial\Omega} fu$ (see Figure 1).

For an *existence theory* one uses the space of equilibrated BV-functions:

$$BV_e(\Omega) = \left\{ v \in L^1(\Omega): Dv \text{ is a measure}, \int_\Omega v = 0 \right\}.$$

For $v \in BV$, we may decompose Dv into its absolutely continuous and singular parts: $Dv = v_x dx + D_s v$. Here v_x denotes the Radon–Nikodym derivative of Dv with respect to Lebesgue measure and $D_s v$ denotes the singular part of Dv. Associated with Ω is a Sobolev constant S_Ω so that

$$\int_{\partial\Omega} |u| \leqslant S_\Omega \int_\Omega |Du| \quad \text{for } u \in BV_e(\Omega).$$

1980 *Mathematics Subject Classification.* Primary 73E99; Secondary 35J50
[1] Research partially supported by the National Science Foundation.

© 1986 American Mathematical Society
0082-0717/86 $1.00 + $.25 per page

One may show that the formula,

$$\int_U \varphi(Du) = \int_U \varphi(u_x)\, dx + |D_s u|(U) \quad \text{for } U \text{ open in } \Omega,$$

is an appropriate *extension of the functional* to $u \in BV(\Omega)$ and that, with this definition, $\int_\Omega \varphi(Du)$ is weakly lower semicontinuous in u. If f satisfies the "safe load condition" $\|f\|_\infty \leqslant S_\Omega$, then

$$I(v) \geqslant \int_\Omega \left[|Dv| - 1/2 \right] - \|f\|_\infty \int_{\partial\Omega} |v| - |\Omega|/2 \quad \text{for } v \in BV_e(\Omega),$$

and the *existence* of a minimizer follows from the lower semicontinuity of $\int_\Omega \varphi(Du)$ as well as the compactness and trace theories for BV functions. In several papers these facts are understood in the context of duality. Here one shows that the solution u above gives the *stress function*

$$\sigma = \varphi_p(u_x) = \begin{cases} u_x & \text{whenever } |u_x| < 1, \\ u_x/|u_x| & \text{whenever } |u_x| \geqslant 1. \end{cases}$$

It is the unique solution of the *dual problem*:

Minimize $\{ \frac{1}{2}\int |\tau|^2 : \tau \in (L^\infty)^n, \ \tau \cdot \nu = f \text{ on } \partial\Omega, \ \operatorname{div}\tau = 0, \ |\tau| \leqslant 1 \}$ where ν is the exterior normal vector field for Ω. Here the strain Du determines the stress σ, but not vice versa (as in linear elasticity).

Let us consider a very elementary example: Suppose $n = 1$, $\Omega = (-1, 1)$, $f_\delta(\pm 1) = \pm\delta$. For $\delta < 1$, $u(t) = \delta t$ is the unique solution. The stress $\equiv \delta$. For

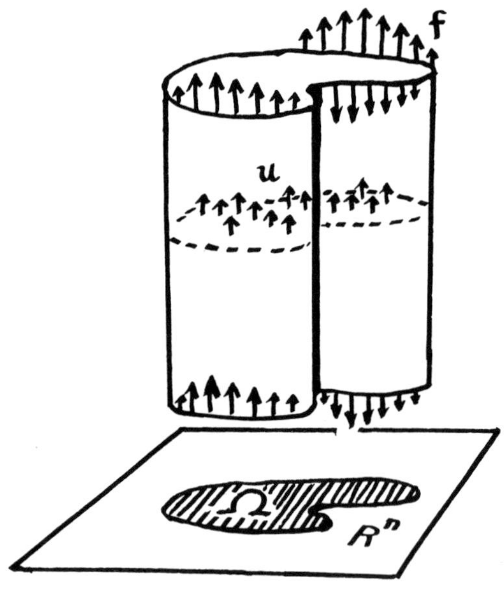

FIGURE 1

$\delta > 1$, there is no solution. For $\delta = 1$, the function

$$u_c(t) = \begin{cases} t + c & \text{for } t > 0, \\ t - c & \text{for } t < 0, \end{cases}$$

is a solution for *any* c, and the stress $\equiv 1$.

In fact *any* monotone u with $\int_{-1}^{1} u = 0$ and $u_x \geq 1$ is a solution because

$$I(u) = \int_{-1}^{1} |Du| - 1/2 - [u(1) - u(0)] = -1.$$

Returning to the general problem, we define (up to a set of Lebesgue measure zero)

$$\text{Elastic region: } E = \left\{ x \colon |\sigma(x)| < 1 \right\} = \left\{ x \colon |u_x(x)| < 1 \right\}$$

$$\text{Plastic region: } P = \left\{ x \colon |\sigma(x)| = 1 \right\} = \left\{ x \colon |u_x(x)| \geq 1 \right\}.$$

Note the extreme cases:

If $\Omega = E$, then u is harmonic.

If $\Omega = P$, then u is a function of least gradient.

There is a useful *first variation formula*: For $\zeta \in BV_e(\Omega)$ with $|D_s\zeta| << |D_s u|$, we easily show that

$$0 = \frac{d}{dt}\bigg|_{t=0} I(u + t\xi) = \int_{\Omega} \varphi_p(u_x) \cdot \zeta_x \, dx + \int_{\Omega} D_s\zeta \cdot \frac{D_s u}{|D_s u|} |D_s u| - \int_{\partial\Omega} f \cdot \zeta dx.$$

This readily gives a sufficient condition for the presence of an elastic region:

If f satisfies safe-load, then $E \neq \varnothing$.

PROOF. If $\Omega = P$, then with $\zeta = u$ above, $\varphi_p(u_x) \cdot \zeta_x = |u_x|$, and

$$\int_{\Omega} |Du| = \int_{\Omega} |u_x| dx + |D_s u|(\Omega) - \int_{\partial\Omega} fu \leq S_\Omega^{-1} \int_{\partial\Omega} |u| dx,$$

contradicting the choice of S_Ω.

It is interesting to note that coexistent with an elastic region may be a nonempty plastic region. In fact

A local jump in f may imply that $P \neq \varnothing$.

PROOF. Suppose $n = 2$, $\Omega = $ Upper half disc, and $f(x) = \delta \operatorname{sgn} x$. If $\Omega = E$, then *near* 0

$$\Delta u = 0 \quad \text{on } \Omega \qquad \text{and} \qquad \partial u/\partial \nu = f \quad \text{on } \partial\Omega.$$

These imply that $u \sim \delta \operatorname{Im}(z \log z) + $ harmonic. This contradicts the condition $|Du| \leq 1$. (Actually f does not have to have a jump; it need only oscillate sufficiently to force $|u_y| > 1$.)

There is an elementary *monotonicity formula*.

For $0 < \rho < r < \operatorname{dist}(a, \partial\Omega)$,

$$\rho^{1-n} \int_{B_\rho(a)} |Du| \leq r^{1-n} \int_{B_r(a)} |Du| + (n - 1)(r - \rho)\omega_n.$$

Hence the $n - 1$ density $\Theta(a) = \lim_{\rho \to 0} \rho^{1-n} \int_{B_\rho(a)} |Du|$ exists.

PROOF. Choose $\zeta = w - u$ with

$$w = \begin{cases} u\left(r\dfrac{x}{|x|}\right) & \text{in } B_r, \\[2mm] u(x) & \text{in } \Omega \sim B_r. \end{cases}$$

In [7] we obtain a variety of mild regularity results.

(1) A global bound. $\|u\|_\infty \leqslant c(\|f\|_\infty, \Omega)$ for $\|f\|_\infty < S_\Omega^{-1}$.

(2) $\Theta(a) = 0 \Leftrightarrow u$ is continuous at a. In fact

$$\|u - u(a)\|_{L^\infty(B_r(a))} \leqslant c\left(r^{1-n}\int_{B_{2r}(a)} |Du| + r\right).$$

(3) For $\mathcal{H}^{n-1}$ almost all a,

$$\Theta(a) > 0 \Leftrightarrow u - \varepsilon < u < u^+ + \varepsilon \quad \text{near } a,$$

where $u^\pm$ are the approximate upper and lower limits of u at a.

(4) Fractures (discontinuities) along which $u^\pm$ are continuous are formally minimal surfaces.

The proofs of these use only convexity properties (e.g. Jensen's inequality), the first variation formula, Sobolev's inequality, and truncation (i.e. one estimates the Lebesgue measure of $\{x \in B_r(a)\colon u(x) > \lambda\}$).

Some open questions. (1). Does $u(x)$ stay near $\{u^-, u^+\}$ in 3 when x is sufficiently near a?

(2) What is the regularity of σ? (In two dimensions it is continuous.)

(3) What is the regularity of the free boundary between E and P?

(4) What is the regularity of the fractures?

(If the fracture divided Ω into 2 regions $\Omega_+ \cup \Omega_-$, then one could use the first variation formula with ζ being the characteristic function of Ω_+ to show that fractures do not occur under safe-load.)

We now turn to the general problem of the infinitesimal displacement of a plastic body subjected to an equilibrated boundary force.

Given a smooth bounded $\Omega \subset \mathbf{R}^n$, $\kappa > 0$, and a function $f\colon \partial\Omega \to \mathbf{R}^n$ which is equilibrated (i.e. $\int_{\partial\Omega} f \cdot R = 0$ for all infinitesimal rigid motions R of $\mathbf{R}^n$), one tries to find a function $u\colon \Omega \to \mathbf{R}^n$, that is equilibrated, minimizing

$$I(u) = \int_\Omega \varphi\big(\varepsilon^D(u)\big) + \frac{\kappa}{2}\int_\Omega |\operatorname{div} u|^2 - \int_{\partial\Omega} f \cdot u,$$

where $\varepsilon(u) = (Du + Du^t)/2$ and $\varepsilon^D(u) = \varepsilon(u) - (1/n)\operatorname{trace} \varepsilon(u) - \text{id}$.

There is again an existence theory for $\|f\|_\infty$ small. One now uses the space

$$P_e(\Omega) = \{u \in BD(\Omega)\colon \operatorname{div} u \in L^2, u \text{ is equilibrated}\},$$

where

$$BD(\Omega) = \{u \in L^1(\Omega)^n\colon \varepsilon(u) \text{ is a measure}\}.$$

The appropriate extension of the functional to BD involves

$$\int_\Omega \varphi(\varepsilon^D) = \int_\Omega \varphi\left(|\varepsilon_a^D|\right) dx + |\varepsilon_s^D|(\Omega),$$

where $\varepsilon^D = (\varepsilon_a^D)\, dx + \varepsilon_s^D$ as before. (However, verifying the relevant regularization properties is more difficult here.) There is again weak lower semicontinuity and the existence theorem involves using the constant T_Ω provided by the Korn–Sobolev inequality:

$$\int_{\partial\Omega} |u| \leqslant T_\Omega \int_\Omega |\varepsilon(u)| \quad \text{for } u \in P_e(\Omega).$$

For duality, the stress σ equals $\sigma^D + (\operatorname{div} u)\,\mathrm{id}$ where

$$\sigma^D = \begin{cases} \varepsilon^D & \text{for } |\varepsilon^D| \geqslant 1, \\ \varepsilon^D/|\varepsilon^D| & \text{for } |\varepsilon^D| \geqslant 1. \end{cases}$$

Then σ minimizes

$$\left\{ \frac{1}{2}\int_\Omega |\tau^D|^2 + \frac{1}{2\kappa n}\int_\Omega (\operatorname{tr}\tau)^2\, dx \colon \tau \in (L^\infty)^{n^2}, \right.$$

$$\left. \tau\cdot\nu = f,\ \operatorname{div}\tau = 0,\ \tau = \tau^t,\qquad |\tau^D| \leqslant 1 \right\}.$$

Freeing oneself from the usual "special form" assumption for solutions, one finds that, for antiplanar shear, there are solutions with horizontal (as well as vertical) fractures. For torsion, there are *non*-St. Venant solutions. In [6], our regularity results for this general displacement problem are somewhat weaker than before. The proofs use convexity, first variation, and the Korn–Sobolev inequality.

(1) $\sigma \in L^s_{\mathrm{loc}}$ for $1 \leqslant s < \infty$. This follows because $\nabla(\operatorname{div} u) = \operatorname{div}\sigma - \operatorname{div}\sigma^D = -f - \operatorname{div}\sigma^D \in H^{-1,\infty}$.

(2) $u \in L^p$ for some $p > n/(n-1)$. Here one shows that

$$\left(\rho^{-n}\int_{B_\rho} |u|^{n/n-1}\right)^{n-1/n} \leqslant A + B\rho^{-n}\int_{B_{2\rho}} |u|$$

and uses a reverse Hölder inequality.

(3)
$$\int_E |\varepsilon(u)^s| - 0 \text{ and } \rho^{-n}\int_{B_\rho(a)} |\varepsilon(u) - \upsilon(u)| \to 0$$

for any Lebesgue point $a \in E$ of σ.

(4) (Anzellotti's Theorem) $\rho^{-n}\int_{B_\rho}\sigma \to \hat\sigma$ in $L^p(dx + |\varepsilon(u)|^s)$ where $\hat\sigma$ is σ extended $dx + |\varepsilon(u)|^s$ almost everywhere.

We have many questions as before.

What is the regularity of σ, u, the free elastic-plastic boundary, and the fracture?

References

1. G. Anzellotti, *On the extremal stress and displacement in Hencky plasticity* (to appear).

2. G. Anzellotti and M. Giaquinta, *Existence of the displacement field for an elasto-plastic body*, Manuscripta Math. **32** (1980), 101–136.

3. ______, *On the existence of fields of stress and displacement for an elastic-plastic body in static equilibrium*, J. Math. Pures Appl. **61** (1982), 219–244.

4. H. Brezis, *Multiplicateur de Lagrange entorsion elastoplastique*, Arch. Rat. Mech. Anal. **49** (1972), 32–40.

5. L. Caffarelli and N. Riviere, *The Lipschitz character of the stress tensor when twisting an elastic-plastic body*, Arch. Rat. Mech. Anal. **69** (1979).

6. R. M. Hardt and D. Kinderlehrer, *Elastic-plastic deformation*, Appl. Math. Optim. **10** (1983), 203–246.

7. ______, *Plastic anti-planar shear*, in preparation.

8. R. Kohn and R. Temam, *Dual spaces of stresses and strains with applications to Hencky plasticity*, Appl. Math. Optim. **10** (1983).

9. G. Strang, *A minimax problem in plasticity theory*, Lecture notes in Math., vol. 701, Springer-Verlag, 1979, pp. 319–333.

10. P. Suquet, *Un espace fonctionnel pour les equations de la plasticité*. Ann. Fac. Sci. Toulouse Math. **1** (1979), 77–87.

11. R. Temam, *Mathematical problems in plasticity theory*, Complementarity Problems and Variational Inequalities, Wiley, New York, 1979.

12. ______, *Existence theorems for variational problems in plasticity*, Nonlinear problems of Analysis in Geometry and Mechanics, Pitman, London, 1981.

13. R. Temam and G. Strang, *Duality and relaxation in variational problems of plasticity*, J. Méc. Théor. Appl. **19** (1980a), 493–527.

14. T. W. Ting, *Elastic-plastic torsion of simply-connected cylindrical bars*, Indiana Univ. Math. J. **20** (1971), 1047–1076.

University of Minnesota

Proceedings of Symposia in Pure Mathematics
Volume **44** (1986)

Tangential Regularity near the $\mathscr{C}^1$-Boundary

ROBERT M. HARDT[1] AND FANG-HUA LIN[2]

1. Introduction. The purpose of this note is to initiate the study of the boundary behavior of an n-dimensional area-minimizing rectifiable current T in $\mathbf{R}^{n+1}$ whose boundary is an oriented submanifold Γ of class $\mathscr{C}^1$. In case Γ is of class $\mathscr{C}^{1,1}$ and extreme (that is, Γ lies on the boundary of its convex hull), then the arguments of W. K. Allard in [**A1** or **A2**, 4, 5.2] show that, for any $0 < \alpha < 1$, $U \cap \operatorname{spt} T$ is a $\mathscr{C}^{1,\alpha}$ submanifold with boundary for some open neighborhood U of Γ in $\mathbf{R}^{n+1}$. In case Γ is of class $\mathscr{C}^{1,\alpha}$, such complete boundary regularity was obtained in [**H-S**, 11.2] without the extremality hypothesis. However, if Γ is only of class $\mathscr{C}^1$, even the existence of a tangent cone at a boundary point is unknown. The discussions of the existence and structure of tangent cones at the boundary in [**A1**, **A2**, **Br** and **H-S**] require a sort of control on the first variation near the boundary or a version of the monotonicity formula at boundary points. Since the boundary term in this formula is infinite for a general $\mathscr{C}^1$ boundary, this approach leads to the need for a higher smoothness hypothesis on Γ, say $\mathscr{C}^{1,\alpha}$ or more generally, $\mathscr{C}^{1,\mathrm{Dini}}$ (see e.g. [**L**] or 5.1).

Here, for a general $\mathscr{C}^1$ extreme boundary Γ, we obtain, at any boundary point, the existence of a tangent cone which is an oriented half n-plane. For such a Γ, we then establish an Excess Improvement Theorem (3.6). The geometric consequence of this in §4 is the absence of interior singularities near Γ and the control of the orienting normal vector field $\mathbf{v}^T$ of T along directions *tangential to the boundary* Γ. In particular, for any continuous vector field v on $\mathbf{R}^{n+1}$ whose restriction to Γ is tangent to Γ, the expression $v \cdot \mathbf{v}^T$ defines a continuous function on some neighborhood of Γ in $\operatorname{spt} T$. This is analogous to a tangential gradient bound in the nonparametric context (as in [**S**]).

1980 *Mathematics Subject Classification.* Primary 49F22, 49F10; Secondary 49F20, 53A10.

[1] Research partially supported by the National Science Foundation.

[2] Research partially supported by the Alfred P. Sloan Foundation.

© 1986 American Mathematical Society
0082-0717/86 $1.00 + $.25 per page

This tangential regularity near a $\mathscr{C}^1$ extreme boundary continues to hold for a codimension one minimizer of the integral of a constant coefficient elliptic integrand. The only proof used here that does not readily carry over is the estimate (3.1). However a corresponding estimate is obtained (using an argument similar to [**SS**, pp. 426–428 or **Bo**, §V]) in [**H-L**] where there is a general study of boundary regularity for a class of parametric obstacle problems.

The extremality hypothesis is used only to obtain the existence at boundary points of tangent cones which are oriented half n-planes. In particular, one easily establishes the tangential boundary regularity in the unoriented Plateau problem where one considers a 2-dimensional area-minimizing flat chain modulo 2 in $\mathbf{R}^3$ which has a (not necessarily extreme) $\mathscr{C}^1$ boundary.

What prevents the full boundary regularity (normal as well as tangential) in the above cases, is the possible nonuniqueness of the tangent half n-planes at a boundary point. In case Γ is $\mathscr{C}^{1,\mathrm{Dini}}$, then these tangent half n-planes will be unique, the excess improvements may be summed as usual to obtain excess decay, and the minimizer will be, near Γ, a $\mathscr{C}^1$ manifold with boundary.

2. Tangent cones at an extreme $\mathscr{C}^1$ boundary.

2.1. LEMMA. *If T is an n-dimensional area-minimizing rectifiable current in $\mathbf{R}^{n+1}$ and $\Gamma = \partial T$ is an oriented extreme submanifold of class $\mathscr{C}^1$, then, at each point of Γ, there exists an oriented tangent cone, which is an oriented half n-plane.*

PROOF. Let H be the convex hull of Γ. If $\mathrm{Int}\, H = \varnothing$, then H is planar and the lemma follows easily. We now assume that $\mathrm{Int}\, H \neq \varnothing$. Then $\mathrm{Bdry}\, H$ may be represented as the polar graph of a Lipschitz function on an n sphere; in particular, the n density $\Theta^n(\mathrm{Bdry}\, H, \cdot)$ is bounded on $\mathrm{Bdry}\, H$. Moreover, by [**F**, 4.4.1, 4.4.7], $\partial T = \partial N$ for some oriented relatively open subset N of $\mathrm{Bdry}\, H$. Then

$$T - N = \pm \partial(\mathbf{E}^{n+1} \llcorner Y)$$

for some bounded open (possibly empty) $Y \subset \mathbf{R}^{n+1}$. For $a \in \Gamma$ and $r > 0$,

$$\partial[T \llcorner \mathbf{U}(a, r)] = \partial\big[N \llcorner \mathbf{U}(a, r) \pm \partial(\mathbf{E}^{n+1} \llcorner \mathbf{U}(a, r)) \llcorner Y\big],$$

hence, by the minimality of $T \llcorner \mathbf{U}(a, r)$, the upper density,

$$\Theta^{n*}(\|T\|, a) \quad \text{is finite.}$$

After a suitable change of coordinates, we assume $a = 0$ and infer, from the closure theorem, the existence of a sequence $\rho_i \to 0+$ such that $\mu_{1/\rho_i \#} T \to C$ in $\mathscr{F}_n^{\mathrm{loc}}(\mathbf{R}^{n+1})$ for some area-minimizing rectifiable current C with $\partial C = \mathbf{E}^{n-1} \times \delta_0 \times \delta_0$. Since Γ is extreme, we can assume, by the convex hull property, that $\mathrm{spt}\, C$ is contained in the closed half-space $\{x \in \mathbf{R}^{n+1}: x_{n+1} \geq 0\}$. By [**H**, 4.6, 3.6], either C is an oriented half n-plane or C is an embedded submanifold with boundary in some open neighborhood of the origin. In the latter case we can choose a sequence $k_i \to \infty$ and an oriented half n-plane P so that

$$\mu_{k_i/\rho_i \#} T \to P \quad \text{in } \mathscr{F}_n^{\mathrm{loc}}(\mathbf{R}^{n+1}) \quad \text{as } i \to \infty.$$

This proves the lemma.

3. Excess comparison. Here we modify the computations of [**H-S**, 3.2] to reduce the question of excess improvement to estimating the average squared height, a quantity easily treated in "blowing-up". Combining this with the harmonic function estimate 3.5 provides some simplification in the usual proofs of excess improvement. We employ, as much as possible, the notations of [**H-S**, §1]. In particular, we study a rectifiable current T that will be in a class $\mathscr{F}$ described in [**H-S**, 1.6] with $m = 1$ (because of 2.1), $\alpha = 0$, $\kappa_T(2) \leqslant 1$, and $\lim_{r \downarrow 0} \kappa_T(r) = 0$ where, for $0 < r \leqslant 2$, $\kappa_T(r)$ is defined by

$$\kappa_T(r) = \sup_{z \neq w, |z| \leqslant r, |w| \leqslant r} \left[\left| D\varphi_T(z) - D\varphi_T(w) \right|^2 + \left| D\psi_T(z) - D\psi_T(w) \right|^2 \right]^{1/2}.$$

3.1. SQUASH LEMMA. *Suppose* $A = \mathbf{C}_2 \sim \mathbf{C}_1$ *and*

$$\begin{aligned}
F(x_1, \ldots, x_{n+1}) &= (x_1, \ldots, x_n, \mu(x_1, \ldots, x_n) X_{n+1} \\
&\quad + [1 - \mu(x_1, \ldots, x_n)] \Psi_T(x_1, \ldots, x_{n-1}))
\end{aligned}$$

where

$$\mu(y) = \begin{cases} 0 & \text{for } |y| \leqslant 1, \\ |y| - 1 & \text{for } 1 \leqslant |y| \leqslant 2, \\ 1 & \text{for } 2 \leqslant |y|. \end{cases}$$

Then

$$\mathbf{M} F_{\#}(T \llcorner A) - \mathbf{M}(T \llcorner A) \leqslant \lambda \int_A \left[1 - \left(v_{n+1}^T \right)^2 \right] d\|T\|$$

$$+ \left(\frac{\beta}{2^{n+1}\lambda} \right) \left[\kappa_T^2(1) + \int_A X_{n+1}^2 \, d\|T\| \right]$$

for any $0 < \lambda < 1$ *where* β *is a constant depending only on* n.

PROOF. As in [**H-S**, 3.2],

$$\mathbf{M} F_{\#}(T \llcorner A) - \mathbf{M}(T \llcorner A) = \int_A \left(|v^T \Delta| - 1 \right) d\|T\| \leqslant \int_A \left(|v^T \Delta|^2 - 1 \right) d\|T\|,$$

where $\Delta(x)$ is the $(n + 1) \times (n + 1)$ matrix whose (k, l)th entry is $(-1)^{k+l}$ times the determinant of the $n \times n$ matrix obtained by deleting the kth row and lth column from the matrix representing $DF(x)$. Then

$$|v^T \Delta|^2 - 1 = \left(v_{n+1}^T \right)^2 - 1 + \sum_{i=1}^{n} \left(\mu v_i^T - a_i v_{n+1}^T \right)^2$$

where $a_i = (1 - \mu) D_i \psi_T + (X_{n+1} - \psi_T) D_i \mu$ for $i = 1, \ldots, n$. Thus we may estimate, for any $0 < \lambda < 1$, that

$$|v^T \Delta|^2 - 1 = (\mu^2 - 1)\left(1 - \left(v_{n+1}^T \right)^2 \right) + \sum_{i=1}^{n} a_i^2 \left(v_{n+1}^T \right)^2 - \sum_{i=1}^{n} 2\mu a_i v_i^T v_{n+1}^T$$

$$\leqslant \sum_{i=1}^{n} a_i^2 \left(v_{n+1}^T \right)^2 + \lambda \sum_{i=1}^{n} \left(v_i^T \right)^2 + \left(\frac{1}{\lambda} \right) \sum_{i=1}^{n} \mu^2 \left(v_{n+1}^T \right)^2 a_i^2.$$

From the inequality

$$a_i^2 \leqslant 2(1-\mu)^2(D_i\psi_T)^2 + 2(X_{n+1}-\psi_T)^2(D_i\mu)^2,$$

we now conclude that

$$\mathbf{M}F_\#(T\llcorner A) - \mathbf{M}(T\llcorner A) \leqslant \lambda\int_A \left(1-\left(v_{n+1}^T\right)^2\right)d\|T\| + \left(\frac{4}{\lambda}\right)\int_A \left[(1-\mu)^2|D\psi_T|^2\right.$$

$$\left. + (X_{n+1}-\psi_T)^2|D\mu|^2\right]d\|T\|$$

$$\leqslant \lambda\int_A \left(1-\left(v_{n+1}^T\right)^2\right)d\|T\| + \left(\frac{\beta}{2^{n+1}\lambda}\right)\left[\kappa_T^2(1) + \int_A X_{n+1}^2\,d\|T\|\right]$$

for some β depending only on n.

3.2. COROLLARY.

$$\mathbf{E}_C(T,1) \leqslant \lambda\mathbf{E}_C(T,2) + \left(\frac{\beta}{\lambda}\right)\left[\kappa_T^2(1) + \int_A X_{n+1}^2\,d\|T\|\right].$$

PROOF. By the area-minimality of $T\llcorner\mathbf{C}_2$,

$$\mathbf{E}_C(T,1) = \mathbf{M}(T\llcorner\mathbf{C}_2) - \mathbf{M}(T\llcorner A) - \mathbf{M}p_\#(T\llcorner\mathbf{C}_1)$$

$$\leqslant \mathbf{M}F_\#(T\llcorner\mathbf{C}_2) - \mathbf{M}(T\llcorner A) - \mathbf{M}F_\#(T\llcorner\mathbf{C}_1)$$

$$= \mathbf{M}F_\#(T\llcorner A) - \mathbf{M}(T\llcorner A)$$

and we may apply 3.1 with λ replaced by $\lambda/2$, the inequality $1-(v_{n+1}^T)^2 \leqslant 2(1-v_{n+1}^T)$, and the formula

$$\mathbf{E}_C(T,r) = r^{-n}\int_{\mathbf{C}_r} \left[1-v_{n+1}^T\right]d\|T\|.$$

3.3. *Scaling.* As in [H-S, 2.3], $T_\rho = (\mu_{1/\rho\#}T)\llcorner\mathbf{U}_3$ belongs to $\mathscr{T}$ and satisfies

$$\mathbf{E}_C(T_\rho,r) = \mathbf{E}_C(T,\rho r) \leqslant \rho^{-n}\mathbf{E}_C(T,r) \quad\text{and}\quad \kappa_{T_\rho} = \kappa_T(\rho r)$$

whenever $T \in \mathscr{T}$, $0 < \rho \leqslant 1/3$, and $\mathbf{E}_C(T,1) + \kappa_T(1)$ is sufficiently small. Applying 3.2 to T_ρ then gives

$$\mathbf{E}_C(T,\rho) \leqslant \lambda\mathbf{E}_C(T,2\rho) + \left(\frac{\beta}{\lambda}\right)\left[\kappa_T^2(\rho) + \rho^{-n-2}\int_{\mathbf{C}_{2\rho}\sim\mathbf{C}_\rho} X_{n+1}^2\,d\|T\|\right].$$

3.4. *Rotating.* Recall from [H-S, 1.1] that γ_ω denotes rotation about $\mathbf{R}^n \times \{(0,0)\}$ through an angle ω. From the formula of 3.2 and the equation

$$1 - v_{n+1}^T = \tfrac{1}{2}\left|v^T - \mathbf{e}_{n+1}\right|^2,$$

we readily find that

$$\mathbf{E}_C\left(\gamma_{\omega\#}T,\tfrac{1}{2}\right) \leqslant 4^n\left[|\omega|^2 + \mathbf{E}_C(T,1)\right]$$

whenever $T \in \mathscr{T}$, $\sup_{\mathbf{C}_1\cap\,\mathrm{spt}\,T}X_{n+1}^2 \leqslant 1$, and $|\omega| \leqslant \pi/4$.

3.5. *Harmonic function estimates.* If h is a harmonic function on $\mathbf{B}^n(0,2)$ and

$$\int_{\mathbf{B}^n(0,2)} |\text{grad } h|^2 \, d\mathscr{L}^n \leqslant 1,$$

then $|\text{grad } h|(0) \leqslant \gamma$ and

$$r^{-n-2} \int_{\mathbf{B}^n(0,r)} \left[h(x) - h(0) - x \cdot \text{grad } h(0) \right]^2 \, d\mathscr{L}^n x \leqslant r^{3/2}$$

whenever $0 \leqslant r \leqslant \delta$ for some constants γ and δ depending only on n.

PROOF. Combine, for example, the inequalities [**H-S**, 8.2(1) (2), p. 470] with the L^2 Poincaré inequality [**G-T**, (7.45)].

3.6. EXCESS IMPROVEMENT THEOREM. *There is a constant ε depending only on n so that for any $T \in \mathscr{T}$ with*

$$\max\left\{ \mathbf{E}_C(T,1), \varepsilon^{-1}\kappa_T^2(1) \right\} \leqslant \varepsilon,$$

there exists an $\omega \in \mathbf{R}$ for which $|\omega| \leqslant \gamma \mathbf{E}_C(T,1)^{1/2}$ and

$$\mathbf{E}_C(\gamma_{\omega\#}T, \theta) \leqslant \max\left\{ \theta \mathbf{E}_C(T,1), \varepsilon^{-1}\kappa_T^2(1) \right\}$$

where $\theta = \delta/2^J$ and J is an integer large enough so that

$$4^{-J}(2/\delta)^n(\gamma^2 + 1) < \theta/2 \quad and \quad 32\beta\theta^{1/2} < 1/2.$$

PROOF. Assuming the theorem false, we may choose, for $v \in \{1,2,\dots\}$, a current $T_v \in \mathscr{T}$ satisfying the hypothesis, with ε replaced by $1/v$, but not the conclusion. We may assume that

$$\lim_{v \to \infty} \kappa_v/\varepsilon_v = 0 \quad \text{where } \kappa_v = \kappa_{T_v}(1) \quad \text{and} \quad \varepsilon_v = \mathbf{E}_C(T_v,1)^{1/2}.$$

(Otherwise $\omega = 0$ would work for some T_v.) Let T_v'' be constructed from T_v as in [**H**, 2.2] by retracting ∂T_v onto $\mathbf{R}^{n-1} \times \{(0,0)\}$ and reflecting about $\mathbf{R}^{n-1} \times \{(0,0)\}$. Then the convergence of κ_v/ε_v to zero implies that

$$\mathbf{E}_C(\gamma_{\omega\#}T_v'', r)/\mathbf{E}_C(\gamma_{\omega\#}T_v, r) \to 2 \quad \text{as } v \to \infty$$

for all ω and r. Approximating as in [**S-S**, Lemma 3 or **F**, 5.3.5(4)] $T_v'' \llcorner \mathbf{C}_1$ by a Lipschitz function f_v we find that

$$\limsup_{v \to \infty} \int |\text{grad}(f_v/\varepsilon_v)|^2 \, d\mathscr{L}^n < \infty.$$

By passing to a subsequence (without changing notation) the functions f_v/ε_v converge strongly in L^2_{loc} to a function h which (by the H^1 bound) has zero trace on $\mathbf{R}^{n-1} \times \{0\}$, is an odd function in x_n, and is harmonic on the unit ball. Choosing $\omega_v = \arctan(\varepsilon_v \zeta)$ where $(\text{grad } h)(0) = (0,\dots,0,\zeta)$, we let g be the Lipschitz function, whose graph approximates $\gamma_{\omega_v\#}T_v''\llcorner\mathbf{C}_{1/2}$ as in [**S-S**, Lemma 3]. From 3.4 we infer that g_v/ε_v converges strongly in L^2_{loc} to $(h - l)|\mathbf{U}^n(0,\tfrac{1}{2})$ where $l(x) = x \cdot \text{grad } h(0)$. Using the estimates for the graphical approximation and 3.5 we deduce that, for all $\theta \leqslant r \leqslant \lambda$ and v sufficiently large,

$$r^{-n-2} \int_{\mathbf{C}_r} X_{n+1}^2 \, d\|\gamma_{\omega_v\#}T_v\| \leqslant 2r^{3/2}\varepsilon_v^2.$$

Applying the squash estimate (3.3) J times with $\lambda = 1/4$, then 3.4, the choice of θ, and the convergence of $\kappa_\nu/\varepsilon_\nu$ to 0, we conclude, for ν sufficiently large, that

$$\mathbf{E}_C\big(\gamma_{\omega_\nu\#}T_\nu, \theta\big) \leqslant \lambda \mathbf{E}_C\big(\gamma_{\omega_\nu\#}T_\nu, 2\theta\big) + \left(\frac{\beta}{\lambda}\right)\left[\theta^{-1}\kappa_{T_\nu}^2(\theta) + \theta^{-n-2}\int_{C_{2\theta}} X_{n+1}^2 \, d\|\gamma_{\omega_\nu\#}T_\nu\|\right]$$

$$\leqslant \cdots \leqslant \lambda^J \mathbf{E}_C\big(\gamma_{\omega_\nu\#}T_\nu, \delta\big) + \left(\frac{\beta}{\lambda}\right)\sum_{j=0}^{J-1}\lambda^j\left[\theta^{-1}\kappa_\nu^2 + 2(2^j\theta)^{3/2}\varepsilon_\nu^2\right]$$

$$\leqslant \lambda^J(2\delta)^{-n}\mathbf{E}_C\left(\gamma_{\omega_\nu\#}T_\nu, \frac{1}{2}\right) + \left(\frac{\beta}{\lambda}\right)\left[\sum_{j=0}^{J-1}(2^{3/2}\lambda)^j\right]\left[\theta^{-2}(\kappa_\nu/\varepsilon_\nu)^2 + 2\theta^{1/2}\right]\theta\varepsilon_\nu^2$$

$$\leqslant \left(\frac{1}{2} + \frac{1}{2}\right)\theta\varepsilon_\nu^2,$$

contradicting the choice of T_ν.

4. Tangential regularity at an extreme $\mathscr{C}^1$ boundary. Here we assume that T and Γ are as in §2.

4.1. *Excess decay.* By the compactness of Γ, there is a positive r_Γ and a function $K_\Gamma \colon [0, r_\Gamma] \to [0, 1]$ so that $\lim_{r\downarrow 0} K_\Gamma(r) = 0$ and, for each point $a \in \Gamma$, the set $\Gamma \cap \mathbf{B}(a, r_\Gamma)$ may (after a rigid motion) be represented as the graph of a $\mathscr{C}^1$ function Φ with

$$\sup_{z\neq\omega,|z|\leqslant r,|\omega|\leqslant r} |D\Phi(z) - D\Phi(w)| \leqslant K_\Gamma(r) \quad \text{for all } 0 \leqslant r < r_\Gamma.$$

For each such a, let R_a be the supremum of the numbers $r \leqslant r_\Gamma$ such that, for some rigid motion Ω of $\mathbf{R}^{n+1}$,

$$\tilde{T} \in \mathscr{T} \quad \text{and} \quad \max\big\{\mathbf{E}_c(\tilde{T}, 1), \varepsilon^{-1}\kappa_{\tilde{T}}^2(1)\big\} \leqslant \varepsilon$$

where $\tilde{T} = (\mu_{3/r\#}\Omega_\# T)\mathbf{L}\mathbf{U}(0, 3)$ and ε is the constant in 3.6. Then, as in [**H-S**, p. 476] the number R_a is positive by 2.1 and the strong convergence to the tangent cone provided by [**F**, 5.4.2]. Moreover the formula in 3.2 and the compactness of Γ imply that

$$R = \inf\{R_a \colon a \in \Gamma\} > 0.$$

Defining inductively λ_j so that $\lambda_0 = \varepsilon$ and

$$\lambda_{j+1} = \max\big\{\theta\lambda_j, \varepsilon^{-1}K_\Gamma^2(R/2^{j+1})\big\},$$

we deduce from 3.3 and 3.6, that $\lim_{j\to\infty}\lambda_j = 0$ and, for each $a \in \Gamma$,

$$\mathbf{E}_C\big((\mu_{3\cdot 2^j/R\#}\Omega_{a\#}^j T)\mathbf{L}\mathbf{U}(0, 3), R/2^j\big) \leqslant \lambda_j$$

for some rigid motion Ω_a^j of $\mathbf{R}^{n+1}$ which maps a to 0 and $\mathrm{Tan}(\Gamma, a)$ to $\mathbf{R}^{n-1} \times \{(0, 0)\}$. Letting

$$\Lambda(r) = \theta^{-n}\lambda_j \quad \text{and} \quad T_{a,r} = \big(\mu_{3/r\#}\Omega_{a\#}^j T\big)\mathbf{L}\mathbf{U}(0, 3)$$

whenever $R/2^{j+1} < r \leqslant R/2^j$ we finally obtain, using 3.3, the uniform decay

$$\mathbf{E}_C(T_{a,r}, r) \leqslant \Lambda(r) \quad \text{for } 0 < r \leqslant R.$$

4.2. THEOREM. $M = \{ x \in \operatorname{spt} T : 0 < \operatorname{dist}(x, \Gamma) < \rho \}$ *is an* (*open*) *real analytic submanifold for any* ρ *sufficiently small enough so that for all* $0 < \sigma \leqslant \rho$,

$$\Lambda(12\sigma) < \inf\left\{ (8c_6)^{-2n}, 8^{-n}\varepsilon_I \right\}$$

where c_6 *and* ε_I *are the height bound and interior regularity constants of* [**S-S**, pp. 418, 421].

PROOF. Suppose $x \in M$ and $\sigma = \operatorname{dist}(x, \Gamma)$. Choose any point $a \in \Gamma$ with $|x - a| = \sigma$ and let $\tilde{x} = (\tilde{x}_1, \ldots, \tilde{x}_{n+1})$ be the point corresponding to x in $\operatorname{spt} T_{a,12\sigma}$, where $T_{a,r}$ is defined as in 4.1. Then $|\tilde{x}| = (3/12\sigma)\sigma = 1/4$, $\tilde{x}_1 = \cdots = \tilde{x}_{n-1} = 0$, because Γ is extreme, and

$$|\tilde{x}_{n+1}| \leqslant c_6 \mathbf{E}_C(T_{a,12\sigma})^{1/2n} \leqslant 1/8$$

by the height bound [**S-S**, p. 421]. Thus $\tilde{x}$ lies in the cylinder

$$p^{-1}\mathbf{B}^{n-1}((0,\ldots,0,1/4), 1/8).$$

The formula in 3.2 gives the interior excess estimate

$$\operatorname{Exc}(T_{a,12\sigma}, (0,\ldots,0,1/4), 1/8) \leqslant 8^n \mathbf{E}_C(T_{a,12\sigma}, 1) \leqslant \varepsilon_I$$

which implies that x is a regular point of $\operatorname{spt} T_{a,12\sigma}$; hence $\tilde{x}$ is a regular point of $\operatorname{spt} T$.

4.3. THEOREM. *For any continuous vector field* v *on* $\mathbf{R}^{n+1}$ *with*

$$v(a) \in \operatorname{Tan}(\Gamma, a) \quad \text{for each } a \in \Gamma,$$

the function on $M \cup \Gamma$ *which is* $v \cdot \mathbf{v}^T$ *on* M *and* 0 *on* Γ *is continuous* (*where* M, Γ *and* $\mathbf{v}^T$ *are as above*).

PROOF. By the compactness of Γ,

$$\zeta(\sigma) = \sup\left\{ |v(x) - v(a)| : x \in \mathbf{R}^{n+1}, a \in \Gamma, |x - a| = \sigma \right\} \to 0 \quad \text{as } \sigma \to 0.$$

For x, a, σ, and $\tilde{x}$ as in the proof of 4.2, we infer from the interior regularity theorem [**S-S**, p. 412] that

$$\left| (y,0,0) \cdot \mathbf{v}_{T_{a,12\sigma}}(x) \right| \leqslant c_1 \Lambda(12\sigma) \quad \text{whenever } y \in \mathbf{S}^{n-2}.$$

This gives the boundary estimate

$$|v \cdot \mathbf{v}_T(x)| \leqslant \left|[v - v(a)] \cdot \mathbf{v}^T(x)\right| + |v(a) \cdot \mathbf{v}^T(x)|$$

$$< \zeta(\sigma) + c_1 \sup_1 |v| \cdot \Lambda(12\sigma),$$

which, combined with the continuity of $\mathbf{v}^T$ on M (4.2) gives the theorem.

5. Nontangential regularity at an extreme $\mathscr{C}^1$ boundary. Suppose T, Γ, and v^T are as in §§2 and 4.

5.1. LEMMA. *If $a \in \Gamma$ and if there is only one tangent cone $C(a)$ (see 2.1) of T at a, then*

$$\sup\left\{\left|v^T(x) - v^T(a)\right| : x \in V \cap \operatorname{spt} T, |x - a| \leqslant \sigma\right\} \to 0 \quad as\ \sigma \to 0$$

where $v^T(a)$ is a normal vector of $C(a)$ and V is some $\mathscr{C}^1$ region in $\mathbf{R}^{n+1} \sim \Gamma$ with $a \in \operatorname{Bdry} V$.

PROOF. Using the formula of 3.2 and the interior regularity theorem [S-S, p. 418] in the wedge

$$W_a = \left\{x - a : \operatorname{dist}[x, \operatorname{Tan}(\Gamma, a)] > \tfrac{1}{2}|x - a|\right\},$$

we infer that any subsequence of a sequence of rotations leading to excess decay as in §4 must have a subsequence convergent to a rotation which determines a tangent cone of T at a. By uniqueness of this cone, any excess decaying family of rotations converges. From 3.4, we infer that the excess measured with respect to the fixed n plane of the unique tangent cone approaches zero. A suitable V may now be chosen using the formula of 3.2 and the interior regularity theorem (as in [H-S, 9.3]).

5.2. THEOREM. *Suppose Γ is of class $\mathscr{C}^{1,\mathrm{Dini}}$; that is (in the notation of 4.1)*

$$\int_0^{r_\Gamma} r^{-1} K_\Gamma(r)\, dr < \infty.$$

Then $U \cap \operatorname{spt} T$ is a $\mathscr{C}^1$ submanifold with boundary for some open neighborhood U of Γ.

PROOF. Recalling 4.1, $\sum_{j=1}^\infty \lambda_j^{1/2}$ is finite because $\sum_{j=1}^\infty K_\Gamma(r_\Gamma/2^j)$ is. For a fixed point $a \in \Gamma$, the estimate on $|\omega|$ in 3.6 and the argument of 4.1 now shows that

$$\limsup_{j \to \infty} \lambda_{j+1}^{-1/2} \left\|\Omega_a^{j+1} - \Omega_a^j\right\| < \infty;$$

hence, the series $\sum_{j=1}^\infty (\Omega_a^{j+1} - \Omega_a^j)$ converges absolutely to a rigid motion Ω_a. Then, by 3.4 and 3.6,

$$\limsup_{r \to 0} \mathbf{E}_C\left(\mu_{3/4\#}\Omega_{a\#}T, r\right)/\Lambda(r) < \infty.$$

Using the interior regularity theorem as in 5.1, we infer that $W_a \cap \tau_a[\operatorname{Tan}(\operatorname{spt} T, a)]$ is contained in the n plane $\Omega_a^{-1}(\mathbf{R}_2^{n-1} \times \{0\})$. Thus the tangent cone at a is unique and 5.1 applies. Noting that the estimates on excess decay and normal convergence in 5.1 depend only on $K_\Gamma(r)$ and not on a, the theorem follows.

REFERENCES

[A1] W. K. Allard, *On boundary regularity for Plateau's problem*, Thesis, Brown University, 1968 and Bull. Amer. Math. Soc. **75** (1969), 522–523.

[A2] ______, *On the first variation of a varifold: boundary behavior*, Ann. of Math. (2) **101** (1975), 418–446.

[Bo] E. Bombieri, *Regularity theory for almost minimal currents*, Arch. Rat. Mech. Anal. **78** (1982), 99–130.

[Br] J. Brothers, *Existence and structure of tangent cones at the boundary of an area minimizing integral current*, Indiana Univ. Math. J. **26** (1977), 1027–1044.

[F] H. Federer, *Geometric measure theory*, Springer-Verlag, 1969.

[G-T] D. Gilbarg and N. Trudinger, *Elliptic partial differential equations of second order*, Springer-Verlag, 1977.

[H] R. Hardt, *On boundary regularity for integral currents or flat chains modulo two minimizing the integral of an elliptic integrand*, Comm. Partial Differential Equations **2** (1977), 1163–1232.

[H-L] R. Hardt and F.-H. Lin, *On the boundary regularity of a class of parametric obstacle problems*, in preparation.

[H-S] R. Hardt and L. Simon, *On the boundary regularity and embedded solutions of the oriented Plateau problem*, Ann. of Math. (2) **110** (1979), 439–486.

[L] F.-H. Lin, *Solutions of the minimal surface system with small boundary data*, preprint.

[S] L. Simon, *Boundary regularity for solutions of the non-parametric least area problem*, Ann. Math. **103** (1976), 429–455.

[S-S] R. Schoen and L. Simon, *A new proof of the regularity theorem for rectifiable currents which minimize parametric elliptic functionals*, Indiana Univ. Math. J. **31** (1982), 415–434.

UNIVERSITY OF MINNESOTA

Proceedings of Symposia in Pure Mathematics
Volume **44** (1986)

Solving Plateau's Problem for Hypersurfaces Without the Compactness Theorem for Integral Currents

ROBERT M. HARDT AND JON T. PITTS[1]

For any $(k - 1)$-dimensional current B in $\mathbf{R}^n$ having finite mass, compact support, and boundary zero, one easily obtains, using the weak compactness of measures, a k-dimensional current S having boundary B and minimizing mass among all currents with boundary B. In 1960, H. Federer and W. Fleming introduced the geometrically interesting subclass of integral currents and proved the deep and fundamental Compactness Theorem for them [**F**, 4.2.17]. This implied the existence, for B integral, of a mass minimizing *integral* current having boundary B. The purpose of this paper is to give a simpler proof for the important hypersurface case $k = n - 1$.

MAIN RESULT. *For any $(n - 1)$-dimensional current N in $\mathbf{R}^n$ having finite mass, compact support, and integral boundary, there exists an integral current T with* $\partial T = \partial N$ *and* $\mathbf{M}(T) \leqslant \mathbf{M}(N)$. This follows without recourse to the Compactness Theorem for Integral Currents.

Our method of proof employs basic facts about functions of bounded variation in $\mathbf{R}^n$ ([**F**, 4.5.9 or **G**]) and general position arguments. Existence, in the context of oriented frontiers, follows from the BV compactness theorem [**F**, 4.2.17 or **G**]. It is well known how this leads to the codimension one integral current existence, *provided the given boundary is extreme*. W. Allard has also shown [**A**] how existence (in all codimensions) is related to the Compactness Theorem for Stationary Integral Varifolds.

1980 *Mathematics Subject Classification*. Primary 49F22.

[1]Research of the first author was supported in part by a grant from the National Science Foundation. Research of the second author was supported in part by grants from the National Science Foundation and the Alfred P. Sloan Foundation.

© 1986 American Mathematical Society
0082-0717/86 $1.00 + $.25 per page

We use the notation of [F]. Throughout n denotes an integer $\geqslant 2$.

LEMMA. *If* $m \in \{1, \ldots, n-1\}$ *and* B *is a countably* $(\mathscr{H}^m, m)$ *rectifiable subset of* $\mathbf{R}^n$, *then the set* Z *of all* $z \in \mathbf{S}^{n-1}$ *such that*

$$\mathscr{H}^m \llcorner B\big(\{x\colon z \in \operatorname{Tan}^m(\mathscr{H}^m \llcorner B, x)\}\big) > 0$$

is contained in the countable union of great $(m-1)$-*spheres.*

PROOF. For each linear subspace $P \in \bigcup_{j=1}^{m} \mathbf{G}(n, j)$, define

$$S(P) = B \cap \{x\colon P \subset \operatorname{Tan}^m(\mathscr{H}^m \llcorner B, x) \in \mathbf{G}(n, m)\},$$
$$G_m = \mathbf{G}(n, m) \cap \{P\colon \mathscr{H}^m(S(P)) > 0\},$$

and

$$G_j = \mathbf{G}(n, j) \cap \big\{ P\colon \mathscr{H}^m(S(P)) > 0 \text{ and } P \text{ is not contained in } Q$$
$$\text{for any } Q \in \bigcup_{k=j+1}^{m} G_k \big\}$$

for $j = 1, 2, \ldots, m-1$.

G_j is countable for each $j \in \{1, \ldots, m\}$, since $S(P)$ and $S(Q)$ are $\mathscr{H}^m$ essentially disjoint sets whenever P and Q are distinct elements of G_j. (This is trivial if $j = m$, and if $j < m$, we otherwise would have $\mathscr{H}^m(S(R)) > 0$, where R is that unique element of $\bigcup_{k=j+1}^{m} G_k$ spanned by $P \cup Q$.) The conclusion follows since $Z = \mathbf{S}^{n-1} \cap \bigcup_{j=1}^{m}[\bigcup G_j]$.

THEOREM. *If* $N \in \mathbf{N}_{n-1}(\mathbf{R}^n)$, N *has compact support, and* $\partial N \in \mathscr{R}_{n-2}(\mathbf{R}^n)$, *then there exists* $R \in \mathbf{I}_{n-1}(\mathbf{R}^n)$ *with* $\partial N = \partial R$ *such that the following three statements are true.*

(1) $\|N - R\| = \|N\| + \|R\|$.

(2) $\overrightarrow{N - R}(x) + \vec{R}(x) = 0$ *for* $\|R\|$ *almost all* $x \in \mathbf{R}^n$.

(3) *If* f *is a real valued* $\mathscr{L}^n$ *measurable function such that* $\partial(\mathbf{E}^n \llcorner f) = N - R$, *and if* λ *and* μ *are the approximate lower and upper limits of* f [F, 4.5.9], *then*

$$0 < \Theta^{n-1}(\|N - R\|, x) = \Theta^{n-1}(\|R\|, x) = \mu(x) - \lambda(x) \in \mathbf{Z}$$

for $\|R\|$ *almost all* $x \in \mathbf{R}^n$.

PROOF. Let $C = \llbracket 0 \rrbracket \times \partial N \in \mathscr{R}_{n-1}(\mathbf{R}^n)$. Let U be a bounded open set containing $\operatorname{spt} N \cup \operatorname{spt} C$, and choose a positive number r sufficiently large that

$$\operatorname{dist}\big(\operatorname{spt}(\tau_{rz \#} C), U\big) > 0$$

for all $z \in \mathbf{S}^{n-1}$.

For each $z \in \mathbf{S}^{n-1}$, let

$$C_z = h_{\#}(\llbracket 0, 1 \rrbracket \times \partial N) + \tau_{rz \#} C,$$

where h is the affine homotopy from τ_{rz} to the identity map. By [F, 4.1.9, 4.1.11, 4.1.28] we have that $\partial C_z = \partial N$, $C_z \in \mathscr{R}_{n-1}(\mathbf{R}^n)$,

$$z \wedge \vec{C}_z(x) = 0 \quad \text{for } \|C_z\| \text{ almost all } x \in U,$$

and

$$z \in \operatorname{Tan}^{n-1}\left(\mathscr{H}^{n-1} \llcorner B_z, x\right) \quad \text{for } \|C_z\| \text{ almost all } x \in U.$$

Here $B_z = \{x: \Theta^{n-1}(\|C_z\|, x) > 0\}$ is an $(\mathscr{H}^{n-1}, n-1)$ rectifiable, $\mathscr{H}^{n-1}$ measurable set, as is $B = \{x: \Theta^{n-1}(\|C\|, x) > 0\}$ [**F**, 4.1.28]. There is a function g of bounded variation such that $\partial(\mathbf{E}^n \llcorner g) = N - C$. By [**F**, 4.5.9(16)], the set

$$F = \left\{x: (\mathscr{L}^n) \operatorname*{ap\,lim\,inf}_{z \to x} g(z) < (\mathscr{L}^n) \operatorname*{ap\,lim\,sup}_{z \to x} g(z)\right\}$$

is an $(\mathscr{H}^{n-1}, n-1)$ rectifiable Borel set. By the Lemma and [**F**, 3.2.19], we may choose $z \in \mathbf{S}^{n-1}$ such that $B \cup F$ and B_z are $\mathscr{H}^{n-1}$ essentially disjoint. Let $R = C_z$.

Whenever $\mathscr{H}^{n-1}(W) < \infty$, we observe that $\|N - C\|(W) = 0$ if $W \subset \mathbf{R}^n \sim F$, and $\|N\|(W) = 0$ if $W \subset \mathbf{R}^n \sim (F \cup B)$. To see this, it suffices to assume that W is $\mathscr{H}^{n-1}$ measurable, in which case the first conclusion follows from [**F**, 4.5.9(14)] since

$$\mathscr{H}^{n-1}\left[W \cap \left\{x: (\mathscr{L}^n) \operatorname*{ap\,lim\,inf}_{z \to x} g(z) = s = (\mathscr{L}^n) \operatorname*{ap\,lim\,sup}_{z \to x} g(z)\right\}\right] > 0$$

for only countably many numbers s. The second conclusion follows from the first, since $\|N\|\llcorner \mathbf{R}^n \sim B = \|N - C\|\llcorner \mathbf{R}^n \sim B$. One consequence of this observation is that

$$\|N\|(B_z) = 0,$$

hence that for any $A \subset \mathbf{R}^n$,

$$\|N - C_z\|(A) - \|N - C_z\|(A \sim B_z) + \|N - C_z\|(A \cap B_z)$$
$$\geqslant \|N\|(A \sim B_z) - \|C_z\|(A \sim B_z)$$
$$+ \|C_z\|(A \cap B_z) - \|N\|(A \cap B_z)$$
$$= \|N\|(A \sim B_z) + \|C_z\|(A \cap B_z)$$
$$= \|N\|(A) + \|C_z\|(A).$$

This proves (1).

A second consequence of our observation is that $\Theta^{n-1}(\|N\|, x) = 0$ for $\mathscr{H}^{n-1}$ almost all $x \in \mathbf{R}^n \sim (F \cup B)$. If not, then there would exist a number t and a compact set $K \subset \mathbf{R}^n \sim (F \cup B)$ such that $\mathscr{H}^{n-1}(K) > 0$ and $\Theta^{n-1}(\|N\|, x) > t > 0$ for all $x \in K$. Since $\|N\|$ is a Radon measure, one infers from [**F**, 2.10.19(3), 2.10.6] that $\|N\|(K) \geqslant t \mathscr{H}^{n-1}(K) > 0$, which is a contradiction. Furthermore, (1) and arguments similar to above also imply that $\mathscr{H}^{n-1}(B_z \sim E) = 0$, where $E = \{x: \lambda(x) < \mu(x)\}$. Thus, for $x \in \mathbf{R}^n$ and $r \in \mathbf{R}$,

$$\|N - C_z\|\mathbf{B}(x, r) - \|N\|(\mathbf{B}(x, r) \sim E) + \|N - C_z\|(\mathbf{B}(x, r) \cap E),$$

and (3) follows from (1), [**F**, 4.1.28, 4.5.9(15)] and our density estimate.

To prove (2), we refer to [F, p. 357] and notice that for $\|C_z\|$ and $\|N - C_z\|$ almost all $x \in B_z$ and $y \in \Lambda^{n-1}\mathbf{R}^n$,

$$
\begin{aligned}
\langle \overrightarrow{N - C_z}(x), y \rangle &= \lim_{r \to 0+} \frac{(N - C_z)(b_{x,r}y)}{\|N - C_z\|\mathbf{B}(x, r)} \\[2mm]
&= \lim_{r \to 0+} \frac{N(b_{x,r}y) - C_z(b_{x,r}y)}{\|N\|\mathbf{B}(x, r) + \|C_z\|\mathbf{B}(x, r)} \\[2mm]
&= \lim_{r \to 0+} \frac{-C_z(b_{x,r}y)}{\Theta^{n-1}(\|C_z\|, x)\alpha(n - 1)r^{n-1}} \\[2mm]
&= -\langle \vec{C_z}(x), y \rangle.
\end{aligned}
$$

(2) follows. This completes the proof of the Theorem.

PROOF OF MAIN RESULT. Let N be as stated and choose R, f, λ, and μ as in the Theorem. By [F, 4.5.9(13)] we have

$$
\begin{aligned}
N - R &= \int_{-\infty}^{\infty} \partial\left[\mathbf{E}^n \llcorner \{ x: f(x) \geqslant s \}\right] d\mathscr{L}^1 s \\[2mm]
&= \int_{\{s: 0 \leqslant s < 1\}} \sum_{j \in \mathbf{z}} \partial\left[\mathbf{E}^n \llcorner \{ x: f(x) \geqslant s + j \}\right] d\mathscr{L}^1 s
\end{aligned}
$$

and

$$
\|N - R\| = \int_{\{s: 0 \leqslant s < 1\}} \sum_{j \in \mathbf{Z}} \left\|\partial\left[\mathbf{E}^n \llcorner \{ x: f(x) \geqslant s + j \}\right]\right\| d\mathscr{L}^1 s.
$$

It follows that one may choose a number $0 < s < 1$ such that

$$
\begin{aligned}
S_j &= \partial\left[\mathbf{E}^n \llcorner \{ x: f(x) \geqslant s + j \}\right] \in \mathscr{R}_{n-1}(\mathbf{R}^n) \quad \text{for all } j \in \mathbf{Z}, \\[2mm]
\mathbf{M}(N - R) &\geqslant \sum_{j \in \mathbf{Z}} \mathbf{M}(\partial\left[\mathbf{E}^n \llcorner \{ x: f(x) \geqslant s + j \}\right]), \\[2mm]
0 &= \|R\|\left(\mu^{-1}(s + j) \cup \lambda^{-1}(s + j)\right) \quad \text{for all } j \in \mathbf{Z},
\end{aligned}
$$

since the set of all such numbers s has positive $\mathscr{L}^1$ measure. We apply [F, 4.5.9(3, 17), 4.5.6(2, 3)], our Theorem, and our choice of s to conclude that for $\|R\|$ almost all $x \in \mathbf{R}^n$:

$$
\lambda(x) \neq s + j \text{ and } \mu(x) \neq s + j \quad \text{for any } j \in \mathbf{Z};
$$

$$
\text{if } \lambda(x) < s + j < \mu(x), \text{ then } \Theta^{n-1}(\|S_j\|, x) = 1;
$$

$$
\text{if } s + j < \lambda(x) \text{ or } \mu(x) < s + j, \text{ then } \Theta^{n-1}(\|S_j\|, x) = 0;
$$

$$
0 < \Theta^{n-1}(\|R\|, x) = \mu(x) - \lambda(x) \in \mathbf{Z};
$$

$$
-\vec{R}(x) = \overrightarrow{N - R}(x) = -\vec{S_j}(x) \quad \text{if } \lambda(x) < s + j < \mu(x).
$$

Let $S = \sum_{j \in \mathbf{Z}} S_j$. For $\|R\|$ almost all $x \in \mathbf{R}^n$,

$$\Theta^{n-1}(\|S\|, x) = \sum_{j \in \mathbf{Z}} \Theta^{n-1}(\|S_j\|, x)$$

$$= \operatorname{card}\{ j \in \mathbf{Z} : \lambda(x) < s + j < \mu(x)\}$$

$$= \mu(x) - \lambda(x)$$

$$= \Theta^{n-1}(\|R\|, x),$$

and $\vec{S}(x) = \vec{R}(x)$. We infer that $\|S\| = \|R\| + \|S - R\|$, hence $\partial(R - S) = \partial N$ and $\mathbf{M}(N) \geq \mathbf{M}(S - R)$. Let $T = R - S$.

References

[A] W. K. Allard, *On the first variation of a varifold*, Ann. of Math. (2) **95** (1972), 417–491.

[F] H. Federer, *Geometric measure theory*, Springer-Verlag, 1969.

[G] E. Giusti, *Minimal surfaces and functions of bounded variation*, Notes on Pure Mathematics, Australian National University, 1977.

University of Minnesota

Texas A & M University

Proceedings of Symposia in Pure Mathematics
Volume **44** (1986)

Complex Analytic Geometry and Measure Theory

F. REESE HARVEY AND H. BLAINE LAWSON, JR.

Abstract. The methods of geometric measure theory have played a time-honored role in the study of complex geometry and the analysis of holomorphic functions of several complex variables. In this article we would like to present in a unified way some of the developments of recent years. Since detailed expositions of this material are already in print, we shall make our discussion concise and aim at presenting the essential ideas and themes of the subject.

1. Local geometry. In this section we shall focus attention on subvarieties of codimension one. Questions in higher codimension are generally reduced to this case.

Let us suppose then that Ω is a complex manifold of dimension $n + 1$ and that $f \colon \Omega \to \mathbf{C}$ is a nonconstant holomorphic function. We would like to study the *divisor* of f which is, roughly speaking, the set of zeros of f together with their multiplicities. We begin with the classical fact that $\log |f|$ is subharmonic and so both it and its first derivatives are locally integrable functions. We then define the *divisor of f* to be the current of bidimension n, n (and therefore bidegree $1, 1$) given by the *Poincaré–Lelong formula*

$$(1.1) \qquad Z_f = (i/\pi)\, \partial\bar{\partial} \log |f|.$$

Specifically, this is the current which to any smooth $2n$-form φ with compact support in Ω assigns the number

$$Z_f(\varphi) = \frac{i}{\pi} \int_\Omega \log |f|\, \partial\bar{\partial}\varphi.$$

REMARK 1.1. Note that Z_f is supported in the zero-locus of f. Furthermore, if $(z_1, \ldots, z_n, w)$ are local coordinates on an open set $\mathscr{O} \subset \Omega$, and if $f = w^m$ in $\mathscr{O}$, then $Z_f = m[w = 0]$ in $\mathscr{O}$ where $[w = 0]$ is defined to be the current given by integration over the canonically oriented hyperplane $w = 0$. Note finally that the current Z_f is flat. This is because $d^c \log |f| \in L^1_{\mathrm{loc}}$ and $dd^c = 2i\partial\bar{\partial}$. (Recall that $d^c = i(\bar{\partial} - \partial)$.)

1980 *Mathematics Subject Classification*. Primary 49F20, 32C25.

© 1986 American Mathematical Society
0082-0717/86 $1.00 + $.25 per page

The following classical result is the fundamental theorem of local analytic geometry.

THEOREM 1.2 (THE WEIERSTRASS PREPARATION THEOREM). *The divisor Z_f is locally the graph of a multivalued holomorphic function. More precisely, at any point $p \in \Omega$ with $f(p) = 0$ there exist local coordinates $(z_1, \ldots, z_n, w)$ defined on a product neighborhood $\mathcal{O} = \Delta' \times \Delta'' \subset \mathbf{C}^n \times \mathbf{C}$, and there exist holomorphic functions $\sigma_1(z), \ldots, \sigma_k(z)$ vanishing at p such that $Z_f \lrcorner \mathcal{O} = Z_p$ where*

$$(1.2) \qquad P(z, w) = w^k + \sigma_1(z)w^{k-1} + \cdots + \sigma_k(z).$$

Note. The function (1.2) is called the Weierstrass polynomial. It has the property that $P|f$ in $\mathcal{O}$, that is, $f = uP$ where u is an invertible element in the ring of holomorphic functions on $\mathcal{O}$.

Note. The divisor $Z_f = Z_p$ is essentially a geometric object, and it is the geometric aspects that we will emphasize when we pass later to higher codimension. To keep a uniform and useful notation throughout the paper, we shall from now on denote Z_f by the symbol V.

PROOF. Fix any point $p \in \Omega$, where $f(p) = 0$, and choose local coordinates $\zeta = (\zeta_1, \ldots, \zeta_n)$, $|\zeta| \leq 1$, in a neighborhood of this point. For almost all ζ with $|\zeta| = 1$, the holomorphic function $\varphi_\zeta(t) \equiv f(t\zeta)$ is nonconstant and therefore has isolated zeros in the disk $|t| < 1$. Thus, there are a unitary change of coordinates to say $(z_1, \ldots, z_n, w)$ and numbers $\rho > 0$ and $\varepsilon > 0$ such that $|f(0, \ldots, 0, w)| \geq 2\varepsilon$ for any $|w| = \rho$. By continuity there is an $r > 0$ such that

$$(1.3) \qquad |f(z, w)| \geq \varepsilon \quad \text{for } |z| \leq r \text{ and } |w| = \rho.$$

Set $\Delta' = \{z: |z| < r\}$ and $\Delta'' = \{w: |w| < \rho\}$, and let $\pi: \Delta' \times \Delta'' \to \Delta'$ denote the projection. The number of zeros of f, counted to multiplicity, in the disk $\{z\} \times \Delta' = \pi^{-1}(z)$ is

$$(1.4) \qquad k(z) = \frac{1}{2\pi i} \int_{|w| = \rho} f^{-1}(z, w) \frac{\partial f}{\partial w}(z, w) \, dw.$$

This integer varies continuously in z and is therefore constant. Let $a_1(z), \ldots, a_k(z)$ denote the zeros of $f_z(w) \equiv f(z, w)$ in the disk $|w| < \rho$. We want to analyze the function $P(z, w) = (w - a_1(z)) \cdots (w - a_k(z))$. (See Figure 1.)

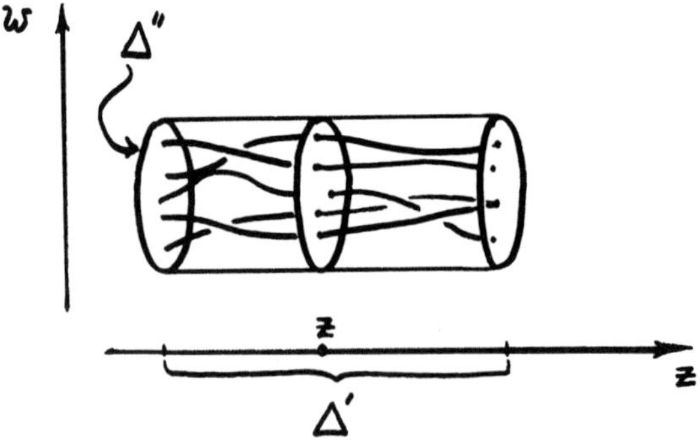

FIGURE 1.

To begin, we notice that the projection π when restricted to supp V is *proper* (since the zeros of f are bounded away from $|w| = \rho$). Consequently, we can define the currents

$$(1.5) \qquad \pi_{\#}(w^p V) \quad \text{for } p = 0, 1, 2, \ldots$$

and observe that

$$(1.6) \qquad \bar{\partial}\left(\pi_{\#} w^p V\right) = 0,$$

$$(1.7) \qquad \pi_{\#}(w^p V) = \sum_{j=1}^{k} a_j(z)^p [\Delta'].$$

This second equation follows from integration over the fibre and the classical fact that, for $z \in \Delta'$, the z-slice $\langle V, \pi, z \rangle = (i/\pi)(\partial/\partial w)(\partial/\partial \bar{w}) \log|f(z, w)| = \Sigma[a_j(z)]$. An immediate consequence of these formulas is that the *power functions*

$$(1.8) \qquad \tau_p(z) \equiv \sum_{j=1}^{k} a_j(z)^p$$

are *holomorphic*.

Recall now that the coefficients of $P(z, w) = \prod(w - a_j(z))$ are the elementary symmetric functions in the $a_j(z)$'s which can be expressed as Newton polynomials in the $\tau_p(z)$'s. Directly, we have that, for $|w| \geqslant \rho$,

$$(1.9) \qquad \log P - \log w^k + \sum_{j=1}^{k} \log\left(1 - \frac{a_j(z)}{w}\right)$$

$$= \log w^k + \sum_{j=1}^{k} \sum_{p=1}^{\infty} \frac{1}{p}\left(\frac{a_j(z)}{w}\right)^p$$

$$= \log w^k + \sum_{p=1}^{\infty} \frac{1}{p}\tau_p(z)\frac{1}{w^p},$$

and so

$$(1.10) \qquad P(z, w) = w^k \exp\left(\sum_{p=1}^{\infty} \frac{1}{p}\tau_p(z)w^{-p}\right) = \sum_{q=-\infty}^{k} \sigma_q(z)w^q$$

where convergence is uniform for $|z| \leqslant r$ and $|w| \geqslant \rho$. Hence, the $\sigma_q(z)$'s are holomorphic. $\square$

It remains to show $Z_f = Z_P$ in $\mathcal{O}$. Note that $u \equiv f/P$ is a well-defined never vanishing function in $\mathcal{O}$ which is holomorphic for each fixed z. The formula

$$u(z, w) = \frac{1}{2\pi i} \int_{|w|=\rho} \frac{f(z, \zeta)/P(z, \zeta)}{\zeta - w} \, d\zeta$$

shows that u is jointly holomorphic and the theorem is proved.

The reader probably noticed that we could have argued more classically by using the argument principle (cf. (1.4)) to construct the functions $\tau_p(z)$. Reasons for presenting the argument above will soon be clear.

The reader may also have noted that $\log P(z, w_0)$ can be defined directly by the formula

$$(1.11) \qquad \pi_{\#}(\log(w - w_0)V) = \log P(z, w_0)[\Delta']$$

for $|w_0| > \rho$.

There are several important consequences of our discussion above. Let $V = Z_f$ be as before.

COROLLARY 1.3. *The support of V has locally finite Hausdorff $2n$-measure. Furthermore, there is a closed subset $\Sigma(V)$ of Hausdorff dimension $2n - 2$ in Ω such that* $\operatorname{supp} V - \Sigma(V)$ *is a complex submanifold of dimension n properly embedded in* $\Omega - \Sigma(V)$.

PROOF. The result is local so we restrict to the neighborhood $\Delta' \times \Delta''$ and consider the Weierstrass polynomial P constructed above. For each $z \in \Delta'$, we let $m(z)$ denote the number of *distinct* roots of $P_z(w) \equiv P(w, z)$. Set $m = \max\{m(z): z \in \Delta'\}$ and consider the subset $U = \{z \in \Delta': m(z) = m\}$. Suppose $z_0 \in U$ and that $a(z_0)$ is a root of P_{z_0}. Then, applying the arguments above and the maximality of m shows that in a neighborhood of $(z_0, a(z_0))$ we can write P as $P(z, w) = (w - a(z))^l u(z, w)$ where u is nonvanishing. It follows that U is open and that the m distinct roots of P are given locally in U by holomorphic functions $A_1(z), \ldots, A_m(z)$. Define the function $D: \Delta' \to \mathbf{C}$ by

$$D(z) \equiv \begin{cases} \displaystyle\prod_{i \neq j} (A_i(z) - A_j(z)) & \text{for } z \in U, \\ 0 & \text{otherwise.} \end{cases}$$

Since this expression is symmetric in the $A_j(z)$'s, it is well defined and holomorphic in U. It is easily seen to be continuous in all of Δ. (Recall that the $A_j(z)$'s are uniformly bounded by ρ and at each point of ∂U some difference $A_i(z) - A_j(z)$ must go to zero.) It now follows from a classical result of Rado (cf. [**Wh**, App. III]) that $D(z)$ is holomorphic in all of Δ'. By induction on dimension, we see that $\operatorname{supp}(Z_D)$ has locally finite $\mathfrak{H}_{2n-2}$-measure. Since $\Sigma(V)$ is finitely sheeted over $\operatorname{supp}(Z_D)$, it also has dimension $2n - 2$.

It remains to prove the first statement. Given $p \in \operatorname{supp} V$ we can choose linear coordinates $(z_1, \ldots, z_{n+1})$, centered at p, and $\delta > 0$, so that the restriction of each projection $\pi_j(z) = (z_1, \ldots, z_{j-1}, z_{j+1}, \ldots, z_{n+1})$ to the support of $V_\delta \equiv V \lrcorner \{z: |z_k| < \delta$ for all $k\}$ is proper. (See the first step of the proof of Theorem 1.2.) Set $\Omega_j = (i/2)^n \, dz_1 \wedge d\bar{z}_1 \wedge \cdots \wedge (\widehat{dz_j \wedge d\bar{z}_j}) \wedge \cdots \wedge d\bar{z}_{n+1}$. By Wirtinger's inequality [**F$_1$**] we have $\mathfrak{H}_{2n}(\operatorname{supp} V_\delta) \leq \Sigma V_\delta(\Omega_j) = \Sigma m_j(\pi\delta^2)^n < \infty$ where m_j is the sheeting number of π_j. This completes the proof. $\square$

The smallest set $\Sigma(V)$ as in Corollary 1.3 is called the *singular set* of V. Its complement $\mathfrak{R}(V) \equiv \operatorname{supp} V - \Sigma(V)$ is called the *regular set* of V and consists of all the n-dimensional complex manifold points of $\operatorname{supp} V$. The theorem above shows that any point $p \in \operatorname{supp} V$ has a neighborhood $\mathcal{O}$ such that $\mathfrak{R}(V) \cap \mathcal{O}$ has only a finite number of connected components. If $\mathfrak{R}(V)$ is itself connected, then V is said to be *irreducible*.

Let V_1, V_2, $V_3, \ldots$ denote the connected components of $\Re(V)$. Each V_j has locally finite $\mathfrak{H}_{2n}$-measure in Ω and so, using the canonical orientation of a complex submanifold, we can define a locally rectifiable current $[V_j] \in R^{\mathrm{loc}}_{2n}(\Omega)$.

COROLLARY 1.4. $d[V_j] = 0$ for all j.

PROOF. $d[V_j]$ is a flat $(2n-1)$-current with support in $\Sigma(V)$ where $\mathfrak{H}_{2n-1}(\Sigma(V)) = 0$. Hence, $d[V_j] = 0$ by the Federer Support Theorem [**F$_2$**, 4.1.20]. $\square$

Note that f vanishes to constant order, say m_j, on V_j.

COROLLARY 1.5. *The divisor V is a d-closed, rectifiable $2n$-current which can be written as a locally finite sum*

$$(1.12) \qquad\qquad V = \sum m_j [V_j]$$

where $m_j \in \mathbf{Z}^+$ and where V_j is a connected component of $\Re(V)$ for each j.

2. Holomorphic chains. Currents of type (1.12) arise naturally in all codimensions and we would like to characterize them. For this we need some definitions. Fix a complex manifold Ω of dimension n. By a *complex analytic subset* of Ω we mean a closed subset $Z \subset \Omega$ which is defined locally by the vanishing of a finite number of holomorphic functions. The results of §1 generalize to the study of Z. In particular, Z is locally the graph of a multivalued holomorphic function. Furthermore, there is an integer $p \geq 0$ (called the *dimension* of Z) and a closed subset $\Sigma(Z)$ of Hausdorff dimension $\leq 2p - 2$ (called the *singular set* of Z), such that the *regular set* $\Re(V) \equiv Z - \Sigma(Z) \subset \Omega - \Sigma(Z)$ is a nonempty properly embedded complex submanifold of dimension p, whose Hausdorff $2p$-measure is locally finite in Ω. Consequently, integration over $Z - \Sigma(Z)$ defines a d-closed current $[\mathbf{Z}] \in R^{\mathrm{loc}}_{2p}(\Omega)$.

DEFINITION 2.1. A *holomorphic p-chain* in Ω is a locally finite sum $V = m_j[Z_j]$ where each m_j is an integer and each Z_j is a p-dimensional complex analytic subvariety of Ω. The chain V is called *positive* if $m_j > 0$ for all j.

Our aim in this section will be to characterize the holomorphic p-chains among the locally rectifiable currents. To this end we recall the Dolbeault decomposition

$$\mathscr{E}'_k(\Omega) = \bigoplus_{p+q=k} \mathscr{E}'_{p,q}(\Omega)$$

of the k-currents on Ω.

DEFINITION 2.2. A current $V \in \mathscr{E}'_{2p}(\Omega)$ is said to be of *type p, p* if $V = V_{p,p}$ in the Dolbeault decomposition, i.e., if $V(\varphi) = 0$ for all r, s-forms $\varphi \in \mathscr{E}^{r,s}(\Omega)$ for $r \neq s$.

We let $R^{\mathrm{loc}}_{p,p}(\Omega) = R^{\mathrm{loc}}_{2p}(\Omega) \cap \mathscr{E}'_{p,p}(\Omega)$ denote the group of locally rectifiable p, p-currents. A current $V \subset R^{\mathrm{loc}}_{2p}(\Omega)$ is of type p, p if and only if its approximate tangent plane $\vec{V}_x$ is a complex subspace $\|V\|$-a.e. The current V is called *positive* if, in addition, the orientation on $\vec{V}_x$ is $\|V\|$-a.e., the canonical one induced by the complex structure. We can now state the characterization theorem.

THEOREM 2.3. (HARVEY AND SHIFFMAN [**HS**]). *Any d-closed current $V \in R^{\mathrm{loc}}_{p,p}(\Omega)$ such that $\mathfrak{H}_{2p+1}(\operatorname{supp} V) = 0$ is a holomorphic p-chain.*

COROLLARY 2.4. (J. KING [**K**$_1$]). *Any positive d-closed current $V \in R^{\mathrm{loc}}_{p,p}(\Omega)$ is a positive holomorphic p-chain.*

This corollary follows from 2.3 because monotonicity for positive currents (see [**H**]) shows that the support has locally finite $\mathfrak{H}_{2p}$-measure.

PROOF OF 2.3 (SKETCH). We restrict attentions to the case of codimension one and we fix a point $p \in \operatorname{supp} V$. Since $\mathfrak{H}_{2p+1}(\operatorname{supp} V) = 0$ we can find coordinates $(z_1, \ldots, z_{n-1}, w)$, as in §1, on a product $\Delta' \times \Delta'' = \{(z, w): |z| < r \text{ and } |w| < \rho\}$ so that the z-projection $\Delta' \times \Delta'' \to \Delta'$ is proper on $\operatorname{supp} V$. The idea then is to construct the rational function

$$(2.1) \qquad R(z, w) = \frac{\prod_{i=1}^{k}(w - a_i(z))}{\prod_{j=1}^{l}(w - b_j(z))} = \frac{P(z, w)}{Q(z, w)}$$

so that $V = \mathrm{Div}_R = Z_P - Z_Q$. We note that for $|w| \geq \rho$ we have

$$(2.2) \quad \log R = \log w^{k-l} + \sum_{i=1}^{k} \log\left(1 - \frac{a_i(z)}{w}\right) - \sum_{j=1}^{l} \log\left(1 - \frac{b_j(z)}{w}\right)$$

$$= \log w^{k-l} + \sum_{i=1}^{k} \sum_{p=1}^{\infty} \frac{1}{p}\left(\frac{a_i(z)}{w}\right)^p - \sum_{j=1}^{l} \sum_{p=1}^{\infty} \frac{1}{p}\left(\frac{b_j(z)}{w}\right)^p$$

$$= \log w^{k-l} + \sum_{p=1}^{\infty} \frac{1}{p} \tau_p(z) \frac{1}{w^p},$$

and R can be reconstructed as

$$(2.3) \qquad R(z, w) = w^{k-l} \exp\left(\sum_{p=1}^{\infty} \frac{1}{p} \tau_p(z) \frac{1}{w^p}\right)$$

where the functions $\tau_p(z) = \Sigma a_i(z)^p - \Sigma b_j(z)^p$ are given, as in §1, by

$$(2.4) \qquad \pi_{\#}(w^p V) = \tau_p(z)[\Delta'].$$

Now the functions τ_p given by (2.4) are well defined and holomorphic and the power series (2.2) converges uniformly for $z \in \Delta'$ and $|w| \geq \rho$. The question is whether the function $R(z, w)$ defined by (2.3) is in fact a rational function of w. Note that $R(z, w)$ does have a Laurent expansion of the form

$$(2.5) \qquad R(z, w) = \sum_{q=-\infty}^{N} \sigma_q(z) w^q$$

where the σ_q's are given by certain polynomial expressions in the τ_p's and where $N = \tau_0 = $ constant on Δ'.

We now invoke a classical condition of Hadamard for determining when such a Laurent series represents a rational function. Consider the series $R = \sum_{-\infty}^{N} \sigma_q w^q$ and observe that:

(i) $\exists$ polynomials $Q = \sum_{j=1}^{l} d_j w^j$ and P such that $QR = P$.

$\Leftrightarrow$ (ii) $\exists (d_0, \ldots, d_l) \in \mathbf{C}^{l+1}$ such that $\sum_{j=0}^{l} d_j \sigma_{q-j} = 0$ for all $q < 0$.

$\Leftrightarrow$ (iii) $\det(\Sigma_{i_0} \cdots \Sigma_{i_l}) = 0$ for all multi-indices $0 > i_0 > i_1 > \cdots > i_l$ where $'\Sigma_i \equiv (\sigma_i, \sigma_{i-1}, \ldots, \sigma_{i-l})$.

By rectifiability and slicing we see that $R(z, w)$ is a rational function with, say, $l(z)$ poles for almost all $z \in \Delta$. The Hadamard condition (iii) must therefore be satisfied for some l on an essential set in Δ'. Since these determinants are holomorphic in Z, the condition then holds everywhere in Δ', and $R(z, w)$ is of the form (2.1). $\square$

This theorem has some nice consequences. We say that a complex analytic subset Z has *pure* dimension if $Z = \mathfrak{R}(V)$.

THE REMMERT–STEIN–SHIFFMAN THEOREM. *Let A be a closed subset with $\mathfrak{H}_{2p-1}(A) = 0$ in a complex manifold Ω. Then if $Z \subset \Omega - A$ is a complex analytic subset of pure dimension p, its closure $\overline{Z}$ is a complex analytic subset of Ω.*

PROOF. By a lemma of Shiffman [S], Z has locally finite $\mathfrak{H}_{2p}$-measure in Ω and so integration defines a current $[Z] \in R_{p,p}^{\mathrm{loc}}(\Omega)$. Since $d[Z]$ is flat and supp $d[Z] \subset A$ we conclude that $d[Z] = 0$ by the Federer Support Theorem. Theorem 2.3 now applies. $\square$

As an immediate consequence of this theorem we see, for example, that given any complex analytic subvariety V of Ω, the closure of each connected component V_j of $\mathfrak{R}(V)$ is *itself* a complex analytic subvariety of Ω. These subvarieties $\overline{V}_j$ are called the *irreducible components* of V.

Another important corollary is the following:

CHOW'S THEOREM. *Every complex analytic subset of $\mathbf{P}^n(\mathbf{C})$ is algebraic.*

PROOF. Let $\pi\colon \mathbf{C}^{n+1} - \{0\} \to \mathbf{P}^n(\mathbf{C})$ denote the canonical map, and given $Z \subset \mathbf{P}^n(\mathbf{C})$ consider the complex cone $\pi^{-1}(Z) \subset \mathbf{C}^{n+1} - \{0\}$. By the theorem above, the closure of $\pi^{-1}(Z)$ is analytic at 0, and hence is defined in a neighborhood of 0 by the vanishing of finitely many holomorphic functions, say, $f_1(z), \ldots, f_r(z)$. Write $f_j(z) = \sum_{k=1}^{\infty} p_{jk}(z)$ where $p_{jk}(z)$ is a homogeneous polynomial of degree k. Then since $\pi^{-1}(z)$ is invariant under scalar multiplication by complex numbers, we see that

$$f_j(z) = 0 \Rightarrow f_j(tz) = \sum t^k p_{jk}(z) = 0 \text{ for all } t \text{ with } |t| \text{ small}$$

$$\Rightarrow p_{jk}(z) = 0 \text{ for all } k.$$

It follows from the Noether theorem that finitely many p_{jk}'s will suffice to define $\pi^{-1}(Z)$. $\square$

3. Boundaries of holomorphic chains. We now focus attention on "pieces" of analytic subsets and try to characterize their boundaries. One could ask, for example, the following question: Given a compact oriented real submanifold $M \subset \mathbf{C}^n$ of dimension $2p - 1$, can one determine whether M is the boundary of a complex submanifold (or, more generally, of a holomorphic p-chain)?

There are some natural conditions necessary for M to be such a boundary.

DEFINITION 3.1. The manifold M is said to be *maximally complex* if (any of) the following three equivalent conditions holds:

(i) $\dim_{\mathbf{R}}(T_x M \cap iT_x M) = 2p - 2$ for all $x \in M$.

(ii) $[M] = [M]_{p,p-1} + [M]_{p-1,p}$ in the Dolbeault decomposition
$$\mathscr{E}'_{2p-1}(\mathbf{C}^n) = \bigoplus_{r+s=2p+1} \mathscr{E}'_{r,s}(\mathbf{C}^n).$$

(See Figure 2.)

(iii) M is locally the graph of a function $f: H \to \mathbf{C}^{n-p}$ on a real hypersurface $H \subset \mathbf{C}^p$ which satisfies the tangential Cauchy–Riemann equations $\bar{\partial}_b f = 0$. (See Figure 3.)

DEFINITION 3.2. The manifold M satisfies the *moment condition* if
$$\int_M \alpha = 0 \quad \text{for all } \alpha \in \mathscr{E}^{p,p-1}(\mathbf{C}^n) \text{ with } \bar{\partial}\alpha = 0.$$

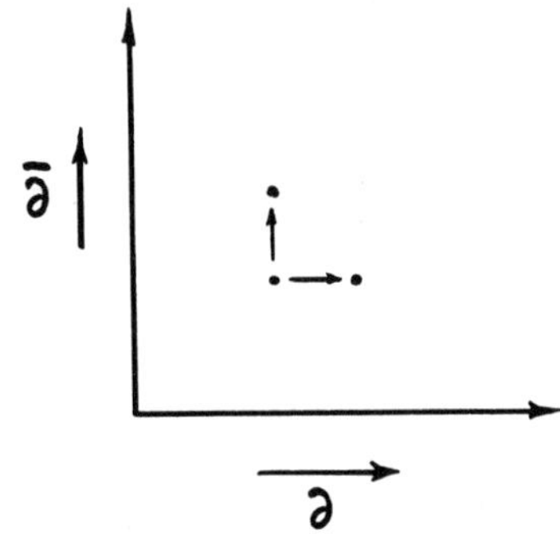

FIGURE 2.

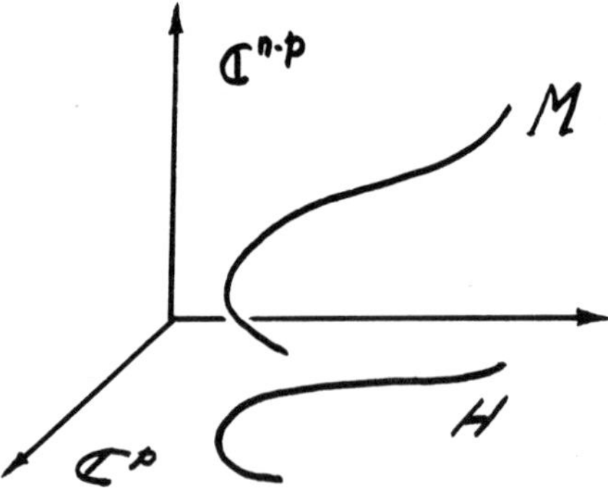

FIGURE 3.

Note. If $p = 1$, then M is automatically maximally complex.

Note. If $p > 1$, then M automatically satisfies the moment condition (since $\bar{\partial}$-closed forms are $\bar{\partial}$-exact in these bidimensions).

Both maximal complexity and the moment condition clearly make sense for general currents on a complex manifold. A d-closed current of dimension $2p - 1$ which satisfies these conditions will be called an *MC-cycle*.

One of the main results we shall discuss is the following theorem proved in $[\mathbf{HL}_1]$. Let Ω be a Stein manifold ($\mathbf{C}^n$, for example).

THEOREM 3.3. (ROUGH STATEMENT). *Every rectifiable MC-cycle in Ω (which is C^1 a.e.) bounds a holomorphic chain.*

This is an interesting contrast with the following result for general MC-cycles, also proved in $[\mathbf{HL}_1]$.

THEOREM 3.4. *There is an isomorphism*

$$H_{2p}(\Omega; \mathbf{R}) \cong \frac{\{ \textit{MC-cycles of dimension } 2p - 1 \}}{d\mathscr{E}_{p,p}(\Omega)_{\mathbf{R}}}.$$

We will now elaborate the main theorem above.

THEOREM 3.3. (PRECISE STATEMENT). *Let M be a compact oriented $(2p - 1)$-dimensional submanifold (with a possible singular set of $\mathfrak{H}_{2p-1}$-measure zero) in a Stein manifold Ω. Assume that M is maximally complex if $p > 1$ or that M satisfies the moment condition if $p - 1$. Then there exists a unique holomorphic p chain V in $\Omega - M$ with $\operatorname{supp} V \Subset \Omega$ and $\mathfrak{M}(V) < \infty$ so that $dV = [M]$.*

Furthermore, there is a compact subset $A \subset M$ of $\mathfrak{H}_{2p-1}$-measure zero so that each point of $M - A$ near which M is of class C^k, $1 \leq k \leq \omega$, has a neighborhood in which $\operatorname{supp} V \cup M$ is a regular class-C^k submanifold with boundary.

Before discussing the proof of this result we make some remarks. Suppose $D \subset \Omega$ is a bounded domain with ∂D of class C^1 and let $f\colon \partial D \to \mathbf{C}$ be a C^1-function. Then f satisfies the tangential Cauchy–Riemann equations $\bar{\partial}_b f = 0$ if and only if the graph of f in $\Omega \times \mathbf{C}$ is maximally complex. Hence, we have the following:

COROLLARY 3.5. (THE BOCHNER EXTENSION THEOREM). *Assume D is a bounded domain with ∂D connected and of class C^1 in a Stein manifold of dimension > 1. Then any $f \in C^1(\partial D)$ which satisfies $\bar{\partial}_b f = 0$ extends to a holomorphic function $F \in \mathcal{O}(D) \cap C^1(\overline{D})$ with $F|_{\partial D} = f$.*

When $p = 1$, the result above retrieves J. Wermer's results on holomorphic approximation on curves $[\mathbf{W}]$.

Concerning boundary regularity in Theorem 3.3 we point out the following. Examples show that in general one does not have boundary regularity everywhere. However, each maximally complex submanifold has an intrinsically defined *Levi form*, and wherever this form has at least one positive eigenvalue, boundary regularity is guaranteed. In particular, if M is pseudoconvex, then M

bounds a holomorphic chain V where supp V has only *isolated* singularities. These singularities are related to the Kohn–Rossi cohomology of M (see Steven Yau [**Y**]).

For an idea of the proof of Theorem 5.3 we consider the case of complex hypersurfaces in $\mathbf{C}^{n+1}$ (where dim $M = 2n - 1$). To begin we choose a good linear projection π of M into $\mathbf{C}^n$. (See Figure 4.)

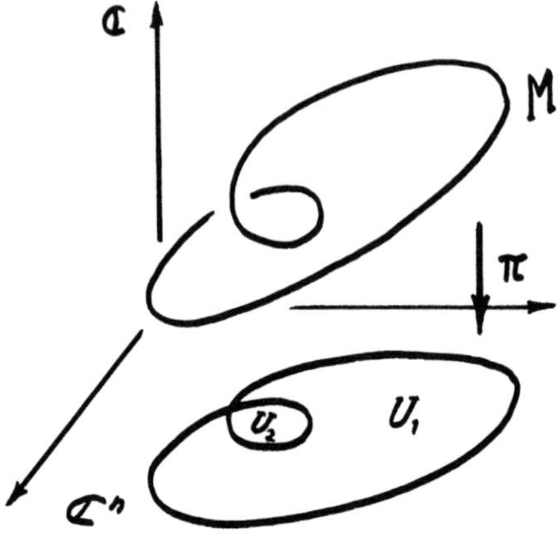

FIGURE 4.

The idea now is to construct a rational function $R(z, w) = \pi(w - a_i(z))/\pi(w - b_j(z))$ over each bounded component U_k of $\mathbf{C}^n - \pi(M)$ so that the divisors of these functions give the desired holomorphic n-chain. To do this, we shall find the holomorphic functions

$$\tau_p(z) = \sum a_i(z)^p - \sum b_j(z)^p$$

and proceed as in §§1 and 2 to construct R. To this end, we observe that if V were the n-chain we are seeking, then we would have $\tau_p[\mathbf{C}^n] = \pi_{\#}(w^p V)$. Taking $\bar{\partial}$ of this equation gives us

$$(3.1) \qquad \bar{\partial}\left(\tau_p[\mathbf{C}^n]\right) = \pi_{\#}(w^p[M])^{0,1}.$$

Now the right-hand side is given data, and so we are faced with finding a solution to the $\bar{\partial}$-equation $\bar{\partial}T_p = \pi_{\#}(w^p[M])^{0,1}$ with *compact support* in $\mathbf{C}^n$.

When $n > 1$, this solution exists and is unique provided that the compatibility condition

$$(3.2) \qquad \bar{\partial}\pi_{\#}(w^p[M])^{0,1} = 0$$

is satisfied. This condition is an immediate consequence of the maximal complexity of M. When $n = 1$, the cohomology group $H^{0,1}_{\text{compact}}(\mathbf{C}) \cong H^{1,0}(\mathbf{C})'$ is not zero, and we must invoke the moment condition to guarantee a compactly supported solution.

It turns out to be useful to solve the equations (3.1) by means of explicit kernels K which satisfy a $\bar{\partial}$-homotopy formula (i.e. "fundamental solution" kernels)

$$(3.3) \qquad \bar{\partial}K + K\bar{\partial} = \bar{\partial}iv(K),$$

and give compatibility with slicing and boundary regularity properties. An in-depth study of such "fundamental solution" kernels is made in [**HP**].

Once the functions τ_p have been given, one can construct the function R by the uniformly convergent series (2.2). The Hadamard conditions for the rationality of R propagate across regular parts of $\pi(M)$ because of boundary regularity for τ_p and the fact that R changes there by a multiplicative factor $(w - f(z))^{\pm 1}$. (Here f is the graphing function for M over $\pi(M)$.) It follows that R is a rational function in each U_k with the desired boundary values and boundary regularity.

Theorem 3.3 generalizes to "larger" complex manifolds. Recall that a Stein manifold Ω is one which admits a proper embedding

$$\Omega \hookrightarrow \mathbf{C}^N = \mathbf{P}^N(\mathbf{C}) - \mathbf{P}^{N-1}(\mathbf{C})$$

for some N, where $\mathbf{P}^N(\mathbf{C})$ denotes complex projective N-space. We now consider complex manifolds Ω which admit proper holomorphic embeddings

$$(3.4) \qquad \Omega \hookrightarrow \mathbf{P}^N(\mathbf{C}) - \mathbf{P}^{N-q}(\mathbf{C})$$

for some N and q. Such manifolds are, in particular, strictly q-convex and even "q-complete" (in that $H^k(\Omega, \mathscr{S}) = 0$ for all $k \geq q$ and all coherent analytic sheaves $\mathscr{S}$.)

THEOREM 3.6. *Let Ω be a complex manifold which admits a proper embedding of type* (3.4). *Suppose that $M \subset \Omega$ is a compact, C^1, oriented real submanifold (with a possible singular set of $\mathring{\mathscr{H}}_{2p-1}$ measure 0) of dimension $2p - 1$ where $p \geq q$. If $p > q$ assume that M is maximally complex, and if $p = q$ assume M satisfies the moment condition. Then all the conclusions of Theorem 3.3 hold for M, i.e., $[M]$ bounds a unique holomorphic p-chain with bounded support in Ω and with good boundary regularity.*

REMARK. The restriction to $p \geq q$ here is important. Note that $\Omega_q \equiv \mathbf{P}^N(\mathbf{C}) - \mathbf{P}^{N-q}(\mathbf{C})$ contains compact complex submanifolds of dimension $q - 1$. Let X be a submanifold of this type which admits a real hypersurface $M \subset X$ whose homology class $[M] \neq 0$ in $H_{2q-3}(X)$. Then M is maximally complex and satisfies the moment condition (which is vacuous since $H^{q-1,q-2}(\Omega_q) = 0$.) Nevertheless, boundary regularity and uniqueness show that M bounds no holomorphic chain in Ω_q. (See Figure 5.)

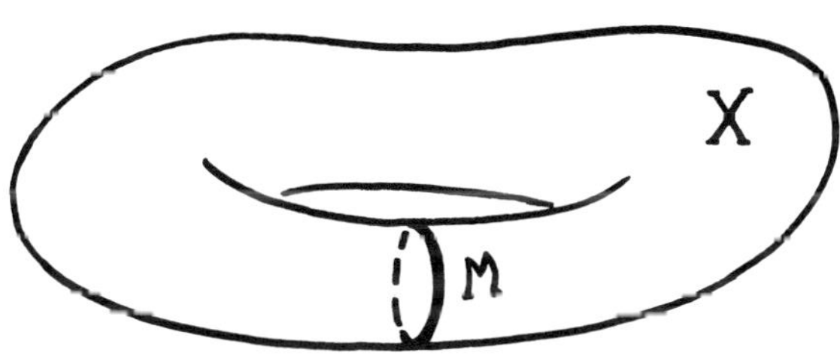

FIGURE 5.

To find such examples of X, consider any algebraic manifold $X \subset \mathbf{P}^N(\mathbf{C})$ with $H^1(X) \neq 0$, e.g., let $X = \mathbf{C}^{q-1}/\Lambda$ where Λ is a lattice satisfying the classical Riemann conditions. A generic $(N - q)$-plane $\mathbf{P}^{N-q}(\mathbf{C})$ will not touch X and can be removed to give $\Omega_q \supset X$. From the isomorphism $H^1(X) = [X, S^1]$ choose a smooth $f: X \to S^1$ with a nonzero class, and let $M \equiv f^{-1}$ (a regular value). Then $[M] \in H_{2q-3}(X)$ is the Poincaré dual of f and is, therefore, not zero.

The question of MC-cycles in $\mathbf{P}^N(\mathbf{C})$ is fascinating. Interesting progress has been made by J. King, P. Dolbeault [D], J. Besnault [BD], and B. Bramaud du Boucheron-Mariaux [Br].

The main results of this section apply to give a Bochner-type extension theorem for meromorphic maps between, say, singular Stein spaces. For example, let $V \subset \mathbf{C}^n$ be an analytic subset in a bounded domain with the property that ∂V is a C^1 submanifold with possible singularities as before. By a *CR-map* of ∂V into a Stein space W, we mean a compact, maximally complex submanifold $\Gamma \subset \mathbf{C}^n \times W$ (with singularities) such that the projection $\pi: \mathbf{C}^n \times W \to \mathbf{C}^n$ restricts to give a *degree-one* map $\pi: \Gamma \to M$. A *meromorphic map* of V into W is defined analogously.

COROLLARY 3.6 [$\mathbf{HL}_{1,2}$]. *Suppose ∂V is connected and of dimension > 1. Then every CR-map of ∂V into a reduced Stein space W extends to a meromorphic map of V into W with boundary regularity almost everywhere.*

4. Projective algebraic cycles. We have seen that complex analytic subsets of $\mathbf{P}^N(\mathbf{C})$ are algebraic, and so the holomorphic chains in $\mathbf{P}^N(\mathbf{C})$ are just the classical algebraic cycles. It is a well-known result of Federer that the *positive* holomorphic chains are homologically mass minimizing in $\mathbf{P}^N(\mathbf{C})$ (and, in fact, also in any Kähler manifold). Of course, the general holomorphic chain, with coefficients of both signs, is not in general minimizing. However, it is *stable* in the sense that it is stationary and the second variation of mass is ≥ 0. The interesting fact is the following.

THEOREM 4.1. *The stable rectifiable currents in $\mathbf{P}^N(\mathbf{C})$ are exactly the algebraic cycles.*

Note. This implies in particular that all odd-dimensional rectifiable currents are unstable in $\mathbf{P}^N(\mathbf{C})$.

PROOF. Let $\mathscr{K}$ denote the space of Killing vector fields (infinitesimal isometries) on $\mathbf{P}^N(\mathbf{C})$. To each $V \in \mathscr{K}$, let φ_t^V be the flow on $\mathbf{P}^N(\mathbf{C})$ generated by the vector field JV where J is the almost complex structure. Suppose now that $T \in \mathbf{R}_p(\mathbf{P}^N(C))$ is stationary and consider the associated quadratic form q_T on $\mathscr{K}$ given by setting

$$q_T(V) = \frac{d^2}{dt^2} \mathfrak{M}\left(\varphi_{t\sharp}^V T\right)\big|_{t=0}.$$

It is a result of Lawson and Simons [**LS**] that

$$\text{trace}(q_T) = -\int_{\mathbf{P}^N(\mathbf{C})} \left\| J\vec{T}_x \right\|^2 d\|T\|(x)$$

where

$$J(v_1 \wedge \cdots \wedge v_p) = \sum_{j=1}^{p} v_1 \wedge \cdots \wedge v_{j-1} \wedge Jv_j \wedge v_{j+1} \wedge \cdots \wedge v_p.$$

It follows that if T is stable, then $p = 2k$ for some k and T is of type k, k. Since T is stationary, $\mathfrak{H}_{2k}(\text{supp } T) < \infty$. Hence, Theorem 2.3 applies directly to prove that T is a holomorphic chain. $\square$

This result has an interesting interpretation. Given $T \in R_{2k}(\mathbf{P}^N(\mathbf{C}))$, consider the function

$$f_T: \text{SL}_{N+1}(\mathbf{C})/\text{SU}_N \to \mathbf{R}^+$$

given by

$$f_T(g) = \mathfrak{M}(g_\sharp T).$$

Then $-\Delta f_T \geqq 0$ and furthermore there exists a point x where $\Delta f_T(x) = 0$ if and only if T is a holomorphic chain.

Similar statements hold if one replaces T by a homology class $\alpha \in H_{2k}(X, \mathbf{Z})$ on an algebraic subvariety $X \subset \mathbf{P}^N(\mathbf{C})$ and defines

$$f_\alpha(g) = \inf\left\{ \mathfrak{M}(g_\sharp T): T \in \alpha \right\} = \inf\{ f_T(g): T \in \alpha \}.$$

(See [**L**] for more details.)

<h3 align="center">REFERENCES</h3>

[**BD**] J. Besnault and P. Dolbeault, *Sur les bords d'ensembles analytiques complexes dans* $\mathbf{P}^n(\mathbf{C})$, Sympos. Math. **24** (1981), 205–213.

[**Bo**] S. Bochner, *Analytic and meromorphic continuation by means of Green's formula*, Ann. of Math. **44** (1943), 652–673.

[**Br**] B. Bramaud du Boucheron-Mariaux, *Chaînes sous-analytiques maximalement complexes*, Thèse de l'Université de Poitiers, France, 1984.

[**C**] W. L. Chow, *On compact complex analytic varieties*, Amer. J. Math. **71** (1949), 893–914.

[**D**] P. Dolbeault, *Sur les chaînes maximalement complexes de bord donné*, (these proceedings).

[**F₁**] H. Federer, *Some theorems on integral currents*, Trans. Amer. Math. Soc. **117** (1965), 43–67.

[**F₂**] H. Federer, *Geometric measure theory*, Springer-Verlag, 1969.

[**H**] R. Harvey, *Holomorphic chains and their boundaries*, Proc. Sympos. Pure Math., Vol. 30, Part 1, Amer. Math. Soc., Providence, R. I., 1977, pp. 309–382.

[**HL₁**] R. Harvey and H. B. Lawson, Jr., *On boundaries of complex analytic varieties, I*, Ann. of Math. **102** (1975) 233–290.

[**HL₂**] R. Harvey and H. B. Lawson, Jr., *On boundaries of complex analytic varieties, II*, Ann. of Math. **106** (1977), 213–238.

[**HP**] R. Harvey and J. C. Polking, *Fundamental solutions in complex analysis I, II*, Duke Math. J. **46** (1979), 253, 300, 301–340.

[**HS**] R. Harvey and B. Shiffman, *A characterization of holomorphic chains*, Ann. of Math. **99** (1974), 553–587.

[**K₁**] J. King, *The currents defined by analytic varieties*, Acta Math. **127** (1971), 185–220.

[L] H. B. Lawson, Jr., *The question of holomorphic carriers*, Proc. Sympos. Pure Math. Vol. 30, Part 2, Amer. Math. Soc., Providence, R. I., (1977), pp. 115–124.

[LS] H. B. Lawson, Jr. and J. Simons, *On stable currents and their applications to global problems in real and complex geometry*, Ann. of Math. **98** (1973) 427–450.

[RS] R. Remmert and K. Stein, *Über die wesentlichen Singularitäten analytischer Mengen*, Math. Ann. **126** (1953), 263–306.

[S] B. Shiffman, *On the removal of singularities of analytic sets*, Michigan Math. J. **15** (1968), 111–120.

[W] J. Wermer, *The hull of a curve in* $\mathbf{C}^n$, Ann. of Math. **68** (1958), 550–561.

[Wh] H. Whitney, *Complex analytic varieties*, Addison-Wesley, Reading, Mass., 1972.

[Y] Stephen S. T. Yau, *Kohn–Rossi cohomology and its application to the complex Plateau problem*, I, Ann. of Math. (2) **113** (1981), 67–110.

RICE UNIVERSITY

STATE UNIVERSITY OF NEW YORK AT STONY BROOK

Proceedings of Symposia in Pure Mathematics
Volume **44** (1986)

Mean Curvature Contraction of Convex Hypersurfaces

GERHARD HUISKEN

We study compact convex hypersurfaces M^n without boundary which are smoothly immersed in $\mathbf{R}^{n+1}$ or some Riemannian manifold N^{n+1}. Let $M^n = M_0$ be given locally by some diffeomorphism

$$F_0 : U \subset \mathbf{R}^n \to F_0(U) \subset M_0 \subset N^{n+1}.$$

We want to move M_0 in the direction of its mean curvature vector, that is, we want to find a whole family $F(\cdot, t)$ of diffeomorphisms corresponding to surfaces M_t such that the evolution equation

$$(1) \qquad \frac{\partial}{\partial t} F(\vec{x}, t) = \vec{H}(\vec{x}, t) \qquad \vec{x} \in U,$$

$$F(\cdot, 0) = F_0,$$

is satisfied. Here $\vec{H}$ is the mean curvature vector of the hypersurface and (1) becomes a quasilinear parabolic system with a smooth solution at least on some short time interval (see Lemma 1). If for example M_0 is a sphere in $\mathbf{R}^{n+1}$, M_t is a family of concentric spheres which shrink towards the centre of the initial sphere in finite time. It was shown in [5], that this behaviour is very typical.

THEOREM 1. *Let $n \geqslant 2$ and assume that $M_0^n \subset \mathbf{R}^{n+1}$ is uniformly convex, i.e. the eigenvalues of its second fundamental form are strictly positive everywhere. Then the evolution equation (1) has a smooth solution M_t on a finite time interval $0 \leqslant t < T$ and the M_t's converge to a single point θ as $t \to T$. Homothetic expansions of the M_t's around θ with fixed total area $|M_0|$ converge to a sphere of area $|M_0|$ in the C^∞-topology.*

REMARKS. (i) The corresponding one-dimensional problem for convex curves in $\mathbf{R}^2$ was solved by Gage and Hamilton [2].

1980 *Mathematics Subject Classification*. Primary 58G11.

© 1986 American Mathematical Society
0082-0717/80 $1.00 + $.25 per page

(ii) Motion by mean curvature is used to model the behaviour of grain boundaries in annealing pure metal. Brakke [1] studied grain boundaries and motion by mean curvature from the view point of geometric measure theory.

If the ambient space N is a general Riemannian manifold, the curvature of N will interfere with the motion of the surfaces M_t. It turns out that the conclusion of Theorem 1 is still valid if we only assume that the initial surface is convex enough to overcome the obstructions imposed by the curvature of N.

In the following Latin indices range from 1 to n, Greek indices range from 0 to n and the summation convention is understood. We denote the induced metric and the second fundamental form on M by $g = \{g_{ij}\}$ and $A = \{h_{ij}\}$. The mean curvature of M is the trace of the second fundamental form, $H = g^{ij}h_{ij}$. We write $\overline{Rm} = \{\overline{R}_{\alpha\beta\gamma\delta}\}$ and $\overline{\nabla}\overline{Rm} = \{\overline{\nabla}_\sigma \overline{R}_{\alpha\beta\gamma\delta}\}$ for the curvature tensor of N and its covariant derivative. Let us denote by $\sigma_x(P)$ the sectional curvature of a plane P at $x \in N$ and let $i_x(N)$ be the injectivity radius of N at x. Let us also agree to write $T_{ij} \geqslant 0$ if all eigenvalues of a symmetric tensor $T = \{T_{ij}\}$ are nonnegative.

THEOREM 2. *Let $n \geqslant 2$ and N^{n+1} be a smooth complete Riemannian manifold which satisfies uniform bounds*

$$-K_1 \leqslant \sigma_x(P) \leqslant K_2, \qquad K_1, K_2 \geqslant 0,$$
$$\left|\overline{\nabla}\overline{Rm}\right|^2 \leqslant L^2, \qquad L \geqslant 0,$$
$$i_x(N) \geqslant i(N) > 0.$$

Let M_0 be a compact connected hypersurface without boundary which is smoothly immersed in N, and suppose that for some $0 < \varepsilon_1 < 1/n$ we have on M_0

(2)
$$H > nK_1^{1/2},$$

(3)
$$h_{ij} \geqslant \varepsilon_1 H g_{ij} + \frac{n(1 - n\varepsilon_1)}{H} K_1 g_{ij} + \frac{n^2}{H^2} L g_{ij}.$$

Then (1) has a smooth solution M_t on a finite time interval $0 \leqslant t < T$ and the M_t's converge to a single point in N as $t \to T$. The surfaces M_t become spherical as $t \to T$ in the sense that all eigenvalues of the second fundamental form approach each other.

REMARKS. (i) Assumption (3) implies $H^2 \geqslant n^2 K_1$, so if (3) holds as a strict inequality, the first condition (2) is automatic. In the case of a locally symmetric ambient space ($\overline{\nabla}\overline{Rm} \equiv 0$) we can choose $L = 0$ and obtain a simpler form of condition (3). In particular, if N is locally symmetric and has nonnegative sectional curvature, then Theorem 2 takes exactly the same form as Theorem 1. In other words, negative sectional curvature puts an obstruction to mean curvature contraction, which has to be overcome by a sufficiently strong convexity assumption.

(ii) Conditions (2), (3) are natural in the sense that they are just strong enough to force all eigenvalues of the intrinsic Ricci tensor of M_0 to be positive (see Lemma 5). Thus M_0 has finite fundamental group $\pi_1(M_0)$ and in the two-dimensional case it follows from the Gauss–Bonnet theorem that M_0 is the immersion of a sphere. In higher dimensions however, this is a nontrivial consequence of Theorem 2: Since M_t is spherical for $t \to T$, already M_0 must have been a sphere. We have

COROLLARY 1. *Any isometric immersion* $M^n \to N^{n+1}$ *with* N *and* M *satisfying the conditions in Theorem 2, is the immersion of a sphere.*

If M_0 is imbedded in N, then it follows from the strong parabolic maximum principle that M_t is imbedded for all $0 \leqslant t < T$. Thus we have

COROLLARY 2. *If* $M^n \to N^{n+1}$ *is an isometric imbedding satisfying the assumptions of Theorem 2, then* M *bounds a region in* N, *and the region is diffeomorphic to a ball.*

The approach to Theorem 1 and Theorem 2 is inspired by Hamilton's paper on *Three-manifolds with positive Ricci curvature* [3]. There the metric of a compact three-dimensional Riemannian manifold with positive Ricci curvature was evolved in direction of the Ricci tensor and a metric of constant positive curvature was obtained in the limit. The evolution equations for the curvature quantities have some similarities, which makes it possible to use some of the ideas in [3].

In the following we give an outline of the proof of Theorem 1 and Theorem 2. Details for the euclidean case can be found in [5] and will appear elsewhere for the general case.

Evolution equations. In local coordinates (1) becomes

$$(1)' \qquad \frac{\partial}{\partial t} F^\alpha(\vec{x}, t) = \vec{H}^\alpha(\vec{x}, t) = \Delta_t F^\alpha(x, t) + \left\{ \overline{\Gamma}^\alpha_{\rho\sigma} \frac{\partial F^\rho}{\partial x_i} \frac{\partial F^\sigma}{\partial x_j} g^{ij} \right\}(\vec{x}, t),$$

where Δ_t is the Laplace–Beltrami operator on M_t and $\overline{\Gamma}$ denotes the connection on N. This is a quasilinear parabolic system and one can show existence at least on some small time interval.

LEMMA 1. *If the initial surface* M_0 *is smooth, then* (1) *has a smooth solution on some maximal open time interval* $0 \leqslant t < T \leqslant \infty$.

Using the Gauss–Weingarten relations one can now derive from (1) evolution equations for all interesting curvature quantities.

LEMMA 2. *Let for a fixed time* $e_0, e_1, e_2, \ldots, e_n$ *be a local orthonormal frame such that restricted to* M, $e_0 = v$ *is normal to* M *and* $e_1, \ldots, e_n$ *are tangential to* M. *Then we have the evolution equations*

(i)
$$\frac{\partial}{\partial t} g_{ij} = -2Hh_{ij},$$

(ii)
$$\frac{\partial}{\partial t} h_{ij} = \nabla_i \nabla_j H - Hh_{il}h^l_j + H\overline{R}_{0i0j},$$

(iii)
$$\frac{\partial}{\partial t} H - \Delta H + H\left(|A|^2 + \overline{\mathrm{Ric}}(v, v) \right).$$

To bring the evolution equation for h_{ij} in a more suitable form, we use Simons'
identity [6] to express $\nabla_i \nabla_j H$ by Δh_{ij}. We obtain

PROPOSITION 1. *We have the evolution equation*

$$\frac{\partial}{\partial t} h_{ij} = \Delta h_{ij} - 2Hh_{il}h_j^l + |A|^2 h_{ij} + h_{ij}\overline{R}_{0l0}{}^l$$

$$- h_{jl}\overline{R}^l{}_{ki}{}^k - h_{il}\overline{R}^l{}_{kj}{}^k + 2h_{kl}\overline{R}^k{}_i{}^l{}_j$$

$$+ \overline{\nabla}_j \overline{R}_{0li}{}^l + \overline{\nabla}_l \overline{R}_{0ij}{}^l.$$

Preserving convexity. In the next step we have to show that the main assumptions on M_0, namely inequalities (2), (3), are preserved during the evolution. In
view of the strict inequality in (2) there is some $\varepsilon_2 > 0$ such that

$$(4) \qquad\qquad\qquad H^2 \geqslant n\varepsilon_2 H^2 + n^2 K_1$$

holds on M_0. Since $|A|^2 \geqslant H^2/n$ and $\overline{\mathrm{Ric}}(\nu, \nu) \geqslant -nK_1$, it follows immediately
from Lemma 2(iii) and the maximum principle that (4) is preserved with the same
ε_2 for all $0 \leqslant t < T$. The harder part, namely that (3) is preserved, follows from a
maximum principle for parabolic systems, [3], when applied to the evolution
equation of A in Proposition 1.

PROPOSITION 2. *If inequalities* (3), (4) *hold on* M_0, *then they remain true on* M_t
for $0 \leqslant t < T$.

The eigenvalues of A. In this step we consider the quantity

$$|A|^2 - \frac{1}{n}H^2 = \left| h_{ij} - \frac{1}{n}Hg_{ij} \right|^2$$

which measures how far the eigenvalues of the second fundamental form diverge
from each other. We show that the eigenvalues approach each other at least at
those points where the mean curvature becomes large.

PROPOSITION 3. *There are constants* $\delta > 0$ *and* $C_0 < \infty$ *depending only on* M_0,
K_1, K_2, L *and* $i(M)$ *such that*

$$|A|^2 - \frac{1}{n}H^2 \leqslant C_0 H^{2-\delta}$$

holds on the whole time interval $0 \leqslant t < T$.

To prove Proposition 3 it is enough to bound the quantity

$$f_\delta = \frac{|A|^2 - H^2/n}{H^{2-\delta}}$$

for sufficiently small $\delta > 0$. The proof of this bound depends heavily on the fact
that inequality (3) is true for all time. One begins with an inequality for the time
derivative of f_δ derived from Lemma 2(iii), Proposition 1 and (3).

LEMMA 3. *We have*

$$\frac{\partial}{\partial t} f_\delta \leqslant \Delta f_\delta + \frac{2(1-\delta)}{H} \langle \nabla_l H, \nabla_l f_\delta \rangle - \frac{1}{2}\varepsilon_1^2 \frac{1}{H^{2-\delta}} |\nabla H|^2$$

$$+ \delta |A|^2 f_\delta + C f_\delta + C \frac{1}{H^{2-\delta}},$$

with a constant C depending only on n, K_1, K_2 and L.

In view of the positive term $\delta |A|^2 f_\delta$ it is not possible to obtain a bound for f_δ by the maximum principle. But the negative term involving $|\nabla H|^2$ can be exploited by the divergence theorem and the Sobolev inequality. At this stage we need the lower bound on the injectivity radius to apply the Sobolev inequality of Hoffmann and Spruck [4], for submanifolds of Riemannian manifolds. An iteration method developed by de Giorgi and Stampacchia leads eventually to Proposition 3.

The maximal time interval. Let again $0 \leqslant t < T$ be the maximal time interval where the smooth solution of (1) exists.

LEMMA 4. *We have $T < \infty$ and $\max_{M_t} |A|^2$ becomes unbounded as $t \to T$.*

PROOF. It follows from a simple comparison argument that the inequality $(\partial/\partial t)H \geqslant \Delta H + \varepsilon_2 H^3/n$ can only be satisfied on a finite time interval since $\min_{M_0} H$ is strictly positive. If $\max_{M_t} |A|^2$ was bounded for $t \to T$, it would be possible to bound higher derivatives of the curvature and to show eventually the existence of a smooth limit surface M_T. Then one could use Lemma 1 to further extend the solution of (1) in contradiction to the maximality of T.

To compare now the maximum value of the mean curvature $H_{\max}$ to the minimum value $H_{\min}$ on M_t, one derives the following gradient bound for the mean curvature from Proposition 3.

PROPOSITION 4. *For any $\eta > 0$ there is a constant C_η depending on C_0, M_0, K_1, K_2 and L such that $|\nabla H|^2 \leqslant \eta H^4 + C_\eta$ holds on $0 \leqslant t < T$.*

Moreover, we have

LEMMA 5. *The intrinsic Ricci curvature R_{ij} of M satisfies*

$$R_{ij} \geqslant (n-1)\varepsilon_1\varepsilon_2 H^2 g_{ij}.$$

PROOF. The Ricci curvature on M is given by Gauss' equation

$$R_{ij} = \overline{R}^l_{ilj} + H h_{ij} - h_{il} h^l_j.$$

Any eigenvalue of $\overline{R}^l_{ilj}$ is the sum of $(n-1)$ sectional curvatures and therefore larger than $-(n-1)K_1$. Any eigenvalue of $H h_{ij} - h_{il} h^l_j$ is larger than $(n-1)H\lambda_1/n$, where λ_1 is the smallest eigenvalue of h_{ij}. But from (3) and (4) we obtain

$$H\lambda_1 \geqslant \varepsilon_1 \left(n^2 K_1 + n\varepsilon_2 H^2 \right) + nK_1 - n^2\varepsilon_1 K_1$$

and the conclusion follows.

In view of Proposition 4 and Lemma 5 we can now use Myer's theorem as in [3 and 5] to compare the mean curvature at different points of the surface and obtain

PROPOSITION 5. $H_{\max}/H_{\min} \to 1$ *as* $t \to T$.

Once this is established it is obvious from Lemma 4 that the diameter of the surfaces M_t tends to zero as $t \to T$. Since the injectivity radius of N is bounded from below, there is $\theta < T$ such that M_θ is contained in some ball $B_\rho(x_0) = \{x \in N \,|\, \mathrm{dist}_N(x_1, x_0) < \rho\}$ where ρ is small compared to $i(N)$ and $(K_1 + K_2)^{-1}$. It is well known that then $B_\rho(x_0)$ is a convex region. Therefore we must have $M_t \subset B_\rho(x_0)$ for all $\theta \leqslant t < T$ and it then follows from $\rho \ll i(N)$, the fact that $H_{\min} \to \infty$ as $t \to T$ and Proposition 3 that for t close to T, M_t is an embedded sphere bounding a convex region. The region enclosed by M_{t_2} is contained in the region enclosed by M_{t_1}, for $t_2 > t_1 \geqslant \theta$ since the surfaces are shrinking. Thus the M_t's converge to a single point as $t \to T$ and it follows from Proposition 3 that the principal curvatures approach each other. This completes the proof of Theorem 2.

REFERENCES

[1] K. A. Brakke, *The motion of a surface by its mean curvature*, Math. Notes, Princeton Univ. Press, Princeton, N.J., 1978.

[2] M. E. Gage, *Curve shortening makes convex curves circular*, preprint.

[3] R. S. Hamilton, *Three-manifolds with positive Ricci curvature*, J. Differential Geom. **17** (1982), 255–306.

[4] D. Hoffman and J. Spruck, *Sobolev and isoperimetric inequalities for Riemannian submanifolds*, Comm. Pure. Appl. Math. **27** (1974), 715–727; **28** (1975), 765–766.

[5] G. Huisken, *Flow by mean curvature of convex surfaces into spheres*, J. Differential Geom. **20** (1984), 237–266.

[6] J. Simons, *Minimal varieties in Riemannian manifolds*, Ann. of Math. (2) **88** (1968), 62–105.

THE AUSTRALIAN NATIONAL UNIVERSITY, AUSTRALIA

$C^{1,\alpha}$ MULTIPLE FUNCTION REGULARITY
AND
TANGENT CONE BEHAVIOUR FOR
VARIFOLDS
WITH SECOND FUNDAMENTAL FORM IN L^P

JOHN E. HUTCHINSON

ABSTRACT. We consider integral varifolds $V \in \mathscr{IV}_n(U)$ where U is an open subset of $\mathbf{R}^N$, having (generalised) second fundamental form in $L^p(V)$ for some $p > n$ in the sense of [**H**] (see also [**H1**]). For such varifolds V we show complete regularity in the following sense. If $x \in \operatorname{spt} \mu_V$ (where μ_V is the measure in $\mathbf{R}^N$ associated with V), then V has a unique tangent cone at x and this tangent cone is a finite union of n-dimensional subspaces P_i with integer multiplicities m_i. Moreover, in some neighbourhood of x we can express V as a finite union of graphs of $C^{1,1-n/p}$, m_i-valued functions defined over the respective affine spaces $x + P_i$.

1. Introduction. In a previous paper [**H**] (see also [**H1**] for a summary) we introduced a notion of second fundamental form for varifolds; more precisely, if U is an open subset of $\mathbf{R}^N$ and V is an n-dimensional integral varifold in $\mathbf{R}^N$ we defined what it means for V to have second fundamental form B in $L^1_{\mathrm{loc}}(\mathbf{G}_n(U), V)$ (cf. 2(14)). In particular, B is a vector-valued function defined V a.e. in $\mathbf{G}_n(U)$, and one should think of B as a function of both position *and* tangent plane direction. If V corresponds to an immersed C^2 submanifold of $\mathbf{R}^N$, then B as above corresponds to the classical second fundamental form in U.

For technical reasons it is convenient to work not with B as above, but with a related function $\mathscr{C}_U V$ also defined V a.e. in $\mathbf{G}_n(U)$ (see 2(13)). One can define B in terms of $\mathscr{C}_U V$ and conversely.

1980 *Mathematics Subject Classification*. Primary 49F22; Secondary 49F20, 58-XX.
Key words and phrases. Curvature varifold, generalised second fundamental form, multiple function regularity.

© 1986 American Mathematical Society
0082-0717/86 $1.00 + $.25 per page

Suppose now that $\mathscr{C}_U V \in L^p(\mathbf{G}_n(U), V)$, where $p > n$, which is equivalent to $B \in L^p(\mathbf{G}_n(U), V)$. We are interested in examining the behaviour of V, and in particular of spt μ_V, where μ_V is the measure in $\mathbf{R}^N$ obtained by projecting V onto $\mathbf{R}^N$.

Suppose further that $x \in$ spt μ_V. We show in 3.4 that V has a varifold tangent cone at x and that this cone is of the form $\sum_{i=1}^J \mathbf{v}(P_i, m_i)$ for some positive integer J and for certain $P_i \in \mathbf{G}(n, N)$ and certain positive integers m_i. Moreover, in some neighbourhood N of x we can write $V = \sum_{i=1}^J V_i$ where each V_i corresponds to the graph of an m_i-valued $C^{1,1-n/p}$ function defined over the n-dimensional affine space $x + P_i$ (see 3.9).

It is interesting to compare our results with the results of Allard [**AW**] concerning the case where V has mean curvature $\mathbf{H} \in L^p(U, \mu_V)$ (one easily shows that this is implied if $\mathscr{C}_U V \in L^p(\mathbf{G}_n(U), V)$ (see [**H**, 5.3.3])). Allard shows that in an open dense subset of $U \cap$ spt μ_V, spt μ_V corresponds locally to the graph of a single-valued $C^{1,1-n/p}$ function. This result cannot be improved to μ_V a.e. regularity. Indeed there is an example in [**B**, 6.1] of a varifold $V \in \mathscr{IV}_2(\mathbf{R}^3)$ with bounded mean curvature $\mathbf{H}$ and a set $A \subset$ spt μ_V with $\mu_V(A) > 0$ such that if $a \in A$ then V does not correspond to the graph of even a multiple-valued function in any neighbourhood of a. Nonetheless, at each $a \in A$ there is a unique varifold tangent cone to V and it is of the form $\mathbf{v}(\mathbf{R}^2, 2)$. Such behaviour is not possible if $\mathscr{C}_{\mathbf{R}^2} V \in L^p_{\mathrm{loc}}(V)$.

The first result we establish is a "monotonicity" type, Formula 3.1, for functions $\psi(P)$ of tangent plane direction. Using an appropriate ψ we then show in 3.3 that in any neighbourhood of each $x \in$ spt μ_V all tangent plane directions are close in a Hölder way (depending on distance from x) to a finite set of tangent plane directions. From this we establish in 3.4 the tangent cone behaviour of V.

It then follows that to analyse the local behaviour of V we can work in a neighbourhood of a point $x \in$ spt μ_V at which V has a varifold tangent *plane P*. It follows from an argument similar to that in [**AW**, §6] that near x, spt μ_V projects orthogonally onto any plane near P with multiplicity which is approximately continuous at x. By means of the Hölder decay result above and a result of Goffman and Waterman [**GW**], one now shows in 3.7(A), (B), that the projection multiplicity is constant. It follows that V locally corresponds to a multiple-valued Lipschitz function.

To prove (3.7(C)) that this Lipschitz function is actually $C^{1,\alpha}$ is technically somewhat involved. One needs a further consequence 3.8 of the monotonicity formula and then one proves the result by induction on multiplicity (see 3.9).

We remark that the present paper is independent of [**H**] in so far as we recall the relevant theorems and definitions in §2. In particular, we do not need any of the convergence, compactness, or existence results from [**H**].

I would like to thank Leon Simon for his continuing interest and advice in this work.

2. Preliminaries and notation. For unexplained notation we refer to [**AW**, **F** or **S**]. In particular, [**AW** and **S**] are the references for material on varifolds.

We will often write $\mathbf{R}^N = \mathbf{R}^n \times \mathbf{R}^{N-n}$, and we denote by

$$\tag{1} \pi, \pi^{\perp}$$

orthogonal projection onto $\mathbf{R}^n$ and $\mathbf{R}^{N-n}$, respectively.

We denote by $\mathbf{G}(n, N)$ the set of n-dimensional subspaces of $\mathbf{R}^N$. There is a natural one-to-one correspondence between members P of $\mathbf{G}(n, N)$ and the corresponding orthogonal projection which we also denote by P. Thus $\mathbf{G}(n, N) \subset \mathbf{R}^{N^2}$ in a natural way, and accordingly the "coordinates" of P, which we denote by $P_{11}, \ldots, P_{1N}, P_{21}, \ldots, P_{NN}$, are the components of the $N \times N$ matrix corresponding to P.

The *Grassmannian* corresponding to an open subset U of $\mathbf{R}^N$ is defined by

$$\mathbf{G}_n(U) = U \times \mathbf{G}(n, N).$$

If $\psi = \psi(P) \in C^0(\mathbf{R}^{N^2})$ or $\psi = \psi(x, P) \in C^0(U \times \mathbf{R}^{N^2})$, then ψ induces a function in $C^0(\mathbf{G}_n(U))$ which we also denote by ψ. In particular, if $V \in \mathscr{IV}_n(U)$ (i.e. V is an integral n-varifold in U, see later) and $E \subset \mathbf{G}(n, N)$, then

$$\tag{2} \int_{U \times E} \psi \, dV$$

is well defined.

If $\psi = \psi(P) \in C^1(\mathbf{R}^{N^2})$ we define

$$\tag{3} D_{ij}^* \psi(P) = \frac{\partial \psi}{\partial P_{ij}}(P),$$

$$\tag{4} |D^*\psi| = \left[\sum_{i,j} \left(D_{ij}^* \psi \right)^2 \right]^{1/2}.$$

If $\psi = \psi(x, P) \in C^1(\mathbf{R}^N \times \mathbf{R}^{N^2})$ we define

$$\tag{5} D_{ij}^* \psi(x, P) = \frac{\partial \psi}{\partial P_{ij}}(x, P),$$

$$\tag{6} D_i \psi(x, P) = \frac{\partial \psi}{\partial x_i}(x, P).$$

Usually $\psi(P)$ and $\psi(x, P)$ as above will be an arbitrary C^1 extension of a C^1 function defined over $\mathbf{G}(n, N)$ or $\mathbf{G}_n(U)$, respectively.

If $T \in \mathrm{Hom}(\mathbf{R}^a, \mathbf{R}^b)$ we define the two norms

$$|T| = \left[\sum_{i,j} T_{ij}^2 \right]^{1/2}, \qquad \|T\| = \sup\{|T(x)| : |x| \leqslant 1\}.$$

It follows

$$\tag{7} \|T\| \leqslant |T| \leqslant \|T\| \min\{\sqrt{a}, \sqrt{b}\}.$$

If $L \in \mathrm{Hom}(\mathbf{R}^n, \mathbf{R}^{N-n})$ then $\mathrm{graph}(L) \in \mathbf{G}(n, N)$. If we denote $\mathrm{graph}(L)$ by P and $\|P - \pi\| < 1$ then

$$(8) \qquad \|L\| \leqslant \|P - \pi\| \left(1 - \|P - \pi\|^2\right)^{-1/2}.$$

(To see this let $Lx = y$; then $(P - \pi)(x, y) = (0, y)$, hence $|y|^2 \leqslant \|P - \pi\|^2(x^2 + y^2)$, and the result follows.) Conversely, if $|L|$ is sufficiently small then

$$(9) \quad |P - \pi| = \sqrt{2}\,|L| +$$

(convergent power series of 2nd and higher order terms in L^{ij});
see [**A**, 1.8] for the calculation.

For $\xi \in \mathbf{R}^N$, $a \in \mathbf{R}^n \subset \mathbf{R}^N$, $\rho > 0$, we will use the notation

$$B_\rho(\xi) = \left\{ z \in \mathbf{R}^N \colon |z - \xi| < \rho \right\}, \qquad B_\rho = B_\rho(0),$$

$$B_\rho^n(a) = \left\{ x \in \mathbf{R}^n \colon |x - a| < \rho \right\}, \qquad B_\rho^n = B_\rho^n(0),$$

$$C_\rho(a) = B_\rho^n(a) \times \mathbf{R}^{N-n}, \qquad C_\rho = C_\rho(0).$$

We next recall some material concerning varifolds.

Following [**S**], we say $M \subset \mathbf{R}^N$ is *countably n-rectifiable* if $\mathrm{M} \subset \mathrm{M}_0 \cup \bigcup_{j=1}^\infty F_j(\mathbf{R}^n)$ where $\mathscr{H}^n(M_0) = 0$ and $F_j \colon \mathbf{R}^n \to \mathbf{R}^N$ are Lipschitz functions for $j = 1, 2, \ldots\,$. If M is countably n-rectifiable and $\mathscr{H}^n$ measurable, and if θ is a positive locally $\mathscr{H}^n$-integrable function on M, we define the *rectifiable n-varifold* $\mathbf{v}(M, \theta)$ to be the equivalence class of all pairs $(\tilde{M}, \tilde{\theta})$, where $\tilde{M}$ is countably n-rectifiable with $\mathscr{H}^n((M \sim \tilde{M}) \cup (\tilde{M} \sim M)) = 0$ and where $\tilde{\theta} = \theta$, $\mathscr{H}^n$ a.e. on $M \cap \tilde{M}$. If θ is an integer $\mathscr{H}^n$ a.e., we say V is an *integral n-varifold*. If moreover U is an open subset of $\mathbf{R}^N$ and $M \subset U$, we say V is an *integral n-varifold* in U, and write $V \in \mathscr{IV}_n(U)$. We will only have need to consider integral varifolds.

If $V \in \mathscr{IV}_n(U)$ we also regard V as a *Radon measure* $\mathbf{G}_n(U)$ in the usual way, see [**AW**, §3 or **S**, §38]. For $\varphi \in C_c^0(\mathbf{G}_n(U))$ we write

$$V(\varphi) = \int \varphi(x, P)\, dV(x, P).$$

If $E \subset U$ we abbreviate

$$\int_E \varphi\, dV = \int_{E \times \mathbf{G}(n, N)} \varphi\, dV.$$

If $V = \mathbf{v}(M, \theta) \in \mathscr{IV}_n(U)$, we denote by $\mu_{\mathbf{v}}$ the corresponding measure on U (or $\mathbf{R}^N$) given by

$$\mu_V(\varphi) = \int \varphi(x)\, dV(x, P) = \int_M \varphi(x)\theta(x)\, d\mathscr{H}^n(x)$$

for $\varphi \in C_c^0(U)$.

If $V, \{V_k\}_{k=1}^\infty \in \mathscr{IV}_n(U)$ we say $V_k \to V$ (in the varifold sense) if $V_k \to V$ in the sense of Radon measures on $\mathbf{G}_n(\mathbf{R}^N)$.

If $P \in \mathbf{G}(n, N)$ and m is a positive integer, we say $V\ (\in \mathscr{IV}_n(U))$ has *varifold tangent* at ξ given by $\mathbf{v}(P, m)$ if

$$\mathbf{v}(P, m) = \lim_{\rho \downarrow 0} V_{\xi, \rho}$$

in the varifold sense. Here we define the varifold $V_{\xi,\rho}$ by

$$V_{\xi,\rho}(\varphi) = \rho^{-n} \int \varphi(\rho x + a, Q)\, dV(x, Q)$$

for all $\varphi \in C_c^0(\mathbf{G}_n(\mathbf{R}^N))$. We denote the varifold tangent at ξ, if it exists, by $T_\xi V$.

Whenever the varifold tangent exists as above we define the *varifold affine approximation* to V at ξ by translation:

$$(10) \qquad\qquad A_\xi V = \mathbf{v}(P + \xi, m).$$

If $V \in \mathscr{IV}_n(U)$ then for μ_V a.e. ξ we have that $T_\xi V$ exists and

$$T_\xi V = \mathbf{v}(\operatorname{Tan}^n(\mu_V, \xi), \theta^n(\mu_V, \xi)),$$

where $\operatorname{Tan}^n(\mu_V, \xi)$ is the approximate tangent plane to μ_v at ξ and $\theta^n(\mu_v, \xi)$ is the n-dimensional density of μ_v at ξ (see [**AW**, 3.5(3)]).

We will be interested in those $V \in \mathscr{IV}_n(U)$ which have generalised mean curvature $\mathbf{H} \in L^P(\mu_V)$ for some $p > n$, and in particular the mean curvature has zero singular part with respect to μ_V. This means (cf. [**AW**, §4] or [**S**, §16])

$$\delta V(X) = \int_U X \cdot \mathbf{H}\, d\mu_V$$

whenever $X \in C_c^0(U; \mathbf{R}^N)$, and

$$(11) \qquad\qquad \int_U |\mathbf{H}|^P d\mu_V < \infty.$$

For such V we can write $V = \mathbf{v}(M, \theta)$, where $M = \operatorname{spt} \mu_V \cap U$ and $\theta(x) = \theta^n(\mu_V, x)$ for $x \in U$ (where $\theta^n(\mu_V, x)$ is the n-dimensional density of μ_V at x). Moreover, M is then relatively closed in U, and θ exists and is finite, is upper semicontinuous, and $\theta(x) \geqslant 1$ for $x \in M$; see [**AW**, 3.5, 8.6 or **S**, 17.9].

In the future, we will always suppose such V are represented in this manner.

It follows that if $T_\xi V = \mathbf{v}(P, m)$ exists, then P is a classical tangent plane to M at ξ, in the sense that

$$(12) \qquad \limsup_{\rho \downarrow 0} \left\{ \rho^{-1} \operatorname{dist}(y, P + \xi);\, y \in M \cap B_\rho(\xi) \right\} = 0,$$

see [**AW**, 8.4 or **S**, 17.11].

We next recall from [**H**, 5.3.1] that if $V \in \mathbf{IV}_n(U)$ then V has *generalised curvature in* U if there exist functions $A_{ijk} \in L^1_{\text{loc}}(V)$ $(1 \leqslant i, j, k \leqslant N)$ such that

$$(13) \qquad 0 = \int \left[P_{ij} D_j \varphi + A_{ijk} D_{jk}^* \varphi + A_{ji} \varphi \right] dV(x, P)$$

for all $i = 1, \ldots, N$ (repeated indices are summed), and all $\varphi \in C_c^1(\mathbf{G}_n(U))$. In this case we write

$$V \in \mathscr{CV}_n(U), \qquad \mathscr{C}_U V = \left\lfloor A_{ijk} \right\rfloor_{i,j,k=1}^N.$$

From [**H**, 5.3.2] we have that the A_{ijk}, if they exist, are determined $V \llcorner U$ almost uniquely.

Recall from [**H**, §5] that if $V = \mathbf{v}(M, \theta)$ where M is a smooth imbedded submanifold of U and $\theta \equiv 1$ on M, then $V \in \mathscr{CV}_n(U)$ and $A_{ijk} = \nabla_{e_i}^M P_{jk}$ (the covariant derivative on M of P_{jk} in the direction e_i^T which is the projection of the ith unit coordinate vector onto the tangent space to M at the appropriate point). Recall also [**H**, 5.2.11] that in this classical case one can define the second fundamental form $B_{ij}^k = \langle B(e_i, e_j), e_k \rangle$ of M in terms of the A_{ijk} and conversely. Namely,

$$(14) \qquad\qquad B_{ij}^k = P_{lj} A_{ikl}, \qquad A_{ijk} = B_{ij}^k + B_{ik}^j.$$

We use (14) to define the *generalised second fundamental form of V in U.*

Notice that the generalised curvature and generalised second fundamental form are defined over $\mathbf{G}_n(U)$, whereas the generalised mean curvature is usually defined over U.

If V has generalised curvature $\mathscr{C}_U V = [A_{ijk}]$ in U, then V has generalised mean curvature $\mathbf{H}$ in U given by

$$(15) \qquad\qquad \mathbf{H}_i(x) = A_{jij}(x, P)$$

for V a.e. (x, P) (see [**H**, 5.3.3]). In particular,

$$(16) \qquad\qquad \int_U |\mathbf{H}|^p \, d\mu_V \leqslant \int_U |\mathscr{C}_U V|^p \, dV.$$

It follows that if

$$\int_U |\mathscr{C}_U V|^p \, dV < \infty,$$

then we can write $V = \mathbf{v}(M, \theta)$ where M and θ are as in the paragraph following (11).

We will always suppose V is so represented.

3. Regularity results. The following monotonicity formula is analogous to a standard result for varifolds V having mean curvature in L^p; see for example [**S**, 17.7]. It is basic to our subsequent results.

3.1. *Monotonicity formula.* Suppose $V \in \mathscr{C}\mathscr{V}_n(B_R(\xi))$, $p > n$, and

$$\left(\int_{B_R(\xi)} |\mathscr{C}_{B_R(\xi)} V|^p \right)^{1/p} \leqslant \Gamma R^{n/p - 1} < \infty.$$

Suppose also that $\psi \in C^1(\mathbf{R}^{N^2})$ and that for all P we have $0 \leqslant \psi(P) \leqslant 1$ and $|D\psi(P)| \leqslant \lambda\psi(P)$.

Then

$$\left[\sigma^{-n} \int_{B_\sigma(\xi)} \psi \, dV \right]^{1/p} \leqslant \left[\rho^{-n} \int_{B_\rho(\xi)} \psi \, dV \right]^{1/p}$$
$$+ (1 + \lambda) \frac{\Gamma}{p - n} \left[\left(\frac{\rho}{R} \right)^{1 - n/p} - \left(\frac{\sigma}{R} \right)^{1 - n/p} \right],$$

whenever $0 < \sigma \leqslant \rho \leqslant R$.

PROOF. Let $\varphi(x, P) = \gamma(r)(x_i - \xi_i)\psi(P)$, where $r = |x - \xi|$, $0 < \rho \leqslant R$, $\gamma \in C^1(\mathbf{R})$, $\gamma(t) = 0$ for $t \geqslant \rho$, $\gamma(t) = 1$ for $t < \rho/2$, and $\gamma'(t) \leqslant 0$ for all t. Let

$A = \mathscr{C}_{B_R} V$. Then from (2.13) we have

$$0 = \int \left[P_{ij}\gamma'(r)(x_j - \xi_j)(x_i - \xi_i)r^{-1}\psi + P_{ij}\gamma(r)\delta_{ij}\psi \right.$$
$$\left. + A_{ijk}\gamma(r)(x_i - \xi_i)D^*_{jk}\psi + A_{jij}\gamma(r)(x_i - \xi_i)\psi \right] dV(x, P).$$

Now $P_{ij}(x_j - \xi_j)(x_i - \xi_i) = r^2 - |(x - \xi)^\perp|^2$, where $(x - \xi)^\perp$ is the projection of $(x - \xi)$ onto the subspace orthogonal to P. Also $P_{ii} = n$. Hence we obtain

$$\int \left[n\gamma(r) + r\gamma'(r) \right] \psi \, dV$$

$$= \int \left[\gamma'(r)|(x - \xi)^\perp|^2 r^{-1} - A_{ijk}\gamma(r)(x_i - \xi_i)D^*_{jk}\psi - A_{jij}\gamma(r)(x_i - \xi_i)\psi \right] dV$$

$$\leqslant (1 + \lambda) \int |A|\gamma(r)|x - \xi|\psi \, dV.$$

Next let $\gamma(r) = \varphi(r/\rho)$ where $\varphi \in C^1(\mathbf{R})$, $\varphi(t) = 1$ if $t \leqslant \frac{1}{2}$, $\varphi(t) = 0$ if $t \leqslant 1$, and $\varphi'(t) \leqslant 0$ for all t. Since

$$\gamma'(r) = \rho^{-1}\varphi'(r/\rho) = -r^{-1}\rho \frac{\partial}{\partial \rho}(\varphi(r/\rho)),$$

we obtain on multiplying through by ρ^{-n-1} that

$$\frac{d}{d\rho}\left[\rho^{-n}\int \varphi(r/\rho)\psi \, dV \right] \geqslant -(1 + \lambda)\rho^{-n-1}\int |A|r\varphi(r/\rho)\psi \, dV$$

$$\geqslant -(1 + \lambda)\rho^{-n}\int |A|\varphi(r/\rho)\psi \, dV$$

$$\geqslant -(1 + \lambda)\rho^{-n}\Gamma R^{n/p-1}\left[\int \varphi(r/\rho)\psi \, dV \right]^{1-1/p}$$

(using Hölder's inequality and noting $0 \leqslant \varphi(r/\rho) \leqslant 1$)

$$\geqslant -(1 + \lambda)\Gamma R^{n/p-1}\rho^{-n/p}\left[\rho^{-n}\int \varphi(r/\rho)\psi \, dV \right]^{1-1/p}.$$

Thus

$$\frac{d}{d\rho}\left[\rho^{-n}\int \varphi(r/\rho)\psi \, dV \right]^{1/p} \geqslant -(1 + \lambda)p^{-1}\Gamma R^{n/p-1}\rho^{-n/p}$$

$$= -\frac{d}{d\rho}\left[(1 + \lambda)\frac{\Gamma}{p - n}(\rho/R)^{1-n/p} \right]$$

If we now integrate from σ to ρ and let φ increase to the characteristic function of $(-\infty, 1)$, the required result follows. $\square$

3.2. *Tangent Plane Measures.* Suppose $V = \mathbf{v}(M, \theta) \in \mathscr{C}\mathscr{V}_n(U)$ where U is an open subset of $\mathbf{R}^N$, and that

$$\int_U |\mathscr{C}_U V|^p \, dV < \infty.$$

We define *tangent plane measures* for $\xi \in U$, $\rho \leqslant d(x, \partial U)$ by

$$(1) \qquad \beta_{\xi,\rho}(\psi) = \int_{B_\rho(\xi)} \psi(P) \, dV/\omega_n \rho^n,$$

$$(2) \qquad \beta_\xi(\psi) = \beta_{\xi,0}(\psi) = \lim_{\rho \downarrow 0} \beta_{\xi,\rho}(\psi),$$

for each $\psi \in C_c^1(\mathbf{G}(n, N))$.

It is clear that $\beta_{\xi,\rho}$ is a Radon measure for $\rho > 0$. Moreover, from 3.1 we see that $\beta_\xi = \beta_{\xi,0}$ is well defined and that β_ξ is then a Radon measure.

For μ_V a.e. $\xi \in U$ we have that $T_\xi V$ exists and then for such ξ we have

$$(3) \qquad \beta_\xi = \theta(\xi)\delta_{T_\xi V},$$

where $\delta_{T_\xi V}$ is the Dirac measure concentrated at $T_\xi V$.

3.3. LEMMA. *Suppose* $V \in \mathscr{C}\mathscr{V}_n(B_R)$ *and* $\mu_V(B_R) \leqslant MR^n$. *Suppose also* $p > n$ *and*

$$\left(\int_{B_R} |\mathscr{C}_{B_R} V|^p \, dV \right)^{1/p} \leqslant \Gamma R^{n/p - 1}.$$

Then there exist $\Gamma^* = \Gamma^*(n, p, M) > 0$, $c^* = c^*(n, p, M)$, *an integer* $J = J(n, p, M)$, *and* $P_1, \ldots, P_J \in \mathbf{G}(n, N)$, *such that if* $0 \leqslant \Gamma \leqslant \Gamma^*$ *and* $(\xi, P) \in \mathrm{spt}(V \llcorner B_{R/2})$, *then* $|P - P_i| \leqslant c^*\Gamma(|\xi|/R)^{1 - n/p}$ *for some* $i = 1, \ldots, J$.

PROOF. Suppose

$$(1) \qquad \xi \in B_{R/2} \quad \text{and} \quad \beta_\xi(P) \geqslant 1$$

(this is true for $V \llcorner B_{R/2}$ a.e. (ξ, P) by 3.1(3)). Our first goal is to prove (13).

For $\varepsilon, \delta \in (0, 1]$, but otherwise yet to be chosen, define $\psi \in C^1(\mathbf{R}^{N^2})$ so that

$$\psi(P) = 1, \qquad \delta \leqslant \psi(Q) \leqslant 1 \quad \text{for all } Q,$$

$$\psi(Q) = \delta \quad \text{if } |Q - P| \geqslant \varepsilon, \text{ and}$$

$$(2) \qquad |D\psi(Q)| \leqslant 2/\varepsilon \quad \text{for all } Q,$$

where

$$|D\psi(Q)| = \left(\sum_{i,j} |D_{ij}^*\psi(Q)|^2 \right)^{1/2}.$$

Henceforth suppose

$$(3) \qquad 0 < \rho < R/2.$$

From 3.1 on letting $\sigma \downarrow 0$, replacing R by $R/2$, Γ by $2^{n/p-1}\Gamma$, and letting $\lambda = 2/(\varepsilon\delta)$ (since $|D\psi(Q)| \leqslant 2|\psi(Q)|/(\varepsilon\delta)$ for all Q), we obtain

$$(4) \qquad \omega_n^{1/p} \leqslant \left(\omega_n \beta_{\xi,\rho}(\psi) \right)^{1/p} + \frac{3}{\varepsilon\delta} \frac{\Gamma}{p - n} \left(\frac{\rho}{R} \right)^{1 - n/p}$$

$$\leqslant \left[\omega_n \beta_{\xi,\rho}\{Q: |Q - P| < \varepsilon\} + \delta\omega_n \beta_{\xi,\rho}(\mathbf{G}(n, N)) \right]^{1/p}$$

$$+ \frac{3}{\varepsilon\delta} \frac{\Gamma}{p - n} \left(\frac{\rho}{R} \right)^{1 - n/p},$$

from (2).

From the mass monotonicity formula [**S**, 17.7], or equivalently 3.1 with $\lambda = 0$ and $\psi \equiv 1$, and from (3), we have

$$
(5) \quad \left[\omega_n \beta_{\xi,\rho}(\mathbf{G}(n, N))\right]^{1/p} = \left[\rho^{-n}\mu_V\left(B_\rho(\xi)\right)\right]^{1/p}
$$

$$
\leqslant \left[(R/2)^{-n}\mu_V\left(B_{R/2}(\xi)\right)\right]^{1/p} + \frac{\Gamma}{p-n}\left(\frac{1}{2}\right)^{1-n/p}
$$

$$
\leqslant (2^n M)^{1/p} + \frac{\Gamma}{p-n}\left(\frac{1}{2}\right)^{1-n/p}.
$$

We now fix $\delta = \delta(n, p, M)$, $c_0 = c_0(n, p, M)$, and $\Gamma^* = \Gamma^*(n, p, M)$ such that

$$
(6) \qquad 0 < \delta \leqslant 1, \qquad \delta\left[(2^n M)^{1/p} + \frac{1}{p-n}\left(\frac{1}{2}\right)^{1-n/p}\right]^p \leqslant \frac{1}{4}\omega_n,
$$

$$
(7) \qquad\qquad c_0 = \left(1 - \left(\frac{3}{4}\right)^{1/p}\right)^{-1} \omega_n^{-1/p} \frac{3}{\delta(p-n)},
$$

$$
(8) \qquad\qquad 0 < \Gamma^* \leqslant 1, \qquad c_0 \Gamma^*\left(\frac{1}{2}\right)^{1-n/p} \leqslant 1.
$$

We impose the restriction

$$
(9) \qquad\qquad\qquad 0 \leqslant \Gamma \leqslant \Gamma^*
$$

and define

$$
(10) \qquad\qquad\qquad \varepsilon = c_0 \Gamma (\rho/R)^{1-n/p}.
$$

Notice ε, $\delta \in (0, 1]$ and so (4) is valid since $0 < \delta \leqslant 1$ by (6) and since $0 < \varepsilon \leqslant 1$ by (10), (3), (9), (8).

We can now estimate two of the terms in (4). From (5), (8), (9), (6), we have

$$
(11) \qquad\qquad \delta\omega_n \beta_{\xi,\rho}(\mathbf{G}(n, N)) \leqslant \tfrac{1}{4}\omega_n.
$$

From (10), (7), we have

$$
(12) \qquad\qquad \frac{3}{\varepsilon\delta} \frac{\Gamma}{p-n}\left(\frac{\rho}{R}\right)^{1-n/p} \leqslant \left(1 - \left(\frac{3}{4}\right)^{1/p}\right)\omega_n^{1/p}.
$$

Substituting (11) and (12) in (4) we obtain

$$
(13) \qquad\qquad \beta_{\xi,\rho}\{Q\colon |Q - P| < \varepsilon\} \geqslant \tfrac{1}{2},
$$

provided $\xi \in B_{R/2}$, $\beta_\xi(P) \geqslant 1$, $0 < \rho < R/2$, $0 \leqslant \Gamma \leqslant \Gamma^* = \Gamma^*(n, p, M)$, $\varepsilon = c_0\Gamma(\rho/R)^{1-n/p}$, and $c_0 = c_0(n, p, M)$.

For $0 < \rho < R/2$ we next define

$$
S_\rho = \left\{ P \in \mathbf{G}(n, N)\colon \beta_\xi(P) \geqslant 1 \text{ for some } \xi \in B_\rho \right\}.
$$

Clearly S_ρ is a decreasing family of sets as $\rho \downarrow 0$.

Now suppose that for some integer K and for $i = 1,\ldots,K$ we have $\beta_{\xi_i}(Q_i) \geqslant 1$, $\xi_i \in B_\rho$, and $|Q_i - Q_j| > 2\varepsilon$ if $i \neq j$ where ε is as in (10). From (13) we have

$$
\text{(14)} \qquad \frac{\mu_V(B_{2\rho})}{\omega_n(2\rho)^n} = \beta_{0,2\rho}(\mathbf{G}(n, N))
$$

$$
\geqslant \sum_{i=1}^{K} \beta_{0,2\rho}\{Q : |Q - Q_i| < \varepsilon\}
$$

$$
\geqslant 2^{-n} \sum_{i=1}^{K} \beta_{\xi_i,\rho}\{Q : |Q - Q_i| < \varepsilon\}
$$

$$
\geqslant 2^{-(n+1)}K.
$$

As in (5) we have

$$
\text{(15)} \qquad \left[\frac{\mu_V(B_{2\rho})}{(2\rho)^n}\right]^{1/p} \leqslant M^{1/p} \leqslant M^{1/p} + \frac{\Gamma}{p-n}\left(\frac{1}{2}\right)^{1-n/p}.
$$

From (14), (15), (9), (8), we see $K \leqslant J = J(n, p, M)$.

Thus for $0 < \rho < R/2$ we have that S_ρ is a subset of the union of J sets each of diameter $\leqslant 4\varepsilon = 4c_0\Gamma(\rho/R)^{1-n/p}$. Since S_ρ is a decreasing family of sets as $\rho \downarrow 0$, it follows there exist $P_1,\ldots,P_J$ such that if $P \in S_\rho$ then $|P - P_i| \leqslant 4c_0\Gamma(\rho/R)^{1-n/p}$ for some $i = 1,\ldots,J$.

Since $\beta_\xi(P) \geqslant 1$ for $V \llcorner B$ a.e. (ξ, P), it follows that for every $(\xi, P) \in \mathrm{spt}(V \llcorner B_\rho)$ we have $|P - P_i| \leqslant 4c_0\Gamma(\rho/R)^{1-n/p}$ for some $i = 1,\ldots,J$. The required result follows with $c^* = 4c_0$. $\square$

Suppose $V \in \mathscr{C}\mathscr{V}_n(U)$ for some open subset U of $\mathbf{R}^N$, and suppose

$$
\int_U |\mathscr{C}_U V|^p < \infty \quad \text{for some } p > n.
$$

Then the following theorem completely describes the tangent cone behaviour of V at each $\xi \in \mathrm{spt}\,\mu_V$.

We remark that the r obtained in the theorem *cannot* be characterised in terms of R, $\theta^n(\mu_V, 0)$, and local curvature and mass bounds alone. The r which we obtain depends in an essential way on $\min_{i\neq j}|P_i - P_j|$, where $P_1,\ldots,P_J$ are as in the theorem. This is the main reason why the proof of $C^{1,1-n/p}$ regularity in Theorem 3.7(C) is somewhat involved.

We also wish to make the following observation at this point. The argument used to establish (i) in the following theorem shows that if $V \in \mathscr{C}\mathscr{V}_n(U)$ and $V = \sum_{i=1}^{J}V_i$ where $\mathrm{spt}\,V_i$ are mutually disjoint, then each $V_i \in \mathscr{C}\mathscr{V}_n(U)$ and $\mathscr{C}_U V_i(x, P) = \mathscr{C}_U V(x, P)$ for V_i a.e. (x, P). However, no such "separation" result is true if we merely require V to have zero first variation, as we easily see by letting V correspond to 3 half-rays in $\mathbf{R}^2$ meeting at the origin at equal angles $2\pi/3$.

3.4. THEOREM. *Suppose $V \in \mathscr{C}\mathscr{V}_n(B_R)$, $0 \in \mathrm{spt}\,\mu_V$, $p > n$, and*

$$\int_{B_R} \left| \mathscr{C}_{B_R} V \right|^p dV < \infty.$$

Then there exist $r \in (0, R)$, an integer J, elements $P_1, \ldots, P_J \in \mathbf{G}(n, N)$, positive integers $m_1, \ldots, m_J$, and varifolds $V_1, \ldots, V_J \in \mathscr{C}\mathscr{V}_n(B_r)$, such that the following are true:

(i) *$V \llcorner B_r = \sum_{i=1}^J V_i$, where $\mathrm{spt}\,V_i$ are mutually disjoint and where $\mathscr{C}_{B_r} V_i(x, P) = \mathscr{C}_{B_r} V(x, P)$ for V_i a.e. (x, P);*

(ii) *$\mathbf{v}(P_i, m_i)$ is the unique varifold tangent to V_i at 0 for $i = 1, \ldots, J$, and hence $\sum_{i=1}^J \mathbf{v}(P_i, m_i)$ is the unique varifold tangent to V at 0;*

(iii) *P_i is a classical tangent plane to $\mathrm{spt}\,\mu_{V_i}$ at 0 for $i = 1, \ldots, J$, in the sense that*

$$\lim_{\rho \to 0} \left\{ \rho^{-1} d(x, P_i) : x \in \mathrm{spt}\,\mu_{V_i} \cap B_\rho \right\} = 0.$$

PROOF. From the previous lemma there exist an integer J, elements $P_1, \ldots, P_J \in \mathbf{G}(n, N)$, and a constant c, such that

$$(1) \qquad \left| P - P_i \right| \leqslant c |x|^{1 - n/p}$$

whenever $(x, P) \in \mathrm{spt}(V \llcorner B_{R/2})$, for some $i = 1, \ldots, J$.

We now choose $r \subset (0, R/2)$ such that

$$(2) \qquad cr^{1-n/p} \leqslant \tfrac{1}{3} \min_{i \neq j} \left| P_i - P_j \right|,$$

and define

$$(3) \qquad V_i = V \llcorner \left\{ (x, P) : x \in B_r \text{ and } |P - P_i| \leqslant \tfrac{1}{3} \min_{i \neq j} |P_j - P_j| \right\}.$$

It is clear that the $\mathrm{spt}\,V_i$ are mutually disjoint, and that $V \llcorner B_r = \sum_{i=1}^J V_i$. Moreover, by choosing r smaller if necessary, we can and will assume that $0 \in \mathrm{spt}\,\mu_{V_i}$ for each $i = 1, \ldots, J$.

We next claim that $V_i \in \mathscr{C}\mathscr{V}_n(B_r)$ and

$$(4) \qquad \mathscr{C}_{B_r} V_i(x, P) = \mathscr{C}_{B_r} V(x, P)$$

for V_i a.e. (x, P). To see this we first recall that for V a.e. (x, P), $\mathbf{v}(P, j)$ is the unique varifold tangent to V at (x, P) where $j = \theta^n(\mu_V, x)$. For such (x, P) it follows from the fact that the $\mathrm{spt}\,V_i$ are mutually disjoint that there is a unique i such that $\mathbf{v}(P, j)$ is a varifold tangent to V_i at (x, P). In particular it follows that each $V_i \in \mathscr{I}\mathscr{V}_n(\mathbf{R}^N)$.

Now suppose $\varphi \in C_c^1(\mathbf{G}_n(B_r))$ is a test function in 2(1). Let χ be a suitable cut-off function with the property that $\chi = 1$ in a neighbourhood of $\mathrm{spt}\,V_i$ and $\chi = 0$ in a neighbourhood of $\mathrm{spt}\,V_j$ for each $j \neq i$. From the fact that 2(1) determines $\mathscr{C}_U V$ almost uniquely with respect to V, it follows that $V \in \mathscr{C}\mathscr{V}_n(B_r)$ and that (4) is true. Thus (1) is established.

We next recall that $\theta^n(\mu_{v_i}, 0)$ exists and is finite and note that moreover $\theta^n(\|\delta V_i\|, 0) = 0$. The latter follows from the fact that for $0 < \rho < r$,

$$\|\delta V_i\|(B_\rho) \leqslant \int_{B_\rho} |\mathscr{C}_{B_r} V_i| dV_i \quad \text{(from 2(15))}$$

$$\leqslant \left[\mu_{V_i}(B_\rho)\right]^{1-1/p} \left[\int_{B_\rho} |\mathscr{C}_{B_r} V|^p dV\right]^{1/p}$$

$$\leqslant c\rho^{n-n/p} \quad \text{(by monotonicity of mass)}.$$

It follows from the scaling properties of μ_{V_i} and $\|\delta V_i\|$, the fact $0 \in \operatorname{spt} \mu_{V_i}$ and so $\theta^n(\mu_{V_i}, 0) \geqslant 1$, and the compactness theorem for integral varifolds, that the set of varifold tangents to V_i at 0 is nonempty and consists of integral varifolds.

If C is a varifold tangent to V_i at 0, it follows from (1) and (3) that $P = P_i$ for C a.e. (x, P). However, from standard results for varifold tangent cones [**AW**, 6.5, 5.2(2)(a) or **S**, 19.3] it follows that $x \in P$ for C a.e. (x, P). Hence $x \in P_i$ for C a.e. (x, P) and so $\operatorname{spt} \mu_C \subset P_i$. Since C is stationary it follows from the constancy theorem [**AW**, 4.6(3) or **S**, 41.1] that $C = \mathbf{v}(P_i, m_i)$ where $m_i = \theta^n(\mu_{V_i}, 0)$. This establishes (ii).

Finally (iii) is just 2(12). $\square$

We now show that if a varifold has zero generalised second fundamental form then it corresponds to a finite sum of affine spaces with positive integer multiplicities.

COROLLARY. *Suppose $V \in \mathscr{CV}_n(U)$ and $\mathscr{C}_U V(x, P) = 0$ for V a.e. $(x, P) \in U$, where U is an open subset of $\mathbf{R}^N$.*

Then

$$v = \sum_{i=1}^{k} \mathbf{v}(M_i, m_i) \llcorner U$$

for some integer k, affine n-dimensional subspaces $M_1, \ldots, M_k$, and nonnegative integers $m_1, \ldots, m_k$.

PROOF. Suppose $a \in \operatorname{spt} \mu_V \cap U$ and without loss of generality let $a = 0$. Now choose r, J, and P_i, m_i, V_i (for $i = 1, \ldots, J$) as in the previous theorem. From (i) of this theorem we see $\mathscr{C}_{B_r} V_i = 0$ V_i a.e. in $\mathbf{G}_n(B_r)$, and in particular we see from 2(15) that V_i is stationary in B_r. From the proof of Lemma 3.3 we see that $|P - P_i| = 0$ for V_i a.e. (x, P) in $\mathbf{G}_n(B_{r/2})$.

We can now apply a height bound type argument. In particular, it follows from [**S**, 19.5] that if $y, z \in \operatorname{spt} \mu_{V_i} \cap B_{r/3}$ and $q(y) \neq q(z)$, where $q(x)$ is the orthogonal projection of x onto the subspace orthogonal to P_i, then we have a contradiction. Thus $\operatorname{spt} \mu_{V_i} \cap B_{r/3} \subset P_i$ (as $0 \in P_i$). By the constancy theorem [**AW**, 4.6(3) or **S**, 41.1] we have that $V \llcorner B_{r/3} = \mathbf{v}(P_i, m_i) \llcorner B_{r/3}$ for some nonnegative integer m_i.

The required result now follows by covering the compact sets spt $\mu_V \cap K$ (for each compact $K \subset U$) with a finite number of open balls corresponding to the $B_{r/3}(0) = B_{r/3}(a)$ of the above argument. $\square$

The following lemma follows fairly easily from the arguments used in [**AW**, 6] to prove the compactness theorem for integral varifolds. See also [**S**, 42.9] for similar arguments.

The conclusion cannot be strengthened to the assertion that f_η is continuous at 0, as one sees by considering the example in [**B**, 6.1].

LEMMA. *Suppose* $V \in \mathscr{IV}_n(B_R)$, $0 \in$ spt μ_V, $\|\delta V\|$ *has no singular part in* B_R, $p > n$, *and*

$$\int_{B_R} |H(V, x)|^p \, d\mu_V(x) < \infty.$$

Suppose also that $\mathbf{v}(P, j)$ *is the unique varifold tangent to* V *at* 0. *Finally suppose that* $\|\pi - P\| < \frac{1}{3}$ *where* π *is projection onto* $\mathbf{R}^n \subset \mathbf{R}^N$.

For each $\eta > 0$ *and* $x \in B_R^n$ *let*

$$f_\eta(x) = \sum \left\{ \theta^m(\mu_V, z) : z \in \pi^{-1}\{x\} \cap B_R \cap \text{spt}\,\mu_V, |\pi^\perp(z)| < \eta \right\}.$$

Then there exists $\eta > 0$ *such that* $f_\eta(0) = j$ *and* f_η *is approximately continuous at* 0.

PROOF. We first check that if $\pi(z_1) = \pi(z_2)$ then

$$|z_1 - z_2| \leqslant |P(z_1 - z_2)| + |P^\perp(z_1 - z_2)|$$
$$\leqslant \tfrac{1}{3}|z_1 - z_2| + |P^\perp(z_1 - z_2)|,$$

and so

$$(1) \qquad |z_1 - z_2| \leqslant \tfrac{3}{2}|P^\perp(z_1 - z_2)|.$$

Let $V_\rho = \mu_{\rho^{-1}\#}V$. By the monotonicity formula [**AW**, 5.1(3) or **S**, 17.7] or 3.1 with $\psi \equiv 1$ we have that for some $r > 0$, $\mu_{V_\rho}(B_1) \leqslant \omega_m(j + 1)$ for all $\rho \in (0, 2r)$. From (1), [**AW**, 6.2], the monotonicity formula again, and the fact that $\theta^n(\mu_V, \xi)$ is an integer for each $\xi \in B_R$ (by 3.4), we see that there exists $\gamma > 0$ such that the following is true:

If $x \in B_\gamma^n$, $\rho \in (0, r)$, and if

$$(2) \qquad \|\delta V_r\|(B_\tau(z)) \leqslant \gamma \mu_{V_r}(B_\tau(z)),$$

$$(3) \qquad \int_{B_\tau(z)} \|P - Q\| dV_\rho(x, P) \leqslant \gamma \mu_{V_r}(B_\tau(z)),$$

whenever $z \in \pi^{-1}\{x\} \cap B_\gamma \cap \text{spt}\,\mu_{V_\rho}$ and $0 < \tau < 1$, then

$$(4) \qquad \sum \left\{ \theta^m(\mu_{V_\rho}, z) : z \in \pi^{-1}\{x\} \cap B_\gamma \cap \text{spt}\,\mu_{V_\rho} \right\} \leqslant j.$$

Let

$$G_\rho = \left\{ z \subset B_\gamma : (2) \text{ and } (3) \text{ are true for all } \tau \in (0, 1) \right\},$$
$$C_\rho = B_\gamma \sim G_\rho.$$

By the Besicovitch covering theorem there exists $c = c(N)$ so that

$$\mu_{V_\rho}(C_\rho) < \frac{c}{\gamma}\left[\|\delta V_\rho\|(B_{1+\gamma}) + \int_{B_{1+\gamma}} \|P - Q\|dV_\rho(x, Q)\right].$$

Since both terms in the right side approach zero as $\rho \to 0$, it follows $\mu_{V_\rho}(C_\rho) \to 0$ as $\rho \to 0$.

From (4) we now have that if

$$g_\rho(x) = \sum\left\{\theta^n(\mu_{V_\rho}, z): z \in \pi^{-1}(x) \cap B_\gamma \cap \operatorname{spt}\mu_{V_\rho}\right\}$$

then

$$(5) \qquad\qquad\qquad g_\rho(x) \leqslant j$$

for all $x \in B_\gamma^n \sim A_\rho$, where $\mathscr{L}^n(A_\rho) \to 0$ as $\rho \to 0$.

However, since $V_\rho \to \mathbf{v}(P, j)$ as $\rho \to 0$, it follows $\pi_{\#}V_\rho \to \mathbf{v}(P, j)$. In other words,

$$\mathbf{v}(\mathbf{R}^n, g_\rho(x)) \to \mathbf{v}(B_\gamma^n, j)$$

and hence $g_\rho(x) \to j$ weakly in B_γ^n. Since g_ρ is integer-valued on B_γ^n, it follows from (5) that $g_\rho(x) = j$ for $x \in B_\gamma^n \sim A_\rho'$, where $\mathscr{L}^n(A_\rho') \to 0$ as $\rho \to 0$.

Since P is a classical tangent plane to $\operatorname{spt}\mu_V$ at 0 and since $\|P - \mathbf{R}^n\| < \frac{1}{3}$ it follows easily that for some $\eta > 0$ and all sufficiently small ρ, $f_\eta(x) = g_\rho(x/\rho)$ for all $x \in B_{\rho\gamma/2}^n$. Hence $f_\eta(x) = j$ for $x \in B_{\rho\gamma/2}^n \sim \rho A_\rho'$, and $\rho^{-n}\mathscr{L}^n(\rho A_\rho') = \mathscr{L}^n(A_\rho') \to 0$ as $\rho \to 0$. Clearly $f_\eta(0) = j$, by choice of η, and so the proof is complete. $\square$

The following proposition is essentially proved in a paper of Goffman and Waterman [**GW**]. Since the proof is short and elegant we give it here for the reader's convenience.

3.6. PROPOSITION. *Suppose U is an open connected subset of $\mathbf{R}^n$ and that f: $U \to \mathbf{R}$ where f is approximately continuous. Then the range of f is connected.*

PROOF. If $\mathscr{R}(f)$, the range of f, is not connected, then there exists a partition $\{A^1, A^2\}$ of $\mathscr{R}(f)$ into nonempty sets both of which are relatively open and closed.

Choose $p_1^i \in U^i = f^{-1}\{A^i\}$ for $i = 1, 2$, and connect p_1^1 and p_1^2 by a continuous path $\varphi(t)$, $0 \leqslant t \leqslant 1$, with $\varphi(0) = p_1^1$ and $\varphi(1) = p_1^2$. Choose $\delta_1 > 0$ so that $B_{\delta_1}(\varphi(t)) \subset U$ for each $t \in [0, 1]$ and so that $\omega_n^{-1}\delta_1^{-n}\mathscr{L}^n[U^i \cap B_{\delta_1}(p_1^i)] \geqslant \frac{1}{2}$ for $i = 1, 2$. Since $\omega_n^{-1}\delta_1^{-n}\mathscr{L}^n[U^i \cap B_{\delta_1}(\varphi(t))]$ is a continuous function of t for $i = 1, 2$, it follows there exists $\xi_1 \in U$, $\xi_1 = \varphi(t)$ for some $t \in [0, 1]$, such that $\omega_n^{-1}\delta_1^{-n}\mathscr{L}^n[U^i \cap B_{\delta_1}(\xi_1)] = \frac{1}{2}$ for both $i = 1, 2$.

Choose $p_2^i \in U^i \cap B_{\delta_1}(\xi_1)$ for $i = 1, 2$. Repeating the above argument with p_1^i replaced by p_2^i ($i = 1, 2$) and U replaced by $B_{\delta_1}(\xi_1)$ we can find $\xi_2 \in B_{\delta_1}(\xi_1)$ and $\delta_2 > 0$ such that $\delta_2 \leqslant \delta_1/2$, $B_{\delta_2}(\xi_2) \subset B_{\delta_1}(\xi_1)$, and $\omega_n^{-1}\delta_2^{-n}\mathscr{L}^n[U^i \cap B_{\delta_2}(\xi_2)] = \frac{1}{2}$ for both $i = 1, 2$.

In this way we obtain a Cauchy sequence $\{\xi_j\}$ with limit ξ, say, and a sequence $\{\delta_j\}$ such that $\{B_{\delta_j}(\xi_j)\}$ is a decreasing sequence of sets with $\omega_n^{-1}\delta_j^{-n}\mathcal{L}^n\{U^i \cap B_{\delta_j}(\xi_j)\} = \frac{1}{2}$ for $i = 1, 2$. Thus $\xi \in \overline{B}_{\delta_j}(\xi_j)$, $B_{\delta_j}(\xi_j) \subset B_{2\delta_j}(\xi)$, and so

$$\omega_n^{-1}(2\delta_j)^{-n}\mathcal{L}^n\{U^i \cap B_{2\delta_j}(\xi)\} \geqslant 2^{-n}2^{-1} = 2^{-n-1}$$

for $i = 1, 2$ and each j. This contradicts the approximate continuity of f at ξ.

Hence $\mathcal{R}(f)$ is connected. $\square$

The following is our main theorem. In conjunction with Theorem 3.4 it implies that if V has second fundamental form in $L^p_{\mathrm{loc}}(V)$ for some $p > n$ then V is locally representable by a finite "union" of graphs of multiple-valued $C^{1,1-n/p}$ functions.

The relevant definitions and material on multiple-valued functions is in the Appendix—see also 2(10) for the definition of the affine varifold approximation $A_z V$ to V at z.

The proof of (A) is almost immediate from 3.5 and 3.6; the proof of (B) is essentially a matter of the relevant definitions. The proof of (C) is more involved; as mentioned before this is caused by the nonexistence of a suitable uniform estimate on r in Theorem 3.4 (see the discussion preceding 3.4).

3.7. THEOREM. *Suppose* $V \in \mathcal{CV}_n(C_R)$ *and* $\mathrm{spt}\,\mu_V \cap C_R$ *is bounded. Suppose* $p > n$ *and*

$$\left[\int_{C_R} |\mathscr{C}_{C_R}V|^p \, dV\right]^{1/p} \leqslant \Gamma R^{n/p-1} < \infty.$$

Finally suppose $\|P - \pi\| \leqslant \theta < \frac{1}{3}$ *for all* $(\xi, P) \in \mathrm{spt}\,V \cap (\mathbf{G}_n(C_R))$.

Conclusions. (A) *For* $a \in B_R^n$ *if*

$$h(a) = \sum\{\theta^n(\mu_V, z): z \in \pi^{-1}\{a\} \cap \mathrm{spt}\,\mu_V\}$$

then h is finite and constant on B_R^n. Let $h(a) = Q$ for $a \in B_R^n$.

(B) V is represented by the graph of a Q-valued function f. Moreover f is affinely approximatable at each $a \in B_R^n$ and $\Sigma\{A_z V: z \in \pi^{-1}\{a\} \cap \mathrm{spt}\,\mu_V\}$ is represented by the graph of $Af(a)$.

(C) For $a \in B_R^n$ let

$$f(a) = \sum_{i=1}^{Q} [\![f_i(a)]\!]$$

and

$$Af(a)(x) = \sum_{i=1}^{Q} [\![f_i(a) + L_i(a)(x)]\!].$$

Then there exist $\Gamma_0 = \Gamma_0(n, p, Q) > 0$, $\theta_0 = \theta_0(n, Q) > 0$, $c_0 = c_0(n, p, Q)$, and $\beta = \beta(n, Q) \in (0, 1)$, such that if $0 \leqslant \Gamma \leqslant \Gamma_0$ and $0 \leqslant \theta \leqslant \theta_0$, then the following is true for any $a_1, a_2 \in B^n_{\beta R}$:

There exists a permutation π of $(1, \ldots, Q)$ such that for $i = 1, \ldots, Q$ both

(i) $|f_i(a_1) - f_{\pi(i)}(a_2)| \leqslant \theta(1 - \theta^2)^{-1/2}|a_1 - a_2|$, and

(ii) $\|L_i(a_1) - L_{\pi(i)}(a_2)\| \leqslant c_0 \Gamma(|a_1 - a_2|/R)^{1-n/p}$.

In particular, f is a $C^{1,1-n/p}$ Q-valued function on $B^n_{\beta R}$.

PROOF OF (A). Suppose $a \in B^n_r$. Since $\pi^{-1}\{a\} \cap \operatorname{spt} \mu_V$ is compact and since at each $z \in \pi^{-1}\{a\} \cap \operatorname{spt} \mu_V$, there is a classical tangent cone which is a finite union of $P \in \mathbf{G}(n, N)$ satisfying $\|P - \pi\| < \frac{1}{3}$ (from 3.4 and the hypothesis), we see that $h(a)$ is always finite.

If we apply 3.4 to each $z \in \pi^{-1}\{a\} \cap \operatorname{spt} \mu_V$ we see there exists $\delta = \delta(a) > 0$ and $j = j(a)$ such that $V \llcorner C_\delta(a) = \sum_{i=1}^{j} V_i$, the $\operatorname{spt} V_i$ are mutually disjoint (although this need not of course be true for $\operatorname{spt} \mu_{V_i}$), for each i, $\pi^{-1}\{a\} \cap \operatorname{spt} \mu_{V_i} = \{z_i\}$ for some z_i (so the z_i need not be all distinct), and for each i there is a unique varifold tangent to V_i at z_i and it is of the form $\mathbf{v}(P_i, m_i)$. Since $(z_i, P_i) \in \operatorname{spt} V$ it follows $\|P_i - \pi\| < \frac{1}{3}$ for each i.

For each i we now define

$$h^i(u) = \sum \left\{ \theta^n(\mu_V, z) : z \in \pi^{-1}\{u\} \cap \operatorname{spt} \mu_{V_i} \right\}$$

for $u \in B^n_\delta(a)$.

By applying 3.5 with the origin at z_i, we see that each h^i is approximately continuous at a (note that from 3.4(iii) there is for each $\eta > 0$ an $\varepsilon > 0$ such that $|\pi(z) - a| < \varepsilon$ and $z \in \operatorname{spt} \mu_{V_i}$ together imply $|\pi^\perp(z - z_i)| < \eta$). Since $h = \sum_{i=1}^{j} h^i$ on $B^n_\delta(a)$ it follows that h is approximately continuous at a and hence approximately continuous in B^n_R.

Since h is integer-valued, as follows from 3.4, it now follows from 3.6 that h is constant on B^n_R. $\square$

PROOF OF (B). From (A) and 4.2 we see that V is represented by the graph of f, where

$$f(x) = \sum_{\pi(z)=x} \theta^n(\mu_V, z) [\![\pi^\perp(z)]\!].$$

Suppose $a \in B^n_R$ and $z \in \operatorname{spt} \mu_V \cap \pi^{-1}\{a\}$. Let $\sum_{i=1}^{J^z} \mathbf{v}(P^z_i, m^z_i)$, where $P_i \neq P_j$ if $i \neq j$ and where the m^z_i are positive integers, be the varifold tangent to V at z and in a neighbourhood E of z let $V \llcorner E = \sum_{i=1}^{J^z} v^z_i$ be the decomposition as in 3.4. It follows from (A) that each V^z_i is represented by the graph of an m^z_i-valued function h^z_i defined in some neighbourhood of a. Moreover, since P^z_i is a classical tangent plane to $\operatorname{spt} \mu_{V_i}$ at z by 3.4(iii), we see that h^z_i is differentiable at a and from 4.2 we see trivially that $A_z V^z_i$ is represented by the graph of $Ah^z_i(a)$.

It now follows by summing that in a neighbourhood of a in $\mathbf{R}^n$ we can represent $V = \sum_{i,z} V^z_i$ by the graph of $\oplus_{i,z} h^z_i$, and so in particular $f = \oplus_{i,z} h^z_i$ in this neighbourhood. Moreover,

$$\sum \left\{ A_z V : z \in \pi^{-1}\{a\} \cap \operatorname{spt} \mu_V \right\}$$

$$= \sum \left\{ A_z V^z_i : z \in \pi^{-1}\{a\} \cap \operatorname{spt} \mu_V, 1 \leqslant i \leqslant J^z \right\},$$

and so is represented by the graph of

$$\bigoplus_{i,z} Ah_i^z(a) = A\left(\bigoplus_{i,z} h_i^z\right)(a)$$

(see 4(10)) $= Af(a)$. $\square$

PROOF OF (C). This follows from Lemma 3.9 by means of a change of origin and rescaling argument, together with the inequalities 2(7), 2(8). $\square$

We first need a technical lemma.

3.8. LEMMA. *Suppose* $V \in \mathscr{C}\mathscr{V}_n(B_R), p > n,$

$$\left[\int_{B_R} |\mathscr{C}_{B_R} V|^p \, dV\right]^{1/p} \leqslant \Gamma R^{n/p-1}.$$

Suppose $E \subset \mathbf{G}(n, N),\ 0 \leqslant \sigma < R,\ \beta_{0,\sigma}(E) \geqslant m.$ *Suppose* $\omega_n^{-1} R^{-n} \mu_V(B_R) \leqslant m + \frac{1}{5}.$

Then there exist $\Gamma^* = \Gamma^*(n, p, m) > 0, c^* = c^*(n, p, m),$ *and* $\alpha = \alpha(n) \in (0, 1)$ *such that if* $0 \leqslant \Gamma \leqslant \Gamma^*,\ (\xi, P) \in \operatorname{spt} V \llcorner B_R,$ *and* $\alpha\sigma \leqslant |\xi| \leqslant \alpha R,$ *then* $d(P, E) \leqslant c^*\Gamma(|\xi|/R)^{1-n/p}.$

PROOF. For $\varepsilon, \delta \in (0, 1]$ but otherwise yet to be chosen, define $\psi \in C^1(\mathbf{R}^{N^2})$ so that

$$\psi(Q) = 1 \quad \text{if } Q \in E, \qquad \delta \leqslant \psi(Q) \leqslant 1 \quad \text{for all } Q,$$
$$\psi(Q) = \delta \quad \text{if } d(Q, E) \geqslant \varepsilon, \text{ and}$$

(1) $$|D\psi(Q)| \leqslant 2/\varepsilon \quad \text{for all } Q,$$

where $|D\psi(Q)| = (\Sigma_{i,j} |D_{ij}^*\psi(Q)|^2)^{1/2}.$

We now apply 3.1 to ψ with $\lambda = 2/(\varepsilon\delta)$ (notice $|D\psi(Q)| \leqslant 2|\psi(Q)|/(\varepsilon\delta)$ for any Q). Since $1 \leqslant 1/(\varepsilon\delta)$ we obtain for $0 \leqslant \sigma \leqslant \rho \leqslant R$ that

(2) $$\left[\omega_n\beta_{0,\sigma}(E)\right]^{1/p} \leqslant \left[\omega_n\beta_{0,\rho}(E_\varepsilon) + \delta\omega_n\beta_{0,\beta}(\mathbf{G}(n, N))\right]^{1/p}$$
$$+ \frac{3}{\varepsilon\delta}\frac{\Gamma}{p-n}\left(\frac{\rho}{R}\right)^{1-n/p}.$$

From 3.1 with $\psi \equiv 1$ and $\lambda = 0$, we have

$$\left[\omega_n\beta_{0,\rho}(\mathbf{G}(n, N))\right]^{1/p} \leqslant \left[\omega_n\beta_{0,R}(\mathbf{G}(n, N))\right]^{1/p} + \frac{\Gamma}{p-n}.$$

But $\beta_{0,R}(\mathbf{G}(n, N)) \leqslant m + \frac{1}{5}$ and hence

(3) $$\beta_{0,\rho}(\mathbf{G}(n, N)) \leqslant m + \frac{2}{5}$$

provided $\Gamma \leqslant c_1$ for some $c_1 = c_1(n, p, m) > 0.$

Since $\beta_{0,\sigma}(E) \geqslant m$, we see from (2) and (3) that if we choose $\delta = \delta(n, p, m)$ sufficiently small and we further require that Γ and ε satisfy

(4) $$c_2\Gamma(\rho/R)^{1-n/p} \leqslant \varepsilon \leqslant 1$$

for suitable $c_2 = c_2(n, p, m) > 0$, then we have

(5) $$\beta_{0,\rho}(E_\varepsilon) \geqslant m - \frac{1}{5},$$

provided $\sigma \leqslant \rho \leqslant R$ (cf. the argument used to prove 3.3(13)).

Next suppose $(\xi, P) \in \operatorname{spt} V \llcorner B_{\alpha\rho}$, where $\alpha \in (0, \frac{1}{2}]$ but α is otherwise yet to be chosen. Notice $\beta_\xi(\{P\}) \geq 1$, as follows from the paragraph after 2(11). Arguing as in the proof of 3.3(4) we obtain

$$\omega_n^{1/p} \leq \left[\omega_n \beta_{\xi,(1-\alpha)\rho}(\{P\})_\varepsilon) + \delta\omega_n\beta_{\xi,(1-\alpha)\rho}(\mathbf{G}(n, N))\right]^{1/p}$$
$$+ \frac{3}{\varepsilon\delta}\frac{\Gamma}{p-n}\left(\frac{(1-\alpha)\rho}{R}\right)^{1-n/p},$$

provided $0 < \varepsilon, \delta \leq 1$.

Now arguing as in the proof of (5) i.e. as for (3.3(13)), we obtain

$$(6) \qquad\qquad \beta_{\xi,(1-\alpha)\rho}(\{P\}_\varepsilon) \geq \tfrac{4}{5},$$

provided $0 \leq \Gamma \leq c_3$ and $c_4\Gamma(\rho/R)^{1-n/p} \leq \varepsilon \leq 1$ for suitable $c_3 = c_3(n, p, m) > 0$ and $c_4 = c_4(n, p, m) > 0$. Under these same conditions it then follows, since $\xi \in B_{\alpha\rho}$, that

$$(7) \qquad\qquad \beta_{0,\rho}(\{P\}_\varepsilon) \geq \tfrac{4}{5}(1-\alpha)^n.$$

Suppose $d(P, E) > 2\varepsilon$. Then we obtain a contradiction from (3), (5), (7), for suitable $\alpha = \alpha(n)$. If we look again at the restrictions necessary for (3), (5), (6), we thus see that if $(\xi, P) \in \operatorname{spt} V \llcorner B_{\alpha\rho}$ and $\sigma \leq \rho \leq R$ then

$$(8) \qquad\qquad d(P, E) \leq 2\max\{c_2, c_4\}\Gamma(\rho/R)^{1-n/P},$$

provided $0 \leq \Gamma \leq \max\{c_1, c_2^{-1}, c_3, c_4^{-1}\}$.

The required result follows from (8) by letting $\rho \to |\xi|/\alpha$ and choosing $c^* = 2\max\{c_2, c_4\}\alpha^{n/p-1}$ and $\Gamma^* = \max\{c_1, c_2^{-1}, c_3, c_4^{-1}\}$. $\square$

Theorem 3.7(C) is implied by the following lemma by means of a change of origin and rescaling argument, together with the inequalities 2(7), 2(8).

3.9. LEMMA. *Suppose* $V \in \mathscr{CV}_n(C_R), p > n,$ *and*

$$\left(\int_{C_R} |\mathscr{C}_{C_R}V|^p\, dV\right)^{1/p} \leq \Gamma R^{n/p-1}.$$

Suppose also that V corresponds to the graph of a Q-valued affinely approximatable function f defined over B_R^n. For each $a \in B_R^n$ and $x \in \mathbf{R}^n$ let $f(a) = \sum_{i=1}^Q [\![f_i(a)]\!]$,

$$Af(a)(x) = \sum_{i=1}^Q [\![f_i(a) + L_i(a)(x)]\!]$$

where $L_1, \ldots, L_Q$ are linear, and let $P_i(a) \in \mathbf{G}(n, N)$ be the graph of $L_i(a)$ for $i = 1, \ldots, Q$. Finally suppose that $\|L_i(a)\| \leq \zeta$ for all $a \in B_R^n$ and all $i = 1, \ldots, Q$.

Then there exist $\Gamma_0 = \Gamma_0(n, p, Q) > 0, \zeta_0 = \zeta_0(n, Q) > 0, c_0 = c_0(n, p, Q),$ and $\beta = \beta(n, Q) \in (0, 1)$ such that if $0 \leq \Gamma \leq \Gamma_0, 0 \leq \zeta \leq \zeta_0,$ and $a \in B_R^n$ then for some permutation π of $(1, \ldots, Q)$ the following are simultaneously true:

$$(i) \qquad\qquad |f_i(0) - f_{\pi(i)}(a)| \leq \zeta|a|,$$

$$(ii) \qquad |P_i(0) - P_{\pi(i)}(a)| \leq c_0\Gamma(|a|/R)^{1-n/p}, \quad \text{for } i = 1, \ldots, Q.$$

PROOF. We already know (i) is true for some π from 4.1 (with $\beta = 1$ and any finite Γ_0 and ζ_0); we need to find a π such that (i) and (ii) hold simultaneously. Naturally π will depend on a.

The proof will be by induction on Q. We first make some observations.

If $J\pi(z, P)$ is the Jacobian of π relative to the pair $(z, P) \in \mathbf{G}_n(\mathbf{R}^N)$ where P is the graph of the linear map $L: \mathbf{R}^n \to \mathbf{R}^{N-n}$ then $(1 + \|L\|^2)^{-n/2} \leqslant J\pi(z, P) \leqslant 1$. It follows from the area formula for mapping varifolds [**AW**, 3.2 or **S**, 15.7] that for any $\rho \in [0, R]$

$$(1) \qquad Q \leqslant \omega_n^{-1}\rho^{-n}\mu_V(C_\rho) \leqslant Q(1 + \zeta^2)^{n/2}.$$

We next let

$$(2) \qquad \operatorname{diam} f(0) = d,$$

and without loss of generality we henceforth suppose by performing a translation in the $\mathbf{R}^{N-n}$ coordinates that

$$(3) \qquad f_1(0) = 0.$$

If $a \in B_R^n$ it follows from 4.1 that

$$(4) \qquad |f_i(a)| \leqslant d + \zeta|a|$$

for $i = 1, \ldots, Q$. Hence

$$\operatorname{spt} \mu_V \cap C_{\rho(1-(\zeta+d/\rho)^2)^{1/2}} \subset B_\rho$$

for $0 < \rho \leqslant R$, provided $\zeta + d/\rho \leqslant 1$. It follows from (1) that

$$(5) \qquad Q\left(1 - \left(\zeta + \frac{d}{\rho}\right)^2\right)^{n/2} \leqslant \omega_n^{-1}\rho^{-n}\mu_V(B_\rho) \leqslant Q(1 + \zeta^2)^{n/2}$$

for $0 < \rho \leqslant R$, provided $\zeta + d/\rho \leqslant 1$ for the first inequality.

We will now prove the lemma by induction on Q, with

$$(6) \qquad \beta(n, Q) = \left(\tfrac{1}{2}\alpha(n)\right)^Q,$$

$$(7) \qquad \Gamma_0(n, p, Q) = \min\left\{\Gamma^*\left(n, p, k - \tfrac{1}{10}\right) : k = 1, \ldots, Q\right\},$$

where α and Γ^* are as in Lemma 3.8. The values of c_0 and ζ_0 will be required to satisfy certain restrictions obtained as we proceed with the proof.

First suppose $Q = 1$. Let $E = \{P_1(0)\}$ and $\sigma = 0$ in 3.8. We henceforth impose the restriction

$$(8) \qquad \left(1 + \zeta_0^2\right)^{1/2} \leqslant 1 + \tfrac{1}{10}.$$

Since $\beta_{0,0}(E) = 1 \geqslant \tfrac{9}{10}$ (from Theorem 3.4, as $f(0) = 0$) and since $\omega_n^{-1}R^{-n}\mu_V(B_R) \leqslant 1 + \tfrac{1}{10} = \tfrac{9}{10} + \tfrac{1}{5}$ from (8) and (5) provided $0 \leqslant \zeta \leqslant \zeta_0$, it follows from 3.8, (4) (with $d = 0$), and (8) (notice that $|(a, f(a))|^2 \leqslant |a|^2 + \zeta^2|a|^2 \leqslant 4|a|^2$ if $0 \leqslant \zeta \leqslant \zeta_0$), that

$$d(P_1(a), P_1(0)) \leqslant c^*\Gamma 2^{1-n/p}(|u|/R)^{1-n/p}$$

provided $0 \leqslant |a| < \alpha R/2, 0 \leqslant \zeta \leqslant \zeta_0, 0 \leqslant \Gamma \leqslant \Gamma^*(n, p, \tfrac{9}{10})$.

Thus, the case $Q = 1$ is true with β and Γ_0 as in (6) and (7), $c_0 = 2^{1-n/p}c*$, ζ_0 subject to (8).

Next assume the lemma is true for any $Q < J$ (where $J \geqslant 2$) with β and Γ_0 as in (6), (7), and with suitable ζ_0 and c_0.

We will prove the lemma for $Q = J$ with β and Γ_0 again as in (6), (7), and with ζ_0 and c_0 subject to certain restrictions obtained as we proceed with the proof. The proof will proceed by defining real numbers $\sigma,\ \tau \in (0, R\,]$ and then successively showing for any $a \in B_{\beta R}^n$ (where $\beta = \beta(n, J)$) with $|a| \in [\,0, \sigma)$, $|a| \in [\,0, \tau)$, or $|a| \in [\,\tau, \beta R)$, the existence of a permutation π of $(1,\ldots,J)$ satisfying (i) and (ii) (assuming $0 \leqslant \Gamma \leqslant \Gamma_0$ and $0 \leqslant \zeta \leqslant \zeta_0$).

Henceforth we define β and Γ_0 by (6) and (7). Moreover, we *fix* $M = M(n, J) > 0$ and *restrict* $\zeta_0 = \zeta_0(n, J)$ so that

$$(9) \qquad J\left(1 - \left(\zeta_0 + \frac{1}{M}\right)^2\right)^{n/2} \geqslant J - \frac{1}{10},$$

$$(10) \qquad J\left(1 + \zeta_0^2\right)^{n/2} \leqslant J + \tfrac{1}{10},$$

ζ_0 sufficiently small that

$$(11) \qquad \|P_i - \pi_i\| < \tfrac{1}{3} \quad \text{for } i = 1,\ldots,Q,$$

and

$$(12) \qquad M \geqslant 1.$$

The reason for these particular inequalities will become clear when we consider the interval $[\,0, \tau)$ (in the case $\tau > \sigma$, see (25) and (26) in particular). We always assume

$$(13) \qquad 0 \leqslant \Gamma \leqslant \Gamma_0, \qquad 0 \leqslant \zeta \leqslant \zeta_0.$$

Unless otherwise clear from context, we will always evaluate our parametrised constants at n, p, J, i.e. $\beta = \beta(n, J)$, $\zeta_0 = \zeta_0(n, p, J)$, etc.

From the definition of d (see (2)) and the fact $J \geqslant 2$, we can find sets S_1, S_2, such that

$$(14) \qquad S = \{f_1(0),\ldots,f_Q(0)\}, \qquad S = S_1 \cup S_2,$$

$$\operatorname{card}(S_j) < J(j = 1, 2), \qquad d(S_1, S_2) \geqslant J^{-1}d.$$

We also define

$$(15) \qquad T = \{P_1(0),\ldots,P_J(0)\}, \qquad d* = \operatorname{diam} T,$$

and consider sets T_1, T_2 such that

$$(16) \qquad T = T_1 \cup T_2, \quad \operatorname{card}(T_j) < J(j = 1, 2), \quad d(T_1, T_2) \geqslant J^{-1}d*$$

The interval $[\,0, \sigma)$. Define

$$(17) \qquad \sigma = \min\{Md, \beta R\}$$

and impose the further restriction

$$(18) \qquad \zeta_0 \leqslant \tfrac{1}{3}J^{-1}\beta M^{-1}.$$

Since $\sigma = 0$ (i.e. $d = 0$) implies $[\,0, \sigma) = \varnothing$, we only consider here the case $\sigma > 0$ (i.e. $d > 0$).

It follows from (13), (18), (14), (17), and 4.1 that $V \llcorner C_{\beta^{-1}\sigma} = V_1 + V_2$ where $\operatorname{spt}\mu_{V_1}$, $\operatorname{spt}\mu_{V_2}$ are mutually disjoint. It follows from the definition 2(13) and the uniqueness [**H**, 5.3.7] of generalised curvature and the use of suitable cut-off functions that for $j = 1, 2$ we have $V_j \in \mathscr{CV}_n(C_{\beta^{-1}\sigma})$,

$$\left[\int_{C_{\beta^{-1}\sigma}} \left|\mathscr{C}_{C_{\beta^{-1}\sigma}} V\right|^p\right]^{1/p} \leqslant \Gamma R^{n/p-1}$$

$$= \Gamma\left(\beta^{-1}\sigma/R\right)^{1-n/p}\left(\beta^{-1}\sigma\right)^{n/p-1}.$$

From (11) and 3.7(B) we see $\|P - \pi\| < \frac{1}{3}$ for all $(\xi, P) \in \operatorname{spt} V_j$, and so we can apply 3.7(B) in $C_{\beta^{-1}\sigma}$ to deduce that each V_j corresponds to a Q_j-valued function $f^{(j)}$. Moreover $Q_j = \operatorname{card}\{k: f_k(0) \in S_j\} < J$ and so we apply the inductive hypothesis to each V_j with R replaced by $\beta^{-1}\sigma$ and Γ by $\Gamma(\beta^{-1}\sigma/R)^{1-n/p}$. Clearly $f(a) = f^{(1)}(a) \oplus f^{(2)}(a)$ for $a \in B_{\beta^{-1}\sigma}^n$.

We now make the further restrictions

(19)$\qquad\qquad \zeta_0 = \zeta_0(n, p, J) \leqslant \min\{\zeta_0(n, p, Q): Q < J\},$

(20)$\qquad\qquad c_0 = c_0(n, p, J) \leqslant \max\{c_0(n, p, Q): Q < J\}.$

Since $\beta(n, p, Q) > \beta(n, p, J) = \beta$ if $Q < J$, it follows from (20) and the inductive hypothesis that to each $a \in B_\sigma^n$ there correspond permutations of $(1, \ldots, Q_j)$ $(j = 1, 2)$ which we can combine to give a permutation π of $(1, \ldots, J)$ such that for $i = 1, \ldots, J$,

(21)$\qquad\qquad\qquad \left|f_i(0) - f_{\pi(i)}(a)\right| \leqslant \zeta|a|,$

(22)$\qquad\qquad \left|P_i(0) - P_{\pi(i)}(a)\right| \leqslant c_0\Gamma\left(\beta^{-1}\sigma/R\right)^{1-n/p}\left(|a|/\beta^{-1}\sigma\right)^{1-n/p}$

$$= c_0\Gamma\left(|a|/R\right)^{1-n/p},$$

provided $0 \leqslant \Gamma(\beta^{-1}\sigma/R)^{1-n/p} \leqslant \min\{\Gamma_0(n, p, Q): Q < J\}$ and $0 \leqslant \zeta \leqslant \min\{\zeta_0(n, p, Q): Q < J\}$. Since $\beta^{-1}\sigma/R \leqslant 1$ from (17), $\Gamma_0 \leqslant \Gamma_0(n, p, Q)$ if $Q < J$ from (7), and $\zeta_0 \leqslant \min\{\zeta_0(n, p, Q): Q < J\}$ from (19), we can replace these two conditions in Γ and ζ by $0 \leqslant \Gamma \leqslant \Gamma_0$ and $0 \leqslant \zeta \leqslant \zeta_0$.

If $\sigma = \beta R$ then we are done; so we henceforth assume

(23)$\qquad\qquad\qquad\qquad \sigma = Md < \beta R.$

The interval $[\,0, \tau)$. We begin by defining

(24)$\qquad\qquad E = \left\{P \in \mathbf{G}(n, N): d(P, T) \leqslant c_0\Gamma(\sigma/R)^{1-n/p}\right\}.$

It follows from (22) that if $\sigma > 0$ then $\operatorname{spt} V \cap \mathbf{G}_n(c_\sigma) \subset C_\sigma \times E$ and so $\beta_{0,\sigma}(E) = \omega_n^{-1}\sigma^{-n}\mu_V(B_\sigma)$, while if $\sigma = 0$ (i.e. $d = 0$) then $E = T$ and so $\beta_{0,\sigma}(E) = J$. If $\sigma > 0$ then from (5), (9), and (23), while if $\sigma = 0$ it is trivial, it follows that

(25)$\qquad\qquad\qquad\qquad \beta_{0,\sigma}(E) > J - \tfrac{1}{10}.$

From (5) and (10) we have

$$(26) \qquad \omega_n^{-1} R^{-n} \mu_V(B_R) \leqslant J + \tfrac{1}{10}.$$

From (4), (12), (23), we see

$$(27) \qquad \left|(a, f_i(a))\right|^2 \leqslant d^2 + (1 + \zeta^2)|a|^2 = \frac{\sigma^2}{M^2} + (1 + \zeta^2)|a|^2$$

$$\leqslant \frac{19}{9}|a|^2 \leqslant 4|a|^2$$

for $\sigma \leqslant |a| < R$.

It follows from Lemma 3.8, (25)–(27), that for $i = 1, \ldots, Q$,

$$d(P_i(a), E) \leqslant 2c^* \Gamma(|a|/R)^{1 - n/p},$$

provided $\sigma \leqslant |a| \leqslant \tfrac{1}{2}\alpha R$, $0 \leqslant \Gamma \leqslant \Gamma^*$, where in the notation of 3.8, $\alpha = \alpha(n)$, $c^* = c^*(n, p, J - \tfrac{1}{10})$, $\Gamma^* = \Gamma^*(n, p, J - \tfrac{1}{10})$. Under these same conditions it then follows from (24) that

$$d(P_i(a), T) \leqslant (2c^* + c_0)\Gamma(|a|/R)^{1 - n/p}.$$

If we now combine this result with (22), we obtain

$$(28) \qquad d(P_i(a), T) \leqslant c_0 \Gamma(|a|/R)^{1 - n/p},$$

provided $a \in B_{\beta R}^n$ (since $\beta \leqslant \tfrac{1}{2}\alpha$), $i = 1, \ldots, Q$, $0 \leqslant \Gamma \leqslant \Gamma_0$ (since $\Gamma_0 \leqslant \Gamma^* = \Gamma^*(n, p, J - \tfrac{1}{10})$ from (7)), $0 \leqslant \zeta \leqslant \zeta_0$, where we now replace the restriction (20) (so far the only restriction on $c_0 = c_0(n, p, J)$) by

$$(29) \qquad c_0 \geqslant 2c^* + \max\{c_0(n, p, Q): Q < J\}.$$

We now define τ' so that

$$(30) \qquad c_0 \Gamma(\beta^{-1}\tau'/R)^{1 - n/p} = \tfrac{1}{3}J^{-1}d^*,$$

and define

$$(31) \qquad \tau = \min\{\tau', \beta R\}.$$

It follows from (16), (28), (30), (31) that $V \llcorner C_{\beta^{-1}\tau} = V_1' + V_2'$ where spt V_1', spt V_2' are mutually disjoint. It then follows from the inductive hypothesis precisely as in the case of the interval $[0, \sigma)$, with σ replaced everywhere in that argument by τ, that for each $a \in B_\tau^n$ there exists a permutation π of $(1, \ldots, J)$ such that for $i = 1, \ldots, J$,

$$(32) \qquad |f_i(0) - f_{\pi(i)}(a)| \leqslant \zeta|a|,$$

$$(33) \qquad |P_i(0) - P_{\pi(i)}(a)| \leqslant c_0 \Gamma(|a|/R)^{1 - n/p},$$

provided $0 \leqslant \Gamma \leqslant \Gamma_0$, $0 \leqslant \zeta \leqslant \zeta_0$, with no further restrictions on ζ_0 or c_0.

The interval $[\tau, \beta R)$. Suppose $a \in B_{\beta R}^n \sim B_\tau^n$. Then $\tau < \beta R$, hence $\tau = \tau'$ from (31), and so $c_0 \Gamma(\beta^{-1}\tau/R)^{1 - n/p} = \tfrac{1}{3}J^{-1}d^*$ from (30). Let π be a permutation of $(1, \ldots, J)$ such that (i) of the lemma is true with $Q = J$ (we know such a π exists from 4.1). From (28) and (15) it follows

$$(34) \qquad \left| P_i(0) - P_{\pi(i)}(a) \right| \leqslant c_0 \Gamma\left(|a|/R \right)^{1-n/p} + 3Jc_0\Gamma\left(\beta^{-1}\tau/R \right)^{1-n/p}$$

$$\leqslant c_0 \Gamma\left(|a|/R \right)^{1-n/p},$$

provided $a \in B_{\beta R}^n \sim B_\tau^n$, $i = 1, \ldots, Q$, $0 \leqslant \Gamma \leqslant \Gamma_0$, $0 \leqslant \zeta \leqslant \zeta_0$, and where we now finally define (compare with the previous restrictions (20), (29))

$$c_0 = \left(1 + 3J\beta^{n/p-1} \right)\left(2c^* + \max\{ c_0(n, p, Q): Q < J \} \right).$$

The required conclusion for $Q = J \geqslant 2$ now follows from (32)–(34), with $\Gamma_0, \zeta_0, c_0, \beta$, as obtained in the proof. $\square$

4. Appendix on multiple-valued functions. We give here the basic facts about multiple-valued functions which are needed in this paper. For an extensive treatment of such functions with applications to regularity for mass-minimising currents see [A].

We define a Q-*tuple* T in $\mathbf{R}^k$ to be an unordered set of Q (not necessarily distinct) elements in $\mathbf{R}^k$. More precisely, if whenever $p \in \mathbf{R}^k$ we write $[\![p]\!]$ for $\mathbf{v}(\{ p \}, 1)$, then a Q-tuple in $\mathbf{R}^k$ is a 0-varifold of the form

$$(1) \qquad T = \sum_{i=1}^{Q} [\![p_i]\!],$$

for some $p_1, \ldots, p_Q \in \mathbf{R}^k$. We sometimes write

$$(2) \qquad p_i \in T,$$

for $i = 1, \ldots, Q$. The set of Q-tuples in $\mathbf{R}^k$ is denoted by

$$(3) \qquad \mathbf{Q}_Q(\mathbf{R}^k) \text{ or } \mathbf{Q}(\mathbf{R}^k).$$

If $T_i \in \mathbf{Q}_{Q_i}(\mathbf{R}^k)$ for $i = 1, \ldots, J$ and $m_1, \ldots, m_J$ are nonnegative integers, then we use usual varifold addition to define

$$(4) \qquad T_1 \oplus T_2 \in \mathbf{Q}_{Q_1+Q_2}(\mathbf{R}^k),$$

$$(5) \qquad \bigoplus_{i=1}^{J} m_i T_i \in \mathbf{Q}_Q(\mathbf{R}^k),$$

where $Q = \sum_{i=1}^{J} m_i Q_i$.

The metric $\mathscr{F}$ on $\mathbf{Q}(\mathbf{R}^k)$ is defined by

$$(6) \quad \mathscr{F}\left(\sum_{i=1}^{Q} p_i, \sum_{i=1}^{Q} q_i \right) = \inf\left\{ \sum_{i=1}^{Q} |p_i - q_{\pi(i)}|: \pi \text{ a permutation of } (1, \ldots, Q) \right\}.$$

If $E \subset \mathbf{R}^n$ and $f\colon W \to \mathbf{Q}(\mathbf{R}^k)$, then we say f is a Q-*valued function on E* (with values in $\mathbf{Q}(\mathbf{R}^k)$).

If f_i are Q_i-valued functions for $i = 1, \ldots, J$ and if $m_1, \ldots, m_J$ are nonnegative integers, then we denote the $Q_1 + Q_2$ and $\sum_{i=1}^{J} m_i Q_i$-valued functions respectively, defined in the natural way, by

$$(7) \qquad f_1 \oplus f_2, \qquad \bigoplus_{i=1}^{J} m_i f_i.$$

We say the Q-valued function f is *continuous* (*continuous at* $a \in E$) if f is *continuous* (*continuous at* $a \in E$) with respect to the $\mathscr{F}$ metric.

If f is a Q-valued function it will often be convenient to write

$$(8) \qquad f(a) = \sum_{i=1}^{Q} [\![f_i(a)]\!],$$

where f_i are real-valued functions with the same domain as f, for $i = 1, \ldots, Q$. *It is important to realise* that even if f is continuous it is not necessarily the case that one can choose the f_i to be continuous functions, even locally. For example, let f: $\mathbf{R}^2 \to \mathbf{R}^2$ be defined by $f(0,0) = (0,0)$ and otherwise $f(x_1, x_2)$ is the 2-tuple consisting of those elements $(y_1, y_2) \in \mathbf{R}^2$ such that $(y_1 + iy_2)^2 = (x_1 + ix_2)$. Then one clearly cannot choose continuous f_1 and f_2 as above in any open set containing the origin.

We say the Q-valued function f: $U \to \mathbf{Q}(\mathbf{R}^k)$, where U is an open subset of $\mathbf{R}^n$, is *affinely approximatable* (or *differentiable*) at $a \in U$ with *affine approximation* $Af(a)$, if there exist affine maps g_i: $\mathbf{R}^n \to \mathbf{R}^k$ for $i = 1, \ldots, Q$ such that

$$f(a) = Af(a)(a), \quad Af(a)(x) = \sum_{i=1}^{Q} [\![g_i(x)]\!]$$

for all $x \in \mathbf{R}^n$, and

$$(9) \qquad \lim_{x \to a} \frac{\mathscr{F}(f(x), Af(x)(x))}{|x - a|} = 0.$$

It follows that if $Af(a)$ exists it is unique; also f is then continuous at a.

If f_1 and f_2 are affinely approximatable at a, then so is $f_1 \oplus f_2$, and clearly

$$(10) \qquad A(f_i \oplus f_2)(a) = Af_1(a) \oplus Af_2(a).$$

We say f is *affinely approximatable* (or *differentiable*) in U if f is affinely approximatable at each $a \in U$.

Suppose f is affinely approximatable in $U \subset \mathbf{R}^n$. At each $a \in U$ we can clearly write

$$(11) \qquad Af(a)(x) = \sum_{i=1}^{Q} [\![f_i(a) + L_i(a)(x - a)]\!]$$

for all $x \in \mathbf{R}^n$, where L_i is linear for $i = 1, \ldots, Q$. We say f is $C^{1,\alpha}$ in U, where $0 < \alpha \leqslant 1$, if there exists a constant c such that

$$(12) \qquad \inf_{\pi} \left\{ \sum_{i=1}^{Q} \left(\frac{|f_i(a_1) - f_{\pi(i)}(a_2)|}{|a_1 - a_2|} + \frac{L_i(a_1) - L_{\pi(i)}(a_2)}{|a_1 - a_2|^{\alpha}} \right) \right.$$

$$\left. : \pi \text{ a permutation of } (1, \ldots, Q) \right\} \leqslant c$$

for all $a_1 \neq a_2$, where $a_1, a_2 \in U$. (Appropriate modifications of the following proposition can be used to show that, in case U is bounded, one obtains an equivalent notion if one replaces the summand by $\|L_i(a_1) - L_{\pi(i)}(a_2)\| / |a_1 - a_2|^{\alpha}$.)

We will need the following elementary result.

4.1. PROPOSITION. *Suppose $U \subset \mathbf{R}^n$ is open and convex and $f: U \to \mathbf{Q}(\mathbf{R}^k)$ is affinely approximatable in U. For each $a \in U$ let*

$$Af(a)(x) = \sum_{i=1}^{Q} [\![f_i(a) + L_i(a)(x - a)]\!] \quad \text{for all } x \in \mathbf{R}^n,$$

where L_i is linear for $i = 1, \ldots, Q$. Suppose $\|L_i(a)\| \leqslant \lambda$ for all $a \in U$ and each i.

Then for each $a_1, a_2 \in U$, if we write $f(a_j) = \sum_{i=1}^{Q} [\![f_i(a_j)]\!]$ for $j = 1, 2$, there exists a permutation π of $(1, \ldots, Q)$ such that

$$\left| f_i(a_1) - f_{\pi(i)}(a_2) \right| \leqslant \lambda |a_1 - a_2|$$

for $i = 1, \ldots, Q$.

PROOF. Suppose $\varepsilon > 0$ and $a_1, a_2 \in U$. Define

$$S = \{ t \in [0, 1] : \text{for each } s \in [0, t] \text{ there exists } \pi \text{ such that}$$

$$\left| f_i(a_1) - f_{\pi(i)}((1 - s)a_1 + sa_2) \right| \leqslant (\lambda + \varepsilon)s|a_1 - a_2| \text{ for all } i = 1, \ldots, Q \}.$$

Clearly $0 \in S$. Moreover S is closed, essentially since f is continuous at $(1 - t)a_1 + ta_2$ and using the fact that the number of permutations of $(1, \ldots, Q)$ is finite.

It follows from the definition of Af that for each $u \in U$ there is a $\delta > 0$ such that if $v \in U$ and $|u - v| < \delta$ then for some permutation π we have $|f_i(u) - f_{\pi(i)}(v)| \leqslant (\lambda + \varepsilon)|u - v|$ whenever $i = 1, \ldots, Q$. It follows from the definition of S that S is open.

Thus $S = [0, 1]$ and so $|f_i(a_1) - f_{\pi(i)}(a_2)| \leqslant (\lambda + \varepsilon)|a_1 - a_2|$ for some π and all $i = 1, \ldots, Q$. If we let $\varepsilon \downarrow 0$ the result follows (noting there exist only a finite number of permutations of $(1, \ldots, Q)$). $\square$

4.2. DEFINITION. Suppose $V \in \mathscr{IV}_n(U \times \mathbf{R}^{N-n})$, where U is an open subset of $\mathbf{R}^n$. Suppose also that $\theta^n(\mu_V, z)$ is a positive integer for each $z \in \operatorname{spt} \mu_V$ and that $\Sigma\{ \theta^n(\mu_V, z) : \pi(z) = x \} = Q$ for all $x \in U$. Then we say V *is represented by the graph of the Q-valued function f* where

$$(13) \qquad f: U \to \mathbf{Q}(\mathbf{R}^{N-n}), \qquad f(x) = \sum_{\pi(z) = x} [\![\theta^n(\mu_V, z) \pi^{\perp}(z)]\!]$$

for all $x \in U$.

Notice that under the assumptions of the above definition we can recover V from f by

$$V = \mathbf{v}(M, \theta), \qquad M = \{ (x, y) : y \in f(x) \},$$

$$(14) \qquad \theta(x, y) = \text{multiplicity of } y \text{ in } f(x).$$

REFERENCES

[A] F. J. Almgren, Jr., *Q valued functions minimising Dirichlet's integral and the regularity of area minimising rectifiable currents up to codimension two*, preprint

[AW] W. K. Allard, *On the first variation of a varifold*, Ann. of Math. (2) **95** (1972), 417–491.

[B] K. A. Brakke, *The motion of a surface by its mean curvature*, Math. Notes, No. 20., Princeton Univ. Press, Princeton, N.J., 1978.

[F] H. Federer, *Geometric measure theory*, Springer-Verlag, 1969.

[GW] C. Goffman and D. Waterman, *Approximately continuous transformations*, Proc. Amer. Math. Soc. **12** (1961), 116–126.

[H] J. E. Hutchinson, *Second fundamental form for varifolds and the existence of surfaces minimising curvature*, preprint.

[H1] ______, *Minimising curvature—a higher dimensional analogue of the Plateau problem*, Proc. Centre for Mathematical Analysis, (N. Trudinger and G. Williams, eds.) Australian National University, vol. 8, 1984.

[S] L. Simon, *Lectures on geometric measure theory*, Proc. Centre for Mathematical Analysis, vol. 3, 1984.

AUSTRALIAN NATIONAL UNIVERSITY

Proceedings of Symposia in Pure Mathematics
Volume **44** (1986)

Pointwise Pinched Manifolds are Space Forms

CHRISTOPHE MARGERIN

The Riemann curvature tensor field is sort of a monster which turns out not to be well understood. In today's Riemannian geometry, a main trend consists in finding connections between local assumptions on the metric and the topology of the underlying manifold, in other words in finding obstructions (mainly topological) to nice properties of the curvature.

In this spirit, R. Hamilton brought forth an important paper [**H**] in which he showed that it was definitely possible to give a positive answer to the following conjecture: Any compact 3-manifold carrying a metric of positive Ricci curvature is diffeomorphic to a space form of positive constant curvature (hence its universal cover is a sphere).

This was done by means of deformation of the initial metric. More than an encouragement to hope for an analogous statement holding in higher dimensions, the paper gave a fundamental tool, namely the "long time existence theorem" which we want to restate [**H**, 14.1]:

"The evolution equation $\partial g / \partial t = -2r$ has a unique solution on a maximal time interval $0 \leqslant t < T \leqslant \infty$. If $T < \infty$ then $\max|R|$ blows up as $t \to T$. This holds for every compact manifold, whatever its dimension may be."

An inviting problem is to try and generalize the three-dimensional result of Hamilton to higher dimensions. Unfortunately, the simple example of $S^2 \times S^2$, with the canonical product metric, shows there is no hope for a dumb mimic to do the job: this metric has positive Ricci curvature but $S^2 \times S^2$, though simply connected, is not diffeomorphic to S^4!

Close to the previous problem is the question of metrics with positive curvature operators (see further for a definition). Is any manifold carrying a metric of positive curvature operator diffeomorphic to a space form (of positive curvature)?

1980 *Mathematics Subject Classification.* Primary 58G11, 53C20.

© 1986 American Mathematical Society
0082-0717/86 $1.00 + $.25 per page

The process of deforming the initial metric according to the Ricci flow ($\partial g/\partial t = -2r$) does not enable us to show the result. Nevertheless, we prove that this flow preserves positiveness for the curvature operator, [Theorem 1], which is a decisive step. This first theorem is joint work with J. P. Bourguignon and J. Lafontaine.

Computation indicates it is worth trying to prove the statement if one imposes a suitable pinching on the curvature (a pointwise estimate suffices). We will show the following:

THEOREM 2. *Any compact n-manifold carrying a metric whose curvature is at least $C(n)$-pinched, is diffeomorphic to a space of constant curvature.*

By a $C(n)$-pinched metric we mean that $u > 0$ and $|\mathscr{D}| < [C(n)]^{1/2}u$, where u is the scalar curvature and

$$\mathscr{D} = R - \frac{u}{2n(n-1)}g\bigcirc\!\!\!\!\!\wedge\, g$$

is the total deviation from being of constant curvature [see further notation]. We show that a possible value for $C(n)$ is $1/(2n(n-1)(n-2))$. Because this is asymptotically strictly larger than $4/((n-2)(n-1)n(n+1))$ (in fact as soon as $n > 6$), our hypothesis does not trivially imply the curvature operator is positive. So, the statement is closely related to the study of bounded curvature ratios on Riemannian manifolds, and we have found it useful to restate the theorem in terms of an a priori estimate on the sectional curvatures [Theorem 3]:

Any compact n-manifold M whose sectional curvature is pointwise $\delta(n)$-pinched (i.e. $\sigma_{\min}(x) \geqslant \delta(n) \cdot \sigma_{\max}(x)$, $\forall x \in M$) is diffeomorphic to a space form providing $\delta(n)$ satisfies an awkward inequality we give on page 327. Here is a simpler, though weaker, translation of it

$$1 - \delta(n) \leqslant \frac{3\sqrt{2n+34} - 20}{4n^2 - 11n - 12} \quad \text{for } n \geqslant 6.$$

This shows $(1 - \delta(n))$ is equivalent to $(3\sqrt{2}\,n^{-3/2}/4)$, for large n. Though we did not pay special attention to the small values of n, we get for cheap $\delta_4 = 0.972$ for $n = 4$ and $\delta_5 = 0.971$ for $n = 5$.

0. Notations. Let us give some indications about our notation. In order to make the most of the irreducible decomposition of curvature type tensors we try to forget as much as possible about indices (a really disgusting aspect of Riemannian geometry).

So, given any compact smooth manifold M, it is always possible to make it Riemannian—i.e. to equip it with a smooth metric g, using partitions of unity (paracompactness is enough in fact). Associated with g is a canonical torsion free connection D, called the Levi–Civita connection, and it is relevant to consider the following second order quantity:

$$R(U, V, W, T) = \big(D_V(D_U W) - D_U(D_V W) + D_{[U,V]}W, T\big),$$

where U, V, W, T are vector fields on M and $(X, Y) \equiv g(X, Y)$. It can be shown that $R(U, V, W, T)(x)$ does not depend on the whole vector fields U, V, W, T, but only on their values at $x \in M$, i.e., R is a 4-tensor: $R \in \otimes^4 TM$. Furthermore, R has the following symmetry properties:

$$R(U, V, W, T) = -R(V, U, W, T) = R(W, T, U, V),$$

i.e., $R \in S^2(\Lambda^2(TM))$ or, in other words, R can be viewed as an endomorphism of $\Lambda^2(TM)$, which is called the "curvature operator" and denoted by $\hat{R}$. Another basic symmetry property of the curvature is the following: $R_{U,V,W,T} + R_{V,W,U,T} + R_{W,U,V,T} = 0$. This is the so-called first Bianchi identity. The second Bianchi identity (which is true for the curvature of any bundle) involves the covariant derivation of R: $D_U R_{X,Y,Z,T} + D_X R_{Y,U,Z,T} + D_Y R_{U,X,Z,T} = 0$. This 4-tensor can be traced in only one way to give a nontrivial 2-tensor, r, which is symmetric and called the Ricci curvature: $\gamma(R) = r$. In its turn, r can be traced to give a scalar, called the scalar curvature: $u = \gamma(r)$.

It is well known that the curvature splits into an orthogonal decomposition which is in fact irreducible under $O(n)$.

$$R = \frac{u}{2n(n-1)} g \textcircled{\wedge} g + \frac{z \textcircled{\wedge} g}{(n-2)} + W.$$

Here: (a) $z = r - u \cdot g/n$ is the traceless part of the Ricci curvature;

(b) if s and t are (symmetric) 2-tensors, we set

$$\left(s \textcircled{\wedge} t\right)(U, V, W, T) = s(U, W)t(V, T) + s(V, T)t(U, W)$$
$$-s(U, T)t(V, W) - s(V, W)t(U, T);$$

which is easily checked to live in $S^2(\Lambda^2(TM))$.

(c) W, which is "what is left", is called the Weyl tensor. $W \equiv 0$ for 3-manifolds.

(d)

$$\mathcal{D} = \frac{z \textcircled{\wedge} g}{n-2} + W.$$

1. Preliminaries.

(A) *Evolution equation for the curvature tensor.* Given any symmetric 2-tensor h, it is always true that, for $\partial g/\partial t = h$, the evolution of the curvature can be written: cf. [**Be**]

$$2\frac{\partial R}{\partial r}(X, Y, Z, T) = D^2_{Y,Z}h(X, T) + D^2_{X,T}h(Y, Z) - D^2_{X,Z}h(Y, T)$$

$$- D^2_{Y,T}h(X, Z) + h(R(X, Y)Z, T) - h(R(X, Y)T, Z).$$

Let us now consider h as an element in $\Omega^1(\Lambda^1(T^*M))$, i.e., as a T^*M-valued differential 1-form (use duality given by g). Then $d^D h$ (the exterior differential associated with D) is in $\Omega^2(\Lambda^1(TM))$, and $d^D h_{X,Y,Z} = (D_X h)_{Y,Z} - (D_Y h)_{X,Z}$. Because of the trivial isomorphism between $\Omega^2(T^*M)$ and $\Omega^1(\Lambda^2(T^*M))$, let us consider $d^D h$ as an element of $\Omega^1(\Lambda^2(T^*M))$, so that $d^D(d^D h) \in \Omega^2(\Lambda^2(T^*M))$.

One gets

$$(d^D d^D h)_{X,Y,Z,T} = D_X(d^D h)_{Y;Z,T} - D_Y(d^D h)_{X;Z,T}$$
$$= (D^2_{X,Z} h)_{T,Y} - (D^2_{X,T} h)_{Z,Y} - (D^2_{Y,Z} h)_{T,X} + (D^2_{Y,T} h)_{X,Z}.$$

Now, if $h = \gamma(H)$, where $H \in \mathscr{S}^2(\Lambda^2(T^*M))$, and if $d^D H = 0$ (which is nothing but the second Bianchi identity: $(D_X H)_{Y,Z} + (D_Y H)_{Z,X} + (D_Z H)_{X,Y} = 0$) we have

$$((\delta^D H)_X Y, Z) = - \sum_{i=1}^{n} \left((D_{e_i} H)_{e_i, X} Y, Z\right)$$
$$= -\left\{ \sum_{i=1}^{n} \left((D_Y H)_{e_i, X} e_i, Z\right) + \left((D_Z H)_{e_i, X} Y, e_i\right) \right\}$$
$$= -((d^D h)_{Y,Z} X).$$

Here δ^D is the adjoint of d^D for the global scalar product (i.e., after integration on the manifold (cf. [**B**, 3.2] for more details)) and $(e_i)_{i=1}^{n}$ is an orthonormal basis of TM for g. The previous computation means that, as a 1-form with values in $\Lambda^2 T^*M$, i.e., as an element of $\Omega^1(\Lambda^2(T^*M))$, $-\delta^D H = d^D h$. We can now compute

$$\left[h \widehat{\wedge} g \circ R \right]_{X,Y} = \sum_{k,l} R_{X,Y,e_k,e_l} \left(h \widehat{\wedge} g \right) e_k \otimes e_l,$$

i.e.,

$$\left[h \widehat{\wedge} g \circ \hat{R} \right]_{X,Y,Z,T} = 2\{ h(R(X,Y)Z,T) - h(R(X,Y)T,Z) \}$$

and conclude

$$2\frac{\partial R}{\partial t} = d^D \delta^D H + \tfrac{1}{2}\left(h \widehat{\wedge} g \right) \circ \hat{R},$$

whenever $h = \gamma(H)$, with $H \in \mathscr{S}^2(\Lambda^2(T^*M))$ satisfying the second Bianchi identity. Because we want to give it the shape of a heat equation, let us now force the Laplace operator into, using the following Weitzenböck formula ([**B**], §4.2):

$$(d^D \delta^D + \delta^D d^D)A = D^*DA + \hat{A} \circ \left(\tfrac{1}{2} r \widehat{\wedge} g - \hat{R}\right) - K(\mathring{A} \circ \mathring{R}).$$

Here, $A \in \mathscr{S}^2(\Lambda^2(T^*M))$, and K is the obvious symmetrization operator which fits $\mathring{A} \circ \mathring{R}$ into $\mathscr{S}^2(\Lambda^2(T^*M))$

$$K(A)_{X,Y,Z,T} = A_{X,Y,Z,T} - A_{Y,X,Z,T} - A_{X,Y,T,Z} + A_{Y,X,T,Z}.$$

By $\mathring{A} \circ \mathring{R}$ we mean

$$(\mathring{A} \circ \mathring{R})_{X,Y,Z,T} = \sum_{i,j} A_{X,e_i,Z,e_j} R_{Y,e_i,T,e_j}.$$

Using once more $d^D H = 0$ (the second Bianchi identity) gives

$$2\frac{\partial R}{\partial t} = D^*DH + \tfrac{1}{2}\left(h \widehat{\wedge} g - \hat{H} \right) \circ \hat{R} + \tfrac{1}{2}\hat{H} \circ \left(r \widehat{\wedge} g - \hat{R} \right) - K(\mathring{H} \circ \mathring{R}).$$

We thus get the following

LEMMA 1. *For $\partial g/\partial t = -2r$, we get the following evolution equation for the curvature tensor:*

$$\frac{\partial R}{\partial t} = -D^*DR - \tfrac{1}{2}\left(r\,\widehat{\wedge}\,g - \hat{R}\right)\circ \hat{R} - \tfrac{1}{2}\hat{R}\circ\left(r\,\widehat{\wedge}\,g - \hat{R}\right) + K(\mathring{R}\circ\mathring{R}).$$

PROOF. Plug $H = -2R$, $h = -2r = -2\gamma(R)$ into the previous computation. Happily, one can go back to indices and check it coincides with 7.1 in [**H**]. ∎

(B) *A generalized maximum principle for tensors.* Let ζ be a symmetric 2-tensor on a vector bundle $\mathscr{B}$ over a compact manifold M, and let $v = p(\zeta, g)$ be a polynomial in ζ formed by contracting products of ζ with itself using the metric. We say that $p(\zeta, g)$ satisfies the null-eigenvector condition if, whenever v is a null eigenvector of ζ, then $(\hat{v}(v), v) \geq 0$.

LEMMA 2. *Suppose that for $0 \leq t \leq t_0$*

$$\partial \zeta/\partial t = \Delta \zeta + (w, D\zeta) + v$$

(where $w \in TM$ and $v = p(\zeta, g)$ satisfies the null-eigenvector condition). If $\zeta_{t=0} \geq 0$, then it remains so for $0 \leq t \leq t_0$.

Proof. Such a statement is given in [**H**, 9.1] in the particular case of the tangent bundle. We give a general statement because we need it for $\Lambda^2(T^*M)$, the bundle of the 2-forms, and because the proof, in the general case, is in no way different. Although the statement in [**H**, 9.1] is correct, there is a mistake in Hamilton's proof (which is not difficult to fix). So we choose to give some details. Note it is sufficient to prove the lemma for $t_0 < \infty$. First we chop the time interval into small pieces, the lengths of which will be fixed later (they only depend on $\max_{x\in M,\ t\in[t_i,t_{i+1}]}|\zeta|$). So let us show $\zeta \geq 0$ for $0 \leq t \leq \delta$, for some small δ depending only on $\max_{x\in M}|\zeta|$. The lemma will be proved by iterating as many times as necessary. Now let us introduce $\tilde{\zeta} = \zeta + \varepsilon(\delta + t)\mathrm{Id}_{\mathscr{B}}$. Here we have

$$\mathrm{Id}_{\mathscr{B}} \equiv g \quad \text{if } \mathscr{B} = TM,$$

$$\mathrm{Id}_{\mathscr{B}} \equiv \tfrac{1}{4}g\,\widehat{\wedge}\,g \quad \text{if } \mathscr{B} = \Lambda^2 M,$$

so $\mathrm{Id}_{\mathscr{B}}$, as a tensor field, may depend on t. We claim $\tilde{\zeta} > 0$ on $0 < t < \delta$, for every $\varepsilon > 0$—which is enough to conclude by taking $\varepsilon \to 0$.

Were our claim false, there would be a first time θ, $0 < \theta \leq \delta$, where $\tilde{\zeta}$ gets a null eigenvector v of unit length at some point $x \in M$. If $\tilde{v} = P(\tilde{\zeta}, g)$, then, by hypothesis, $(\tilde{v}(v), v) \geq 0$ at (x, θ).

We also note that $|\tilde{v} - v| \leq C|\tilde{\zeta} - \zeta|$, C being a constant depending only on $\max(|\tilde{\zeta}| + |\zeta|)$, since p is a polynomial. From this, we get

$$(v(v), v) \geq -C_1\varepsilon\delta$$

for a constant C_1, depending only on $|\zeta|$ (w.l.o.g. set $\sup(\varepsilon, \delta) \leqslant 1$). There is another constant C_2 for which $\partial(v, v)_t / \partial t \leqslant C_2$ for all t (since $t_0 < \infty$). The only quantity depending on t is g, the metric, and $(\ ,\ )_s$ is obtained from g by inversing and tensoring. For example, in the case of $\Lambda^2 T^*M$ we get

$$\left| \frac{\partial}{\partial t}(\omega, \omega) \right| = \left| -2\omega_{ij}\omega_{kl}\frac{\partial}{\partial t}g_{pq}g^{ip}g^{kq}g^{jl} \right| \leqslant 2|\omega|^2 \left| \frac{\partial g}{\partial t} \right|.$$

Set $C = \mathrm{Sup}\{C_1, C_2\}$, and extend v as a vector field $\tilde{v}$ on M with values in $\mathscr{B}$, such that $Dv = 0$ at x, $\tilde{v}$ independent of t, and let $f = (\tilde{\zeta}\tilde{v}, \tilde{v})$, so that $f \geqslant 0$ on $0 \leqslant t \leqslant \theta$, and all over M, i.e. $\partial f/\partial t \leqslant 0$ and $df = 0$ and $\Delta f \geqslant 0$ at (x, θ). But

$$\frac{\partial f}{\partial t} = \left(\frac{\partial}{\partial t}\zeta\right)(\tilde{v}, \tilde{v}) + \varepsilon(\delta + t)\frac{\partial}{\partial t}(\tilde{v}, \tilde{v}) + \varepsilon(\tilde{v}, \tilde{v}),$$

and at (x, θ), $D\tilde{v} = 0$, i.e. $df = ((D\zeta)v, v)$, $\Delta f = (\Delta\zeta \cdot v, v)$, but

$$\left(\frac{\partial}{\partial t}\zeta\right)(v, v) = (\Delta\zeta \cdot v, v) + ((w \cdot D\zeta)(v), v) + (\nu(v), v).$$

This implies $-C\varepsilon\delta \leqslant (\nu(v), v) \leqslant \varepsilon(-1 + 2\delta C)$ and the contradiction if $C\delta < \frac{1}{3}$. ∎

2. The Ricci flow preserves positive curvature operator metrics.

More precisely, we will show the following statement:

THEOREM 1. *Let M be a compact n-dimensional manifold with a metric g_0 of positive curvature operator $\hat{R}(0)$. Then, the metric $g(t)$, obtained by following the flow of $\partial g/\partial t = -2r$, remains positively curved. In fact the infimum of the spectrum of $\hat{R}(t)$ is always greater than or equal to the infimum of the spectrum of $\hat{R}(0)$.*

Proof. Apply the maximum principle, cf. §1(B), to the heat equation describing the evolution of the curvature operator under the flow $\partial g/\partial t = -2r$, (this has been computed in §1(A)) with $\zeta = R$, $w = 0$, and

$$\nu = -\tfrac{1}{2}\left(\left(r\widehat{\wedge}g - \hat{R}\right)\circ \hat{R}\right) - \tfrac{1}{2}\left(\hat{R}\circ\left(r\widehat{\wedge}g - \hat{R}\right)\right) + K(\mathring{R}\circ\mathring{R}).$$

If $\omega \in \Lambda^2 T^*M$ is such that $\hat{R}(\omega) = 0$, we get

$$0 = \tfrac{1}{2}\left(\left(r\widehat{\wedge}g - \hat{R}\right)\circ \hat{R}(\omega), \omega\right) = \tfrac{1}{2}\left(\hat{R}\circ\left(r\widehat{\wedge}g - \hat{R}\right)\omega, \omega\right),$$

and

$$(K(\mathring{R}\circ\mathring{R})\omega, \omega) = 4(\mathring{R}\cdot\mathring{R}(\omega), \omega),$$

since $\omega \in \Lambda^2 T^*M$, with

$$(\mathring{R}\circ\mathring{R}(\omega), \omega) = \sum R_{I,X,J,Y}R_{I,X',J,Y'}\omega_{X,X'}\omega_{Y,Y'}.$$

Let us now write R in its polar form, i.e., $R = \sum \lambda_\alpha \omega_\alpha \otimes \omega_\alpha$ where $\{\lambda_\alpha\}$ is the spectrum of $\hat{R}$—this is always possible since $R \in S^2\Lambda^2 TM$. We get

$$(\mathring{R}\circ\mathring{R}(\omega), \omega) = \sum_{\alpha, \beta} \lambda_\alpha\lambda_\beta\left(\sum_{X,X'} (\omega_\beta \circ \omega_\alpha)(X, X')\omega(X, X')\right)^2,$$

so, starting with $R \geqslant 0$ implies it remains so for all time. ∎

Let us recall that, as an operator on $\Lambda^2 TM$, $g\bigcirc\!\!\!\!\wedge\, g = 4 \cdot \mathrm{Id}_{\Lambda^2 TM}$. So, if $\inf_\alpha \lambda_{\alpha(t=0)} \geq 4C_0$, then $(R - C_0 g\bigcirc\!\!\!\!\wedge\, g)_{(t=0)} \geq 0$. Now, for $R - C_0 g\bigcirc\!\!\!\!\wedge\, g = H$, we get the following evolution equation

$$\frac{\partial H}{\partial t} = \Delta H - \tfrac{1}{2}\left(r\bigcirc\!\!\!\!\wedge\, g - \hat{R}\right) \circ \hat{R} - \tfrac{1}{2}\left(\hat{R} \circ \left(r\bigcirc\!\!\!\!\wedge\, g - \hat{R}\right)\right)$$

$$+ K(\mathring{R} \circ \mathring{R}) + 4C_0 r\bigcirc\!\!\!\!\wedge\, g.$$

So, we can apply the maximum principle with $w = 0$ and

$$v = -\tfrac{1}{2}\left(\left(r\bigcirc\!\!\!\!\wedge\, g - \hat{R}\right)\circ\hat{R}\right) - \tfrac{1}{2}\left(\hat{R}\circ\left(r\bigcirc\!\!\!\!\wedge\, g - \hat{R}\right)\right) + K(\mathring{R}\circ\mathring{R}) + 4C_0 r\bigcirc\!\!\!\!\wedge\, g$$

so that, if $H(\omega) = 0$, i.e. $R(\omega) = 4C_0\omega$, we get

$$\tfrac{1}{2}\left(\left(r\bigcirc\!\!\!\!\wedge\, g - \hat{R}\right)\hat{R}(\omega), \omega\right) = 2C_0\left(\left(r\bigcirc\!\!\!\!\wedge\, g\right)(\omega), \omega\right) - 8C_0^2(\omega, \omega)$$

$$= \tfrac{1}{2}\left(\hat{R}\circ\left(r\bigcirc\!\!\!\!\wedge\, g - \hat{R}\right)\omega, \omega\right).$$

I.e., $(v(\omega, \omega) = 16C_0^2(\omega, \omega) + (K(\mathring{R}\circ\mathring{R})\omega, \omega) > 0$, which shows that $(R - C_0 g\bigcirc\!\!\!\!\wedge\, g)_t \geq 0$ for all t. This is what is stated in Theorem 1. ∎

Unfortunately, we are sort of stuck at this point, since we are unable to prove the same result for the normalized evolution equation—in other words, since we cannot show that, if $\varepsilon > 0$ is such that $(\hat{R} - \varepsilon \cdot u \cdot g\bigcirc\!\!\!\!\wedge\, g)_{t=0} \geq 0$, then, this remains true for all t. We now show why it is impossible to use the maximum principle—at least in a naive way (i.e., using only the symmetries of $S^2\Lambda^2(TM)$ and the first Bianchi identity) to get the result.

Let us first write down the evolution equation for $\hat{H} = \hat{R} - \varepsilon u g\bigcirc\!\!\!\!\wedge\, g$,

$$\frac{\partial \hat{H}}{\partial t} = +\Delta(\hat{H}) - \tfrac{1}{2}\left\{\left(r\bigcirc\!\!\!\!\wedge\, g - \hat{R}\right)\circ\hat{R} + \hat{R}\circ\left(r\bigcirc\!\!\!\!\wedge\, g - \hat{R}\right)\right\} + K(\mathring{R}\circ\mathring{R})$$

$$+ 4\varepsilon u r\bigcirc\!\!\!\!\wedge\, g - \varepsilon\left(-\Delta u + \frac{\partial u}{\partial t}\right)g\bigcirc\!\!\!\!\wedge\, g.$$

Now, if $\omega \in \Lambda^2 TM$ is such that $(\hat{R} - \varepsilon u g\bigcirc\!\!\!\!\wedge\, g)(\omega) = 0$, we get

$$\tfrac{1}{2}\left(\left\{\left(r\bigcirc\!\!\!\!\wedge\, g\circ\hat{R}\right) + \left(\hat{R}\circ r\bigcirc\!\!\!\!\wedge\, h\right)\right\}\omega, \omega\right) = 4\varepsilon u\left(r\bigcirc\!\!\!\!\wedge\, g(\omega), \omega\right),$$

and, as already seen, $(K(\mathring{R}\circ\mathring{R})\omega, \omega) = 8\Sigma_{\alpha<\beta}\lambda_\alpha\lambda_\beta(\omega_\alpha\circ\omega_\beta, \omega)^2$. Finally,

$$(\hat{R}\circ\hat{R}(\omega), \omega) = 16\varepsilon^2 u^2\|\omega\|^2 \quad \text{and} \quad -\Delta u + \frac{\partial u}{\partial t} = 2|r|^2 \quad \left(\text{where } |r|^2 = (r, r)\right).$$

PROPOSITION 1. *It is possible to construct a tensor R in* $S^2\Lambda^2(TM)$, *satisfying the first Bianchi identity, for which* $\langle(\partial\hat{H}/\partial t - \Delta(\hat{H}))\omega, \omega\rangle \leq 0$, *i.e.,*

$$8 \sum_{\alpha<\beta} \lambda_\alpha\lambda_\beta(\omega_\alpha\circ\omega_\beta, \omega)^2 + 16\varepsilon^2 u^2\|\omega\|^2 - 8\varepsilon|r|^2|\omega|^2 < 0$$

as soon as $n \geq 4$.

PROOF. Consider $R = \sum_{i<j} \lambda_{ij}(e_i \wedge e_j) \otimes (e_i \wedge e_j)$.

(a) First, let us check it satisfies the 1st Bianchi identity

$$R(U, V, W, T) + R(V, W, U, T) + R(W, U, V, T) = 0.$$

(b) Now, remark

$$u = 2 \sum_{k<l} R_{klkl} = 2 \sum_{k<l} \lambda_{kl} \quad \text{and} \quad R(e_i \wedge e_j) = 2\lambda_{ij}(e_i \wedge e_j).$$

Let us require $\lambda_{01} \leqslant \lambda_{ij}$ whenever $i < j$. In this case, $\omega = e_0 \wedge e_1$ and $\hat{R}(\omega) = 4\varepsilon u \omega = 2\lambda_{01}\omega$, i.e.,

$$\varepsilon = \frac{\lambda_{01}}{2u} = \frac{\lambda_{01}}{4\sum_{k<l} \lambda_{kl}},$$

so that

$$\nu(\omega, \omega) = 16\varepsilon^2 u^2 \omega^2 + 8 \sum_{\alpha<\beta} \lambda_\alpha \lambda_\beta (\omega_\alpha \circ \omega_\beta, \omega)^2 - 8\varepsilon |r|^2 |\omega|^2$$

$$= 8\left\{ \lambda_{01}^2 + \sum_{p=2}^{n-1} \lambda_{0p}\lambda_{1p} - \lambda_{01}\left(\frac{\sum_j (\sum_i \lambda_{ij})^2}{\sum_{k,l} \lambda_{k,l}} \right) \right\}.$$

Since $(e_i \wedge e_j) \circ (e_k \wedge e_l) = 0$ if $\{i, j\} \cap \{k, l\} = \varnothing$, and $(e_0 \wedge e_p) \circ (e_1 \wedge e_p) = -e_0 \otimes e_1$. Moreover, we have set $\lambda_{kl} = \lambda_{lk}$. If $n = 3$, we can check

$$\nu(\omega, \omega) = a^2 + bc - a\frac{\left((a + b)^2 + (b + c)^2 + (c + a)^2\right)}{2(a + b + c)}$$

with $a \leqslant b$, $a \leqslant c$, i.e. $\nu(\omega, \omega) \geqslant 0$, and $\nu(\omega, \omega) = 0$ iff $\{(b = a$ or $c = 0)$ and $(c = a$ or $b = 0)\}$. If $n \geqslant 4$ let us choose λ_{23} very large compared to any λ_{ij}, $\{i, j\} \neq \{2, 3\}$. If $\lambda_{23}/\lambda_{ij}$ is sufficiently large for $\{i, j\} \neq \{2, 3\}$, then $M(w, w) < 0$. ∎

This is the main motivation for the second part of the work: to try the same problem with an a priori pinching estimate on the curvature tensor. As a matter of fact, our result forgets about the original question (is a manifold carrying a metric with positive curvature operator covered by a sphere?) and seems to be more closely related to the study of Riemannian manifolds with bounded curvature ratios than we expected it to be.

3. Sufficiently pinched manifolds are covered by spheres. Let us restate the main theorem.

THEOREM 2. *Any compact n-manifold carrying a metric with $C(n)$-pinched curvature is diffeomorphic to a space of constant curvature. $C(n)$ can be chosen to equal $\gamma/(n(n - 1)(n - 2))$, where $\gamma = \frac{1}{2}$ for all $n \geqslant 6$, $\gamma = 6/25$ for $n = 4$, $\gamma = 48/125$ for $n = 5$.*

We want to point out that one of the interests of the result is that $C(n)$ is greater (at least for $n > 6$) than the a priori constant which implies—for algebraic reasons—the positivity of the curvature operator. Let us first determine this philosophical barrier. We claim

$$\frac{4}{(n-2)(n-1)(n+1)n} \geq \frac{|\mathscr{D}|^2}{u^2}$$

implies

$$R = \frac{u}{2n(n-1)} g \bigcirc\!\!\!\wedge g + \frac{z \bigcirc\!\!\!\wedge g}{n-2} + W$$

is positive as an operator on $\Lambda^2 TM$. We only need to show that under the pinching assumption any eigenvalue λ of $(z \bigcirc\!\!\!\wedge g)/(n-2) + W$, which is a tracefree endomorphism of $\Lambda^2 TM$, is smaller than $2u/(n(n-1))$ (since $g \bigcirc\!\!\!\wedge g = 4 \cdot \mathrm{Id}_{\Lambda^2 TM}$). But this is a consequence of the following trivial algebraic manipulation. If

$$\sum_{i=1}^{N} \lambda_i = 0,$$

then

$$\lambda_p^2 = \left(\sum_{i \neq p} \lambda_i \right)^2 \leq (N-1) \sum_{i \neq p} \lambda_i^2 \quad \text{(for every } p\text{)},$$

i.e.,

$$\lambda_p^2 \leq \frac{N-1}{N} \sum_{i=1}^{N} \lambda_i^2.$$

Here, this gives

$$\lambda^2 \leq \frac{(n-2)(n+1)}{n(n-1)} \frac{4u^2}{(n-2)(n-1)(n+1)} = \frac{4}{n^2(n-1)^2} u^2,$$

i.e.,

$$\lambda^2 \leq \left[\frac{2u}{n(n-1)} \right]^2. \quad \text{q.e.d.} \quad \blacksquare$$

The proof of the main theorem is based upon a basic estimate which can be stated as follows:

PROPOSITION 2. *For any n manifold, which is $C_0(n)$-pinched, there exist $1 < \alpha < 2$ and a constant A such that $|\mathscr{D}|^2/u^\alpha \leq A$ is satisfied for all t, $0 < t < T$. Here we can choose $C(n)$ as in Theorem 2.*

Note 1. We can always take $A = \sup_{M, t=0}(|\mathcal{D}|^2/u^\alpha)$.

Note 2. The value we give for $C(n)$, although better than the a priori reasonable barrier that we already discussed, is not sharp. For example we give here a very rough upper bound for $|(\hat{W} \circ \hat{W}, W) + 4(\mathring{W} \circ \mathring{W}, W)|$. We hope to be able to discuss this point in a forthcoming paper.

Proof. Using the evolution equation for the curvature operator and contractions of it, we compute

$$\frac{\partial}{\partial t} \frac{|\mathcal{D}|^2}{u^\alpha} = \frac{\partial}{\partial t}|\mathcal{D}|^2 \frac{1}{u^\alpha} - \frac{|\mathcal{D}|^2}{u^{\alpha+1}} \frac{\partial u}{\partial t}$$

and

$$\frac{\partial u}{\partial t} = \Delta u + 2|r|^2, \qquad \frac{\partial |\mathcal{D}|^2}{\partial t} = 2\left(\frac{\partial \mathcal{D}}{\partial t}, \mathcal{D}\right) + 8(\gamma(\hat{D} \circ \hat{D}), r).$$

Moreover, we get

$$\frac{\partial}{\partial t}\mathcal{D} = \frac{\partial}{\partial t}\left(R - \frac{u}{2n(n-1)}g \bigwedge g\right) = \Delta\mathcal{D} - \tfrac{1}{2}\left(r\bigwedge g \circ \hat{R}\right) - \tfrac{1}{2}\left(\hat{R} \circ r\bigwedge g\right)$$

$$+ \hat{R} \circ \hat{R} + K(\mathring{R} \circ \mathring{R})$$

$$+ \frac{2u}{n(n-1)}z\bigwedge g + g\bigwedge g\left(-\frac{|z|^2}{n(n-1)} + \frac{u^2}{n^2(n-1)}\right).$$

Using

$$\Delta|\mathcal{D}|^2 = 2|D\mathcal{D}|^2 + 2(\Delta\mathcal{D}, \mathcal{D})$$

and

$$\Delta\left(\frac{|\mathcal{D}|^2}{u^\alpha}\right) = \frac{\Delta|\mathcal{D}|^2}{u^\alpha} - \frac{\alpha|\mathcal{D}|^2\Delta u}{u^{\alpha+1}} - 2\alpha\frac{\left(D|\mathcal{D}|^2, Du\right)}{u^{\alpha+1}} + \frac{\alpha(\alpha+1)|\mathcal{D}|^2|Du|^2}{u^{\alpha+2}},$$

we get

$$\frac{\partial}{\partial t}\left(\frac{|\mathcal{D}|^2}{u^\alpha}\right) = \Delta\left(\frac{|\mathcal{D}|^2}{u^\alpha}\right) + \left\{-2\frac{|D\mathcal{D}|^2}{u} + \frac{2\alpha}{u^{\alpha+1}}\left(D|\mathcal{D}|^2, Du\right) - \frac{\alpha(\alpha+1)}{u^{\alpha+2}}|\mathcal{D}|^2|\Delta u|^2\right\}$$

$$- \frac{2}{u^\alpha}\left(\frac{4u}{n}\hat{\mathcal{D}} + \frac{2u}{n(n-1)}z\bigwedge g + z\bigwedge g \circ \hat{\mathcal{D}}, \hat{\mathcal{D}}\right)$$

$$+ \frac{2}{u^\alpha}\left(\hat{\mathcal{D}} \circ \hat{\mathcal{D}} + \frac{4u}{n(n-1)}\hat{\mathcal{D}}, \hat{\mathcal{D}}\right)$$

$$+ \frac{8}{u^\alpha}\left((\mathring{\mathcal{D}} \circ \mathring{\mathcal{D}}, \mathcal{D}) + \frac{u}{2n(n-1)} \cdot 4 \cdot |\gamma(D)|^2 - \frac{4u}{2n(n-1)}(\mathring{\mathcal{D}} \cdot \mathring{\mathcal{D}})_{jllj}\right)$$

$$+ \frac{4u}{n(n-1)} \cdot \frac{1}{u^\alpha} \cdot \frac{|z\bigwedge g|^2}{n-2} + \frac{8}{u^\alpha}(\gamma(\mathring{\mathcal{D}} \circ \mathring{\mathcal{D}}), r) - \frac{2\alpha}{u^{\alpha+1}}|\mathcal{D}|^2|r|^2.$$

First order terms can be rearranged in the following way,

$$\{\cdots\} = \frac{2(\alpha - 1)}{u}\left(Du, D\left(\frac{|\mathscr{D}|^2}{u^\alpha}\right)\right) - 2\frac{|uD\mathscr{D} - Du \cdot \mathscr{D}|^2}{u^{\alpha+2}}$$

$$- \frac{(2 - \alpha)(\alpha - 1)}{u^{\alpha+2}}|\mathscr{D}|^2 \cdot |Du|^2.$$

Polynomial terms of order zero we can deal with in the following way,

$$\frac{2}{u^\alpha}\left\{+4(\mathring{\mathscr{D}} \circ \mathring{\mathscr{D}}, \mathscr{D}) + (\hat{\mathscr{D}} \circ \hat{\mathscr{D}}, \mathscr{D}) + \frac{8u|z|^2}{n(n-1)} - \frac{\alpha|\mathscr{D}|^2|r|^2}{u}\right\} = \mathscr{A}_\alpha,$$

where we used $(\mathring{\mathscr{D}} \circ \mathring{\mathscr{D}})_{ijji} = \frac{1}{2}|\mathscr{D}|^2$ (1st Bianchi identity). As a matter of fact, we can use this step for both a cheap (but rough) a priori estimate (a)—and to check the quantity $\mathscr{A}_2$ is identically 0 (b) for $\mathbf{C}P^m$ equipped with its canonical metric (such a metric is Einstein, so $\partial g/\partial t = -(1/(2m)) \cdot 2u_{\mathbf{C}P^m} \cdot g$ and $|\mathscr{D}|^2/u^2$ is a constant).

(a) Since we want to apply the (classical) maximum principle to $|\mathscr{D}|^2/u^\alpha$, we need to exhibit a constant K such that $|\mathscr{D}|^2/u^2 < K$ implies $\mathscr{A}_\alpha \leqslant 0$ for some $\alpha < 2$.

A rough estimate. If $|\mathscr{D}|/u < 2/(5n(n - 1))$, then it remains so for all $t < T$. In fact, there exist $1 < \alpha < 2$ and a constant $\mathscr{B}$ such that $|\mathscr{D}|^2/u^\alpha < \mathscr{B}$ for all $t < T$.

This cheap statement is proved in the following lines:

$$\frac{|\mathscr{D}|}{u} < \frac{2}{5n(n-1)} \quad \text{implies} \quad 5|\mathscr{D}|^3 < \left(\frac{\alpha}{n} - \frac{2(n-2)}{n(n-1)}\right)|\mathscr{D}|^2 u$$

for some $1 < \alpha < 2$, so that

$$5|\mathscr{D}|^3 + 2u\frac{|\mathscr{D}|^2(n-2)}{n(n-1)} < \alpha|\mathscr{D}|^2\frac{|r|^2}{u}.$$

But

$$|z|^2 = \frac{|z \bigcirc\!\!\!\!\wedge\, g|^2}{4(n-2)} \leqslant \frac{(n-2)}{4}|\mathscr{D}|^2.$$

(b) Let us compute $\mathscr{A}_2$ for $\mathbf{C}P^m$ equipped with its canonical metric s.t.: $R_{\mathbf{C}P^m} = \frac{1}{2}g \bigcirc\!\!\!\!\wedge\, g + J \bigcirc\!\!\!\!\wedge\, J + 4J \otimes J$ (the sectional curvature ranges between 1 and 4, and J denotes the complex structure). First,

$$\mathscr{R} = \frac{1}{2}\left(g \bigcirc\!\!\!\!\wedge\, g + J \bigcirc\!\!\!\!\wedge\, J + 4J \otimes J\right) - \frac{m+1}{2m-1}g \bigcirc\!\!\!\!\wedge\, g$$

$$= -\frac{3}{2(2m-1)}g \bigcirc\!\!\!\!\wedge\, g + \frac{1}{2}J \bigcirc\!\!\!\!\wedge\, J + 2J \otimes J,$$

then, computing $(s \bigwedge s) \circ (t \bigwedge t)$ for s and $t \in S^2(TM)$, for s and $t \in \Lambda^2(TM)$, and for $s \in S^2(TM)$ and $t \in \Lambda^2(TM)$, we get

$$(\mathring{\mathscr{D}} \circ \mathring{\mathscr{D}}, \mathscr{D}) = \frac{16m(m^2 - 1)}{(2m - 1)^2}(2m^2 + m + 8).$$

As far as $(\hat{\mathscr{D}} \circ \hat{\mathscr{D}}, \mathscr{D})$ is concerned, one can use [**BK**, 5.2] to get

$$(\hat{\mathscr{D}} \circ \hat{\mathscr{D}}, \mathscr{D}) = \frac{4^3}{(2m - 1)^2}m(m^2 - 1)(4m^2 + 2m - 11).$$

So

$$\mathscr{A}_2 = \frac{4^3}{(2m - 1)^2}m(m^2 - 1)\{(4m^2 + 2m - 11)$$

$$+ (2m^2 + m + 8) - 3(m + 1)(2m - 1)\},$$

i.e.,

$$\mathscr{A}_2 \equiv 0. \quad \text{q.e.d.} \quad \blacksquare$$

Because we claim in Proposition 1 a sharper result than in "a rough estimate", we need to work a bit more. Let us rewrite $\mathscr{A}_\alpha$ as

$$\mathscr{A}_\alpha = \frac{2}{u^\alpha}\left(4(\mathring{\mathscr{D}} \circ \mathring{\mathscr{D}}, \mathscr{D}) + (\hat{\mathscr{D}} \circ \hat{\mathscr{D}}, \mathscr{D}) + 8|z|^2 u\left(\frac{1}{n(n - 1)} - \frac{1}{n(n - 2)}\right)\right.$$

$$\left. - \frac{2u}{n}|W|^2 - \frac{8|z|^4}{(n - 2)u} - \frac{2|W|^2|z|^2}{u} + (2 - \alpha)|\mathscr{D}|^2|r|^2\right).$$

By expanding $(\mathring{\mathscr{D}} \circ \mathring{\mathscr{D}}, \mathscr{D})$ and $(\hat{\mathscr{D}} \circ \hat{\mathscr{D}}, \mathscr{D})$, we get

$$(\mathring{\mathscr{D}} \circ \mathring{\mathscr{D}}, \mathscr{D}) = (\mathring{W} \circ \mathring{W}, W) - \frac{6}{n - 2}z_{ij}W_{iabc}W_{jcba}$$

$$+ \frac{3(n - 4)}{(n - 2)^2}(\mathring{W}(z), z) + \frac{2(8 - 3n)}{(n - 2)^3}(z \circ z, z);$$

and

$$(\hat{\mathscr{D}} \circ \hat{\mathscr{D}}, \mathscr{D}) = (\hat{W} \circ \hat{W}, W) + \frac{12}{n - 2}z_{ij}W_{iabc}W_{jabc}$$

$$+ \frac{24}{(n - 2)^2}(\mathring{W}(z), z) + \frac{8(n - 4)}{(n - 2)^3}(z \circ z, z).$$

This finally gives

$$\mathscr{A}_\alpha = \frac{4}{u^{\alpha+1}}\left(2u(\mathring{W} \circ \mathring{W}, W) + \frac{u}{2}(\hat{W} \circ \hat{W}, W)\right.$$

$$- \frac{8u}{(n - 2)^2}(z \circ z, z) + \frac{6u}{n - 2}(\mathring{W}(z), z)$$

$$\left. - \frac{4|z|^2u^2}{(n - 2)(n - 1)n} - \frac{u^2}{n}|W|^2 - |z|^2|W|^2 - \frac{4|z|^4}{(n - 2)} + \frac{(2 - \alpha)}{2}|\mathscr{D}|^2|r|^2\right),$$

where we have used

$$W_{abci}W_{abcj}z_{ij} - 2W_{iabc}W_{jcba}z_{ij}$$
$$= \left(W_{abci}W_{abcj} + 2W_{bcai}W_{abcj}\right)z_{ij}$$

(using the 1st Bianchi identity)

$$= \left(-W_{bcai}W_{abcj} + W_{acbi}W_{abcj} + 2W_{bcai}W_{abcj}\right)z_{ij}$$
$$= \left(W_{abci} + W_{bcai}\right)W_{abcj}z_{ij} = 0.$$

We now have to estimate $(z \circ z, z)$. We want to make the most of the tracefree property of z.

LEMMA 3. *If P is a symmetric endomorphism whose eigenvalues are $\lambda_1,\ldots,\lambda_N$, and if P is tracefree (i.e., $\sum_{i=1}^{N}\lambda_i = 0$) then*

$$\sum_{i=1}^{N}\lambda_i^3 \leqslant \frac{N-2}{\sqrt{N(N-1)}}|P|^3.$$

PROOF.

$$(P \circ P, P) = \gamma(P \circ P \circ P) = \sum_{i=1}^{N}\lambda_i^3,$$

which is homogeneous. So we have to maximize $\sum_{i=1}^{N}\lambda_i^3$ under the following constraints: $\sum \lambda_i = 0$ and $\sum_{i=1}^{N}\lambda_i^2 = 1$ (or any constant). Introducing the two Lagrange multipliers α and β we get $3\lambda_i^2 + 3\alpha\lambda_i + \beta = 0$ (for all i) so $\beta = -3/N$ (since $\sum \lambda_i = 0$) and there are only two possible values for the λ_i

$$\left(= -\frac{\alpha}{3} \pm \sqrt{\frac{\alpha^2}{9} + \frac{1}{N}}\right).$$

So we have to maximize

$$\left(k - (N-k)\frac{k^3}{(N-k)^3}\right)\frac{(N-k)^{3/2}}{k^{3/2}N^{3/2}}, \qquad k \in \{1,\ldots,N-1\},$$

i.e., to maximize

$$\frac{1-2x}{x\sqrt{1-x}} \quad \text{on} \left\{\frac{1}{N}, \frac{2}{N},\ldots,1-\frac{1}{N}\right\}: x = \frac{1}{N}.$$

So we get

$$\sum_{i=1}^{N}\lambda_i^3 < \frac{N-2}{\sqrt{N(N-1)}}|P|^3 \quad \blacksquare$$

As a corollary, we get here

$$\left|\frac{8}{(n-2)^2}u(z \circ z, z)\right| \leqslant \frac{8u}{(n-2)}\frac{|z|^3}{\sqrt{n(n-1)}}.$$

A rough estimate for $(\mathring{W}(z), z)$ would be $|W|\,|z|^2$. Rewriting

$$(\mathring{W}(z), z) = \frac{(n-2)^2}{8}\left(W, \frac{z\,\widehat{\circledwedge}\,g}{(n-2)} \circ \frac{z\,\widehat{\circledwedge}\,g}{(n-2)} \right),$$

we can study how $(z\,\widehat{\circledwedge}\,g \circ z\,\widehat{\circledwedge}\,g)$ splits under the original decomposition of $S^2(\Lambda^2(T^*M))$.

LEMMA 4.

$$(\mathring{W}(z), z) \leqslant \frac{1}{\sqrt{2}} \sqrt{\frac{(n-2)}{(n-1)}}\, |W|\,|z|^2.$$

PROOF.

$$(\mathring{W}(z), z) \leqslant \frac{(n-2)^2}{8}|W|$$

$$\cdot \left\{ \left| \frac{z\,\widehat{\circledwedge}\,g}{n-2} \circ \frac{z\,\widehat{\circledwedge}\,g}{n-2} \right|^2 - \gamma\left(\gamma\left(\frac{z\,\widehat{\circledwedge}\,g}{n-2} \,\hat{\circ}\, \frac{z\,\widehat{\circledwedge}\,g}{n-2} \right) \right)^2 \frac{2}{n(n-1)} \right.$$

$$\left. - \frac{1}{n-2}\left(\gamma\left(\frac{\left(z\,\widehat{\circledwedge}\,g\right)}{n-2} \circ \frac{\left(z\,\widehat{\circledwedge}\,g\right)}{n-2} \right)^2 - \frac{1}{n}\left(\gamma\left(\gamma\frac{\left(z\,\widehat{\circledwedge}\,g\right)}{n-2} \circ \frac{\left(z\,\widehat{\circledwedge}\,g\right)}{n-2} \right) \right)^2 \right) \right\}^{1/2},$$

i.e.,

$$(\mathring{W}(z), z) \leqslant \frac{(n-2)^2}{8}|W|\left\{ \frac{16}{(n-2)^4}\left((n-8)|z\circ z|^2 + 3|z|^4 \right) \right.$$

$$\left. - \frac{32|z|^4}{n(n-1)(n-2)^2} \leqslant -16\frac{(n-4)^4}{(n-2)^5}\left(|z\circ z|^2 - \frac{1}{n}|z|^4 \right) \right\}^{1/2}$$

$$\leqslant \frac{(n-2)^2}{8}|W|\left\{ \frac{32}{(n-2)^5}\left(-n|z\circ z|^2 + \frac{(n^2-3n+3)|z|^4}{n-1} \right) \right\}^{1/2}$$

$$\leqslant \frac{(n-2)^2}{8}|W|\left(\frac{32}{(n-2)^3}\frac{z^4}{(n-1)} \right)^{1/2} = \frac{1}{\sqrt{2}}\sqrt{\frac{(n-2)}{(n-1)}}\,|W|\,|z|^2. \quad \blacksquare$$

In order to apply the maximum principle, we need to show the following is negative for some $\alpha < 2$:

$$\frac{(2-\alpha)}{2}|\mathscr{D}|^2|r|^2 + 2u(\mathring{W}\circ\mathring{W}, W) + \frac{u}{2}(\hat{W}\circ\hat{W}, W)$$

$$- \frac{1}{n}u^2|W|^2 - \frac{4u^2|z|^2}{n(n-1)(n-2)} - \frac{4|z|^4}{n-2}$$

$$+ \frac{8u}{n-2}\frac{1}{\sqrt{n(n-1)}}|z|^3 - |W|^2|z|^2 + \frac{3\sqrt{2}}{(n-1)(n-2)}u|W||z|^2.$$

From this point on, we can either look for a sharp estimate for $C(n)$, n small—or search for an optimal power in n—i.e. an asymptotic formula for $C(n)$. The last choice turns out to be a good one, since the value we get for $C(n)$ is greater than $4/((n-2)(n-1)n(n+1))$, at least for $n > 6$, as already mentioned. To do so, we do not need to worry too much about $|4(\mathring{W}\circ\mathring{W}, W) + \hat{W}\circ\hat{W}, W)|$, which we roughly overestimate by $5|W|^3$. (This can be improved: using the trick on page 319

$$|(\hat{W}\circ\hat{W}, W)| \leqslant \frac{n^2-n-4}{\sqrt{n(n-1)(n^2-n+2)}}|W|^3 \Bigg).$$

We are now looking for the largest $C(n)$ such that $|\mathscr{D}|^2/u^2 < C(n)$ implies the existence of an $\varepsilon > 0$ with

$$Q - \frac{5}{2}u|W|^3 - \frac{1}{n}u^2|W|^2 - \frac{4}{n(n-1)(n-2)}u^2|z|^2 - \frac{4}{n-2}|z|^4$$

$$+ \frac{8u}{n-2}\frac{|z|^3}{\sqrt{n(n-1)}} - |W|^2|z|^2 + \frac{3\sqrt{2}\,u|W||z|^2}{\sqrt{(n-1)(n-2)}} \leqslant -\varepsilon|\mathscr{D}|^2u^2.$$

First, let us estimate Q in terms of $|z|^2 \cdot |W|^2$, $|z|^2 \cdot u^2$, or $u^2 \cdot |W|^2$, using

$$\begin{cases} 2xy \leqslant x^2/a^2 + a^2y^2 & \forall\, a \neq 0, \\ |\mathscr{D}|^2 = |W|^2 + 4|z|^2/(n-2) < C(n) \cdot u^2, \end{cases}$$

where $C(n)$ is the constant we have to determine. We obtain

(a)

$$\frac{5}{2}|W|^3u \leqslant \frac{5u^2}{2}|W|^2d + \frac{5}{8d}|W|^4$$

$$\leqslant \frac{5u^2}{2}|W|^2d + \frac{5}{8d}|W|^2u^2C(n) - \frac{5}{2}\frac{|W|^2|z|^2}{d(n-2)},$$

(b)

$$\frac{8u}{(n-2)}\frac{|z|^3}{\sqrt{n(n-1)}} \leqslant \frac{4u^2|z|^2}{(n-2)n(n-1)a} + \frac{4|z|^4}{(n-2)} \cdot a,$$

(c)

$$\frac{4}{(n-2)}|z|^4(a-1) \leqslant C(n)|z|^2u^2(a-1) - |W|^2|z|^2(a-1) \quad (\text{if } a \geqslant 1),$$

(d)

$$\frac{3\sqrt{2}\,u|W|z^2}{\sqrt{(n-1)(n-2)}} \leqslant \frac{9|z|^2|W|^2}{4b} + \frac{2b|z|^2u^2}{(n-1)(n-2)},$$

i.e.,

(1)

$$Q \leqslant u^2|W|^2\left\{-\frac{1}{n} + \frac{5C(n)}{8d} + \frac{5}{2}d\right\}$$

$$+ |z|^2|W|^2\left\{-\frac{5}{2d(n-2)} + \frac{9}{4b} - 1 - (a-1)\right\}$$

$$+ |z|^2u^2\left\{C(n)(a-1) + \right.$$

$$\left. \frac{2b}{(n-1)(n-2)} + \frac{4}{n(n-1)(n-2)a} - \frac{4}{n(n-1)(n-2)}\right\}.$$

This formula can be used to get estimates on $C(n)$, better than $4/((n-2)(n-1)n(n+1))$, at least as soon as $n > 6$. Because the estimate for $4(\mathring{W} \circ \mathring{W}, W) + (\hat{W} \circ \hat{W}, W)$ we gave here is not sharp, there is no chance to get a sharp value for $C(n)$. We have to solve simultaneously

(i)

$$-\frac{1}{n} + \frac{5}{8d}C(n) + \frac{5d}{2} < 0,$$

(ii)

$$-\frac{5}{2d(n-2)} + \frac{9}{4b} - a < 0,$$

(iii)

$$C(n)(a-1) + \frac{2b}{(n-1)(n-2)} + \frac{4}{n(n-1)(n-2)}\left(\frac{1}{a} - 1\right) < 0.$$

First, we choose to show it is possible to find $K \in \mathbf{R}^+$ such that

$$C(n) \geqslant \frac{K}{(n-2)(n-1)n}.$$

Let us set $a = 2$ and $x = C(n) \cdot (n-1)(n-2)n$ and rewrite (iii)

$$0 < b < \left(1 - \frac{x}{2}\right)\frac{1}{n}.$$

Then, (ii) is equivalent to

$$0 < d < \frac{5}{(n-2)} \cdot \frac{(2-x)}{9n - 8 + 4x},$$

and (iii) to (at least if $x < 2$ and $n \geq 2$)

$$-8(2-x)(n-1)(n-2)(9n-8+4x) + (n-2)(9n-8+4x)^2 x$$
$$+ 100(2-x)^2 n(n-1) < 0.$$

This means

$$f(x) = x^3 16(n-2) + x^2(204n^2 - 404n + 192)$$
$$+ x(153n^3 - 1050n^2 + 1280n - 384)$$
$$- 144n^3 + 960n^2 - 1072n + 256 < 0.$$

All the coefficients of $f(x)$, but the last one, are positive for $n \geq 6$. Considering the case $n \geq 6$, we get $f(x)$ is increasing, and we can check $f(\frac{1}{2}) = -135n^3/2 + 486n^2 - 531n + 108$ which is negative for $n \geq 6$.   q.e.d.   ∎

Note 1. It is clear we can improve the value of x by dealing with any specific value of n. Here, we are essentially concerned in giving a value of $C(n)$ which does not imply a priori the positiveness of $\hat{R}$, for almost all n (all $n > 6$). Looking for an equivalent of $C(n)$, n large, the previous computation gives $16/(17n^3)$ (which can be compared with $4/n^4$).

Note 2. If one is interested in low dimensions ($n = 4, 5$), not included in the previous computation, it is not hard to check from (1) that the following does the job:

$$\text{for } n = 4, \quad x_4 = \frac{6}{25} \quad \left(a = 10, d = \frac{1}{20}, b = \frac{3}{20}\right),$$

$$\text{for } n = 5, \quad x_5 = \frac{48}{125} \quad \left(a = 4, d = \frac{1}{25}, b = \frac{27}{298}\right).$$

This can be improved according to the remark on page 321. To go further into the proof of Theorem 2, we need another basic estimate on the gradient of the scalar curvature. We have the following generalization of Theorem 11.1 of Hamilton (cf. **[H]**).

PROPOSITION 3. *Under the hypothesis of Theorem 2, for every $\eta > 0$ we can find a constant $C(\eta)$ depending only on η and the initial value of the metric, such that, on $0 \leq t < T$, we have $|Du| \leq \frac{1}{2}\eta^2 u^{3/2} + C(\eta)$.*

Proof. We have
(a)

$$\frac{\partial}{\partial t}|Du|^2 = \Delta|Du|^2 - 2|DDu|^2 + 4\left(Du, D|r|^2\right);$$

(b)

$$\frac{\partial}{\partial t}\left(\frac{|Du|^2}{u}\right) = \Delta\left(\frac{|Du|^2}{u}\right) - \frac{2}{u^3}|Du \cdot Du|^2 + \frac{4}{u^3}(uDDu, DuDu) - \frac{2}{u^3}|uDDu|^2$$
$$+ \frac{4}{u}\left(Du, D|r|^2\right) - \frac{2}{u^2}|Du|^2|r|^2;$$

(c)

$$\frac{\partial}{\partial t}u^2 = \Delta u^2 - 2|Du|^2 + 4u|r|^2,$$

$$\frac{\partial}{\partial t}|r|^2 = \Delta|r|^2 - 2|Dr|^2 + 4(\mathring{R}(r), r),$$

$$\frac{\partial}{\partial t}|z|^2 = \Delta|z|^2 - 2|Dr|^2 + 4(\mathring{R}(r), r) + \frac{2}{n}|Du|^2 - \frac{4u}{n}|r|^2.$$

Once more, we need to make the most of the irreducible decomposition of Dr. Set $Dr = E + F$ with $\gamma(Dr) = \gamma(E)$ for every available trace. If

$$E_{ijk} = \frac{1}{A}\left(g_{ij}D_k u + g_{ik}D_j u\right) + \frac{1}{B}g_{jk}D_i u,$$

we get $2((n + 1)/A + 1/B) = 1$ and $2/A + n/B = 1$, which, in turn, implies

$$|Dr|^2 \geqslant |E|^2 = \frac{|Du|^2}{(n + 1)(n - 1)}\frac{(3n - 2)}{2}.$$

Finally, we get

$$-|Dr|^2 + \frac{1}{n}|Du|^2 \leqslant -\frac{(n - 2)^2}{n(3n - 1)}|Dr|^2,$$

and rewrite

$$\frac{\partial}{\partial t}|z|^2 \leqslant \Delta|z|^2 - \frac{2(n - 2)^2}{(3n - 2)n}|Dr|^2 + 4(\mathring{R}(r), r) - \frac{4u}{n}|r|^2.$$

At this point, we need the following remark.

LEMMA 5. *Under the pinching hypothesis of Theorem 2, the Ricci curvature is always positive definite.*

If $|\mathscr{D}|^2/u^2 \leqslant 1/(2n(n - 1)(n - 2))$, then

$$|z|^2 \leqslant \frac{1}{4(n - 2)}\left|z\bigcirc g\right|^2 \leqslant \frac{(n - 2)}{4}|\mathscr{D}|^2,$$

and the eigenvalues of z satisfy

$$\lambda^2 \leqslant \frac{n - 1}{n}|z|^2 \leqslant \frac{(n - 1)(n - 2)}{4n}|\mathscr{D}|^2,$$

i.e.,

$$r > \frac{u}{n}g\left(1 - \frac{1}{2\sqrt{2}}\right) > 0. \quad\blacksquare$$

We can evaluate $(Du, D|r|^2)$ by

$$\left(Du, D|r|^2 \right) \leqslant 2|Du|\,|r|\,|Dr| \leqslant 2\sqrt{n}\,|r|\,|Dr|^2 \leqslant 2\sqrt{n}\,u|Dr|^2,$$

since $|r|^2 \leqslant u^2$, which, in turn, follows from the lemma. Now,

$$\frac{\partial}{\partial t}\left(\frac{|Du|^2}{u} - \eta u^2 \right) \leqslant \Delta\left(\frac{|Du|^2}{u} - \eta u^2 \right) + 2\eta|Du|^2 - \frac{2|r|^2|Du|^2}{u^2}$$

$$+ \frac{8\sqrt{n}}{u}|r|\,|Dr|^2 - 4\eta u|r|^2$$

$$\leqslant \Delta\left(\frac{|Du|^2}{u} - \eta u^2 \right) + 8\sqrt{n}\,|Dr|^2 - \frac{4\eta u^3}{n},$$

since $|r|^2 \geqslant u^2/n$ implies $|r|^2/u^2 > \eta$ for $\eta \leqslant 1/n$. We want to get rid of $8\sqrt{n}\,|Dr|^2$. Let us consider

$$F = \frac{|Du|^2}{u} - \eta u^2 + \frac{8n\sqrt{n}\,(3n-2)}{2(n-2)^2}|z|^2,$$

which gives

$$\frac{\partial F}{\partial t} \leqslant \Delta F - \frac{4\eta u^3}{n} + \frac{8n\sqrt{n}\,(3n-2)}{2(n-2)^2} \cdot 4 \cdot \left(\mathring{R}(r,r) - \frac{u}{n}|r|^2 \right).$$

Let us remark

$$\left((\mathring{R}(r), r) - \frac{u}{n}|r|^2 \right) \leqslant \frac{u|z|^2}{n}\left(\frac{(2n-3)(n+2) + (n-1)(n-2)}{(n-1)(n-2)} \right)$$

$$+ \frac{2}{n-2}|z|^3 + |W|\,|z|^2.$$

Using Proposition 2, we conclude that there exists a constant $C(\eta)$, for every $0 < \eta < 1/n$, depending only on η, and the initial metric such that

$$\frac{\partial F}{\partial t} \leqslant \Delta F + C(\eta).$$

At this point, we need a cheap generalization of an argument of Hamilton to show that, if $u \geqslant \rho$ at $t = 0$ for $\rho > 0$, then $T < n/2\rho$. The proof of this fact is as follows:

$$\frac{\partial u}{\partial t} = u + 2|r|^2 \geqslant \Delta u + \frac{2u^2}{n}.$$

Now, $f(t) = n\rho/(n - 2\rho t)$ satisfies $f(0) = \rho$, and $\partial f/\partial t = 2f/n$ so that,

$$\frac{\partial}{\partial t}(u - f) \geqslant \Delta(u - f) + \frac{2}{n}(u^2 - f^2),$$

and, by the maximum principal, $u \geqslant f$, i.e., $T < n/2\rho$. ∎

Now we can use once more the maximum principle to show $\max F_t \leqslant \max F_0 + C(\eta)t$ and the previous remark to conclude $\max F_t \leqslant \tilde{C}(\eta)$, where $\tilde{C}(\eta)$ depends only on η and the initial metric. I.e.,

$$|Du|^2 \leqslant \eta u^3 + \tilde{C}(\eta)u \leqslant 2\eta u^3 + \frac{\tilde{C}(\eta)}{4},$$

which establishes Proposition 2. ∎

To end the proof of Theorem 2 we can follow [**H**]. Since $|R|_{\max}$ blows up as t goes to T, we can use Proposition 2 to show $u_{\max}$ blows up as well as t goes to T. Let us remark that, using Myer's theorem, we can show

$$\lim_{t \to T} \frac{u_{\max}}{u_{\min}} = 1.$$

There exists a time $\theta > 0$ such that $C(\eta) < \frac{1}{2}\eta^2 u_{\max}^{3/2}$ for all $t \geqslant \theta$. Then, if $x \in M$ satisfies $u(x) = \sup_M u$, on any geodesic out of x, γ, of length $\leqslant 1/\eta u_{\max}^{1/2}$, we have $u \geqslant (1 - \eta)u_{\max}$. By Myer's theorem, since $r > \varepsilon ug$, all these geodesic arcs contain a conjugate point as soon as $\eta < \sqrt{\varepsilon}(1 - \eta)/\pi\sqrt{n - 1}$ i.e., as soon as η is sufficiently small. So, $u_{\min} \geqslant (1 - \eta)u_{\max}$ as soon as t is large enough.

Now applying the maximum principle to $u - f$, where $f(0) = u_{\max}$ and $df/dt = 2u_{\max}f$, we get $u \leqslant f$, i.e., f blows up as $t \to T$. But

$$\frac{1}{2} \log\frac{f(t)}{f(0)} = \int_0^t u_{\max} \, dt,$$

which means $\int_0^T u_{\max} \, dt$ diverges as $t \to T$. Now, if $\bar{u} = \int u \, dV / \int dV$ is the average scalar curvature, we get

$$\int_0^T \bar{u} \, dt \geqslant \int_t^T u_{\min} \, dt \geqslant \frac{1}{2} \int_t^T u_{\max} \, dt = \infty,$$

where we choose t such that $u_{\min}/u_{\max} \geqslant \frac{1}{2}$ on $[t, T[$.

Last step: back to the normalized evolution equation. We are now ready to conclude. We want to go back to the normalized evolution equation (such that the volume of $\tilde{g}$ is constant)

$$\frac{\partial}{\partial t} \tilde{g} = \frac{2\tilde{\bar{u}}}{n} \tilde{g} - 2\tilde{r} \quad \text{(i)}.$$

Then, if $\partial g/\partial t = -2r$ has a solution on a maximal interval $0 \leqslant t < T$, (i) has an associated solution on $0 \leqslant \tilde{t} < \tilde{T}$ such that $\tilde{r} = r$, $\tilde{u} = u/\psi$, $\tilde{\bar{u}} = \bar{u}/\psi$, $\tilde{t} = \int \psi(t) \, dt$, where ψ satisfies $\tilde{g} = \psi g$ and $\int d\tilde{V} = 1$. By homogeneity, we then get $\tilde{u}_{\max}/\tilde{u}_{\min} \to 1$ as $\tilde{t} \to \tilde{T}$ and $\tilde{r} \geqslant \varepsilon\tilde{u}\tilde{g}$ for some $\varepsilon > 0$. Using Myer's theorem, we get

$$\tilde{V} \leqslant \left(\sqrt{\frac{n - 1}{\varepsilon u_{\min}}}\right)^n \cdot \sigma_n \quad (\text{since } \tilde{r} \geqslant \varepsilon\tilde{u}_{\min}\tilde{g}),$$

where σ_n is the volume of the canonical sphere of radius 1. I.e.,

$$u_{\min} < \sigma_n^{2/n} \cdot \frac{(n - 1)}{\varepsilon}.$$

Now, because $\tilde{u}_{\max}/\tilde{u}_{\min} \to 1$, we have a constant C for which $\tilde{u}_{\max} \leqslant C$ on $[0, \tilde{T}[$. Since $d\tilde{t}/dt = \psi$ and $\psi\tilde{\tilde{u}} = \bar{u}$, we get

$$\int_0^{\tilde{T}} \tilde{\tilde{u}}\, d\tilde{t} = \int_0^t \bar{u}\, dt = \infty, \quad \text{i.e., } \tilde{T} = \infty.$$

Since $|\mathscr{D}|^2/u^2 \to 0$, according to Proposition 2 (the main step of the proof), $|\tilde{\mathscr{D}}|^2/\tilde{u}^2 \to 0$ and the sectional curvatures become arbitrarily pinched (at least pointwise). But $\tilde{u}_{\min}/\tilde{u}_{\max} \to 1$ means the pinching is global. From this point on, Theorem 2 is a consequence of the differential sphere theorem ([$\mathbf{C} \cdot \mathbf{E}$, §7]). ∎

Note. We could in fact follow Hamilton's last pages (§17) to get an exponential C^∞ convergence of $\tilde{g}(t)$ to $\tilde{g}(\infty)$, a metric of constant sectional curvature. For sake of shortness, and because it is well done in [$\mathbf{H}$], we do not develop this point here.

Using Calderón–Zygmund inequalities, E. Ruh proved in 1982 (cf. [$\mathbf{R}$]) a sphere theorem under an only pointwise pinching estimate on the sectional curvature. Unhappily, his method does not enable him to give any explicit bound on the ratio $\sigma_{\min}/\sigma_{\max}$ he a priori needs to prove the theorem. Here we give a translation of Theorem 2 in terms of a pinching of the sectional curvatures. Doing so, we get for free an improvement of Ruh's statement.

THEOREM 3. *Any compact manifold M^n, $n > 3$, whose sectional curvature is pointwise $c(n)$-pinched, is diffeomorphic to a space form of constant positive curvature (i.e. S^n if simply connected). Here*

$$c(n) = \frac{(4n - 3)(n - 2)/9 - \sqrt{2n/9 + 23/6 + 1/(n - 2)}}{(n - 2)(4n - 3)/9 - 2 - 1/(2(n - 2))} \qquad (n > 5),$$

i.e., $1 - c(n)$ is equivalent to $\frac{3}{4} \cdot \sqrt{2}\, n^{-3/2}$ for n large, ($c(n)$-pinched means $\sigma_{\min}(x) \geqslant c(n) \cdot \sigma_{\max}(x) > 0, \forall x \in M$).

Proof. Remember the sectional curvature determines the whole curvature tensor and let us write down what this means.

(i) $\quad 4R(X, Y, Z, Y) = R(X + Z, Y, X + Z, Y)$
$$- R(X - Z, Y, X - Z, Y) \quad \text{(by linearity)},$$

(ii) $\quad 6R(X, Y, Z, T) = R(X, Y + T, Z, Y + T)$
$$- R(X, Y - T, Z, Y - T) - R(Y, X + T, Z, X + T)$$
$$+ R(Y, X - T, Z, Y - T) \quad \text{(by linearity and the 1st Bianchi identity)}.$$

Now, if σ is δ-pinched and (e_i) is an orthonormal basis,

$$\max|R_{ijkl}| < \tfrac{2}{3}(1 - \delta)\sigma_{\max} \quad \text{if } i \neq j \neq k \neq l \neq i, i \neq k, j \neq l,$$

$$\max R_{ijik} \leqslant \tfrac{1}{2}(1 - \delta)\sigma_{\max} \quad \text{if } i \neq j \neq k \neq i,$$

$$\max R_{ijij} \leqslant \sigma_{\max},$$

so that $u \geqslant \delta \sigma_{\max} n(n-1)$,

$$|R|^2 \leqslant \sigma_{\max}^2 \left\{ \frac{4}{9}(1-\delta)^2 n(n-1)(n-2)(n-3) \right.$$

$$\left. + (1-\delta)^2 n(n-1)(n-2) + 2n(n-1) \right\},$$

and

$$\frac{|\mathscr{D}|^2}{u^2} \leqslant \left(|R|^2 - \frac{2u^2}{n(n-1)} \right) \frac{1}{u^2} \leqslant \frac{(1-\delta)(n-2)}{n(n-1)} \left(\frac{4n-3}{9} \right) \frac{(1-\delta)}{\delta^2}$$

$$+ \frac{2}{n(n-1)} \left(\frac{1-\delta^2}{\delta^2} \right).$$

To apply Theorem 2, we require

$$\frac{|\mathscr{D}|}{u^2} \leqslant \frac{1}{2n(n-1)(n-2)} \qquad (n > 5),$$

i.e.,

$$\left(2(1+\delta) + (1-\delta)(n-2)\frac{(4n-3)}{9} \right) \frac{(1-\delta)}{\delta^2} \leqslant \frac{1}{2(n-2)}$$

which gives

$$\delta \geqslant \frac{(n-2)(4n-3)/9 - \sqrt{2n/9 + 23/6 + 1/(n-2)}}{(n-2)(4n-3)/9 - 2 - 1/2(n-2)}.$$

Asymptotically, we have $1 - \delta$ is equivalent to $\frac{3}{4} \cdot \sqrt{2}\, n^{-3/2}$. ∎ (For $n = 4$, we get $\delta = 0.972$, for $n = 5$, we get $\delta = 0.971$), which establishes Theorem 3. This can, in turn, be improved according to the remarks of pages 321 and 323.

Nota bene. The Institute was an opportunity to discover many people were working on the question we deal with here: G. Huisken independently got a theorem close to Theorem 2 supra although his techniques do not enable him to get rid of a stronger assumption which implies a priori the curvature operator is positive.

REFERENCES

[Be] A. Besse, *Einstein manifolds*, Springer-Verlag (to appear).

[B] J. P. Bourguignon, *Les variétés de dimension 4 à signature non nulle dont la courbure est harmonique sont d'Einstein*, Invent. Math. **63** (1981), 263–268.

[B.K] J. P. Bourguignon and H. Karcher, *Curvature operators: pinching estimates and geometric examples*, Ann. E.N.S. 4ème, t. 11 (1978), 71–92.

[C.E] J. Cheeger and D. G. Ebin, *Comparison theorems in Riemannian geometry*, North-Holland, 1975.

[H] R. Hamilton, *Three-manifolds with positive Ricci curvature*, J. Differential Geom. **17** (1982), 255–306.

[R] E. A. Ruh, *Riemannian manifolds with bounded curvature ratios*, J. Differential Geom. **17** (1982), 643–653.

CENTRE DE MATHÉMATIQUES DE L'ECOLE POLYTECHNIQUE, FRANCE

Proceedings of Symposia in Pure Mathematics
Volume **44** (1986)

The Multiplicity of Generic Projections
of n-Dimensional Surfaces in $\mathbf{R}^{n+k}$ ($n + k \leqslant 4$)

DANA NANCE

Let $M^n \subset \mathbf{R}^{n+k}$ be a compact, imbedded manifold, or manifold-with-boundary. For any $v \in S^{n+k-1}$ we let p_v denote the orthogonal projection with kernel spanned by v, and we define the *multiplicity* of M with respect to the projection p_v as follows:

$$n(M, v, y) = n(v, y) = \mathcal{H}^0\{(p_v|M)^{-1}(y)\} \qquad (y \in \operatorname{Im} p_v),$$

$$n(M, v) = n(v) = \sup\{n(v, y) : y \in \operatorname{Im} p_v\}.$$

We will also make use of the *essential multiplicity*

$$\operatorname{ess} n(v) = \operatorname{ess} \sup\{n(v, y) : y \in \operatorname{Im} p_v\}.$$

In [1] the following result was proved.

THEOREM 1. *If $k = 1$ and M is of class $(n + 1, \alpha)$ for some $\alpha > 0$, then*
(a)

$$\int_{S^1} n(v)\, d\mathcal{H}^1 v \leqslant 2\int_M \|DN\|(x)\, d\mathcal{H}^1 x + \pi\mathcal{H}^0(\partial M) \quad (if\ n = 1);$$

(b)

$$\int_{S^2} n(v)\, d\mathcal{H}^1 v \leqslant 2\int_M \|DN\|^2 d\mathcal{H}^2 + 2\int_M \|D^2 N\| d\mathcal{H}^2$$

$$+ 6\int_{\partial M} \|DN\| d\mathcal{H}^1 + 8\int_{\partial M} \|D\tau\| \quad (if\ n = 2);$$

where $N: M \to S^n$ is the Gauss map and $\tau: \partial M \to S^2$ is a unit tangent vector field.

1980 *Mathematics Subject Classification.* Primary 53C65; Secondary 58C27.

© 1986 American Mathematical Society
0082-0717/86 $1.00 + $.25 per page

It will be important to note that the hypotheses of Theorem 1(a) can be weakened: (a) holds even if M is only immersed (provided that we count the multiplicity of boundary points which bound more than one arc), and if M is of class 1 (rather than $(2, \alpha)$) at the boundary. The proof is identical to that given in [1].

In the present paper we will study $n(\nu)$ for the case $n = 3$, $k = 1$ (and, necessarily, the case $n = 2$, $k = 2$); we will show that if M is of class $q = 4 + \alpha$ (i.e. $(4, \alpha)$) for $\alpha > 0$, then

$$\int_{S^3} [n(\nu)]^\beta \, d\mathscr{H}^3 \nu < \infty \quad \text{for } \beta < (q - 4)/2(q - 1).$$

Special attention should be paid to the use in Theorem 9 of Yomdin's generalization [2] of the Morse–Sard theorem; I believe this to be the first use of his theorem in an integral-geometric context.

DEFINITION. A compact set $M \subset \mathbf{R}^{n+k}$ is a *finite-curvature stratified manifold* (FSM) if $M = M^0 \cup M^1 \cup \cdots \cup M^n$, where:

(1) M^i is an $(n - i)$-dimensional manifold of class $(n - i + 1, \alpha)$;

(2) the points of M^{i+1} are boundary points of M^i in the usual sense;

(3) $N_{(j)}$ (the Gauss map of M^j) is continuous on M^{j+1} (naturally, it is of class $(n - j, \alpha)$ on M^j);

(4)

$$\int_M J_r N_0 \, d\mathscr{H}^n < \infty \quad \text{for } 0 \leqslant r \leqslant n.$$

NOTATION. If $M = M^0 \cup \cdots \cup M^n \subset \mathbf{R}^{n+1}$ is a FSM, let

$$M_*^{j,b} = \left\{ (x, \nu) \in M^j \times S^n: \nu \cdot N_0(x) = \nu \cdot \langle \nu, DN_0(x) \rangle \right.$$

$$= \nu \cdot \langle \nu \otimes \nu, D^2 N_0(x) \rangle = \cdots$$

$$\left. = \nu \cdot \langle (\otimes \nu)^{b-1}, D^{b-1} N_0(x) \rangle = 0, \nu \cdot \langle (\otimes \nu)^b, D^b N_0(x) \rangle \neq 0 \right\};$$

$$E_*^{j,b} = \left\{ (x, \nu) \in M^j \times S^n: \right.$$

$$\left. N_{(j)} \wedge \langle \nu, DN_0 \rangle \wedge \cdots \wedge \langle (\otimes \nu)^{b-1}, D^{b-1} N_0 \rangle = 0 \right\};$$

$$^2 M_*^{j,1} = \left\{ (x, \pi) \in M^j \times G_0(n + 1, 2): N_0 \lrcorner \pi^* = 0, \pi \cdot \langle \pi, \Lambda^2 DN_0(x) \rangle \neq 0 \right\}$$

$$(\pi^* = \text{dual covector to } \pi);$$

$$^2 M_*^{j,2} = \left\{ (x, \pi) \in M^j \times G_0(n + 1, 2): N_0 \lrcorner \pi^* = 0, \pi \cdot \langle \pi, \Lambda^2 DN_0(x) \rangle = 0, \right.$$

$$\left. \langle \pi, \Lambda^2 DN_0(x) \rangle \neq 0 \right\};$$

$$M_x^{b,i} = \left\{ \nu: (x, \nu) \in M_*^{b,i} \right\}; \qquad M_\nu^{b,i} = \left\{ x: (x, \nu) \in M_*^{b,i} \right\};$$

$$E_x^{b,i}, E_\nu^{b,i}, {}^2 M_x^{b,i}, {}^2 M_\pi^{b,i} \text{ defined analogously;}$$

$$M_\nu = \left\{ x \in M: \nu \cdot N_0(x) = 0 \right\} \cup M^1 \cup \cdots \cup M^n;$$

$$\partial M = M^1 \cup \cdots \cup M^n;$$

$$P(x, \nu) = x; \qquad Q(x, \nu) = \nu.$$

It will occasionally be useful to write $N_{(1)}$ as $N_0 \wedge N_1$; the reader should be careful to distinguish between N_1 and $N_{(1)}$.

LEMMA 2. *If $M = M^0 \cup \cdots \cup M^n \subset \mathbf{R}^{n+1}$ is a FSM such that $N_{(j)}$ is of class $(n - j - 1, \alpha)$ in a neighborhood of each $x \in M^{j+1}$, and if ∂M is also a FSM, then for almost all $v \in S^n$, $p_v(M_v)$ is a FSM.*

PROOF. By the implicit function theorem, $(M^{j,b} \cup M^{j,b+1}) \setminus E^{j,b-1}$ is a $(2n - j - b)$-dimensional manifold of class $(n + 1 - j - b, \alpha)$ and $p_v(M_v^{j,b} \setminus E_v^{j,b-1})$ is a $(n - j - b)$-dimensional manifold of class $(n + 1 - j - b, \alpha)$. By calculation, $J_n(Q|M_*^{j,b} \cap E_*^{j,b}) = 0$, hence by the Morse–Sard theorem

$$\dim\left\{ v\colon M_v^{j,b} \cap E_v^{j,b} \neq \varnothing \right\} \leqslant (n - 1) + \frac{(2n - j - b) - (n - 1)}{n + 1 - j - b + \alpha} < n,$$

hence $\mathscr{H}^n\{ v\colon M_v^{j,b} \cap E_v^{j,b} \neq \varnothing\} = 0$. Since $M_v^{j,b} \subset E_v^{j,b}$ if $j + b > n$, it also follows that $\mathscr{H}^n\{ v\colon M_v^{j,b} \neq \varnothing\} = 0$ if $j + b > n$. It then follows from a simple induction argument that

$$p_v(M_v) = \bigcup_{j+b\leqslant n} p_v\left(M_v^{j,b} \setminus E_v^{j,b-1}\right)$$

for $\mathscr{H}^n$-almost all v.

Assume v is a direction for which the above decomposition holds. If $x_0 \in M_v^{j,b}$, let $n' - n - j - b + 1$ and let $\psi\colon \mathbf{R}^{n'} \to M_v^{j,b-1} \cup M_v^{j,b}$ be a parametrization of $M_v^{j,b-1} \cup M_v^{j,b}$ such that $\varphi(0) = x_0$; $\varphi(x_1,\ldots,x_{n'-1}, 0) \in M_v^{j,b}$ for all $x_1,\ldots,x_{n'-1}$; and $\partial\varphi(0, 0)/\partial x_{n'} = v$. Let $\varphi'_\pm(x_1,\ldots,x_{n'-1}, x_n) = p_v \circ \varphi(x_1,\ldots,x_{n'-1}, \pm(x_{n'})^{1/2})$. The partial derivatives of $\varphi'_\pm$ are continuous at 0; in particular,

$$\frac{\partial\varphi'_\pm}{\partial x_{n'}}(x, 0) = \frac{1}{2} p_v\left[\frac{\partial^2\varphi}{\partial x_{n'}^2}(x, 0) \right].$$

The partial derivatives $\partial\varphi'_\pm/\partial x_i$ ($i \leqslant n' - 1$) are independent, since $v \notin \mathrm{Tan}(M_v^{j,b}, x_0)$; to show that $\partial\varphi'_\pm/\partial x_{n'}$ is independent of these we note that

$$\frac{\partial\varphi}{\partial x_i} \cdot \langle (\otimes v)^{b-1}, D^{b-1}N_0(x_0)\rangle = 0, \qquad v \cdot \langle (\otimes v)^{b-1}, D^{b-1}N_0(x_0)\rangle = 0,$$

$$\to \frac{\partial\varphi'}{\partial x_i} = p_v\left(\frac{\partial\varphi}{\partial x_i} \right) \perp \langle (\otimes v)^{b-1}, D^{b-1}N_0(x_0)\rangle,$$

while

$$\frac{\partial^2\varphi}{\partial x_{n'}^2} \cdot \langle (\otimes v)^{b-1}, D^{b-1}N \rangle = -\frac{\partial\varphi}{\partial x_n'} \cdot \langle (\otimes v)^b, D^bN \rangle$$

$$= -v \cdot \langle (\otimes v)^b, D^bN \rangle \neq 0.$$

Hence φ'_+ and φ'_- parametrize a neighborhood of $p_v(x_0) \in p_v(M_v^{j,b-1} \cup M_v^{j,b})$ as the union of two C^1 manifolds-with-boundary. This implies that the Gauss map for $p_v(M_v^{j,b-1})$ extends continuously to $p_v(M_v^{j,b})$.

Finally, the finite-curvature condition for $p_\nu(M_\nu^{0,1} \cup M_\nu^{1,0})$ follows from the co-area formula. For future reference, we define a (nonunit) normal to $M_\nu^{j,b}$ by $N^{j,b} \equiv N_{(j)} \wedge \langle \nu, DN_0 \rangle \wedge \cdots \wedge \langle (\otimes \nu)^b, D^b N_0 \rangle$ and a (nonunit) normal to $p_\nu(M_\nu^{j,b})$ by

$$Z^{j,b} \equiv \left(N_{(j)} \llcorner \nu^* \right) \wedge \left(\bigwedge_{1 \leqslant k \leqslant b} \langle (\otimes \nu)^k, D^k N_0 \rangle \right)$$
$$+ (-1)^{b+j+1} \nu \cdot \langle (\otimes \nu)^b, D^b N_0 \rangle N^{j,b-1}.$$

Then

$$(*) \quad \int_{\nu \in S^n} \int_{p_\nu(M_\nu^{j,b})} J_r\left(\frac{Z^{j,b}}{|Z^{j,b}|} \right)(y) \, d\mathcal{H}^{n-j-b} y \, d\mathcal{H}^n \nu$$

$$= \int_{\nu \in S^n} \int_{M_\nu^{j,b}} J_r\left(\frac{Z^{j,b}}{|Z^{j,b}|} \right) [p_\nu(x)] J_{n-j-b}(p_\nu|M_\nu^{j,b})(x) \, d\mathcal{H}^{n-j-b} x \, d\mathcal{H}^n \nu$$

$$= \int_{M_*^{j,b}} J_r\left(\frac{Z^{j,b}}{|Z^{j,b}|} \right) [p_\nu(x)] J_{n-j-b}(p_\nu|M_\nu^{j,b})(x) J_n Q(x,\nu) \, d\mathcal{H}^{2n-j-b}(x,\nu)$$

$$= \int_{M^j} \int_{M_x^{j,b}} J_r\left(\frac{Z^{j,b}}{|Z^{j,b}|} \right) [p_\nu(x)] J_{n-j-b}(p_\nu|M_\nu^{j,b})(x) \frac{J_n Q(x,\nu)}{J_{n-j} P(x,\nu)}$$
$$\cdot d\mathcal{H}^{n-b}\nu \, d\mathcal{H}^{n-j} x.$$

When $j = 0, b = 1, J_n Q / J_n P = |\langle \nu, DN_0 \rangle|$,

$$J_{n-1}(p_\nu|M_\nu^{0,1}) = \frac{|\nu \cdot \langle \nu, DN_0 \rangle|}{|\langle \nu, DN_0 \rangle|},$$

and the integrand is $\leqslant J_{r+1} N_0(x)$ (equal if $r = n - 1$). When $j = 1, b = 0$, it is trivial that $(*)$ is bounded if $r = 0$. If $r = n - 1$, the integral $(*)$ becomes

$$\int_{\nu \in S^n} \int_{e \cdot \nu = 0} \mathcal{H}^0\{ x : e \wedge N_0 \wedge N_1(x) = 0 \} \, d\mathcal{H}^{n-1} d\mathcal{H}^n$$

$$= \int_{e \in S^n} \int_{\nu \cdot e = 0} \mathcal{H}^0\{ x : e \wedge N_0 \wedge N_1(x) = 0 \} \, d\mathcal{H}^{n-1} d\mathcal{H}^n$$

$$= (n-1)\alpha(n-1) \int_{M^1} \int_{e \wedge N_0 \wedge N_1(x) = 0} \langle \pi, \Lambda^{n-1}[(e \cdot N_0)DN_0 + (e \cdot N_1)DN_1] \rangle$$
$$\cdot d\mathcal{H}^1 d\mathcal{H}^{n-1}$$

$$\leqslant 2\pi(n-1)\alpha(n-1) \int_{M^1} J_{n-1}(N_0 \wedge N_1)(x) \, d\mathcal{H}^{n-1} x.$$

Since for almost all directions ν,

$$\int_{p_\nu(M_\nu^{j,b})} J_r\left(\frac{Z^{j,b}}{|Z^{j,b}|} \right)(y) \, d\mathcal{H}^{n-1} y < \infty$$

for $r = 0$ and $n - 1$, the same is true for $0 \leqslant r \leqslant n - 1$ by Schwarz's inequality.

LEMMA 3. *If $M \subset \mathbf{R}^4$ is a 3-dimensional manifold-with-boundary, then for almost all $\nu \in S^3$ and almost all $\sigma \in S^3$ such that $\sigma \cdot \nu = 0$, $p_\sigma\{[p_\nu(M_\nu)]_\sigma\}$ is a FSM.*

PROOF. The curvature estimate follows directly from an application of Lemma 2 to $p_\nu(M_\nu)$, since it does not require smoothness of order (n, α) at the boundary. However, the remaining conclusions of Lemma 2 do require this degree of smoothness. To circumvent this difficulty, one observes that many of the singular sets of $p_\sigma | p_\nu(M_\nu)$ can also be described as projections of curves *in* M which have the requisite degree of smoothness. For example: if $y = p_\nu(x)$,

$$y \in \left[p_\nu\left(M_\nu^{0,1} \right) \right]_\sigma^1 \to x \in {}^2M_{\nu \wedge \sigma}^1,$$

$$y \in \left[p_\nu\left(M_\nu^{0,1} \right) \right]_\sigma^2 \to x \in {}^2M_{\nu \wedge \sigma}^2,$$

$$y \in \left[p_\nu\left(M_\nu^{0,2} \right) \right]_\sigma^1 \to x \in {}^2M_{\nu \wedge \sigma}^1 \quad \text{or} \quad {}^2M_{\nu \wedge \sigma}^2.$$

The last example is particularly important: the projection p_σ, in a certain sense, does not even "see" the boundary set $p_\nu(M_\nu^{0,2})$. (See Figure 1.)

Once the singular set $[p_\nu(M_\nu)]_\sigma$ is properly decomposed, the proof that $p_\sigma[p_\nu(M_\nu)]_\sigma$ is a FSM proceeds in a substantially similar fashion to Lemma 2. I omit the details for reasons of space and tediousness.

REMARK. The fact that $p_\nu(M_\nu)$ is not smooth enough at the boundary to permit a pure inductive argument is not an artifact of the proof of Lemma 2, but a real phenomenon. (Consider, for example, the curvature of the standard cusp $y^2 = x^3$ near $(0, 0)$). Moreover, the fact that the curvature is, in general, infinite at cusp points is responsible for the behavior illustrated in Figure 1.

The next lemma provides the key "inductive" step in the proof of the Multiplicity Estimate.

LEMMA 4 (WEAK CROSSING LEMMA). *If* $M = M^0 \sqcup \cdots \sqcup M^n \subset \mathbf{R}^{n+1}$ *is a FSM, and if* ∂M *is a FSM, then for almost all* $\nu \in S^n$, $\sigma \in S^n$, *such that* $\sigma \cdot \nu = 0$,

$$\operatorname{ess} n(M, \nu) \leqslant \operatorname{ess} n(p_\nu(M^\nu), \sigma).$$

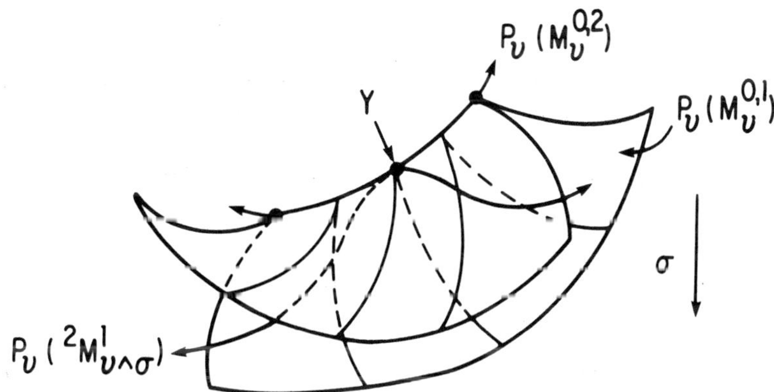

FIGURE 1. Here $y \in [p_\nu(M_\nu^{0,2})]_\sigma^1$. The curve $p_\nu({}^2M_{\nu \wedge \sigma}^1)$ is the set of points at which σ is tangent to $p_\nu(M_\nu^{0,1})$; note that this curve crosses *smoothly* from the front to the back sheet at y. The space in which the figure is drawn is $\operatorname{Im} p_\nu \approx \mathbf{R}^3$.

PROOF. By the finite-curvature condition,

$$\int_{G_0(n+1,2)} \mathcal{H}^{n-2}\{x \in M: \pi \subset \mathrm{Tan}(M, x)\}\, d\mathcal{H}^{2n-2}\pi$$

$$\leq C(n)\int_M |\Lambda^2 DN_0| < \infty;$$

$$\int_{G_0(n+1,2)} \mathcal{H}^{n-2}\{x \in M^1: \pi \cdot N_{(1)}(x) = 0\}\, d\mathcal{H}^{2n-2}\pi$$

$$\leq C(n)\int_{M^1} |\Lambda^1 DN_{(1)}| < \infty.$$

Hence for almost all $v \in S^n$, almost all $\sigma \perp v$, and almost all $y \in \mathrm{Im}(p_\sigma \circ p_v)$,

$$(p_\sigma \circ p_v)^{-1}(y) \cap \Big[\{x \in M: N_0(x) \lrcorner (\sigma \wedge v)^* = 0\}$$

$$\cup \{x \in M^1: (\sigma \wedge v) \cdot N_{(1)}(x) = 0\}\Big] = \varnothing.$$

Also we can assume that $(p_\sigma \circ p_v)^{-1}(y) \cap (M^2 \cup \cdots \cup M^n) = \varnothing$, since the Hausdorff dimension of the latter set is at most $(n - 2)$. For any such y, $C = (p_\sigma \circ p_v)^{-1}(y) \cap M$ is a 1-manifold with boundary $\partial C = (p_\sigma \circ p_v)^{-1}(y) \cap M^1$. Let $L_\sigma^y = p_\sigma^{-1}(y) \cap \mathrm{Im}\, p_v$. By a counting argument like that in Lemma 9 of [1], for any $x \in L_\sigma$,

$$n(M, x) \leq \sharp \text{ of fold points of } C + \tfrac{1}{2}\sharp(\partial C)$$
$$= \mathcal{H}^0\{p_v(M_v^{0,1}) \cap L_\sigma\} + \tfrac{1}{2}\mathcal{H}^0\{p_v(M^{1,0}) \cap L_\sigma\}$$
$$\leq \mathcal{H}^0\{p_v(M_v) \cap L_\sigma\} = n[p_v(M_v), \sigma, y]$$
$$\leq \mathrm{ess}\, n[p_v(M_v), \sigma] \quad \text{for almost all } y.$$

Since almost all points of $\mathrm{Im}\, p_v$ are contained in $\cup L_\sigma^y$, this proves the lemma.

PROPOSITION 5. *If $M \subset \mathbf{R}^4$ is a 3-dimensional manifold-with-boundary of class $(4, \alpha)$, then for $\mathcal{H}^3$-almost all $v \in S^3$, $\mathrm{ess}\, n(M, v) = n(M, v)$.*

To prove this we need a lemma on the multiplicity of projections of the boundary.

LEMMA 6. *If $M^n \subset \mathbf{R}^{n+2}$ is a manifold of class (n, α), $n = 1$ or 2, then for almost all $v \in S^{n+1}$, $n(M, v) \leq n + 1$.*

PROOF. Let $f_2: M \times M \to S^{n+1}$ be given by $f_2(x, y) = (x - y)/|x - y|$. By the Morse–Sard theorem, $\mathcal{H}^{n+1}[f_2(\Sigma)] = 0$ where $\Sigma = \{(x, y): J_{n+1}f_2(x, y) = 0\}$. Σ includes all pairs (x, y) for which the self-intersection of $p_v(M)$ at $p_v(x) = p_v(y)$ is not transverse $(v = (x - y)/|x - y|)$; in the case $n = 2$, it also includes all

pairs (x, y) for which both $\mathrm{Tan}(M, x)$ and $\mathrm{Tan}(M, y)$ contain $(x - y)/|x - y|$.
Let $P^{3,0} = \{(x, y, z) \in M \times M \times M : (x - y) \wedge (x - z) = 0\}$;

$$P^{4,0} = \{(x, y, z, w) \in M \times M \times M \times M : (x - y) \wedge (x - z)$$
$$= (x - y) \wedge (x - w) = 0\};$$
$$P^{3,1} = \{(x, y, z) \in P^{3,0} : [(x - y) \lrcorner N(x)^*] \wedge [(x - y) \lrcorner N(y)^*]$$
$$\wedge [(x - y) \lrcorner N(z)^*] = 0\}.$$

Let $f_{i,j} \colon P^{i,j} \to S^{n+1}$ be given by $f_{i,j}(x, y, z) = (x - y)/|x - y|$ [or $f_{i,j}(x, y, z, w) = (x - y)/|x - y|$].

In the case $n = 1$, $P^{3,0}$ is a 1-manifold at (x, y, z) except in case $(x, y) \in \Sigma$ (or $(y, z) \in \Sigma$, or $(x, z) \in \Sigma$); hence $\mathscr{H}^2[f_{3,0}(P^{3,0})] = 0$. In the case $n = 2$, observe that $P^{3,1} \subset \{(x, y, z) \in P^{3,0} : J_3 f_{3,0}(x, y, z) = 0\}$, so that $\mathscr{H}^3[f_{3,1}(P^{3,1})] = 0$. $P^{4,0}$ is a 2-manifold at (x, y, z, w) except in case (without loss of generality) $(x, y) \in \Sigma$ or (w.l.o.g.) $(x, y, z) \in P^{3,1}$. Hence $\mathscr{H}^3[f_{4,0}(P^{4,0})] = 0$. This proves the Lemma.

LEMMA 7. *Given* $M = M^0 \cup M^1$ *as in Proposition 5. Let* $x_0 \in M_\nu^{j,b}$ *and let* $y_0 = p_\nu(x_0)$. *If* $j = 1$, *assume* $\nu \notin \mathrm{Tan}(M^1, x_0)$. *There exists a neighborhood* $U \subset M$ *of* x_0 *such that*:

(a) $\mathscr{H}^0\{(p_\nu|U)^{-1}(y_0)\} = 1$;

(b) *For any* $\sigma \perp \nu$ *such that* $[\sigma \cdot N_0(x_0)][\nu \cdot \langle (\otimes \nu)^b, D^b N_0(x_0) \rangle] > 0$ *there exists* $\varepsilon > 0$ *such that*

$$\mathscr{H}^0\{(p_\nu|U)^{-1}(y_0 + t\sigma)\} \geqslant 2 \quad \textit{for all } 0 < t < \varepsilon \ (b \textit{ odd}, j = 0)$$

or

$$\mathscr{H}^0\{(p_\nu|U)^{-1}(y_0 + t\sigma)\} \geqslant 1 \quad \textit{for all } -\varepsilon < t < \varepsilon \ (b \textit{ even}, j = 0)$$

or

$$\mathscr{H}^0\{(p_N|U)^{-1}(y_0 + t\sigma)\} \geqslant 1 \quad \textit{for all } 0 < t < \varepsilon \textit{ or for all } -\varepsilon < t < 0 \ (j = 1).$$

PROOF. Choose a smooth parametrization φ of the curve $(p_\nu|U)^{-1}(L_\sigma)$, where $L_\sigma = \{y_0 + t\sigma\}$, such that $\varphi(0) = x_0$ and show that

$$\varphi^{(b+1)}(0) \cdot \sigma = \left[-\nu \cdot \langle (\otimes \nu)^b, D^b N_0(x_0) \rangle \right] / [\sigma \cdot N_0(x_0)] \neq 0.$$

LEMMA 8 (ASYMMETRIES OF THE SPHERE). *If* $\nu_1, \ldots, \nu_s \in S^n$ *and any* $(n + 1)$ *of the vectors* $\nu_1, \ldots, \nu_s$ *are linearly independent, then there exists* $w \in S^n$ *such that*

$$\mathscr{H}^0\{\nu_i \colon w \cdot \nu_i > 0\} - \mathscr{H}^0\{\nu_i \colon w \cdot \nu_i < 0\} \geqslant \min\{s, n\}.$$

PROOF. The result for $s \leqslant n$ follows by linear algebra. Denote the above difference by $d(w)$. Suppose $n = 1$. Consider any point w_i such that $w_i \cdot \nu_i = 0$. By hypothesis, $\nu_j \cdot w_i \neq 0$ if $j \neq i$. For sufficiently small t,

$$d[(\cos t)w_i + (\sin t)\nu_i] - d[(\cos t)w_i - (\sin t)\nu_i] = 2.$$

The conclusion follows immediately.

Suppose the lemma holds for $n \leqslant N - 1$. Let $v_1, \ldots, v_s \in S^N$. Let $V_1 = \{w: w \cdot v_1 = 0\}$. The points $p_{v_1}(v_2)/|p_{v_1}(v_2)|, \ldots, p_{v_1}(v_s)/|p_{v_1}(v_s)|$ satisfy the hypothesis of the lemma, with n replaced by $N - 1$; hence there exists $w_1 \in V_1$ such that $d(w_1) \geqslant N - 1$. For small enough t,

$$d\big[(\cos t)w_1 + (\sin t)v_1\big] \geqslant (N - 1) + 1 = N.$$

This proves the lemma.

PROOF OF PROPOSITION 5. If there exists $y_0 \notin p_\nu(M_\nu)$ such that $n(M, \nu, y_0) = \operatorname{ess} n(M, \nu) < \infty$, then $\operatorname{ess} n(M, \nu) = n(M, \nu)$, because there is an open neighborhood U of y_0 such that $y \in U \to n(M, \nu, y) = n(M, \nu, y_0)$.

Let ν be chosen so that the stratification of $p_\nu(M_\nu)$ of Lemma 2 holds. Suppose $y_0 \in p_\nu(M_\nu)$, with

$$(p_\nu|M)^{-1}(y_0) = \big\{x_1^1, \ldots, x_r^1, x_1^2, \ldots, x_s^2, x_1^3, \ldots, x_t^3\big\},$$

where $\{x_1^1, \ldots, x_r^1\} \subset \bigcup\{M_\nu^{0,b}: b \text{ even}\}$; $\{x_1^1, \ldots, x_s^1\} \subset \bigcup\{M_\nu^{0,b}: b \text{ odd}\}$; $\{x_1^1, \ldots, x_t^1\} \subset \bigcup M_\nu^{1,b} = \partial M$. By Lemma 6, we can assume that $t \leqslant 3$, and if $t = 3$ that

$$\big(\nu \lrcorner N_{(1)}(x_1^3)*\big) \wedge \big(\nu \lrcorner N_{(1)}(x_2^3)*\big) \wedge \big(\nu \lrcorner N_{(1)}(x_3^3)*\big) \neq 0.$$

By arguments similar to those of Lemma 6, it can be shown that, in fact, *any* three of the vectors $\{N_0(x_j^2), (\nu \lrcorner N_{(1)}(x_k^3)*)\}$ are linearly independent (i.e. if we assume ν lies outside a particular set of measure 0). Corresponding to this set of vectors we choose w in Lemma 8. Then by Lemma 7, there exists $\varepsilon > 0$ such that for $0 < u < \varepsilon$,

$$\mathscr{H}^0\big\{(p_\nu|M)^{-1}(y_0 + uw)\big\} \geqslant r + \left(\frac{s}{2}\right)2 + 3 = \mathscr{H}^0\big\{(p_\nu|M)^{-1}(y_0)\big\}.$$

(If $t = 2$ then we may have $\nu \lrcorner N_{(1)}(x_2^3)* = 0$, and we apply Lemma 8 to $\{N_0(x_j^2)\} \cup \{\nu \lrcorner N_{(1)}(x_1^3)*\}$ instead.) Also, for small enough ε, $y_0 + uw \notin p_\nu(M_\nu)$. Hence

$$\sup\{n(\nu, y): y \in p_\nu(M_\nu)\} \leqslant \sup\{n(\nu, y): y \notin p_\nu(M_\nu)\},$$

and the proposition follows.

THEOREM 9. *If* $M \subset \mathbf{R}^4$ *is a 3-dimensional manifold-with-boundary of class* $(q + 1) > 4$, *then*

$$\int_{S^3} [n(M, \nu)]^\beta \, d\mathscr{H}^3 \nu \leqslant C < \infty$$

for any $\beta < (q - 3)/2q$, $\beta \geqslant 0$, *where* C *is a constant depending only on* q, β, $\|DN_0\|_q$, $\|DN_{(1)}\|_{q-1}$, $\mathscr{H}^3(M)$, *and the injectivity radius of* ∂M.

PROOF. By Lemmas 2 and 3, $p_\nu(M_\nu)$ and $p_\sigma[p_\nu(M_\nu)_\sigma]$ are FSM, for almost all $\nu, \sigma \in \{\nu\}^\perp$. That their boundaries are also FSM follows from the curvature estimates in this proof. Hence by Lemma 4 and Proposition 5,

$$n(M, \nu) \leqslant \operatorname{ess} n[p_\nu(M_\nu), \sigma] \leqslant \operatorname{ess} n[p_\sigma[p_\nu(M_\nu)_\sigma], \tau]$$

for almost all ν, $\sigma \in \{\nu\}^{\perp}$, and $\tau \in \{\nu, \sigma\}^{\perp}$. Moreover, we have stratifications

$$p_{\nu}(M_{\nu}) = \hat{M}^2 \cup \hat{M}^1 \cup \hat{M}^0,$$

$$\left(\hat{M}^i = \bigcup_{j+b=3-i} p_{\nu}\left(M_{\nu}^{j,b} \setminus E_{\nu}^{j,b}\right) \right),$$

$$p_{\sigma}\left[p_{\nu}(M_{\nu})\sigma \right] = \hat{\hat{M}}^1 \cup \hat{\hat{M}}^0.$$

By Theorem 1(a), there exists τ such that

$$\mathrm{ess}\, n\left[p_{\sigma}\left[p_{\nu}(M_{\nu}) \right]_{\sigma}, \tau \right] \leqslant \frac{1}{\pi} \int_{\hat{\hat{M}}^1} \|DN_{\hat{\hat{M}}^1}\| d\mathcal{H}^1 + \frac{1}{2} \mathcal{H}^0(\hat{\hat{M}}^0).$$

By integral estimates identical to those used in proving Theorem 1(b),

$$\int_{\sigma \perp \nu} \mathrm{ess}\, n\left[p_{\nu}(M_{\nu}), \sigma \right] d\mathcal{H}^2\sigma \leqslant I_1(\nu) + I_2(\nu) + I_3(\nu) + I_4(\nu) + I_5(\nu)$$

$$\equiv 2\int_{\hat{M}^2} J_2 N_{\hat{M}^2}\, d\mathcal{H}^2$$

$$+ \frac{1}{2} \int_{\hat{M}^2} \sum_{e \in (\hat{M}^2)_y^2} |e \cdot \langle e \otimes e, D^2 N_{\hat{M}^2}(y) \rangle| d\mathcal{H}^2 y$$

$$+ \int_{\hat{M}^1} 6\|D(N_{\hat{M}^2}|\hat{M}^1)\| d\mathcal{H}^1 + \int_{\hat{M}^1} 8\|DN_{\hat{M}^1}\| d\mathcal{H}^1$$

$$+ 4\pi \mathcal{H}^0(\hat{M}^0).$$

(One can show that, for almost all ν, $D(N_{\hat{M}^2}|\hat{M}^1)$ is defined even though $DN_{\hat{M}^2}$ may not be defined at points of $\hat{M}^1$; see p. 340.) Hence there exists σ such that $\mathrm{ess}\, n[\, p_{\nu}(M_{\nu}), \sigma]$ is less than $(4\pi)^{-1}$ times the above quantity.

Denote by $I_1^{j,b}$, $I_2^{j,b}$, $I_3^{j,b}$, $I_4^{j,b}$, the integrals I_1, I_2, I_3, I_4, restricted to the submanifold $p_{\nu}(M_{\nu}^{j,b} \setminus E_{\nu}^{j,b})$, so that $I_k = \sum_{j,b} I_k^{j,b}$.

By the calculations in Lemma 2,

$$\int_{\nu} \int_{\hat{M}^2(\nu)} J_2 N_{\hat{M}^2}\, d\mathcal{H}^2 \leqslant 4\pi \int_M J_3 N_M\, d\mathcal{H}^3 + 8\pi^2 \int_{\partial M} J_2 N_{\partial M}\, d\mathcal{H}^2.$$

However, it is not possible to obtain such an *a priori* estimate for $\int_{\nu} I_2(\nu)\, d\mathcal{H}^3\nu$. Indeed, if $y \in p_{\nu}(M_{\nu}^{0,1} \setminus E_{\nu}^{0,1}) \cap \hat{M}^2$, $y = p_{\nu}(x)$, $x \in M_{\nu}^{0,1} \setminus E_{\nu}^{0,1}$, and $e \in \mathrm{Tan}(M_{\nu}^{0,1}, x)$, then

$$\langle p_{\nu}(e) \otimes p_{\nu}(e), D^2 N_{\hat{M}^2}(y) \rangle$$

$$= \langle e \otimes e, D^2 N_0(x) \rangle - \frac{\nu \cdot \langle e \otimes e, D^2 N_0(x) \rangle}{\nu \cdot \langle \nu, DN(x) \rangle} \langle \nu, DN(x) \rangle$$

and hence

$$\int_{\nu \in S^3} \int_{p_\nu(M^{0,1})} \sum_{e \in (\hat{M}^2)^2_y} |e \cdot \langle e \otimes e, D^2 N_{\hat{M}^2}(y)\rangle| \, d\mathcal{H}^2 y \, d\mathcal{H}^3 \nu$$

$$= \int_{x \in M^0} \int_{M^{0,1}_x} \sum_{e \in (\hat{M}^2)^2_{p(x)}} |e \cdot \langle e \otimes e, D^2 N_{\hat{M}^2}[p_\nu(x)]\rangle \nu|$$

$$\cdot \langle \nu, DN(x)\rangle | \, d\mathcal{H}^2 \nu \, d\mathcal{H}^3 x$$

$$= \int_{M^0} \int_{M^{0,1}_x} \sum_{\substack{e \in M^{0,2}_x \\ e \cdot \langle \nu, DN(x)\rangle = 0}} \left| \frac{p_\nu(e)}{|p_\nu(e)|} \cdot \langle e \otimes e, D^2 N(x)\rangle \nu \cdot \langle \nu, DN(x)\rangle \right.$$

$$\left. - \nu \cdot \langle e \otimes e, D^2 N(x)\rangle \frac{p_\nu(e)}{|p_\nu(e)|} \cdot \langle \nu, DN(x)\rangle \right| |p_\nu(e)|^{-2} \, d\mathcal{H}^2 \nu \, d\mathcal{H}^3 x$$

$$= \int_{M^0} \int_{M^{0,1}_x} \sum_e |e \cdot \langle e \otimes e, D^2 N(x)\rangle \nu \cdot \langle \nu, DN(x)\rangle| |p_\nu(e)|^{-3} \, d\mathcal{H}^2 \nu \, d\mathcal{H}^3 x \quad (\dagger)$$

$$= 2\pi \int_{x \in M} \int_{e \in M^{0,2}_x} |e \cdot \langle e \otimes e, D^2 N(x)\rangle| |\langle e, DN(x)\rangle|^{-1}$$

$$\cdot (e_3(e, x) \cdot \langle e_3(e, x), DN(x)\rangle)^2 \, d\mathcal{H}^1 e \, d\mathcal{H}^3 x \quad (\dagger\dagger)$$

(in the last integral, $e_3(x)$ is a unit vector in $\mathrm{Tan}(M, x)$ orthogonal to e and $\langle e, DN(x)\rangle$). Note that there is no reason either of the last two integrals should be bounded: either $|\langle e, DN(x)\rangle|$ or $|p_\nu(e)|$ may be zero, and $\nu \cdot \langle \nu, DN(x)\rangle$ is only of the order of $|p_\nu(e)|^2$. [Thus at this stage the equalities $(\dagger)$ and $(\dagger\dagger)$ should be considered formal.]

To get around this problem, we appeal to Yomdin's theorem (1.1 or 4.1 in [2]). According to his terminology, let $\Delta(\varepsilon) = \Delta(Q, 1, 1, \varepsilon) = Q\{(x, \nu) \in M^{0,1}_*: J_3 Q(x, \nu) < \varepsilon\}$. Observe that if

$$J_3 Q(x, \nu) \left(= |\langle \nu, DN(x)\rangle| / \left[1 + |\langle \nu, DN(x)\rangle|^2\right]^{1/2} \right) \geq \varepsilon$$

and $e \in M^2_x$ (i.e. $e \cdot \langle e, DN(x)\rangle = 0$), then

$$|p_\nu(e)| > \varepsilon/2 \|DN(x)\| \quad \text{or} \quad |\langle e, DN(x)\rangle| > \varepsilon(\sqrt{2} - 1)/2.$$

By applying both of the integrals $(\dagger)$ and $(\dagger\dagger)$, one finds that

$$\int_{\nu \in S^3 \setminus \Delta(\varepsilon)} I_2^{0,1}(\nu) \, d\mathcal{H}^3 \nu \leq (C_1/\varepsilon) \int_{M^0} \|DN(x)\|^2 \|D^2 N(x)\| d\mathcal{H}^3,$$

where C_1 is an *a priori* constant. [Note: This constant includes the upper bound $\mathcal{H}^1(M^1_x) \leq 4\pi$, derived from the fact that (assuming $J_3 N(x) \neq 0$) the intersection of M^1_x with any plane through 0 contains at most four points.] As observed before, we have no hope of bounding

$$\int_{\Delta(\varepsilon)} I_2^{0,1}(\nu) \, d\mathcal{H}^3 \nu,$$

but from Yomdin's theorem

$$\mathcal{H}^3[\Delta(\varepsilon)] \leqslant C_2\left\{ \sum_{i=0}^{2} \overline{A}(5,3,i)\left(\frac{r}{\delta}\right)^i \left(\frac{R}{\delta}\right)^{(5-i)/q} \delta^3 + \overline{A}(5,3,3) r^3 \left(\frac{R}{\delta}\right)^{2/q} \varepsilon \right\},$$

where δ is chosen at our discretion; $r = \frac{1}{2} \sup\{\|DN(x)\|: x \in M\}$; $R = \|N\|_q$; $\overline{A}(5,3,i)$ are *a priori* constants; and $C_2(M)$ is the number of balls of radius $1/r$ needed to cover M (C_2, in fact, depends only on r, $\mathcal{H}^3(M)$, and the injectivity radius of ∂M). Let $\delta = \varepsilon^{q/(q-1)}$. Then

$$\mathcal{H}^3\left\{ v: I_2^{0,1}(v) \geqslant C/\varepsilon^{1+(q-3)/(q-1)} \right\}$$
$$\leqslant \mathcal{H}^3[\Delta(\varepsilon)] + \mathcal{H}^3\left\{ v \in S^3 \setminus \Delta(\varepsilon): I_2^{0,1}(v) \geqslant C/\varepsilon^{1+(q-3)/(q-1)} \right\}$$
$$\leqslant C_2'\varepsilon^{(q-3)/(q-1)} + C_1'\varepsilon^{(q-3)/(q-1)} = C\varepsilon^{(q-3)/(q-1)}.$$

Hence

$$\mathcal{H}^3\left\{ v: I_2^{0,1}(v) \geqslant C/\varepsilon^{(2q-4)/(q-3)} \right\} \leqslant C\varepsilon$$

and

$$\int_{S^3} \left(I_2^{0,1}(v) \right)^\beta d\mathcal{H}^3 < \infty \quad \text{for } \beta < \frac{1}{2}\left(\frac{q-3}{q-2}\right).$$

Having dealt with $I_2^{0,1}(v)$ in great detail, we will give the remaining bounds more briefly; the methods for deriving them are highly similar to the above argument.

$$\int_{v \in S^3} I_2^{1,0}(v) \, d\mathcal{H}^3 v$$

(§)
$$\leqslant \int_{v \in S^3} \int_{\partial M} \sum_{\substack{e:\, (e \wedge v)\cdot \langle e, DN_{(1)}(x)\rangle = 0 \\ e \lrcorner N_{(1)}(x)^* = 0}}$$

$$\cdot \left\{ \|D^2 N_{(1)}(x)\| + \|DN_{(1)}(x)\|^2 \right] |v \wedge e|^{-2} \, d\mathcal{H}^2 x \, d\mathcal{H}^3 v$$

$$= \int_{\partial M} \left\{ \|D^2 N_{(1)}(x)\| + \|DN_{(1)}(x)\|^2 \right] \int_{v \in S^3} \sum_e |v \wedge e|^{-2} \, d\mathcal{H}^3 v \, d\mathcal{H}^2 x.$$

If we set

$$E(e) = e \cdot \langle e, DN_1(x)\rangle N_0(x) - e \cdot \langle e, DN_0(x)\rangle N_1(x)$$
$$= e \lrcorner \langle e, DN_{(1)}(x)\rangle^*,$$

the interior integral is bounded (formally) by

(§§)
$$8\pi \int_{e \in \operatorname{Tan}(\partial M,\, x) \cap S^3} \|DN_{(1)}(x)\|/|E(e)|^2 \, d\mathcal{H}^1 e.$$

Let $\Delta(\varepsilon) = \{v: |v \wedge e| < \varepsilon$ for some $x \in \partial M$, $e \in \operatorname{Tan}(\partial M, x)$ s.t. $|e \lrcorner \langle e, DN_{(1)}(x)\rangle^*| < \varepsilon r\}$. Then, as before,

$$\int_{S^3 \setminus \Delta(\varepsilon)} I_2^{1,0}(v) \, d\mathcal{H}^3 v \leqslant \varepsilon^{-1} r^{-1} \int_{\partial M} \left(\|D^2 N_{(1)}\| + \|DN_{(1)}\|^2 \right) \|DN_{(1)}\| d\mathcal{H}^3$$

and $\mathcal{H}^3[\Delta(\varepsilon)] \leqslant C_3 \varepsilon^{(q-2)/(q-1)}$ (consider $Q|\{(x, e) \in \partial M \times S^3 : e \lrcorner N_{(1)}(x)^* = 0\})$

$$\to \int_{S^3} \{I_2^{1,0}(\nu)\}^\beta \, d\mathcal{H}^3 \nu < \infty \quad \text{for } \beta < \frac{q-2}{2q-3}.$$

The bound for $I_4^{j,b}$ is a special case of an estimate that applies for all n. If $(x, e) \in M^{j,b}$ then

$$J_n(Q|M^{j,b})(x, \varepsilon) \leqslant |N^{j,b}(x, e)|/b! |N^{0,b}(x, e)|$$

(see p. 331 for notation), so that if

$$|N^{j,b}(x, \varepsilon)| \leqslant \varepsilon^b$$

then $J_n(Q|M^{j,b}) \leqslant \frac{1}{b!}\varepsilon$ or $|N^{0,b}(x, \varepsilon)| \leqslant \varepsilon^{b-1}$. By a simple inductive argument, we prove that if $|N^{j,b}(x, \varepsilon)| \leqslant \varepsilon^b$, then $J_n(Q|M^{j,b} \cup \cdots \cup M^{j,i}) \leqslant \frac{1}{i!}\varepsilon$ for some $i \leqslant b$.

Let

$$\Delta_i(\varepsilon) = Q\left(\left\{(x, e): J_n(Q|M^{j,b} \cup \cdots \cup M^{j,i}) \leqslant \frac{1}{i!}\varepsilon\right\}\right).$$

Then

$$\varepsilon^{-1+b(n-b-j)} \int_{\nu \in S^n \setminus \bigcup_{i \leqslant b} \Delta_i(\varepsilon)} \int_{M_\nu^{j,b}} J_{n-j-b}\left[\frac{Z^{j,b}}{|Z^{j,b}|}\right] d\mathcal{H}^{n-j-b} \, d\mathcal{H}^n$$

$$\leqslant \varepsilon^{-1+b(n-b-j)} \int_{M^j} \int_{M_x^{j,b} \setminus \bigcup \Delta_i(\varepsilon)} |N^{j,b}|^{-(n-b-j)} J_{n-j-b}(N^{j,b}|M_\nu^{j,b}).$$

$$\cdot \frac{|Z^{j,b}|}{|N^{j,b-1}|} d\mathcal{H}^{n-b}\nu \, d\mathcal{H}^{n-j}x$$

$$\leqslant \int_{M^j} \int_{M_x^{j,b}} J_{n-j-b}(N^{j,b}|M_\nu^{j,b})(x) \, d\mathcal{H}^{n-b}\nu \, d\mathcal{H}^{n-j}x$$

(since $|Z^{j,b}| \leqslant |N^{j,b}|$, $|N^{j,b}| \geqslant \varepsilon^b$ and $|N^{j,b-1}| \geqslant \varepsilon^{b-1}$)

$$\leqslant C(n, j, b) \int_{M^j} P(DN_0, \ldots, DN_{(j)}, D^2 N_0, \ldots, D^b N_0) \, d\mathcal{H}^{n-j}$$

for some continuous function P. By Yomdin's theorem,

$$\mathcal{H}^n\left[\bigcup_{i \leqslant b} \Delta_i(\varepsilon)\right] \leqslant C\varepsilon^{1-((n-j)/(q-j))}.$$

Hence

$$\int_{S^n} \left[\int_{M_\nu^{j,b}} J_{n-j-b}\left(\frac{Z^{j,b}}{|Z^{j,b}|}\right) d\mathcal{H}^{n-j-b}\right]^\beta d\mathcal{H}^n < \infty$$

for $\beta < (q-n)/(b(n-b-j)(q-j))$. If $n = 3$, this holds for $\beta < \frac{1}{2}(q-3)/q$.

The remaining two integrals,

$$\int_{S^3} I_3^{j,b}(\nu) \, d\mathcal{H}^3\nu \quad \text{and} \quad \int_{S^3} I_5^{j,b}(\nu) \, d\mathcal{H}^3\nu$$

are, as it turns out, bounded *a priori*, so that we need not resort to Yomdin's theorem for them. The reader is invited to work out the details; it is helpful to note, in particular, that for $y = p_\nu(x) \in \hat{M}^1$, $N_{\hat{M}^2}(y) = N_0(x)$. Also note that

$$\int_{S^3} I_5^{0,3}(\nu) \, d\mathcal{H}^3 \nu \leqslant C_4 \int_M \|D^3 N_0\| d\mathcal{H}^3,$$

where C_4 is an *a priori* constant; this is the only estimate in which third derivatives of N_0 appear "naturally" (i.e. without applying Yomdin's theorem).

Putting all the above estimates together, we have

$$\int_\nu [n(\nu)]^\beta \, d\mathcal{H}^3 \leqslant \int_\nu \left[\sum_{j=1}^5 I_j(\nu) \right]^\beta d\mathcal{H}^3 < \infty$$

for $\beta < (q - 3)/2q$. In fact, since at each step it would be possible to bound $\int_\nu [I_j(\nu)]^\beta \, d\mathcal{H}^3$ by an *a priori* constant

$$C\left[n, \beta, q, \|DN_0\|_1, \|DN_0\|_q, \|DN_{(1)}\|_1, \|DN_{(1)}\|_{q-1}, \mathcal{H}^3(M), \partial M \right],$$

the same holds for $\int_\nu [n(\nu)]^\beta \, d\mathcal{H}^3$.

FINAL REMARKS. Although certain aspects of the proof of a general multiplicity theorem for n-dimensional surfaces in $\mathbf{R}^{n+1}$ are visible, it is still too early to tell if such a theorem can be proved. The stratification in Lemma 2 seems to take care of the problem of classifying the singularities of the projection map. However, a difficulty which may be significant is that Yomdin's theorem requires supremum estimates as well as integral estimates on the derivatives of N_0, while the dimension-2 case involved only integral estimates. Since the step up to the next dimension presently requires integral estimates, it may be necessary to develop new methods to attack the problem when $n \geqslant 4$. Even if Nature has no objections to the truth of a higher-dimensional version, the number and complexity of the integral estimates is a substantial technical problem. The integrals that depend only on curvature terms can be handled in a uniform fashion, as in Lemma 2 and Theorem 9; however, so far I have no way of dealing with the integrals of third and higher derivatives except on a case-by-case basis. If, in fact, those integrals turn out to be "no worse" than the curvature integrals, one would have the following result (the best possible with my current techniques).

CONJECTURE. If $M \subset \mathbf{R}^{n+1}$ is an n-dimensional manifold-with-boundary of class $(n + 1, \alpha)$, then

$$\int_{S^n} [n(\nu)]^\beta \, d\mathcal{H}^n \nu < \infty \quad \text{for } 0 < \beta < C(n, \alpha);$$

if M is smooth then the result holds for $0 < \beta < ([n^2/4])^{-1}$.

REFERENCES

1. D. Nance, *A multiplicity estimate for projections of surfaces*, J. Differential Geometry **20** (1984).

2. Y. Yomdin, *The geometry of critical and near-critical values of differentiable mappings*, Math. Ann. **264** (1983), 495–515.

DUKE UNIVERSITY

Proceedings of Symposia in Pure Mathematics
Volume **44** (1986)

Deformation of Riemannian Metrics and Manifolds
with Bounded Curvature Ratios

SEIKI NISHIKAWA

The aim of this paper is to show that a Riemannian metric on a compact manifold can be deformed into a metric of constant curvature if it is sufficiently near a metric of constant positive curvature (Theorem 2). This proves that any compact Riemannian manifold of positive sectional curvature is diffeomorphic to a spherical space form if the ratios of the sectional curvature are close to one at each point (Theorem 3).

Our method of proof has stemmed from the variational problem on the total scalar curvature functional first studied by D. Hilbert, and is inspired by the recent work of R. Hamilton [1] in which he proved that a Riemannian metric of strictly positive Ricci curvature on a compact 3-manifold can be deformed into a metric of constant positive curvature.

1. The evolution equation. Let M be a compact n-manifold and $\mathcal{M}(M)$ be the set of Riemannian metrics $g = (g_{ij})$ on M. Let R denote the scalar curvature of g and consider the total scalar curvature of g,

$$I(g) = \int_M R \, d\mu(g)$$

as a functional defined on $\mathcal{M}(M)$. Let $g(t) = (g_{ij}(t))$ be a smooth variation of $g = g(0)$ and set $h = (h_{ij}) = (d/dt)g(t)|_{t=0}$ the variation vector at g. Then it is known that the first variation of the functional $I: \mathcal{M}(M) \to \mathbf{R}$ is given by

$$(1) \qquad \frac{d}{dt} I(g(t))\Big|_{t=0} = \int_M \left(-R_{ij} + \frac{1}{2}Rg_{ij}\right) h^{ij} \, d\mu,$$

where (R_{ij}) denotes the Ricci curvature of g and $h^{ij} = g^{ik}g^{jl}h_{kl}$ (cf. [5]).[1]

1980 *Mathematics Subject Classification.* Primary 53C20; Secondary 58D17, 58E11, 58G11.

[1] We use the index-notation for tensors. Indices are raised and lowered in the usual fashion and the Einstein summation convention is always used.

© 1986 American Mathematical Society
0082-0717/86 $1.00 + $.25 per page

Let $\mathscr{H}(M)$ denote the subset of $\mathscr{M}(M)$ consisting of those metrics g with a fixed total volume. Then for the variation of the functional $I|\mathscr{H}(M)$, the restriction of I to $\mathscr{H}(M)$, we have the volume constraint

$$(2) \qquad \frac{d}{dt}\operatorname{Vol}(M, g(t))\bigg|_{t=0} = \frac{1}{2}\int_M g_{ij}h^{ij}\,d\mu = 0.$$

From (1) and (2) we obtain a classical result due to D. Hilbert that $g \in \mathscr{H}(M)$ is an Einstein metric, namely g satisfies $R_{ij} = Rg_{ij}/n$, if and only if g is a critical point of the functional $I|\mathscr{H}(M)$.

Note that from (1) we may consider the tensor field $-R_{ij} + \frac{1}{2}Rg_{ij}$ as the gradient field of the functional I on $\mathscr{M}(M)$. Then, projecting it onto the subset $\mathscr{H}(M)$, with respect to the L^2 inner product on tensor fields on M, we obtain the gradient field of the functional $I|\mathscr{H}(M)$

$$\left[\operatorname{grad}(I|\mathscr{H}(M))\right]_{ij}(g) = -R_{ij} + rg_{ij}/n + (R - r)g_{ij}/2,$$

where r is the average of the scalar curvature R,

$$r = \int_M R\,d\mu \bigg/ \int_M d\mu.$$

Consider the evolution equation

$$\partial g_{ij}/\partial t = -R_{ij} + rg_{ij}/n + (R - r)g_{ij}/2,$$

which would define the gradient flow of the functional $I|\mathscr{H}(M)$. This evolution equation, however, will not in general have solutions even for a short time, since it implies a backward heat equation for R (cf. Remark 1).

On the other hand, let α and β be real numbers and consider the evolution equation

$$(3) \qquad \partial g_{ij}/\partial t = \alpha\left(-R_{ij} + rg_{ij}/n\right) + \beta(R - r)g_{ij},$$

which also would define a flow on $\mathscr{H}(M)$, since the right side of (3) satisfies the constraint (2) for any α and β. Then the following holds.

THEOREM 1. *If $\alpha > 0$ and $\beta < \alpha/2(n - 1)$, then the evolution equation (3) has a unique solution for a short time on any compact n-manifold M with any initial metric g at $t = 0$.*

To prove Theorem 1, we first note that as in [1, §3] there exists a change of variables which transforms (3) into the evolution equation

$$(4) \qquad \partial g_{ij}/\partial t = -\alpha R_{ij} + \beta Rg_{ij}.$$

Namely, beginning with (4), choose for instance the normalization factor $\psi = \psi(t)$ so that $\tilde{g} = \psi g$ has $\operatorname{Vol}(M, \tilde{g}) = 1$ for all t. Then, taking the new time scale $\tilde{t} = \int\psi(t)\,dt$, an elementary calculation yields that $\tilde{g}$ satisfies the evolution equation (3). Consequently, to prove the short-time existence of solutions we can deal with the unnormalized equation (4) which is easier to handle. Then by rescaling we can obtain the result for the normalized equation (3).

Now denote the right side of (4) as

$$E(g)_{ij} = -\alpha R_{ij} + \beta Rg_{ij},$$

which is a nonlinear partial differential operator of order 2 in g. For any symmetric tensor $T = (T_{ij})$ we define the linear partial differential operator $L(g)$, depending on the metric g and its connection, by

$$[L(g)T]_k = g^{pq}\left(\nabla_p T_{qk} - \frac{\alpha - 2\beta}{2(\alpha - n\beta)}\nabla_k T_{pq}\right),$$

which is of order 1 in g and T. Then it follows from the contracted second Bianchi identity $\nabla_k R = 2\nabla_p R_k{}^p$ that

(5) $$L(g)E(g) = 0.$$

We will see that the identity (5) is in fact the integrability condition for the evolution equation (4). To do this, let $DE(g)$ denote the linearization of E at g, and $\sigma(L(g), \zeta)$, $\sigma(DE(g), \zeta)$, denote respectively the symbols of the linear operators $L(g)$, $DE(g)$, in the direction ζ. Then from (5) we have

$$\sigma(L(g), \zeta) \circ \sigma(DE(g), \zeta) = 0,$$

which shows that the image of $\sigma(DE(g), \zeta)$ must lie in the null space Null $\sigma(L(g), \zeta)$ of $\sigma(L(g), \zeta)$, and therefore $\sigma(DE(g), \zeta)$ must have a null eigenspace.

On the other hand, by the same analysis done in [1, §4], we can see without difficulty the following result.

LEMMA 1. *The eigenvalues of the symbol $\sigma(DE(g), \zeta)$ restricted to the null space* Null $\sigma(L(g), \zeta)$ *are* $(\alpha - 2(n - 1)\beta)/2$ *with multiplicity* 1 *and* $\alpha/2$ *with multiplicity* $n(n - 1)/2$.

This shows that when $\alpha > 0$ and $\beta < \alpha/2(n - 1)$ all the eigenvalues of the eigenspaces of $\sigma(DE(g), \zeta)$ in Null $\sigma(L(g), \zeta)$ are positive. Theorem 1 is then an immediate consequence of the existence theorem of Hamilton [1, Theorem 5.1] for evolution equations with integrability condition.

REMARK 1. If $\alpha < 2(n - 1)\beta$, then the evolution equation (3) will not in general have solutions even for a short time, since it implies, for the scalar curvature R, the evolution equation

$$\partial R/\partial t = \tfrac{1}{2}(\alpha - 2(n - 1)\beta)\Delta R + \alpha R_{ij}R^{ij} - \alpha rR/n - \beta(R - r)R,$$

which is a backward heat equation when $\alpha < 2(n - 1)\beta$.

REMARK 2. Einstein metrics are critical points of the functional $I|\mathcal{H}(M)$, but in general they are not locally maxima nor minima (cf. [3, 5]).

2. Deformation of Riemannian metrics. Let M be a compact n-manifold ($n \geq 3$) and $g = (g_{ij})$ be a given Riemannian metric on M. Recall that g is a metric of constant curvature if and only if the Riemannian curvature tensor (R_{ijkl}) of g satisfies

$$R_{ijkl} = \frac{R}{n(n - 1)}(g_{il}g_{jk} - g_{ik}g_{jl}),$$

R being the scalar curvature of g. We denote by $Z = (Z_{ijkl})$ the so-called concircular curvature tensor of g, that is

$$Z_{ijkl} = R_{ijkl} - \frac{R}{n(n-1)}(g_{il}g_{jk} - g_{ik}g_{jl}),$$

which measures the deviation of g from being a metric of constant curvature. Let $G = (G_{ij})$ denote the traceless Ricci tensor of g, that is

$$G_{ij} = R_{ij} - Rg_{ij}/n,$$

which measures the deviation of g from being an Einstein metric.

We consider the evolution equation

$$(6) \qquad\qquad \partial g_{ij}/\partial t = -2R_{ij} + 2rg_{ij}/n,$$

where $r = \int R\, d\mu / \int d\mu$ is the average of the scalar curvature R, and try to improve the given metric by means of the deformation defined by the solution of this equation. Note that on account of Theorem 1 (with $\alpha = 2$, $\beta = 0$) the initial value problem for (6) has a unique solution for a short time with any initial metric at $t = 0$. Recall that the factor r serves to normalize the equation so that the deformation preserves the volume.

In the following, we are concerned with the long-time existence of solutions for the initial value problem for (6). Our goal is to prove the following result.

THEOREM 2. *Let g_0 be a Riemannian metric on a compact n-manifold M. Let R_0 and Z_0 denote the scalar curvature and the concircular curvature tensor of g_0 respectively. Suppose that $R_0 > 0$ and $|Z_0| < R_0/4n(n-1)$, where $|Z_0|$ denotes the norm of Z_0 with respect to g_0. Then the following hold.*

(1) *The initial value problem for the evolution equation* (6) *has a unique solution $g(t, x)$ with $g(0, x) = g_0(x)$, $x \in M$ for all time $t \in [0, \infty)$.*

(2) *As $t \to \infty$ the metrics $g_t(x) = g(t, x)$ converge to a metric of constant positive curvature in the C^∞ topology.*

It should be remarked that the assumption $R_0 > 0$ combined with $|Z_0| < R_0/4n(n-1)$ implies that the sectional curvature of the initial metric g_0 is strictly positive everywhere.

The proof of Theorem 2 requires several a priori estimates which control the evolution of curvatures under the deformation. To derive these estimates, we can deal with, as in the previous section, the unnormalized evolution equation

$$(7) \qquad\qquad \partial g_{ij}/\partial t = -2R_{ij},$$

which is easier to handle, and then rescale them to apply to the normalized equation (6). However, it will be seen that the solution of (7) will blow up (become singular) after some finite time T. A crucial point in proving the long-time existence for (6) on $[0, \infty)$ is to show that $\int_0^T \psi(t)\, dt = \infty$, where $\psi(t)$ is the normalization factor which transforms the equation (7) into (6).

Now, let g_0 be a Riemannian metric on M and assume that the evolution equation (7) has a solution $g = g(t, x)$ with initial metric $g(0, x) = g_0(x)$ on the interval $0 \leqslant t < T$. Then the following evolution equations for curvature tensors are known in [1, §7].

LEMMA 2. *We have the evolution equations*

$$\partial R_{ijkl}/\partial t = \Delta R_{ijkl} - 2\left(B_{ijkl} - B_{ijlk} + B_{ikjl} - B_{iljk} \right)$$
$$-\left(R_i{}^p R_{pjkl} + R_j{}^p R_{ipkl} + R_k{}^p R_{ijpl} + R_l{}^p R_{ijkp} \right).$$

$$\partial R_{ij}/\partial t = \Delta R_{ij} + 2 R_{pq} R^p{}_{ij}{}^q - 2 R_{ip} R^p{}_j,$$

$$\partial R/\partial t = \Delta R + 2\left| R_{ij} \right|^2;$$

where (B_{ijkl}) *denotes the tensor defined by* $B_{ijkl} = R_{pijq} R^p{}_{kl}{}^q$ *and* $|R_{ij}|^2 = R_{ij} R^{ij}$.

In particular, the scalar curvature R of g satisfies the parabolic inequality

$$\partial R/\partial t \geqslant \Delta R + 2R^2/n,$$

for $R^2 \leqslant n|R_{ij}|^2$. Consequently, it follows from the maximum principle that if $R > 0$ at $t = 0$, then it remains so for all time. Moreover, we have a bound on T, the maximal time for which the solution of (7) exists: If $R \geqslant \rho$ at $t = 0$ for some constant $\rho > 0$, then $R \geqslant \rho$ on $0 \leqslant t < T$ and hence $T \leqslant n/2\rho$.

Our a priori estimates are based on the following

LEMMA 3. *The norm* $|Z|$ *of the concircular curvature tensor* Z *of* g *satisfies the evolution equation*

$$\partial|Z|^2/\partial t = \Delta|Z|^2 - 2|\nabla Z|^2 + P,$$

where

$$P = \frac{16}{n(n-1)} R|G|^2 - 8\left(Z_{pijq} Z^p{}_{kl}{}^q Z^{ijkl} + Z_{pikq} Z^p{}_{jl}{}^q Z^{ijkl} \right).$$

PROOF. With the aid of evolution equations in Lemma 2 together with the identities $Z_{pij}{}^p = G_{ij}$, $G_p{}^p = 0$, we obtain by straightforward calculation

$$\frac{\partial}{\partial t}|Z|^2 = 2\left(\frac{\partial}{\partial t} Z_{ijkl} \right) Z^{ijkl} + 4\left(\frac{\partial}{\partial t} g^{ip} \right) Z_p{}^{jkl} Z_{ijkl}$$

$$= \Delta|Z|^2 - 2\left| \nabla_m Z_{ijkl} \right|^2 - 8\left(B_{ijkl} Z^{ijkl} + B_{ikjl} Z^{ijkl} \right)$$

$$= \Delta|Z|^2 - 2\left| \nabla_m Z_{ijkl} \right|^2 - 8\left(Z_{pijq} Z^p{}_{kl}{}^q Z^{ijkl} + Z_{pikq} Z^p{}_{jl}{}^q Z^{ijkl} \right)$$

$$+ \frac{16}{n(n-1)} R|G|^2.$$

From Lemma 3 we obtain the following estimates which enable us to control the evolution of $|Z|$ compared to R^γ.

LEMMA 4. *If* $R > 0$ *at* $t = 0$, *then for any* γ *with* $1 \leqslant \gamma \leqslant 2$ *we have*

$$\frac{\partial}{\partial t}\left(\frac{|Z|^2}{R^\gamma} \right) \leqslant \Delta\left(\frac{|Z|^2}{R^\gamma} \right) + \frac{2(\gamma - 1)}{R} \nabla_k R \cdot \nabla^k\left(\frac{|Z|^2}{R^\gamma} \right) + Q,$$

where

$$Q = \frac{16}{R^{\gamma-1}} \left(\frac{|Z|}{R} - \frac{1}{4n(n-1)} + \frac{2-\gamma}{8n} \right) |Z|^2.$$

PROOF. First note that $R > 0$ for all time. Then, for any γ, a computation analogous to that in [1, Lemma 10.3] yields

$$\frac{\partial}{\partial t}\left(\frac{|Z|^2}{R^\gamma} \right) = \Delta\left(\frac{|Z|^2}{R^\gamma} \right) + \frac{2(\gamma-1)}{R} \nabla_k R \cdot \nabla^k\left(\frac{|Z|^2}{R^\gamma} \right) + A,$$

where

$$A = -\frac{2}{R^{\gamma+2}} |R \cdot \nabla Z - \nabla R \cdot Z|^2 - \frac{(2-\gamma)(\gamma-1)}{R^{\gamma+2}} |\nabla R|^2 |Z|^2$$
$$- \frac{8}{R^\gamma}\left(Z_{pijq} Z^p{}_{kl}{}^q Z^{ijkl} + Z_{pikq} Z^p{}_{jl}{}^q Z^{ijkl} - \frac{2}{n(n-1)} R|G|^2 \right)$$
$$- \frac{2\gamma}{R^{\gamma+1}} |R_{ij}|^2 |Z|^2.$$

We will estimate A from above.[2] When $1 \leqslant \gamma \leqslant 2$, it is easy to see

$$A \leqslant \frac{2}{R^{\gamma+1}}\left(8R|Z|^3 + \frac{8}{n(n-1)} R^2|G|^2 - \frac{\gamma}{n} R^2|Z|^2 \right),$$

for $(2-\gamma)(\gamma-1) \geqslant 0$ and $R^2 \leqslant n|R_{ij}|^2$. On the other hand, we have

$$|G|^2 \leqslant \frac{n-2}{4} |Z|^2,$$

which can be seen for instance from the identity

$$|Z|^2 = |W|^2 + \frac{4}{n-2} |G|^2,$$

where W denotes the Weyl conformal curvature tensor. This completes the proof.

LEMMA 5. *If $R > 0$ and $|Z| < R/4n(n-1)$ at $t = 0$, then both conditions continue to hold on $0 \leqslant t < T$. Moreover, there exists a γ with $1 < \gamma < 2$ such that on $0 \leqslant t < T$ we have*

$$|Z|^2 < CR^\gamma,$$

where $C = 1/16n^2(n-1)^2$.

PROOF. Since $Q < 0$ whenever $|Z| < R/4n(n-1)$ in Lemma 4 (with $\gamma = 2$), the first assertion follows immediately from the maximum principle for the parabolic inequality.

To the second assertion, first note that, since the manifold is compact, there is a constant c_1 with $0 < c_1 < 1/4n(n-1)$ such that $|Z| < c_1 R$ holds at $t = 0$. Then choose a constant γ with $1 < \gamma < 2$ which satisfies

$$|Z|^2 < CR^\gamma, \qquad C = 1/16n^2(n-1)^2$$

---

[2] Our estimate is quite rough. A better estimate can be seen in [2, 4].

at $t = 0$ and

$$c_1 < \frac{1}{4n(n-1)} - \frac{2-\gamma}{8n}.$$

Then the second assertion follows also from the maximum principle applied to Lemma 4 (with $1 < \gamma < 2$).

In order to compare the scalar curvature R at distant points, we need to estimate the evolution of the gradient of R. We begin with the following

LEMMA 6. *If $R > 0$ and $|Z| < R/4n(n-1)$ at $t = 0$, then for any η with $0 \leqslant \eta \leqslant 1/n$ we have*

$$\frac{\partial}{\partial t}\left(\frac{|\nabla_k R|^2}{R} - \eta R^2\right) \leqslant \Delta\left(\frac{|\nabla_k R|^2}{R} - \eta R^2\right) + \frac{24}{n-1}|\nabla_k R_{ij}|^2 - \frac{4}{n}\eta R^3.$$

PROOF. By a direct computation as in [1, Lemmas 11.3, 11.4], it is verified for any η that

$$\frac{\partial}{\partial t}\left(\frac{|\nabla_k R|^2}{R} - \eta R^2\right) = \Delta\left(\frac{|\nabla_k R|^2}{R} - \eta R^2\right) + \frac{4}{R}\nabla_k R \cdot \nabla^k |R_{ij}|^2 + A,$$

where

$$A = -\frac{2}{R^3}|R \cdot \nabla_k \nabla_l R - \nabla_k R \cdot \nabla_l R|^2 - \frac{2}{R^2}|R_{ij}|^2 |\nabla_k R|^2$$

$$+ 2\eta|\nabla_k R|^2 - 4\eta R|R_{ij}|^2.$$

Since $0 \leqslant \eta \leqslant 1/n$ and $R > 0$ for all time, it is easy to see

$$A \leqslant 4\eta R^3/n,$$

for $R^2 \leqslant n|R_{ij}|^2$. On the other hand, since the assumption $|Z| < R/4n(n-1)$ continues to hold for all time by Lemma 5, it follows that

$$|R_{ijkl}|^2 < \frac{3}{n(n-1)}R^2,$$

for $|Z|^2 = |R_{ijkl}|^2 - 2R^2/n(n-1)$, and therefore

$$|R_{ij}|^2 < \frac{6}{n(n-1)^2}R^2,$$

for $|Z|^2 = |R_{ijkl}|^2 - 2R^2/n(n-1)$, and therefore

$$|R_{ij}|^2 < \frac{6}{n(n-1)^2}R^2,$$

for $2|R_{ij}|^2 \leqslant (n-1)|R_{ijkl}|^2$. Consequently, since $|\nabla_k R|^2 \leqslant n|\nabla_k R_{ij}|^2$, we get by the Cauchy–Schwarz inequality

$$\frac{4}{R}\nabla_k R \cdot \nabla^k |R_{ij}|^2 \leqslant \frac{8}{R}|\nabla_k R||\nabla_k R_{ij}||R_{ij}| < \frac{24}{n-1}|\nabla_k R_{ij}|^2,$$

and this proves the lemma.

350 SEIKI NISHIKAWA

We note also the following result.

LEMMA 7. *If $R > 0$ at $t = 0$, then we have*

$$\frac{\partial}{\partial t}|Z|^2 \leqslant \Delta|Z|^2 - c_2|\nabla_k R_{ij}|^2 + \frac{4(n-2)}{n(n-1)}R|Z|^2 + 16|Z|^3,$$

where $c_2 = 4(n-2)^2/n(n-1)(3n-2)$.

PROOF. From Lemma 3 we have

$$\frac{\partial}{\partial t}|Z|^2 \leqslant \Delta|Z|^2 - 2|\nabla Z|^2 + \frac{4(n-2)}{n(n-1)}R|Z|^2 + 16|Z|^3,$$

for $4|G|^2 \leqslant (n-2)|Z|^2$ and $R > 0$ for all time. On the other hand, by the same argument as in [**1**, Lemma 11.6], we can obtain the estimate

$$|\nabla_k R|^2 \leqslant \frac{2(n-1)(n+2)}{3n-2}|\nabla_k R_{ij}|^2.$$

Therefore we have

$$|\nabla Z|^2 = |\nabla_m R_{ijkl}|^2 - \frac{2}{n(n-1)}|\nabla_k R|^2$$

$$\geqslant \frac{2(n-2)^2}{n(n-1)(3n-2)}|\nabla_k R_{ij}|^2,$$

for $2|\nabla_k R_{ij}|^2 \leqslant (n-1)|\nabla_m R_{ijkl}|^2$, and the result follows.

As an immediate consequence of these lemmas, the following gradient estimate for R is obtained.

LEMMA 8. *If $R > 0$ and $|Z| < R/4n(n-1)$ at $t = 0$, then for any η with $0 < \eta \leqslant 1/n$ there exists a constant $C(\eta)$ depending only on η, n, and g_0 such that*

$$|\nabla_k R|^2 \leqslant \eta R^3 + C(\eta).$$

PROOF. Let $0 < \eta \leqslant 1/n$ and set

$$F = \frac{|\nabla_k R|^2}{R} - \eta R^2 + c_3|Z|^2,$$

where $c_3 = 24n(3n-2)/(n-1)^2(n-2)^2$. Then it follows from Lemma 6 and Lemma 7 that

$$\partial F/\partial t \leqslant \Delta F + c_4 R|Z|^2 + c_5|Z|^3 - 4\eta R^3/n,$$

c_4 and c_5 being constants depending only on n. Recall that by Lemma 5 we can find a constant c_6 and some $\delta > 0$ depending only on n and the initial metric g_0, such that

$$|Z|^2 \leqslant c_6 R^{2-\delta}.$$

Then with a constant $C(\eta)$ depending only on $\eta > 0$, n and δ, we get

$$c_4 R|Z|^2 + c_5|Z|^3 - 4\eta R^3/n \leqslant C(\eta),$$

and therefore

$$\partial F/\partial t \leqslant \Delta F + C(\eta).$$

Since the maximal existence time T for (7) is bounded, this implies by the maximum principle that

$$F \leqslant C(\eta) \quad \text{on } 0 \leqslant t < T,$$

for some (possibly larger) constant $C(\eta)$ depending only on η, n and g_0. This proves the result, since $0 < \eta \leqslant 1/n$ is arbitrary.

As remarked before, the curvature assumptions of Theorem 2 together with Lemma 5 imply that all the sectional curvatures of the metric $g(t, x)$ are bounded below by some constant $\varepsilon > 0$ independent of time t. Therefore, as in [1, Theorem 15.1] we can apply Meyer's theorem on the diameter of manifolds of strictly positive Ricci curvature to Lemma 8 and prove that

$$R_{\max}/R_{\min} \to 1 \quad \text{as } t \to T,$$

where $R_{\max}$ and $R_{\min}$ denote respectively the maximum value and the minimum value of the scalar curvature R.

Once this is established, we can proceed in the same way as in the corresponding part of Hamilton [1] to prove Theorem 2, and the result is obtained. For example, since the ratio $R_{\max}/R_{\min}$ is unchanged under the rescaling, we can see that for the solution of the evolution equation (6) the scalar curvature is bounded above, while it is unbounded for (7), and in consequence the solution exists for all time $t \in [0, \infty)$.

The convergence of the initial metric g_0 to the desired metric of constant positive curvature is exponential. In fact, as in [1, Lemma 17.2], it is easy to see

LEMMA 9. *If $R > 0$ and $|Z| < R/4n(n - 1)$ at $t = 0$, then for the initial value problem for (6) we can find constants $C < \infty$ and $\tau > 0$ such that*

$$|Z|^2 \leqslant Ce^{-\tau t}.$$

3. Manifolds with bounded curvature ratios. Let δ be a positive number with $0 < \delta \leqslant 1$. A Riemannian manifold (M, g) is said to be *locally δ-pinched* if there exists a positive function $A: M \to R$ such that at every point $x \in M$ its sectional curvature K satisfies $\delta A(x) \leqslant K \leqslant A(x)$. Then, since a manifold of constant positive curvature is isometric to a spherical space form S^n/Γ (cf. [8]), as an immediate consequence of Theorem 2 we have the following differentiable pinching theorem, which has been proved by E. Ruh [6] by a different method.

THEOREM 3. *There exists $\delta = \delta(n)$ with $\frac{1}{4} < \delta < 1$ such that any compact locally δ-pinched Riemannian n-manifold M is diffeomorphic to a spherical space form S^n/Γ.*

Note that, since the evolution equation (6) is invariant under the action of the diffeomorphism group of M, any isometries which exist in the initial metric in Theorem 2 are preserved as the metric evolves. Therefore, in Theorem 3, if G is the isometry group of M, then G is isomorphic to a closed subgroup of the isometry group of the spherical space form $\overline{M} = S^n/\Gamma$, and the diffeomorphism is equivariant with respect to the actions of G on M and $\overline{M}$ respectively.

REMARK 3. Our pinching condition of Theorem 2 in fact requires that $\delta(n) > 0.99$ and $\delta(n) \to 1$ as $n \to \infty$. However, it should be remarked that well-known differentiable pinching theorems (cf. [7]) do not apply under the local pinching assumptions of Theorem 3.

ACKNOWLEDGMENT. I would like to express my gratitude to Mr. K. Akutagawa for many helpful discussions and to Professor G. Huisken and Dr. C. Margerin for explaining their ideas.

REFERENCES

1. R. S. Hamilton, *Three-manifolds with positive Ricci curvature*, J. Differential Geometry **17** (1982), 255–306.

2. G. Huisken, *Ricci deformation of the metric on a Riemannian manifold*, preprint.

3. N. Koiso, *On the second derivative of the total scalar curvature*, Osaka J. Math. **16** (1979), 413–421.

4. C. Margerin, *Some results about the positive curvature operators and point-wise $\delta(n)$-pinched manifolds*, informal notes.

5. Y. Mutō, *On Einstein metrics*, J. Differential Geometry **9** (1974), 521–530.

6. E. A. Ruh, *Riemannian manifolds with bounded curvature ratios*, J. Differential Geometry **17** (1982), 643–653.

7. T. Sakai, *Comparison and finiteness theorems in Riemannian geometry*, Advanced Studies in Pure Mathematics **3** (1984), 125–181.

8. J. A. Wolf, *Spaces of constant curvature*, McGraw-Hill, New York, 1967.

KYUSHU UNIVERSITY, JAPAN

Proceedings of Symposia in Pure Mathematics
Volume **44** (1986)

A Regularity Condition at the Boundary
for Weak Solutions
of Some Nonlinear Elliptic Systems

GEORGE PAULIK

1. Introduction. This note considers boundary regularity of weak solutions to certain diagonal elliptic systems. More precisely, given a domain $\Omega \subset \mathbf{R}^n$, $n \geqslant 2$, let $u \in W^{1,2} \cap L^\infty(\Omega, \mathbf{R}^m)$ be a vector-valued function such that

$$-A(u^i) = f^i(x, u, \nabla u),$$

i.e.

$$(1) \qquad \int_\Omega a^{\alpha\beta}(x, u, \nabla u) D_\alpha u^i D_\beta \phi^i \, dx = \int_\Omega f^i(x, u, \nabla u) \phi^i \, dx$$

$$\text{for all } \phi \in W_0^{1,2} \cap L^\infty(\Omega, \mathbf{R}^m).$$

The convention of summing over repeated indices is used. The following structure is assumed: $f^i(x, u, \nabla u)$ and $a^{\alpha\beta}(x, u, \nabla u) \, (= a^{\alpha\beta})$ are measurable and

(a) $\|u\|_{L^\infty} \leqslant M$;

(b) $\lambda|\xi|^2 \leqslant a^{\alpha\beta}\xi^\alpha\xi^\beta \leqslant \Lambda|\xi|^2$, for some constants $0 < \lambda \leqslant \Lambda < \infty$ and all $\xi \in \mathbf{R}^n$;

(c) $|f(x, u, p)| \leqslant a(a^{\alpha\beta}p_\alpha^i p_\beta^i) + b$, for some constants a, b and all $p = (p_\alpha^i) \in \mathbf{R}^{n \times m}$;

(d) $u(x) \cdot f(x, u, p) \leqslant a^*(a^{\alpha\beta}p_\alpha^i p_\beta^i) + b^*$, for some constants a^*, b^* and all $p = (p_\alpha^i) \in \mathbf{R}^{n \times m}$.

Systems of this type occur in differential geometry and physics. A good example is the system corresponding to a harmonic mapping between two Riemannian manifolds; for this example and others, see [**H**].

A weak solution u is considered regular if it is locally Hölder continuous; generally this is the best that can be expected assuming only measurability of $a^{\alpha\beta}$ and f. A point $x_0 \in \partial\Omega$ is regular for the system (1) if solutions of the Dirichlet

---

1980 *Mathematics Subject Classification.* Primary 35J65

© 1986 American Mathematical Society
0082-0717/86 $1.00 + $.25 per page

problem with continuous boundary data are continuous at x_0. In [**W1, W2, HW,** and **GH**] the regularity of u is established under certain smallness conditions. L. A. Caffarelli recently gave a new proof of this regularity ([**C**], see also [**T**]). The techniques of [**C**] and [**GT**] (Chapter 8) enable one to prove the sufficiency of the Wiener criterion for the regularity of $x_0 \in \partial\Omega$; this is the main result of this paper. The smallness condition $a^* + aM < 2$ is assumed; this is known to be optimal for interior regularity, see [**H**].

2. Preliminaries and statement of theorem. Let $x_0 \in \partial\Omega$ and let $B_R(x_0)$ denote the open ball of radius R about x_0. Write $\Omega_R = B_R(x_0) \cap \Omega$ and $\partial\Omega_R = B_R(x_0) \cap \partial\Omega$. For $R < 1$ the capacity of $B_R(x_0) - \Omega$ is

$$\operatorname{cap}\{B_R(x_0) - \Omega\} = \inf\left\{\int |\nabla v|^2 : v \in C_0^1(B_1(x_0)), v = 1 \text{ on } B_R(x_0) - \Omega\right\}.$$

Write $\chi(R) = R^{2-n}\operatorname{cap}\{B_R(x_0) - \Omega\}$ and consider the series $\sum_{j=0}^{\infty}\chi(R^j)$. The Wiener criterion states that x_0 is regular for the Laplacian if and only if the series diverges, see [**W**]. Similar results have been obtained for much broader classes of second order elliptic equations in divergence form, see [**LSW, S, GZ**].

For $w \in W^{1,2}(\Omega, \mathbf{R})$ define

$$M_{\partial R} = \sup_{\partial\Omega_R} w = \inf\{k \in \mathbf{R}: (w - k)^+ \text{ is the limit in } W^{1,2}(\Omega, \mathbf{R})$$

$$\text{of a sequence of functions in } C_0^1(\overline{\Omega} - \partial\Omega_R)\}.$$

The quantity $m_{\partial R} = \inf_{\partial\Omega_R} w$ is defined similarly and $\operatorname{osc}_{\partial\Omega_R} w = M_{\partial R} - m_{\partial R}$. Write $M_R = \sup_{\Omega_R} w$, $m_R = \inf_{\Omega_R} w$, and $\operatorname{osc}_{\Omega_R} w = M_R - m_R$.

THEOREM. *Let $u \in W^{1,2} \cap L^{\infty}(\Omega, \mathbf{R}^m)$ be a solution of* (1) *satisfying the given structural conditions and assume $a^* + aM < 2$. Suppose the Wiener condition is satisfied at $x_0 \in \partial\Omega$, i.e., $\sum_{j=0}^{\infty}\chi(R^j)$ diverges, and that*

$$\lim_{R \to 0} \operatorname{osc}_{\partial\Omega_R} u^i = 0, \qquad 1 \leqslant i \leqslant m.$$

Then $\lim_{R \to 0}\operatorname{osc}_{\Omega_R} u^i = 0, 1 \leqslant i \leqslant m$.

REMARK. If $w, f \in W^{1,2}(\Omega, \mathbf{R})$, $f \in C^0(\overline{\Omega}, \mathbf{R})$, and $w - f \in W_0^{1,2}(\Omega, \mathbf{R})$, then

$$\inf_{\Omega_R} f \leqslant m_{\partial R} \leqslant M_{\partial R} \leqslant \sup_{\Omega_R} f,$$

and so $\lim_{R \to 0}\operatorname{osc}_{\partial\Omega_R} w = 0$ with $\lim_{R \to 0} m_{\partial R} = \lim_{R \to 0} M_{\partial R} = f(x_0)$. Furthermore, it is also true that

$$m_R \leqslant m_{\partial R} \leqslant M_{\partial R} \leqslant M_R.$$

Thus, if $\lim_{R \to 0}\operatorname{osc}_{\Omega_R} w = 0$ then w is continuous at x_0 and $\lim_{x \to x_0, x \in \Omega} w(x) = f(x_0)$. In particular, suppose u is a solution of (1) with $a^* + aM < 2$, and $u - g \in W_0^{1,2}(\Omega, \mathbf{R}^m)$, $g \in W^{1,2}(\Omega, \mathbf{R}^m) \cap C^0(\overline{\Omega}, \mathbf{R}^m)$. If the Wiener condition is satisfied at $x_0 \in \partial\Omega$, then u is continuous at x_0 with $\lim_{x \to x_0, x \in \Omega} u(x) = g(x_0)$. In this sense the theorem is an extension of classical results.

3. Proof of theorem. The function au is a solution of a system similar to (1) but with $a = 1$. Therefore, assume $a = 1$, $a^* + M < 2$. The theorem is a consequence of the following lemma.

LEMMA. *Let $\theta \in \mathbf{R}^m$ be the vector such that*

$$\lim_{R \to 0} \sup_{\partial \Omega_R} |u - \theta| = \lim_{R \to 0} \inf_{\partial \Omega_R} |u - \theta| = 0.$$

Fix $\eta \in (0,1)$. Then there are constants $C = C(n, (2 - a^* - M)^{-1}, \Lambda/\lambda) > 0$, $C' = C'(\max\{b, b^*\}, (2 - a^* - M)^{-1}) > 0$ such that for $R < 4^{-1}$ one of the following must hold:

(i)
$$M_{4R} \leqslant m(1 - \eta)^{-1} \max_i \operatorname*{osc}_{\partial \Omega_{4R}} u^i + C'(1 - \eta)^{-1} \operatorname*{osc}_{\Omega_R} v;$$

(ii)
$$|u(x) - C^{-1}\chi(R)\theta| \leqslant M_{4R}(1 - C^{-1}\chi(R)\eta), \qquad x \in \Omega_R.$$

Here M_{4R} refers to the function $|u|$ and v is an auxiliary function which is continuous at x_0.

To prove the theorem assume the lemma is true and fix $R < 4^{-1}$. Fix $\eta \in (0,1)$ at a value to be determined later. Consider the following iteration. Apply the lemma to $u_0 = u$. If (i) holds, apply the lemma to $u_1 = u$ at the radius $4^{-1}(4R) = R$. If (ii) holds, let $u_1 = u - C^{-1}\chi(R)\theta$. Note that u_1 is a solution of a system similar to (1), and on Ω_R,

$$a_1^* + M_1^* \leqslant \left(a_0^* + C^{-1}\chi(R)M_0\right) + M_0\left(1 - C^{-1}\chi(R)\eta\right)$$
$$= a_0^* + M_0 + M_0(1 - \eta)C^{-1}\chi(R).$$

If η is close to 1, $a_1^* + M_1^* < 2$, and the lemma can be applied to u_1 at the radius $4^{-1}(4R) = R$. The following induction argument shows that the lemma can be applied to u_k at the radius $4^{1-k}R$, $k = 0, 1, 2, \ldots$, if η is chosen appropriately.

After k applications of the lemma, suppose constant vectors have been subtracted at steps $k_1, \ldots, k_j$. Write $d_{k_i} = C^{-1}\chi(4^{1-k_i}R)\eta$. A straightforward calculation shows

$$a_k^* + M_k \leqslant a_0^* + M_0 + M_0(1 - \eta)\eta^{-1}$$
$$\cdot \left[d_{k_1} + (1 - d_{k_1})d_{k_2} + \cdots + (1 - d_{k_1}) \cdots (1 - d_{k_{j-1}})d_{k_j}\right]$$
$$= a_0^* + M_0 + M_0(1 - \eta)\eta^{-1}\left[1 - \prod_{i=1}^{j}(1 - d_{k_i})\right]$$
$$< a_0^* + M_0 + M_0(1 - \eta)\eta^{-1}, \qquad x \in \Omega_{4^{1-k}R}.$$

Choose η such that $(1 - \eta)\eta^{-1}M_0 < 2 - a_0^* - M_0$ and the argument is finished. Note that $a_k^* + M_k$ is bounded independently of k by a value less than 2, so the constants C and C' can be chosen to be independent of k.

Using the notation of the above argument, one of the following must hold for each k:

$$\text{(I)} \qquad M_k \leqslant m(1-\eta)^{-1} \max_i \operatorname*{osc}_{\partial\Omega_{4^{1-k}R}} u^i + C' \operatorname*{osc}_{\Omega_{4^{-k}R}} v;$$

$$\text{(II)} \qquad M_{k+1} \leqslant M_0 \prod_{i=1}^{j} (1 - d_{k_i}).$$

This implies $\lim_{k\to\infty} M_k = 0$. For, either infinitely many steps of the iteration provide an estimate (I), or the infinite product $\prod_{i=1}^{\infty}(1-d_{k_i})$ is zero. The M_k provide estimates for $\operatorname{osc}_{\Omega_{4^{1-k}R}} u^i$, so the theorem is proved assuming the proof of the lemma.

PROOF OF LEMMA. Suppose (i) does not hold, so

$$M_{4R} > m(1-\eta)^{-1} \max_i \operatorname*{osc}_{\partial\Omega_{4R}} u^i + C'(1-\eta)^{-1} \operatorname*{osc}_{\Omega_R} v.$$

Write $l = (a^* + M_{4R})/2$ and let $\xi \in \mathbf{R}^m$ be any vector with $|\xi| \leqslant 1 - l$. Let $v \in W_0^{1,2}(\Omega, \mathbf{R})$ be the solution to the Dirichlet problem

$$A(v) = 1 \qquad \text{on } \Omega,$$
$$v = 0 \qquad \text{on } \partial\Omega.$$

The solution v is bounded and is continuous at x_0 with $\lim_{x\to x_0, x\in\Omega} v(x) = 0$, see [**GT**, Chapter 8]. Without loss of generality let $b = \max\{b, b^*\}$. Proceeding exactly as in [**C**], it follows from the structural assumptions (a), (c), (d) and the smallness condition $a^* + M < 2$ that

$$h = 2^{-1}\left(M_{4R}^2 - |u|^2 \right) + (1-l)M_{4R} - \xi \cdot u - 2bv$$

is a nonnegative supersolution in Ω_{4R}, i.e., $A(h) \leqslant 0$. Let $s = \inf_{\partial\Omega_{4R}} h$ and define

$$h_s^-(x) = \begin{cases} \inf\{h(x), s\}, & x \in \Omega_{4R}, \\ s, & x \in B_{4R}(x_0) - \Omega_{4R}. \end{cases}$$

Then from inequality (8.80), p. 208 in [**GT**],

$$s\chi(R) \leqslant C \inf_{B_R} h_s^- \leqslant C \inf_{\Omega_R} h.$$

Here $C = C(n, \Lambda/\lambda)$. Using the fact that $\sup_{\Omega_R} v = 0$, the inequality becomes

$$C^{-1}\chi(R)\left[(1-l)M_{4R} - \xi \cdot \theta\right] + C^{-1}\chi(R)\left[\xi \cdot \theta - \sup_{\partial\Omega_{4R}} \xi \cdot u\right]$$

$$\leqslant 2^{-1}\left(M_{4R}^2 - |u(x)|^2 \right) + (1-l)M_R - \xi \cdot u(x) + 2b \operatorname*{osc}_{\Omega_R} v, \qquad x \in \Omega_R.$$

Choosing $\xi = (1-l)u(x)/|u(x)|$ and again arguing as in [**C**], the inequality becomes

$$\left| u - C^{-1}\chi(R)\theta \right|^2 \leqslant M_{4R}^2\left(1 - C^{-1}\chi(R)\right)^2$$

$$+ (1-l)^{-1}M_{4R}C^{-1}\chi(R) \sup_{\partial\Omega_{4R}} \xi \cdot (u - \theta) + C' \operatorname*{osc}_{\Omega_R} v$$

$$\leqslant M_{4R}^2\left(1 - C^{-1}\chi(R)\right)^2 + M_{4R}C^{-1}\chi(R) m \max_i \operatorname*{osc}_{\partial\Omega_{4R}} u^i + C' \operatorname*{osc}_{\Omega_R} v,$$

for all $x \in \Omega_R$. Here $C' = C'(\max\{b, b^*\}, (2 - a^* - M)^{-1})$, and $C = C(n, (2 - a^* - M)^{-1}, \Lambda/\lambda)$ is large such that $C^{-1}\chi(R) < 2^{-1}$. Now write the right side of the inequality as

$$M_{4R}^2\left[1 - C^{-1}\chi(R)(1 - \delta)\right]^2.$$

Solve for δ and note that $1 - \delta > \eta$ if $C^{-1}\chi(R) < 2^{-1}$, $0 < \eta < 1$, and $M_{4R} > m(1 - \eta)^{-1}\max_i \mathrm{osc}_{\partial\Omega_{4R}} u^i + (1 - \eta)^{-1}C'\mathrm{osc}_{\Omega_R} v$. Thus,

$$\left|u - C^{-1}\chi(R)\theta\right|^2 \leqslant M_{4R}^2\left(1 - C^{-1}\chi(R)\eta\right)^2, \qquad x \in \Omega_R,$$

and the proof is finished.

REMARK. It is not known if the Wiener condition is necessary for boundary regularity. That is, if $\sum_{j=0}^{\infty}\chi(R^j)$ converges for $x_0 \in \partial\Omega$, does there exist a solution u of (1), satisfying the structural conditions and $a^* + aM < 2$, such that u is not continuous at x_0? The main difficulty is the inability to prove that the Dirichlet problem has a solution for given boundary data. However, for systems corresponding to harmonic mappings the Dirichlet problem is solvable ([HKW]) and it is possible to prove the necessity of the Wiener criterion, see [P]. In this setting geometric restrictions on the size of the image of the harmonic map replace the hypothesis $a^* + aM < 2$.

REFERENCES

[C] L. A. Caffarelli, *Regularity theorems for weak solutions of some nonlinear systems*, Comm. Pure Appl. Math. **35** (1983), 833–838.

[GH] M. Giaquinta and S. Hildebrandt, *Estimation à priori des solutions faibles de certains systèmes non linéaires elliptiques*, Séminaire Goulaouic-Meyer-Schwartz, 1980–1981.

[GT] D. Gilbarg and N. S. Trudinger, *Elliptic partial differential equations of second order*, 2nd ed., Springer-Verlag, 1983.

[GZ] R. Gariepy and W. P. Ziemer, *A regularity condition at the boundary for solutions of quasilinear elliptic equations*, Arch. Rat. Mech. Anal. **67** (1977), 25–39.

[H] S. Hildebrandt, *Quasilinear elliptic systems in diagonal form*, Vorlesungsreihe SFB 72, no. 11, Bonn, 1982.

[HKW] S. Hildebrandt, H. Kaul and K. O. Widman, *An existence theorem for harmonic mappings of Riemannian manifolds*, Acta Math. **138** (1977), 1–16.

[HW] S. Hildebrandt and K. O. Widman, *On the Hölder continuity of weak solutions of quasilinear elliptic systems of second order*, Ann. Scuola Norm. Sup. Cl. Pisa (4) **4** (1977), 145–178.

[LSW] W. Littman, G. Stampacchia and H. Weinberger, *Regular points for elliptic equations with discontinuous coefficients*, Ann. Scuola Norm. Sup. Cl. Pisa (3) **17** (1963), 43–77.

[P] G. Paulik, Ph.D. thesis, Indiana University (1985).

[S] G. Stampacchia, *Le problème de Dirichlet pour les équations elliptiques du second ordre à coefficients discontinus*, Ann. Inst. Fourier (Grenoble) **15** (1965), 189–258.

[T] P. Tolksdorf, *A new proof of a regularity theorem*, Invent. Math. **71** (1983), 43–49.

[W1] M. Wiegner, *Ein optimaler Regularitätssatz für schwache Lösungen gewisser elliptischer Systeme*, Math. Z. **147** (1976), 21–28.

[W2] ______, *A-priori Schranken für Lösungen gewisser elliptischer Systeme*, Manuscripta Math. **18** (1976), 279–297.

[W] N. Wiener, *The Dirichlet problem*, J. Math. Phys. **3** (1924), 127–146.

INDIANA UNIVERSITY

Proceedings of Symposia in Pure Mathematics
Volume **44** (1986)

Solutions to the Navier–Stokes Inequality
With Singularities on a Cantor Set

VLADIMIR SCHEFFER[1]

1. Introduction. This paper gives an outline of the proof of Theorem 1.1 below.

DEFINITION. If $u\colon R^3 \times (a, b) \to R^n$ is a function then we set

$$\Delta u = \sum_{i=1}^{3} \frac{\partial^2 u}{\partial x_i^2}, \quad \nabla u = \left(\frac{\partial u}{\partial x_1}, \frac{\partial u}{\partial x_2}, \frac{\partial u}{\partial x_3} \right), \quad |\nabla u|^2 = \sum_{i=1}^{n} \sum_{j=1}^{3} \left(\frac{\partial u_i}{\partial x_j} \right)^2.$$

THEOREM 1.1. *There exist* $S \subset R^3$ *and functions* $u\colon R^3 \times [0, \infty) \to R^3$, $p\colon R^3 \times [0, \infty) \to R$ *with the following properties*:

(1.1)   *there is a compact set* $K \subset R^3$ *such that* $u(x, t) = 0$ *for all* $x \notin K$;

(1.2)   *for fixed t, the function* $u_t\colon R^3 \to R^3$ *defined by* $u_t(x) = u(x, t)$ *is* C^∞;

(1.3)
$$\sum_{i=1}^{3} \frac{\partial u_i}{\partial x_i}(x, t) = 0;$$

(1.4)
$$p(x, t) = \int_{R^3} \sum_{i=1}^{3} \sum_{j=1}^{3} \frac{\partial u_j}{\partial x_i}(y, t) \frac{\partial u_i}{\partial x_j}(y, t)\big(4\pi|x - y|\big)^{-1} \, dy;$$

(1.5)   *there exists* $M < \infty$ *such that* $\|u_t\|_2 \leqslant M$ *for all* t;

(1.6) $|\nabla u|^2$, $|u|^3$, *and* $|u||p|$ *are integrable*;

1980 *Mathematics Subject Classification.* Primary 76D05.
[1] This research was supported in part by a Sloan Foundation Fellowship.

© 1986 American Mathematical Society
0082-0717/80 $1.00 + $.25 per page

(1.7) *if g: $R^3 \times (0, \infty) \to R$ is a C^∞ function with compact support and $g \geqslant 0$ then*

$$\int_0^\infty \int_{R^3} |\nabla u|^2 g \leqslant \int_0^\infty \int_{R^3} \left(2^{-1}|u|^2 + p \right) u \cdot \nabla g$$

$$+ \int_0^\infty \int_{R^3} 2^{-1}|u|^2 \left(\frac{\partial g}{\partial t} + \Delta g \right);$$

(1.8) *if $x \in S$ and U is a neighborhood of $(x, 1)$ then u is not essentially bounded on U;*

(1.9) S is homeomorphic to the Cantor ternary set.

In [2], it is proved that (1.3)–(1.8) imply $\dim(S) \leqslant 1$, where dim is the Hausdorff dimension. Reference [1] is an extension of [2]. Theorem 1.1 is an example that shows that S can be nontrivial.

The Navier–Stokes equations for a viscous incompressible fluid in R^3 are

$$(1.10) \qquad \frac{\partial u_i}{\partial t} = -\sum_{j=1}^3 u_j \frac{\partial u_i}{\partial x_j} - \frac{\partial p}{\partial x_i} + \Delta u_i + f_i, \qquad \sum_{i=1}^3 \frac{\partial u_i}{\partial x_i} = 0,$$

where f is an external force and p is the pressure. One may impose the additional conditions

$$(1.11) \qquad \sum_{i=1}^3 f_i u_i \leqslant 0, \qquad \sum_{i=1}^3 \frac{\partial f_i}{\partial x_i} = 0$$

on the force f. In [3], it is shown that conditions (1.3), (1.4), (1.7) imply a weak form of (1.10) and (1.11). This says that Theorem 1.1 can be written informally as follows: There exists a solution to (1.10) and (1.11) with internal singularities on all the points of a Cantor set.

Reference [3] exhibits an example where S is a single point.

2. Definitions and preliminaries. The following definitions are taken from [3]. We let $C_c^\infty(U, R^n)$ be the set of all C^∞ functions f: $U \to R^n$ with compact support. If $\{f_1, f_2\} \subset C_c^\infty(R^3, R)$, $f_i \geqslant 0$, $0 < b$ and $\nu > 0$, then the 4-tuple (f_1, f_2, b, ν) is called *admissible* when there exist functions u: $R^3 \times [0, b] \to R^3$, p: $R^3 \times [0, b] \to R$, satisfying (1.1)–(1.6) and the condition

$$\int_{R^3} 2^{-1}(f_2(x))^2 g(x, b)\, dx - \int_{R^3} 2^{-1}(f_1(x))^2 g(x, 0)\, dx + \int_0^b \int_{R^3} \nu |\nabla u|^2 g$$

$$\leqslant \int_0^b \int_{R^3} \left(2^{-1}|u|^2 + p \right) u \cdot \nabla g + \int_0^b \int_{R^3} 2^{-1}|u|^2 \left(\frac{\partial g}{\partial t} + \nu \Delta g \right)$$

for every $g \in C_c^\infty(R^3 \times R, R)$ with the property $g \geqslant 0$. We set $P = \{(x_1, x_2) \in R^2: x_2 > 0\}$. If $c = (c_1, c_2) \in R^2$ and $c_1^2 + c_2^2 = 1$ then R_c: $R^3 \to R^3$ is the rotation

$$R_c(x_1, x_2, x_3) = (x_1, c_1 x_2 - c_2 x_3, c_1 x_3 + c_2 x_2)$$

about the x_1 axis. If $f \in C_c^\infty(P, R)$ then $R(f)$ is defined by

$$R(f)(x_1, x_2, x_3) = f\left(x_1, \left(x_2^2 + x_3^2\right)^{1/2}\right) \quad \text{for } x_2^2 + x_3^2 \neq 0,$$

$$R(f)(x_1, 0, 0) = 0.$$

If $\{g_1, g_2\} \subset C_c^\infty(P, R)$, $g_i \geq 0$, $0 < b$ and $\nu > 0$, then (g_1, g_2, b, ν) is called P-admissible if $(R(g_1), R(g_2), b, \nu)$ is admissible.

LEMMA 2.1. *If (f_1, f_2, b, ν) is admissible (P-admissible), (f_3, f_4, b', ν) is admissible (P-admissible), and $f_2 \geq f_3$, then $(f_1, f_4, b + b', \nu)$ is admissible (P-admissible).*

If $f \in C_c^\infty(P, R)$, $v = (v_1, v_2) \in C_c^\infty(P, R^2)$, $f \geq 0$, and $f(x) > |v(x)|$ holds for all $x \in \mathrm{spt}(v)$ then $p^*[v, f]$ is the C^∞ function from R^3 into R defined by

$$p^*[v, f](x) = \int_{R^3} \sum_{i=1}^{3} \sum_{j=1}^{3} \left(\frac{\partial}{\partial x_i} u_j[v, f]\right)(y)$$

$$\cdot \left(\frac{\partial}{\partial x_j} u_i[v, f]\right)(y)(4\pi|x - y|)^{-1} dy,$$

where $u[v, f]$ is given by

$$u[v, f](x_1, x_2, 0) = \left(v_1(x_1, x_2), v_2(x_1, x_2), \left((f(x_1, x_2))^2 - |v(x_1, x_2)|^2\right)^{1/2}\right)$$

$$\text{if } (x_1, x_2) \in P,$$

$$u[v, f](R_c(x_1, x_2, 0)) = R_c(u[v, f](x_1, x_2, 0)) \quad \text{if } c \in R^2, |c| = 1,$$

$$(x_1, x_2) \in P, u[v, f](x_1, 0, 0) = 0.$$

We also set $p[v, f](x_1, x_2) = p^*[v, f](x_1, x_2, 0)$ if $(x_1, x_2) \in P$. If $f \in C_c^\infty(P, R)$ we define

$$L(f)(x_1, x_2) = \Delta f(x_1, x_2) + x_2^{-1}\left(\frac{\partial f}{\partial x_2}(x_1, x_2)\right) - x_2^{-2} f(x_1, x_2)$$

$$\text{if } (x_1, x_2) \in P.$$

3. The construction. The concepts below are taken from [3].

DEFINITION. An *initial structure* is an 11-tuple

$$I = (T, K_1, K_2, U_1, U_2, g_1, g_2, f_1, f_2, v_1, v_2)$$

satisfying (3.1)–(3.8):

(3.1) $K_i \subset U_i \subset P$, K_i is compact, U_i is open, closure $(U_i) \subset P$;

(3.2) $g_i \in C_c^\infty(P, R)$, $\mathrm{spt}(g_i) \subset U_i, 0 \leq g_i \leq 1$,

$$K_i \subset \mathrm{interior}\{x: g_i(x) = 1\};$$

(3.3) $f_i \in C_c^\infty(P, R)$, $v_i = (v_{i1}, v_{i2}) \in C_c^\infty(P, R^2)$;

(3.4)   closure(U_1) and closure(U_2) are disjoint compact sets, $\mathrm{spt}(v_i) \subset K_i$;

(3.5) $f_i \geq 0$, $f_i(x) = 0$ if $x \notin U_i$,

$$f_i(x) > |v_i(x)| \quad \text{if } x \in U_i;$$

$$(3.6) \qquad x_2 \frac{\partial}{\partial x_1} v_{i1}(x_1, x_2) + x_2 \frac{\partial}{\partial x_2} v_{i2}(x_1, x_2) + v_{i2}(x_1, x_2) = 0;$$

$$(3.7) \quad (f_2(x))^2 - Tv_2(x) \cdot \nabla(p[v_1, f_1] - p[0, f_1])(x) > |v_2(x)|^2 \quad \text{if } x \in U_2;$$

$$(3.8) \qquad L(f_i) \geqslant 0 \quad \text{if } x \notin K_i, \qquad L(f_i) > 0 \quad \text{if } x \in U_i \sim K_i.$$

If I is an initial structure then a *resolution* of I is a triple (δ, h_1, h_2) satisfying (3.9)–(3.13):

$$(3.9) \qquad \delta > 0, \qquad h_i \colon P \times (-\delta, T + \delta) \to R \text{ is } C^\infty;$$

$$(3.10) \qquad h_i \geqslant 0, \qquad h_i(x, t) > |v_i(x)| \quad \text{if } x \in \mathrm{spt}(g_i);$$

$$(3.11) \qquad (h_1(x, t))^2 = (f_1(x))^2 - 2t\delta g_1(x);$$

$$(3.12) \quad (h_2(x, t))^2 = (f_2(x))^2 - 2t\delta g_2(x)$$
$$- \int_0^t v_2(x) \cdot \nabla(p[v_1, h_{1, r}] - p[0, h_{1, r}])(x) \, dr$$

where $h_{i, r}(x) = h_i(x, r)$;

$$(3.13) \qquad L(h_{i, t})(x) \geqslant 0 \quad \text{if } g_i(x) < 1.$$

In [3], we prove

LEMMA 3.1. *If I is an initial structure then there exists $\delta_0 > 0$ such that there is a resolution (δ, h_1, h_2) of I for every $\delta \in (0, \delta_0)$.*

In [3], we also prove

LEMMA 3.2. *If (δ, h_1, h_2) is a resolution of*
$$I = (T, K_1, K_2, U_1, U_2, g_1, g_2, f_1, f_2, v_1, v_2)$$
then there exists $v_0 > 0$ such that $(f_1 + f_2, k, T, v)$ is P-admissible when $k(x) = (h_1 + h_2)(x, T)$ and $0 < v < v_0$.

We conclude

LEMMA 3.3. *If $I = (T, K_1, K_2, U_1, U_2, g_1, g_2, f_1, f_2, v_1, v_2)$ is an initial structure and $\varepsilon > 0$ then there exist $v_0 > 0$ and a function $k \in C_c^\infty(P, R)$ such that $k \geqslant 0$, $(f_1 + f_2, k, T, v)$ is P-admissible when $0 < v < v_0$, and*
$$\left\| f_1 + \left(f_2^2 - Tv_2 \cdot \nabla(p[v_1, f_1] - p[0, f_1]) \right)^{1/2} - k \right\|_\infty < \varepsilon.$$

DEFINITION. If $f \colon P \to R$, $X \subset P$, and $a \in R$, then $f^a \colon P \to R$ and X^a are defined by $f^a(x) = f(x - (a, 0))$ and $X^a = \{ x + (a, 0) \colon x \in X \}$.

LEMMA 3.4. *If $I = (T, K_1, K_2, U_1, U_2, g_1, g_2, f_1, f_2, v_1, v_2)$ is an initial structure, $\varepsilon > 0$ and $a \in R$ satisfy the condition*

$$(3.14) \qquad U_1, U_2, U_1^a, U_2^a \text{ have disjoint closures,}$$

then there exist $v_0 > 0$ and $k \in C_c^\infty(P, R)$ such that $k \geqslant 0$,

$$(f_1 + f_2 + f_1^a + f_2^a, k, 2T, v) \text{ is P-admissible} \quad \text{if } 0 < v < v_0,$$

and

$$\left\| f_1 + \left(f_2^2 - Tv_2 \cdot \nabla(p[v_1, f_1] - p[0, f_1]) \right)^{1/2} \right.$$
$$\left. + f_1^a + \left[\left(f_2^2 - Tv_2 \cdot \nabla(p[v_1, f_1] - p[0, f_1]) \right)^{1/2} \right]^a - k \right\|_\infty < \varepsilon.$$

In addition, v_0 is independent of a.

Outline of proof. From (3.14) and $\mathrm{spt}(v_i) \subset \mathrm{spt}(f_i) \subset \mathrm{closure}(U_i)$ we obtain

$$p[v_1, f_1 + f_1^a + f_2^a] - p[0, f_1 + f_1^a + f_2^a]$$

(3.15)
$$= p[v_1, f_1] + p[0, f_1^a] + p[0, f_2^a]$$
$$- (p[0, f_1] + p[0, f_1^a] + p[0, f_2^a])$$
$$= p[v_1, f_1] - p[0, f_1].$$

This implies that

$$I' = (T, K_1 \cup K_1^a \cup K_2^a, K_2, U_1 \cup U_1^a \cup U_2^a, U_2, g_1 + g_1^a + g_2^a, g_2,$$
$$f_1 + f_1^a + f_2^a, f_2, v_1, v_2)$$

is an initial structure. We set

$$f_3 = \left(f_2^2 - Tv_2 \cdot \nabla(p[v_1, f_1] - p[0, f_1]) \right)^{1/2}.$$

Let $\eta > 0$. Lemma 3.3 and (3.15) yield $v_1 > 0$ and $k_1 \in C_c^\infty(P, R)$ such that $k_1 \geqslant 0$,

(3.16) $(f_1 + f_1^a + f_2^a + f_2, k_1, T, v)$ is P-admissible   if $0 < v < v_1$,

(3.17) $\| f_1 + f_1^a + f_2^a + f_3 - k_1 \|_\infty < \eta$.

From (3.14), $\mathrm{spt}(v_i) \subset \mathrm{spt}(f_i)$ for $i = 1, 2$, and $\mathrm{spt}(f_3) \subset \mathrm{closure}(U_2)$ we conclude

$$p[v_1^a, f_1^a + f_1 + f_3] - p[0, f_1^a + f_1 + f_3]$$

(3.18)
$$= p[v_1^a, f_1^a] + p[0, f_1] + p[0, f_3]$$
$$- (p[0, f_1^a] + p[0, f_1] + p[0, f_3])$$
$$= p[v_1^a, f_1^a] - p[0, f_1^a].$$

This implies that

$$I'' = (T, K_1^a \cup K_1 \cup K_2, K_2^a, U_1^a \cup U_1 \cup U_2, U_2^a, g_1^a + g_1 + g_2, g_2^a,$$
$$f_1^a + f_1 + f_3, f_2^a, v_1^a, v_2^a)$$

is an initial structure. Once again, Lemma 3.3 and (3.18) yield $v_2 > 0$ and $k_2 \in C_c^\infty(P, R)$ such that $k_2 \geqslant 0$,

(3.19) $(f_1^a + f_1 + f_3 + f_2^a, k_2, T, v)$ is P-admissible   if $0 < v < v_2$

and

$$\left\| f_1^a + f_1 + f_3 + \left((f_2^a)^2 - Tv_2^a \cdot \nabla(p[v_1^a, f_1^a] - p[0, f_1^a]) \right)^{1/2} - k_2 \right\|_\infty < \eta,$$

which can be rewritten as

(3.20) $\| f_1^a + f_1 + f_3 + f_3^a - k_2 \|_\infty < \eta$.

Properties (3.16), (3.17), say that

$$(f_1 + f_1^a + f_2^a + f_2, f_1 + f_1^a + f_2^a + f_3, T, v)$$

is very close to being P-admissible for small ν. Properties (3.19), (3.20) say that

$$\left(f_1^a + f_1 + f_3 + f_2^a, f_1^a + f_1 + f_3 + f_3^a, T, \nu \right)$$

is very close to being P-admissible for small ν. A suitable modification of Lemma 2.1 shows that

$$\left(f_1 + f_1^a + f_2 + f_2^a, f_1 + f_1^a + f_3 + f_3^a, 2T, \nu \right)$$

is very close to being P-admissible for small ν. A perturbation argument yields the conclusion of the lemma.

DEFINITION. If $f \colon R^3 \to R$, $a \in R$, $b \in R$, $c > 0$, then $f^{a,\,b,\,c} \colon R^3 \to R$ is defined by $f^{a,\,b,\,c}(x) = c^{-1} f(c^{-1}(x - (a, b, 0)))$.

The next lemma is also proved in [3].

LEMMA 3.5. *There exists an initial structure*

$$I = (T, K_1, K_2, U_1, U_2, g_1, g_2, f_1, f_2, v_1, v_2)$$

and there exist $a \in R$, $b \in R$, $0 < c < 1$, such that the following holds: If $G = R(f_1 + f_2)$ and

$$G' = R\left(f_1 + \left(f_2^2 - Tv_2 \cdot \nabla (p[v_1, f_1] - p[0, f_1]) \right)^{1/2} \right)$$

then $\mathrm{spt}(G^{a,\,b,\,c}) \subset \mathrm{spt}(G)$ and

$$G'(x) > G^{a,\,b,\,c}(x) \quad \text{for all } x \in \mathrm{spt}(G^{a,\,b,\,c}).$$

An examination of the construction in [3] shows that the conclusion of Lemma 3.5 can be strengthened to

$$\mathrm{spt}(G^{a,\,b,\,c}) \cup \mathrm{spt}(G^{a',\,b,\,c}) \subset \mathrm{spt}(G),$$

$$\mathrm{spt}(G^{a,\,b,\,c}) \cap \mathrm{spt}(G^{a',\,b,\,c}) = \varnothing,$$

$$G'(x) > G^{a,\,b,\,c}(x) + G^{a',\,b,\,c}(x) \quad \text{if } x \in \mathrm{spt}(G^{a,\,b,\,c}) \cup \mathrm{spt}(G^{a',\,b,\,c}),$$

where a' is a certain constant such that $a' \neq a$.

Combining the above, Lemma 3.3 and Lemma 3.4, we obtain

LEMMA 3.6. *There exist $G \in C_c^\infty(R^3, R)$, $a \in R$, $a' \in R$, $b \in R$, $0 < c < 1$, $T > 0$, $\nu > 0$, such that $G \geqslant 0$, $G \neq 0$, $a \neq a'$,*

$$\mathrm{spt}(G^{a,\,b,\,c}) \cup \mathrm{spt}(G^{a',\,b,\,c}) \subset \mathrm{spt}(G),$$

$$\mathrm{spt}(G^{a,\,b,\,c}) \cap \mathrm{spt}(G^{a',\,b,\,c}) = \varnothing,$$

$$(G, G^{a,\,b,\,c} + G^{a',\,b,\,c}, T, \nu) \text{ is admissible,}$$

$$(G + G^{s,\,0,\,1}, G^{a,\,b,\,c} + G^{a+s,\,b,\,c}, 2T, \nu) \text{ is admissible}$$

provided $s \in R$ satisfies $\mathrm{spt}(G) \cap \mathrm{spt}(G^{s,\,0,\,1}) = \varnothing$.

DEFINITION. If $f \colon R^3 \to R$ is a function, $a \in R$, and $c > 0$, then $f^{a,\,c} \colon R^3 \to R$ is defined by $f^{a,\,c} = f^{a,\,0,\,c}$.

Applying a translation to Lemma 3.6 we obtain

LEMMA 3.7. *There exist* $H \in C_c^\infty(R^3, R)$, $d \in R$, $0 < c < 1$, $T > 0$, $\nu > 0$, *such that* $H \geqslant 0$, $H \neq 0$, $d \neq 0$,

$$\operatorname{spt}(H^{0,c}) \cup \operatorname{spt}(H^{d,c}) \subset \operatorname{spt}(H), \qquad \operatorname{spt}(H^{0,c}) \cap \operatorname{spt}(H^{d,c}) = \varnothing,$$

$$(H, H^{0,c} + H^{d,c}, T, \nu) \text{ is admissible},$$

$$(H + H^{s,1}, H^{0,c} + H^{s,c}, 2T, \nu) \text{ is admissible}$$

provided $s \in R$ *satisfies* $\operatorname{spt}(H) \cap \operatorname{spt}(H^{s,1}) = \varnothing$.

When eddies in a fluid are far apart, they have little influence on each other. The following generalization of Lemma 3.7 is a precise statement of this fact.

LEMMA 3.8. *There exist* $H \in C_c^\infty(R^3, R)$, $d \in R$, $0 < c < 1$, $T > 0$, $\nu > 0$, $M < \infty$, *such that* $H \geqslant 0$, $H \neq 0$, $d \neq 0$,

$$\operatorname{spt}(H^{0,c}) \cup \operatorname{spt}(H^{d,c}) \subset \operatorname{spt}(H), \qquad \operatorname{spt}(H^{0,c}) \cap \operatorname{spt}(H^{d,c}) = \varnothing,$$

and the following hold:

(1) *If* $\{a_1, \ldots, a_p\} \subset R$ *and* $|a_i - a_j| \geqslant M$ *for* $i \neq j$ *then*

$$\left(\sum_j H^{a_j, 1}, \sum_j H^{a_j, c} + \sum_j H^{d + a_j, c}, T, \nu \right) \text{ is admissible};$$

(2) *if, in addition,* $s \in R$, $\operatorname{spt}(H) \cap \operatorname{spt}(H^{s,1}) = \varnothing$ *and* $|a_i - (a_j + s)| \geqslant M$ *for* $i \neq j$ *then*

$$\left(\sum_j H^{a_j, 1} + \sum_j H^{s + a_j, 1}, \sum_j H^{a_j, c} + \sum_j H^{s + a_j, c}, 2T, \nu \right) \text{ is admissible}.$$

LEMMA 3.9. *If* (f_1, f_2, t, ν) *is admissible then* $(f_1^{a, c^m}, f_2^{a, c^m}, tc^{2m}, \nu)$ *is admissible.*

LEMMA 3.10. $(f^{a, c})^{a', c'} = f^{ac' + a', cc'}$.

DEFINITION. We fix H, d, c, T, ν, M, as in Lemma 3.8 and set $[\alpha, \sigma] = H^{\alpha, \sigma}$ for $\alpha \in R$, $\sigma > 0$.

Combining the last three lemmas we conclude

LEMMA 3.11. *If* $m \in \{0, 1, 2, \ldots\}$, $\{a_1, \ldots, a_p\} \subset R$, *and* $|a_i - a_j| \geqslant Mc^m$ *for* $i \neq j$ *then*

$$\left(\sum_j [a_j, c^m], \sum_j [a_j, c^{m+1}] + \sum_j [c^m d + a_j, c^{m+1}], c^{2m}T, \nu \right)$$

is admissible. If, in addition, $s \in R$, $\operatorname{spt}(H) \cap \operatorname{spt}(H^{c^{-m}s, 1}) = \varnothing$, *and* $|a_i - (a_j + s)| \geqslant Mc^m$ *for* $i \neq j$ *then*

$$\left(\sum_j [a_j, c^m] + \sum_j [s + a_j, c^m], \sum_j [a_j, c^{m+1}] + \sum_j [s + a_j, c^{m+1}], 2c^{2m}T, \nu \right)$$

is admissible.

DEFINITION. We set $E_k = \{e = (e_0, e_1, \ldots, e_k) \colon e_i \in \{0, 1\}\}$. Let n be a positive integer such that $c^n M < |d|(1 - c)$, $c^n < 1/3$.

A consequence of Lemma 3.8 and Lemma 3.11 is the following

LEMMA 3.12. *If $k \in \{0, 1, 2, \ldots\}$ then*

$$\left(\sum_{e \in E_k} \left[e_0 d + e_1 c^n d + e_2 c^{2n} d + \cdots + e_k c^{kn} d, c^{(k+1)n} \right], \right.$$

$$\left. \sum_{e \in E_{k+1}} \left[e_0 d + e_1 c^n d + e_2 c^{2n} d + \cdots + e_{k+1} c^{(k+1)n} d, c^{(k+1)n+1} \right], c^{2(k+1)n} T, \nu \right)$$

is admissible. If, in addition, $i \in \{1, 2, \ldots, n-1\}$ then

$$\left(\sum_{e \in E_{k+1}} \left[e_0 d + e_1 c^n d + e_2 c^{2n} d + \cdots + e_{k+1} c^{(k+1)n} d, c^{(k+1)n+i} \right], \right.$$

$$\left. \sum_{e \in E_{k+1}} \left[e_0 d + e_1 c^n d + e_2 c^{2n} d + \cdots + e_{k+1} c^{(k+1)n} d, c^{(k+1)n+i+1} \right], 2 c^{2((k+1)n+i)} T, \nu \right)$$

is admissible.

PROOF. We set $m = (k+1)n$ and

$$\{ a_1, \ldots, a_p \} = \{ e_0 d + e_1 c^n d + \cdots + e_k c^{kn} d : e \in E_k \}.$$

Observe that $c^n < 1/3$ yields $|a_i - a_j| \geq c^{kn} |d|$ for $i \neq j$. This fact and $c^n M < |d|$ imply that the first part of Lemma 3.11 can be used to obtain the first conclusion of Lemma 3.12.

To obtain the second conclusion, we keep the same a_j but set $m = (k+1)n + i$. We also set $s = c^{(k+1)n} d$. If $i \neq j$ then the property $c^n < 1/3$ implies

$$\left| a_i - \left(a_j + c^{(k+1)n} d \right) \right| \geq c^{kn} |d| - c^{(k+1)n} |d|.$$

This fact and $c^n M < |d|(1 - c^n)$ imply $|a_i - (a_j + s)| \geq M c^m$ for $i \neq j$. We will be able to apply Lemma 3.11 and reach the desired conclusion as soon as we show

$$\mathrm{spt}(H) \cap \mathrm{spt}(H^{c^{-m}s, 1}) = \varnothing.$$

Since we have $c^{-m} s = c^{-i} d$, this is equivalent to

$$\mathrm{spt}(H^{0, c^i}) \cap \mathrm{spt}(H^{c^{-i}d, 1})^{0, c^i} = \varnothing,$$

and Lemma 3.10 says that this is the same as

$$\mathrm{spt}(H^{0, c^i}) \cap \mathrm{spt}(H^{d, c^i}) = \varnothing.$$

The above can be proved by induction on i. One uses

$$\mathrm{spt}(H^{0, c^{i+1}}) = \mathrm{spt}(H^{0, c})^{0, c^i} \subset \mathrm{spt}(H^{0, c^i}),$$

$$\mathrm{spt}(H^{d, c^{i+1}}) = \mathrm{spt}(H^{0, c})^{d, c^i} \subset \mathrm{spt}(H^{d, c^i}),$$

which follow from Lemma 3.10 and $\mathrm{spt}(H^{0, c}) \subset \mathrm{spt}(H)$ (see Lemma 3.8).

From Lemma 3.12 and Lemma 2.1 we get

LEMMA 3.13. *If $k \in \{0, 1, 2, \ldots\}$ then*

$$\left(\sum_{e \in E_k} \left[e_0 d + e_1 c^n d + e_2 c^{2n} d + \cdots + e_k c^{kn} d, c^{(k+1)n} \right], \right.$$

$$\sum_{e \in E_{k+1}} \left[e_0 d + e_1 c^n d + e_2 c^{2n} d + \cdots + e_{k+1} c^{(k+1)n} d, c^{(k+2)n} \right],$$

$$\left. c^{2(k+1)n} T + \sum_{i=1}^{n-1} 2 c^{2((k+1)n+i)} T, \nu \right)$$

is admissible.

The set S of all infinite sums $e_0 d + e_1 c^n d + e_2 c^{2n} d + \cdots$, where $e_i \in \{0, 1\}$, is a Cantor set. Lemma 3.13 and Lemma 2.1 give us a weak solution to (1.10) and (1.11) with the viscosity term Δu_i replaced by $\nu \Delta u_i$ such that there are singularities on all the points of $S \times \{(0, 0)\}$ at time

$$\sum_{k=0}^{\infty} \left(c^{2(k+1)n} T + \sum_{i=1}^{n-1} 2 c^{2((k+1)n+i)} T \right) < \infty.$$

A simple scaling of the equations gives us Theorem 1.1.

REFERENCES

1. L. Caffarelli, R. Kohn and L. Nirenberg, *Partial regularity of suitable weak solutions of the Navier-Stokes equations* (to appear).

2. V. Scheffer, *Hausdorff measure and the Navier-Stokes equations*, Comm. Math. Phys. **55** (1977), 97–112.

3. ______, *A solution to the Navier-Stokes inequality with an internal singularity*, Comm. Math. Phys. (to appear).

RUTGERS UNIVERSITY

Proceedings of Symposia in Pure Mathematics
Volume **44** (1986)

Asymptotic Behaviour
of Minimal Submanifolds and Harmonic Maps

LEON SIMON

Abstract. This article summarizes recent work concerning behaviour of minimal submanifolds and harmonic maps near isolated singular points, and asymptotic behaviour near infinity of solutions of the minimal surface equation over the exterior of bounded domains in $\mathbf{R}^n$.

1. Isolated singular points. It is actually convenient to begin our discussion by looking at a general class of functionals defined over *cylindrical domains* $\Sigma \times \mathbf{R}$, where Σ is a compact Riemannian manifold. Specifically, we look at functionals

$$(1.1) \qquad \mathscr{G}(u) = \mathscr{F}(u) + \mathscr{R}(u),$$

with

$$\mathscr{F}(u) = \int_a^b \int_\Sigma e^{-mt} F(\omega, u, \nabla u, \dot{u}) \, d\omega \, dt \qquad (-\infty \leqslant a < b \leqslant \infty),$$

$$\mathscr{R}(u) = \int_a^b \int_\Sigma e^{-mt} R(\omega, t, u, \nabla u, \dot{u}) \, d\omega \, dt,$$

where $m \in \mathbf{R}$, $\nabla =$ gradient on Σ, $u = \partial u/\partial t$, F, R are given smooth functions of their arguments, and R has exponential decay as $t \uparrow \infty$. The precise conditions assumed about F, R are given below. The function u may be vector-valued; it is assumed to take values in $\mathbf{R}^N$, $N \geqslant 1$, or in some vector subbundle of $\mathbf{R}^N$ (in the sense that $u(\omega, t) \in V_\omega$, where $\{V_\omega\}_{\omega \in \Sigma}$ is a smoothly varying family of subspaces of $\mathbf{R}^N$, each V having fixed dimension k, $1 \leqslant k \leqslant N$). In the latter case $\nabla u(\omega, t)$ is taken to mean $\Sigma_{i,j=1}^N p(\omega)^{ij} (\nabla u_j) \otimes e_i$, where $e_1, \ldots, e_N$ is the standard orthonormal basis for $\mathbf{R}^N$, $u_j = e_j \cdot u$ is the jth component of u, and where $(p(\omega)^{ii})$ is the matrix of the orthogonal projection of $\mathbf{R}^N$ onto the subspace V_ω. We shall in fact always assume we are in this "vector bundle" setting, since we can always specialize back to the simpler case by taking $V_\omega = \mathbf{R}^N$, $\omega \in \Sigma$.

1980 *Mathematics Subject Classification*. Primary 49F22.

© 1986 American Mathematical Society
0082-0717/86 $1.00 + $.25 per page

$\mathcal{M}(u)$, $\mathcal{N}(u)$ (with values at (ω, t) in $V_\omega \subset \mathbf{R}^N$) denote e^{mt} times the Euler–Lagrange operators associated with the functionals $\mathcal{F}(u)$ and $\mathcal{G}(u)$ respectively. Then $\mathcal{M}$, $\mathcal{N}$ are second order quasilinear operators—i.e. linear in the second derivatives, with coefficients depending on ω, u, ∇u (and on t in the case of $\mathcal{N}(u)$).

To describe the exact conditions on F, R, assume (without loss of generality by the Nash embedding theorem) that Σ is a submanifold of some $\mathbf{R}^P$. Then we require that F, R are smooth on $\Sigma \times \mathbf{R}^N \times \mathbf{R}^{PN} \times \mathbf{R}^N$ and $\Sigma \times \mathbf{R} \times \mathbf{R}^N \times \mathbf{R}^{PN} \times \mathbf{R}^N$ respectively and that (writing $F = F(\omega, z, p, q)$, $(\omega, z, p, q) \in \Sigma \times \mathbf{R}^N \times \mathbf{R}^{PN} \times \mathbf{R}^N$, and denoting partial derivations by subscripts) for $\omega \in \Sigma$, $(z, p, q) \in V_\omega \times (T_\omega \Sigma \otimes V_\omega) \times V_\omega$,

$$(1.2) \qquad F_{p_i^\alpha p_j^\beta}(\omega, z, p, q)\lambda^\alpha \lambda^\beta \xi_i \xi_j > 0, \qquad \lambda \neq 0, \xi \neq 0,$$

$$(1.3) \qquad q \cdot F_q(\omega, z, p, q) > 0, \qquad q \neq 0,$$

$$q \cdot F_q(\omega, z, p, q) \geq c(p, z)|q|^2, \qquad |q| \leq 1,$$

$$(1.4) \quad \sup_{t > 0} e^{\varepsilon_0 t}(|R| + |DR| + |D^2 R|) \qquad (\omega, t, z, p, q) \leq c(z, p, q) < \infty,$$

where $\varepsilon_0 > 0$ and where D is the full gradient on the product space $\Sigma \times \mathbf{R} \times \mathbf{R}^N \times \mathbf{R}^{PN} \times \mathbf{R}^N$.

Notice that (see [SL1, I, §1] for discussion) the conditions (1.2), (1.3) guarantee that the linearization of the Euler–Lagrange operator $\mathcal{M}(u)$ at $u = \varphi$, where φ is a t-independent C^2 solution of $\mathcal{M}(\varphi) = 0$, gives a *strongly elliptic* second order linear operator $\mathcal{L}_\varphi$ on $\Sigma \times \mathbf{R}$ of the form

$$\mathcal{L}_\varphi u = a(\ddot{u} - m\dot{u}) + L_\varphi u,$$

where $a(\omega)$: $V_\omega \to V_\omega$ is a positive isomorphism (in fact $a(\omega)$ is the restriction to V_ω of the linear transformation $\mathbf{R}^N \to \mathbf{R}^N$ represented by the matrix $(F_{q^\alpha q^\beta}(\omega, \varphi(\omega), \nabla\varphi(\omega), 0))$) and where L_φ is a second order elliptic operator on Σ (operating on smooth functions v with $v(\omega) \in V_\omega$ and with $(L_\varphi v)(\omega) \in V_\omega$), and where L_φ is selfadjoint relative to the $L^2(\Sigma; \mathbf{R}^N)$ inner product.

To motivate consideration of such functionals we look at two key examples, the *energy functional* for maps between Riemannian manifolds and the *area functional* over a cone, extrema for which give harmonic maps and minimal submanifolds.

To describe the energy functional, consider the unit ball $B_1(0)$ in $\mathbf{R}^n$, equipped with a smooth metric $g = \sum_{i,j=1}^n g_{ij}(x)\, dx^i\, dx^j$ with $g_{ij}(0) = \delta_{ij}$, $\partial g_{ij}/\partial x^k(0) = 0$, $i, j, k = 1, \ldots, n$, and a map $h: B_1(0) \to \mathbf{R}^N$, $N \geq 1$, where $\mathbf{R}^N$ is equipped with a smooth metric $\gamma = \sum_{i,j=1}^N \gamma_{ij}(y)\, dy^i\, dy^j$. The "energy" of such a map h is defined to be

$$\mathcal{E}(h) = \int_{B_1(0)} \gamma_{ij}(h) \frac{\partial h^i}{\partial x^k} \frac{\partial h^j}{\partial x^l} g^{kl}(x)\sqrt{g}\, dx.$$

(More intrinsically, $\mathcal{E}(h) = \int_M |dh|^2\, dv$, where $M = (B_1(0), g)$, dv is the volume element of M, dh is the linear map induced by h, and $|dh|$ is the inner product norm $(\text{trace}(dh \circ \text{adjoint}(dh)))^{1/2}$ computed relative to the metrics g, γ.)

Introducing new variables $\omega = x/|x|$, $t = -\log|x|$, $u(\omega, t) = h(x)$, one easily checks that (after writing $g^{kl}(x) = \delta_{kl} + (g^{kl}(x) - \delta_{kl})$ and noting that $|g^{kl}(x) - \delta_{kl}| \leq c|x|^2 \equiv ce^{-2t}$) the energy functional $\mathscr{E}(h)$ takes the form (1.1)–(1.4) with $m = n - 2$, $\varepsilon_0 = 2$, $a = 0$, $b = \infty$, and the asymptotic behaviour as $t \uparrow \infty$ of $u(\omega, t)$ gives exactly the asymptotic behaviour as $|x| \downarrow 0$ of $h(x)$.

To describe our second key example—the area functional over a cone—suppose Σ is a compact $(n - 1)$-dimensional minimal submanifold of the sphere $S^{N-1} \subset \mathbf{R}^N$ ($N > n$), so that the cone $C = \{\lambda y \colon \lambda > 0, \, y \in \Sigma\}$ is an n-dimensional minimal submanifold of $\mathbf{R}^N$. For $0 < \rho_1 < \rho_2$ write $C_{\rho_1, \rho_2} = C \cap (B_{\rho_2} \sim \bar{B}_{\rho_1})$ ($B_\rho = \{x \in \mathbf{R}^N \colon |x| < \rho\}$), and consider submanifolds M which can be expressed as "graphs" over C_{ρ_1, ρ_2} in the sense that

$$M = G_{\rho_1, \rho_2}(h) \equiv \left\{ \frac{r\omega + h(r\omega)}{\sqrt{1 + |h(r\omega)/r|^2}} \colon r\omega \in C_{\rho_1, \rho_2} \right\},$$

where $h \in C^2(C_{\rho_1, \rho_2}; \mathbf{R}^N)$ and $h(x)$ is normal to C at x for each $x \in C_{\rho_1, \rho_2}$. Then the area of M is given by the functional $\mathscr{A}(h)$,

$$\mathscr{A}(h) = \mathscr{H}^n\left(G_{\rho_1, \rho_2}(h)\right) = \int_{C_{\rho_1, \rho_2}} JH \, d\mathscr{H}^n,$$

where $H(x) = (1 + |h(x)/|x||^2)^{-1/2}(x + h(x))$ and where JH is the Jacobian of H. Introducing variables $\omega = x/|x|$, $t = -\log|x|$, $u(\omega, t) = h(r\omega)/r$ ($r = |x|$), one checks (see [**SL1**, I, §3]) that $\mathscr{A}(h)$ has the same form as $\mathscr{F}(u)$ in (1.1) above with $m = n$, $a = \log \rho_2^{-1}$, $b = \log \rho_1^{-1}$. (The reader should keep in mind that for the area functional both $u(\omega, t)$ ($- h(r\omega)/r$) and $\mathscr{M}(u)$ (ω, t) take values in $V_\omega \equiv (T_\omega C)^\perp$; here $T_\omega C$ is the tangent space of C at ω, which is the same as the tangent of C at $r\omega$, $r > 0$, and $(\,)^\perp$ is orthogonal complement taken in $\mathbf{R}^N$.) For an explicit formula for the function F in this case, and for the check that F does satisfy (1.2) and (1.3) we refer the reader to [**SL1**, I, §3].

We now proceed with our discussion of asymptotic limit results. First of all we note that the positivity condition (1.3) easily implies the existence of "limit points" for $u(t)$ as $t \uparrow \infty$ in case u has finite C^2 norm on $\Sigma \times (0, \infty)$ (and in some more general cases also); the main result is given in the following lemma, the straightforward proof of which is based on the appropriate "monotonicity" formulae, which are a consequence of the condition (1.3). We refer to [**SL1**, II, §2] for details.

1.5. **Lemma.** *Suppose $u \in C^2(\Sigma \times (0, \infty))$ with $|u|_{C^2} < \infty$ and $\mathscr{N} u = 0$, and suppose that (1.2)–(1.4) hold. Then for each sequence $\{t_j\} \uparrow \infty$, there is a subsequence $\{t_{j'}\}$ and a t-independent C^2 function $\varphi = \varphi(\omega)$ with $\mathscr{M}\varphi = 0$ and*

$$\lim_{j' \to \infty} |u - \varphi|_{C^2(\Sigma \times (t_{j'} - K, \, t_{j'} + K))} = 0$$

for each fixed $K > 0$.

In view of this lemma one is tempted to conjecture that there is actually a unique asymptotic limit φ for any such u at ∞; that is, that there is a $\varphi = \varphi(\omega)$ with $\mathcal{M}\varphi = 0$ and

$$\lim_{t \uparrow \infty} |u - \varphi|_{C^2(\Sigma \times (t, \infty))} = 0.$$

(This is evidently equivalent to proving that there is a unique limit point φ, independent of which sequences $\{t_j\}$, $\{t_{j'}\}$ we choose in Lemma 1.5 above.)

In fact this turns out to be correct, provided we make an additional assumption. Specifically, we need to assume, in addition to (1.2)–(1.4), that $F(\omega, z, p, 0)$, $D_\omega F(\omega, z, p, 0)$, $D_\omega^2 F(\omega, z, p, 0)$ are real analytic functions of z, p, uniformly for $\omega \in \Sigma$. Thus we assume

$$(1.6) \qquad \sup_{\substack{\omega \in \Sigma \\ j = 0, 1, 2}} \left| D_z^\alpha D_p^\beta D_\omega^j F(\omega, z, p, 0) \right| \leqslant c^{|\alpha| + |\beta|} \alpha! \beta!$$

for all $\alpha \in \mathbf{Z}_+^N$, $\beta \in \mathbf{Z}_+^{PN}$, $(z, p) \in \mathbf{R}^N \times \mathbf{R}^{PN}$, where $c = c(z, p) < \infty$. ($\mathbf{Z}_+$ is the set of nonnegative integers.)

Then we have the following theorem, the proof of which is rather lengthy and given in [**SL1**, II, §4]. (See also [**SL2**].)

THEOREM 1. *Suppose* $u \in C^2(\Sigma \times (0, \infty))$ *with* $|u|_{C^2} < \infty$ *and* $\mathcal{N} u = 0$, *and suppose that* (1.2)–(1.4), *and* (1.6) *all hold. Then there is a t-independent C^2 function* $\varphi = \varphi(\omega)$ *with* $\mathcal{M}\varphi = 0$ *and*

$$\lim_{t \uparrow \infty} |u - \varphi|_{C^2(\Sigma \times (t, \infty))} = 0.$$

One shortcoming of Theorem 1 as far as applications are concerned is the assumption that $|u|_{C^2(\Sigma \times (0, \infty))} < \infty$. In the case of the energy functional $\mathcal{E}$ discussed above (writing $u(\omega, t) = h(r\omega)$, $t = -\log r$), the assumption $|u|_{C^2} < \infty$ requires that $h \in C^2(B_1(0); \mathbf{R}^N)$ with

$$(*) \qquad |Dh(x)| \leqslant c/r, \qquad |D^2 h| \leqslant c/r^2,$$

which, as an *a priori* assumption, limits the applicability of Theorem 1. One case when $(*)$ holds automatically (at least after a suitable change of scale) occurs when $B_1(0) \subset \mathbf{R}^3$ (i.e. $n = 3$) and h is a *minimizing* harmonic map $(B_\rho(0), g) \to (\mathbf{R}^N, \gamma)$ for some $\rho > 0$. (h minimizing means that $\mathcal{E}(h) \leqslant \mathcal{E}(\tilde{h})$ whenever h and $\tilde{h}$ differ by a $W^{1,2}$ function with compact support in $B_\rho(0)$.) In this case the regularity theory of [**SU** or **GG**] guarantees automatically that either h is smooth in a neighbourhood of 0 or $h \in C^2(B_\sigma(0) \sim \{0\})$ and that $(*)$ holds for $0 < r < \sigma$ for some $\sigma > 0$. Then by changing scale ($x \mapsto \sigma^{-1} x$) we have a harmonic map h satisfying $(*)$ on $B_1(0)$, and hence Theorem 1 guarantees that (provided the metric γ is real-analytic) there is a C^2 harmonic map $\varphi: S^2 \to (\mathbf{R}^N, \gamma)$ such that $\lim_{r \downarrow 0} h(r\omega) = \varphi(\omega)$, $\omega \in S^2$, where the convergence is relative to the C^2 norm on S^2. Notice that this gives a rather precise description of the singularity at 0.

Theorem 1 also has a more general version which does not require $|u|_{C^2} < \infty$ or even that u be *a priori defined* over all of $\Sigma \times (0, \infty)$. To describe this more general version, let ε, δ, $\beta \in (0, 1)$, $T_0 \geq 1$, φ any C^2 t-independent solution of $\mathcal{M}\varphi = 0$, and let $\mathcal{P} = \mathcal{P}(\varepsilon, \delta, \beta, T_0, \varphi)$ be the class of all C^2 solutions u of $\mathcal{N}u = 0$ on $\Sigma \times (0, b)$ $(b = b(u) > 0)$ such that:

(i)

      b is *maximal* for u; that is, u cannot be extended to be a C^2

      solution of $\mathcal{N}u = 0$ on $\Sigma \times (0, b')$ with $b' > b$;

(ii)

$$|u - \varphi|_{C^2(\Sigma \times (0, T_0))} \leq \delta;$$

(iii)

$$\text{if } m > 0, \quad \mathcal{F}_\Sigma(u(t)) \geq \mathcal{F}_\Sigma(\varphi) - \delta e^{-\varepsilon t}, \quad 0 < t < b,$$
$$\text{if } m < 0, \quad \mathcal{F}_\Sigma(u(t)) \leq \mathcal{F}_\Sigma(\varphi) + \delta e^{-\varepsilon t}, \quad 0 < t < b;$$

(iv)

$$|u - \varphi|_{C^2(\Sigma \times (0, \tau))} \leq \delta^\beta, \quad \tau \geq T_0 \Rightarrow b \geq \tau + T_0, \quad |u - \varphi|_{C^2(\Sigma \times (\tau, \tau + T_0))} \leq 1.$$

Subject to these assumptions we have

THEOREM 2. *There are* δ, $\beta \in (0, 1)$ (*depending on* ε, T_0, φ, F, R) *so that any* $u \in \mathcal{P}$ *is* C^2 *on* $\Sigma \times (0, \infty)$ (*i.e.* $b = \infty$) *and there is a* t-*independent* $\tilde{\varphi}$ *with* $\mathcal{M}\tilde{\varphi} = 0$ *and*

$$\lim_{t \uparrow \infty} |u - \tilde{\varphi}|_{C^2(\Sigma \times (t, \infty))} = 0.$$

For the proof we again refer to [**SL1**, Part II] or to [**SL2**]. (The latter reference contains a somewhat more abstract, and slightly more general version of the above result; however the result of [**SL2**] is less convenient to apply in practice.)

It is rather straightforward to apply Theorem 2 to the special case of the energy and area functionals (discussed above) in order to deduce results about asymptotic behaviour of harmonic maps and minimal submanifolds. For example if h: $(B_1(0), g) \to (\mathbf{R}^N, \gamma)$ is a minimizing harmonic map (see discussion after Theorem 1 for the terminology), if the metric γ is real analytic, and if at least one of the tangent maps φ: $S^{n-1} \to (\mathbf{R}^N, \gamma)$ of h at 0 (see [**SU**] for the terminology) is smooth, then

$$\lim_{r \downarrow 0} h(r\omega) = \varphi(\omega), \quad \omega \in S^{n-1},$$

where the limit is in the C^2 sense on S^{n-1}. Notice that this gives a good picture of the singularity of $r = 0$. It is an open question whether or not an analogous result holds if the *a priori* assumption that φ be smooth is dropped.

For the area functional, the general result above enables us to prove the following: If M is a minimal submanifold of a complete Riemannian manifold N with N isometrically embedded in $\mathbf{R}^P$, if $p \in \overline{M} \sim M$ is a singular point of M, and if one of the tangent cones C of M at p is multiplicity one and *regular* (i.e. sing $C \subset \{0\}$), then C is the unique tangent cone of M at p. In fact, relative to normal coordinates for N with origin corresponding to p and with $T_p N$ identified with $\mathbf{R}^P$, we can write (for suitably small $\rho > 0$)

$$M \cap B_\rho(0) = \left\{ x + h(x) : x \in C \cap B_\rho(0) \right\}$$

where h is a C^2 function on $C \cap B_\rho(x)$ with values normal to C and with $\lim_{|x| \to 0} h(x)/|x| = 0$.

For a detailed discussion of how these results follow from Theorem 2 given above, we refer to [**SL1**, II, §5] and [**SL2**, §7].

We wish to make one further point here concerning Theorem 2. Namely, Theorem 2 does not say anything about the *rate* at which $|u - \tilde{\varphi}|_{C^2(\Sigma \times (t, \infty))}$ converges to zero. Examples show (see the discussion of [**SL2**]) that the convergence need not be exponential in t in general[1]—additional assumptions are necessary. One interesting result in this direction is the following

THEOREM 3. *Suppose* (1.2)–(1.4), (1.6) *hold, and suppose* φ *is a solution of* $\mathcal{M}\varphi = 0$ *which is t-independent, and such that the operator* L_φ *(as above) has the property that all* $C^2(\Sigma)$ *solutions* ψ *of* $L_\varphi \psi = 0$ *can be expressed*

$$\psi = \lim_{s \downarrow 0} s^{-1}(\varphi_s - \varphi)$$

(limit relative to $C^2(\Sigma)$*) for some 1-parameter family* $\{\varphi_s\}_{s \in (0,1)}$ *of t-independent* C^2 *solutions of* $\mathcal{M}\varphi_s = 0$. *Then for any* $\beta, \varepsilon, \sigma \in (0,1)$ *there is* $\eta \in (0,1)$ *such that* $u \in \mathscr{P}(\varepsilon, \sigma, \beta, T_0, \varphi)$ *and*

$$e^{-\varepsilon T_0} + |u - \varphi|_{C^2(\Sigma \times (0, T_0))} < \eta$$

implies $u \in C^2(\Sigma \times (0, \infty))$ *and there are* $\mu > 0$ *and a t-independent* $\tilde{\phi}$ *satisfying*

$$\mathcal{M}\tilde{\varphi} = 0, \ |u - \tilde{\varphi}|_{C^2(\Sigma \times (t, \infty))} \leqslant c\eta e^{-\mu t}.$$

For the proof of this result we refer the reader to [**SL1**, II, §6]. Note that the result is a generalization of a technique applied to minimal submanifolds in [**AA**]. (The proof of [**SL1**] is somewhat simplified in addition to being applicable in the general setting described above.)

The main disadvantage of Theorem 3 is that it is in general difficult to check the condition (∗) (and it does not always hold).

In conclusion we mention that there is a straightforward implicit function theorem method for constructing examples of solutions u of $\mathcal{N}u = 0$ ($\mathcal{N}$ the Euler–Lagrange operator for the functional $\mathscr{F}$ of (1.1)) with exponential decay at ∞. The precise results are described in detail in [**SL1**, Part I]. The method is a

---

[1] D. Adams [**AD**] has recently obtained a general result about existence of examples where the convergence is like t^{-1}; in particular, his result implies the existence of many examples of minimal submanifolds which decay to their tangent cones at *logarithmic* rate.

natural extension of that proposed for the area functional in [**CHS**]. The results of [**SL1**, Part I], when applied to the energy and area functionals, give very large classes of harmonic maps and minimal submanifolds with isolated singularities of prescribed type. For example, given a minimal cone C of dimension n in $\mathbf{R}^P$ with $\operatorname{sing} C = \{0\}$, it is possible to construct a rich class of 1-parameter families $\{M_\alpha\}_{\alpha \in (-1,1)}$ of minimal submanifolds which are expressible in the form

$$M_\alpha = \{x + h_\alpha(x): x \in C \cap B_1(0)\},$$

where each $h_\alpha(x)$ takes values normal to C and satisfies $|h_\alpha(x)| \leqslant c|\alpha|\,|x|^{1+q}$, $c, q > 0$, independent of α. (For details, including a discussion of the extent to which the boundary values $h_\alpha|C \cap \partial B_1(0)$ can be controlled, and the relation of this to q, we refer to [**CHS** and **SL1**, Part I].) In the codimension 1 case ($P = n + 1$) one can prove that all these examples, for $|\alpha|$ small enough, are actually *minimizing* in $B_1(0)$ provided that the cone C is *strictly* minimizing. (C is said to be strictly minimizing if there is a $\Theta > 0$ such that $\mathbf{M}(C \cap B_1(0)) \leqslant \mathbf{M}(T)$ $- \Theta\varepsilon^n$ whenever $\varepsilon \in (0,1)$ and T is an integer-multiplicity current with $\partial T = \partial(C \llcorner B_1(0))$ and $\operatorname{spt} T \subset \overline{B}_1(0) \sim B_\varepsilon(0)$; notice that all the codimension 1 minimizing cones given in [**L**] are strictly minimizing. For discussion we refer to [**HS**].)

2. Asymptotic behaviour of minimal graphs over exterior domains. Let u be a C^2 solution of the minimal surface equation over $\mathbf{R}^n \sim \Omega$, where Ω is bounded and open in $\mathbf{R}^n$. Thus u satisfies

$$\sum_{i,j=1}^n \left(\delta_{ij} - \left(1 + |\operatorname{grad} u|^2\right)^{-1} (D_i u)(D_j u)\right) D_i D_j u = 0$$

in $\mathbf{R}^n \sim \Omega$, where $D_i = \partial/\partial x^i$.

An interesting question is: What can one say about the asymptotic behaviour of u as $|x| \to \infty$? In contrast to solutions of linear elliptic equations (e.g. the equation $\Delta u = 0$), it turns out that the possible behaviour is quite restricted. The first result in this direction was established by L. Bers [**BL**] in case $n = 2$. In this case $Du(x)$ automatically is bounded and has a limit as $|x| \to \infty$. Recently Bers' result was shown to also be true in case $3 \leqslant n \leqslant 7$ in [**SL3**]. Indeed for $3 \leqslant n \leqslant 7$ one gets the slightly stronger conclusion that there is an $a \in \mathbf{R}^n$ and a constant $c \in \mathbf{R}$ such that

$$(*) \qquad\qquad \lim_{\rho \uparrow \infty} |u - l|_{C^2(\mathbf{R}^n \sim B_\rho(0))} = 0,$$

where l is the affine function given by $l(x) = a \cdot x + c$. This is in fact a special case of a result established in [**SL3**] for arbitrary $n \geqslant 3$. Namely, if $n \geqslant 3$ then either there is an affine function l such that $(*)$ holds or else all tangent cones for graph u at ∞ (see [**SL3**, §2] for terminology) are vertical cylinders $C \times \mathbf{R}$, where C is an $(n - 1)$-dimensional multiplicity one minimizing cone in $\mathbf{R}^n$ with $0 \in \operatorname{sing} C$. In particular none of these "tangent cylinders at ∞" can be a hyperplane because $\operatorname{sing} C = \varnothing$ if C is a hyperplane. Since there are no minimizing cones C

with dimension$(C) \leqslant 6$ and sing $C \neq \varnothing$ (by the codimension 1 regularity theory —see e.g. [**SL5**, Chapter 7 or **FH**, Chapter 5]), the result $(*)$ for $3 \leqslant n \leqslant 7$ thus follows from this general result.

Notice that the above result also implies that (in arbitrary dimension n) if u is bounded on a set of rays of the form $\{ \lambda v\colon \lambda > \lambda_0, v \in K \}$ where $\lambda_0 \geqslant 0$ and K is a subset of S^{n-1} with $\mathscr{H}^{n-2}(K) > \sup \mathscr{H}^{n-2}(\mathrm{spt}\, C \cap S^{n-1})$, where the sup is taken over all multiplicity 1 minimizing cones C of dimension $n - 1$ in $\mathbf{R}^n$ (with $\partial C = 0$). This follows rather easily, because if $C \times \mathbf{R}$ is a tangent cylinder for graph u at ∞, then there is a sequence $\lambda_j \downarrow 0$ such that graph u_j $(u_j(x) = \lambda_j u(\lambda_j^{-1} x))$ converges weakly to $C \times \mathbf{R}$, and in particular $\lambda_j u(\lambda_j^{-1} v) \to \infty$ for any $v \in S^{n-1} \sim \mathrm{spt}\, C$, because $\mathscr{H}^n(\mathrm{graph}\, u_j \cap B_\rho(y)) \geqslant c\rho^n$ for $y \in \mathrm{graph}\, u_j$ and $\rho > 0$, where $c = c(n) > 0$.

A natural question to ask is whether or not tangent cylinders at ∞ are *unique*: notice that an affirmative answer to this question would give quite a good picture of the asymptotic behaviour of u at ∞. It is proved in [**SL4**] that if there is at least one tangent cylinder $C \times \mathbf{R}$ at infinity with C *regular* (i.e. sing $C = \{0\}$), then $C \times \mathbf{R}$ is the unique tangent cylinder at ∞. Since all 7-dimensional minimizing cones C in $\mathbf{R}^8$ are automatically regular (by the codimension 1 regularity theory) it follows that tangent cylinders at ∞ are always unique in case $n = 8$. Whether or not this is true in general for $n \geqslant 9$ is an open question.

Another question which has so far been only partially settled is the question of the *rate of growth* of solutions u as $|x| \to \infty$ in case $n \geqslant 8$. In [**BG**] it was suggested that all nonlinear *entire* solutions (i.e. solutions u as above with $\Omega = \varnothing$) could have growth bounded by some power of $|x|$ as $|x| \to \infty$. (This is sometimes referred to as the polynomial growth conjecture.) Indeed it seems reasonable to make the same conjecture for all solutions u which are C^2 on $\mathbf{R}^n \sim \Omega$, Ω bounded. Some progress has been made on this recently; in [**SL4**] it is proved that if u is C^2 on $\mathbf{R}^n \sim \Omega$, Ω bounded, and if one of the tangent cylinders $C \times \mathbf{R}$ at ∞ is such that C is *regular, strictly minimizing,* and *strictly stable,* then there are indeed positive c, γ such that $|u(x)| \leqslant c(1 + |x|^\gamma)$ for all $x \in \mathbf{R}^n \sim \Omega$. Using related estimates it is proved in [**SL4**] that there exist a rather large class of nonlinear entire solutions of the minimal surface equation. (The first examples of such solutions were given in [**BDG**], where it was proved that there is at least one entire solution defined over $\mathbf{R}^m \times \mathbf{R}^m$ for each $m \geqslant 4$.)

It is tempting to conjecture that for any regular minimizing cone C in $\mathbf{R}^n$ there is an entire solution u having $C \times \mathbf{R}$ as tangent cylinder at ∞. (In [**SL4**] this is shown to be true for each of the minimizing cones C in the list of Lawson [**L**].)

References

[**AA**] F. J. Almgren and W. K. Allard, *On the radial behaviour of minimal surfaces and the uniqueness of their tangent cones*, Ann. of Math. (2) **113** (1981), 215–265.

[**AD**] D. Adams (to be submitted as part of Australian National University Ph.D. thesis).

[**BL**] L. Bers, *Isolated singularities of minimal surfaces*, Ann. of Math. (2) **53** (1951), 364–386.

[**BG**] E. Bombieri and E. Giusti, *Harnack's inequality for elliptic differential equations on minimal surfaces*, Invent. Math. **15** (1972), 24–46.

[**BDG**] E. Bombieri, E. De Giorgi and E. Giusti, *Minimal cones and the Bernstein problem*, Invent. Math. **7** (1969), 243–268.

[**CHS**] L. Caffarelli, R. Hardt and L. Simon, *Minimal surfaces with isolated singularities*, Manuscripta Math. **48** (1984), 1–18.

[**FH**] H. Federer, *Geometric measure theory*, Springer-Verlag, 1969.

[**GG**] M. Giaquinta and E. Giusti, *The singular set of the minima of certain quadratic functionals*, Ann. Scuola Norm. Sup. Cl. Pisa (4) **11** (1984), 45–55.

[**HS**] R. Hardt and L. Simon, *Area minimizing hypersurfaces with isolated singularities*, J. Reine Angew. Math. (to appear).

[**L**] H. B. Lawson, *The equivariant Plateau problem and interior regularity*, Trans. Amer. Math. Soc. **173** (1972), 231–250.

[**SU**] R. Schoen and K. Uhlenbeck, *A regularity theory for harmonic maps*, J. Differential Geometry **17** (1982), 307–336.

[**SL1**] L. Simon, *Isolated singularities of extrema of geometric variational problems*, C.M.A. Research Report, 1984.

[**SL2**] ______, *Asymptotics for a class of non-linear evolution equations, with applications to geometric problems*, Ann. of Math. (2) **118** (1983), 525–571.

[**SL3**] ______, *Asymptotic behaviour of minimal graphs over exterior domains*, C.M.A. Research Report, 1984.

[**SL4**] ______, *Non-linear entire solutions of the minimal surface equation*, in preparation.

[**SL5**] ______, *Lectures on Geometric Measure Theory*, Proc. of the Centre for Mathematical Analysis, Vol. 3, Australian National University, 1983.

AUSTRALIAN NATIONAL UNIVERSITY, AUSTRALIA

Proceedings of Symposia in Pure Mathematics
Volume **44** (1986)

Complete Catalog of Minimizing Embedded
Crystalline Cones

JEAN E. TAYLOR[1]

Grain boundaries and surfaces of crystalline materials have a surface free energy which in general depends on the normal direction of the interface relative to the crystal lattice(s). Determining the surface energy minimizing configurations of such interfaces, for a given surface free energy function, is an interesting mathematical problem; it reduces in the case of isotropic (i.e. constant) surface energy to the minimal surface problem. A first step is to classify minimizing cones, since they can arise as tangent cones to minimizing or asymptotically minimizing surfaces. In the isotropic case for two-dimensional surfaces in $\mathbf{R}^3$, the only minimizing embedded cones are planes. For anisotropic surface energy functions, we give here a catalog of 12 types of embedded minimizing cones, and prove that it is a complete catalog among embedded minimizing crystalline cones (see below for definitions). This catalog is being published in the metallurgy literature as well [TC1], with an incomplete proof of completeness but with an emphasis on its new techniques, its place in the history of the subject and its predictions.

The surface free energy function Φ is a positive real-valued continuous function defined on the space of unit vectors in $\mathbf{R}^3$ (corresponding to oriented normal vectors to a surface). In physical applications, the temperature, pressure, and the orientation of each crystalline region are fixed. One of the regions is arbitrarily denoted by I (crystal) and the other by II (matrix); the interface is oriented so that the unit normal points from I to II. In such physical problems there are normally restrictions on possible Φ's due to symmetry groups and statistical-mechanical computations; mathematically, one can choose either to abide by these restraints or to ignore them. (Recent research on "quasicrystals" suggests that the classical symmetry constraints need not always be obeyed anyway [SBGC, LS].)

1980 *Mathematics Subject Classification.* Primary 49B34, 82A55, 82A60; Secondary 49F10, 49F22.
[1] This research was partially supported by NSF Grant MCS-8301869.

© 1986 American Mathematical Society
0082-0717/86 $1.00 + $.25 per page

To any such Φ there is associated a positively oriented convex body

$$W = \left\{ x \in \mathbf{R}^3: x \cdot n \leqslant \Phi(n) \text{ for each unit vector } n \right\},$$

the Wulff shape; it is the equilibrium shape of the crystal of fixed volume embedded in the matrix [**W**, **D**, **T1**]. Its central inversion WI is the equilibrium shape of the matrix of fixed volume embedded in the crystal and has all its normal vectors pointing inwards. (Note: if Φ is extended by positive homogeneity of degree 1 to a function on $\mathbf{R}^3$ and if it is then a convex functional, it is the support function of the convex body W.)

We consider here a type of prescribed boundary problem. Physically, one can think of prescribing a boundary curve as a way of isolating part of a larger interface. This way of regarding a prescribed boundary curve is particularly apt for the description of local features of surfaces, such as edges and corners. Minimizing cones can also arise as tangent cones to surfaces which are minimizing or solve other variational problems.

We catalog here all the embedded minimizing cones which are composed of a finite number of planar wedges whose normals are normals of W. More precisely, let **D** be the unit disk $\{x \in \mathbf{R}^2: |x| \leqslant 1\}$; we classify the continuous maps s: $\mathbf{D} \to \mathbf{R}^3$ with $s(\partial \mathbf{D}) \subset \mathbf{S}^2$ satisfying (i)–(iv):

(i) s is positively homogeneous of degree 1 (i.e. $s(r, \theta) = rs(1, \theta)$) for all points in the disk (given by polar coordinates (r, θ)). A mapping with this property is called a *cone*.

(ii) For some smallest positive integer K, on each open interval

$$(2\pi(k - 1)/K, 2\pi k/K) \qquad (k = 1, \ldots, K)$$

of θ's, $s(1, \cdot)$ is parametrized proportional to arc length, and the oriented unit normal of the embedding $s|[(0, 1) \times (2\pi(k - 1)/K, 2\pi k/K)]$ is constant and is a normal of W or a limit of normals of W (so that s maps the disk onto a distinguished collection of oriented planar wedges). Such a mapping is called *crystalline*.

(iii) s has Properties (1)–(2) of §4; we show there that any minimizing surface must have these properties wherever it is locally representable by a continuous mapping having properties (i) and (ii) above. Such a mapping is called *locally minimizing*.

(iv) s is an *embedding*.

We show that there are up to 12 types of such cones (depending on the nature of W); by means of barriers, we show that each of these cones is in fact minimizing, both as a mapping taking a prescribed boundary and as an integral current. Thus we have classified the embedded minimizing crystalline cones. *This analysis is not restricted to polyhedral W. In particular, the n-diagram defined below need not have a finite number of elements.*

These cones when "saddle-shaped" may have more than three wedges meeting at the center, even if only three smooth parts of ∂W meet at any corner of W. All of these cones remain minimizing under perturbations in Φ which do not affect the n-diagram. Some cones provide examples disproving several previous expectations in metallurgy; this is discussed at length in [**CT** and **TC1**].

The partial differential equation approach is of little use here, since all the derivatives in the weighted mean curvature formula are either zero or undefined. Rather, the primary technique is the use of "labelled cycles" on the sphere $\mathbf{S}^2$ to describe the cones (see §1). The labelled cycle is sort of a generalized Gauss map of the outer part of the corresponding cone, specifying its normal directions and its "curvature"; one of the mappings of the disk $\mathbf{D}$ into $\mathbf{S}^2$ with $\partial\mathbf{D}$ mapping to the edges of the labelled cycle then gives the generalized Gauss map of the center of the cone. (Which equivalence class of mappings to choose will be obvious for each cone of the catalog, though not in general for nonembedded cones.) There is a natural way to associate to each vertex $n(k)$ of the labelled cycle an oriented wedge $P(k)$ of the disk with normal $n(k)$, such that the terminal ray of $P(k)$ is the initial ray of $P(k + 1)$. By defining a positively homogeneous homeomorphism of $(0, 1] \times [2\pi(k - 1)/K, 2\pi k/K]$ onto $P(k)$, we will construct a continuous mapping which has properties (i)–(ii) above and Property 1 of §4; since crystalline cones with Property 1 define labelled cycles, we thus in fact classify the labelled cycles with Property 2 of §4 that give rise to embedded cones.

We do not consider minimizing cones having other normals. If Φ is sufficiently nonconvex, then any minimizing piecewise smooth cone with other normals exists only as an "infinitesimally corrugated" surface, i.e. varifold, not as a classical surface [**T1**].

This catalog considers only cones which are topologically disks. There are additional features of minimizing surfaces such as cusps in interfaces that are not reflected in their tangent cone behavior [**TC2**]. Other important questions remain open, such as the structure of junctions of interfaces in polyphase and polycrystalline materials, analogous to the singularities in soap films and compound soap bubbles.

The author would like to thank Dr. John Cahn for providing the impetus to complete this catalog and Professor F. J. Almgren, Jr., for numerous helpful suggestions on its exposition.

1. Descriptive tools.

1.1. *The n-Diagram.* Edges and corners of W are particularly important. (The tangent cones at such points are automatically minimizing [**T1**] and so can be used as parts of comparison surfaces or barriers as in §3 below.) For this reason, it is convenient to describe the Wulff shape in terms of an n-diagram. The n-diagram is a decomposition of the unit sphere $\mathbf{S}^2$ into various sets induced by the generalized Gauss map of ∂W.

1.1.1. *Definitions.* Following [**F**, 3.1.21], we define $\mathrm{Nor}(W, a)$, the normal cone of W at a, for any a in $\mathbf{R}^3$ as

$$\mathrm{Nor}(W, a) = \mathbf{R}^3 \cap \left\{ u\colon u \cdot v \leqslant 0 \text{ for every } v \in \mathrm{Tan}(W, a) \text{ with } |v| = 1 \right\},$$

where

$$\mathrm{Tan}(W, a) \cap \left\{ v\colon |v| = 1 \right\} = \bigcap_{\varepsilon > 0} \mathrm{Clos}\left\{ \frac{x - a}{|x - a|} \colon a \neq x \in W \text{ and } |x - a| < \varepsilon \right\}.$$

$\mathrm{Nor}(W, a)$ is always closed and convex [**F**, 3.1.21]. We consider $\mathrm{Nor}(W, a) \cap \mathbf{S}^2$ for a's which are in ∂W. If $\partial\,\mathrm{Tan}(W, a)$ is a plane, then $\mathrm{Nor}(W, a) \cap \mathbf{S}^2$ is a single point. If $\partial\,\mathrm{Tan}(W, a)$ consists of two halfplanes, then $\mathrm{Nor}(W, a) \cap \mathbf{S}^2$ is the geodesic connecting the (oriented) normals of the halfplanes. If $\mathrm{Nor}(W, a) \cap \mathbf{S}^2$ is neither a point nor a single geodesic, then W has a corner at a, and $\mathrm{Nor}(W, a) \cap \mathbf{S}^2$ is two-dimensional and is bounded by the normals to $\mathrm{Tan}(W, a)$ at points which have tangent planes, and by the shortest geodesics connecting these normal vectors. We let

$$N^0 = \mathrm{Clos}\left\{ v \in \mathbf{S}^2 \colon \{v\} = \mathrm{Nor}(W, a) \cap \mathbf{S}^2 \text{ for some } a \in \partial W \right\},$$

$$N^1 \doteq \mathrm{Clos}\left\{ C\colon C = \mathrm{Nor}(W, a) \cap \mathbf{S}^2, \text{ some } a \in \partial W, \text{ and } \dim(C) = 1 \right\}$$
$$\sim \left\{ C\colon \dim(C) = 0 \right\}$$

(here the closure is in the Hausdorff distance on compact sets [**F**, 2.10.21]),

$$N^2 = \left\{ \mathbf{S}^2 \cap \mathrm{Nor}(W, a)\colon a \in \partial W \text{ and } \dim(\mathbf{S}^2 \cap \mathrm{Nor}(W, a)) = 2 \right\}.$$

We call the elements of N^0 the *vertices* of the *n*-diagram, the elements of N^1 the *tie lines* of the *n*-diagram, and the elements of N^2 the *tie regions* of the *n*-diagram.

If W is itself a polyhedron (in which case Φ will be called *crystalline*) then the *n*-diagram contains only isolated vertices, each representing a facet of W, connected by isolated tie lines and tie polygons, representing respectively edges and corners of W. The *n*-diagram was introduced and called the *crystal graph* in that case in [**T2**]. It was introduced in general independently in [**C**]. A polyhedral W and its corresponding *n*-diagram are shown in Figure 1. On the other hand, if W is a right circular cone, its *n*-diagram consists of one tie region (corresponding to the apex of the cone), a circle of vertices (the normals of the curved part of W), one isolated vertex (the normal to the base of the cone), and a continuum of tie lines, each joining a vertex on the circle to the isolated vertex. There are also W such that there are tie lines but no tie regions, or tie regions but no tie lines, or a countable dense set of tie regions.

1.1.2. *Properties of the n-diagram.* These statements follow from the definitions above and the convexity of W.

(1) Each tie region is convex within $\mathbf{S}^2$ and is contained in an open hemisphere.

(2) No interior point of one tie line is a point of any other tie line.

(3) The endpoints of each tie line are vertices. (There are examples of W's such that all interior points of a tie line are also vertices.)

(4) Let $R = \mathrm{Nor}(W, a) \cap \mathbf{S}^2$ be in N^2. Then the oriented unit normals to $\mathrm{Tan}(W, a)$ at points of $\mathrm{Tan}(W, a)$ which have tangent planes are in N^0, and the great circle segments (if any) in ∂R are tie lines.

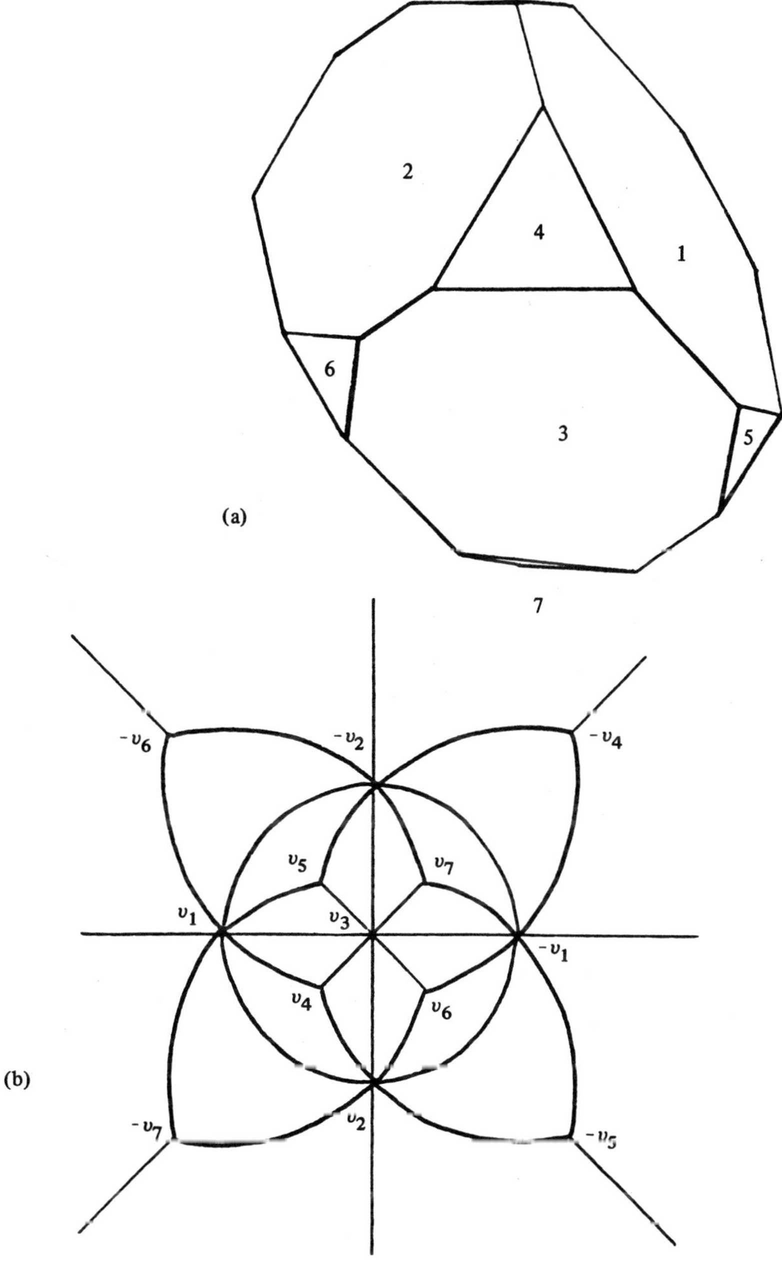

FIGURE 1. (a) a Wulff shape W. (b) stereographic projection from the direction $-v_3$ of the n-diagram corresponding to this W.

(5) If $n \in N^0$ then $\Phi(n) = x \cdot n$ for some $x \in \partial W$; however, if Φ is sufficiently nonconvex, $\Phi(u) > x \cdot u$ for all $x \in \partial W$ and $u \in \mathbf{S}^2 \sim N^0$.

1.2. *Labelled cycles and corresponding cones, angles, etc.* If all the oriented unit normals of a minimizing cone are vertices of the *n*-diagram, it is convenient to describe the cone in terms of a "labelled cycle on the *n*-diagram"; such labelled cycles are the fundamental objects of this paper.

1.2.1. *Labelled cycles.* Suppose K is a positive integer. By a *labelled cycle of length K* we mean a pair of functions

$$n: \{0, 1, \ldots, K\} \to N^0, \qquad \mathrm{Lbl}: \{(k-1, k): k = 1, \ldots, K\} \to \{-1, 1\},$$

such that $n(0) = n(K)$ and for each $k = 1, \ldots, K$, $n(k-1)$ and $n(k)$ are both contained in one element of N^1 or of N^2.

For each $k = 1, \ldots, K$, $n(k)$ is called a *vertex* of the labelled cycle. Since each element of N^1 and N^2 is contained in an open hemisphere, there is a unique shortest geodesic with endpoints $n(k-1)$ and $n(k)$; this geodesic, oriented to go from $n(k-1)$ to $n(k)$, is denoted $e(k-1, k)$, and is called an *edge* of the cycle. $\mathrm{Lbl}(k-1, k)$ is called the *label* of $e(k-1, k)$; when $\mathrm{Lbl}(k-1, k) = 1$ the edge is called *regular* and otherwise it is called *inverse*. See the discussion at the end of 1.2.4 for the relevance of the terms regular and inverse.

We will always work "modulo K" when dealing with indices of a labelled cycle of length K. In particular, $e(i, i+1) = e(0, 1)$ when $i = K$. We also do not distinguish between labelled cycles (n, Lbl) and (n', Lbl') which differ only by a cyclic permutation of all their indices (*i.e. such that $n'(i) = n(i + j)$ and $\mathrm{Lbl}'(i, i+1) = \mathrm{Lbl}(i + j, i + j + 1)$ for all i and some j*).

Finally, when we say two edges *cross* we always mean *transversally on their interiors* (and that can only happen inside tie regions).

1.2.2. *Labelled fragments.* In the proofs to follow, it will be useful to localize and extend the idea of a labelled cycle. We thus define a *labelled fragment* to be a pair of functions

$$n: \{0, 1, 2\} \to \mathbf{S}^2, \qquad \mathrm{Lbl}: \{(k-1, k): k = 1, 2\} \to \{-1, 1\},$$

with only the restriction that $n(k) \neq \pm n(k-1)$ for $k = 1, 2$. Again, we let the shortest geodesic with endpoints $n(k-1)$ and $n(k)$, oriented to go from $n(k-1)$ to $n(k)$, be denoted $e(k-1, k)$, and call it an *edge* of the fragment; it need not be contained in an element of N^1 or N^2. A labelled fragment is understood to have $K = 2$ in what follows.

1.2.3. *Points, etc.* Associated with each labelled cycle of length K or any labelled fragment (n, Lbl) and each $k = 1, \ldots, K$ are

$$\mathrm{Pole}(k-1, k) = n(k-1) \times n(k) / |n(k-1) \times n(k)| \in \mathbf{S}^2,$$

$$\mathrm{Point}(k-1, k) = -\mathrm{Lbl}(k-1, k)\mathrm{Pole}(k-1, k) \in \mathbf{S}^2,$$

$$\mathrm{Ray}(k-1, k) = \{t\,\mathrm{Point}(k-1, k): 0 \leqslant t \leqslant 1\},$$

and

$$E(k; k - 1) \quad [\text{resp.,} \ E(k - 1; k)],$$

defined to be the limiting unit tangent vector at $n(k)$ [resp., $n(k - 1)$] to $e(k - 1, k)$.

Note that for each $k = 1, \ldots, K$,

$$\text{Pole}(k - 1, k) = n(k) \times E(k; k - 1) = n(k - 1) \times E(k - 1; k).$$

1.2.4. *Wedges and cones.* Associated with a labelled cycle of length 1 is the embedded oriented disk with deleted origin $P(1) = n(1)^{\perp} \sim \{O\}$, where $n^{\perp}$ denotes the oriented unit disk centered at the origin whose oriented unit norm is n. One can choose any embedding which is positively homogeneous and parametrized by arc length around the boundary.

Associated with each labelled cycle (n, Lbl) of length K is the sequence of oriented wedges

$$P(n, \text{Lbl}; k) \equiv P(k): \qquad k = 1, \ldots, K,$$

and associated with each labelled fragment is the oriented wedge

$$P(n, \text{Lbl}; k) \equiv P(k): \qquad k = 1,$$

defined by requiring $P(k)$ to be the part of $n(k)^{\perp}$ obtained by rotating $\text{Ray}(k - 1, k)$ into $\text{Ray}(k, k + 1)$ in the positive (counterclockwise) direction and then deleting the origin. In case $\text{Ray}(k - 1, k) = \text{Ray}(k, k + 1)$ we rotate through the angle 2π rather than 0. $P(k)$ is called the *wedge* of the cone coresponding to $n(k)$. We define $I_k = [2\pi(k - 1)/K, 2\pi k/K]$, and define

$$s_k: (0, 1] \times I_k \to \mathbf{R}^3$$

to be the positively homogeneous homeomorphism onto $P(k)$ which induces the correct orientation, is parametrized proportional to arc length along $\{1\} \times I_k$, and satisfies $s(r, \theta) = rs(1, \theta)$ for all (r, θ).

In case (n, Lbl) is a labelled cycle, we also let

$$s(n, \text{Lbl}) \equiv s: \mathbf{D} \to \mathbf{R}^3$$

be the positively homogeneous mapping of the disk which is given (in polar coordinates) by s_k on the domain of s_k for $k = 1, \ldots, K$.

Note that the sequence of edges $e(k - 1, k)$, $k = 1, \ldots, K$, gives a generalized Gauss map of $s|(r_0, 1] \times [0, 2\pi]$ for any $0 < r_0 < 1$.

Note that if $\text{Lbl}(k - 1, k) = 1$, then $P(k)$ bends down from $P(k - 1)$ along $\text{Ray}(k - 1, k)$ (i.e. the exterior angle of $s|(I_{k-1} \cup I_k)$ in the oriented tangent plane to $\mathbf{S}^2$ at $\text{Point}(k - 1, k)$ is negative), whereas if $\text{Lbl}(k - 1, k)$ is -1, then $P(k)$ bends up from $P(k - 1)$. The terminology of "regular" and "inverse" was adopted since the property of bending down is a property of tangent cones to W and the property of bending up is a property of tangent cones to WI, the central inversion of W with negative orientation (so that its exterior normals point inwards).

1.2.5. *Angles.* Associated to each labelled cycle of length $K \geq 2$ are the

$$\text{edge angle function } \theta \colon \{0, \ldots, K-1\} \to (-\pi, \pi]$$

and the

$$\text{wedge angle function } \gamma \colon \{0, \ldots, K-1\} \to (0, 2\pi]$$

defined by requiring $\theta(k)$ to be the exterior angle at $n(k)$ of the labelled cycle (i.e. the angle from $E(k; k-1)$ to $E(k; k+1)$ in $n(k)^{\perp}$), and by requiring γ to be the central angle of $P(k)$. Note that θ does not depend on the edge labels, and that γ is unchanged when all the edge labels are multiplied by -1.

The edge angle $\theta(1)$ and the wedge angle $\gamma(1)$ are similarly defined for any labelled fragment. Additionally, given a labelled cycle (n, Lbl), we denote by $\gamma_F(i, j; v)$, the wedge angle for a labelled fragment (n_F, Lbl_F) with vertices $n_F(0) = n(i)$, $n_F(1) = n(j)$, $n_F(2) = v$, $\mathrm{Lbl}_F(0, 1) = \mathrm{Lbl}(j-1, j) \neq \mathrm{Lbl}_F(1, 2)$, and by $\gamma_F(v; j, k)$, the wedge angle for a labelled fragment with $n(0) = v$, $n(1) = n(j)$, $n(2) = n(k)$, but with $\mathrm{Lbl}_F(1, 2) = \mathrm{Lbl}(j, j+1) \neq \mathrm{Lbl}_F(0, 1)$. We denote by

$$\gamma_G(i, j; v) \quad \text{and} \quad \gamma_G(v; j, k),$$

the wedge angles for the labelled fragments defined as above except that each " $\neq$ " is replaced by " $=$ " (so that all the labels are equal). We denote by $\theta(i, j, v)$ and $\theta(v, j, k)$ the edge angles for the appropriate labelled fragments.

1.2.6. *Convex, even, etc.* In view of the following proposition, the precise values of $\mathrm{Lbl}(k-1, k)$ and $\mathrm{Lbl}(k, k+1)$ are often not as important as whether they are the same or different. We therefore say that $n(k)$ and $P(k)$ are *even* if these labels are the same and that they are *odd* if the labels are different. We also define $n(k)$ and $P(k)$ to be *convex* if $\gamma \leq \pi$ and to be *nonconvex* otherwise.

1.3. PROPOSITION. *Suppose (n, Lbl) is a labelled cycle of length K and $k \in \{1, \ldots, K\}$, or that (n, Lbl) is a labelled fragment and $k = 1$.*

(1) If $n(k)$ is odd, then $\gamma(k) = \theta(k) + \pi$.

(2) If $n(k)$ is even, then $\gamma(k) = \theta(k) + 2\pi$ in case $\theta(k) \leq 0$ (here $P(k)$ is nonconvex), and $\gamma(k) = \theta(k)$ in case $\theta(k) > 0$ (here $P(k)$ is convex).

PROOF. By the definitions of $\gamma(k)$ and $P(k)$, $\gamma(k)$ is the angle from $\mathrm{Point}(k-1, k)$ to $\mathrm{Point}(k, k+1)$. From 1.2.3 we obtain

$$\mathrm{Lbl}(k-1, k)\mathrm{Point}(k-1, k) \times E(k; k-1) = n(k)$$

$$= \mathrm{Lbl}(k, K+1)\mathrm{Point}(k; k+1) \times E(k; k+1),$$

so that when $n(k)$ is odd, the angle from $\mathrm{Point}(k-1, k)$ to $\mathrm{Point}(k, k+1)$ equals π plus the angle from $E(k; k-1)$ to $E(k; k+1)$, whereas if $n(k)$ is even, then one angle differs from the other by 0 or 2π; whether one adds 2π or not is dictated by the ranges of θ and γ.

2. The catalog. We give here a catlog $\mathscr{C}$ of twelve types of labelled cycles (or more precisely, of ten types of cycles and two types of modifications of these ten types). In the following sections, we will prove first that all cones in the catalog

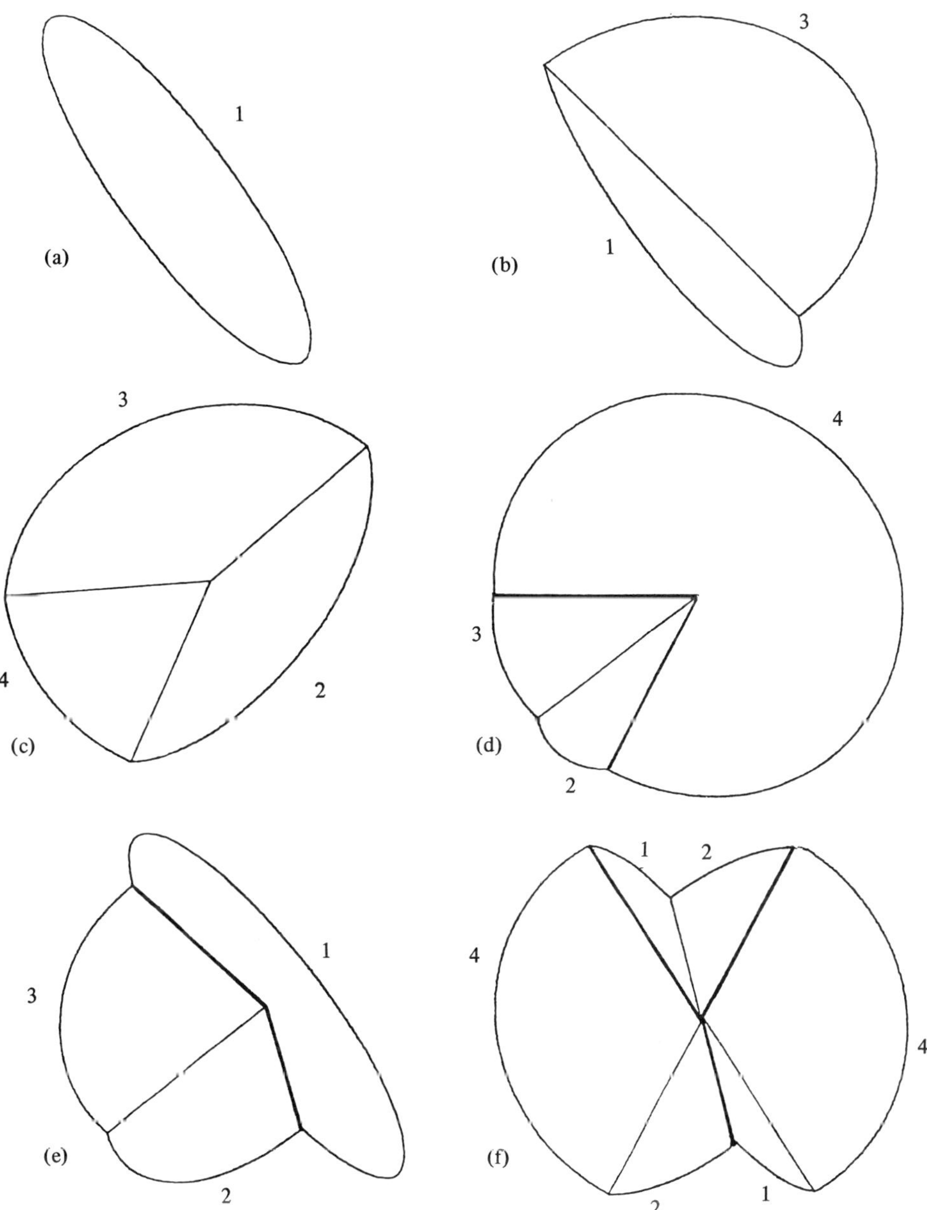

FIGURE 2. (a) through (1) are examples of cones of types (1) through (12). Cones (a) through (k) are minimizing for the W shown in Figure 1, but (1) requires a different W such as the octahedron where planes 4, 5, 6, and 7 meet at a single corner.

are minimizing and embedded, and then that every embedded minimizing crystal-line cone is defined by a labelled cycle in the catalog. Examples of cones corresponding to types (1)–(12) are shown in Figure 2(a)–(l) respectively.

For ease of description, we sometimes specify the indices, but (as stated in 1.2.1) we do not distinguish between labelled cycles which differ only by a cyclic permutation of all their indices.

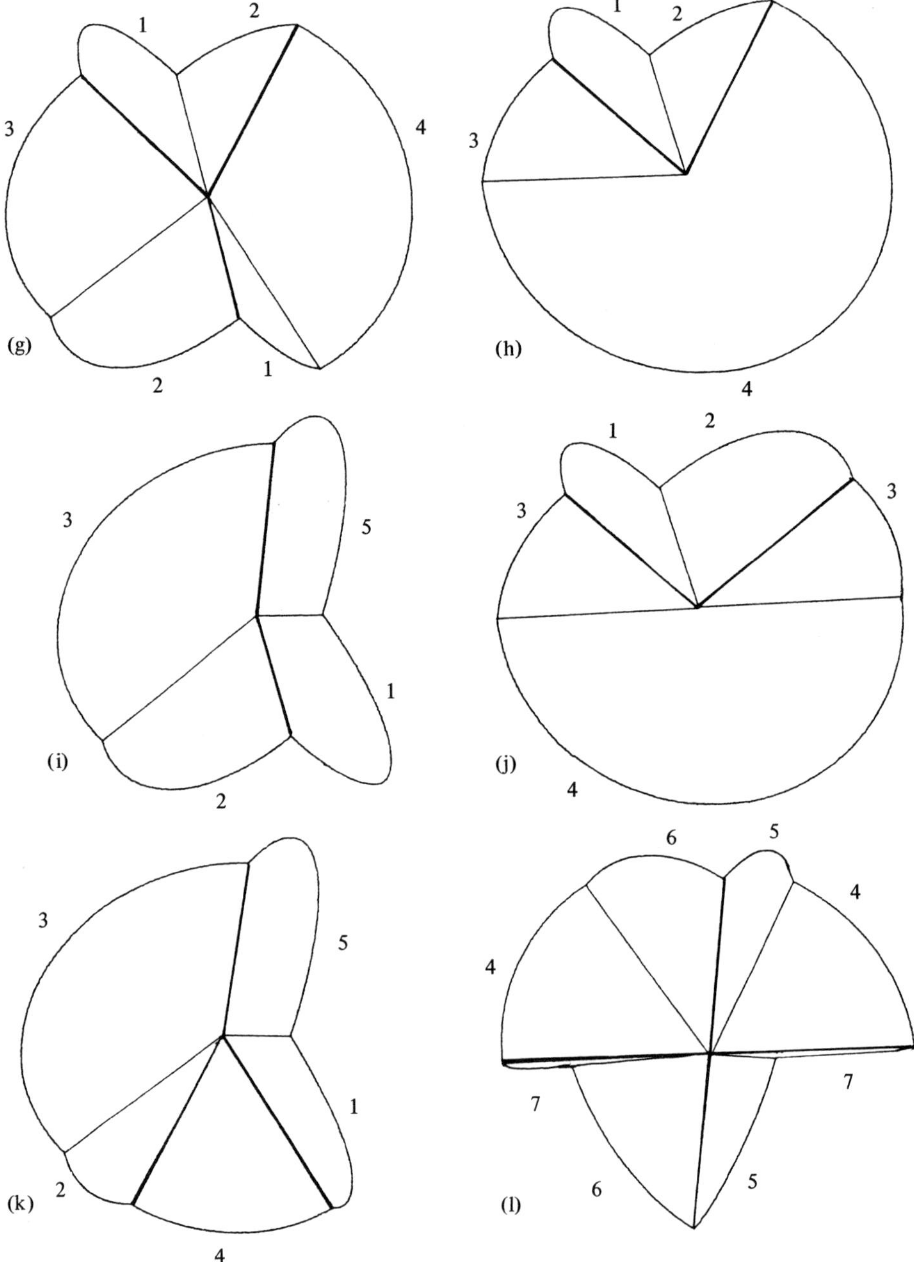

2.1.$\mathscr{C}$. (1) Any labelled cycle of length 1 belongs to $\mathscr{C}$.

(2) Any labelled cycle of length 2, with $\mathrm{Lbl}(0,1) = \mathrm{Lbl}(1,2)$ (so that $e(0,1)$ is $e(1,2)$ with the opposite orientation), belongs to $\mathscr{C}$.

(3) Any labelled cycle of length K, whose edge labels are all equal and whose vertices are distinct vertices of a single tie region, ordered so that $\theta(i) > 0$ for each i and the edges of the labelled cycle form a simple closed curve, belongs to $\mathscr{C}$. Thus the vertices are taken in the positive (counterclockwise) direction around the tie region and are all convex even.

(4) Any labelled cycle of length 3 whose vertices are in a single tie region, ordered so that each $\theta(i) < 0$ and labelled so that two vertices are odd (and hence convex) and one even (and hence nonconvex), belongs to $\mathscr{C}$.

(5) Any other labelled cycle of length 3 whose vertices are in one open hemisphere of the n-diagram, ordered so that each $\theta(i) < 0$ and labelled so that two vertices are odd and one is even, belongs to $\mathscr{C}$. (Cycles of length 3 whose vertices are not in a single tie region occur whenever W has a triangular facet with straight edges.) More generally they occur whenever ∂W contains a triangle not in a single facet nor bounding three triangular wedges meeting at a single corner.)

(6) Any labelled cycle of length 6 with $n(i) = n(i + 3)$, $i = 0, 1, 2$, and all $n(i)$ in one open hemisphere, ordered so that each $\theta(i) < 0$ and labelled so that edge labels alternate between $+1$ and -1, belongs to $\mathscr{C}$.

(7) Any labelled cycle of length 6, with $n(1) = n(4)$, and all vertices contained in an open hemisphere, ordered so that $\theta(i) < 0$ if $i \neq 1$ and labelled so that edge labels alternate, belongs to $\mathscr{C}$.

(8) Any labelled cycle of length 4 with edge labels alternating belongs to $\mathscr{C}$ if $n(1)$ and $n(3)$ are endpoints of a tie line and $\theta(2), \theta(4) < 0$.

(9) Any other labelled cycle of length 4, with edge labels alternating, belongs to $\mathscr{C}$ if $\theta(2), \theta(4) < 0$.

(10) Any labelled cycle of length 5, with $n(0) = n(2)$, $n(1)$ even, and $n(i)$ convex odd for $i \neq 1$, belongs to $\mathscr{C}$.

(11) Suppose (n, Lbl) is a labelled cycle of type (4)–(12) (having perhaps been previously modified as indicated here or below) and for some i, some $R \in N^2$ containing $e(i - 1, i)$, and some vertex $v \in R \sim e(i - 1, i)$, either $\theta(i - 2, i - 1, v) \leqslant \theta(i - 1)$ or $\theta(i - 2, i - 1, v) = \pi$ and $P(i - 1)$ is nonconvex, and either $\theta(v, i, i + 1) \leqslant \theta(i)$ or $\theta(v, i, i + 1) = \pi$ and $P(i)$ is nonconvex. Then the labelled cycle (n', Lbl') belongs to $\mathscr{C}$, where $\mathrm{Lbl}'(k - 1, k) = \mathrm{Lbl}(k - 1, k)$ for $k = 1, \ldots, i$, $\mathrm{Lbl}'(k, k + 1) = \mathrm{Lbl}(k, k - 1)$ for $k = i, \ldots, K$, $n'(k) = n(k)$ for $k = 0, \ldots, i - 1$; $n'(k + 1) = n(k)$ for $k = i, \ldots, K$, and $n'(i) = v$.

(12) Let (n, Lbl) be a labelled cycle of type (1)–(12) (having perhaps been previously modified). Suppose that for some i and some $R \in N^2$ there are two vertices v_1, v_2, in $R \sim e(i - 1, i)$ with the properties that $\theta(v_2, i, i + 1) < \theta(i)$, $\theta(i - 2, i - 1, v_1) < \theta(i - 1)$, and $\theta(i - 1, i, v_2) < \theta(i - 1, i, v_1) < 0$ (i.e. the $n(i - 1) - v_1$ and $n(i) - v_2$ geodesics in R cross); if $n(i - 1)$ [resp. $n(i)$] is convex

even, then we also require that $n(i - 2)$ [resp., $n(i + 1)$] be in R. Then the labelled cycle (n', Lbl') where $n'(k) = n(k)$, $k = 0, \ldots, i - 1$, $n'(i) = v_1$, $n'(i + 1) = v_2$, $n'(k + 2) = n(k)$, $k = i, \ldots, K$, $\text{Lbl}'(k - 1, k) = \text{Lbl}(k - 1, k)$ for $k = 1, \ldots, i - 1$, $\text{Lbl}'(k + 1, k + 2) = \text{Lbl}(k - 1, k)$ for $k = i + 1, \ldots, K$, belongs to $\mathscr{C}$ provided $\text{Lbl}'(i - 1, i)$ and $\text{Lbl}'(i, i + 1)$ are chosen so that the number of nonconvex vertices is not increased.

2.2. *Comments*. Note that not all types of labelled cycle may be possible for any given n-diagram (a trivial example is that when there is a tangent plane to each point on ∂W, only labelled cycles of type (1) are possible).

Distinctions are made between type (4) and (5) cycles because type (4) cycles exist whenever there is a tie region, but for there to be type (5) cycles requires special structure in the n-diagram (also, type (5) cones provide counterexamples to expectations previously held in metallurgy [**CT**]). Similarly, type (6) cycles can be derived from any type (4) or (5) cycle, and type (8) cycles exist whenever two tie regions share a tie line; type (7) and (9) cycles require more structure in the n-diagram. Type (10) cycles are also philosophically akin to type (11) modifications of type (4) or (5) cycles, in that they both involve inserting convex even vertices. Thus there are in some sense only four types of minimizing embedded crystalline cones, those of type (5), (7), (9) (all of which are saddle-shaped) and the "convex" ones (types (1)–(3)), together with certain modifications of these basic types.

In case W is a polyhedron where only three facets meet at any corner, all edges of a labelled cycle have to be oriented tie lines; furthermore, there are at most 3 vertices in type 3 cycles and no type (11) modifications to edges previously modified. There are no type (12) cycles and no type (11) modifications of cycles of type (4) or (8). Thus there are at most 12 vertices in any labelled cycle of the catalog, and there are a finite number of such labelled cycles. But if the n-diagram contains a tie region that is not a tie polygon (e.g. the tie region corresponding to the apex of W if W is a circular cone), there is no upper bound to the number of vertices in a type (3), (11), or (12) labelled cycle. Thus for such W, the number of wedges in a cone in this catalog need have no upper bound and the number of labelled cycles (of these and other types) in the catalog is infinite. However, if there is an upper bound to the number of vertices in the boundary of each tie region, then there is an upper bound to the number of vertices in any labelled cycle in the catalog, and if additionally there are a finite number of vertices in N^0 then there are a finite number of cycles in the whole catalog.

3. Proof that cones in the catalog are embedded and minimizing. Cones of types (1) and (2) are single planes and two halfplanes with a common diameter respectively and hence are embedded disks. For labelled cycles of type (3), one notes that all normals $n(i)$ are in a single open hemisphere and all points $\text{Point}(i - 1, i)$ are in a single open hemisphere (the same one if all edges are labelled inverse, the opposite one if all are labelled regular). Thus the mapping s

can fail to be an embedding only if it has a generalized branch point at the origin. But the edges of the labelled cycle form a simple closed curve, so there can be no such branch point.

Cones of types (4) and (5) consist of one wedge with central angle greater than π and two wedges with central angle less than π, and are hence embedded. If (n, Lbl) is a labelled cycle of type (6) or (7), then for each $k = 2, \ldots, 6$, $n(k)$ is odd and $\theta(k)$ is negative, so $\gamma < \pi$ and $P(k - 1)$ intersects $P(k)$ only along $\text{Ray}(k - 1, k)$. If $\text{Lbl}(1, 2) = 1$, then $\text{Lbl}(3, 4) = 1$ and $\text{Lbl}(4, 5) = \text{Lbl}(0, 1) = -1$; otherwise, all signs are the opposite of these. Thus either $P(2)$ and $P(3)$ are above $n(1)^{\perp}$, and $P(4)$ and $P(5)$ are below it, or vice versa; thus $P(2)$ and $P(3)$ cannot intersect $P(4)$ or $P(5)$ and vice versa.

Cones of types (8) and (9) consist of four odd wedges, with $\gamma(2) < \pi$ and $\gamma(4) < \pi$. Thus $P(k - 1)$ intersects $P(k)$ only along $\text{Ray}(k - 1, k)$, for each $k = 1, 2, 3, 4$. Since $P(1)$ is odd, either $P(2)$ lies above $P(1)$ and $P(4)$ lies below $P(1)$, or vice versa; assume the former. Thus $P(2)$ and $P(4)$ do not intersect. $P(3)$ crosses $n(1)^{\perp}$ in one ray, since $\text{Ray}(2, 3)$ and $\text{Ray}(3, 4)$ are on opposite sides of $n(1)^{\perp}$. This ray is below the plane of $n(2)^{\perp}$ and thus in the half disk on the other side of $\text{Ray}(0, 1)$ from $P(1)$. Similarly it is above the plane of $n(4)^{\perp}$ and thus on the other side of $\text{Ray}(1, 2)$ from $P(1)$. Therefore $P(3)$ does not intersect $P(1)$.

Labelled cycles of type (10) have $\gamma(1) = \pi$ and $\gamma(k) < \pi$ for $k = 0, 2, 3, 4$. Furthermore, either $P(1)$ is above $n(0)^{\perp}$ and $P(3)$ and $P(4)$ are below it, or vice versa. Therefore such cones are embedded.

Note that cones of types (1)–(7) and (10) project 1-to-1 onto a subset of a plane.

Suppose (n, Lbl) is a labelled cycle of length K of one of the types (4)–(10) (so $K = 3, 4, 5$, or 6). If we modify its cycle as described in type (11) to produce a new labelled cycle (n', Lbl'), we insert a convex even vertex, which we may assume is $n'(K)$, leaving $n'(i) = n(i)$, and $\text{Lbl}'(i - 1, i) = \text{Lbl}(i - 1, i)$ for $i = 0, \ldots, K - 1$. Note that by the definition of a type (11) cone,

$$0 < \gamma(n', \text{Lbl}'; K - 1) < \gamma(n, \text{Lbl}; K - 1), \qquad 0 < \gamma(n', \text{Lbl}'; 0) < \gamma(n, \text{Lbl}; 0).$$

Thus the cone corresponding to (n', Lbl') is formed from that corresponding to (n, Lbl) by truncating the two wedges containing $\text{Ray}(K - 1, K)$ by that convex even wedge. If $s(n, \text{Lbl})$ projects 1-1, therefore, so does $s(n', \text{Lbl}')$, whereas if (n, Lbl) is type (8) or (9), then the new wedge is in an entirely different region of the ball from $P(1), \ldots, P(K - 2)$ (as was shown in the proof that type (8) and (9) cycles are embedded).

Suppose (n, Lbl) is a labelled cycle of length K of type (1)–(11). A type (12) modification inserts two vertices v_1 and v_2 (which we may assume are $n'(1)$ and $n'(2)$) to produce a labelled cycle (n', Lbl') of length $K + 2$. Again, we see that

$$\gamma(n', \text{Lbl}'; 0) < \gamma(n, \text{Lbl}; 0) \quad \text{and} \quad \gamma(n', \text{Lbl}'; 3) < \gamma(n, \text{Lbl}; 1).$$

Since only these angles are changed and the number of nonconvex vertices is not increased, type (12) modifications of cones which project 1-1 yield cones which project 1-1 and hence are embedded. Type (12) modifications to cones of types (8) and (9) are embedded because all the vertices of such a modification are odd (including the new ones) and the new wedges are again in an entirely different region of the ball from the other wedges of (n, Lbl) as was shown in the proof of embeddedness of type (8) and (9) cones.

Cones of type (1) are uniquely minimizing because they are tangent planes to W or limits of tangent planes to W [**T1**]. The cones of type (2) that are tangent cones to W or WI are uniquely minimizing, as can be seen by translating the origin of W to a point on ∂W having that tangent cone; by Stokes' theorem (as used in [**T3**]), such a translation of the origin does not affect any surface energy comparisons, and a 2-wedge tangent cone is easily seen to be uniquely minimizing when F is zero on both plane directions. (This method of proof clarifies, extends, and simplifies the proof given in [**T1**].) Now uniquely minimizing surfaces may be used as barriers in a proof that another embedded surface S is uniquely minimizing, by showing that every point of S can be contacted from front and back by some of those uniquely minimizing surfaces. Furthermore, if every point of S can be contacted on both sides by the use of known minimizing surfaces, then S is minimizing, though perhaps not uniquely so. Although it is a slight abuse of terminology, we will call the latter type of comparison surfaces "barriers" as well. Finally, not every point need be contacted from both sides: if one can show (using barriers) that all but certain subsets can be so contacted, and that the surface contains the boundaries of those subsets, and that there can be no "tubes" connecting different such subsets, then one can use the fact that those subsets are known *a priori* to be minimizing to conclude that the whole cone is minimizing.

Cones of types (1), (2), and (3) are all minimizing, as can be shown by using only planes and tangent cones to W as barriers. These cones themselves then turn out to give a large enough class of barriers to prove that all other surfaces in the catalog are minimizing. Corresponding to any nonconvex even vertex of the labelled cycle one can use a type (1) barrier (from the front if both labels are 1, from the back if both are -1). If the cone is type (4)–(10), one can use the type (2) surface corresponding to any edge of the labelled cycle as a barrier from the appropriate side. That the surface lies entirely to one side of each such barrier follows directly from the proof of embedding given above if the surface is of type (1)–(10). For type (8) or (9) surfaces with a nonconvex wedge, such barriers do not contact all points of the nonconvex wedge from both sides (all points are contacted from at least one side), but they do pin down its entire boundary and all of the other wedges; one then uses the fact that plane segments are in fact minimizing. For type (11) modifications, one uses as a "barrier" the type (3) surface derived from the pair of edges containing a convex even vertex (or sequence of such edges, if there are a succession of type (11) modifications) instead of the barrier corresponding to the original unmodified edge. None of the other barriers are blocked by the modification. The surface may be able to be contacted by barriers from the other side as well; in any case, the convex even wedge or sequence of convex even wedges has its boundary pinned down and separated from the rest of the cone by barriers and is known to be minimizing, so s is minimizing. Similarly, a pair of wedges inserted in a type (12) modification is known to be minimizing, can be contacted from at least one side by barriers, does not affect the operation of other barriers, and has its boundary pinned down (and isolated from the rest of the cone) by barriers. Thus type (12) modifications are also minimizing.

Provided the edges of their labelled cycles are oriented tie lines, cones of types (4)–(10) can be pinned down on both sides by uniquely minimizing surfaces, and hence they are uniquely minimizing. Type (3) cones, and cones whose labelled cycles contain edges that are not in tie lines, are not uniquely minimizing. It is unknown whether or not type (11) cones are uniquely minimizing, given that all their edges are in tie lines.

4. Properties of minimizing crystalline cones. Suppose that $s: \mathbf{D} \to \mathbf{R}^3$ is a continuous mapping of a disk with properties (i)–(ii) of the Introduction. Denote its wedges by $\{P(k)\}$ and their corresponding oriented normals by $\{n(k)\}$ or, when more convenient, $\{n_k\}$.

Property 1. *The normals of any two adjacent wedges of s are vertices of a single tie line or tie region of the n-diagram. As a consequence, there is a labelled cycle associated to s, and two edges of the labelled cycle can cross each other only if all four endpoints are vertices of a single tie region with at least four vertices.*

PROOF. If the normals of two adjacent wedges were not part of a single tie figure, then either there would be a tie line crossing transversally the great circle segment between the normals, in which case a local barrier argument (using the tangent cone to the edge of W or WI corresponding to that tie line) shows that a neighborhood of the intersection is not minimizing (see Figure 3(a)), or there would be at least one element of N^0 on the interior of that great circle segment (and no tie lines crossing it transversally), in which case a plane segment with that normal can be hung just below the intersection, with triangular patches at the ends, decreasing energy proportional to its length times its width at a cost proportional to its width squared and thereby decreasing energy overall if the width is small enough (see Figure 3(b)). Therefore a labelled cycle can be associated to s. The last statement of Property 1 follows from 1.1.2 (1)(2).

PROPERTY 2. *If a wedge $P(i)$ of s has a central angle of less than $180°$ and is even, then $n(i-1)$, $n(i)$, and $n(i+1)$ are among the vertices of a single tie region, in positive (counterclockwise) order around the tie region. Furthermore, if there are two or more adjacent convex even wedges in s, then their normals, together with those of the adjacent wedges, are all part of the same tie region, in counterclockwise order around the tie region.*

PROOF. Suppose that $P(i)$ is even, with central angle less than π, but that n_{i-1}, n_i, and n_{i+1} are not all among the vertices of a single tie region. Without loss of generality, we take $i = 1$ and assume that $\mathrm{Lbl}(i - 1, i) = \mathrm{Lbl}(i, i + 1) = -1$. Let $x \in \mathbf{R}^3$ be such that

$$x \cdot n_i = \Phi(n_i), \qquad i = 0, 1, 2;$$

since there is no tie region containing $n(0)$, $n(1)$, and $n(2)$, there is a v (in fact, a v in N^0) such that $x \cdot v > \Phi(v)$.

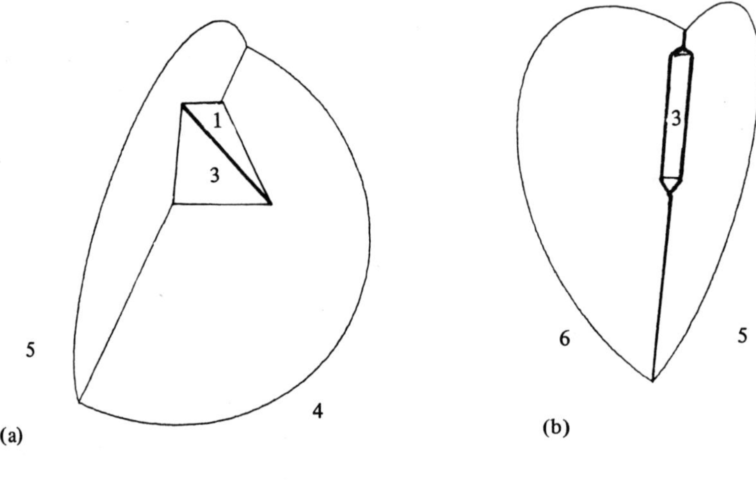

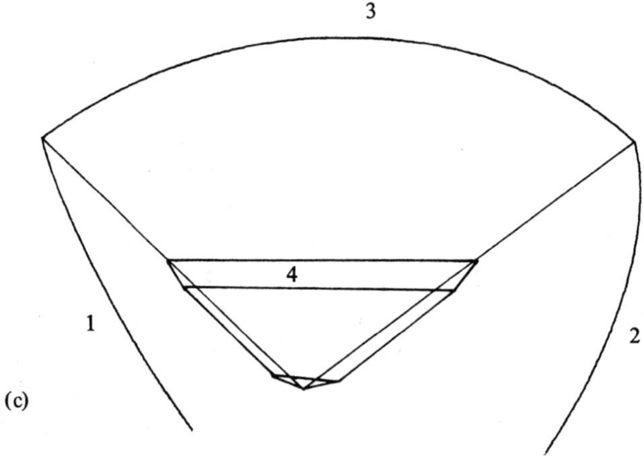

FIGURE 3. (a) A deformation decreasing the energy of part of a cone with adjacent wedges having normals v_4 and v_5 which are separated by a transverse tie line in the n-diagram. (b) A deformation decreasing the energy of the part of a cone having normals v_5 and v_6 which are separated by a vertex v_4 in the n-diagram. (c) A deformation decreasing the energy of the part of a cone corresponding to $n(0) = v_2$, $n(1) = v_3$, $n(2) = v_1$, and $\mathrm{Lbl}(0, 1) = \mathrm{Lbl}(1, 2) = -1$.

We make the deformation illustrated in Figure 3(c), where we cut off a large triangle of the wedge $P(1)$ and raise it a small distance t, patching the raised part (cut off or extended as necessary along its boundary) back into the rest of the cone by cutting out trapezoids in $P(0)$ and $P(2)$ and adding a trapezoid with normal v and a small triangle at the origin. More specifically, we select two fixed points z_1 and z_2 in $P(n_1)$ and four more points z_3, z_4, z_5, z_6 (with all odd numbered points in $P(0)$ and all even numbered ones in $P(2)$) as follows. Given any $t' > t > 0$, let z_1 be the solution of

$$n_1 \cdot z_1 = 0, \quad n_0 \cdot z_1 = 0, \quad v \cdot z_1 = t',$$

and z_3 be the solution of

$$n_1 \cdot z_3 = t, \quad n_0 \cdot z_3 = 0, \quad v \cdot z_3 = t'.$$

Let z_5 be the point of intersection of the line

$$n_1 \cdot z_5 = t, \qquad n_0 \cdot z_5 = 0,$$

with the ray bisecting the central angle of the wedge $P(0)$. Let z_2, z_4 and z_6 be the corresponding points when n_0 is replaced throughout by n_2. Finally, choose t' so that $\max\{|z_1|, |z_2|\} = 3/4$. For small enough t, $|z_5|$ is much less than $|z_3|$, and $|z_6|$ is much less than $|z_4|$ (and all are less than 1). We construct a new surface by removing from $P(1)$ the triangle with vertices O, z_1, z_2, and from $P(0)$ and $P(2)$ the trapezoids with vertices O, z_1, z_3, z_5, and O, z_2, z_4, z_6 respectively and replacing them by the quadrilateral with vertices z_3, z_4, z_6, z_5 (which is in the plane with normal n_1 at distance t from the origin), the trapezoid with vertices z_1, z_2, z_4, z_3 (which is in a plane with normal v), and the small triangle with vertices O, z_5, z_6. (Clearly there are mappings which accomplish this, and even ones which stay in the class of embeddings if s is an embedding.)

Let $L_0 = |z_1|$, $L_2 = |z_2|$, and $L_1 = |z_1 - z_2|$. Let $c_0 = n_0 \cdot n_1$, $c_2 = n_2 \cdot n_1$, and $c_1 = v \cdot n_1$. One can compute the change in integral of the new surfaces compared to the old, to first order in t, by looking at the rectangular approximations to the strips added and removed in forming the new surface. The result is that the energy change is $O(t^2)$ plus t times the following:

$$L_1\big(\Phi(v) - c_1\Phi(n_1)\big)\big(1 - c_1^2\big)^{-1/2} - L_0\big(\Phi(n_0) - c_0\Phi(n_1)\big)\big(1 - c_0^2\big)^{-1/2}$$
$$-L_2\big(\Phi(n_2) - c_2\Phi(n_1)\big)\big(1 - c_2^2\big)^{-1/2}.$$

(This formula first appeared, for more general deformations involving raising and lowering plane segments, in [**T3**, Proposition 5, part (4)]; in that paper there was a hypothesis that Φ was blunt, which was in fact unnecessary for this first-variation-type formula.)

We show that this first order change in energy is negative by computing and reassembling the components of the trivial vector equation

$$z_2 + (z_1 - z_2) + (-z_1) = 0.$$

The computations are simplified if we assume (without loss of generality) that

$$n_1 = (0,0,1), \quad n_0 = (a_0, b_0, c_0), \quad n_2 = (a_2, b_2, c_2), \quad v = (a_1, b_1, c_1)$$

(note that c_i is as defined before for $i = 0, 1, 2$). One then obtains the two formulas

$$L_1 a_1 (1 - c_1^2)^{-1/2} - L_0 a_0 (1 - c_0^2)^{-1/2} - L_2 a_2 (1 - c_2^2)^{-1/2} = 0$$

and

$$L_1 b_1 (1 - c_1^2)^{-1/2} - L_0 b_0 (1 - c_0^2)^{-1/2} - L_2 b_2 (1 - c_2^2)^{-1/2} = 0.$$

These are the first two components of the vector equation

$$0 = L_1 (v - c_1 n_1)(1 - c_1^2)^{-1/2} - L_0 (n_0 - c_0 n_1)(1 - c_0^2)^{-1/2}$$
$$- L_2 (n_2 - c_2 n_1)(1 - c_2^2)^{-1/2};$$

the third component is identically zero. In particular, the right side, dotted with any vector in $\mathbf{R}^3$, is zero. Therefore, dotting this vector equation with the x defined at the beginning of this proof and substituting in the three equalities and one inequality there, we obtain exactly that the first order change in energy is negative, as desired.

The second statement of Property 2 then follows easily from the first.

5. Proof of completeness of catalog.

5.1. *Ideas of the proof.* If we could know a priori that the cone had all its normals in one hemisphere of the unit sphere, then there would be a straightforward proof of the completeness of the catalog, based on angle sums of the stereographic projections of the labelled cycle, and of the cone intersected with the unit sphere; this "proof" was given in [TC1]. However, there is no way to know a priori that a cone has this property (and in fact a minimizing cone of type (8) need not have it if W is a tetrahedron, for example). There may be a way to extend that proof to the general case, but it has not yet been found. We therefore give here a messier proof which works in general.

The proof assumes that an embedded crystalline minimizing cone is not one of the cones in the catalog, and derives a contradiction. The fundamental observation is that for a wedge $P(i)$ to have a small central angle, either $n(i)$ must be odd and the labelled cycle must make a sharp right turn at $n(i)$ (i.e. $\theta(i)$ must be close to $-\pi$), or $n(i)$ must be even and the cycle must make a gradual left turn at $n(i)$ (i.e. $\theta(i)$ must be positive and close to 0). More specifically, for $P(i)$ to avoid intersecting the disk $n(k)^\perp$, we must have

$$\gamma(i) < \min\{\gamma_F(i - 1, i; n(k)), \gamma_G(i - 1, i; n(k))\}$$

(see subsection 1.2.5 for the definition of γ_F and γ_G) since $\text{Ray}(i, i + 1)$ is in $n(k)^\perp$ when $e(i, i + 1)$ is in the great circle containing $n(k)$ and $n(i)$. By Proposition 1.3, then, either $n(i)$ is convex odd and $e(i - 1, i)$ and $e(i, i + 1)$ are on the same side of the great circle containing $n(i)$ and $n(k)$, or $n(i)$ is convex even and $e(i - 1, i)$ and $e(i, i + 1)$ are on opposite sides of that great circle; in the latter case, both edges are contained in a single tie region, by Property 2. Since edges of the cycle can cross each other only if they also are both inside a single tie region, such crossing and convex even vertices can be effectively deleted, and part of the cycle will get trapped by sharp right turns into smaller and smaller triangles and not be able to join up with the rest of the cycle, as it must. (For any particular n-diagram having a finite number of vertices, a contradiction will also arise because at some point there simply will not be any vertices of the n-diagram in the part of $\mathbf{S}^2$ where the next vertex must be.)

As a useful shorthand notation, given nonantipodal vertices $n(j)$, $n(k)$, of the labelled cycle, we denote the point where the great circle through $n(j)$ and $n(k)$ crosses $e(i, i + 1)$, if it does, by $n(i, i + 1; j, k)$.

5.2. *s has a nonconvex even wedge implies s is* (*a modification of*) *type* (4) *or* (5). Suppose that s is a minimizing embedded crystalline cone, having a nonconvex even wedge (which we may assume is $P(0)$); by Property 1 there is a labelled cycle (n,Lbl), of length K, associated to s. We assume that (n, Lbl) is not type (4) or (5) or a type (11) or (12) modification thereof and derive a contradiction.

We distinguish six cases.

Case 1 (*see Figure* 4). *Either* $e(1, 2)$ *does not cross* $e(K - 1, K)$, $e(2, 3)$ *does not cross* $e(0, 1)$ *and*

$$\gamma(K - 1) \geqslant \max\{\gamma_F(n(1); K - 1, K), \ldots, \gamma_F(n(j); K - 1, K)\},$$

where j is such that $n(j)$ *is the first vertex following* $n(0)$ *which is not convex even, or* $e(K - 2, K - 1)$ *does not cross* $e(0, 1)$, $e(K - 3, K - 2)$ *does not cross* $e(K - 1, K)$ *and*

$$\gamma(1) \geqslant \max\{\gamma_F(0, 1; n(K - 1)), \ldots, \gamma_F(0, 1; n(j'))\},$$

where j' is such that $n(j')$ *is not convex even but* $n(k)$ *is convex even for* $j' < k < K$.

Note that j and j' are defined: if all vertices but $n(0)$ were convex even, then (by Property 2) they would all be from the same tie region and (by Proposition 1.3 and the convexity of tie regions) $P(1)$ and $P(K - 1)$ would intersect along a ray. Because the arguments are symmetrical, we need consider the former

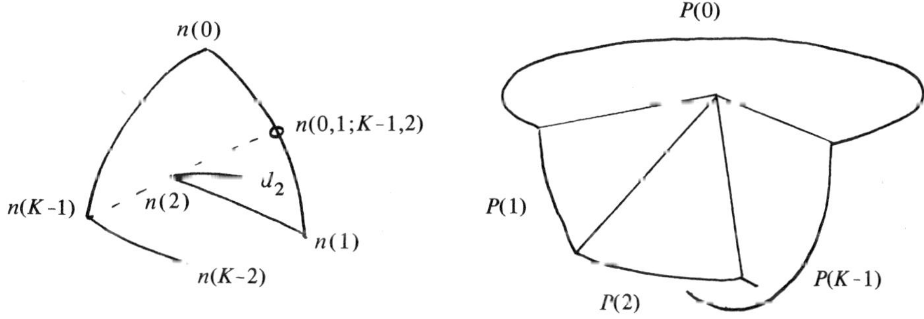

FIGURE 4. Typical labelled cycle and cone for the Case 1 argument.

possibilty of Case 1 only. Thus $\gamma(K-1)$ is "too big," and we will show that some wedge must intersect $P(K-1)$ (unless the situation reduces to another case). Since $P(1)$ cannot intersect $P(K-1)$ (recall s is not type (4) or (5) by hypothesis), $\gamma(1) < \gamma_F(0, 1; n(K-1))$. If $n(1)$ is convex even, then by Proposition 1.3, $0 < \theta(1) < \theta(0, 1, n(K-1)) + \pi$, and in the arguments to follow, this situation is indistinguishable from the one with the labelled cycle (n', Lbl'), where $n'(0) = n(0)$, and $n'(i) = n(i+1)$, $\mathrm{Lbl}'(i-1, i) = \mathrm{Lbl}(i, i+1)$ for $1 \leqslant i \leqslant K - 1$ (except that we may now be in Case 5 or 6 instead of Case 1). See figure 5(a), (b). Thus we may assume without loss of generality that $j = 1$, and so that $n(1)$ is odd and $\theta(1) < \theta(0, 1, n(K-1))$. Since $e(1, 2)$ is assumed not to cross $e(K - 1, K)$, we must then have $n(2)$ inside the triangle with corners $n(K-1), n(0)$, and $n(1)$.

To avoid $P(2)$ intersecting $P(K-1)$, $\gamma(2) < \gamma_F(1, 2; n(0, 1; K - 1, 2)) = \gamma_G(1, 2; n(K-1))$. If $n(2)$ is even, it is convex and again the situation is equivalent to that for the labelled cycle (n', Lbl') where vertex $n(2)$ is deleted, the remaining vertices are renumbered, and the remaining edge labels are moved up. See Figure 5(c), (d). By the condition on $\gamma(2)$, the final alternative is that $n(2)$ must be odd and $n(3)$ must be inside the triangle with corners $n(1)$, $n(2)$, and $n(0, 1; K - 1, 2)$ (recall that $e(2, 3)$ does not cross $e(0, 1)$). In order that $P(3)$ does not intersect $P(K-1)$, $\gamma(3) < \gamma_F(2, 3; n(K-1))$. Suppose that $e(3, 4)$ crosses $e(1, 2)$. Then $n(3)$ is odd, and $n(4)$ is inside the $n(K - 1, K; 2, 3)$-$n(3)$-$n(K - 1)$

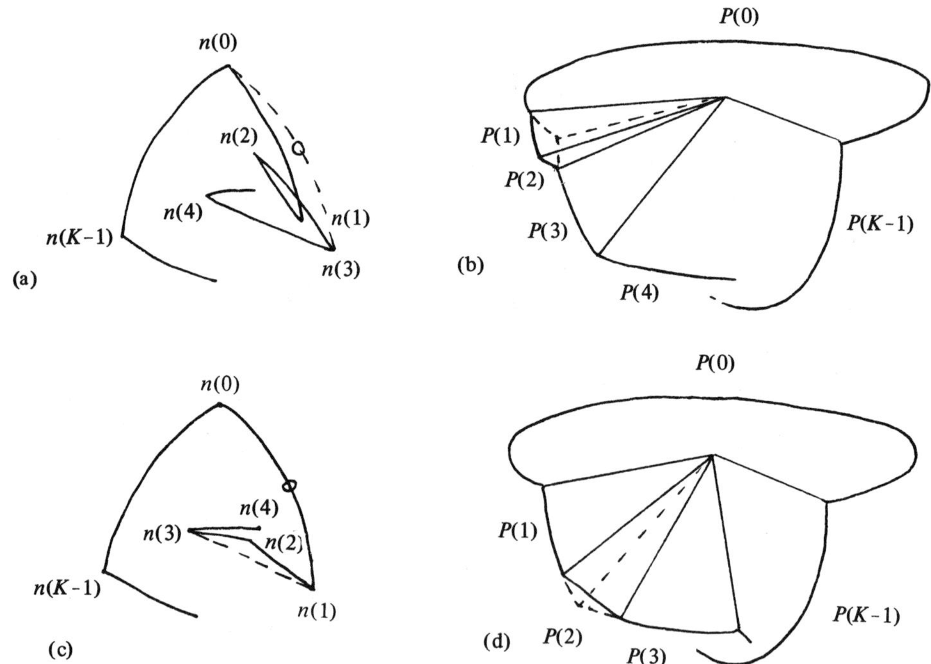

FIGURE 5. The effect of removing edges and vertices and relabelling in the crossing-edge and convex-even situations.

triangle. The situation is equivalent to the one with the labelled cycle (n', Lbl') of length $K' = K - 2$ where $n'(i) = n(i)$ for $i = 0, 1$, $\mathrm{Lbl}'(0, 1) = \mathrm{Lbl}(0, 1)$, and $n'(i) = n(i + 2)$ and $\mathrm{Lbl}'(i - 1, i) = \mathrm{Lbl}(i + 1, i + 2)$ for $2 \leqslant i \leqslant K'$; we then go back to the beginning of this paragraph (or to Case 5 or 6). If $n(3)$ is convex even, we can remove $n(3)$ as before. Thus $n(3)$ must be odd and $n(4)$ must be inside the $n(2)\text{-}n(3)\text{-}n(1, 2; K - 1, 3)$ triangle. Successive vertices are similarly trapped by sharp right turns in ever decreasing triangles; if there are a finite number of vertices inside the $n(K - 1)\text{-}n(0)\text{-}n(1)$ triangle, then one soon runs out of possible vertices; but in any case, since the edges must somehow join up with $n(K - 1)$ to make a cycle, this is impossible. Thus we have reached a contradiction and Case 1 does not happen.

Case 2. Suppose $j' \leqslant K - 1$. Let

$$\gamma_M = \max\left\{ \gamma_F\big(j' - 1, j'; n(K)\big), \gamma_G\big(k - 1, k; n(K)\big) : j' < k < K \right\},$$

and let k' be the lowest value of the index $(j' \leqslant k' \leqslant K - 1)$ at which the maximum is attained. Suppose

$$\gamma_M \geqslant \max\left\{ \gamma_F\big(n(i); k', K\big) : 0 < i \leqslant j \right\}.$$

(Or equivalently, suppose all the symmetric statements hold with 1 replacing $K - 1$, etc.)

Again, we need only consider the former alternative. The argument to rule out this case is essentially equivalent to that in Case 1. That is, one can consider the situation to be as if $n(k' + 1), \ldots, n(K - 1)$ were deleted to form a new labelled cycle (n', Lbl') with length $K' = k' + 1$, but it is not in fact $P(n', \mathrm{Lbl}', K' - 1)$ that must not be intersected but $P(n, \mathrm{Lbl}; k') \cup \cdots \cup P(n, \mathrm{Lbl}; K - 1)$. The arguments with respect to the positions of $n(2)$, $n(3), \ldots$, however, are identical, since crossing any edge $e(k, k + 1)$ of (n, Lbl) for $k \geqslant k'$ implies crossing the edge $e(K' - 1, K)$ of (n', Lbl') (by Property 1 and the convexity of tie regions). Thus Case 2 cannot happen.

Case 3 (see Figure 6). $n(j + 1)$ is inside the $n(0)\text{-}n(j)\text{-}n(j')$ triangle but not the $n(j')\text{-}n(K)\text{-}n(0, j; j', j' - 1)$ triangle, and the corresponding statement also holds for the other end of the labelled cycle.

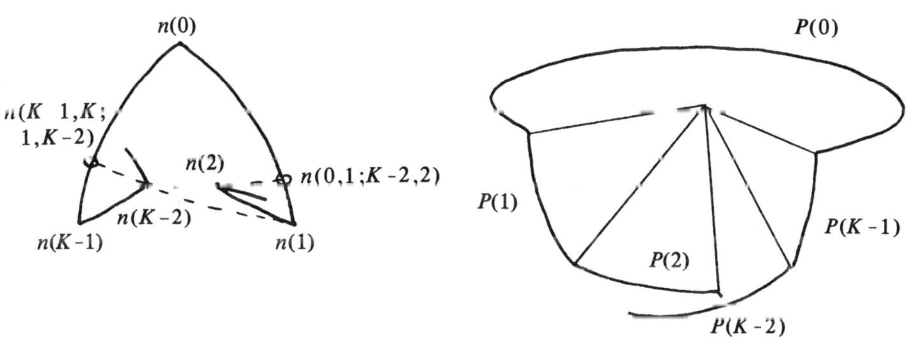

FIGURE 6. Typical labelled cycle and cone for the Case 3 argument.

As in Case 1, we may (by eliminating vertices and renumbering) assume without loss of generality that $j = 1$ and $j' = K - 1$. Then, to avoid intersections of their wedges with $P(0)$, both $n(K - 1)$ and $n(1)$ are odd. $n(K - 2)$ cannot follow $n(2)$ in the labelled cycle, since otherwise both would be convex even (if one were odd, both would be and both would have $\gamma > \pi$, making the cone not embedded) and thus s would be a type 11 modification of a type 5 surface, contrary to hypothesis. Since $P(K - 2)$ and $P(2)$ cannot intersect, at least one of them—say $P(2)$—must end before it reaches the plane of the other. Then either we are in one of the edge-crossing or convex-even situations again (which can be handled as before by deleting one or two vertices and renumbering the rest, possibly thereby becoming another case), or $n(2)$ is odd and $n(3)$ must be inside the $n(1)$-$n(2)$-$n(0, 1; 2, K - 2)$ triangle. If

$$\gamma(K - 2) \geqslant \gamma_F(n(K - 1, k; 1, K - 2); K - 2, K - 1),$$

then all the vertices following $n(2)$ are again trapped by right turns into successively smaller triangles by the necessity of their wedges avoiding intersections with $P(K - 2)$. If $n(K - 2)$ is convex even, we can delete it as in Case 1. If $e(K - 3, K - 2)$ crosses $e(K - 1, K)$, we are in Case 5. Thus we are reduced to $n(K - 3)$ being inside the $n(K - 1)$-$n(K - 2)$-$n(K - 1, K; 1, K - 2)$ triangle. But then some odd wedge $P(k)$ must have $\gamma(k) > \pi$ in order for the cycle to escape that triangle other than by $e(i + 1, i + 2)$ crossing $e(i - 1, i)$ for some i or by having convex-even vertices (both of which can be handled as before). As in Case 2, the wedges $P(k), \ldots, P(K - 2)$ will block the wedges $P(3)$, $P(4), \ldots,$ preventing that part of the cycle from escaping the $n(1)$-$n(2)$-$n(0, 1; 2, K - 2)$ triangle. Once again, this is impossible, since the cycle must join up.

Case 4 (*see Figure* 7). $n(K - 2)$ *and* $n(2)$ *are both inside the* $n(0)$-$n(1)$-$n(K - 1)$ *triangle, and* $n(2)$ *is inside the* $n(K - 1)$-$n(K)$-$n(0, 1; K - 1, K - 2)$ *triangle* (*or the symmetric statement holds for* $n(2)$ *in place of* $n(K - 2)$).

Note that for $P(2)$ to intersect $P(K - 2)$, $\gamma(2)$ would have to be well over 180°, and in fact $P(2)$ would intersect $P(K - 1)$ first. Now since $P(2)$ does not intersect $P(K - 1)$, either we are in the edge-crossing or convex-even situation,

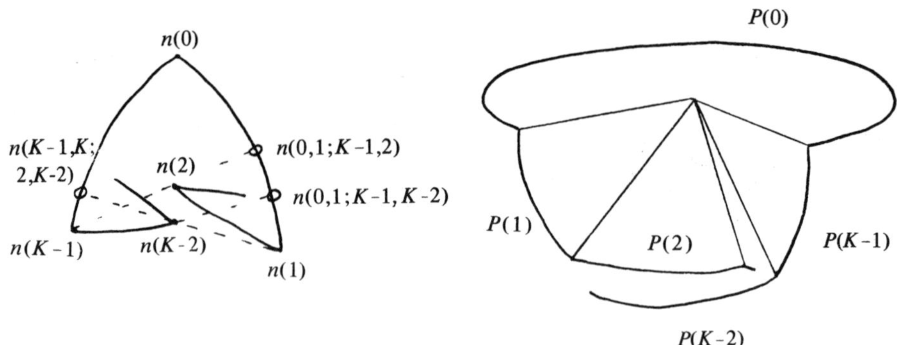

FIGURE 7. Typical labelled cycle and cone for the Case 4 argument.

which can be handled by deleting vertices as before, or $n(2)$ is odd and $n(3)$ must be inside the $n(1)$-$n(2)$-$n(0, 1; K - 1, 2)$ triangle. If $n(3)$ lies on the same side of the $n(K - 1)$-$n(K - 2)$ great circle as $n(2)$, then the same argument as in Case 1 (i.e. trying to avoid intersections with $P(K - 1)$) produces an eventual contradiction. If $n(3)$ is on the other side of the $n(K - 1)$-$n(K - 2)$ great circle, then the plane segments must avoid intersecting $P(K - 2)$ as well as $P(K - 1)$, and we proceed as in Case 3.

Case 5. $e(K - 3, K - 2)$ *crosses* $e(K - 1, K)$ *or* $e(2, 3)$ *crosses* $e(0, 1)$.

To avoid intersections with $P(0)$, $n(K - 1)$ and $n(K - 2)$ must both be odd. As before, the situation is essentially equivalent to the one of the labelled cycle (n', Lbl') of length $K' = K - 2$, where $n'(i) = n(i)$ and $\mathrm{Lbl}'(i, i + 1) = \mathrm{Lbl}(i, i + 1)$ for $0 \leqslant i \leqslant K - 1$ (the "essentially" meaning here the same as in Case 2, i.e. that intersections with $P(n', \mathrm{Lbl}'; K' - 1)$ must be interpreted as intersections with

$$P(n, \mathrm{Lbl}; K - 1) \cup P(n, \mathrm{Lbl}; K - 2) \cup P(n, \mathrm{Lbl}; K - 3)).$$

Thus we are back in Cases 1 to 4.

Case 6. $e(1, 2)$ *crosses* $e(K - 1, K)$ *or* $e(K - 2, K - 1)$ *crosses* $e(0, 1)$, *but* $e(2, 3)$ *does not cross* $e(0, 1)$ *and* $e(K - 3, K - 2)$ *does not cross* $e(K - 1, K)$.

Again, we need consider only the former alternative. We may assume that $n(2)$ is not convex even, since if it were, one could delete it as before, arriving in another case or back in this case. By Property 1 and the assumption that $e(2, 3)$ does not cross $e(0, 1)$, therefore $e(2, 3)$ does not cross $e(K - 1, K)$, and $\theta(1, 2, n(0)) < \theta(2)$. The situation is then equivalent to the one with the labelled cycle of length $K - 1$ with $n'(i) = n(i)$, $i = 1$, 2, $\mathrm{Lbl}'(0, 1) = \mathrm{Lbl}(2, 3)$, and $n'(i) = n(i + 1)$ and $\mathrm{Lbl}'(i - 1, i) = \mathrm{Lbl}(i, i + 1)$ for $i = 2, \ldots, K - 1$. Note that $n'(1)$ is now nonconvex even and $n'(0)$ is either convex even or nonconvex odd. Renumbering so that $n''(0) = n'(1)$, we are back in one of the first four cases.

Thus we conclude that if the labelled cycle corresponding to s contains a nonconvex even vertex, that cycle must be of type (4), (5), or a modification of such involving reinserting convex even vertices or pairs of vertices whose adjacent edges cross. Properties 1 and 2 and nonintersection of wedges corresponding to any string of deleted vertices and their immediately adjacent vertices limit the number of ways convex even vertices and crossing edges can be reinserted to type (11) and type (12) modifications.

5.3 *Other types.* Since we have classified embedded minimizing crystalline cones with a nonconvex even wedge we now assume s has no nonconvex even wedges. By Property 1 again there is a labelled cycle (n, Lbl) associated to s. If it is of length 1, then it is type (1). If it is of length 2, then the vertices must be even (otherwise, $\gamma(0)$ would be 2π, $\mathrm{Ray}(0, 1)$ would equal $\mathrm{Ray}(1, 2)$, and s would not be an embedding) and hence the cycle is type (2). If it is of length 3, then (since we assumed that there are no nonconvex even vertices) all vertices are convex even. In general, when all vertices are convex even and the length is at least 3, all vertices are part of the same tie region (Property 2) which lies in one hemisphere (1.1.2); since the labelled cycle gives a generalized Gauss map of the outer part of the cone, and the cone has no branch point at the origin (in particular, it has a 1-1 projection), the edges of the labelled cycle must form a simple closed curve and thus the cycle must be type (3).

Next, suppose that the length K of (n, Lbl) is at least 4 and that $n(K - 2) = n(K)$. Then $n(K)$ is even and $\gamma(K - 1) = \pi$ and (to avoid $P(K - 2)$ intersecting $P(K)$) $\gamma(K - 2) + \gamma(K) < \pi$. If $n(K - 2)$ and $n(K)$ are both odd, then as far as the arguments of the previous section are concerned, (n, Lbl) is equivalent to (n', Lbl'), the labelled cycle of length $K - 2$ with $n'(0)$ nonconvex even defined by $n'(i) = n(i)$ and $\text{Lbl}'(i, i + 1) = \text{Lbl}(i, i + 1)$ for $i = 0, \ldots, K - 2$. Therefore (n', Lbl') is type (4) or (5) or a type (11) or (12) modification thereof, and (n, Lbl) is type (10) or a type (11) or (12) modification thereof. If $n(K - 2)$ and $n(K)$ are both even, then they are both convex (as stated above) and hence (n, Lbl) is the result of two type (11) modifications to the labelled cycle (n', Lbl') with $n'(0) = n(K - 1)$, $n'(i) = n(i)$, and $\text{Lbl}'(i - 1, i) = \text{Lbl}(i - 1, i)$ for $i = 1, \ldots, K - 3$. Since (n', Lbl') has $n'(0)$ nonconvex even, it is type (4) or (5) or a modification thereof, and thus (n, Lbl) is a type (11) modification of type (4) or (5). Finally, if $n(K - 2)$ is even and $n(K)$ is odd, then (n, Lbl) is a type (11) modification of the labelled cycle (n', Lbl') with $n(K - 2)$ deleted, which in turn is a type (12) modification of a type (4) or (5) cycle (we know (n', Lbl') is type (12) since $e(0, 1)$ is at least partly in the triangle with vertices $n(K - 3)$, $n(K - 2)$, and $n(K - 1)$, which is in a tie region by Property 2).

We are reduced to the case where the length K of the labelled cycle is at least 4, there are no nonconvex even vertices, and $n(i) \neq n(i + 2)$ for any i. For ease of description, we will further assume that there are no convex even vertices and no pairs of edges $e(i - 1, i)$, $e(i + 1, i + 2)$, which cross as in type (12) modifications; if there were, we could equivalently consider the cycles where they were deleted, as in the previous section and above. Thus in particular all vertices are odd. We will show that only type (6)–(9) cycles are possible. (Again, Properties (1) and (2) and nonintersection (in a comparatively local way) will then limit the ways in which such convex even vertices and pairs of crossing edges can occur in the original cycle to those specified in type (11) and (12) modifications of types (6)–(9).)

First suppose that $n(0) = n(3)$. Then to avoid intersections, $n(i)$ must be convex for $i = 1, 2$. As far as the arguments of the previous section are concerned, the situation is equivalent to that for the labelled cycle with $n'(i) = n(i + 3)$ and $\text{Lbl}'(i - 1, i) = \text{Lbl}(i + 2, i + 3)$ for $i = 1, \ldots, K - 3$. Either $n'(0)$ is nonconvex even, in which case the arguments of the previous section apply directly, predicting that the number of vertices in the labelled cycle (n', K') is 3 and thus that $K = 6$ and s is type (6) or (7), or $n'(0)$ is convex even. But $n'(0)$ cannot be convex even, given the assumptions of no crossing edges and no convex even vertices; again, at least one side of the cycle is forced into an infinite cycle, which is impossible.

Finally, suppose additionally that there is no i such that $\dot{n}(i) = n(i + 3)$. Then for some i, it must be the case that both $n(i + 3)$ and $n(i - 1)$ are outside the triangle to the right of $e(i, i + 1)$ with vertices $n(i)$, $n(i + 1)$, and $n(i + 2)$, since the contrary assumption (and the embeddedness of s) would force an infinite spiral of vertices in the labelled cycle. Then one can apply the arguments of the previous section to the labelled cycle with $n(i + 1)$ deleted (and $n'(i)$ thus nonconvex even), since the cones corresponding to the two labelled cycles differ only in regions away from where the intersections that cause the contradictions of the previous section would take place. The conclusion is that the cycle (n', Lbl') has only three vertices, and thus the cycle (n, Lbl) has four vertices, all odd, and is therefore type (8) or (9).

This completes the proof of the completeness of the catalog.

REFERENCES

[C] J. W. Cahn, *Transitions and phase equilibria among grain boundary structures*, J. Physique **43** (1982), 199–213.

[CT] J. W. Cahn and J. E. Taylor, *A contribution to the theory of surface energy minimizing shapes*, Scripta Met. (1984).

[D] A. Dinghas, *Über einen geometrischen Satz von Wulff für die Gleichgewichtsform von Kristallen*, Zeitschrift für Kristallographie **105** (1944), 304–314.

[LS] D. Levine and P. J. Steinhardt, *A new class of ordered structures*, Phys. Rev. Lett. **53** (1984), 2477–2480.

[SBGC] D. Schechtman, I. Blech, D. Gratias and J. W. Cahn, *Metallic phase with long-range orientational order and no translational symmetry*, Phys. Rev. Lett. **53** (1984), 1951–1953.

[T1] J. E. Taylor, *Crystalline variational problems*, Bull. Amer. Math. Soc. **84** (1978), 568–588.

[T2] ______, *Constructing crystalline minimal surfaces*, Ann. of Math. Studies **105** (1983), 271–288.

[T3] ______, *F-minimal hypersurfaces characterize F up to a point*, Indiana Univ. Math. J. **31** (1982), 789–799.

[TC1] J. E. Taylor and J. W. Cahn, *Catalog of saddle shaped surfaces in crystals*, Acta Metallurgica (to appear, fall 1985).

[TC2] ______, *The cusp: a new singularity in anisotropic energy minimizing surfaces*, preprint.

[W] G. Wulff, *Zur Frage der Geschwindigkeit des Wachsthums und der Auflösung der Krystallflachen*, Zeitschrift für Krystallographie und Mineralogie **34** (1901), 499–530.

RUTGERS UNIVERSITY

Proceedings of Symposia in Pure Mathematics
Volume **44** (1986)

Liouville Theorems for Stable Harmonic Maps into Either Strongly Unstable, or δ-Pinched, Manifolds

S. WALTER WEI

1. Introduction. In [5], a compact Riemannian manifold N is defined to be *strongly unstable* if it is not the domain or target of any nonconstant smooth stable harmonic map (into or from an arbitrary *compact* Riemannian manifold); also, the homotopy class of any smooth map, to or from N, contains a map of arbitrarily small energy. The stable harmonic map in our context means a local minimum of the energy functional.

Professor Ralph Howard and the author then gave a useful geometric criterion for N with Riemannian metric $\langle\ ,\ \rangle_N$ to be strongly unstable, i.e., a selfadjoint linear map, Q_y^N, from the tangent space, $T_y(N)$ to N at y, into itself, given by

$$(1) \qquad \langle Q_y^N(X), X\rangle_N = \sum_{i=1}^{n} 2|h(X, \alpha_i)|^2 - \langle h(X, X), h(\alpha_i, \alpha_i)\rangle_{R^q},$$

is negative definite for every $y \in N$, where $\alpha_1, \alpha_2, \ldots, \alpha_n$ is an orthonormal basis for $T_y(N)$ and h is the second fundamental form of N isometrically embedded into R^q.

As an application of this criterion, we showed, in particular, that a compact, connected symmetric space N is strongly unstable if and only if Q_y^N is negative definite for every y in N; if and only if N is one of the following: (A) a sphere S^n, $n > 2$; (B) Grassman manifolds $\mathrm{Sp}(p + q)/\mathrm{Sp}(p) \times \mathrm{Sp}(q)$ (including the quaternionic projective spaces $\mathrm{QP}(n)$, where $p = n$, and $q = 1$); (C) the Cayley plane $F_4/\mathrm{Spin}(9)$; (D) $\mathrm{SU}(2n)/\mathrm{Sp}(n)$, $n \geq 2$; (E) any compact simply-connected irreducible group manifolds other than the exceptional group E_8; or (F) an arbitrary finite product of any manifolds from (A)–(E) (cf. [5]).

1980 *Mathematics Subject Classification.* Primary 58E15, 58F20, 58F30, 58F35.

© 1986 American Mathematical Society
0082-0717/86 $1.00 + $.25 per page

The purpose of this paper is to derive general extrinsic and intrinsic stability inequalities and use them to establish Liouville theorems for stable harmonic maps, without any growth condition from R^2 (or more generally, from complete Riemannian manifolds), into either a *complete* (not necessarily compact) Riemannian manifold N which is negative definite Q_y^N for every $y \in N$, or δ-pinched for $\frac{1}{4} < \delta < 1$ (i.e., all sectional curvatures are in the half-open interval $(\delta K, K]$ for some $K > 0$).

AN EXTRINSIC VERSION OF LIOUVILLE THEOREM 1.1. *Let M be a complete Riemannian manifold covered by a collection of compact geodesic balls, B_r, of radius $0 < r < \infty$, centered at a fixed point x_0, i.e.*

$$(2) \qquad M = \bigcup_{r \in (0, \infty)} B_r$$

such that for sufficiently large r,

$$(3) \qquad \text{volume}(B_r) < \sigma r^2,$$

where σ is a fixed constant, independent of r. Then any smooth stable harmonic map from M into a complete (not necessarily compact) Riemannian manifold N, with Q_y^N (defined in (1)) negative definite for every y in N, is constant. In particular, any smooth stable harmonic map, from R^2, or any compact Riemannian manifold, into such an N, is constant.

AN INTRINSIC VERSION OF LIOUVILLE THEOREM 1.2. *Any smooth stable harmonic map, from a complete Riemannian manifold M, satisfying the volume growth assumption (3), into a compact, simply-connected, δ_k-pinched Riemannian manifold N^k, is a constant for $\frac{1}{4} < \delta_k < 1$ where δ_k is defined in (11) (hence $k > 2$). In particular, any smooth stable harmonic map, from R^2, or any compact Riemannian manifold, into such an N^k, is constant.*

By a similar technique, the result can be generalized to higher-dimensional Euclidean space or to complete Riemannian manifolds with rather weaker volume growth assumption under a suitable condition on N. Furthermore, it can be applied to the study of minimizing tangent maps, the existence and regularity of minimizing harmonic maps into certain strongly unstable manifolds. We will discuss the latter in a forthcoming paper [11].

We wish to point out that Professors R. Schoen and Karen Uhlenbeck [9] have shown that any smooth stable harmonic map, $u: R^n \to S^k$, is constant, provided $n = 2$ and $k = 3$; or $2 \leqslant n \leqslant \min(4, [[k/2]])$, and $k \geqslant 4$ where $[[t]]$ denotes the greatest integer in t. Our work also generalizes an interesting theorem of Ralph Howard [4] which asserts that no stable harmonic map, from a compact Riemannian manifold into a compact, simply-connected, δ_k-pinched Riemannian manifold N^k for $k > 2$ and $\frac{1}{4} < \delta_k < 1$ (δ_k defined in (11)), exists, unless it is a constant map.

The author wishes to thank Professor S. Y. Cheng for many interesting conversations related to this work.

2. An Extrinsic Version of a Stability Inequality and the Proof of the Liouville Theorem 1.1. Let u be a smooth stable harmonic map from a complete, noncompact n-dimensional Riemannian manifold M into a k-dimensional complete Riemannian manifold N (which is isometrically embedded into R^q with second fundamental form h.) Let ϕ be a smooth, real-valued function on M with compact support, and $e_1, \ldots, e_n$ be a local frame in M; α will be a unit vector in R^q, and α^T, the tangential projection of $\alpha|_N$ onto N. Denote by $u_t^{\phi\alpha^T}$ a variation of u with $u_0^{\phi\alpha^T} = u$, and deformation vector, $\phi\alpha^T \in u^{-1}(TN)$, pull-back bundle from the tangent bundle TN of N. Denote ∇, the Riemannian connection on TN, and ∇', the pull-back connection. The second variational formula (cf. [2]) for energy, $E(u_t^{\phi\alpha^T})$, then yields

$$\frac{d^2}{dt^2} E\left(u_t^{\phi\alpha^T}\right)\bigg|_{t=0} = \int_M \left(\sum_i \left|\nabla'_{e_i}\phi\alpha^T\right|^2 - \langle R^N(\phi\alpha^T, \tilde{e}_i)\phi\alpha^T, \tilde{e}_i\rangle_N \right) dV$$

where $\tilde{e}_i = du(e_i)$, R^N is the curvature tensor in N, and dV is the volume element in M.

$$= \int_M \sum_{i=1}^n |e_i\phi|^2 |\alpha^T|^2 + \phi^2 \left|\nabla'_{e_i}(\alpha - \alpha^\perp)\right|^2 - 2(e_i\phi)\phi\langle \alpha^T, \nabla'_{e_i}\alpha^\perp \rangle_N$$

$$-\phi^2 \sum_i \langle R^N(\alpha^T, \tilde{e}_i)\alpha^T, \tilde{e}_i\rangle_N \, dV$$

$$= \int_M |\nabla\phi|^2 |\alpha^T|^2 + \sum_i \phi^2 \left|A^{\alpha^\perp}\tilde{e}_i\right|^2 + 2(e_i\phi)\phi\langle \alpha^T, A^{\alpha^\perp}\tilde{e}_i\rangle_N$$

$$-\phi^2 \sum_i \langle R^N(\alpha^T, \tilde{e}_i)\alpha^T, \tilde{e}_i\rangle_N \, dV,$$

where the Weingarten map $A^{\alpha^\perp}$ is given by $A^{\alpha^\perp}\tilde{e}_i = -\nabla_{\tilde{e}_i}\alpha^\perp$. Hence $\langle A^{\alpha^\perp}\tilde{e}_i, \tilde{e}_j\rangle_N = \langle h(\tilde{e}_i, \tilde{e}_j), \alpha^\perp \rangle_N$. Now choose an orthonormal basis $\alpha_1, \ldots, \alpha_q$ in R^q such that $\alpha_1, \ldots, \alpha_k$ are tangent to N and $\alpha_{k+1}, \ldots, \alpha_q$ are orthogonal to N. Computing the second variational formula along $\phi\alpha_p$, and summing over p from 1 to q (in a manner similar to [5, 6 or 10]), we obtain by the Gauss curvature equation and the stability assumption:

$$0 \leqslant \sum_{p=1}^q \frac{d^2}{dt^2} E\left(u_t^{\phi\alpha_p}\right)\bigg|_{t=0}$$

$$= \int_M k|\nabla\phi|^2 + \sum_{i=1}^n \sum_{j=1}^k \phi^2\left(2|h(\alpha_j, \tilde{e}_i)|^2 - \langle h(\alpha_j, \alpha_j), h(\tilde{e}_i, \tilde{e}_i)\rangle_N\right) dV.$$

Therefore, we have established

AN EXTRINSIC STABILITY INEQUALITY 2.1. With the same notation as above, for any smooth, real-valued function ϕ defined on M with compact support,

$$(4) \quad 0 \leqslant \int_M k|\nabla\phi|^2 + \sum_{i=1}^n \sum_{j=1}^k \phi^2\left(2|h(\alpha_j, \tilde{e}_i)|^2 - \langle h(\alpha_j, \alpha_j), h(\tilde{e}_i, \tilde{e}_i)\rangle_N\right) dV,$$

408   S. W. WEI

or

$$0 \leqslant \int_M k|\nabla\phi|^2 + \sum_{i=1}^n \langle Q_y^N(\tilde{e}_i), \tilde{e}_i \rangle_N \phi^2 \, dV.$$

Furthermore, if we assume Q_y^N is negative definite, then there exists a positive function $\psi(x)$ defined on M, such that $\langle Q_{u(x)}^N(X), X \rangle_N < -k\psi(x)|X|^2$, where $X = du(Z)$ for every Z in $T_x(M)$, and x in M. (4) therefore becomes

PROPOSITION 2.2. *Let u be a stable harmonic map from a complete Riemannian manifold M into a complete Riemannian manifold N, with Q_y^N negative definite. Then there exists a function $\psi(x) > 0$ defined on M, such that*

$$(5) \qquad \int_M \psi\phi^2|du|^2 \, dV \leqslant \int_M |\nabla\phi|^2 \, dV$$

for every smooth real-valued function ϕ defined on M with compact support. In particular, if N is compact, then there exists a constant $c > 0$, such that

$$(5') \qquad \int_M \phi^2|du|^2 \, dV \leqslant c\int_M |\nabla\phi|^2 \, dV$$

for every smooth real-valued function ϕ defined on M with compact support.

On the other hand, for $r > 1$ we can define ϕ_r as follows

$$\phi_r(x) = \begin{cases} 1 & \text{on } B_r, \\ \log(r^2/|x|)/\log r & \text{on } B_{r^2} \setminus B_r, \\ 0 & \text{off } B_{r^2}, \end{cases}$$

when $|x|$ denotes the geodesic distance of x from x_0.

Substituting ϕ_r into (5), we have

$$\int_{B_r} \psi|du|^2 \, dV \leqslant \int_{B_{r^2} \setminus B_r} \frac{1}{(\log r)^2|x|^2} \, dV$$

$$= \int_r^{r^2} \left(\frac{d}{dt} \int_{B_t} \frac{1}{(\log r)^2|x|^2} \, dV \right) dt$$

$$= \int_r^{r^2} \frac{1}{(\log r)^2 t^2} \left(\frac{d}{dt} \mathrm{vol}(B_t) \right) dt.$$

We can integrate by parts and use (3) to obtain

$$\int_{B_r} \psi|du|^2 \, dV \leqslant \left(\frac{\mathrm{vol}(B_{r^2})}{(\log r)^2 r^4} - \frac{\mathrm{vol}(B_r)}{(\log r)^2 r^2} - \int_r^{r^2} \frac{2}{(\log r)^2 t^3} \mathrm{vol}(B_t) \, dt \right)$$

$$\leqslant \sigma \left(\frac{1}{(\log r)^2 r^2} + \frac{1}{(\log r)^2} + \int_r^{r^2} \frac{2}{(\log r)^2 t} \, dt \right)$$

for sufficiently large r. It follows that

$$\int_{B_r} \psi |du|^2 \, dV \leqslant \sigma \left(\frac{1}{(\log r)^2 r^2} + \frac{1}{(\log r)^2} + \frac{2}{\log r} \right)$$

for some constant σ and sufficiently large r.

As r tends to ∞, the limit concludes that u is constant and proves the extrinsic Liouville theorem.

Combining the result in [**5**], we obtain the following immediate

COROLLARY 2.3. (i) *If the principal curvatures of an ovaloid N^k satisfy*

$$\lambda_k < \lambda_1 + \cdots + \lambda_{k-1} \quad (assume \ 0 < \lambda_i \leqslant \cdots \leqslant \lambda_k),$$

or (ii) *if the ovaloid N^k is pointwise, δ-pinched for some $\delta > 1/(k-1)$, then any smooth stable harmonic map, from a complete Riemannian manifold M satisfying* (3) *(for example, R^2, or any compact Riemannian manifold), into N^k, is constant.*

COROLLARY 2.4. *Let N^k be a compact submanifold of the unit sphere S^{q-1} with the second fundamental form B satisfying*

$$|B|^2 < \begin{cases} \dfrac{4(k-2)}{\sqrt{k}+5} & \text{for } 3 \leqslant k \leqslant 8 \\[2mm] \dfrac{2(k-2)}{\sqrt{k}+1} & \text{for } 9 \leqslant k. \end{cases}$$

Then any smooth stable harmonic maps, from M as in Theorem 1.1 or, in particular, R^2, or any compact Riemannian manifold, into N^k, are constant.

As a complementary theorem to Corollary 2.4, we consider the polar image of the unit normal bundle of an immersed submanifold and obtain

COROLLARY 2.5. *Let $\hat{N}(M)$ be the bundle of unit normal vectors to a compact n-dimensional manifold M in S^{q-1} with the smallest squared principal curvature $\kappa(M)$ of M over all x in M and all normal directions satisfying*

$$\kappa(M) > \begin{cases} n(\sqrt{q-2}+5)/4(q-4), & \text{for } 5 \leqslant q \leqslant 10, \\[2mm] n(\sqrt{q-2}+1)/2(q-4), & \text{for } 11 \leqslant q. \end{cases}$$

Then a smooth, stable harmonic map from a complete Riemannian manifold with quadratic volume growth (3) *into $\hat{N}(M)$ is constant.*

COROLLARY 2.6. *Let N^k be one of the symmetric spaces, (A)–(F), listed in the introduction. Then any smooth stable harmonic map, from M as in Theorem 1.1 or, in particular, any compact Riemannian manifold, or R^2, into N^k, is constant.*

COROLLARY 2.7. *Let a complete Riemannian manifold, $N_i^{K_i}$, be isometrically embedded into R^{q_i}, with $Q_{x_i}^{N_i^{K_i}}$ negative definite for every $x_i \in N_i^{K_i}$, and $1 \leqslant i \leqslant t$. Then, any smooth harmonic map, from M as above (for example, R^2), into $N_1^{K_1} \times N_2^{K_2} \times \cdots \times N_t^{K_t}$, is constant. In particular, any smooth harmonic map, from such M or R^2, into an arbitrary finite product of any manifolds N^k, in Corollaries 2.3(i), (ii), 2.4, 2.5, 2.6, is constant.*

We can also study stable harmonic maps into compact Riemannian manifolds with boundary. For example

COROLLARY 2.8. *Let $\bar{S}^k_+$ be a closed upper k-hemisphere. Then any stable harmonic map, u into $\bar{S}^k_+$, from an arbitrary compact Riemannian manifold M, is constant for $k > 3$.*

PROOF. We assume $u = (u_1, \ldots, u_{k+1})$; then harmonicity implies

$$\Delta u_{k+1} + |du|^2 u_{k+1} = 0.$$

By the maximum principle and assumption $u_{k+1} \geqslant 0$, either $u_{k+1} = 0$ or u is constant. Now the corollary follows from the fact that S^{k-1} is not the target of any nonconstant, stable harmonic map defined on a compact manifold for $k > 3$ (cf. [**8** or **10**]).

3. An Intrinsic Version of a Stability Inequality and a Liouville Theorem. In this section we use the same notations M^n, N^k, TN, ∇, ϕ, $e_1, \ldots, e_n$, R^N, $\langle\,,\,\rangle_N$, dV, as in §2. Let f be a smooth stable harmonic map from M into N, and $V_y(g)$ be the gradient of $g \circ \rho_y$, where $\rho_y(y')$ is the geodesic distance of y' from y in N, and g is a real-valued C^1-map defined on R.

Denote $f_t^{\phi V_y(g)}$, a variation of f with $f_0^{\phi V_y(g)} = f$, and deformation vector $\phi V_y(g) \in f^{-1}(TN)$. Denote $\tilde{\nabla}'$, the pull back connection; then

$$\frac{d^2}{dt^2} E\left(f_t^{\phi V_y(g)}\right)\bigg|_{t=0} = \int_M \sum_{i=1}^n \left|\tilde{\nabla}'_{e_i}\phi V_y(g)\right|^2 - \langle R^N(\phi V_y(g), \hat{e}_i)\phi V_y(g), \hat{e}_i\rangle_N \, dV$$

$$(\text{where } \hat{e}_i = df(e_i))$$

$$= \int_M |\nabla\phi|^2 |V_y(g)|^2 + \sum_{i=1}^n \phi^2 \left(\left|\tilde{\nabla}'_{e_i} V_y(g)\right|^2 - \langle R^N(V_y(g), \hat{e}_i)V_y(g), \hat{e}_i\rangle_N\right)$$

$$+ \phi e_i(\phi)\hat{e}_i\left(\left|V_y(g)\right|^2\right) dV.$$

Integrating the second variational formula along $\phi V_y(g)$ over N, with volume element dW, and applying the Fubini theorem and the stability assumption, we acquire

$$(6) \quad 0 \leqslant \int_N \frac{d^2}{dt^2} E\left(f_t^{\phi V_y(g)}\right)\bigg|_{t=0} dW = \int_M |\nabla\phi|^2 \, dV \int_N |V_y(g)|^2 \, dW$$

$$+ \int_M \phi^2 \, dV \int_N \sum_{i=1}^n \left(\left|\tilde{\nabla}'_{e_i} V_y(g)\right|^2 - \langle R^N(V_y(g), \hat{e}_i)V_y(g), \hat{e}_i\rangle_N\right) dW$$

$$+ \sum_{i=1}^n \int_M (e_i\phi)\phi \, dV \int_N \hat{e}_i\left(\left|V_y(g)\right|^2\right) dW.$$

Now we assume N is simply-connected, δ-pinched for $\frac{1}{4} < \delta < 1$, (therefore, the radius of injectivity $> \pi$ by a theorem of Klingenberg [**7**]), and let

$$g(t) = \begin{cases} -\cos t, & |t| \leqslant \pi, \\ 1, & |t| \geqslant \pi. \end{cases}$$

Then

$$V_y(g) = \begin{cases} \sin(\rho_y)\nabla\rho_y, & \text{at } y' \text{ with } \rho_y(y') \leqslant \pi, \\ 0, & \text{otherwise.} \end{cases}$$

It follows from the Cauchy-Schwarz inequality and the compactness of N that

$$(7) \quad \sum_{i=1}^{n} \int_M (e_i\phi)\phi \, dV \int_N \hat{e}_i\left(|V_y(g)|^2\right) dW$$

$$= \sum_{i=1}^{n} \int_M (e_i\phi)(\phi|\hat{e}_i|) \, dV \int_N \bar{e}_i\left(|V_y(g)|^2\right) dW$$

$$\left(\text{where } \bar{e}_i = \begin{cases} \hat{e}_i/|\hat{e}_i| & \text{if } \hat{e}_i \neq 0 \\ 0 & \text{otherwise} \end{cases}\right)$$

$$\leqslant \sum_{i=1}^{n} \int_M c \operatorname{vol}(N)\left((e_i\phi)^2/2\varepsilon + \phi^2|\hat{e}_i|^2\varepsilon/2\right) dV$$

$$\left(\text{where } \varepsilon > 0 \text{ is a constant and } c = \max_N |\nabla|V_y(g)|^2|\right)$$

$$= (c \operatorname{vol}(N)/2\varepsilon)\int_M |\nabla\phi|^2 \, dV + (\varepsilon c \operatorname{vol}(N)/2)\int_M \phi^2|df|^2 \, dV.$$

Furthermore, an estimate of Howard [4], followed from careful manipulation of the volume comparison theorem of Bishop and Crittenden [1], and the Hessian comparison theorem of Greene and Wu [3], implies

$$(8) \quad \int_M \phi^2 \, dV \int_N \sum_{i=1}^{n} \left|\tilde{\nabla}'_{e_i}V_y(g)\right|^2 - \langle R^N(V_y(g), \hat{e}_i)V_y(g), \hat{e}_i\rangle_N dW$$

$$< \operatorname{vol}(S^{k-1})F_k(\delta)\int_M \phi^2|df|^2 \, dV,$$

where

$$(9) \quad F_k(\delta) = \int_0^\pi \max\left\{\cos^2 t, \, \delta \sin^2 t \cos^2\sqrt{\delta}\, t/\sin^2\sqrt{\delta}\, t\right\}$$

$$\times \left(\sin(\sqrt{\delta}\, t)/\sqrt{\delta}\right)^{k-1} - (k-1)\delta\cos^2 t \sin^{k-1} t \, dt.$$

Combining (6)–(9), we obtain

$$(10) \quad \left(-\operatorname{vol}(S^{k-1})F_k(\delta) - \varepsilon c \operatorname{vol}(N)/2\right)\int_M \phi^2|df|^2 \, dV$$

$$\leqslant (1 + c/2\varepsilon)\operatorname{vol}(N)\int_M |\nabla\phi|^2 \, dV.$$

Now set

$$(11) \quad \delta_k = \inf\{\delta: 1/4 < \delta \text{ and } F_k(\delta) < 0 \text{ (hence } k > 2)\}.$$

Choose $0 < \varepsilon < -2\operatorname{vol}(S^{k-1})F_k(\delta_k)/c \operatorname{vol}(N)$ and let

$$C = (1 + c/2\varepsilon)\operatorname{vol}(N)/\left(-\operatorname{vol}(S^{k-1})F_k(\delta_k) - \varepsilon c \operatorname{vol}(N)/2\right).$$

Then (10) becomes

An Intrinsic Stability Inequality 3.1. With the same notations as above, for any smooth, real-valued function ϕ defined on M with compact support,

$$\int_M \phi^2 |df|^2 \, dV \leqslant C \int_M |\nabla \phi|^2 \, dV.$$

The proof of the intrinsic Liouville theorem now follows as in §2.

References

1. R. L. Bishop and R. J. Crittenden, *Geometry of manifolds*, Academic Press, New York, 1966.

2. J. Eells and L. Lemaire, *A report on harmonic maps*, Bull. London Math. Soc. **10** (1978), 1–68.

3. R. E. Greene and H. Wu, *Function theory on manifolds which possess a pole*, Lecture Notes in Math, vol. 699, Springer-Verlag, 1979.

4. R. Howard, *The non-existence of stable submanifolds, varifolds, and harmonic maps in sufficiently pinched simply connected Riemannian manifolds*, Michigan J. Math. (to appear).

5. R. Howard and S. W. Wei, *Non-existence of stable harmonic maps to and from certain homogeneous spaces and submanifolds of Euclidean space*, preprint.

6. ______, *On the existence and non-existence of stable submanifolds and currents in positively curved manifolds and the topology of submanifolds*, preprint.

7. W. Klingenberg, *Contributions to Riemannian geometry in the large*, Ann. Math. **69** (1959).

8. P. F. Leung, *On the Stability of harmonic maps*, Lecture Notes on Harmonic Maps, Springer-Verlag, vol. 949, 1982.

9. R. Schoen and K. Uhlenbeck, *Regularity of minimizing harmonic maps into the sphere*, preprint.

10. S. W. Wei, *An average process in the calculus of variations and the stability of harmonic maps*, Bull. Inst. Math. Acad. Sinica **11** (1983).

11. ______, *Existence and regularity of minimizing harmonic maps into strongly unstable manifolds*, in preparation.

University of California, Los Angeles

Proceedings of Symposia in Pure Mathematics
Volume **44** (1986)

A Regularity Theorem
for Minimizing Hypersurfaces Modulo p

BRIAN WHITE

Introduction. Given any $(m - 1)$-dimensional rectifiable cycle S in $\mathbf{R}^{m+1}$, and any integer $p \geqslant 2$, there exists an m-dimensional rectifiable surface (current) T that minimizes area (or any other even parametric elliptic integrand Φ) subject to the condition that $\partial T \equiv S \pmod{p}$. In case S is smooth, we may also (by virtue of our decomposition theorem) regard T as the solution of a free boundary problem: minimize area among all surfaces T with boundary $S + pS'$ where S' is a free boundary (any integral flat chain).

Such Φ-minimizing hypersurfaces modulo p are interesting because, unlike solutions to other formulations of Plateau's problem, they exhibit singularities even in low dimensions. Also, recently surfaces mod p have been used to prove regularity results about ordinary (i.e., integral current) solutions to variational problems [**W2**].

In this paper we show (§3) that if a Φ-minimizing hypersurface modulo p is, in some cylinder, near a disk with multiplicity $k < p/2$, then it is regular in a smaller cylinder. This implies (by standard arguments, §4) the almost everywhere regularity of T in case p is odd. That in turn implies (§5), by some results of Almgren, that the singular set has dimension $< m$ for arbitrary integrands Φ and $\leqslant m - 1$ for the area integrand.

The idea of the proof is as follows. First $T \llcorner C_1$ (where C_r is the solid cylinder of radius r) is decomposed into p pieces $T_1, T_2, \ldots, T_p$ such that

$$(\partial T)\llcorner \operatorname{int} C_1 = p(\partial T_i)\llcorner \operatorname{int} C_1 \qquad (1 \leqslant i \leqslant p).$$

Then $(p - k)$ of these pieces are singled out as bad. We show, by a comparison, that the area of the bad pieces in a cylinder of radius r decays so rapidly as $r \to 0$ that it is already 0 when $r = \frac{1}{2}$. Consequently $(\partial T)\llcorner \operatorname{int} C_{1/2} = 0$, so that $T \llcorner C_{1/2}$

1980 *Mathematics Subject Classification.* Primary 49F22.

© 1986 American Mathematical Society
0082-0717/86 $1.00 + $.25 per page

is Φ-minimizing as an ordinary rectifiable current. Hence the usual regularity theory (cf. [**SS**]) applies to $T \llcorner C_{1/2}$.

Very precise descriptions of the singular set have been given when $p = 4$ [**W1**] and when $p = 3$, $m = 2$, and $\Phi = $ area [**T**].

I would like to thank Esther Gokhale for her constant encouragement.

1. Preliminaries. In general we follow the conventions of [**F**]. In particular

1.1. Φ is a parametric integrand of degree m on $\mathbf{R}^{m+1}$. That is, Φ: $(\mathbf{R}^{m+1} \sim \{0\}) \times \Lambda_m \mathbf{R}^{m+1} \to (0, +\infty)$ is a continuous function such that

$$\Phi(x, t\omega) \equiv t\Phi(x, \omega), \quad \text{for } t > 0.$$

We also assume that Φ is *even*, i.e. in that $\Phi(x, \omega) \equiv \Phi(x, -\omega)$.

1.2. If T is a rectifiable current, we say

$$T \equiv 0 \pmod{p}$$

provided $T = pS$ for some rectifiable current S. For integral flat chains (such as boundaries of rectifiable currents) the definition of congruence modulo p is slightly more complicated: $T \equiv 0 \pmod{p}$ provided T is the limit (in the $\mathscr{F}$ topology) of a sequence pS_i (where the S_i are integral flat chains). It is not known whether that implies that $T = pS$ for some integral flat chain S, though it happens to be so in the cases that arise in our proofs.

1.3. If T is an m-dimensional rectifiable current in $\mathbf{R}^{m+1}$ and Φ is an even parametric integrand, we define

$$\Phi(T) = \int \Phi(x, \vec{T}x) \, d\|T\|x.$$

We say that T is Φ-minimizing modulo p provided $\Phi(T) \leqslant \Phi(T')$, whenever $\partial T \equiv \partial T' \pmod{p}$.

1.4. $\mathbf{M}(T)$ is the mass of T; $\mathbf{M}^p(T)$ is the mod p mass of T:

$$\mathbf{M}^p(T) = \inf\{\mathbf{M}(T + S) \colon S \equiv 0 \pmod{p}\}.$$

$\mathscr{F}(T)$ is the integral flat norm of T:

$$\mathscr{F}(T) = \inf\{\mathbf{M}(T + \partial R) + \mathbf{M}(R) \colon R \text{ rectifiable}\}.$$

$\mathscr{F}^p(T)$ is the corresponding mod p (semi)norm:

$$\mathscr{F}^p(T) = \inf\{\mathbf{M}(T + \partial R + pS) + \mathbf{M}(R) \colon R, S \text{ rectifiable}\}.$$

Also, $\mathrm{spt}^p(T)$ is the mod p support of T:

$$\mathrm{spt}^p(T) = \bigcap\{\mathrm{spt}\, T' \colon T' \equiv T \pmod{p}\}.$$

If $\mathbf{M}^p(T) < \infty$, then T is congruent modulo p to a *representative modulo p*, i.e., a rectifiable current T' with densities $\leqslant p/2$ almost everywhere, so that:

$$\mathbf{M}(T') = \mathbf{M}^p(T'), \qquad \mathrm{spt}(T') = \mathrm{spt}^p(T')$$

[**F**, pp. 430–431]. Note that if p is odd, then this T' is unique. Note also that if T is Φ-minimizing modulo p, then it is a representative modulo p.

1.5. As in [**W1**], we denote the tangent cone to T at x (if it exists and is unique) by $\mathrm{TAN}(T, x)$.

1.6. If $f\colon \mathbf{R}^m \to \mathbf{R}$ is $\mathscr{L}^1$ with compact support, then $\mathbf{E}^m \llcorner f$ is the dual m-dimensional current.

1.7. LEMMA. *Let T be a rectifiable current. Then for every x and $R > 0$ there is an $r \in (R/2, R)$ such that*

(1) $$\mathbf{M}\big[\partial(T \llcorner \mathbf{B}(x, r)) - (\partial T)\llcorner \mathbf{B}(x, r)\big] \leqslant (R/2)\mathbf{M}(T \llcorner \mathbf{B}(x, R))$$

and an $r' \in (R/2, R)$ such that

(2) $$\mathscr{F}^p(T \llcorner \mathbf{B}(x, r')) \leqslant (2 + R)\cdot\mathscr{F}^p(T).$$

REMARK. Note that if $(\partial T)\llcorner \operatorname{int} \mathbf{B}(x, R) \equiv 0 \pmod p$, then (1) implies

$$\mathbf{M}^p\big[\partial(T \llcorner \mathbf{B}(x, r))\big] \leqslant (R/2)\mathbf{M}(T \llcorner \mathbf{B}(x, R)).$$

PROOF. Note that for almost all r

$$\frac{d}{dr}\mathbf{M}(T \llcorner \mathbf{B}(x, r)) \geqslant \mathbf{M}\big[\partial(T \llcorner \mathbf{B}(x, r)) - (\partial T)\llcorner \mathbf{B}(x, r)\big].$$

$$\therefore \quad \mathbf{M}(T \llcorner \mathbf{B}(x, R)) \geqslant \int_{R/2}^{R} \mathbf{M}\big[\partial(T \llcorner \mathbf{B}(x, r)) - (\partial T)\llcorner \mathbf{B}(x, r)\big]\, dr.$$

(1) follows immediately.

To see (2), note that for every $\varepsilon > 0$ there exist rectifiable currents Q and S such that

$$\mathbf{M}(T + \partial Q + pS) + \mathbf{M}(Q) \leqslant \mathscr{F}^p(T) + \varepsilon.$$

On the other hand,

$$\begin{aligned}
\mathscr{F}^p(T \llcorner \mathbf{B}(x, r)) &\leqslant \mathbf{M}(T \llcorner \mathbf{B}(x, r) + \partial(Q \llcorner \mathbf{B}(x, r)) \\
&\quad + pS \llcorner \mathbf{B}(x, r)) + \mathbf{M}(Q \llcorner \mathbf{B}(x, r)) \\
&\leqslant \mathbf{M}\big[T \llcorner \mathbf{B}(x, r) + (\partial Q)\llcorner \mathbf{B}(x, r) + pS \llcorner \mathbf{B}(x, r)\big] \\
&\quad + \mathbf{M}\big[\partial(Q \llcorner \mathbf{B}(x, r)) - (\partial Q)\llcorner \mathbf{B}(x, r)\big] + \mathbf{M}\big[Q \llcorner \mathbf{B}(x, r)\big] \\
&\leqslant \mathbf{M}\big[T + \partial Q + pS\big] + \mathbf{M}\big[\partial(Q \llcorner \mathbf{B}(x, r)) - (\partial Q)\llcorner \mathbf{B}(x, r)\big] + \mathbf{M}(Q) \\
&\leqslant \mathscr{F}^p(T) + \varepsilon + \mathbf{M}\big[\partial(Q \llcorner \mathbf{B}(x, r)) - (\partial Q)\llcorner \mathbf{B}(x, r)\big].
\end{aligned}$$

But now by applying (1) to Q, we see we can choose r' so that

$$\mathscr{F}^p(T \llcorner \mathbf{B}(x, r')) \leqslant \mathscr{F}^p(T) + \varepsilon + (R/2)\mathbf{M}(Q)$$

$$\leqslant (1 + R/2)(\mathscr{F}^p(T) + \varepsilon). \qquad \square$$

1.8. ROCKING HORSE APPROXIMATION THEOREM.

Hypotheses: $\varnothing$,   *Conclusions*: $\varnothing$.

PROOF (ALLARD). By conclusions (1), (7), and (256) of F6.8.3, together with 1.2.3, 5.1.7, 2.8.8, and 1.9.19b applied to the equality 7.0.7, we see that the hypotheses of 4.8.2 are satisfied. $\square$

2. A decomposition theorem.

DECOMPOSITION THEOREM. *Let U be an open convex subset of $\mathbf{R}^{m+1}$ and let T be an m-dimensional rectifiable current such that*

$$\operatorname{spt} T \subset \overline{U},$$

$$\mathbf{M}(T \llcorner \partial U) = 0,$$

$$(\partial T) \llcorner U \equiv 0 \quad (\operatorname{mod} p).$$

Then there exist rectifiable currents $T_1, T_2, \ldots, T_p$ such that

$$T = \sum T_i,$$

$$\|T\| = \sum \|T_i\|,$$

$$(\partial T) \llcorner U = p\left[(\partial T_i) \llcorner U\right], \qquad 1 \leqslant i \leqslant p.$$

REMARK. More generally $\overline{U}$ can be replaced by any compact Riemannian $(m + 1)$-dimensional manifold M with (possibly empty) boundary such that $H_m(M, \partial M; \mathbf{Z}_p) = 0$.

PROOF. Since $\operatorname{spt} T$ is compact, we may assume $\overline{U}$ is compact. By hypothesis,

$$T(\operatorname{mod} p) \in Z_m^p(\overline{U}, \partial U).$$

But $H_m(\overline{U}, \partial U; \mathbf{Z}_p) = 0$, so there exists an $(m + 1)$-dimensional rectifiable current R in $\overline{U}$ such that

(1) $$\operatorname{spt}\left[(T - \partial R)(\operatorname{mod} p)\right] \subset \partial U.$$

Since R is an $(m + 1)$-dimensional rectifiable current in $\mathbf{R}^{m+1}$, we can write R as

$$R = \mathbf{E}^{m+1} \llcorner r,$$

where $r: U \to \mathbf{Z}$ is an $\mathscr{L}^1$ function. Note that we may choose R so that $-p/2 < r(x) \leqslant p/2$ for all x: this insures that ∂R is rectifiable ([**W1**, 2.3]).

By (1),

$$(T - \partial R) \llcorner U = pT_p$$

for some rectifiable current T_p. Hence

$$(T - T_p - \partial R) \llcorner U = (p - 1)T_p,$$

so

$$\left[\partial(T - T_p)\right] \llcorner U \equiv 0 \quad (\operatorname{mod} p - 1).$$

Now $T - T_p$ satisfies the hypotheses of the theorem with p replaced by $p - 1$. So by induction we have $T_1, T_2, \ldots, T_{p-1}$ such that

$$T - T_p = \sum_{i=1}^{p-1} T_i, \qquad \|T - T_p\| = \sum_{i=1}^{p-1} \|T_i\|.$$

Thus it suffices to prove that $\|T\| = \|T - T_p\| + \|T_p\|$. This is proved exactly as in the decomposition theorem of [**W2**]. $\quad\square$

3. The key result.

3.1. THEOREM. *Let Φ be an even parametric integrand of degree m on $\mathbf{R}^{m+1}$ such that (for all ω)*

$$\lambda^{-1}|\omega| \leqslant \Phi(x,\omega) \leqslant \lambda|\omega|.$$

Then there is an $\varepsilon = \varepsilon(\lambda, p, m) > 0$ with the following property. If T is an m-dimensional rectifiable current in $C_R = \mathbf{B}^m(0, R) \times \mathbf{R}$ such that T is Φ-minimizing modulo p and if

$$(\partial T)\llcorner\operatorname{int}(C_R) \equiv 0 \quad (\operatorname{mod} p),$$

$$\Pi_{\#}(T) \equiv k(\mathbf{E}^m\llcorner\mathbf{B}^m(0, R)) \quad (\operatorname{mod} p),$$

$$\mathbf{M}(T) \leqslant (k\alpha_m + \varepsilon)R^m,$$

where $0 < k < p/2$, then

$$(\partial T)\llcorner\operatorname{int} C_{R/2} = 0.$$

(Here $\Pi\colon \mathbf{R}^m \times \mathbf{R} \to \mathbf{R}^m$ is orthogonal projection and α_m is the volume of the unit m-dimensional ball $\mathbf{B}^m$.)

PROOF. By scaling, we may assume without loss of generality that $R = 1$. By the decomposition theorem, we have rectifiable currents T_i ($1 \leqslant i \leqslant p$) such that

$$
\begin{gathered}
T - \sum T_i, \qquad \|T\| = \sum \|T_i\|, \\
(\partial T_i)\llcorner\operatorname{int} C_1 = (\partial T_j)\llcorner\operatorname{int} C_1.
\end{gathered}
$$
(0)

Since $\Pi_{\#}(T_i)$ is an m-dimensional rectifiable current in $\mathbf{B}^m$,

$$\Pi_{\#}(T_i) = \mathbf{E}^m\llcorner t_i$$

for some $\mathscr{L}^1$ function $t_i\colon \mathbf{B}^m \to \mathbf{Z}$, where

$$\operatorname{TAN}(\Pi_{\#}(T_i), x) = t_i(x)\mathbf{E}^m \quad (\text{a.e. } x \in \mathbf{B}^m).$$
(1)

Since $\Pi_{\#}(T) = \sum \Pi_{\#}(T_i) \equiv k(\mathbf{E}^m\llcorner\mathbf{B}^m) \pmod{p}$,

$$\sum t_i(x) \equiv k \quad (\operatorname{mod} p) \quad (\text{a.e. } x \in \mathbf{B}^m).$$
(2)

On the other hand,

$$k\alpha_n + \varepsilon \geqslant \mathbf{M}(T) = \sum \mathbf{M}(T_i)$$

$$\geqslant \sum \mathbf{M}(\Pi_{\#}(T_i)) = \sum \int_{\mathbf{B}^m} |t_i(x)|\, dx.$$

Thus,

$$\int_{\mathbf{B}^m} \left[\sum_{i=1}^{p} |t_i(x)| - k \right] dx \leqslant \varepsilon.$$

Since the integrand is integer-valued, if $\varepsilon < \alpha_m$, then we can find an $a \in \operatorname{int} \mathbf{B}^m$ such that

$$\sum |t_i(a)| \leqslant k$$

and such that (1) and (2) hold. It follows that

$$\sum_{i=1}^{p} t_i(a) = \sum_{i=1}^{p} |t_i(a)| = k.$$

In particular, $t_i(a) = 0$ for all but k values of i. Without loss of generality,

$$t_i(a) = 0 \quad \text{for } i > k,$$

(4)
$$\therefore \quad \sum_{i=1}^{k} t_i(a) = k.$$

(A typical example is shown in Figure 1. The surfaces $T_1, T_2, \ldots, T_k$ cover a portion of $\mathbf{B}^m$ (including the point a) with multiplicity k. The remaining surfaces $T_{k+1}, \ldots, T_p$ are "upside-down" in that they cover the rest of $\mathbf{B}^m$ ($p - k$) times with the reverse orientation.)

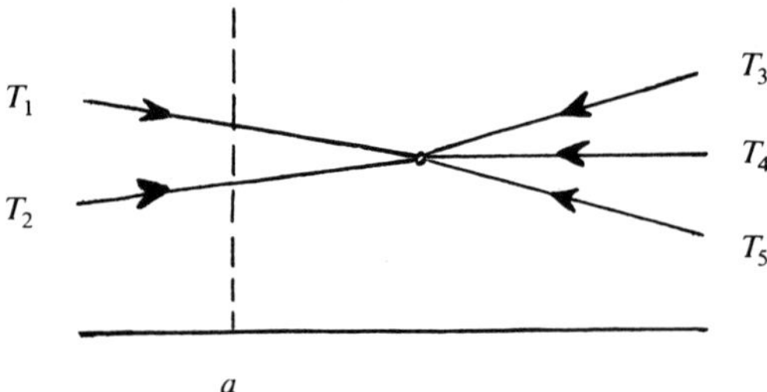

FIGURE 1. $k = 2$, $p = 5$.

The first step is to show that the upside-down pieces $T_{k+1}, T_{k+2}, \ldots, T_p$ have small mass. Let

$$R = \sum_{i=1}^{k} (T_i - T_{i+k}).$$

Then by (0),

$$(\partial R) \llcorner \operatorname{int} C_1 = 0.$$

Thus by the constancy theorem [F, 4.1.7], $\Pi_\#(R)$ is a constant multiple of $\mathbf{E}^m \llcorner \mathbf{B}^m$. Now,

$$\text{TAN}(\Pi_\# R, a) = \text{TAN}\left(\Pi_\#\left(\sum_{i=1}^{k} (T_i - T_{i+k})\right), a\right)$$

$$= \sum_{i=1}^{k} \left(\text{TAN}(\Pi_\# T_i, a) - \text{TAN}(\Pi_\# T_{i+k}, a)\right)$$

$$= \mathbf{E}^m \llcorner \sum_{i=1}^{k} \left(t_i(a) - t_{i+k}(a)\right)$$

$$= \mathbf{E}^m \llcorner k \quad (\text{by } (4)).$$

Hence $\Pi_{\#}R = k(\mathbf{E} \llcorner B^m)$. Thus, $\mathbf{M}(R) \geqslant k\alpha_m$. On the other hand,

$$\mathbf{M}(R) = \mathbf{M}\left(\sum_{i=1}^{m} (T_i - T_{i+k}) \right)$$

$$\leqslant \sum_{i=1}^{k} (\mathbf{M}(T_i) + \mathbf{M}(T_{i+k}))$$

$$= \mathbf{M}(T) - \sum_{i>2k} \mathbf{M}(T_i)$$

$$< k\alpha_m + \varepsilon - \sum_{i>2k} \mathbf{M}(T_i)$$

so $\sum_{i>2k} \mathbf{M}(T_i) < \varepsilon$. In particular, $\mathbf{M}(T_p) < \varepsilon$.

Likewise, the same argument (with a suitably chosen R) shows that

$$(5) \qquad \mathbf{M}(T_i) < \varepsilon \quad \text{if } i > k.$$

The next step is to estimate $\mathbf{M}(T_i \llcorner C_r)$ $(i > k)$. To do this, we construct comparison surfaces as follows. For $j > k$ let

$$(6) \qquad Q_j = \sum_{i=1}^{k} T_i - kT_j.$$

Notice by (0) that $(\partial Q_j)\llcorner \text{int } C_1 = 0$, so

$$(7) \qquad \left(\partial(T - Q_j) \right)\llcorner \text{int } C_1 \equiv 0 \quad (\bmod\, p).$$

For $0 < r < 1$, let $S_j = S_j(r)$ be a mass-minimizing rectifiable current such that

$$(8) \qquad \partial S_j \equiv \partial\left[(T - Q_j)\llcorner C_r \right] \quad (\bmod\, p).$$

Then $T \llcorner C_r$ and $(Q_j \llcorner C_r) + S_j$ have the same boundary modulo p, so

$$\Phi(T \llcorner C_r) \leqslant \Phi(Q_j \llcorner C_r) + \Phi(S_j).$$

$$\therefore \quad \sum_{i=1}^{p} \Phi(T_i \llcorner C_r) \leqslant \Phi\left(\sum_{i=1}^{k} T_i \llcorner C_r - kT_j \llcorner C_r \right) + \Phi(S_j)$$

$$\leqslant \sum_{i=1}^{k} \Phi(T_i \llcorner C_r) + k\Phi(T_j \llcorner C_r) + \Phi(S_j).$$

$$\therefore \quad \sum_{i=k+1}^{p} \Phi(T_i \llcorner C_r) \leqslant k\Phi(T_j \llcorner C_r) + \Phi(S_j).$$

Sum from $j = k + 1$ to p,

$$(p - k) \sum_{i>k} \Phi(T_i \llcorner C_r) \leqslant k \sum_{j>k} \Phi(T_j \llcorner C_r) + \sum_{j>k} \Phi(S_j).$$

$$\therefore \quad (p - 2k) \sum_{i>k} \Phi(T_i \llcorner C_r) \leqslant \sum_{j>k} \Phi(S_j),$$

so

$$(p - 2k) \sum_{i>k} \mathbf{M}(T_i \llcorner C_r) \leqslant \lambda^2 \sum_{j>k} \mathbf{M}(S_j).$$

Since $p - 2k \geqslant 1$,

$$(9) \qquad\qquad E(r) \leqslant \lambda^2 \sum_{j>k} \mathbf{M}(S_j),$$

where

$$E(r) = \sum_{i>k} \mathbf{M}(T_i \mathbin{\llcorner} C_r).$$

On the other hand, for almost all $r \in (0,1)$,

$$\frac{d}{dr}\mathbf{M}\big[(T - Q_j)\mathbin{\llcorner} C_r\big] \geqslant \mathbf{M}\big\{\partial\big[(T - Q_j)\mathbin{\llcorner} C_r\big] - \big[\partial(T - Q_j)\mathbin{\llcorner} C_r\big]\big\}$$

$$\geqslant \mathbf{M}^p\big\{\partial\big[(T - Q_j)\mathbin{\llcorner} C_r\big] - \big[\partial(T - Q_j)\mathbin{\llcorner} C_r\big]\big\}$$

$$= \mathbf{M}^p(\partial S_j)$$

by (7) and (8). Hence by the isoperimetric inequality [**F**, 4.2.10$^\nu$ (p. 431)],

$$\mathbf{M}(S_j) \leqslant K\big[(d/dr)\mathbf{M}((T - Q_j)\mathbin{\llcorner} C_r)\big]^{m/(m-1)}$$

(where K depends only on m). By definition of Q_j (6), $T - Q_j = \sum_{i>k} T_i + kT_j$, so

$$\mathbf{M}(S_j) \leqslant K\left[\frac{d}{dr}\Big\{\sum_{i>k}\mathbf{M}(T_i\mathbin{\llcorner} C_r) + k\mathbf{M}(T_j\mathbin{\llcorner} C_r)\Big\}\right]^{m/(m-1)}.$$

Thus (summing),

$$\sum_{j>k}\mathbf{M}(S_j) \leqslant K\big(pE'(r)\big)^{m/(m-1)}.$$

Hence by (9)

$$E(r) \leqslant \lambda^2 K\big(pE'(r)\big)^{m/(m-1)}$$

or

$$E'(r) \geqslant p^{-1}(\lambda^2 K)^{(1/m)-1}E(r)^{1-(1/m)}.$$

Thus if $E(1/2) > 0$, then

$$E'(r)E(r)^{(1/m)-1} \geqslant \beta = p^{-1}(\lambda^2 K)^{(1/m)-1}$$

for almost all $r \geqslant \tfrac{1}{2}$, so

$$(d/dr)\big[E(r)^{1/m}\big] \geqslant \beta.$$

Integrate from $r = \tfrac{1}{2}$ to 1,

$$E(1)^{1/m} \geqslant 2^{-1}\beta, \qquad E(1) \geqslant 2^{-m}\beta^m.$$

But $E(1) < (p - k)\varepsilon$ (by (5)), so $\varepsilon > (p - k)^{-1}2^{-m}\beta^m$. Now this holds if $E(1/2) > 0$. Conversely, if $\varepsilon < p^{-1}2^{-m}\beta^m$, then $E(1/2) = 0$. But then

$$\mathbf{M}(T_p\mathbin{\llcorner} C_{1/2}) = 0$$

so

$$\mathbf{M}\big((\partial T)\mathbin{\llcorner} C_{1/2}\big) = p\mathbf{M}\big((\partial T_p)\mathbin{\llcorner} C_{1/2}\big) = 0. \quad \square$$

4. $\mathscr{H}^m$-Almost everywhere regularity. For the remainder of this paper, Φ will be Lipschitz in its first argument, $C^{2,\alpha}$ in its second, and elliptic modulo p. Specifically,

(H1) $$\lambda^{-1}|\omega| \leqslant \Phi(x,\omega) \leqslant \lambda|\omega|,$$

(H2) $$|\Phi(x,\omega) - \Phi(y,\omega)| \leqslant L|x-y| \quad \text{for } |\omega| \leqslant 1,$$

(H3) $$\Phi(x,\cdot)|\{\omega\colon |\omega|=1\} \quad \text{has } C^{2,\alpha} \text{ norm } \leqslant L$$

(H4) $$\Phi_x(T) - \Phi_x(R) \geqslant \Lambda[\mathbf{M}(T) - \mathbf{M}(R)],$$

provided R is supported in an m-dimensional plane, has density $\leqslant p/2$ almost everywhere, and $\partial T \equiv \partial R \pmod{p}$. Here Φ_x is the constant coefficient integrand $\Phi_x(\omega) = \Phi(x,\omega)$. Note that (H2) implies

4.1. $$|\Phi(T) - \Phi_x(T)| \leqslant Lr\mathbf{M}(T) \quad \text{if spt } T \subset \mathbf{B}(x,r).$$

We begin with three basic, elementary propositions about Φ-minimizers.

4.2. PROPOSITION. *There exists $\Gamma = \Gamma(\lambda, p, m)$ such that if T is Φ-minimizing modulo p and if $(\partial T)\llcorner\mathrm{int}\mathbf{B}(x,R) \equiv 0 \pmod{p}$, then*

(1) $$\mathbf{M}(T\llcorner\mathbf{B}(x,R)) \leqslant \Gamma R^m.$$

Furthermore, there exists $r \in (R/2, R)$ such that

(2) $$\mathbf{M}^p(\partial(T\llcorner\mathbf{B}(x,r))) \leqslant \Gamma r^{m-1}.$$

If also $x \in \mathrm{spt}\, T$, then

(3) $$\mathbf{M}(T\llcorner\mathbf{B}(x,R)) \geqslant \Gamma^{-1}R^m.$$

PROOF. Note that there is an S with support in $\partial\mathbf{B}(x,R)$, with densities $\leqslant p/2$, and such that $\partial S \equiv \partial(T\llcorner\mathbf{B}(x,R)) \pmod{p}$. Hence,

$$\begin{aligned}
\Phi(T\llcorner\mathbf{B}(x,R)) &\leqslant \Phi(S) \\
&\leqslant \lambda\mathbf{M}(S) \\
&\leqslant \lambda(p/2)\mathscr{H}^m(\partial\mathbf{B}^{m+1}(x,R)) \\
&= \lambda(p/2)(m+1)\alpha_{m+1}R^m.
\end{aligned}$$

$$\therefore \quad \mathbf{M}(T\llcorner\mathbf{B}(x,R)) \leqslant \lambda^2(p/2)(m+1)\alpha_{m+1}R^m.$$

This proves (1); (2) follows by Lemma 1.7. (3) is proved using the isoperimetric inequality [**F**, 4.2.10^p, p. 431]; see [**F**, 5.1.6, p. 523].

4.3. PROPOSITION. *If T is Φ-minimizing mod p, then*

$$\Phi(T) \leqslant \Phi(S) + \lambda\mathscr{F}^p(T-S)$$

for every S. Hence also

$$\mathbf{M}(T) \leqslant \lambda^2\mathbf{M}(S) + \lambda^2\mathscr{F}^p(T-S).$$

PROOF. Let R and Q be rectifiable currents of dimension $m + 1$ and m, respectively. Then since T and $T + \partial R + pQ$ have the same boundary (modulo p),

$$\Phi(T) \leqslant \Phi(T + \partial R + pQ)$$
$$\leqslant \Phi(S) + \Phi(T - S + \partial R + pQ)$$
$$\leqslant \Phi(S) + \lambda \mathbf{M}(T - S + \partial R + pQ).$$

But $\mathscr{F}^p(T - S) = \inf\{\mathbf{M}(T - S + \partial R + pQ) + \mathbf{M}(R)\}$, so we are done.

4.4. PROPOSITION. *Let T be Φ-minimizing mod p, $x \in \operatorname{spt} T$, and S be any other rectifiable m-current. Then*

$$\operatorname{dist}(x, \operatorname{spt} S \cup \operatorname{spt}^p(\partial T)) < c \cdot \mathscr{F}^p(T - S)^{1/m}$$

where $c = c(\lambda, p, m)$.

PROOF. Let R be $\operatorname{dist}(x, \operatorname{spt} S \cup \operatorname{spt}^p(\partial T))$ or 1, whichever is smaller. By Lemma 1.7, there is an $r \in (R/2, R)$ such that

$$\mathscr{F}^p((T - S)\llcorner \mathbf{B}(x, r)) \leqslant 3\mathscr{F}^p(T - S).$$
$$\therefore \quad \mathscr{F}^p(T \llcorner \mathbf{B}(x, r)) \leqslant 3\mathscr{F}^p(T - S).$$

By Propositions 4.3 (applied to $T\llcorner \mathbf{B}(x, r)$ and 0 in place of T and S, respectively) and 4.2,

$$\Gamma^{-1} r^m \leqslant \mathbf{M}(T\llcorner \mathbf{B}(x, r)) \leqslant 3\lambda^2 \mathscr{F}^p(T - S).$$
$$\therefore \quad r^m \leqslant 3\Gamma\lambda^2 \cdot \mathscr{F}^p(T - S).$$
$$\therefore \quad R^m \leqslant 2^m \cdot 3\Gamma\lambda^2 \cdot \mathscr{F}^p(T - S).$$

4.5. THEOREM. *For every $\varepsilon > 0$ there is a $\delta = \delta(\lambda, \Lambda, L, \alpha, p, m) > 0$ such that if $r < \delta$,*

$$T \text{ is } \Phi\text{-minimizing mod } p,$$
$$(\partial T)\llcorner \operatorname{int} C_r \equiv 0 \quad (\operatorname{mod} p),$$
$$\operatorname{spt} T \subset \mathbf{B}^m(0, r) \times [-\delta r, \delta r]$$

and

$$\Pi_\sharp(T) \equiv k(\mathbf{E}^m \llcorner \mathbf{B}^m(0, r)) \quad (\operatorname{mod} p),$$

then

$$(4) \qquad \mathbf{M}(T) \leqslant (|k|\alpha_m + \varepsilon)r^m$$

and if $|k| < p/2$, then $\operatorname{spt}(T\llcorner C_{r/8})$ is a $C^{1,1/2}$ submanifold.

PROOF. We may assume $0 \leqslant k \leqslant p/2$. Note that there is a rectifiable current S in $\partial \mathbf{B}^m(0, r) \times [-\delta r, \delta r]$ such that

$$\partial S \equiv \partial T - \partial(kD) \quad (\operatorname{mod} p)$$

where $D = \mathbf{E}^m \llcorner \mathbf{B}^m(0, r)$. Note we can choose S to have densities $\leqslant p/2$, so that

$$\mathbf{M}(S) \leqslant (p/2)\mathscr{H}^m(\operatorname{spt} S)$$
$$\leqslant (p/2)\mathscr{H}^m(\partial \mathbf{B}^m(0, r) \times [-\delta r, \delta r])$$
$$\leqslant \delta c r^m.$$

Since $\partial(kD) \equiv \partial(T - S)$, ellipticity implies

$$(5) \quad \Lambda[\mathbf{M}(T - S) - \mathbf{M}(kD)] \leqslant \Phi_0(T - S) - \Phi_0(kD)$$
$$\leqslant \Phi_0(T) - \Phi_0(kD) + \Phi_0(S)$$
$$\leqslant \Phi(T) - \Phi(kD) + 2Lr[\mathbf{M}(T) + \mathbf{M}(kD)] + \Phi_0(S)$$

(by 4.1). Because $\partial(kD + S) \equiv \partial T$,

$$\Phi(T) \leqslant \Phi(kD + S)$$
$$\leqslant \Phi(kD) + \Phi(S).$$

So

$$\mathbf{M}(T) \leqslant \lambda^2(\mathbf{M}(kD) + \mathbf{M}(S))$$
$$\leqslant \lambda^2(k\alpha_m + \delta c)r^m$$

and (5) becomes

$$\Lambda[\mathbf{M}(T - S) - \mathbf{M}(kD)] \leqslant \Phi(S) + 2Lr[\mathbf{M}(T) + \mathbf{M}(kD)] + \Phi_0(S)$$
$$\leqslant \Phi(S) + 2Lr(c'r^m) + \Phi_0(S).$$
$$\therefore \quad \Lambda[\mathbf{M}(T) - \mathbf{M}(kD)] \leqslant \mathbf{M}(S) + \Phi(S) + \Phi_0(S) + 2Lc'r^{m+1}$$
$$\leqslant (1 + 2\lambda)\mathbf{M}(S) + 2Lc'r^{m+1}$$
$$\leqslant c''\delta r^m.$$
$$\therefore \quad \mathbf{M}(T) \leqslant \left(k\alpha_m + \Lambda^{-1}c''\delta\right)r^m.$$

Thus we get (4) if $\delta \leqslant \varepsilon\Lambda/c''$.

Note that if δ is small enough, then (by Theorem 3.1) $T \llcorner C_{r/2}$ has no boundary in int $C_{r/2}$ and satisfies the hypotheses of [**SS**], according to which $\mathrm{spt}(T \llcorner C_{r/8})$ is then a $C^{1,1/2}$ submanifold. $\square$

REMARK. In case $p = 2$, the last conclusion of Theorem 4.5 is true without the hypothesis $k < p/2$. To see this, let $T = T_1 + T_2$ be the decomposition of §2. Then

$$T \equiv T' \pmod 2$$

where $T' = T_1 - T_2$. Now T' is also Φ-minimizing mod 2 (hence Φ-minimizing) and $(\partial T')\llcorner \mathrm{int}\, C_r = 0$, so T' satisfies the hypotheses of [**SS**].

4.6. THEOREM. *Let T be an m-dimensional rectifiable current in $\mathbf{R}^{m+1}$ that is Φ-minimizing mod p, where p is odd and Φ is an even integrand satisfying the hypotheses* (H1–H4). *Then T is regular $\mathcal{H}^m$-almost everywhere in $\mathbf{R}^{m+1} \sim \mathrm{spt}^p(\partial T)$.*

Indeed, if d is the Hausdorff dimension of the singular set of T, then $d < m - \varepsilon$ (where $\varepsilon > 0$ does not depend on T). If Φ is m-dimensional area (with respect to a Riemannian metric), then $d \leqslant m - 1$.

PROOF. For $\mathcal{H}^m$-almost every $x \in \mathrm{spt}\, T \sim \mathrm{spt}^p(\partial T)$, T has a tangent cone that is a plane with some multiplicity k, $|k| \leqslant p/2$. Since p is odd, $|k| < p/2$. One can show (cf. [**F**, 5.1.6, p. 523]), using the lower density bound of Proposition

4.2, that this plane is the ordinary (not just measure theoretic) tangent cone to spt T at x. Hence the hypotheses of Theorem 4.5 hold in some neighborhood of x, so that x is a regular point of T. This proves the $\mathscr{H}^m$-almost everywhere regularity.

The $\mathscr{H}^{m-\varepsilon}$ almost everywhere regularity follows from a general argument of Almgren. Since his argument has not yet appeared in print, we explain it in the next section.

The last statement ($d \leqslant m - 1$) follows from Almgren's "stratification of singularities" theorem. Define the index of a stationary cone C to be the dimension of the set of points y at which $\Theta^m(C, y) = \Theta^m(C, 0)$ (where $\Theta^m(C, y)$ is the m-dimensional density of C at y). Almgren's first observation is that this set is a subspace (so that the index is an integer) and that C is the Cartesian product of this subspace with a cone of dimension $(m - \text{index}(C))$. Now define the index $i(x)$ of a point x in any stationary varifold V to be the maximum index of all tangent cones to V at x. Then Almgren's theorem asserts that the set of points in V of index $\leqslant k$ has Hausdorff dimension $\leqslant k$. (Although the theorem appears in the enormous [A, 2.27], its proof is short, elegant, and independent of the rest of that paper.)

Note that if x is of index m in T, then x must have a tangent cone that is an m-dimensional plane. In our case this implies that x is a regular point. Hence a singular point can have index at most $m - 1$. $\square$

5. $\mathscr{H}^{m-\varepsilon}$-Almost everywhere regularity.

5.1. WORK-RACCOON THEOREM. *Let $\mathscr{S}$ be a family of compact subsets of $\mathbf{B}(0, 1)$ such that*

(1) $\qquad$ *if $K \in \mathscr{S}$, $\quad x \in \mathbf{B}(0, 1)$ and $0 < R \leqslant 1$,*

then there is an $r \in [R/2, R]$ such that
$$(K \cap \mathbf{B}(x, r) - x)/r \in \mathscr{S}.$$

(2) $\qquad$ *$\mathscr{S}$ is compact in the Hausdorff metric on sets.*

Then $\{s: \mathscr{H}^s(K) = 0 \text{ for all } K \in \mathscr{S}\}$ is open.

PROOF. Define $h^s(K)$ to be the infimum of $\sum r_i^s$ among all coverings
$$K \subset \bigcup_i \text{int } \mathbf{B}(x_i, r_i), \qquad r_i \leqslant 1/2.$$

Note that for K compact we can restrict ourselves to finite covers. Note also that $h^s(K) = 0$ if and only if $\mathscr{H}^s(K) = 0$.

Suppose $h^s(K) = 0$ for all $K \in \mathscr{S}$. We must show that there is a $t < s$ such that $h^t(K) = 0$ for all $K \in \mathscr{S}$.

By hypothesis, each $K \in \mathscr{S}$ has a finite cover $\mathscr{C}(K)$ by open balls such that
$$\sum_{\text{int } \mathbf{B}(x, r) \in \mathscr{C}(K)} r^s < 4^{-s}.$$

Let $\mathscr{U}(K)$ be the set of all $R \in \mathscr{S}$ such that
$$R \subset \bigcup \{\text{int } \mathbf{B}(x, r): \text{int } \mathbf{B}(x, r) \in \mathscr{C}(K)\}.$$

Then $\mathscr{U}(K)$ is an open subset of $\mathscr{S}$ that contains K. Since $\mathscr{S}$ is compact, we can find $K_1, K_2, \ldots, K_N$ such that

$$\mathscr{S} \subset \mathscr{U}(K_1) \cup \mathscr{U}(K_2) \cup \cdots \cup \mathscr{U}(K_N).$$

Let the open balls in $\mathscr{C}(K_i)$ be int $\mathbf{B}(x_{ij}, r_{ij})$ $(1 \leqslant j \leqslant M)$.

Since

$$\max_{i \leqslant N} \sum_{j=1}^{M} r_{ij}^s < 4^{-s},$$

there exists a $t < s$ such that

$$\max_{i \leqslant N} \sum_{j=1}^{M} r_{ij}^t < 4^{-s}.$$

Now let K be any member of $\mathscr{S}$. Then there exists $x_k \in \mathbf{B}(0,1)$ and $R_k \in (0, 1/2)$ such that

$$K \subset \bigcup_k \text{int } \mathbf{B}(x_k, R_k), \qquad \sum R_k^t < h^t(K) + \varepsilon.$$

Hence we can find r_k with $R_k \leqslant r_k \leqslant 2R_k$ such that

$$K \subset \bigcup_k \text{int } \mathbf{B}(x_k, r_k), \qquad \sum r_k^t < 2^t h^t(K) + 2^t \varepsilon$$

and

$$(K \cap \mathbf{B}(x_k, r_k) - x_k)/r_k \in \mathscr{S}.$$

Hence

$$(K \cap \mathbf{B}(x_k, r_k) - x_k)/r_k \in \mathscr{U}(K_{i(k)})$$

for some $i(k) \leqslant N$. Thus,

$$(K \cap \mathbf{B}(x_k, r_k) - x_k)/r_k \subset \bigcup_j \text{int } \mathbf{B}(x_{i(k)j}, r_{i(k)j}).$$

$$\therefore \quad K \cap \mathbf{B}(x_k, r_k) \subset \bigcup_j \text{int} \mathbf{B}(r_k x_{i(k)j} + x_k, r_k r_{i(k)j}).$$

$$\therefore \quad K \subset \bigcup_k \bigcup_j \text{int } \mathbf{B}(z_{kj}, r_k r_{i(k)j}).$$

$$\therefore \quad h^t(K) \leqslant \sum_{k,j} (r_k r_{i(k)j})^t$$

$$= \sum_k r_k^t \left(\sum_j r_{i(k)j}^t \right)$$

$$< \sum_k r_k^t \cdot 4^{-s}$$

$$< 2^t (h^t(K) + \varepsilon) 4^{-s}.$$

$$\therefore \quad h^t(K) < 2^{-s} (h^t(K) + \varepsilon)$$

for all $\varepsilon > 0$. Hence $h^t(K) \leqslant 2^{-s} h^t(K)$.

$$\therefore \quad h^t(K) = 0.$$

We are now ready to complete the proof of Theorem 4.6.

5.2. THEOREM. *Let Φ satisfy the hypotheses (H1–H4) of §4, and let p be odd or $p = 2$. Let G be the set of $s \in (0, +\infty)$ such that*

$$\mathscr{H}^s(\operatorname{sing}(T) \sim \operatorname{spt}^p(\partial T)) = 0$$

for every T that is Φ-minimizing modulo p. Then G is open.

REMARK. $\operatorname{sing}(T)$ is defined to be the set of points $x \in \operatorname{spt} T$ for which $(\operatorname{spt} T) \cap \operatorname{int} \mathbf{B}(x, r)$ is *not* a C^1 submanifold-with-boundary for any $r > 0$.

PROOF. For $x \in \mathbf{R}^{m+1}$ and $0 \leqslant r \leqslant 1$, let $\Phi_{x,r}$ be the integrand

$$\Phi_{x,r}(z, \omega) = \Phi(rz + x, \omega).$$

Note that each $\Phi_{x,r}$ satisfies the hypotheses (H1–H4) (with the same bounds λ, Λ, α, and L that Φ does). Let $\mathscr{T}$ be the set of rectifiable m-currents T in $\mathbf{B}(0, 2)$ such that

$$T \text{ is } \Phi_{x,r} \text{ minimizing mod } p \text{ for some } |x| \leqslant 1, \ r \leqslant 1,$$

$$(\partial T) \llcorner \operatorname{int} \mathbf{B}(0, 2) \equiv 0 \quad (\operatorname{mod} p),$$

$$\mathbf{M}^p(\partial T) \leqslant 2^{m-1}\Gamma$$

(where Γ is as in 4.2). Standard arguments (along the lines of 4.3, using [F, 4.2.17^p]) show that $\mathscr{T}$ is $\mathscr{F}^p$-compact.

Now let $\mathscr{S}$ be the set of compact sets K such that

$$K \subset \operatorname{sing}(T) \cap \mathbf{B}(0, 1) \quad \text{for some } T \in \mathscr{T}.$$

We claim that $\mathscr{S}$ satisfies the hypotheses (1) and (2) of Theorem 5.1. To see (1), let $K \in \mathscr{S}$ and $0 \leqslant R \leqslant 1$. By Proposition 4.2, there is an $r \in (R/2, R)$ such that

$$\mathbf{M}^p(\partial(T \llcorner \mathbf{B}(x, 2r))) < 2^{m-1}\Gamma r^{m-1}$$

so

$$\mathbf{M}^p(\partial T') \leqslant 2^{m-1}\Gamma$$

where T' is the image of $T \llcorner \mathbf{B}(x, 2r)$ under the map $z \to (z - x)/r$. Thus $T' \in \mathscr{T}$ (since it minimizes $\Phi_{x,r} \operatorname{mod} p$) and

$$(K \cap \mathbf{B}(x, r) - x)/r \in \mathscr{S}.$$

To see (2), let K_i be a sequence in $\mathscr{S}$. Let T_i be the corresponding sequence in $\mathscr{T}$ and Φ_{x_i, r_i} be the corresponding sequence of integrands. By passing to a subsequence, we may assume K_i converges to some set K, T_i converges to some T (in $\mathscr{F}^p$), and $x_i \to x$, $r_i \to r$. Then $T \in \mathscr{T}$. If $\mathscr{U}$ is an open set containing $\operatorname{spt} T \cup \partial \mathbf{B}(0, 2)$, then (by Proposition 4.4) $\operatorname{spt} T_i \subset \mathscr{U}$ for sufficiently large i. In particular, each regular point z of T has a neighborhood in which (by Theorem 4.5) T_i is regular for large enough i. That is, a sequence of singular points from the T_i cannot converge to a regular point of T. Hence $K \subset \operatorname{sing}(T)$, so $K \in \mathscr{S}$.

Hence $\mathscr{S}$ satisfies the hypotheses of the Work-Raccoon Theorem. The conclusion is the desired result. $\square$

References

[A] F. J. Almgren, Jr., *Q-valued functions minimizing Dirichlet's integral and the regularity of area minimizing rectifiable currents up to codimension two*, preprint.

[ASS] F. J. Almgren, Jr., R. Schoen and L. Simon, *Regularity and singularity estimates on hypersurfaces minimizing parametric elliptic variational integrals*, Acta Math. **139** (1977), 217–265.

[F] H. Federer, *Geometric measure theory*, Springer-Verlag, 1969.

[SS] R. Schoen and L. Simon, *A new proof of the regularity theorem for rectifiable currents which minimize parametric elliptic functionals*, Indiana Univ. Math. J. **31** (1982), 415–434.

[T] J. E. Taylor, *Regularity of the singular sets of two-dimensional area minimizing flat chains modulo 3 in $\mathbf{R}^3$*, Invent. Math. **22** (1973), 119–159.

[W1] B. White, *The structure of minimizing hypersurfaces* mod 4, Invent. Math. **53** (1979), 45–58.

[W2] B. White, *Regularity of area minimizing hypersurfaces at boundaries with multiplicity*, Ann. of Math. Studies **103** (1983), 293–301.

STANFORD UNIVERSITY

Proceedings of Symposia in Pure Mathematics
Volume **44** (1986)

Regularity of Quasi-Minima and Obstacle Problems

WILLIAM P. ZIEMER[1]

1. Introduction. We recall a concept that has been recently introduced by Giaquinta and Giusti [**GG**] which is very useful in unifying the $C^{0,\alpha}$-regularity theory of elliptic differential equations. Let $\Omega \subset R^n$ be a bounded, open set and consider the functional

$$(1) \qquad J(u, E) = \int_E f(x, u, \nabla u),$$

where $u = (u^1, \ldots, u^N)$ is a mapping from Ω into R^N, $E \subset \Omega$, and where $f(x, z, p)$ is a real-valued Carathéodory function; that is, measurable in x for every (z, p) and continuous in (z, p) for almost all $x \in \Omega$. We assume that the function f also satisfies the following structural inequalities:

$$(2) \qquad |p|^m - b|z|^m - g(x) \leqslant f(x, z, p)$$
$$\leqslant \alpha|p|^m + b|z|^m + g(x)$$

where $\alpha, m \geqslant 1$, and b, q are functions in $L^q(\Omega)$ for $q > N/m$ if $m \leqslant N$, and $q = 1$ for $m > N$. If f is convex in p, then it is well known that the functional (1) is lower semicontinuous in the weak topology of $W^{1,q}_{\mathrm{loc}}(\Omega, R^N)$.

The essence of the work introduced in [**GG**] is to determine the regularity of extremals by working directly with the functional J in (1) instead of dealing with its Euler–Lagrange equation. Thus, no differentiability assumptions are placed on the integrand in (1). In [**GG**], it was shown that minima of the functional J in (1) are Hölder continuous directly without passing through the Euler–Lagrange equation. Moreover, it was observed that this result along with some others also hold for quasi-mimina. The concept of a quasi-minimum is somewhat similar in spirit to the notion of almost minimal currents introduced by Almgren, [**A**]. See also [**B**].

1980 *Mathematics Subject Classification.* Primary 35J65, 49A29.
[1]Research supported in part by a grant from the NSF.

© 1986 American Mathematical Society
0082-0717/86 $1.00 + $.25 per page

DEFINITION 1.1. A function $u \in W^{1,m}_{\text{loc}}(\Omega, R^N)$ is a sub Q-minimum (super Q-minimum) for the functional J if $Q \geqslant 1$ and

$$J(u, K) \leqslant QJ(u + \varphi, K)$$

whenever $\varphi \in W^{1,m}(\Omega)$ is nonpositive (nonnegative), with spt $\varphi = K \subset \Omega$.

In this paper, we will review some recent contributions on the regularity of quasi-minima as well as include some new results concerning boundary regularity. This will then lead directly to a new treatment of obstacle problems for elliptic operators in divergence form.

2. Preliminaries and examples. Before discussing the regularity properties of quasi-minima, we will first consider some examples.

EXAMPLE 2.1. Consider the system

$$(3) \qquad D_\alpha A_i^\alpha(x, u, \nabla u) = B_i(x, u, \nabla u), \qquad i = 1, 2, \ldots, N,$$

where A_i^α and B_i are Baire functions and satisfy the following structure conditions:

$$(4) \qquad A_i^\alpha(x, u, p)\, p_\alpha^i \geqslant |p|^m - b|u|^m - c_2(x);$$

$$|A(x, u, p)| \leqslant a_0|p|^{m-1} + b|u|^{m-1} + a_2(x);$$

$$|B(x, u, p)| \leqslant b_0|p|^{m-1} + b|u|^{m-1} + b_2(x).$$

Here a_0, b_0, and b are nonnegative constants and a_2, b_2, and c_2 are all elements of a suitable Lebesgue space. It is shown in [GG] that a weak solution $u \in W^{1,m}_{\text{loc}}(\Omega, R^N)$ of (3) with structure (4) is a quasi-minimum of the functional

$$J(u, \Omega) = \int_\Omega \left(|Du|^m + b|u|^m + h(x) \right) dx,$$

where

$$h(x) = a_2(x)^{m/(m-1)} + b_2(x)^{m/(m-1)} + c_2(x).$$

In particular, if $u \in W^{1,2}_{\text{loc}}(\Omega)$ is a weak solution of the equation ($N = 1$)

$$D_i\big(a^{ij}(x)D_j u\big) = 0,$$

where the coefficients a^{ij} are measurable on Ω and satisfy

$$a^{ij}\xi_i\xi_j \geqslant \lambda|\xi|^2, \quad \forall \xi \in R^n, \lambda > 0,$$

$$|a^{ij}| \leqslant \Lambda,$$

then u is a quasi-minimum of the Dirichlet integral. That is, there is a $Q \geqslant 1$ such that

$$\int_\Omega |\nabla u|^2 \leqslant Q \int_\Omega |\nabla v|^2,$$

whenever $v \in W^{1,2}_{\text{loc}}(\Omega)$ with spt$(u - v) \subset \Omega$.

EXAMPLE 2.2. A mapping $u: \Omega \to R^n$, $u \in W^{1,n}_{\text{loc}}(\Omega; R^n)$, is called quasi-regular if there is a constant K such that

$$|Du(x)|^n \leqslant K|Ju(x)|,$$

where Ju denotes the Jacobian of u. If u is a homeomorphism, then u is called quasi-conformal. It is shown in [**GG**] that u is a quasi-minimum for the functional $\int_\Omega |Du|^n$.

EXAMPLE 2.3. Consider the integrand in (2) with $N = 1$ and let $\psi \in W^{1,m}(\Omega)$. Also, let $\beta \in W^{1,m}(R^n)$ be such that $\beta|(R^n - \Omega)$ is continuous. We study the following obstacle problems. Let M be the convex set in $W^{1,m}(\Omega)$ of all v such that $v - \beta \in W^{1,m}_0(\Omega)$ (i.e., $v = \beta$ on $\partial\Omega$) and $v \geqslant \psi$ a.e. on Ω. Let u be the extremal of the problem

$$(5) \qquad \inf\left\{ \int_\Omega f(x, v, \nabla v) : v \in M \right\}.$$

It is a routine matter to verify that

$$(6) \qquad \int_\Omega \sum_{i=1}^n \frac{\partial f}{\partial p_i}(x, u, \nabla u)\left(\frac{\partial v}{\partial x_i} - \frac{\partial u}{\partial x_i} \right) \geqslant 0$$

whenever $v \in K$. This is a special case of the variational inequality

$$(7) \qquad \int_\Omega A(x, u, \nabla u) \cdot \nabla\varphi + B(x, u, \nabla u)\varphi \geqslant 0,$$

where $u \geqslant \psi$ and $\varphi \in W^{1,m}_0(\Omega)$ is a nonnegative test function. If we take $\varphi = v - u$ where $v \geqslant \psi$, it follows easily that there is a $Q \geqslant 1$ such that

$$(8) \qquad J(u, K) \leqslant QJ(v, K),$$

where J is the functional in Example 2.1 and $K = \text{spt}(v - u)$. In order to eliminate the condition $v \geqslant \psi$, let $w = \sup(v, \psi)$ and note that $w \geqslant \psi$, spt $w \subset K$, and (8) is valid with v replaced by w. Because $|\nabla w| \leqslant |\nabla v| + |\nabla \psi|$, it therefore follows that u is a quasi-minimum for the functional

$$(9) \qquad J(u, \Omega) + \int_\Omega |\nabla\psi|^m.$$

3. Regularity of quasi-minima, $N > 1$. One of the basic results in the investigation of regularity of elliptic systems of the type (3), (4) is that weak solutions $u \in W^{1,m}_{\text{loc}}(\Omega, R^N)$ in fact are elements of $W^{1,m+\varepsilon}_{\text{loc}}(\Omega, R^n)$ for some $\varepsilon > 0$, cf. [**BO, M, G, ME, GM, S**]. A similar result is true for quasi-minima [**GG**].

THEOREM 3.1. *Let $u \in W^{1,m}(\Omega, R^N)$ be a quasi-minimum of the functional* (1) *with structure* (2), *where b is a constant and $g \in L^q(\Omega)$, $q > 1$. Then, there is an $\varepsilon > 0$ such that $u \in W^{1,m+\varepsilon}_{\text{loc}}(\Omega, R^N)$. Moreover, there exists $r_0 > 0$ such that for every ball $B(r) \subset \Omega$ with $r < r_0$,*

$$\left\{ \fint_{B(r)} \left(|Du|^m + b|u|^m\right)^{(m+\varepsilon)/m} \right\}^{1/(m+\varepsilon)} \leqslant C\left\{ \fint_{B(r)} \left(|Du|^m + b|u|^m\right) \right\}^{1/m}$$

$$+ \left(\fint_{B(r)} g^{(m+\varepsilon)/m} \right)^{m/(m+\varepsilon)}.$$

Here, $\fint$ denotes the integral average.

Because solutions of elliptic systems are quasi-minima, it is not possible, in general, for quasi-minima to be continuous. We do, however, have the following regularity results for quasi-minima of the Dirichlet integral.

THEOREM 3.2. *If $u \in W^{1,2}(\Omega, R^N)$ has the property that for every ball $B \subset \Omega$,*

$$\int_B |Du|^2 \leqslant \frac{n-1}{n-2+2\alpha} \int_B |\nabla v|^2, \qquad 0 < \alpha < 1/2,$$

whenever $v \in W^{1,2}(\Omega, R^N)$ agrees with u on ∂B, then $u \in C^{0,\alpha}(\Omega)$.

For the proof of this theorem we will need the following

LEMMA 3.3. *If v is harmonic on Ω, then*

$$(10) \qquad \int_{\partial B(r)} |\nabla_t v|^2 = \int_{\partial B(r)} |\nabla_N v|^2 + \frac{n-2}{r} \int_{B(r)} |\nabla v|^2$$

whenever $B(r) \subset\subset \Omega$.

Here ∇_t and ∇_N denote the tangential and normal components of the gradient, respectively. The proof of the lemma is quite simple and will not be given here. It follows immediately from the divergence theorem (consider the vector field $V(x) = (x \cdot \nabla v(x))\nabla v(x)$ and integration using spherical coordinates). We also recall the following well-known property of harmonic functions: If v is harmonic on Ω and $B(r) \subset\subset \Omega$, then

$$(11) \qquad \int_{B(r)} |\nabla v|^2 \leqslant \frac{r}{n} \int_{\partial B(r)} |\nabla v|^2.$$

PROOF OF THEOREM 3.2. For notational simplicity, we consider only the case $N = 1$. Select $x_0 \in \Omega$ and consider the ball $B(x_0, r)$ centered at x_0 and radius r. For brevity write $B(r) = B(x_0, r)$. We consider only those r for which $B(r) \subset\subset \Omega$ and for which u, when restricted to $\partial B(r)$, has a well-defined trace. This latter requirement is satisfied for almost all $r > 0$. Let v be the harmonic function on $B(r)$ such that $v = u$ on $\partial B(r)$. If we let $Q = (n-1)/(n-2+2\alpha)$ then from (10) and (11),

$$\int_{B(r)} |\nabla u|^2 \leqslant Q \int_{B(r)} |\nabla v|^2 \leqslant \frac{Qr}{n} \int_{\partial B(r)} |\nabla v|$$

$$\leqslant \frac{Qr}{n} \int_{\partial B(r)} |\nabla_t v|^2 + |\nabla_N v|^2$$

$$\leqslant \frac{Qr}{n} \int_{\partial B(r)} |\nabla_t u|^2 + |\nabla_t v|^2 - \frac{Q(n-2)}{n} \int_{B(r)} |\nabla v|^2.$$

Therefore,

$$(12) \qquad \left(1 + \frac{n-2}{n}\right) \int_{B(r)} |\nabla u|^2 \leqslant \frac{2Qr}{n} \int_{\partial B(r)} |\nabla u|^2.$$

If we set

$$D(r) = \int_{B(r)} |\nabla u|^2,$$

(12) yields the differential inequality

$$D(r) \leqslant \frac{Qr}{n-1} D'(r)$$

which, in turn, implies

$$\int_{B(r)} |\nabla u|^2 \leqslant Cr^{n-2+2\alpha}.$$

By Morrey's lemma, cf. [**GT**, Theorem 7.19], this implies that $u \in C^{0,\alpha}(\Omega)$.

De Giorgi has shown that weak solutions to systems of the form (3), (4) are not necessarily continuous, [**DG**]. Because elliptic systems of the form

$$(13) \qquad D_\alpha\left(A_{ij}^{\alpha\beta}(x)D_\beta u^j\right) = 0, \quad i = 1, 2, \ldots, N,$$

$A_{ij}^{\alpha\beta}$ bounded, measurable,

$$\Lambda|\xi|^2 \geqslant A_{ij}^{\alpha\beta}\xi_\alpha^i\xi_\beta^j \geqslant \lambda|\xi|^2, \quad \forall \xi \in R^{nN},$$

are quasi-minima it follows from Theorem 3.2 that weak solutions of (13) are Hölder continuous provided Λ/λ is sufficiently close to 1.

A close inspection of the above proof reveals that we did not use the full strength of the definition of quasi-minimum. Indeed, the only requirement on u is that for every ball $B \subset \Omega$,

$$(14) \qquad \int_B |\nabla u|^2 \leqslant \frac{n-1}{n-2+2\alpha} \int_B |\nabla v|^2,$$

whenever v is a harmonic function agreeing with u on $\partial\beta$.

4. Boundary behavior of quasi-minima, $N = 1$. In his celebrated work on the Hölder continuity of linear elliptic equations with bounded measurable coefficients, DeGiorgi employed a delicate analysis of the solution truncated at arbitrary levels, [**DG2**]. This technique was pursued further in [**LU**] where a wide class of quasilinear equations was investigated. Later, Moser established a Harnack inequality for linear elliptic equations by employing an iteration procedure along with the John–Nirenberg Lemma and thus provided an alternate proof of DeGiorgi's theorem [**MO**]. His work was extended to quasilinear equations by Serrin [**S**] and Trudinger [**T**]. Recently DiBenedetto and Trudinger [**DT**] established a Harnack inequality for quasi-minima. Their proof did not rely on the John–Nirenberg Lemma but rather employed a technique of Krylov and Safanov [**KS**] which was used to establish Harnack's inequality for solutions of nondivergence equations. The result in [**DT**] is, in fact, more general than that stated above and is given in the following theorem.

THEOREM 4.1. [**DT**]. *If $u \in W_{\mathrm{loc}}^{1,m}(\Omega)$ $(N = 1)$ is a sub Q-minimum of the functional J with $g \equiv 0$ in the structure* (1), (2), *then for every ball $B(r) \subset\subset \Omega$ and any $\sigma \in (0,1)$, $p > 0$, we have*

$$\sup_{B(\sigma r)} u \leqslant C\left(\fint_{B(r)} u^p\right)^{1/p}$$

where C depends on only σ, m, p, and the structure (2). Alternatively, if u is a super Q-minimum of J then there exists $p > 0$ such that for any $\sigma, \tau \in (0, 1)$, we have

$$\left(\fint_{B(\sigma r)} u^p\right)^{1/p} \leqslant C \inf_{B(\tau r)} u.$$

Thus, if u is a Q-minimum of J, then

$$\sup_{B(r)} u \leqslant C \inf_{B(r)} u.$$

We will now investigate the behavior of quasi-minima at the boundary of Ω. For this purpose let $\beta \in W^{1,m}(R^n)$ have the property that $\beta|(R^n - \Omega)$ is continuous.

DEFINITION 4.2. If $u \in W^{1,m}_{\text{loc}}(\Omega)$, $x_0 \in \partial\Omega$, and $L \in R^1$, we say that

$$(15) \qquad\qquad u(x_0) \leqslant L \text{ weakly}$$

if for every $k > L$ there is an $r > 0$, such that $\eta(u - k)^+ \in W^{1,m}_0(\Omega)$ whenever $\eta \in C_0^\infty[B(x_0, r)]$.

The condition

$$(16) \qquad\qquad u(x_0) \geqslant L \text{ weakly}$$

is defined analogously and $u(x_0) = L$ if both (15) and (16) hold. Observe that if $u - \beta \in W^{1,m}_0(\Omega)$, then $u(x) = \beta(x)$ weakly for each $x \in \partial\Omega$.

THEOREM 4.3. *Let $\Omega \subset R^n$ be an open bounded set, and let $u \in W^{1,m}(\Omega)$ be a sub Q-minimum of J with $g \equiv 0$ in (2). Suppose $x_0 \in \partial\Omega$ and that $u(x_0) \leqslant L$ weakly. If*

$$\Lambda = \limsup_{x \to x_0} u(x) > L$$

then

$$\lim_{r \to 0} r^{-n} \int_{B(x_0, r) \cap \Omega} |u(x) - \Lambda| = 0.$$

PROOF. For $k > L$, let

$$u_k = \begin{cases} (u - k)^+, & \text{on } \Omega, \\ 0, & \text{otherwise,} \end{cases}$$

and let

$$\mu_k(r) = \sup\{u_k(x): x \in B(x_0, 1)\}.$$

Choose $r > 0$ so small that $\eta u_k \in W^{1,m}_0(\Omega)$ whenever $\eta \in C_0^\infty[B(x_0, r)]$. For $0 < h < \mu_k(r)$ let $v = \mu_k(r) - u_k$ and $\varphi = \eta(v - h)^-$. Note that $\varphi \in W^{1,m}_0(\Omega)$. Now using the techniques in [DT] it follows that there is a $p > 0$ such that

$$(17) \qquad \left(\fint_{B(x_0, r/2)} v^p\right)^{1/p} \leqslant C \inf_{B(x_0, r/4)} v \leqslant C[\mu_k(r) - \mu_k(r/4)] \to 0$$

as $r \to 0$. The conclusion readily follows from (17).

Note that (17) also implies

$$(18) \qquad |B(x_0, r) - \Omega| \cdot r^{-n} \to 0$$

as $r \to 0$. Also notice that there is a result analogous to Theorem 4.3 for u a super Q-minimum. That is, if $u(x_0) \geq l$ weakly and if

$$\lambda = \liminf_{x \to x_0} u(x) < l,$$

then

$$\lim_{r \to 0} \fint_{B(x_0, r) \cap \Omega} |u(x) - \lambda| = 0.$$

This follows from Theorem 4.3 applied to $-u$, which is a sub Q-minimum of a functional related to J in a natural way.

Now suppose β is as above and that $u \in W^{1,m}_{\mathrm{loc}}(\Omega)$ is a Q-minimum of J with $u - \beta \in W^{1,m}_{\mathrm{loc}}(\Omega)$. For $x_0 \in \partial\Omega$, it follows from above that if either

$$\limsup_{x \to x_0} u(x) = \Lambda > \beta(x_0) \quad \text{or} \quad \liminf_{x \to x_0} u(x) = \lambda > \beta(x_0),$$

then

$$(19) \qquad \fint_{B(x_0, r)} |u - \Lambda| \to 0 \quad \text{or} \quad \fint_{B(x_0, r)} |u - \lambda| \to 0.$$

Therefore, it is not possible to have

$$(20) \qquad \liminf_{x \to x_0} u(x) < \beta(x_0) < \limsup_{x \to x_0} u(x).$$

Moreover, if

$$(21) \qquad \limsup_{r \to 0} \frac{|B(x_0, r) - \Omega|}{r^n} > 0,$$

then

$$(22) \qquad \lim_{x \to x_0} u(x) = \beta(x_0).$$

Finally, note that Theorem 4.3 also yields the fact that if u is a sub Q-minimum (super Q-minimum) then u is upper semicontinuous (lower semicontinuous) on Ω.

We now proceed to discuss the boundary regularity of quasi-minima and establish continuity at the boundary under an assumption considerably weaker than (21). For this purpose we introduce the notion of capacity. For $1 < p < n$, the p-capacity of a set $E \subset R^n$ is defined as

$$(23) \qquad \gamma_p(E) = \inf\left\{ \int |\nabla v|^p : E \subset \mathrm{int}\{v \geq 1\}, v \in C_0^\infty(R^n) \right\}.$$

A well-known relationship between capacity and Hausdorff measure is as follows, cf. [FZ].

$$\gamma_p(E) = 0 \Rightarrow H^{n-p+\varepsilon}(E) = 0, \quad \forall \varepsilon > 0,$$

$$H^{n-p}(E) < \infty \Rightarrow \gamma_p(E) = 0.$$

Consider $\beta \in W^{1,m}(R^n)$ such that $\beta|(R^n - \Omega)$ is continuous and suppose that $u - \beta \in W_0^{1,m}(\Omega)$ where u is a quasi-minimum. From (19) and (20) we have that every point of $\partial\Omega$ is a Lebesgue point of u. Also, because u satisfies Harnack's inequality on Ω, it is continuous there. This, along with a modification of the techniques in [GZ1] yield the following.

THEOREM 4.4. *If u is a quasi-minimum of J in (1), (2), $u - \beta \in W_0^{1,m}(\Omega)$, $x_0 \in \partial\Omega$, and*

$$(24) \qquad \limsup_{r \to 0} \frac{\gamma_m[B(x_0, r) - \Omega]}{r^{n-m}} > 0,$$

then u is continuous at x_0.

The proof of this appears in [Z1].

It is tempting to conjecture that Theorem 4.4 remains valid with (24) replaced by the following weaker requirement

$$(25) \qquad \int_0 \left\{ \frac{\gamma_m[B(x_0, r) - \Omega]}{r^{n-m}} \right\}^{1/(m-1)} \frac{dr}{r} = \infty.$$

In case $m = 2$, this reduces to the classical Wiener criterion for boundary regularity of harmonic functions. It was shown in [GZ2] that (25) is sufficient for boundary continuity of weak solutions of a wide class of equations in divergence form. For a narrower class of equations it is known that (25) is also necessary for boundary continuity. It appears however, that (25) may not be sufficient for the case of quasi-minimum. The reason is that (17) is known to hold for only some $p > 0$ (presumably small), whereas in the case of solutions of divergence type equations, a result corresponding to (17) is valid for $p > n(m - 1)/(n - m)$. The weak Harnack inequality (17) with a suitable (large) choice of p plays a critical role in proving boundary regularity of solutions of equations under assumption (25). It might be possible perhaps to prove that (25) is sufficient for boundary regularity of quasi-minima with the exponent $1/(m - 1)$ in the integrand replaced by some smaller exponent. But then, of course, (25) could not be a necessary condition for the class of quasi-minima.

5. Obstacle problem, $N = 1$. Let ψ be a function (an obstacle) on Ω with the property that every point in Ω is a Lebesgue point for ψ and that ψ is upper semicontinuous on Ω. For example, as we have seen, a sub Q-minimum has this property. Let

$$M = W^{1,m}(\Omega) \cap \{v: v \geqslant \psi \text{ q.e.}\}$$

(the term q.e. means except for a set of γ_m-capacity 0). Recall that if $u \in W^{1,m}(\Omega)$, then u has a Lebesgue point q.e. on Ω, cf. [FZ]. In this section, we consider the obstacle problem for a Q-minimum u of the functional J in (1), (2) relative to M. That is, consider $u \in W$ with the property

$$(26) \qquad J(u, K) \leqslant QJ(v, K),$$

for all $v \in M$ such that $u - v \in W_0^{1,m}(\Omega)$ and with $K = \{u \neq v\}$.

THEOREM 5.1. *If u is a Q-minimum for* (26) *relative to M, then u is continuous on* Ω.

PROOF. We sketch the proof. Complete details appear in [**Z2**]. First observe that u is a super Q-minimum for (26) because if $\varphi \in W_0^{1,m}(\Omega)$ is nonnegative, then $u + \varphi \in M$ and thus

$$J(u) \leqslant QJ(u + \varphi).$$

As we have observed, Theorem 4.3 implies that u is therefore lower semicontinuous on Ω. We have assumed that ψ is upper semicontinuous and consequently the contact set $C = \{u = \psi\}$ is closed relative to Ω. Of course, $u|C$ is continuous on C.

In order to see that u is continuous on $\Omega - C$ choose $x_0 \in \Omega - C$ and let $B \subset \Omega - C$ be an open ball centered at x_0. Note that

$$J(u) \leqslant QJ(u + \varphi)$$

for all $\varphi \in W^{1,m}(\Omega)$ such that spt $\varphi \subset B$ and $|\varphi|$ sufficiently small. Referring to [**GG**, 4.1] we see that this is sufficient to establish the Hölder continuity of u in B, although the modulus of continuity will depend on the distance from x_0 to C.

Now choose $x_0 \in \Omega \cap C$. Because u is a super Q-minimum, we know that

$$(27) \qquad \liminf_{\substack{x \to x_0 \\ x \in U}} u(x) \geqslant u(x_0) = \psi(x_0).$$

If we regard x_0 as a boundary point of the open set $U = \Omega - C$, we may apply the analysis of Theorem 4.3. To this end assume

$$(28) \qquad \Lambda \equiv \limsup_{\substack{x \to x_0 \\ x \in U}} u(x) > \psi(x_0)$$

and choose k so that $\psi(x_0) < k < \Lambda$. Note that $k > \sup\{\psi(x): x \in B(x_0, r)\}$ for all sufficiently small $r > 0$ because ψ is upper semicontinuous. Let $u_k = (u - k)^+$ and for a fixed small $r > 0$, let $\mu_k = \sup\{u_k(x): x \in B(x_0, r)\}$ and $v = \mu_k - u_k$. Finally, let $\varphi = -\eta(v - h)^-$, where $0 \leqslant h \leqslant \mu_k$ and $\eta \in C_0^\infty[B(x_0, r)]$ is the usual cut-off function. Observe that $\varphi \in W_0^{1,m}(\Omega)$ and that

$$u(x) > \mu_k - h + k, \qquad u(x) + \varphi(x) \geqslant k > \psi(x),$$

for those x for which $\varphi(x) \neq 0$. Therefore $J(u) \leqslant QJ(u + \varphi)$ and thus proceeding as in Theorem 4.3, we again obtain (17) which, in turn, yields

$$\lim_{r \to 0} \fint_{R(x_0, r)} |u(x) - \Lambda| \to 0.$$

But this contradicts the fact that $u(x_0) = \psi(x_0) < \Lambda$, and the proof is complete. The above result and others were obtained by J. H. Michael and the author for equations in divergence form [**MZ**].

We conclude this paper by considering obstacle problems within a smaller class of objects, namely equations of the form (3) with $N = 1$ and structure (4). Following the motivation provided by [**FM1**], we consider obstacles ψ that are bounded above but otherwise quite irregular and investigate conditions under which the associated solution will be continuous at a particular point. For this purpose choose $x_0 \in \Omega$ and let

$$E(\varepsilon.r) = B(x_0, r) \cap \{ y: \psi(y) \geqslant \psi(r) - \varepsilon \text{ for } \gamma_m - \text{q.e. } y \},$$

where $\psi(r) = \sup\{ \psi(y): y \in B(x_0, r)\}$. A condition for continuity of a solution of the variational inequality (7) is given by the following result, [Z2].

THEOREM 5.1. *Let ψ and $u \in W_{\text{loc}}^{1,m}(\Omega)$ be as above and let $x_0 \in \Omega$. If, for every $\varepsilon > 0$,*

$$(29) \qquad \int_0 \left\{ \frac{\gamma_m[E(\varepsilon, r)]}{r^{n-m}} \right\}^{1/(m-1)} \frac{dr}{r} = \infty$$

then u is continuous at x_0.

This result was obtained in [**FM2**] for a special class of elliptic operators and the proof relies on the associated Green function. The proof in [**Z2**] is quite different because no Green function is available. Instead, the proof makes strong use of the following fact, which was first established in [**GZ2**] in a slightly different context:

$$(30) \qquad \int_0 \left[r^{m-n} \int_{B(x_0, r)} |\nabla u|^m \right]^{1/(m-1)} \frac{dr}{r} = \infty.$$

It is also possible to obtain estimates of the modulus of continuity of u in terms of the modulus of continuity of ψ.

If one considers extending Theorem 5.1 to quasi-minima, a difficulty is encountered that is similar to the one that was discussed following Theorem 4.4. In particular, it seems doubtful that (30) remains valid for super Q-minima.

REFERENCES

[A] F. J. Almgren, Jr., *Existence and regularity almost everywhere of solutions to elliptic variational problems with constraints*, Mem. Amer. Math. Soc., No. 165 (1976).

[B] E. Bombieri, *Regularity theory for almost minimal currents*, Arch. Rat. Mech. Anal. **78** (1982), 99–130.

[BO] B. V. Boyarskii, *Homeomorphic solutions of Beltrami systems*, Dokl. Akad. Nauk SSSR **102** (1955), 661–664.

[DG] E. DeGiorgi, *Un esempio di estremali discontinue per un problema variazionale di tipo ellittico*, Boll. Un. Mat. Ital. **4** (1968), 135–137.

[DG2] ______, *Sulla differenziabilita e l'analiticita delle estremali degli integrali multipli regolari*, Mem. Accad. Sci. Torino Cl. Sci. Fis. Mat. Natur. **3** (1957), 25–43.

[DT] E. Di Benedetto and N. Trudinger, *Harnack inequalities for quasi-minima of variational integrals*, Australian National University Research Report, CMA-R47-83, 1983.

[FZ] H. Federer and W. P. Ziemer, *The Lebesgue set of a function whose distribution derivatives are pth power summable*, Indiana Univ. Math. J. **22** (1972), 139–158.

[FM1] J. Frehse and U. Mosco, *Variational inequalities with one-sided irregular obstacles*, Manuscripta Math. **28** (1979), 219–233.

[FM2] ______, *Wiener obstacles*, Sonderforschungsbereich 72, Universität Bonn, # 643, 1984.

[GZ1] R. Gariepy and W. P. Ziemer, *Behavior at the boundary of solutions of quasilinear elliptic equations*, Arch. Rat. Mech. Anal. **56** (1974), 372–384.

[GZ2] ______, *A regularity condition at the boundary for solutions of quasilinear elliptic equations*, Arch. Rat. Mech. Anal. **67** (1977), 25–39.

[G] F. Gehring, *The L^p-integrability of the partial derivatives of a quasi-conformal mapping*, Acta Math. **130** (1973), 265–277.

[GG] M. Giaquinta and E. Giusti, *Quasi-Minima*, Ann. Inst. H. Poincaré Anal. Non Linéaire **1** (1984), (to appear).

[GT] D. Gilbarg and N. Trudinger, *Elliptic partial differential equations of second order*, Springer-Verlag, 1977.

[GM] M. Giaquinta and G. Modica, *Regularity results for some classes of higher order nonlinear elliptic systems*, J. Reine Angew. Math. **311–312** (1979), 145–169.

[KS] N. V. Krylov and M. V. Safonov, *Certain properties of solutions of parabolic equations with measurable coefficients*, Izv. Akad. Nauk SSSR **40** (1980), 161–175.

[LU] O. A. Ladyzhenskaya and Ural'tseva, *Linear and quasilinear elliptic equations*, Academic Press, New York, 1968.

[M] N. G. Meyers, *An L^p-estimate for the gradient of solutions of second order elliptic divergence equations*, Ann. Scuola Norm. Sup. Pisa Cl. Sci. **17** (1963), 189–206.

[ME] N. G. Meyers and A. Elcrat, *Some results on regularity for solutions of nonlinear elliptic systems and quasiregular functions*, Duke Math. J. **42** (1975), 121–136.

[MZ] J. Michael and W. P. Ziemer, *Regularity of solutions of obstacle problems* (to appear).

[MO] J. Moser, *A new proof of DeGiorgi's theorem concerning the regularity problem for elliptic differential equations*, Comm. Pure Appl. Math. **13** (1960), 457–468.

[SE] J. Serrin, *Local behavior of solutions of quasilinear elliptic equations*, Acta Math. **111** (1964), 247–302.

[S] E. Stredulinsk ', *Higher integrability from reverse Hölder inequalities*, Indiana Univ. Math. J. **29** (1980), 408–417.

[T] N. S. Trudinger, *On Harnack type inequalities and their applications to quasilinear elliptic equations*, Comm. Pure Appl. Math. **20** (1967), 721–747.

[Z1] W. P. Ziemer, *Boundary behavior of quasi-minima* (to appear).

[Z2] ______, *Variational inequalities and irregular obstacles* (to appear).

Indiana University

Proceedings of Symposia in Pure Mathematics
Volume **44** (1986)

Some Open Problems in Geometric Measure Theory and its Applications Suggested by Participants of the 1984 AMS Summer Institute

J. E. BROTHERS, EDITOR

1. Existence and examples.

Problem 1.1. Does every smooth compact Riemannian manifold M^n contain infinitely many distinct compact minimal submanifolds N^k without boundary of each (any) dimension k? It is tempting to construct pieces of periodic-like minimal surfaces and to try to fit them together. A. Schoen in [**SA**] has illustrations of a number of possible periodic pieces for $k = 2$ and $n = 3$. Lawson constructed infinitely many distinct N^2's in $\mathbf{S}^3$ in [**LB1**]. For general Riemannian manifolds M^n, $n \in \{3, 4, 5, 6, 7\}$, at least one minimal N^{n-1} exists according to Pitts [**PJ**].   F. ALMGREN

Problem 1.2. Let $\mathbf{R}^3$ be divided into regions of volume one by a rectifiable set S. (a) Does there exist S for which $\lim_{R \to \infty} R^{-3} \mathscr{H}^2(S \cap \mathbf{B}_R)$ is as small as possible? For which S is this true? Is S triply periodic (invariant under some lattice $\mathbf{Z}^3 \simeq \Lambda \subset \mathbf{R}^3$ of translations)? (b) Restrict problem (a) to triply periodic sets S. (c) Restrict to a specific lattice Λ (i.e., find the fundamental domain in a flat torus of smallest boundary area). Lord Kelvin [**TW**] proposed a specific solution to problem (a) which is invariant under the lattice generated by e_1, e_2, $2^{-1/2}e_3$. R. GULLIVER

Problem 1.3. (a) Does there exist a complete $M^2 \subset \mathbf{R}^3$, with constant mean curvature $H \neq 0$, which is asymptotic near infinity to three cylinders of constant mean curvature H? [**WB8**] and [**SL1**] may be useful. (b) Does there exist a complete M of constant mean curvature which has k ends, $3 \leqslant k < \infty$? Katsuei Kenmotsu has a variation of the Weierstrass representation which may help solve this problem. Observe that the Gauss map is harmonic [**RV**]. Examples for $k = \infty$ (doubly periodic and triply periodic surfaces of constant mean curvature) are constructed in [**LB1**].   R. GULLIVER

1980 *Mathematics Subject Classification.* Primary 49F20; Secondary 49F22, 53A10, 53C42

© 1986 American Mathematical Society
0082-0717/86 $1.00 + $.25 per page

Problem 1.4. Suppose that the unit circle $\mathbf{S}^1 \subset \mathbf{R}^2$ is the boundary ∂S of a compact connected constant mean curvature surface S immersed in $\mathbf{R}^3$. Is S a round spherical cap? [**GRH**] uses Alexandrov reflection methods to prove that this is true provided S is embedded and omits the exterior of $\mathbf{S}^1$ in the plane $\mathbf{R}^2$. Can holomorphic differential methods be exploited as in [**HH**, Part 2]? Recall that the Gauss map $g: S \to \mathbf{S}^2$ of a constant mean curvature surface is harmonic [**RV**]. R. Schoen suggests comparing this problem with the following one about a harmonic map $h: S \to \mathbf{S}^2$: If $h|\partial S$ is the identity onto an equator $\mathbf{S}^1 \subset \mathbf{S}^2$, is S a hemisphere, and is h the identity or the identity following a reflection? Jost observes [**JJ**, Chapter 4] that such maps are the only energy minimizing maps with this boundary data.   R. GULLIVER AND R. KUSNER

Problem 1.5. Does there exist a connected codimension one area minimizing integer multiplicity rectifiable current with two isolated singular points? See the discussion in Problem 1.6.   R. HARDT

Problem 1.6. Does there exist a codimension one area minimizing integer multiplicity rectifiable current with zero boundary which has an isolated singularity and is *not* a cone? The minimality of a cone example was first shown in [**BDG**]. Caffarelli, Hardt, and Simon [**CHS**] have shown that the intersection of such a cone with the unit ball can be perturbed to give one parameter families of noncone minimal surfaces with an isolated singularity. The mass minimality of many of these examples was recently established by Hardt and Simon [**HS2**]. However, all of these examples have nonzero boundary and hence do not settle the above question.   R. HARDT

Problem 1.7. Characterize those complete Riemannian manifolds which admit an exhaustion by compact domains $\Omega_1 \Subset \Omega_2 \Subset \Omega_3 \Subset \cdots$ such that each boundary $\partial\Omega_k$ is smooth and of positive mean curvature with respect to the outer normal; e.g., concentric balls in $\mathbf{R}^n$. Some results in this direction are in [**BR**].   B. LAWSON

Problem 1.8. The only known examples of singular two-dimensional area minimizing rectifiable currents in $\mathbf{R}^n$ are complex curves, and the problem is to find other examples. Regularity theory [**AF5**] only says that the singular set must be zero-dimensional. On the other hand, every singularity must be complex to first order [**MF2**]. A simple candidate is the union of two regular minimal surfaces which intersect orthogonally in $\mathbf{R}^4$.   F. MORGAN

Problem 1.9. The only known example of a singular two-dimensional area minimizing flat chain modulo two in $\mathbf{R}^n$ is a sum of orthogonal planes. The problem is to find another. It is known that any singularity is isolated, and that the tangent cone is a sum of orthogonal planes [**MF2**]. A simple candidate is the union of two regular minimal surfaces which intersect orthogonally in $\mathbf{R}^4$. F. MORGAN

Problem 1.10. Can there be a solution u of the minimal surface equation with $u \in C^2(\Omega)$, Ω a connected open subset of $\mathbf{R}^n$, such that $\Omega \neq \mathbf{R}^n$ and the graph of u is a complete submanifold of $\mathbf{R}^{n+1}$ without boundary? (It may be that one should also throw in the hypothesis that the graph of u, as a current, is minimizing in

$\mathbf{R}^{n+1}$, although this might be automatic.) A negative answer to this question leads immediately to a restriction on the growth near infinity of nonlinear entire solutions u of the minimal surface equation, normalized so that $u(0) = 0$, $\sup_{|x|<1}|Du(x)| \leqslant 1$. For a partial result (settling the question in some special cases) see [**SL2**].   L. SIMON

Problem 1.11. Is there an example (other than $\mathbf{R}^2 \subset \mathbf{R}^3$) of a hypercone C in $\mathbf{R}^{n+1}$ which is minimizing but not strictly minimizing? (C *strictly minimizing* means that if $C_1 = C \llcorner \mathbf{B}_1(0)$, then there is a constant $\theta > 0$ such that $\mathbf{M}(C_1) \leqslant \mathbf{M}(S) - \theta \varepsilon^n$ whenever $\varepsilon \in (0,1)$, $\partial S = \partial C_1$, and spt $S \cap \mathbf{B}_\varepsilon(0) = \varnothing$; for further discussion and equivalent characterizations, see [**HS2**].) The analogous question in higher codimension is also of interest.   L. SIMON

Problem 1.12. The only known example of an indecomposable integral n-cycle in $\mathbf{R}^{n+1}$ which is stable but not area minimizing is the cone on $\mathbf{S}^1(1) \times \mathbf{S}^5(\sqrt{5})$ in $\mathbf{R}^8 = \mathbf{R}^2 \times \mathbf{R}^6$ [**SP**]. Are there other examples? Smooth ones? There are smooth minimal hypersurfaces asymptotic to the above cone at infinity, and these share the symmetries of the cone, but they are unstable.   B. SOLOMON

Problem 1.13. Let C be a curve in $\mathbf{R}^n$ and let $\alpha(C)$ denote the least area of any surface (i.e., integer multiplicity current) with boundary C. Young [**YL**] gave an example of a C in $\mathbf{R}^4$ for which $\alpha(2C) < 2\alpha(C)$. Morgan [**MF4**] and White [**WB7**] (by very different methods) constructed for each $p = 2, 3, 4, \ldots$, a curve C such that $p^{-1}\alpha(pC) < k^{-1}\alpha(kC)$ for $1 \leqslant k < p$. Indeed, they constructed curves C (smooth except at one point) such that $\inf\{k^{-1}\alpha(kC): k = 1, 2, \ldots\}$ is not attained for any k. (This quantity is the mass of the mass minimizing *real* current with boundary C [**FH4**].) On the other hand, the following problems are open: (a) Is there a curve C such that $\alpha(2C) < \alpha(C)$? (b) If so, is the ratio of $\alpha(2C)$ to $\alpha(C)$ bounded away from zero? (c) If so, is the ratio of $\inf\{k^{-1}\alpha(kC)\colon k = 1, 2, \ldots\}$ to $\alpha(C)$ bounded away from zero? In codimension one (e.g. if C is a curve in $\mathbf{R}^3$) none of these pathologies is possible: $\alpha(kC) = k\alpha(C)$ for every $k \geqslant 1$. (This is proved for arbitrary elliptic integrands in [**WB4**].)   B. WHITE

Problem 1.14. Does every Riemannian metric on $\mathbf{S}^3$ admit (a) four minimal embedded 2-spheres? (b) five minimal embedded tori? (c) minimal embedded surfaces of every genus (or at least arbitrarily high genus)? Lawson [**LB1**] constructed minimal surfaces of every genus in the standard $\mathbf{S}^3$. Simon and Smith [**SSR**] proved by a minimax argument that every metric admits a minimal embedded $\mathbf{S}^2$. J. Li and White [**WB8**] independently showed that every metric near the constant curvature metric admits four minimal embedded 2-spheres. White's method also gave five minimal tori (and many similar examples in higher dimensions) for metrics near the standard one. By a continuity method (path lifting argument) White concluded that every metric of positive Ricci curvature in $\mathbf{S}^3$ admits at least two minimal 2-spheres and (if one assumes that conjecture that the only minimal embedded torus in the standard $\mathbf{S}^3$ is the Clifford torus) at least one minimal torus. (Positive Ricci curvature comes in only because it is a hypothesis of the Choi–Schoen compactness theorem [**CS**]. Is the Choi–Schoen theorem true for positive scalar curvature?) Of course (a) is analogous to the Lyusternik–Schnirlman theorem (every metric on $\mathbf{S}^2$ has at least three embedded closed geodesics).   B. WHITE

Problem 1.15. Does every embedding of $\mathbf{S}^2$ in $\mathbf{S}^3$ extend to a minimal embedding of $\mathbf{B}^3$ in $\mathbf{R}^4$? In one dimension lower ($\mathbf{B}^2$ in $\mathbf{R}^3$) the answer is "yes" [AS, TT]. Indeed, the area minimizing map must be an embedding [MY]. However, although every embedding of $\mathbf{S}^2$ in $\mathbf{S}^3$ can be extended to an area minimizing map of $\mathbf{B}^3$, there are embeddings for which every such extension is extremely degenerate, so that the image is not a ball [WB6]. B. White has obtained a partial result: Let R be either of the two regions into which the image of $\mathbf{S}^2$ divides $\mathbf{S}^3$. Then there exists a simply connected region U in $\mathbf{R}^4$ such that $\partial U - R$ is a stable embedded minimal surface. The answer is definitely negative for higher codimensions. There is an embedding f of $\mathbf{S}^3$ in $\mathbf{S}^6$ that does not extend to any continuous map of $\mathbf{B}^4$ whose image lies in a stationary varifold with boundary $f(\mathbf{S}^3)$. (The "nonexistence" example of Lawson and Osserman [LO] actually has this property.)   B. WHITE

2. Rigidity, invariance, and classification of minimal surfaces.

Problem 2.1. Suppose M is a smooth imbedded minimal hypersurface in $\mathbf{S}^{n+1}$. Is it true that M is rigid in the sense that any nearby minimal hypersurface in $\mathbf{S}^{n+1}$ is a rigid motion of M? In [AW2] this is shown to be the case if M is a product of two spheres. A preprint of Dajczer and Gromoll shows this to be the case if $n \geqslant 3$ and a nearby hypersurface is isometric to M.   W. ALLARD

Problem 2.2. What conditions on a compact group of isometries G and a G-invariant boundary B will guarantee that every (or some) area minimizing integral current with boundary B is G-invariant? Examples of groups and boundaries that admit *no* invariant area minimizing integral currents can be found in [FH2, 5.4.17; BJ3]. [MF7] contains an example of an $\mathbf{S}^1$-invariant one-manifold which bounds noninvariant area minimizing integral currents. Positive results can be found in [BJ3, LB3, MF7] for codimension one and in [BD2, BD3] for the support of B consisting of an integer sum of orbits. [LJ] contains positive results in case the action of G is polar [DJ] and the dimension of B is arbitrary. Little is known about the case where G is not connected. A related problem is the study of stability of invariant stationary currents. Some results are known for minimal orbits [BJ4, BJ5]. One can also study these questions for non-compact groups [BJ6] and for invariant parametric integrands other than area [BJ3, BJ6, LJ].   D. BINDSCHADLER

Problem 2.3. Show that for each genus there are finitely many (up to rigid motions) compact embedded minimal surfaces of genus g in $\mathbf{S}^3$. [The smooth compactness due to Choi and Schoen [CS] reduces this problem to the following isolation theorem: If there is a sequence of compact embedded minimal surfaces of genus g, then all but finitely many of them are congruent (i.e. differ by rigid motions) to the limit.] See also [LB2].   H. CHOI

Problem 2.4. Are compact embedded minimal hypersurfaces of $\mathbf{S}^n$ topologically unique for $n > 3$? It is known that if Σ_1, Σ_2 are two compact minimal surfaces of the same genus embedded in $\mathbf{S}^3$, then there is a homeomorphism $f: \mathbf{S}^3 \to \mathbf{S}^3$ such that $f(\Sigma_1) = \Sigma_2$. (The surfaces may *not* be congruent, by example.) To what extent is this true in higher dimensions? Something is known. If $M^{n-1} \subset \mathbf{S}^n$ is minimal, compact, and embedded then it is connected [**FT**], and if D_1, D_2 are the connected components of $\mathbf{S}^n - M^{n-1}$ then the homomorphisms $i_*: \pi_1(M^{n-1}) \to \pi_1(D_k)$ are *surjective* for each k. It is further known that the double $\mathscr{D}(D_k)$ of each D_k carries a metric of positive scalar curvature. See [**GL1, GL2, LB2, MSY**].
B. Lawson

3. Foundations and basic tools.

Problem 3.1. Suppose V is a k-dimensional integral varifold in $U = \{x \in \mathbf{R}^n: |x| < R\}$ and V is stationary with respect to some elliptic integrand Φ. Let $d = \limsup_{r \to 0} r^{-k}\|V\|\{x: |x| < r\}$. Does there exist $c > 0$, independent of Φ, such that $r^{-k}\|V\|\{x: |x| < r\} \geq c \cdot d$ when $0 < r < R/2$? This is well known to be the case when Φ is the area integrand; one may take $c = 1$ [**AW2**]. It is shown in [**AW4**] that this is the case if $k = n - 1$ and V is the varifold corresponding to a smooth imbedded hypersurface, provided the second variation of V with respect to Φ is bounded away from $-\infty$. W. Allard

Problem 3.2. According to Allard [**AW2**, §7] and Almgren [**AF1**] there are various pairs of numbers $0 < a(k) < \infty$ and $0 < b(k, n) < \infty$ depending only on $n \in \{3, 4, 5, \cdots\}$ and $k \in \{2, 3, \cdots, n - 1\}$ with the following properties: Suppose (a) M is a compact smooth $(k - 1)$-dimensional submanifold of $\mathbf{R}^n$ without boundary (more general varifold boundaries are also permitted); (b) V is a k-dimensional integral varifold in $\mathbf{R}^n$ (for more general varifolds the formulas contain a density term); (c) $0 \leq K < \infty$ and $\|\delta V\|\llcorner(\mathbf{R}^n \sim M) \leq K\|V\|$ (i.e. V has mean curvatures dominated by K with respect to boundary M); (d) $K^k\mathbf{M}(V) \leq a(k)$. Then $\mathbf{M}(V) \leq b(k, n)\mathscr{H}^{k-1}(M)^{k/(k-1)}$. In the absence of counterexamples, it is plausible to ask whether or not the optimal values for $a(k)$ and $b(k, n)$ are achieved for $K > 0$ when V is a round k-sphere with a spherical cap removed, and for $K = 0$ when V is a flat k-disk. The general bounded mean curvature and stationary cases are completely open. Almgren does, however, show the following in [**AF6**]: (1) If $M = 0$ and $K = k$ (as would be the case for the standard k-sphere $\mathbf{S}^k$), then $\mathbf{S}(V) \geq \mathscr{H}^k(\mathbf{S}^k)$; here $\mathbf{S}(V)$ is the size of V. The inequality is also true for general (real) rectifiable varifolds. (2) Each size and mass bounded (real) rectifiable $(k - 1)$-cycle T with compact support in $\mathbf{R}^n$ is the boundary of a k-dimensional normal current Q with $\mathbf{M}(Q) \leq \gamma(k)\mathbf{M}(T)\mathbf{S}(T)^{1/(k-1)}$, where $\gamma(k) = 1/k^{k/(k-1)}\alpha(k)^{1/(k-1)}$ (i.e. the extreme case is the flat k-disk); the same inequality holds if T and Q are restricted to be integral currents or flat chains modulo ν. F. Almgren

Problem 3.3. Is the subspace $\mathbf{I}_k^\nu$ of the flat k-chains modulo ν in $\mathbf{R}^n$ the quotient of the k-dimensional integral currents in $\mathbf{R}^n$ by ν times those integral currents? The example of a countable sum of two-dimensional real projective spaces given in Federer's treatise [**FH2**, 4.2.26] illustrates one difficulty of a proof; the example is not a counterexample. In codimension zero the answer is "yes" [**WB1**, 2.3]. F. ALMGREN

Problem 3.4. Let $A \subset B \subset \Omega$ be open subsets of an open set Ω with ∂A, ∂B oriented boundaries of least area. Suppose that ∂A, ∂B have a point x_0 of Ω in common. Does it follow that $A = B$? If x_0 is a regular point for ∂A and ∂B then $\partial A = \partial B$ follows from the maximum principle in a neighborhood of x_0 and then generally by analytic continuation. If x_0 is singular then one can show that the tangent cones of ∂A and ∂B at x_0 are the same. If M^n, N^n, have mean curvature 0 and if $M^n = N^n$ on a compact set of positive $\mathscr{H}^n$-measure, does it follow that the restrictions of M^n and N^n to some open set coincide?   E. BOMBIERI

Problem 3.5. Let $M^n \subset \mathbf{R}^{n+k}$ be a minimal submanifold. If $k = 1$, one has Sobolev inequalities of the sort

$$\inf_{c} \|f - c\|_{L^{n/(n-1)}(\mathbf{B}_{\beta_n R}(x_0))} \leqslant \gamma_n \|\nabla f\|_{L^1(\mathbf{B}_R(x_0))},$$

where c denotes a constant, $x_0 \in M^n$, and where $R < \operatorname{dist}(x_0, \partial M^n)$ and β_n, γ_n are positive absolute constants. If $k \geqslant 2$, a similar result need not be true, mainly because M^n need not be connected. The problem is whether such a result holds for *minimizing* $M^n \subset R^{n+k}$, $k \geqslant 2$, in the form

$$\inf_{g} \|f - g\|_{L^{n/(n-1)}(\mathbf{B}_{\beta_{n,k}R}(x_0))} \leqslant \gamma_{n,k} \|\nabla f\|_{L^1(\mathbf{B}_R(x_0))},$$

where now the inf is over all g with $\nabla g = 0\|$ $M^n\|$ almost everywhere. E. BOMBIERI

Problem 3.6. Find a suitable analogue for integral currents of positive codimension of the characterization of sets of finite perimeter in terms of the set of points where the Lebesgue density is neither zero nor one [**FH2**, 4.5.11]. Relevant references are [**BJ1**, **PS1**, **PS2**]. In these papers the concept of measure theoretic exterior normal is extended to integral currents and integral currents modulo ν of all dimensions and codimensions.   J. BROTHERS

Problem 3.7. Is the Cartesian product of two area minimizing surfaces area minimizing? For normal currents, a proof has been given in certain low dimensions and codimensions [**MF6**, 1.3, 5.2]. However, for integral currents (or similarly for oriented manifolds with boundary), Almgren has given a counterexample (see [**MF4**, Introduction]). The question is completely open for flat chains modulo two.   F. MORGAN

Problem 3.8. Can every normal current T in $\mathbf{R}^n$ be decomposed as a convex (integral) combination of integral currents? In codimension one the answer is "yes" if $\partial T = 0$ [**FH2**, 4.5.9(13)]   F. MORGAN

Problem 3.9. What is first variation? The definition of the first variation of a surface requires deformations of the ambient space that move the surface along with it (with an obvious interpretation for immersed surfaces). Essentially, the deformations are C^1 close and cannot separate sheets or decompose singularities.

Regular parts of surfaces with zero first variation are locally area minimizing, but this is not true for singular parts (e.g. two crossing lines in $\mathbf{R}^2$) or for surfaces which are stationary for non-convex integrands (e.g. a line segment may be stationary even when a zig-zag has much less integral). Perhaps some other definition of first variation would be useful. One possibility would be to compare a surface T with a one parameter family of surfaces that are flat-close or mass-close to it, though there is as yet no calculus for computing such a variation. Also, Pitts' defining conditions for almost-minimizing varifolds include a notion of discretized flows of surfaces which one might wish to regard as variations [**PJ**]. A more subtle extension of the same idea is Brakke's construction of motion by mean curvature [**BK**]. Is it worth developing a theory of generalized variations? That is, are there any good encompassing theorems? J. TAYLOR

Problem 3.10. Do there exist extremely unrectifiable sets? That is, for some $r < k < n$, is there a compact set C in $\mathbf{R}^n$ of finite, positive Hausdorff r-dimensional measure, such that C projects in almost every k-plane of $\mathbf{R}^n$ to a set of r-dimensional measure zero? For $r = k$, such sets C are called purely unrectifiable and many examples are known. If r is an integer less than k, the condition above is much stronger than pure unrectifiability. If the answer is "yes" for some noninteger r, one could look for an analogue of the Besicovitch–Federer structure theorems [**FH2**, §3.3] for r-dimensional sets. Of course if C lies in a k-dimensional plane in $\mathbf{R}^n$ it cannot have this property. See N. Falconer's forthcoming book for various pathological examples of r-dimensional sets. Also see [**MP1, MP2, MP3, MP4**] and papers by J. Marstrand listed in the bibliography of [**FH2**]. B. WHITE

4. Regularity and structure.

Problem 4.1. Suppose V is an (area) stationary integral k-dimensional varifold in an open subset of $\mathbf{R}^n$. Let S, the singular set, be the set of points near which $\mathrm{spt}\|V\|$ is not smooth. It follows from [**AW2**] that S is a closed nowhere dense subset of $\mathrm{spt}\|V\|$. Is it true that $\|V\|(S) = 0$? In case $k = 1$, the answer is "yes" and S and V near S are completely understood; see [**AA1**]. To handle $k > 1$ one is tempted to use Almgren's new methods; see [**AF5**]. Much more would be needed. I feel this is an extremely difficult problem. W. ALLARD

Problem 4.2. Is it possible that a k-dimensional stationary integral varifold in $\mathbf{R}^n$ [**AW2**] almost everywhere (many places) has second fundamental form with norm locally summable to the $m + \varepsilon$ power in the sense of Hutchinson [**HJE**]? If so one could apply Hutchinson's multiple-valued function regularity results [**HJE**]. There is nothing in sight which seems powerful enough to give such a result a priori. Is it possible that the existence of tangent cones in such V forces large curvatures to be concentrated on small sets? F. ALMGREN

Problem 4.3. What properties of boundaries are reflected in properties of minimal surfaces spanning those boundaries? To what extent is it possible, for example, to correlate combinatorial complexity or singularity structure in area minimizing currents with boundary properties? Such questions are part of a fundamental area of research in geometric measure theory. Among the present fruits of such research are the boundary regularity results of Allard [**AW1, AW3, AF5**, 5.23], and of Hardt and Simon [**HR, HS1**]. Among varying boundaries the uniqueness of surfaces of least area in general dimensions and codimensions is almost sure according to Morgan [**MF1, AF5**, 5.24]. Everywhere interior regularity of surfaces of least unoriented area of dimension two and general codimensions is similarly almost sure according to White [**WB2**]. We also know the complex boundary criteria of Harvey and Lawson [**HL1**] that boundary tangent planes be maximally complex, and the complex interior regularity criterion of Yau in terms of vanishing Kohn–Rossi $\bar{\partial}_b$ cohomology groups [**YS**]. Some upper bounds on curvature and topological type are given by Hardt and Simon in [**HS1**], and various lower bounds on topological type are given in examples of Almgren, Hubbard, and Thurston in [**AS, AT, HJ**]. Finally we know from Milani [**MA**] how to use A. Tognoli's extension of H. Hironaka's resolution of singularities technique to give examples in which nontrivial boundary cobordism is an obstruction not only to interior regularity of an area minimizing surface spanning such a boundary but also to the possible real analyticity of its singular set. F. ALMGREN

Problem 4.4. It would be of interest to prove some theorems about the Gauss map of a complete minimal graph.   E. BOMBIERI

Problem 4.5. Consider a (stationary) minimal surface M properly immersed in $\mathbf{B}^3 - \{0\}$ in $\mathbf{R}^3$. Must M have finite topological type? [**PJ**, pp. 230 ff.] gives a proof (although not a statement) for stable embeddings. For stable immersions this is proved in [**GLB**].   R. GULLIVER

Problem 4.6. Can a minimizing disk in $\mathbf{R}^3$ with C^∞ boundary Γ have a boundary branch point? [**GLF**] shows that if Γ is real analytic then a surface with a boundary branch point does not minimize area for mappings of the given topological type. In [**GR**] a surface with a boundary branch point is constructed for which a neighborhood of the branch point is isotopic to the graph of $z = \mathrm{Re}\{\exp(-w^{1/7})\}$, $-\pi/2 \leqslant \arg w \leqslant 5\pi/2$, $9\pi/14 \leqslant \arg(-w^{-1/7}) \leqslant 15\pi/14$. For this example the method of [**GLF**] fails.   R. GULLIVER

Problem 4.7. What regularity holds for the two-dimensional mapping problem (as in the classical Plateau problem) in which the area integrand is replaced by a parametric integrand? The existence of a minimizer with square integrable first derivatives was established by Morrey and others; see [**MC2**, Chapter 9] for a discussion.   R. HARDT

Problem 4.8. What can be said about the boundary regularity of a codimension one stationary rectifiable current T having a smooth boundary? In case T is area minimizing the complete boundary regularity was established in [**HS1**]. However, stationary currents in general may have boundary singularities such as self intersections or boundary branch points; an example is in [**GLF**]. How large is the set of boundary branch points near which spt T fails to be an immersed manifold with boundary?   R. HARDT

Problem 4.9. Remove the hypothesis on the support of the current in Theorem 2.7 of [**HSB**]. That is, show that each locally rectifiable, d-closed current of bidimension p, p is a holomorphic chain (i.e. analytic cycle). The support hypothesis requires that the Hausdorff $2p + 2$ measure of the support vanish or, if the codimension is one, that the support should be nowhere dense.   R. HARVEY

Problem 4.10. Consider a surface M of constant mean curvature with convex boundary ∂M lying in a plane. Find conditions on ∂M for M to be convex. The corresponding capillary problem has been treated recently by Chen and Huang [**CH**], Finn [**FR2**], and Korevaar [**KN**].   W. H. HUANG

Problem 4.11. Develop a regularity theory for stationary harmonic maps. In [**SU1**] Schoen and Uhlenbeck demonstrated that energy minimizing maps (which are in particular weak solutions for the Euler–Lagrange equations of harmonic maps) are regular outside a set of codimension at least three in the domain. Furthermore, they characterized the possible singularities of such maps infinitesimally. Similar results were obtained by Giaquinta and Giusti [**GG1** and **GG2**] for minima of a somewhat larger class of quadratic variational problems under the restriction, however, that the image is a priori confined to lie in a homeomorphic image of a ball. Furthermore, minima solving a Dirichlet problem were shown to have no singularities at all in a neighborhood of the boundary [**SU2, JM**]. A crucial ingredient in all proofs was a monotonicity formula, and such a formula holds for a weakly harmonic map (i.e. a critical point of the energy functional), provided it is also stationary with respect to composition with a one parameter family of local diffeomorphisms of the domain [**PP, KW**]. Therefore, it is natural to define a stationary harmonic map as a critical point of the energy functional that is also stationary with respect to these variations of the independent variables, and to ask for regularity properties of such maps. So far, only the case of a two-dimensional domain has been settled, since Schoen [**SR**] could reduce it to the case of a weakly conformal harmonic map where Grüter [**GRM**] previously obtained full regularity. (Note that in the two-dimensional case the regularity of minima was already proved by Morrey [**MC1**].) It should be expected that any partial regularity results will have to make strong use of the variational structure of the problem since for the class of nonlinear elliptic systems with quadratic growth in the gradient a wide range of counterexamples to various regularity questions is available, mainly by the work of M. Meier [**GM**].   J. JOST

Problem 4.12. Which curves C bound a disk D minimally embedded in $\mathbf{R}^n$? Which of these C bound a unique minimal D? Any extreme Jordan curve $C \subset \mathbf{R}^3$ bounds an embedded minimal disk D [**AS, TT**]; in this case the area minimizing D is embedded [**MY**]. This is a key to the construction of various classes of embedded minimal surfaces (not merely disks) in three-manifolds [**HS**]. Nitsche [**NJ**] has shown that if the total absolute curvature $\int_C |\kappa|\, ds$ is less than 4π then the minimal disk D spanning $C \subset \mathbf{R}^3$ is unique. A number of authors have conjectured that this estimate suffices to show D is embedded (see the introduction to [**MY**]) Very recently, L. P. Jorge has discovered other sufficient conditions for embeddedness and uniqueness. For $n \geq 5$ one can show that a generic smooth curve $C \subset \mathbf{R}^n$ bounds some embedded minimal disk D by using the generic regularity of D [**BT**], the smooth dependence of D upon C [**WB3**], and Whitney's ideas. The fascinating case is $n = 4$; here, for simplicity, assume $C \subset \mathbf{S}^3 = \partial \mathbf{B}^4$. The topologists have asked the following [**KIR, RD**]: (i) Does slice imply ribbon? If $C_0 \subset \mathbf{S}^3$ bounds a smooth disk D_0 embedded in $\mathbf{B}^4$, then can C_0 be isotoped in $\mathbf{S}^3$ to C_1 which bounds a minimal disk D_1 embedded in $\mathbf{B}^4$? (ii) Are ribbon disks topologically unique? Suppose $C_i \subset \mathbf{S}^3$ is spanned by an embedded minimal disk $D_i \subset \mathbf{B}^4$, $i = 0, 1$. If C_0 is isotopic to C_1 in $\mathbf{S}^3$, is D_0 properly isotopic to D_1 in $\mathbf{B}^4$? A caution to (i): The slice disk D_0 may not be properly isotopic to *any* minimal D_1 [**HJ**], hence the technique of minimizing area within an isotopy class (so successful in dimension three) can not be naively mimicked in dimension four. A remark to (ii): For *knotted* C_i some topologists expect the conclusion to be false. For *unknotted* C_i the conclusion is known to hold provided $\int_{C_i} |\kappa| \, ds < 8\pi$. If (ii) is true one may be able to answer the following question affirmatively by means of the continuity method in which C is connected to a geodesic $\mathbf{S}^1$ via a generic path of curves: Does every unknotted Jordan curve $C \subset \mathbf{S}^3$ bound some embedded minimal disk $D \subset \mathbf{B}^4$? (See Problem 1.15. Using the "bridge principle" one easily constructs such a C for which the area minimizing disk is *not* embedded.) Presumably minimal surfaces can play an important role in the study of four-manifolds [**AF3**].   R. KUSNER

Problem 4.13. What regularity can be proved for solutions of the minimal surface system? It is known that for $n \geq 4$ and $k \geq 3$ there is a Lipschitz function $f\colon \mathbf{R}^n \to \mathbf{R}^k$ which is a solution of the minimal surface system and is *not* C^1 [**LO**]. The graphs on these cones are absolutely minimizing [**HL3**]. There are examples of Gulliver–Kinderlehrer–Spruck [**OR**] of $W^{1,2}$ weak solutions of the minimal surface system of the form $u\colon \mathbf{B}^4 - \{0\} \to \mathbf{R}^4$ where $u(x) = f(|x|)x$. Some of these are bounded and discontinuous. For $n = 2$ or 3 Lipschitz solutions are analytic, and for $n > 3$ there is a constant C_n (explicit and not so small) such that solutions with Lipschitz constant $\leq C_n$ are analytic [**F-C**].   B. LAWSON

Problem 4.14. Let n and k be integers greater than one, and let M be a smooth compact orientable Riemannian manifold of dimension $n + k$. Does M support a k-dimensional stationary integral varifold such that $\mathrm{spt}\|V\|$ is regular except for a compact singular set of $\|V\|$ measure zero (or, more optimistically, of Hausdorff dimension at most $k - 2$)? The analogous theorem for hypersurfaces ($n = 1$) was established in [**PJ** and **SS**]. The regularity theory in these papers depends heavily on analysis of the second variation of area, and therefore may not generalize to codimensions greater than one.   J. PITTS

Problem 4.15. Is the Gauss map on an area minimizing hypersurface in $\mathbf{R}^{n+1}$ an energy minimizing map to $\mathbf{S}^n$? The answer is "yes" for hypersurfaces which are graphs [**SU3**]. I have shown that the analogous proposition with "minimizing" replaced by "stable" is true.   B. SOLOMON

Problem 4.16. Suppose Φ is a two-dimensional crystalline integrand on $\mathbf{R}^3$, and B is a one-dimensional polyhedral cycle whose edges are all parallel to edges of the Wulff shape for Φ. Suppose S has boundary B, minimizes the integral of Φ, and among all such minimizers it maximizes volume (or, more generally, obeys crystal corner barriers). Is S polyhedral? The question is really one of finiteness of the number of plane segments, since the number of tangent directions should be finite (see [**TJ4**], though the proof of Theorem 6.1 there is incomplete). Furthermore, if the number of plane segments is finite, there is an explicit upper bound to that number in terms of the number of segments in B, the Euler characteristic of S, and the structure of Φ [**TJ8**]. I am not convinced that this problem is difficult; I know only that I have not yet found a proof of it after eight years of trying on and off. Perhaps one should try the multiple-valued function approach (see Problem 4.17)?   J. TAYLOR

Problem 4.17. Are crystalline minimal surfaces multiple-valued functions? The idea here is that if one can control the multiplicity of some projection of such surfaces, then there is a built-in Lipschitz constant to the multiple-valued function (given by the steepest slope in the boundary of the Wulff shape for the crystalline integrand). This could lead to new explicit computational methods.   J. TAYLOR

Problem 4.18. Can ellipticity be localized, and is a minimizing (or stationary) surface regular at elliptic tangent planes? Almgren's definition for ellipticity of an m-dimensional integral Φ in $\mathbf{R}^n$ is that there is a constant $c > 0$ such that whenever D is an m-dimensional surface in an m-dimensional plane and $T \subset \mathbf{R}^n$ is any other surface with $\partial T = \partial D$, then $\Phi(T) - \Phi(D) \geq c(M(T) - M(D))$. One could define the idea of Φ being elliptic at a given plane direction by restricting the above to D's which lie in that plane. Or, more locally, one might define a convex integrand Φ to be elliptic on a plane π if Φ agrees with an elliptic integrand in a neighborhood of π. An integrand is then elliptic if it is elliptic on all planes. The theorems involving ellipticity now hypothesize that Φ be elliptic on all planes. One might conjecture that local ellipticity would be sufficient. This is of interest for more general integrands, whose minimizing surfaces might be regular pieces (in low dimensions) patched together by singularities and/or "infinitesimally corrugated" varifold pieces.   J. TAYLOR

Problem 4.19. How regular is a minimizing integral current at a boundary with multiplicity at least two? (or stationary integral varifold with density greater than one-half at the boundary?) That is, let T be an integral current that minimizes an elliptic functional such that ∂T is, in some neighborhood U of a point x, a smooth manifold with multiplicity $p \geq 2$. Is T regular near x? In codimension one, in U, T can be decomposed into p minimizing currents each of which takes the given boundary with multiplicity one [**WB1**]. Thus one gets as much regularity as one has for the multiplicity one case; in particular, one has boundary regularity for the area functional [**HS1**]. In higher codimensions one does not get complete boundary regularity even in the multiplicity one case. (For example $\{(z^3, z^4):$ Im $z \geq 0\}$ is an area minimizing current in $\mathbf{C}^2 = \mathbf{R}^4$ with a boundary singularity at the origin.) On the other hand, in the case of area almost all tangent cones at boundary points are sums of half-planes [**BJ2**]. See also [**AW3** and **HR**] for the basic known results concerning stationary varifolds and currents minimizing elliptic functionals, respectively, at points of density one-half. B. WHITE

Problem 4.20. Prove almost everywhere regularity for currents modulo p that minimize area or other even elliptic functionals. What do the singular sets look like? For $p = 2$ or 3 almost every point has a tangent plane with multiplicity one and hence is a regular point by the general theory (Allard's regularity theorem for area [**AW2**], and [**AF2**] for other functionals). For hypersurfaces in $\mathbf{R}^{m+1}$ and p odd, the singular set has dimension $\leqslant m - \varepsilon$ and indeed $\leqslant m - 1$ for area [**WB9**]. For two-dimensional area minimizing currents modulo 3 in $\mathbf{R}^3$, the singular set consists of analytic curves along which three minimal surfaces meet at equal angles [**TJ1**]. The support of any minimizing hypersurface modulo 4 is locally the union of the supports of two minimizing integer multiplicity currents and hence has a codimension one singular set that can be fairly precisely described [**WB1**].   B. WHITE

Problem 4.21. Does there exist a sequence of pairs (R_i, S_i) of absolutely area minimizing $(n - 1)$-dimensional integral currents in $\mathbf{R}^n$ with the following properties: (i) $R_i \neq S_i$, (ii) $\partial R_i = \partial S_i$ and spt ∂R_i lies on the boundary of a strictly convex body in $\mathbf{R}^n$, (iii) (spt ∂R_i) $\cap$ (spt ∂R_j) $= \varnothing$ for $i \neq j$, (iv) $R_i \to T$ and $S_i \to T$ in the flat topology. This question is motivated by the study of functions of least gradient in [**PZ**]. Let $u: \Omega \to \mathbf{R}$ be such a function. Whenever t is such that $\mathscr{L}^n\{x: u(x) = t\} > 0$, the currents $\partial(\mathbf{R}^n \llcorner \Omega \cap \{x: u(x) \leqslant t\})$ and $\partial(\mathbf{R}^n \llcorner \Omega \cap \{x: u(x) \geqslant t\})$ are, when restricted to Ω, absolutely area minimizing and have the same boundary. We call the region $\{x: u(x) = t\}$ a *lens*. Clearly, there are only countably many lenses, but it was not decided whether it is possible for them to accumulate. F. Morgan has pointed out that by using decreasing curves in Enneper's surface one can construct examples which satisfy all but condition (iii). W. ZIEMER

5. Singularity problems.

Problem 5.1. Consider an area stationary integral cone C. Is there a way of using C as a parameter space for nearby stationary surfaces? We have in mind that C is smooth away from a well-understood singular set. The object is to extend the methods of [**AA2** or **SL1**] to this case to obtain similar results. W. ALLARD

Problem 5.2. Are the tangent cones to area minimizing surfaces unique? This is one of the most celebrated of the unsolved problems in geometric measure theory. Such uniqueness is not known even when, for example, each possible tangent cone is assumed to be twice the rectifiable current associated with an oriented linear subspace. For rectifiable currents or flat chains modulo two one knows that all singular points, a fortiori all singular points with nonunique tangent cones, are of Hausdorff codimension at least two [**AF5**, **FH3**]. For two-dimensional area minimizing surfaces in $\mathbf{R}^n$, White [**WB5**] showed such uniqueness; one would expect his technique to generalize to manifolds. In general dimensions and codimensions, Allard and Almgren [**AA2**] gave assumptions (sometimes realized) sufficient to guarantee uniqueness at singular points having as one tangent cone the central cone over a minimal submanifold of $\mathbf{S}^{n-1}$ while Simon [**SL1**] subsequently showed by a new method uniqueness at all such points (with no additional assumptions). The tangent cone stratification of [**AF5**, 2.27] is one method by which the tangent cone problem might usefully be subdivided. F. ALMGREN

Problem 5.3. Singularities in m-dimensional area minimizing integer multiplicity currents are closed sets with vanishing $(m - 2 + \varepsilon)$-dimensional Hausdorff measure for each $\varepsilon > 0$ according to Almgren [**AF5**]. In all known examples, however, such singular sets have locally finite $(m - 2)$-dimensional measure. Is this true in general? A weaker (possibly easier) result would be that such singular sets have locally finite $(m - 2)$-dimensional capacity (which is not incompatible with infinite $(m - 2)$-dimensional measure). An added incentive to singularity research is the hope of solving in complete generality the least area mapping problem with domain dimensions at least three following the technique set forth and applied by White in [**WB6**].   F. ALMGREN

Problem 5.4. Is it possible for the singular set of an area minimizing integer (real) rectifiable current to be a Cantor type set with possibly non-integer Hausdorff dimension? The area problem possibly has enough self-similarities to generate such a singular set. Cantor type singular sets of various dimensions between one and two can be realized for two-dimensional surfaces in $\mathbf{R}^3$ minimizing the integral of a convex (but not uniformly so) parametric integrand according to [**TJ4**].   F. ALMGREN

Problem 5.5. Suppose $f: \mathbf{B}^m(0, 1) \to \mathbf{Q}$ is stationary (not minimizing) for Dirichlet's integral in the sense of [**AF5**, Chapter 2]. Does the singular set of f have measure zero? Is it of Hausdorff dimension $m - 2$ as in the minimizing case? Is the graph of f real analytic (unknown even in the minimizing case)? Similar questions apply for $g: \mathbf{B}^m(0, 1) \to \mathbf{Q}$ which is minimizing (or stationary) for the $\mathbf{Q}$ elliptic integrals of Mattila in [**MP5**]; continuity of such g is known only for $m = 2$ in the minimizing case.   F. ALMGREN

Problem 5.6. According to [**AF5**, 2.13(5)(v)] each Dirichlet integral minimizing $\mathbf{Q}$-valued function can be approximated locally by homogeneous functions. Is there only one such homogeneous approximating function at each point or does the function change with the radius within which one is approximating? Possibly Simon's new parabolic estimates [**SL1**] would be relevant.   F. ALMGREN

Problem 5.7. Let C be a stable (or minimizing) hypercone in $\mathbf{R}^n$. Given an $\varepsilon > 0$, can one find a smooth embedded hypersurface of positive mean curvature properly embedded in $\mathbf{B}(0; 1) \cap C_\varepsilon$ (where C_ε is the ε-neighborhood of C in $\mathbf{R}^n$)?

This is related to the following question: Let Ω be a compact domain in a Riemannian manifold such that $\partial\Omega$ is of minimal area for the contained volume. Can $\partial\Omega$ be approximated by a smooth hypersurface of positive mean curvature? B. LAWSON

Problem 5.8. A pivotal question in the classification of area minimizing cones (see Problem 5.9) is when is a pair of planes area minimizing (as a locally rectifiable current)? The geometric relationship between two oriented k-dimensional planes is given by k angles $0 \leqslant \gamma_1 \leqslant \gamma_2 \leqslant \cdots \leqslant \gamma_k$. The conjecture (see [**MF3**]) is that the pair is area minimizing if and only if $\gamma_k \leqslant \gamma_1 + \cdots + \gamma_{k-1}$. The result is proved only for $k = 2$ [**MF2**] and the trivial case $k = 1$. If $\gamma_k = \gamma_1 + \cdots + \gamma_{k-1}$, the planes are simultaneously special Lagrangian and therefore area minimizing. For $\gamma_k < \gamma_1 + \cdots + \gamma_{k-1}$, it has been shown for $k \leqslant 3$ that the planes are calibrated and therefore area minimizing [**MF3**]. For $\gamma_k > \gamma_1 + \cdots + \gamma_{k-1}$, for $k = 2$, a curved comparison surface has been exhibited [**MF2**]; another method would be to show that every area minimizing pair of planes is calibrated by a constant coefficient differential form and to study such forms. All that is known is that for every area minimizing *normal* current there is a dual *measurable coflat* form [**FH4**, 4.10(1)]. The related conjecture for calibrations appears in Problem 6.5. A more general question asks when a sum of l k-dimensional planes is area minimizing.   F. MORGAN

Problem 5.9. Classification of area minimizing cones is the first major component of the classification of singularities in area minimizing surfaces. In the class of k-dimensional locally rectifiable currents in $\mathbf{R}^n$, the area minimizing cones have been classified only for $k = 1$, for $k = 2$ (as a finite sum of complex planes [**MF2**]), and for $k = n - 1 \leqslant 6$ (as a plane with multiplicity). For $k = n - 1 = 7$, a compactness argument shows that there are only finitely many diffeomorphism classes. Other examples are given by certain sums of planes [**MF3**], certain cones over products of spheres [**BD1**], certain equivariant cones [**LB3**], calibrated cones [**HL3**, **BRR**], and certain sums [**MF5**, §5] and cartesian products of previous examples [**MF6**, 5.2]. In the class of k-dimensional locally flat chains modulo ν, the area minimizing cones have been classified for $k = 1$; for $k = \nu = 2$ [**MF2**]; for $k + 1 = n = \nu = 3$ [**TJ1**]; and for $k + 1 = n \leqslant 7$ and $\nu = 4$ [**WB1**]. Also for $k = n - 1$, the problem coincides with the problem for rectifiable currents [**MF5**, §4]. Other examples are given by products of previous examples with planes and by certain sums of previous examples ([**MF5**, §5] applies).   F. MORGAN

Problem 5.10. (i) When are tangent cones of minimal surfaces and tangent maps of harmonic maps unique? This question is fundamental in that it guarantees well-controlled asymptotic behavior on approach to singular points in a wide variety of circumstances. For two-dimensional area minimizing surfaces the question has been settled by White [**WB5**] and for the stationary "multiplicity one regular" case by Simon [**SL1**]. See also Almgren and Allard [**AA2**] for an earlier result using a different method. For the three-dimensional case (and in certain higher dimensional cases) the unique tangent harmonic map question is also settled in [**SL1**]. (ii) When are the "tangent cylinders at infinity" of an entire minimal graph unique? Again this is a basic question in that an affirmative answer gives a very good picture of how the minimal graph behaves asymptotically at infinity. For discussion, and for the proof of uniqueness in certain cases (when at least one of the tangent cylinders at infinity is of the form $C \times R$ with sing $C = \{0\}$), see [**SL2**].   L. SIMON

Problem 5.11. Does a minimal submanifold M, with a unique tangent cone at 0, approach this tangent cone with rate at least $|x|^{\gamma}(\gamma > 1)$ as $|x| \to 0$? This is true in all known cases; see in particular the many examples in [**CHS** and **HS2**]. Almgren and Allard [**AA2**] prove it is true if the tangent cone C has multiplicity one, provided sing $C = \{0\}$ and all Jacobi fields of $C \cap \{x: |x| = 1\}$ are "integrable" (i.e. are generated as initial velocity vector fields of one-parameter families of minimal submanifolds of the sphere). The method of Simon [**SL1**] does not need such an integrability restriction, but does not give much information about the *rate* at which M approaches its tangent cone.   L. SIMON

Problem 5.12. If C is a hypercone in $\mathbf{R}^n$ with sing $C = \{0\}$, is it true that C strictly minimizing (Problem 1.11) implies C strictly stable? Strictly stable means that $\inf_{\Omega \Subset C} \lambda_1(\Omega) > 0$ where $\lambda_1(\Omega)$ is the first eigenvalue of the Jacobi field operator for C on $\Omega \Subset C$ relative to zero boundary values.   L. SIMON

Problem 5.13. Can one give bounds on the growth (near infinity) of an entire solution $u \in C^2(\mathbf{R}^n)$ of the minimal surface equation? One conjecture is that the growth might satisfy $|u(x)| \leqslant c(1 + |x|)^{\gamma}$ for some constants c, γ. This has been proved under certain restrictions on the tangent cylinder at infinity by Simon [**SL2**]; see Problem 5.10. See also [**BG**] for a discussion and [**BDG, SL2**] for the existence of examples of nonlinear entire solutions.   L. SIMON

Problem 5.14. Classify (M, ε, δ) minimal cones in higher dimensions. Two-dimensional (M, ε, δ) minimal cones in $\mathbf{R}^3$ (or a $C^{1,\alpha}$ three-dimensional manifold) have been classified [**TJ2**] (resp., [**TJ3**]). To my knowledge, there are no higher dimensional results. A reasonable conjecture is that the cone over the $(k - 2)$-dimensional skeleton of a standard k-dimensional simplex is (M, ε, δ) minimizing in $\mathbf{R}^k$. Is it true? If so, then products of these with lines and planes are also minimizing; does this exhaust the list of possibilities up to dimension seven? J. TAYLOR

Problem 5.15. What is the structure of minimizing surfaces for nonelliptic (convex) integrands? A two-dimensional surface in $\mathbf{R}^3$ which minimizes the integral of an elliptic integrand is regular everywhere [**ASS**]. This is certainly not the case for nonelliptic integrands. In fact, for integrands that are neither elliptic nor crystalline very little is known. For a survey of what is known in the nonelliptic cases, see [**TJ5**]; see also [**TJ7**]. (It is no loss of generality only to consider convex integrands in the following sense: If Φ^c is the convex integrand with the same Wulff shape as Φ, and if S minimizes the integral of Φ^c, then there is a (nonintegral) varifold V with the same underlying set as S such that V minimizes Φ; one just puts the appropriate infinitesimal corrugation on the bad tangent planes of S.)   J. TAYLOR

Problem 5.16. Do the singularities in an area minimizing current disappear after generic perturbations of its boundary? In some cases the answer is "no" for topological reasons. For example, $\mathbf{CP}^2$ is not the boundary of any manifold; thus any imbedding of $\mathbf{CP}^2$ into Euclidean space can only bound singular area minimizing currents (if the image of $\mathbf{CP}^2$ is in the boundary of a convex set, then by Allard's boundary regularity theorem [**AW3**], the singularities will be interior singularities). Also, if $3 \leqslant k \leqslant n - 3$, the set of pairs of oriented k-planes in $\mathbf{R}^n$ contains a nonempty open subset of pairs that are area minimizing [**MF3**], so the answer may be "no" in these dimensions and codimensions. On the other hand, if $k = 2$ or $n - 2$, then the subset of pairs that are area minimizing is closed and has empty interior [**MF2**]. On the positive side, it is known that the set of $C^{2,\alpha}$ curves in $\mathbf{R}^n$ that bound singular area minimizing currents modulo two is of codimension at least one in the set of all $C^{2,\alpha}$ curves [**WB10**]. The proof depends heavily on the fact that such currents are actually regular as immersed surfaces; the singularities are just self-intersections. Similarly, Böhme and Tromba [**BT**] show that generic curves in $\mathbf{R}^n$ ($n \geqslant 4$) do not bound any minimal disks with branch points. One should also consult the work of Caffarelli, Hardt, and Simon [**CHS**], who show that for various area minimizing cones in $\mathbf{B}^n$, the singularity persists under a large family of perturbations of the boundary. One can also ask whether singularities in a homologically area minimizing cycle in a Riemannian manifold disappear after generic perturbations of the metric. Again, there can be topological obstructions (some integral homology classes of compact manifolds cannot be represented by the image of any manifold). Also, two-dimensional area minimizing cycles modulo two are probably regular for generic metrics by an argument similar to [**WB10**]. For seven-dimensional area minimizing frontiers in $\mathbf{R}^8$ (where there are only isolated singularities) R. Hardt and L. Simon have announced that one sided perturbations of the boundary always eliminate singularities.   B. WHITE

6. Calibrated geometry.

Problem 6.1. This problem concerns properties of holomorphic chains T with $\partial T = M$, where M is a compact real analytic cycle of dimension $2p - 1$ in $\mathbf{C}^n$, and M is maximally complex and satisfies the moment condition. It is conjectured in [**HL1**] that $\overline{\operatorname{spt} T} = \operatorname{spt} T \cup \operatorname{spt} M$ is a finite union of $2p$-dimensional blocks (i.e. is semianalytic). Mariaux [**MB**] has proved that $\overline{\operatorname{spt} T}$ is subanalytic provided M is a subanalytic chain and $|M| = \operatorname{spt} M$ satisfies (C): Each singular point has a neighborhood which is contained in a complex subvariety of pure dimension p. Moreover, (C) is satisfied if M is smooth, or $\dim_{\mathbf{R}} |M| = 1$, or if $|M|$ is semialgebraic (i.e. semianalytic and locally described by Nash functions). The problem is to decide whether, for $p \geqslant 2$, (C) is necessary for $\overline{\operatorname{spt} T}$ to be subanalytic. P. DOLBEAULT

Problem 6.2. What are the obstructions to the construction in a complex manifold X of a holomorphic p-chain T with a given boundary M satisfying the moment condition, or maximally complex? There is no obstruction for $\operatorname{spt} M$ a compact subset of either $X = \mathbf{C}^n$ [**HL1**] or $X = \mathbf{P}^n(\mathbf{C}) - \mathbf{P}^{n-q}(\mathbf{C})$, $p \geqslant q$ [**HL2**].

There is also no obstruction to the construction of a maximally complex $(2p - 1)$-chain with $\partial M = N$, where N is a given $(2p - 2)$-chain with holomorphic dimension $p - 2$, in a real hyperplane of $\mathbf{C}^n$ [**BD**]. For the corresponding "soft problem" (i.e. M and T currents) obstructions are known (Harvey–Lawson). These problems (when they have meaning) are also of interest for M noncompact. P. DOLBEAULT

Problem 6.3. Give an effective description of special Lagrangian cones [**HL3**]. Some elementary facts are already known and there is at least one theorem, due to R. Bryant. It says that a special Lagrangian cone in $\mathbf{R}^6$, whose link in $\mathbf{S}^5$ is an immersed two-sphere, is a three-plane. The corresponding question can be asked for coassociative and Cayley geometries. In the associative geometry the links of cones are compact holomorphic curves in $\mathbf{S}^6$, and by Bryant, many such exist. B. LAWSON

Problem 6.4. Characterize the boundaries of varieties in a calibrated geometry [**HL3**] (in the spirit of [**HL1** and **HL2**]). It is known that in each calibrated geometry the boundaries must satisfy a local condition (analogous to "maximal complexity") and a global "moment condition." Sometimes the local condition is vacuous. B. LAWSON

Problem 6.5. Study of the faces of the Grassmannian $G(k, \mathbf{R}^n)$ of oriented k-dimensional subspaces of $\mathbf{R}^n$ has provided a wealth of examples of interesting area minimizing surfaces and has advanced the understanding of the singularities that occur. What patterns or principles can be found in the general classification of faces? For example, it is conjectured that two k-dimensional planes lie on a common face of the Grassmannian $G(k, \mathbf{R}^n)$ if and only if their characterizing angles $\gamma_1 \leqslant \gamma_2 \leqslant \cdots \leqslant \gamma_k$ satisfy $\gamma_k \leqslant \gamma_1 + \cdots + \gamma_{k-1}$. (It follows that their union is area-minimizing; see Problem 5.8.) This conjecture has been proved only in the cases $k = 2$ or $n - 2$ [**MF2**] and $k = 3$ or $n - 3$ [**DH**, **MF6**, 1.2]. More generally, can geometric criteria be given for a set of l k-planes to lie on a common face? See also [**HM, HAR**]. F. MORGAN

Problem 6.6. Harvey and Lawson [**HL3, HAR**] describe rich calibrated geometries associated with large faces of the Grassmannian $G(k, \mathbf{R}^n)$ of oriented k-planes in $\mathbf{R}^n$, especially for $n = 6, 7, 8$. What are the largest faces in higher dimensions? (The face of special Lagrangian planes has dimension $(k + 2)(k - 1)/2$ in $G(k, \mathbf{R}^{2k})$, which has dimension k^2.) The largest known faces are associated with subgroups of the orthogonal group which act doubly transitively on the sphere, but those have all been found. F. MORGAN

7. Real Analysis.

Problem 7.1. Consider the space of real-valued functionals on the convex bodies in $\mathbf{R}^n$ which are additive, continuous, and invariant under euclidean motions. Hadwiger [**HAH**] showed that the n mixed volumes of Minkowski form a basis of this space. He applied this fact to give an elegant proof of the principal kinematic formula for these sets. More recently, Schneider [**SCR**] was able to prove the analogous result for locally defined measure-valued functionals on the convex bodies; that is, that the curvature measures form a basis for this space. In his classic work on extrinsic curvature Federer [**FH1**] asked for a similar characterization of the curvature measures on sets of positive reach. The fruit of a result of this type would certainly include a greatly improved proof of the kinematic formula in this broader context. One may also ask the same question in general homogeneous spaces. In this connection a recent paper by Ralph Howard contains a proof of the kinematic formula for smooth submanifolds of a Riemannian homogeneous space.   J. BROTHERS AND J. FU

Problem 7.2. If $f: \mathbf{R}^n \to \mathbf{R}$ is of class C^n then Sard's theorem tells us that the set of critical values of f has measure zero. Examples of Whitney [**WH**] show that it is not sufficient merely to assume f to be of class C^{n-1}. However, in case $n = 1$ classical results in real analysis [**SAS**, Chapter VII, Theorems 6.5 and 7.2] imply that if f has bounded variation then the set of $f(x)$ such that there is a sequence $x_n \to x$ with $(f(x_n) - f(x))/(x_n - x) \to 0$ has measure zero. It would be interesting to find analogous extensions of Sard's theorem for functions on $\mathbf{R}^n$ which are not C^n. For example, what can be said if the second distributional derivative $D^2 f$ of $f: \mathbf{R}^2 \to \mathbf{R}$ is a measure? It is not difficult to show that Sard's theorem holds under the much stronger hypothesis $f \in C^{1,1}$.   J. FU

Problem 7.3. It is known that the Hessian $D^2 f$ of a convex function $f: \mathbf{R}^n \to \mathbf{R}$ is a nonnegative semidefinite matrix of measures [**DR**]. It is also possible to show that, in a definite sense, each subdeterminant of $D^2 f$ is a well-defined measure. This fact is related to the existence of curvature measures of all orders on the boundary of a convex body [**SCR**]. Do analogous properties hold for differences of convex functions?   J. FU

Problem 7.4. What is the dual of $BV(\Omega)$, Ω open in $\mathbf{R}^n$? Not much is known. In [**MZ**] Meyers and Ziemer found some measures in the dual of $BV(\Omega)$, but no one knows how small a part this is. One possible application would be the formulation of dual variational principles in free boundary fluid flow.   P. D. SMITH

8. Applications of geometric measure theory.

Problem 8.1. What physical or other phenomena can be modelled in a geometric measure theory setting? There are intriguing engineering problems and physical and biological phenomena in which the geometry of the situation seems best regarded as an independent variable. What is the role of geometric measure theory in such situations? There have already been a number of such applications. Taylor's crystalline variational calculus seems particularly useful as illustrated in [**TJ4, TJ7, ATZ**]. Soap bubbles and films and various immiscible liquid configurations have been modelled within geometric measure theory [**TJ3, ATJ**] as well as the various other minimal partitioning phenomena discussed in [**AF2, AF4**]. Brakke studied mathematical motion by mean curvature [**BK**] in part as a model for grain boundary migration in annealing metals. Kohn and others have used techniques of geometric measure theory in the study of elastic and plastic deformations [**KR**] as have Scheffer and others in the study of singularities in solutions of the Navier–Stokes equations for fluid motion [**CKN, SV**]. F. ALMGREN

Problem 8.2. Do theoretically optimal load bearing structures sometimes have a fractal structure? A paper of Dyson's [**DF**], for example, describes an elaborate iterative structure of beams for use in a space station which is highly resistant to failure by buckling. One can ask whether or not least mass structures for prescribed load bearing ability are sometimes infinitely complicated, e.g. should an optimal railroad bridge have infinitely many trusses?—note [**KS**]. F. ALMGREN

Problem 8.3. What is the regularity of the set of fracture points in an elastic-plastic body subjected to traction forces? See the discussion and references in [**HK**]. R. HARDT

Problem 8.4. Is a sessile drop always convex if the surface energy is not just proportional to area? The usual sessile drop problem, where the surface energy is proportional to surface area and one minimizes surface energy plus gravitational energy for a region of given volume contained in the upper half space, has a long history. Finn has extraordinarily accurate upper and lower bounds for its shape [**FR1**]. Yet when the surface energy is anisotropic (depends on tangent plane direction), it is not even known if the region must be a convex body. Presumably if the Wulff shape (the solution in the absence of gravity) has a vertical axis of infinite rotational symmetry, the body is convex and more or less computable; for very small gravitational constants g, the body can also presumably be shown to be convex. But otherwise the problem is wide open, even for elliptic integrands. See [**ATZ**] for a general discussion of the problem; see [**TJ4**] for the construction of the $g = 0$ solution (the Wulff shape). See [**ASB**] for a related more general open problem; see [**TJ6**] for some of the results that can be proved if one knows the region is convex. J. TAYLOR

BIBLIOGRAPHY

[**AW1**] W. K. Allard, *On boundary regularity for Plateau's problem*, Bull. Amer. Math. Soc. **75** (1969), 522–523.

[**AW2**] ______, *On the first variation of a varifold*, Ann. of Math. (2) **95** (1972), 417–491.

[**AW3**] ______, *On the first variation of a varifold: Boundary behavior*, Ann. of Math. (2) **101** (1975), 418–446.

[**AW4**] ______, *An a priori estimate for the oscillation of the normal to a hypersurface whose first and second variation with respect to an elliptic integrand is controlled*, Invent. Math. **73** (1983), 287–321.

[**AA1**] W. K. Allard and F. J. Almgren, *The structure of stationary one-dimensional varifolds with positive density*, Invent. Math. **34** (1976), 83–97.

[**AA2**] ______, *On the radial behavior of minimal surfaces and the uniqueness of their tangent cones*, Ann. of Math. (2) **113** (1981), 215–265.

[**AF1**] F. J. Almgren, *Three theorems on manifolds with bounded mean curvature*, Bull. Amer. Math. Soc. **71** (1965), 755–756.

[**AF2**] ______, *Existence and regularity almost everywhere of solutions of elliptic variational problems with constraints*, Mem. Amer. Math. Soc., No. 165 (1976).

[AF3] ______ , Xeroxed lecture notes, c. 1980.

[AF4] ______ , *Minimal surface forms*, Math. Intelligencer **4** (1982), 164–172.

[AF5] ______ , **Q** *valued functions minimizing Dirichlet's integral and the regularity of area minimizing rectifiable currents up to codimension two*, preprint.

[AF6] ______ , *Optimal isoperimetric inequalities*, preprint.

[ASS] F. J. Almgren, R. Schoen and L. Simon, *Regularity and singularity estimates on hypersurfaces minimizing parametric elliptic variational integrals*, Acta. Math. **139** (1977), 217–265.

[AS] F. J. Almgren, and L. Simon, *Existence of embedded solutions of Plateau's problem*, Ann. Scuola Norm. Sup. Pisa Cl. Sci. (4) **6** (1979), 447–495.

[ASB] F. J. Almgren and B. Super, *Multiple-valued functions in the geometric calculus of variations*, Astérisque **118** (1984), 13–22.

[ATJ] F. J. Almgren and J. E. Taylor, *The geometry of soap films and soap bubbles*, Scientific American, 1976, pp. 82–93.

[AT] F. J. Almgren and W. P. Thurston, *Examples of unknotted curves which bound only surfaces of high genus within their convex hulls*, Ann. of Math. (2) **105** (1977), 527–538.

[ATZ] J. E. Avron, J. E. Taylor and R. K. P. Zia, *Equilibrium shapes of crystals in a gravitational field: Crystals on a table*, J. Statist. Phys. **33** (1983), 493–522.

[BD] J. Besnault and P. Dolbeault, *Sur les bords d'ensembles analytiques complexes dans* $\mathbf{P}^n(\mathbf{C})$, Sympos. Math. **24** (1981), 205–213.

[BD1] D. Bindschadler, *Absolutely area-minimizing singular cones of arbitrary codimension*, Trans. Amer. Math. Soc. **243** (1978), 223–233.

[BD2] ______ , *Invariant solutions to the oriented Plateau problem of maximal codimension*, Trans. Amer. Math. Soc. **261** (1980), 439–462.

[BD3] ______ , *Finite groups and invariant solutions to one-dimensional Plateau problems*, Proc. Amer. Math. Soc. **261** (1980), 621–626.

[BT] R. Böhme and A. Tromba, *The index theorem for classical minimal surfaces*, Ann. of Math. (2) **113** (1981), 447–499.

[BDG] E. Bombieri, E. De Giorgi and E. Giusti, *Minimal cones and the Bernstein problem*, Invent. Math. **7** (1969), 243–268.

[BG] E. Bombieri and E. Giusti, *Harnack's inequality for elliptic differential equations on minimal surfaces*, Invent. Math. **15** (1972), 24–46.

[BK] K. A. Brakke, *The motion of a surface by its mean curvature*, Math. Notes No. 20, Princeton Univ. Press, Princeton, N.J., 1978.

[BR] R. Brooks, *The fundamental group and the spectrum of the Laplacian*, Comment. Math. Helv. **56** (1981), 581–598.

[BJ1] J. E. Brothers, *Stokes' theorem* I; II, Amer. J. Math. **92** (1970), 657–670; ibid. **93** (1971), 479–484.

[BJ2] ______ , *Existence and structure of tangent cones at the boundary of an area minimizing integral current*, Indiana Univ. Math. J. **26** (1977), 1027–1044.

[BJ3] ______ , *Invariance of solutions to invariant parametric variational problems*, Trans. Amer. Math. Soc. **262** (1980), 159–179.

[BJ4] ______ , *Stability of minimal orbits*, Trans. Amer. Math. Soc. (to appear).

[BJ5] ______ , *Second variation estimates for minimal orbits*, Proc. Sympos. Pure Math. **44** (1985), 139–148.

[BJ6] ______ , *Stability of non-compact minimal orbits*, in preparation.

[BRR] R. Bryant, *Submanifolds and special structures on the octonions*, J. Differential Geometry **17** (1982), 185–232.

[CHS] L. Caffarelli, B. Hardt and L. Simon, *Minimal surfaces with isolated singularities*, Manuscripta Math. **48** (1984), 1–18.

[CKN] L. Caffarelli, R. Kohn and L. Nirenberg, *Partial regularity of suitable weak solutions of the Navier–Stokes equations*, Comm. Pure Appl. Math. **35** (1982), 771–831.

[CH] J. T. Chen and W. H. Huang, *Convexity of capillary surfaces in outer space*, Invent. Math. **67** (1982), 253–259.

[CS] H. Choi and R. Schoen, *The space of minimal embeddings of a surface into a three-dimensional manifold of positive Ricci curvature*, preprint.

[DJ] J. Dadok, *Polar coordinates induced by actions of compact Lie groups*, Trans. Amer. Math. Soc. **288** (1985), 125–137.

[**DH**] J. Dadok and R. Harvey, *Calibrations on* $\mathbf{R}^6$, Duke Math. J. **50** (1983), 1231–1243.

[**DR**] R. M. Dudley, *On second derivatives of convex functions*, Math. Scand. **41** (1977), 159–174; ibid. **46** (1980), 61.

[**DF**] F. Dyson, *The search for extraterrestrial technology*, Perspectives in Modern Physics, (R. E. Marshak, ed.), Interscience, New York, 1966.

[**FH1**] H. Federer, *Curvature measures*, Trans. Amer. Math. Soc. **93** (1959), 418–491.

[**FH2**] ______, *Geometric measure theory*, Springer-Verlag, 1969.

[**FH3**] ______, *The singular sets of area minimizing rectifiable currents with codimension one and of area minimizing flat chains modulo two with arbitrary codimension*, Bull. Amer. Math. Soc. **76** (1970), 767–771.

[**FH4**] ______, *Real flat chains, cochains, and variational problems*, Indiana Univ. Math. J. **24** (1974), 351–407.

[**FR1**] R. Finn, *The sessile liquid drop* I, *Symmetric case*, Pacific J. Math. **88** (1980), 549–587.

[**FR2**] ______, *Existence criteria for capillary free surfaces with gravity*, Indiana Univ. Math. J. **32** (1983), 439–460.

[**F-C**] D. Fischer-Colbrie, *Some rigidity theorems for minimal submanifolds of the sphere*, Acta Math. **145** (1980), 29–46.

[**FT**] T. Frankel, *On the fundamental group of a compact minimal submanifold*, Ann. of Math. (2) **83** (1966), 68–73.

[**GM**] M. Giaquinta, *Multiple integrals in the calculus of variations and nonlinear elliptic systems*, Ann. of Math. Studies No. 105 (1983).

[**GG1**] M. Giaquinta and E. Giusti, *On the regularity of the minima of variational integrals*, Acta Math. **148** (1982), 31–46.

[**GG2**] ______, *The singular set of the minima of certain quadratic functionals*, Ann. Scuola Norm. Sup. Pisa **11** (1984), 45–55.

[**GL1**] M. Gromov and H. B. Lawson, *Spin and scalar curvature in the presence of a fundamental group*, Ann. of Math. (2) **111** (1980), 209–320.

[**GL2**] ______, *Positive scalar curvature and the Dirac operator on complete Riemannian manifolds*, Inst. Hautes Etudes Sci. Publ. Math. **58** (1983).

[**GRM**] M. Grüter, *Regularity of weak H-surfaces*, J. Reine Angew. Math. **329** (1981), 1–15.

[**GR**] R. Gulliver, *A minimal surface with an atypical boundary branch point*, preprint.

[**GLB**] R. Gulliver and H. B. Lawson, *The structure of stable minimal hypersurfaces near a singularity*, preprint.

[**GLF**] R. Gulliver and F. Lesley, *On boundary branch points of minimizing surfaces*, Arch. Rational Mech. Anal. **52** (1973), 20–25.

[**GRH**] R. Gulliver and H. Rosenberg, *Symmetry for hypersurfaces of constant mean curvature*, preprint.

[**HJ**] J. Haas, Ph.D. thesis, Univ. of California, Berkeley, 1981.

[**HS**] J. Haas and P. Scott, *The existence of least area surfaces in three-manifolds*, preprint.

[**HAH**] H. Hadwiger, *Vorlesungen über Inhalt, Oberfläche und Isoperimetrie*, Springer-Verlag, 1957.

[**HR**] R. Hardt, *On the boundary regularity for integral currents or flat chains modulo two minimizing the integral of an elliptic integrand*, Comm. Partial Differential Equations **2** (1977), 1163–1232.

[**HK**] R. Hardt and D. Kinderlehrer, *Some regularity results in plasticity*, Proc. Sympos. Pure Math. **44** (1985), 239–244.

[**HS1**] R. Hardt and L. Simon, *Boundary regularity and embedded solutions for the oriented Plateau problem*, Ann. of Math. (2) **110** (1979), 439–486.

[**HS2**] ______, *Area minimizing hypersurfaces with isolated singularities*, preprint.

[**HAR**] R. Harvey, *Calibrated geometries*, Proc. Int. Cong. Math., 1983.

[**HL1**] R. Harvey and H. B. Lawson, *On boundaries of complex analytic varieties*, Ann. of Math. (2) **102** (1975), 223–290.

[**HL2**] ______, *On the boundaries of complex analytic varieties* II, Ann. of Math. (2) **106** (1977), 213–238.

[**HL3**] ______, *Calibrated geometries*, Acta Math. **148** (1982), 45–157.

[**HM**] R. Harvey and F. Morgan, *The faces of the Grassmannian of three-planes in* $\mathbf{R}^7$, preprint.

[**HSH**] R. Harvey and B. Shiffman, *A characterization of holomorphic chains*, Ann. of Math. (2) **99** (1974), 553–587.

[**HH**] H. Hopf, *Differential geometry in the large*, Lecture Notes in Math., vol. 1000, Springer-Verlag, 1983.

[HJ] J. Hubbard, *On the convex hull genus of space curves*, Topology **19** (1980), 203–208.

[HJE] J. E. Hutchinson, $C^{1,\,\alpha}$ *multiple function regularity and tangent cone behaviour for varifolds with second fundamental form in* L^P, Proc. Sympos. Pure Math. **44** (1985), 281–306.

[JJ] J. Jost, *Harmonic maps between surfaces*, Lecture Notes in Math., vol. 1062, Springer-Verlag, 1984.

[JM] J. Jost and M. Meier, *Boundary regularity for minima of certain quadratic functionals*, Math. Ann. **262** (1983), 549–561.

[KW] H. Karcher and J. C. Wood, *Non-existence results and growth properties for harmonic maps and forms*, SFB-40 preprint, Bonn, 1983.

[KIR] R. Kirby, *Problems in low-dimensional topology*, in preparation.

[KR] R. V. Kohn, *New integral estimates for deformations in terms of their nonlinear strains*, Arch. Rational Mech. Anal. **78** (1982) 131–172.

[KS] R. V. Kohn and G. Strang, *Structural design optimization, homogenization, and relaxation of variational problems*, (Proc. Conf. Disordered Media, N.Y.U., 1981), Lecture Notes in Physics, vol. 154, Springer-Verlag, 1982, pp. 131–147.

[KN] N. Korevaar, *Capillary surface convexity above convex domains*, Indiana Univ. Math. J. **32** (1983), 73–81.

[LJ] J. Lander, *Area minimizing integral currents with boundaries invariant under polar actions*, Ph.D. thesis, M.I.T., Cambridge, Mass., 1984.

[LB1] H. B. Lawson, *Complete minimal surfaces in* $\mathbf{S}^3$, Ann. of Math. (2), **92** (1970), 335–374.

[LB2] ______, *The unknottedness of minimal embeddings*, Invent. Math. **11** (1970), 183–187.

[LB3] ______, *The equivariant Plateau problem and interior regularity*, Trans. Amer. Math. Soc. **173** (1973), 231–249.

[LO] H. B. Lawson and R. Osserman, *Non-existence, non-uniqueness and irregularity of solutions of the minimal surface system*, Acta Math. **139** (1977), 1–17.

[MB] B. Mariaux, *Thèse de* 3e *cycle*, Thèse, Université de Poitiers, 1984; C. R. Acad. Sci. Paris Sér. I Math. (to appear).

[MP1] P. Mattila, *Hausdorff m regular and rectifiable sets in n-space*, Trans. Amer. Math. Soc. **205** (1975), 263–274.

[MP2] ______, *Hausdorff dimension, orthogonal projections and intersections with planes*, Ann. Acad. Sci. Fenn. Ser. A I Math. **1** (1975), 227–244.

[MP3] ______, *Hausdorff dimension and exceptional sets of linear trasnformations*, Ann. Acad. Sci. Fenn. Ser. A I Math. **1** (1975), 387–392.

[MP4] ______, *Integralgeometric properties of capacities*, Trans. Amer. Math. Soc. **266** (1981), 539–554.

[MP5] ______, *Lower semicontinuity and existence theorems for elliptic variational integrals of multiple-valued functions*, Trans. Amer. Math. Soc. **280** (1983), 589–610.

[MSY] W. Meeks, L. Simon and S.-T. Yau, *Embedded minimal surfaces, exotic spheres and manifolds of positive Ricci curvature*, Ann. of Math. (2) **116** (1982), 151–168.

[MY] W. Meeks and S.-T. Yau, *The classical Plateau problem and the topology of three-dimensional manifolds*, Topology **21** (1982), 409–442.

[MZ] N. Meyers and W. P. Ziemer, *Wirtinger and Poincaré inequalities for BV functions*, Amer. J. Math. **99** (1977), 1345–1360.

[MA] A. Milani, *Non-analytic minimal varieties*, Istituto Matematico "Leonida Torielli," 77–13, Università degli Studi di Pisa, 1977.

[MF1] F. Morgan, *Measures on spaces of surfaces*, Arch. Rational Mech. Anal. **78** (1982), 335–359.

[MF2] ______, *On the singular structure of two-dimensional area minimizing surfaces in* $\mathbf{R}^n$, Math. Ann. **261** (1982), 101–110.

[MF3] ______, *On the singular structure of three-dimensional area-minimizing surfaces in* $\mathbf{R}^n$, Trans. Amer. Math. Soc. **276** (1983), 137–143.

[MF4] ______, *Area minimizing currents bounded by higher multiples of curves*, Rend. Circ. Mat. Palermo (2) **33** (1984), 1–10.

[MF5] ______, *Examples of unoriented area-minimizing surfaces*, Trans. Amer. Math. Soc. **283** (1984), 225–237.

[MF6] ______, *The exterior algebra of* $\Lambda^k \mathbf{R}^n$ *and area minimization*, Linear Algebra Appl. **66** (1985), 1–28.

[MF7] ______, *On finiteness of the numbers of stable minimal hypersurfaces with a fixed boundary*, preprint.

[MC1] C. B. Morrey, *The problem of Plateau on a Riemannian manifold*, Ann. of Math. (2) **49** (1948), 807–851.

[MC2] ______, *Multiple integrals in the calculus of variations*, Springer-Verlag, 1966.

[NJ] J. Nitsche, *A new uniqueness theorem for minimal surfaces*, Arch. Rational Mech. Anal. **52** (1973), 319–329.

[OR] R. Osserman, *Properties of solutions to the minimal surface equation in higher codimensions*, U.S.A.-Japan Seminar on Minimal Submanifolds and Geodesics, Tokyo, 1978, pp. 163–172.

[PZ] H. Parks and W. Ziemer, *Jacobi fields and regularity of functions of least gradient*, Ann. Scuola Norm. Sup. Pisa Cl. Sci. (to appear).

[PS1] S. Paur, *Stokes' theorem for integral currents modulo v*, Amer. J. Math. **99** (1977), 379–388.

[PS2] ______, *An estimate of the density at the boundary of an integral current modulo v*, Proc. Amer. Math. Soc. **62** (1977), 319–322.

[PJ] J. Pitts, *Existence and regularity of minimal surfaces on Riemannian manifolds*, Math. Notes, No. 27, Princeton Univ. Press, Princeton, N.J., 1981.

[PP] P. Price, *A monotonicity formula for Yang-Mills fields*, Manuscripta Math. **43** (1983), 131–166.

[RD] D. Rolfsen, *Knots and links*, Publish or Perish, Berkeley, 1976.

[RV] E. Ruh and J. Vilms, *The tension field of the Gauss map*, Trans. Amer. Math. Soc. **149** (1970), 569–573.

[SAS] S. Saks, *Theory of the Integral*, Hafner, New York, 1933.

[SV] V. Scheffer, *A solution to the Navier–Stokes inequality with an internal singularity*, preprint.

[SA] A. H. Schoen, *Infinite periodic minimal surfaces without self-intersections*, NASA Technical Note No. 5541, NASA, Washington, D.C.

[SCR] R. Schneider, *Curvature measures of convex bodies*, Ann. Mat. Pura Appl. **116** (1978), 101–134.

[SR] R. Schoen, *Analytic aspects of the harmonic map problem*, preprint.

[SS] R. Schoen and L. Simon, *Regularity of stable minimal hypersurfaces*, Comm. Pure Appl. Math. **34** (1981), 741–797.

[SU1] R. Schoen and K. Uhlenbeck, *A regularity theory for harmonic maps*, J. Differential Geometry **17** (1982), 307–335.

[SU2] ______, *Boundary regularity and the Dirichlet problem for harmonic maps*, J. Differential Geometry **18** (1983), 253–268.

[SU3] ______, *Regularity of minimizing maps into the sphere*, preprint.

[SP] P. Simoes, *On a class of minimal cones in $\mathbf{R}^n$*, Bull. Amer. Math. Soc. **80** (1974), 488–489.

[SL1] L. Simon, *Asymptotics for a class of nonlinear evolution equations with applications to geometric problems*, Ann. of Math. (2) **118** (1983), 525–571.

[SL2] ______, *Entire solutions of the minimal surface equation* (in preparation).

[SSR] L. Simon and R. Smith, *On the existence of embedded minimal 2-spheres in the 3-sphere, endowed with an arbitrary metric*, preprint.

[TJ1] J. E. Taylor, *Regularity of the singular sets of two-dimensional area minimizing flat chains modulo 3 in $\mathbf{R}^3$*, Invent. Math. **22** (1973), 119–159.

[TJ2] ______, *The structure of singularities in soap-bubble-like and soap-film-like minimal surfaces*, Ann. of Math. (2) **103** (1976), 489–539.

[TJ3] ______, *The structure of singularities in solutions to ellipsoidal variational problems with constraints in $\mathbf{R}^3$*, Ann. of Math. (2) **103** (1976), 541–546.

[TJ4] ______, *Crystalline variational problems*, Bull. Amer. Math. Soc. **84** (1978), 568–588.

[TJ5] ______, *Equilibrium shapes for surfaces and grain boundaries*, Phase Transformations and Material Instabilities in Solids (Proc. Mathematical Research Center Conf.) Madison, 1983.

[TJ6] ______, *Is there gravity-induced facetting in crystals?* (Proc. Variational Methods for Equilibrium Problems of Fluids, Trento, 1983), Astérisque **118** (1984), 243–253.

[TJ7] ______, *Complete catalog of minimizing embedded crystalline cones*, Proc. Sympos. Pure Math. **44** (1985), 379–403.

[TJ8] ______, *An a priori bound on the number of plane segments of an F-minimizing F-crystalline mapping*, in preparation.

[TW] W. Thompson (Lord Kelvin), *On the division of space with minimum possible error*, Acta Math. **11** (1887), 121–134.

[TT] F. Tomi and A. Tromba, *Extreme curves bound embedded minimal surfaces of the type of the disk*, Math. Z. **158** (1978), 137–145.

[WB1] B. White, *The structure of minimizing hypersurfaces* mod 4, Invent. Math. **53** (1979), 45–58.

[WB2] ______, *Singularity structure and generic regularity of two-dimensional area minimizing surfaces*, Ph.D. thesis, Princeton Univ., Princeton, N. J., 1981.

[WB3] ______, *The space of m-dimensional surfaces that are stationary for a parametric elliptic integrand*, preprint, 1982.

[WB4] ______, *Regularity of area-minimizing hypersurfaces at boundaries with multiplicity*, Seminar on Minimal Submanifolds, Ann. of Math. Studies, **103** (1983), 293–301.

[WB5] ______, *Tangent cones to two-dimensional area minimizing integral currents are unique*, Duke Math. J. **50** (1983), 143–160.

[WB6] ______, *The existence of least-area mappings of N-dimensional domains*, Ann. of Math. (2) **118** (1983), 179–185.

[WB7] ______, *The least area bounded by multiples of a curve*, Proc. Amer. Math. Soc. **90** (1984), 230–232.

[WB8] ______, *The space of minimal submanifolds of a manifold with varying metrics*, preprint.

[WB9] ______, *A regularity theorem for minimizing hypersurfaces* mod p, preprint.

[WB10] ______, *The space of m-dimensional stationary surfaces in* $\mathbf{R}^n$, preprint.

[WH] H. Whitney, *A function not constant on a connected set of critical points*, Duke Math. J. **1** (1935), 514–517.

[YS] S.-T. Yau, *Kohn–Rossi cohomology and its application to the complex Plateau problem*, Ann. of Math. (2) **113** (1981), 67–110.

[YL] L. C. Young, *Some extremal questions for simplicial complexes V: The relative area of a Klein bottle*, Rend. Circ. Mat. Palermo (2) **12** (1963), 257–274.

INDIANA UNIVERSITY

ABCDEFGHIJ–AMS/AP–89876